Air Pollution Modeling and Its Application XIV

Previous Volumes in this Mini-Series

Volumes I–XII were included in the NATO Challenges of Modern Society Series.

AIR POLLUTION MODELING AND ITS APPLICATION I
Edited by C. De Wispelaere

AIR POLLUTION MODELING AND ITS APPLICATION II
Edited by C. De Wispelaere

AIR POLLUTION MODELING AND ITS APPLICATION III
Edited by C. De Wispelaere

AIR POLLUTION MODELING AND ITS APPLICATION IV
Edited by C. De Wispelaere

AIR POLLUTION MODELING AND ITS APPLICATION V
Edited by C. De Wispelaere, Francis A. Schiermeier, and Noor V. Gillani

AIR POLLUTION MODELING AND ITS APPLICATION VI
Edited by Han van Dop

AIR POLLUTION MODELING AND ITS APPLICATION VII
Edited by Han van Dop

AIR POLLUTION MODELING AND ITS APPLICATION VIII
Edited by Han van Dop and Douw G. Steyn

AIR POLLUTION MODELING AND ITS APPLICATION IX
Edited by Han van Dop and George Kallos

AIR POLLUTION MODELING AND ITS APPLICATION X
Edited by Sven-Erik Gryning and Millán M. Millán

AIR POLLUTION MODELING AND ITS APPLICATION XI
Edited by Sven-Erik Gryning and Francis A. Schiermeier

AIR POLLUTION MODELING AND ITS APPLICATION XII
Edited by Sven-Erik Gryning and Nadine Chaumerliac

AIR POLLUTION MODELING AND ITS APPLICATION XIII
Edited by Sven-Erik Gryning and Ekaterina Batchvarova

Air Pollution Modeling and Its Application XIV

Edited by

Sven-Erik Gryning
Risø National Laboratory
Roskilde, Denmark

and

Francis A. Schiermeier
U.S. Environmental Protection Agency
Research Triangle Park, North Carolina

Kluwer Academic / Plenum Publishers
New York, Boston, Dordrecht, London, Moscow

Library of Congress Cataloging-in-Publication Data

Air pollution modeling and its application XIV / edited by Sven-Erik Gryning and Francis A. Schiermeier.
p. cm.
Includes bibliographical references and index.
ISBN 0-306-46534-5
1. Air--Pollution--Mathematical models--Congresses. 2. Atmospheric diffusion--Mathematical models--Congresses. I. Gryning, Sven-Erik. II. Schiermeier, Francis A. III. NATO/CCMS International Technical Meeting on Air Pollution Modeling and Its Application (24th : 2000 : Boulder, Colo.)

TD881 .A47524 2001
628.5'3'015118--dc21

00-052737

FRONT COVER

The cover photo depicts Boulder Colorado's Flatirons, which serves as a backdrop to the city. Boulder, Colorado, is located in the western United States, just east of the Rocky Mountain continental divide at an elevation of 1632 meters. The town was laid out in 1859, the same year that gold was discovered nearby. Once supported by mines of coal and precious rare metals, the city later attracted important industrial and research organizations. Today it serves as host to the University of Colorado, the National Center for Atmospheric Research (NCAR), the University Corporation for Atmospheric Research (UCAR), and several U.S. National Oceanic and Atmospheric Administration (NOAA) research laboratories.

Proceedings of the Millennium NATO/CCMS International Technical Meeting on Air Pollution Modeling and Its Application, held May 15–19, 2000, in Boulder, Colorado

ISBN: 0-306-46534-5

233 Spring Street, New York, New York 10013

http://www.wkap.nl/

10 9 8 7 6 5 4 3 2 1

A C.I.P. record for this book is available from the Library of Congress

Printed in the United States of America

PREFACE

In 1969 the North Atlantic Treaty Organization (NATO) established the Committee on the Challenges of Modern Society (CCMS). The subject of air pollution was from the start one of the priority problems under study within the framework of various pilot studies undertaken by this committee. The organization of a periodic conference dealing with air pollution modeling and its application has become one of the main activities within the pilot study relating to air pollution. The first five international conferences were organized by the United States as the pilot country; the second five by the Federal Republic of Germany; the third five by Belgium; the next four by The Netherlands; and the five most recent by Denmark.

This volume contains the papers and poster abstracts presented at the Millennium (24th) NATO/CCMS International Technical Meeting (ITM) on Air Pollution Modelling and Its Application, held in Boulder, Colorado, during May 15-19, 2000. This ITM was jointly organized by the Risø National Laboratory of Denmark (Pilot Country); by the EPA National Exposure Research Laboratory, United States (Host Country); and by the American Meteorological Society, United States (Host Organization).

The ITM was attended by 135 participants representing 30 countries from North and South America, Europe, Asia and Australia. The main topics of this ITM were: Role of Atmospheric Models in Air Pollution Policy and Abatement Strategies; Integrated Regional Modelling; Global and Long-Range Transport; Regional Air Pollution and Climate; New Developments; and Model Assessment and Verification.

Invited papers were presented by P. Builtjes of The Netherlands (Major Twentieth Century Milestones in Air Pollution Modelling and Its Application); W. Klug of Germany (What Did We Learn From the ETEX Experiment?); and A. Venkatram of the United States (Challenges of Air Pollution Modelling and Its Application in the Next Millennium).

On behalf of the ITM Scientific Committee and as organizers and editors, we should like to express our gratitude to all the participants who made the meeting so successful, resulting in the publication of this book. Among the participants, we especially recognize the efforts of the chairpersons and rapporteurs. Thanks are extended to the American Meteorological Society, the local host organization, who so capably assisted us in organizing this Millennium ITM.
Finally, we wish to express our gratitude to the sponsors. The major sponsorship was provided by the United States Environmental Protection Agency. In addition, financial support was received from Risø National Laboratory, the NATO Committee on the Challenges of Modern Society, and EURASAP (European Association for the Science of Air Pollution). A special grant was given by NATO/CCMS to facilitate attendance of scientists from Central and Eastern Europe. The next conference in this series will be held in 2001 in Belgium.

Francis A. Schiermeier
Conference Chairman
United States

Sven-Erik Gryning
Scientific Committee Chairman
Denmark

THE MEMBERS OF THE SCIENTIFIC COMMITTEE FOR THE MILLENNIUM (24th) NATO/CCMS INTERNATIONAL TECHNICAL MEETINGS ON AIR POLLUTION MODELLING AND ITS APPLICATION

G. Schayes, Belgium
D. Syrakov, Bulgaria
D. Steyn, Canada
S. E. Gryning (chairman), Denmark
H. Olesen, Denmark
W. Klug, Germany
N. Chaumerliac, France
G. Kallos, Greece
D. Anfossi, Italy
H. van Dop, Holland
T. Iversen, Norway
C. Borrego, Portugal
R. Salvador, Spain
B. Fisher, United Kingdom
F. Schiermeier, USA
Y. Schiffman, USA

HISTORY OF THE NATO/CCMS AIR POLLUTION PILOT STUDIES

Pilot Study on Air Pollution: International Technical Meetings (ITM) on Air Pollution Modelling and Its Application

Dates of Completed Pilot Studies:

1969 - 1974 Air Pollution Pilot Study (United States Pilot Country)
1975 - 1979 Air Pollution Assessment Methodologies and Modelling (Germany)
1980 - 1984 Air Pollution Control Strategies and Impact Modelling (Germany)

Dates and Locations of Pilot Study Follow-Up Meetings:

Pilot Country - United States (R.A. McCormick, L.E. Niemeyer)

February 1971 - Eindhoven, The Netherlands, First Conference on Low Pollution Power Systems Development
July 1971 - Paris, France, Second Meeting of the Expert Panel on Air Pollution Modelling

---

All following meetings were entitled NATO/CCMS International Technical Meetings (ITM) on Air Pollution Modelling and Its Application

---

October 1972 - Paris, France - Third ITM
May 1973 - Oberursel, Federal Republic of Germany - Fourth ITM
June 1974 - Roskilde, Denmark - Fifth ITM

Pilot Country - Germany (Erich Weber)

September 1975 - Frankfurt, Federal Republic of Germany - Sixth ITM
September 1976 - Airlie House, Virginia, USA - Seventh ITM
September 1977 - Louvain-La-Neuve, Belgium - Eighth ITM
August 1978 - Toronto, Ontario, Canada - Ninth ITM
October 1979 - Rome, Italy - Tenth ITM

Pilot Country - Belgium (C. De Wispelaere)

November 1980 - Amsterdam, The Netherlands - Eleventh ITM
August 1981 - Menlo Park, California, USA - Twelfth ITM
September 1982 - Ile des Embiez, France - Thirteenth ITM
September 1983 - Copenhagen, Denmark - Fourteenth ITM
April 1985 - St. Louis, Missouri, USA - Fifteenth ITM

Pilot Country - The Netherlands (Han van Dop)

April 1987 - Lindau, Federal Republic of Germany - Sixteenth ITM
September 1988 - Cambridge, United Kingdom - Seventeenth ITM
May 1990 - Vancouver, British Columbia, Canada - Eighteenth ITM
September 1991 - Ierapetra, Crete, Greece - Nineteenth ITM

Pilot Country - Denmark (Sven-Erik Gryning)

November 1993 - Valencia, Spain - Twentieth ITM
November 1995 - Baltimore, Maryland, USA - Twenty-First ITM
June 1997 - Clermont-Ferrand, France - Twenty-Second ITM
September 1998 - Varna, Bulgaria - Twenty-Third ITM
May 2000 - Boulder, Colorado - Twenty-Fourth (Millennium) ITM

CONTENTS

ROLE OF ATMOSPHERIC MODELS IN AIR POLLUTION POLICY AND ABATEMENT STRATEGIES

INTEGRATED REGIONAL MODELLING

GLOBAL AND LONG-RANGE TRANSPORT

REGIONAL AIR POLLUTION AND CLIMATE

NEW DEVELOPMENTS

MODEL ASSESSMENT AND VERIFICATION

POSTER SESSION

ROLE OF ATMOSPHERIC MODELS IN AIR POLLUTION POLICY AND ABATEMENT STRATEGIES

chairpersons:	S. E. Gryning F. Schiermeier
rapporteurs:	J. Irwin P. Suppan

MAJOR TWENTIETH CENTURY MILESTONES IN AIR POLLUTION MODELLING AND ITS APPLICATION

Peter J.H. Builtjes

TNO-MEP, Dep. of Environmental Quality,
P.O.Box 342, 7300 AH Apeldoorn, the Netherlands and
University Utrecht, Institute of Marine and
Atmospheric Research, Princetonplein 5,
3584 CC Utrecht, the Netherlands
e-mail: p.j.h.builtjes@mep.tno.nl

INTRODUCTION

Air pollution modelling is an attempt to describe the functional relation between emissions and occurring concentrations and deposition. Air pollution measurements present these occurring concentrations and deposition, but they can only give a snapshot at specific locations and times. In principle air pollution modelling can give a more complete and consistent description, including an analysis of the causes-the emissions sources-which have led to these concentrations/deposition. However, it is often difficult to assess how good the modelling description is.

Air pollution models play an important role in science, because of their capability to investigate the importance of the relevant processes, and they play a major role in application Air pollution models are the only method which can quantify the relation between emissions and concentrations/depositions, including the consequences of future scenario's and the determination of the effectivity of abatement strategies.

Air pollution modelling in this paper will basically be restricted to the troposphere. This choice is made because in general modelling studies in the stratosphere are not called air pollution studies, they belong more to the field of atmospheric chemistry. However, stratospheric models can be important to determine the boundary conditions for tropospheric models.

In this paper a rough attempt is made to detect some milestones in the last century concerning air pollution modelling and its application. A large number of books and review articles exist already which from their own perspective present overviews and highlights. No attempt is made here to harmonise all this existing information. This paper presents my personal view at this moment without being in the position to make a thorough study. But I

Air Pollution Modeling and Its Application XIV, Edited by
Gryning and Schiermeier, Kluwer Academic/Plenum Publishers, New York, 2001

hope to show some of the pleasure and beauty of air pollution modelling. Because modelling is great fun!

TRANSPORT AND DIFFUSION

The concentrations of species in the atmosphere are determined by transport and diffusion. This means that in considering milestones throughout the last century, some remarks should be made concerning transport and diffusion. Transport, so mean velocity, has already been measured and studied for centuries, for example for sailing ships etc. Diffusion, so turbulent motions, is a more modern phenomenon. Although turbulent motions have been observed from the moment people looked to for example rivers and streams, one could mention the paper by Reynolds (1895) as a scientific starting point in which the famous criterion for the change from laminar to turbulent flow in pipes is formulated.

One of the first articles that mention turbulence in the atmosphere was published by Taylor (1915). In later years he also developed the famous Taylor-theory of turbulent diffusion, Taylor (1921). In this theory it is shown that the diffusion from a point source can only be described with a constant eddy diffusivity K for travel times which are much larger than the turbulent integral time scale, the so-called diffusion limit. For smaller time-scales the effective turbulent diffusivity is proportional to the travel time.

Until about 1950 a number of studies were performed on the subject of diffusion in the atmosphere, as Richardson (1925), Sutton (1932), Bosanquet (1936), Church (1949), Thomas e.a. (1949), Inoue (1950), Batchelor (1950). Already the paper by Richardson considered long range aspects upto over 80 km. Bosanquet is one of the first publications about the impact of chimney plumes. A paper by Chamberlain (1953) already considered the deposition of aerosols.

DISPERSION OF EMISSIONS FROM POINT SOURCES

The study of the dispersion from low and high level point sources, especially experimental, was a major topic shortly after 1955. Papers on this subject appeared from Smith (1957), Gifford (1957 a,b), Hay and Pasquill (1957), Record (1958) and Haugen (1959) both devoted to the Prairie grass experiment, Stewart (1958), Monin (1959), Ogura (1959). Maybe the first paper on this subject was by Roberts (1923).

The book by Pasquill, Atmospheric Diffusion, which appeared in 1962 definitely formed a major milestone in summarising the work performed uptill that moment. This book made it also clear that air pollution modelling around the beginning of the sixties was focussed on local dispersion phenomena, mainly from point sources with SO_2 as major component in the application studies.

The Gaussian plume model was borne, in which the horizontal and vertical spread of the plume was determined experimentally. Tables appeared with the famous Pasquill-Gifford sigma-values in the horizontal and vertical direction, and as a function of the atmospheric stability ranging from very stable, class F up to very unstable, class A. The experimental sigma values are in their functions with distance from the source in reasonable agreement with the Taylor-theory. The differences are caused by the fact that the Taylor-theory holds for homogeneous turbulence, which is not the case in the atmosphere.

During the sixties the studies concerning dispersion from a point source continued and were broadening in scope. Major studies were performed by Högstrom (1964), Turner (1964), Briggs (1965) -the famous plume-rise formula's-, Moore (1967), Klug (1968). The use and application of the Gaussian plume model spreaded over the whole globe, and became a standard technique in every industrial country to calculate the stack height required in permits, see for example Beryland (1974) who published a standard work in Russian. The Gaussian plume model concept was soon applied also to line and area-sources. Gradually, the importance of the mixing height was realised (Holzworth, 1967, Deardorff, 1974) and its major influence on the magnitude of ground level concentrations.

The Journal of Atmospheric Environment in which over the years, and uptill now, many papers about modelling and application have been published had its first issue in 1967.

In February 1971 the first NATO/CCMS international technical meeting-ITM- on air pollution modelling and its application was held in Eindhoven, the Netherlands, the meeting was called: First conference on low pollution power systems development. The second meeting, held in Paris in July 1971 was called the second meeting of the expert panel on air pollution modelling, the third meeting was held in October 1972 in Paris, and was called the third ITM.

This start of the ITM-meetings clearly showed the growing interest in -local-air pollution modelling and its application.

The impression from the air pollution modelling papers published in the sixties and seventies is that they are mainly written by meteorologists, specialised in boundary layer meteorology and atmospheric turbulence. Studies focussed often on the effect of atmospheric stability on plume spread. During the next decade, next to the continuos research on local dispersion, see for an nice overview Nieuwstadt and van Dop (1982), the spatial scale of air pollution modelling increased substantially.

MODELLING OF TRANSPORT OF AIR POLLUTION AT ALL SCALES

The Gaussian plume model was the central air pollution model during the sixties. During the seventies the central model was given by the 'full' diffusion equation:

$$\frac{\partial c}{\partial t} + U\frac{\partial c}{\partial x} + V\frac{\partial c}{\partial y} + W\frac{\partial c}{\partial z} =$$

$$\frac{\partial}{\partial x}\left(K_h \frac{\partial c}{\partial x}\right) + \frac{\partial}{\partial y}\left(K_h \frac{\partial c}{\partial y}\right) + \frac{\partial}{\partial z}\left(K_z \frac{\partial c}{\partial z}\right)$$

+ chemistry + emissions - dry deposition - wet deposition.

The first comprehensive discussion of this equation is given by Lamb and Seinfeld (1973). This equation nicely shows by its different terms that other scientists beyond boundary layer meteorologists became part of the science of air pollution modelling.

The terms on the left side describe the impact of the mean transport in the atmosphere on the change in time of a specific species at a specific location, the first three terms on the righthand side present the turbulent transport terms. These terms together give the influence of the meteorological conditions, for the area under consideration. The diffusion equation is valid on all scales, from local scale to the global scale. It should be noted that also the Gaussian plume model can be derived from this equation, provided that the eddy diffusivities reflect the sigma-values, which means that they have to be a function of the distance to the source for small distances, as is given by the Taylor-theory.

Solving the diffusion equation means that adequate meteorological input has to be provided. Meteorology as such is not the subject of this paper, but a number of remarks can be made.

Meteorological input to air pollution models

Air pollution models at scales beyond the local scale- in which case only local meteorological data are needed- require consistent meteorological fields of mean velocity and turbulence as input. During the seventies the method used was to create meteorological input based on observations, the so-called diagnostic method. This method has the advantage that it is-by definition-close to the observations. After about 1985 studies have been performed to use the output of meteorological models-as they are currently used in weather forecast-which are based on the full physical equations, as input to air pollution models. This so-called prognostic method has the advantage that the output is consistent. The disadvantage is that the result can deviate rather substantially from the observations. Because the weather forecast models have clearly improved over the last decade, there is a tendency during the nineties to use prognostic models more and more.

As from the days of the first application of the Gaussian plume model, the weakest point in the meteorological data is still the determination of the mixing height, or in other words the determination of the vertical exchange by both mean convection and turbulence.

The chemistry term in air pollution models

The next term in the diffusion equation are the sources and sinks of a specific species due to the atmospheric chemistry. Although atmospheric chemistry is not the subject of this paper, a number of remarks can be made. In most applications of local models, as the Gaussian plume model, the chemistry term is neglected, the emissions are assumed to be inert. In the first studies of the dispersion of SO_2 at scales beyond the local scale the chemistry term was simplified to a linear decay with time. Already from the beginning of the seventies, attempts were made to use non-linear gas phase chemistry in air pollution models.

Atmospheric chemistry as such is a relatively new science. The first book on the subject was published by Junge (1963). Gasphase chemistry, mainly focussed on ozone made a big step forwards when the OH-radical was 'found', Levy (1971). Photochemical schemes were developed and were used in air pollution models since the beginning of the seventies.

Using indoor and outdoor smogchambers, these schemes were gradually improved over the years. The basic processes of gasphase ozone-chemistry are rather well understood nowadays. Much less is still known about heterogeneous chemistry and aerosols. Major textbooks are by Seinfeld (1986) and Warneck (1988).

The emission input to the models

Again, emission data bases are not the main theme of this paper, but emissions are too vital in air pollution modelling to be neglected. The often used phrase: garbage in, garbage out is very true indeed for the results of air pollution models.

From the beginning of the seventies, countries started to collect information about emissions into the atmosphere. See for an early paper, Roth e.o. (1974). Starting with the acidifying substances as SO_2 and NO_x, the amount of species gradually increased, including NH_3, speciated VOC, CO, CH_4, etc. The structure of these emission data bases are similar for all species. Activity data are collected and multiplied by an emission factor which is, hopefully, based on measurements and is generally applicable. Official emission data were collected in the framework of EMEP and the US-EPA. These official emissions play a large role in international negotiations, but are not necessarily the most accurate or correct ones for use in air pollution models. Over the years the reliability of emission data has gradually improved as new data on emission factors became available as well as better statistics concerning activities/source categories. Also an improvement has been made to come to a total coverage of all relevant source categories.

From the beginning of the eighties attention has been given to biogenic emissions also, especially for VOC's.

Over the last decade studies have been performed to try to improve the accuracy of emissions by making use of field observations and inverse modelling in which way the emissions leading to a certain observation can be determined.

Dry and wet deposition as sink in the models

Next to the decrease of concentrations due to chemical transformation, dry and wet deposition are a major sink term in the diffusion equation.

In the first air pollution models in which dry deposition was incorporated, it was treated by using a dry deposition velocity, which multiplied by the concentration at a certain height gave the vertical flux out of the system. This dry deposition velocity was based on measurements, and the assumption that below a reference height of say 50 m, the concept of a constant flux-layer was valid. Already around 1975 the dry deposition velocity concept was improved by using the resistance analogy. The dry deposition velocity is splitted in 3 resistances, the aerodynamic resistance Ra, the viscous sublayer resistance Rb and the surface resistance Rc. Ra follows directly from boundary layer theory, Rb is a function of the species itself, and Rc is a function of the species and the surface. It can be stated that the knowledge about Ra and Rb is well established, but that the accurate determination of Rc is still a subject of study. For many species Rc is the limiting parameter for the dry deposition process, and is consequently of primary importance. Essential for the determination of Rc is an accurate, and fit-for-the -purpose-landuse data base. The same data-base is important for biogenic emissions and for the surface roughness which is needed in the calculation of Ra.

Wet deposition is important for water soluble species. A distinction is made between in-cloud and below-cloud scavenging. Wet deposition is related to clouds and cloud-formation, and aqueous phase chemistry.

Numerics

The numerical method to solve the transport-diffusion equation is an essential element of the success, or failure, of air pollution modelling. Two elements play a role here, speed and accuracy. The system of coupled differential equations is large, and often a substantial amount of model runs has to be performed. Consequently, research has been performed to come to relatively fast numerical solvers, which is of course related to the computer capacity available. Accuracy is a major point especially for grid models. Grid models have the handicap to introduce numerical diffusion, which can lead to artificial diffusion in the system which can be larger than the physical diffusion. The first method which has addressed this point by introducing counter-diffusion/flux correction is the Shasta-method, Boris (1973).

Numerous studies have been performed over the last decades concerning the numerical problems associated with solving air pollution models. The current solution methods are considered to have sufficient accuracy, research continues to increase the speed of the solutions in combination with the required storage.

Observations

Observations as such are not a term in the diffusion equation. However, they are important as initial and boundary conditions-provided these do not come from an other model. But even more important, observations are used to determine the reliability of the model. Calculated concentrations are compared with observations which leads to a determination of the model performance. Where model evaluation, or testing, validation, uptill about 1990 was basically a graphical comparison between calculated and observed values, more detailed studies have been performed recently, including attempts to come to a model evaluation protocol, quality assurance/quality control and the use of proper statistics.

AIR POLLUTION MODELLING AT URBAN AND LARGER SCALES

Shortly after 1970 the awareness grew that air pollution was not only a local phenomenon. It became clear-first especially in Europe- that the SO_2- and NO_x-emissions from tall stacks could lead to acidification at large distances from the sources. It also became clear-first especially in the US-that ozone was a problem in urbanized and industrialized areas. It became also clear that these situations could not be tackled by simple Gaussian-plume type modelling.

Two different modelling approaches were followed, Lagrangian modelling and Eulerian modelling. In Lagrangian modelling an air parcel is followed along a trajectory, and is assumed to keep its identity during its path along the trajectory. In Eulerian modelling the area under investigation is divided in grids, both in the vertical and in the horizontal direction.

Lagrangian modelling directed to the description of long-range transport of sulfur started with studies by Rohde (1972, 1974), Eliassen (1975) and Fisher (1975). The work by Eliassen was the beginning of the well-known EMEP-trajectory model which has been used over the years, and is still used, to calculate transboundary air pollution of acidifying species and later also photo-oxidants. Lagrangian modelling is often used to cover longer time periods, upto years.

Eulerian modelling started with studies by Reynolds (1973) for ozone in urbanised areas with Shir and Shieh (1974) for SO_2 in urban areas, and Egan (1976) and Carmichael (1979) for regional scale sulfur. From the modelling studies by Reynolds over the Los Angeles basin the well-known Urban Airshed Model-UAM-originated. Eulerian modelling is often used only for specific episodes.

So in general, Lagrangian modelling was mostly performed in Europe, over large distances and longer time-periods, and focussed primarily on SO_2. Eulerian grid modelling was mostly performed in the US, over urban areas and restricted to episodic conditions, and focussed primarily on O_3. Also hybrid approaches were studied, as well as particle in cell methods, see Sklarew e.o. (1971). Early papers on both Eulerian and Lagrangian modelling are by Friedlander and Seinfeld (1969), Eschenroeder and Martinez (1970) and Liu and Seinfeld (1974).

A nice overview of long-range transport modelling in the seventies was presented by Johnson (1980) at the tenth ITM.

The next obvious step in scale is global modelling of the complete troposphere. The first global models were 2-d models in which the global troposphere was averaged in the longitudinal direction, see Isaksen, 1978. The first 3-D global models were developed by Peters, 1979, see also Zimmermann, 1988.

AIR POLLUTION MODELLING FROM 1980-2000

Analysing the studies on air pollution modelling over the last two decades it can be stated that although many studies have been performed, no real new inventions have taken place. In other words, the basic modelling-tools were developed and ready for use by 1980. Obviously the modelling-tools have been substantially improved over this period, a short overview is presented below. The experience in air pollution modelling has increased substantially, much has been gained by taking new developments in other fields into account as in meteorology and atmospheric chemistry. Also the increase in computerpower and-storage has made more complex models possible, and the -coloured-new presentation techniques made the analysis of the overwhelming amount of calculated data possible. The proceedings of the ITM's over these years give a nice and rather complete overview of the progress made.

(The fact of the missing mile-stones after 1980 can also be caused by the short distance to these years, and the fact that the author was part of the process. It is also tempting to think about a relation with the economical cycles of Kondratief, or more likely with our current state of post-modernism. One element of post-modernism is the lack of clear issues and mile-stones, as is for example expressed by Fukuyama (1992) in ' The end of history'. See also Derrida (1991,1992) and Finkelkraut (1996))

A very comprehensive overview concerning Eulerian grid modelling has been presented by Peters e.a., 1995. The concept given in this paper is that models have gone from first generation models -with relatively simple chemistry- to second generation models with expanded chemistry, dry and wet deposition and improved numerics The third generation models contain interactions among chemistry, deposition and meteorology. Current state-of-the-art models are such third generation models. An example is the models-3 system, Dyun & Ching (1999).

This overview by Peters shows the substantial improvement since 1980, but also show that the basic modelling tools were in place by 1980.

Modelling at local scales, point, line and area-sources has improved by taking the new parametrisations of the boundary layer, Monin-Obukov lenght scale and friction velocity into account. This has led to a more physical description of the plume spread than the empirical stability classification of the Pasquill-Gifford type. See for the research on the convective boundary layer Deardorff (1970), Willis and Deardorff (1978). Also, instead of classification into specific wind-directions and stability classes for the calculation of yearly averaged and higher percentile concentrations, more and more hour-by-hour calculations are performed. In 1991 a European initiative started with the aim of harmonisation of atmospheric local scale dispersion models. The 6th international harmonisation workshop was held in Rouen, France in October 1999.

Modelling at urban scales shows a wide variety of approaches. Taking a complete urban-industrialised area into account of say 20 x 20 km^2 upto 100 x 100 km^2 is often done by using an Eulerian grid model with a finest grid-resolution of 1 x 1 km^2 (this lower limit is determined by the Taylor-diffusion limit). Provided the meteorological input, the emissions etc are available, modelling the overall air pollution field over the city is possible. The difficulty arise for modelling concentrations at the street-level, under the influence of local obstacles and buildings. In such cases often empirical approaches are used, or parametrisations based on analyses of field data or results obtained in atmospheric boundary layer windtunnels. Gaussian plume type models are used, dedicated streetcanyon models have been developed, attempts are made to use CFD-modelling. Which approach is most suited depends completely on the situation under consideration.

Modelling at regional to continental scales has shown a number of developments over the last two decades. The episodic photo-oxidant models show a development to modelling over longer time scales on an hour-by-hour basis. Models for ozone and sulfur, which were restricted to the boundary layer and the lower troposphere, now include also the free troposphere and the stratosphere. The species considered now include also aerosols, 'full' aqueous phase chemistry, and also species like heavy metals and pesticides. There is also a tendency to go from using diagnostic meteorological data as input to the use of prognostic meteo-data, and associated with that to go from off-line modelling to on-line modelling with respect to meteo-input. The same tendency holds for going from off-line emission input including time and temperature dependencies to on-line emission handling in which the time and temperature dependencies are treated inside the model. These remarks for continental scale models hold to a large extent also for global models.

There is also a clear tendency to go from Lagrangian modelling to Eulerian modelling. There is of course a place for both model approaches, and which model is best suited depends on the situation. In the case of an analysis of observations at a specific location, or in case of the determination of the impact of a restricted source area, Lagrangian-trajectory models are the obvious choice. In general however, Eulerian grid models are the better modelling approach. The real world, and the meteorological models are 3-D Eulerian. The concept of an airparcel that keeps its identity is questionable. Nearly all observations are Eulerian observations, and Eulerian models give a coherent and consistent overall concentration pattern in one calculation. And over larger areas Eulerian models are more computational efficient for the same spatial concentration pattern.

Modelling studies over the last 20 years have been used extensively for policy applications. The effecivity of proposed abatement strategies were determined, and numerous sensitivity studies were performed, especially on emissions. Models were also used to perform scenario studies, and were used as part of integral assessment studies. But also science-oriented studies were performed to determine for example the relative contribution of different processes to concentration levels, to assess the impact of vertical exchange, to analyse observed trends, etc.

Modelling studies often play the role to integrate knowledge of different areas as atmospheric chemistry, observations, meteorology, etc. A nice example of the use of models in both science and policy can be found in Borrell e.a. (1997).

As has been stated it is difficult to define clear milestones during the last two decades because to a large part the research was further development based on 'milestones' made before 1980. However, two items could be mentioned.

Model evaluation has been part of model application and development always. From about 1990, attempts have been made to come beyond the simple comparison of observed versus calculated concentrations on a graphical basis, which nearly always resulted in 'reasonable agreement'.

Attention is given to evaluation of separate processes, and evaluation of the complete model. The spatial representativeness of the observations is addressed, and also the non-linearity problem and the possibilities of multi-component testing. Also model evaluation protocols are tried out. There is a clear drive to come to a coherent quality assurance/quality control of air pollution models.

The second item is the development of data-assimilation techniques which can handle non-linear chemistry. Data-assimilation, often nudging, is common practice in meteorology for linear processes, but is under development for non-linear chemistry like ozone formation. Two approaches are followed, extended Kalman filtering (van Loon, 1997) and adjoint 4-D var (Elbern, 1997).

Both methods make it possible to integrate observations and calculated concentrations from models. By data-assimilation, observations can be used as input to the model system, by which the model system will adjust the chosen parameters with their given noise in such a way that the model results will be closer to the observations then in the case of the base-run without data-assimilation. Further development of data-assimilation has a large potential to make optimal use of the knowledge contained in both observations and modelling.

Both items, model evaluation and data-assimilation, might lead to the possibility that the current models, with their high degree of sophistication, will be used to their full extent. At the moment only limited use is made of the large capabilities of air pollution models. And both items might also lead to a better insight into the inherent limitations of the models and their limit to predict concentrations with a certain degree of reliability.

Although, as has been said in the introduction, that modelling is great fun, modelling can also be dangerous. In case models are used without sufficient knowledge about their behaviour and limitations, large mistakes can be made. Models that are based on insufficient science can led to erroneous results. Sometimes, modellers have a greater believe in their model results then they can justify. Modelling is great fun, but finally modelling is just a tool with the aim to improve our scientific knowledge and to support application studies. For the fun and danger of modelling, the book by Killus, 1985, is recommended.

AIR POLLUTION MODELLING ASPECTS BEYOND THE MAIN STREAM OF ACTIVITY

A number of interesting aspects of modelling have not been addressed in the above given overview. They will be listed here.

Statistical and empirical modelling

Analysis of observed concentrations , often including the meteorological observations leads to empirical and statistical relations. These relations have as main purpose to complete the observations, or in other words to interpolate between observations. In this way a more complete picture of the air pollution situation is made. Modern techniques like neural networks and fuzzy logic have the same purpose, and can be viewed as modern methods of statistical and empirical modelling. In the definition of air pollution modelling given in the beginning of this paper: to make a functional relation between emissions and concentrations, statistical and empirical modelling should not be considered air pollution modelling in the strict sense.

Atmospheric boundary layer windtunnel modelling

Windtunnels have been used since about 1970 to study local dispersion phenomena under the influence of obstacles. Windtunnels are capable of accurate modelling of flow and dispersion in highly complex situations were mathematical modelling is not possible. Windtunnels are frequently used upto this moment for such specific situations.

Heavy gas dispersion

The research of dispersion of gases heavier than air started around 1970 with the increased use of liquid LPG and LNG. Mathematical models have been developed based on field trials and windtunnel experiments. This type of research has often been addressed at ITM's.

Interaction of turbulence and chemistry

In the situation that the chemical time scale and the turbulent time scale are of the same order of magnitude, an interaction, correlation between turbulence and chemistry will occur. This situation takes place by chemical reactive plumes and by dry deposition processes, both mainly in the case of NO-NO_2-O_3. Over the years a number of studies have been performed into this aspect. The current conclusion is that this phenomenon is only relevant in a limited number of cases.

Large eddy simulation-LES-and Computational fluid dynamics-CFD

Both LES and CFD have been further developed and used over the years to address rather local dispersion phenomena. Although studies are still under way, a major breakthrough has not been made (yet).

CONCLUSIONS

From the above described attempt to define the milestones in air pollution modelling over the last century, the following milestones appeared:

- 1895	Reynolds number
- 1921	Taylor diffusion theory
- around 1930	Diffusion in the atmosphere, Richardson, Sutton, Bosanquet
- around 1955	Dispersion from a point source, Smith, Gifford, Pasquill-Prairie grass experiment
- 1962	Pasquill: Atmospheric diffusion
- around 1965	Further development of the Gaussian plume model and the importance of mixing height, Högstrom, Turner, Briggs, Moore, Klug, Holzworth
- 1967	First issue Atmospheric Environment
- 1971	First ITM
- around 1975	long range transport models: Lagrangian: Rohde, Eliassen, Fisher Eulerian, Reynolds, Roth, Seinfeld Shir, Shieh, Egan, Carmichael
- 1978/1979	2-D global model: Isaksen, Rohde, 3-D global model: Peters
- around 1980	Basic modelling tools established
- after 1980	Gradual improvements, overview Peters 1995
- after 1990	Modern model evaluation studies
- around 1995	Data-assimilation of non-linear chemistry

Acknowledgement

I would like to thank all colleagues for valuable suggestions and additions, especially Phil Roth.

REFERENCES

Batchelor, G.K., 1950, 'The application of the similarity theory of turbulence to atmospheric diffusion' *Quart. J.R.Met.Soc.* 76:133.

Beryland, M.Y., 1975, Contemporary problems of atmospheric diffusion and pollution of the atmosphere. Gidrometezdat, Leningrad, translated into english by NERC, US EPA.

Boris, J.P., and Book, D.L., 1973, 'Flux-corrected transport.I.SHASTA, an algorithm that works' *J.Comp.Phys.* 11:38.

Borrell, P., Builtjes, P.J.H., Grennfelt, P., Hov O., 1997, Photo-oxidants, acidification and tools: policy applications of Eurotrac results. Springer.

Bosanquet, C.H., and Pearson, J.L., 1936, 'The spread of smoke and gases from chimneys' *Trans. Faraday Soc.* 32:1249.

Briggs, G.A., 1965, 'A plume rise model compared with observations' *J.Air Poll. Control Association* 15:433.

Carmichael, G.R., and Peters, L.K., 1979, 'Numerical simulation of the regional transport of SO_2 and sulfate in the eastern United States, Proc. 4 th *Symp. on turbulence, diffusion and air pollution,* AMS 337.

Chamberlain, A.C., 1953, 'Aspects of travel and deposition of aerosol and vapour clouds' A.E.R.E., HP/R 1261, H.M.S.O.

Church, P.E., 1949, 'Dilution of waste stack gases in the atmosphere' *Ind.Eng. Chem.* 41:2753.

Deardorff, J.W., and Willis, G.E., 1975, 'A parameterization of diffusion into the mixed layer' J. *Appl. Met* 14:1451.

Deardorff, J.W., 1970, Convective velocity and temperature scales for the unstable planetary boundary layer and for Rayleigh convection. J. Atm. Sci. 27, 1211-1213.

Derrida, J., 1991, L'autre Cap, Paris.

Derrida, J. 1992, Donne la Mort, Paris.

Dyun, D.W. and J.K.S. Ching, 1999, Science algorithm of the EPA Models-3 Cummunity Multiscale Air Quality (CMAQ) Modelling System., EPA-Dep./600/R-99/030.

Egan, B.A., Rao, K.S., and Bass, A., 1976, 'A three dimensional advective-diffusive model for long range sulfate transport and transformation' *7 th ITM*, 697, Airlie House.

Elbern, H. e.a., 1997, 'Variational data assimilation for tropospheric chemistry modelling' *J. Geoph. Res.* 102:15967.

Eliassen, A., and Saltbones, J., 1975, 'Decay and transformation rates of SO_2 as estimated from emission data, trajectories and measured air concentrations' *Atm. Env.* 9:425.

Eschenroeder, A.Q. and J.R. Martinez, 1970, "Mathematical Modelling of Photochemical Smog", American Institute Aeronautics and Astronautics (Proceedings), Eight Aerospace Sciences Meeting, New York, Jan 19-21.

Finkelkraut, A. 1996, A L'Humanity perdu, Paris.

Fisher, B.E.A., 1975, 'The long range transport of sulfur dioxide, *Atm.Env.* 9,:1063.

Friedlander, S.K. and J.H. Seinfeld, 1969, "A Dynamic Model of Photochemical Smog", Environ., Science Technol., 3, 1175 (1969).

Fukuyama, F. 1992, "The end of history", New York, Free Press.

Gifford, F.A., 1957a, 'Relative atmospheric diffusion of smoke plumes' *J. Met.* 14:410.

Gifford, F.A., 1957b, 'Further data on relative atmospheric diffusion' *J. Met.* 14:475

Haugen, D.A., 1959, 'Project Prairie grass, a field programme in diffusion' *Geographical research paper* 59, vol III, G.R.D.A.F.C., Bedford, Mass.

Hay, J.S., and Pasquill, F., 1957, 'Diffusion from a fixed source at a height of a few hunderd feet in the atmosphere' *J. Fluid Mech.* 2:299.

Högstrom, U., 1964, 'An experimental study on atmospheric diffusion' Tellus, 16:205.

Holzworth, G.C., 1967, 'Mixing depth, wind speed and air pollution potential for selected locations in the U.S.A.', *J.Appl.Met.* 6:1039.

Inoue, E., 1950, 'On the turbulent diffusion in the atmosphere' *J.Met.Soc.* Japan, 28:13.

Isaksen, I.S.A., and Rohde, H., 1978, 'A two-dimensional model for the global distribution of gases and aerosol particles in the troposphere' *Rep. AC*-47, Dep. of Meteor. Univ Stockholm, Sweden.

Junge, C.E., 1963, Air chemistry and radioactivity, *academic press*, New York, London.

Johnson, W.B., 1980, 'Interregional exchange of air pollution: model types and applications' *10 th ITM*, 3, Amsterdam.

Killus, J., 1985, 'Sunsmoke', *Berkely Publ. Group.*

Klug, W., 1968, 'Diffusion in the atmospheric surface layer: comparison of similarity theory with observations' Quart. *J.R. Met.Soc.* 94:555.

Kolmogorov, A.N., 1941, The local structure of turbulence in incompressible viscous fluid for very large Reynolds numbers, C.R. Acad. Sci. U.R.S.S., 30, 301.

Lamb, R.G. and J.H. Seinfeld, 1973. Mathematical modelling of urban air pollution - General theory. Envir. Sci. Technol. 7, 253-261.
Levy, H., 1971, 'Normal atmosphere: large radical and formaldehyde concentrations predicted' *Science* 173:141.
Liu, M.K. and J.H. Seinfeld, 1974, Ön the Validity of Grid and Trajectory Models of Urban Air Pollution", Atmos. Environ., Vol. 9, pp. 555-574.
Loon, M. van, and Heemink, A.W., 1997, 'Kalman filtering for non-linear atmospheric chemistry models: first experiences CWI-Rep. *MAS-R9711*, March.
Monin,A.S., 1959, 'Smoke propagation in the surface layer of the atmosphere' Atmospheric diffusion and air pollution, ed. Frenkiel and Sheppard, advances in *Geophysics*, 6:331, Academic Press.
Monin, A.S., 1955, The equation of turbulent diffusion, Dokl. Akad. Naak., 105, 256.
Moore, D.J., 1967, 'Physical aspects of plume models' *Atm.Env.* 1:411.
Nieuwstadt, F.T.M., and Dop, H. van, 1982, Atmospheric turbulence and air pollution modelling, D.Reidel Publish. Comp.
Obukov, A.M., 1941, Energy distribution in the spectrum of turbulent flow, Izv. Akad. Naak. Geogr. I Gesfiz, 5, 453.
Ogura, Y., 1959, 'Diffusion from a continuous source in relation to a finite observation interval' Atmospheric diffusion and air pollution, ed. Frenkiel and Sheppard, advances in Geophysics 6,149, Academic Press.
Pasquill, F., 1962, Atmospheric diffusion *Van Nostrand*, New York.
Peters, L.K., and Jouvanis, A.A., 1979, 'Numerical simulation of the transport and chemistry of CH_4 en CO in the troposphere', *Atm.Env.* 13:1443.
Peters, L.K., e.a., 1995, 'The current state and future direction of Eulerian models in simulating the tropospheric chemistry and transport of trace species: a review', *Atm.Env.* 29,2:189.
Record, F.A., and Cramer, H.E., 1958, 'Preliminary analysis of Project Prairie grass diffusion *measurements' J.Air Poll.Cont.Ass* 8:240.
Reynolds, S., Roth, P., and Seinfeld, J., 1973, 'Mathematical modelling of photochemical air pollution' *Atm.Env* 7.
Reynolds, O., 1895, 'On the dynamical theory of incompressible viscous fluids and the determination of the criterion' Phil. *Transactions of the Royal Soc. of London.* series A, 186:123.
Richardson, L.F., and Proctor, D., 1925, 'Diffusion over distances ranging from 3 to 86 km'. *Memoirs of the R.Met. Soc*,vol 1:1.
Roberts, O.F.T., 1923, The theoretical scattering of smoke in a turbulent atmosphere Proc. Roy. Soc. A, 104, 640.
Rohde, H., 1972, 'A study of the sulfur budget for the atmosphere over northern Europe' *Tellus*, 24:128.
Rohde, H., 1974, 'Some aspects of the use of air trajectories for the computation of large scale dispersion and fallout patterns' Adv. in *Geophysics* 18B:95, academic press.
Roth, P.M., P.J.W. Roberts, J.K. Liu, S.D. Reynolds and J.H. Seinfeld, 1974, "Mathematical Modelling of Photochemical Air Pollution - Part II. A Model and Inventory of Pollutant Emissions, Atmos. Environ., Vol. 8, pp. 97-130.
Seinfeld, J.H., 1986, 'Atmospheric chemistry and physics of air pollution', *John Wiley & Sons.*
Shir, C.C. and L.J. Shieh, 1974, A generalized urban air pollution model and its application to the study of SO_2-distribution in the St. Louis Metropolitan area, J. Appl. Met. 19, 185-204.
Sklarew, R.C. e.o., 1971, A particle-in-cell method for numerical solution of the atmospheric diffusion equation and application to air pollution problems; Systems, Science and Software, Ca-Reg 35R-844, Vol I.
Smith, F.B., 1957, 'The diffusion of smoke from a continuous elevated piont source into a turbulent atmosphere' *J.Fluid Mech.* 2:49.
Stewart, N.G. e.a., 1958, 'The atmospheric diffusion of gases discharged from the chimney of the Harwell Pile' *Int J.Air Poll.* 1:87.
Sutton, O.G., 1932, 'A theory of eddy diffusion in the atmosphere' *Proc. Roy. soc. A*, 135:143.
Taylor, G.I., 1915, 'Eddy motion in the atmosphere' Phil. *Transactions of the Royal Soc. of London.* Series A, 215:1.
Taylor, G.I., 1921, 'Diffusion by continuous movements' *proc. London Math. soc.* 20:196.
Thomas, M.D., e.a., 1949, 'Dispersion of gases from tall stacks' *Ind. and En. Chemistry*, 41:2409.
Turner, D.B., 1964, 'A diffusion model for an urban area' *J.Appl. Met.* 3:83.
Warneck, P., 1988, 'Chemistry of the natural atmosphere', *Int.Geoph.Series* 41, Academic Press.
Willis and Deardorff, 1978, A laboratory study of dispersion from elevated source within a modelled convection mixed layer, Atm. Env. 12, 1305-1311.
Zimmermann, P.H., 1988, 'Moguntia: a handy global tracer model' *17 th ITM.* 593, Cambridge.

DISCUSSION

S. T. RAO: You have done an excellent job of reviewing the history of air pollution modelling and its applications. I think the review would not be complete without recognizing the significant contributions of Russian scientists, notably Kolmogorov (1941), Monin (1955) and Obukov (1941), etc.

Also I think that data-assimilation is a major milestone since it has contributed towards improving modelling simulations.

P. BUILTJES: Thank you for mentioning the Russian meteorologists. The reason for not mentioning them explicitly was that I tried to restrict myself to air pollution modelling, and give less attention to meteorology. However, I agree that they made important contributions with a clear impact on air pollution modelling.

Further, I fully agree with your remark concerning data-assimilation. In my opinion data-assimilation will grow in importance very quickly.

J. WEIL: A major milestone missed in your overview is the breakthrough in understanding the convective boundary layer by numerical simulation (LES), Deardorff (1970), convective tank experiments, Willis and Deardorff (1978) and field observations. This led to an overhaul of the Gaussian Plume Model with application of convective scaling of dispersion (convective velocity scale and convective boundary layer depth), inclusion of skewness, vertical velocity etc.

This information is now used in state-of-the-art short range dispersion models all over the world. Furthermore, this understanding has been used to improve the vertical exchange in long-range transport models.

P. BUILTJES: Thank you for this useful addition to my presentation. I indeed only mentioned very briefly the improvements made in boundary layer theory and its impact on local modelling. As in my answer to the previous speaker, my reason was an attempt to restrict myself and not put emphasis on meteorology. But I agree that these developments had a major impact on local modelling. However, should I then also have given attention to aspects as stratosphere-troposphere exchange which have a major impact on continental and global scale modelling?

A. ELIASSEN: Isn't it so that the major basic understanding and tools in air pollution modelling were developed before 1980, a period in which you very easily found some basic milestones? After 1980 people have concentrated on filling gaps in the understanding and refining modelling tools. In such a period it is more difficult to find mile-stones. In other words, this science has gone through the normal typical time development and has now reached its maturity.

P. BUILTJES: I basically agree with your comment. However, would this mean that for the future air pollution modelling is 'just' application, and no further development is needed?

B. E. A. FISHER: The problem between 1980-2000 has been that models have become more complex, but the measurements have not matched this, despite the development of new monitoring techniques. Is this the reason it is hard to choose a mile-stone for this period? There is a fundamental mismatch between prediction and validation and this needs to be corrected, and this correction may be the essential mile-stone for modelling post-2000.

P. BUILTJES: It is true that models can calculate much more quantities then can be measured, and that this hampers further developments. Adequate model validation is essential for the future, and should be carried out right now.

R. BORNSTEIN: Could we include as advances in the 80's:

- Met observations via profilers, satellites, etc.
- Use of model outputs for urban planning

P. BUILTJES: I do see a major advantage in satellite observations, especially of chemical species in the troposphere, as they are becoming available. However, their accuracy is still limited. And the role of models in urban planning and more general in integrated assessment studies is of growing importance.

R. SAN JOSE: First question is similar to the one asked by Bornstein about the importance of remote sensing on AQM. The second question is related to the importance of Internet revolution on AQM output to provide real-time and on-line applications in air pollution to the citizens. I think this is a major step forward in AQM development

P. BUILTJES: Yes, if we learn how to use and assess satellite observations, and if we use internet in an tailored way to provide the information really needed and wanted, we can benefit much from these new developments.

WHY AND HOW TO HARMONISE AIR POLLUTION IMPACT ASSESSMENT MODELS?

Jan G. Kretzschmar

Vito
Flemish Institute for Technological Research
Boeretang 200, B-2400 Mol, Belgium
Email : kretzscj@vito.be

INTRODUCTION

Within the European Union (EU) the Council Directives 85/337/EEC and 97/11/EC (European Union, 1985 and 1997), and the corresponding national and/or regional laws in the different Member States (EIA Centre, 1998), regulate the environmental impact assessments (EIAs) required to evaluate the effects of certain public and private projects on the environment.

Inherent to an environmental impact assessment study is the fact that one tries to forecast the future situation, taking into account the present one and the characteristics of, or alternatives for the planned private or public works. The most appropriate way to tackle this problem is trying to simulate the present and future situations by means of a (computer)model. According to the Council Directive on Ambient Air Quality Assessment and Management (frequently indicated by Framework Directive) "Assessment of air quality means any method used to measure, calculate, predict or estimate the level of a pollutant in the ambient air" and therewith air pollution models are recognised as assessment tools (de Leeuw et al., 1997).

Since the early days hundreds of different atmospheric transport and dispersion models have been developed, eventually validated and certainly applied in practice. The obvious consequence is that – within a general regulatory framework without any modelling guidance or control at the EU-level – major discrepancies between the simulated results and the real impact of new sources of air pollution do occur. As 'open market' and 'free competition' are basic rights and obligations within the European Union, it is obvious that not harmonised regulatory models can lead to unfair competition when the location of a new industrial plant or zone has to be decided, as well as to insufficient local air quality management. This problem has up to now not been solved within the EU inventorizing and reviewing steps, despite the fact that model developers and users initiated the "European Initiative on Harmonisation of Regulatory Air Pollution Models" already in 1991. Up to now industry too has shown little interest in the harmonisation topic. This is nevertheless somewhat easier to understand, as explained by M. Tasker (1997).

Air Pollution Modeling and Its Application XIV, Edited by
Gryning and Schiermeier, Kluwer Academic/Plenum Publishers, New York, 2001

SITUATION IN THE EU AND THE US

In his paper "Creating harmony from dissonance in model development: a business approach" at the 5th Harmonisation Conference, Richard H. Schulze (1998) summarises the situations in the EU and the US as follows:

"Well over one hundred governmental bodies regulate air pollution in Europe. In some countries, the regulations are quite centralised, while in others bodies at the provincial, regional or canton levels perform regulation. Individual countries, abetted by industry and labour, are generally reluctant to cede on air quality matters much authority to a central authority such as the European Union. This highly fragmented situation impairs the development of both a uniform approach to air quality regulation, as well as mutually acceptable models for assessing air quality.

The United States, in contrast, has a moderately strong central government and thus most regulations are developed at the national level by the Environmental Protection Agency (EPA) and then delegated to the states. The U.S. EPA models are universally accepted by industry and agencies throughout the country. Still, it took some 15 years after EPA was established for models developed in states like Texas to be supplanted by the EPA models."

Already by the end of the eighties some members of the European modelling community realised that indeed the situation within the European Union was far from what it had to be. Therefore they started the "Initiative on Harmonisation within Atmospheric Dispersion Modelling for Regulatory Purposes" on the 13th of June 1991 at a meeting in the Institute of Prospective Studies of the Joint Research Centre in Ispra, Italy. This was clearly a bottom-up approach on a voluntary basis, without any formal organisational structure – except an International Steering Committee - without any regular external funding but with the following set of ambitious objectives :

- foster the development of "new generation models" based on a more justifiable parameterisation of the dispersion processes, and turn them into practical, generally accepted tools fitting the various and evolving needs of decisionmakers ;
- approach other organisations with direct interests in harmonisation of ATDMs, such as major industrial groups and national or regional governments ;
- increase the pressure and lobbying at the European level to initiate some EU-wide organised action with respect to the harmonisation of models (similar to what was already started some 20 years ago with respect to the intercomparison, harmonisation and standardisation of ambient air pollution monitoring strategies, techniques and data evaluation and interpretation) ;
- work out transnational project proposals to be submitted to DG XII of the Commission, or to the EC's Human Capital and Mobility Programme ;
- and start some kind of visible actions, having a high probability of success and generating a (much more) widespread recognition of, and support to the need of harmonising regulatory models in Europe.

RESULTS OF THE HARMONISATION INITIATIVE

The undoubtedly most visible results over the past 7 years of the Harmonisation Initiative are the six international workshops/conferences successively organised by different members of the Steering Committee (Olesen, 1992; Cuvelier, 1994; Cosemans, 1995; Kretzschmar, 1997; Bartzis, 1998):

From the papers published in the corresponding proceedings, it can be deduced that, as a function of time, there was a clear-cut evolution in the ambitions, or maybe in the sense for reality of the organisers. Indeed the whole Harmonisation Initiative started with the focus on "next generation models", and "harmonise those advanced (and better ?) models during their development and testing phases" in order to avoid the problems one was facing then with what was developed over the past 20 years. But while gradually moving from the first workshop into the fifth international conference, "real life situations", "models frequently used for EIAs" and "officially approved, or at least accepted models" were gaining more and more attention, together with "validation data sets", "evaluation procedures", "meteorological data and their pre-processing", "intercomparisons" and "urban problems". Many practical exercises and intercomparisons illustrated anyway that the most frequently used impact assessment models were still very much of the type already seen before, and that the deviations in the outputs were still frequently much larger than what was acceptable, seen the importance of the conclusions and their consequences.

A second deliverable, generated as an essential tool to make reliable intercomparisons of model validations within the context of the common exercises organised for the workshops, was a model validation toolkit with validated data of various tracer experiments. Presently available in the toolkit – already distributed to more than 150 research groups – are Kincaid, LillestrØm, Copenhagen and Indianapolis data. Full details of the data sets, and the protocols to evaluate the results, are given by H.R. Olesen (1998) and on the website of the Harmonisation Initiative (htpp://www.dmn.dk/AtmosphericEnvironment/harmoni.htm). While this toolkit has proven merits, there are also some weak points, with as major ones :

- the data sets, or the underlying tracer releases, did not have "harmonisation of models" as major objective. So there is a risk that they are not optimal for that purpose ;
- the data sets were in general not obtained from experiments on European sources, projects and/or sites representative for the regulatory purposes aimed at within the context of the EU Directives.

The model validation toolkit is certainly useful to intercompare models, to detect weak points in a given model and to eventually correct them if those weak points (or errors) are due to poor physics, mathematics or software. What has to be avoided as an abuse is using the data sets to calibrate or trim a given model by some kind of best-fit trial and error approach.

As third result – at the level of the European Commission this time – the Harmonisation Initiative was one of the driving forces to start in 1994 the COST 710 action on "Harmonisation in the pre-processing of meteorological data for dispersion models". Fifteen different countries participated and the final report was published in 1998. Its follow-up, the COST 715 action : Meteorology Applied to Urban Pollution Problems, started by the end of the same year.

Last but, in the (very ?) long run hopefully not least, is the result reached by the Harmonisation Initiative at the level of the European Environment Agency. Indeed in the working programme of EEA's European Topic Centre on Air Quality (ETC-AQ) a specific project on "Harmonisation in the use of models for ambient air quality and pollution dispersion/transport" was initiated mid 1994 (EEA, 1994) with the following aims :

MA3-1 : Review requirements on models and model applications ;
MA3-2 : Report on state of art, needs and trends ;
MA3-3 : Establish documentation centre and toolkits for testing relevant models ;
MA3-4 : Report on guidance in selected areas and in application of relevant models.

While very promising, and maybe too ambitious, it looks like those aims are only partly realised by 2000. At http://www.etcaq.rivm.nl we find some descriptive reports and a model documentation system – developed under MA3-3 but without toolkit for testing – with the description of 91 different models : very specific ones and very general ones, simple and sophisticated ones, some used for regulatory purposes but many not, ...
ETC-AQ correctly warns the users on the website as follows :
"There are several sources of inhomogeneities in the Model Documentation System, each of which may constitute a serious loss in comparability between individual models :

- the submissions by the individual modellers differ significantly regarding the information provided, especially concerning model limitations ;
- remarks related to quality assurance/quality control (QA/QC)of the models and the fact that their results are solely based on information provided by the modellers ;
- consult the links to the own WWW pages of the modellers for additional information."

It must thus be concluded that this is maybe the start of a model documentation system, that could be useful within the context of a EU-wide and EEA supported action on the harmonisation of air pollution models used for regulatory purposes. In its present state it is marginally contributing with some risk to be contraproductive as people could claim that their model is "The Model" or "at least as good as all other models" or "recognised by the EEA" or simply that it has been verified/certified because their model is in the EEA's Model Documentation System on the Web (despite the following ETC-AQ and EEA disclaimers on the website: there can be no responsibility of either ETC-AQ or the European Environment Agency for confusions or misinterpretations which may result from (the previously mentioned) inhomogeneities).

POSSIBLE SOLUTIONS

Learning from successful examples, the solution looks very simple at a first glance. The EPA has solved the problem in the US, so take the same approach for the European Union. Seen EPA's experience one can hope to do this within a reasonable number of years. But taking into account the more complex situations in the European Union with its national Member States, each having their own governmental structure, traditions and laws, it could turn out to take 10 to 15 years. Pragmatic approach but there is a fundamental difference between the US EPA and the European Environment Agency (EEA) as relatively new institution under the European Union. The mandate of the EEA is to provide the Member States with information on the environment. Thus, it is not the European Union's counterpart of the US EPA, which is in possession of regulatory power. As conclusion it can be stated that this approach cannot solve the problem.

A second possible approach is described in R.H. Schulze's (1998) paper. He states:

1. Define first of all a hierachy of models in order to efficiently and effectively use the scarce resources, both computer and human:
   - "screening models" for the vast bulk of sites;
   - "advanced models" for multiple source configurations or sites for which the "screening models" give excessive concentrations;
   - "refined models" to improve accuracy of predictions and to check the previous ones;
   - "site specific models" if not covered by the previous ones.
2. The development of a "European refined model" will require the pooling of all available knowledge with respect to more than 20 issues for which algorithms will

have to be tested and validated. This is out of the scope of a single team or a single organisation, so pool the 10 to 15 best European institutes, distribute the tasks with respect to the different issues according to each's proven expertise and manage the developments within a well-defined and systematic way.

3. Do not forget the quality of the input data – as well meteorological as emission data – or the monitoring data needed to validate each of the many nested models in this "European refined model". Problems also highlighted by M. Tasker (1997, 1998).

The creation of such a universally acceptable and applicable model, involving the development and validation of dozens of subroutines each addressing a special topic, would possibly deserve some profounder analysis and discussion, although the feasability of the whole approach seems questionable, if not unrealistic.

In what preceeds we mainly tried to solve the harmonisation problem from the point of view of the modellers, looking for a solution to their worries. Is this the only way to go ? Certainly not. So let us therefore return to the real problem namely "evaluate the environmental impact of certain public and private projects "as stated in the 1985 and 1997 EU Directives, and reverse the harmonisation reasoning by starting from "the projects" towards the "modelling tools" we need to "assess the project's impact" upon the air pollution within a "regulatory framework". This leads to the following tentative scheme of successive steps.

1. *Inventorise the projects* one is dealing with, as well from a conceptual (EU Directives), as from a practical, application oriented point of view (the EIAs already carried out since the Directives became law in the different Member States). The characteristics of the projects and the assessments are the guidelines for the inventory, without nevertheless forgetting to note the model(s) used in the respective EIAs.
2. As this inventory could result into a too broad and diverse collection of projects, *reduce the inventory as a function of the application type e.g. stick to regulatory and compliance purposes.*
3. Look for simularities in the projects in order to *categorise them* and to obtain a limited number of *subsets*. Apart manageability, information on frequency of occurrence and level of priority are generated too.
4. *Define the set of physical parameters relevant to air pollution* transport, dispersion and deposition, and thus *to be taken into account* in the assessment process of the projects such as (Moussiopoulos et al., 1998):
   - meteorology
   - topography and orography
   - spational scale and detail
   - urban or not
   - type of pollutant(s)
   - source characteristics
5. Combining 3. and 4. leads to a *matrix of "type of project" versus governing "physical (and chemical) parameters"*. Look for possible combinations of the matrix elements in order to reduce their number without mixing members of different families (projects) into inconsistent (modelling) populations.
6. *Define the modelling requirements for each* of the remaining *matrix elements* and decide upon the level of complexity of the model(s) needed to assess the air pollution impact of the project: screening-, advanced-, refined- or site specific model.
7. Now turn to the *model validation requirements.* Is a validation required and to what extent ? What data are needed ? Are these data available ? How many independent sets are available ? What is the quality of the data sets ?

8. As the models used in the different projects were already identified in the procedure's first step, each of the final matrix elements resulting from step 5 contains a family of projects and a corresponding population of models. The logical step 8 is thus to *evaluate* those *models w.r.t. the modelling requirements* defined in step 6 *and the validation requirements* specified under 7.
9. Simultaneously with 8 the issue *"reference model"* can be tackled. Is there already one ? Does one need one ? Is the candidate "reference model" present in the available population of models ?
10. With or without "reference model" *intercomparisons* of the models within one matrix element are now possible, and the *harmonisation procedure* can start for as far (independent) *data* (sets) are available to meet the *"model validation requirements"*.

Documentation-, information distribution- and updating procedures are necessarily inherent to all steps of the sketched methodology. A methodology which seems tempting, not simple and certainly open for discussion with respect to its scientific value and its feasibility.

CONCLUSIONS

More than 15 years after the publication of the 85/337/EEC Directive on Environmental Impact Assessment reporting, specific actions for the harmonisation of air pollution regulatory modelling started within the 1994-1999 work programme of the European Environment Agency. This was certainly co-triggered by the preparatory work within, and the pressure by the 1991 Initiative on Harmonisation, a voluntary collaboration between leading air pollution modelling teams in the Member States. A Harmonisation Initiative that over the past years helped to pave the way by organising five successful international workshops where the results of model intercomparisons and validations, based on a newly developed and tested "model validation kit", were presented and discussed. Important deviations in the outputs, and consequently major uncertainties with respect to the decision making, proved that the harmonisation of air pollution models, as well the presently used ones as the next generation, is still needed.

Different approaches have already been proposed, each of them having merits and shortcomings. A clearcut strategy and methodology at the European level is nevertheless still missing, and pessimists or disbelievers even call it "mission impossible". Without minimising the barriers and difficulties to overcome in the coming years, the progress over the past years has proven the will and duty of all involved parties to succeed.

REFERENCES

Bartzis, J.G. & Konte, K., 1998, *Pre-prints of the 5th International Conference on Harmonisation within Atmospheric Dispersion Modelling for Regulatory Purposes*, Rhodes 1998, INTRP/Environmental Research Laboratory.

Cosemans, G. & Maes, G. (eds), 1995, *Proceedings of the third workshop on Harmonisation within Atmospheric Dispersion Modelling for Regulatory Purposes*, Mol 1994, Int. J. Environment and Pollution, Vol. 5, Nos. 4-6.

Cuvelier, D. (ed), 1994, *Workshop on Intercomparison of Advanced Practical Short-range Atmospheric Dispersion Models*, Manno 1993, JRC-Ispra, EUR 15603 EN.

de Leeuw, F., Berge, E., Gronskei, K. & Tombrou, M., 1997, Review of the requirements for models and model applications, *Int. J. Environment and Pollution*, Vol. 8, Nos. 3-6, pp. 391-400.

EIA Centre, 1998, EIA Legislation and Regulations in the EU, *EIA Leaflet Series, Leaflet 5*, (http://www.art.man.ac.uk/EIA/EIAC.htm).

European Environment Agency, 1994, *Annual Workprogramme Mid 1994-1995*, EEA/052/94, Copenhagen.

European Union, 1985, Council Directive 85/337/EEC of 27 June 1985 on the assessment of the effects of certain public and private projects on the environment, *Official Journal of the European Communities*, L175, pp. 40-47.

European Union, 1997, Council Directive 97/11/EC of 3 March 1997 amending Directive 85/337/EEC on the assessment of the effects of certain public and private projects on the environment, *Official Journal of the European Communities*, L73, pp. 5-15.

Kretzschmar, J.G. & Cosemans, G. (eds), 1997, *Proceedings of the 4th Workshop on Harmonisation within Dispersion Modelling for Regulatory Purposes*, Oostende 1996, Int. J. Environment and Pollution, Vol. 8, Nos. 3-6.

Moussiopoulos, N., de Leeuw, F., Karatzas, K. & Bassoukos, A., 1998, The air quality Model Documentation System of the European Environment Agency, *Pre-prints of the 5th International Conference on Harmonisation within Atmospheric Dispersion Modelling for Regulatory Purposes*, eds. J.G. Bartzis & K. Konte, INTRP/Environmental Research Laboratory, pp. 31-37.

Olesen, H.R. & Mikkelsen, T. (eds), 1992, *Proceedings of the Workshop on Objectives for Next Generation of Practical Short-Range Atmospheric Models*, RiSØ 1992, DCAR, Roskilde, Denmark.

Olesen, H.R., 1998, Model Validation Kit-status and outlook, *Pre-prints of the 5th International Conference on Harmonisation within Atmospheric Dispersion Modelling for Regulatory Purposes*, eds. J.G. Bartzis & K. Konte, INTRP/Environmental Research Laboratory, pp. 63-70.

Schulze, R.H., 1998, Creating harmony from dissonance in model development – a business approach, *Pre-prints of the 5th International Conference on Harmonisation within Atmospheric Dispersion Modelling for Regulatory Purposes*, eds. J.G. Bartzis & K. Konte, INTRP/Environmental Research Laboratory, pp. 26-30.

Tasker, M., 1997, Gas dispersion modelling harmonisation : an industrial newpoint, *Int. J. Environment and Pollution*, Vol. 8, Nos. 3-6, pp. 241-249.

Tasker, M., 1998, The harmonisation of gas dispersion modelling for industrial applications. A few thoughts for the future, *Pre-prints of the 5th International Conference on Harmonisation within Atmospheric Dispersion Modelling for Regulatory Purposes*, eds. J.G. Bartzis & K. Konte, INTRP/Environmental Research Laboratory, pp. 18-25.

DISCUSSION

P. BUILTJES: Are these minimum requirements for models, or is it still possible to use highly empirical models in the EIA?

J. KRETZSCHMAR: The minimum requirements for EIA models are:

1. Based on sound physics;
2. PC compatible with sufficient speed;
3. Can accept (yearly) series of successive hourly values as input;
4. Give output statistics compatible with Air Quality Standards (or Guidelines);
5. Model ground level concentration as well as deposition values;
6. Are validated on past situations for which measured values are available.

A. ELIASSEN: What is actually the rationale for harmonisation of models? Isn't it rather diversity we want, and isn't it diversity that supports progress?

J. KRETZSCHMAR: The rationale for harmonisation of EIA models is that the same physico (chemical) situation around (a) given source(s) must be modelled in a comparable way by different models pretending that they simulate reality. The same European Air Quality Standards are to be respected (as a minimum) in all the EU Member States, so all EIAs must use comparable tools (as well for monitoring as for modelling). To make models comparable we need at least some harmonisation and intercomparisons.

HOW SHOULD THE PHOTOCHEMICAL MODELING SYSTEMS BE USED IN GUIDING EMISSIONS MANAGEMENT DECISIONS?

S. T. Rao[1,2], C. Hogrefe[1], H. Mao[1], J. Biswas[1], I. Zurbenko[2], P. S. Porter[3], P. Kasibhatla[4], and D. A. Hansen[5]

[1]Department of Earth and Atmospheric Sciences, State University of New York, University at Albany, Albany, New York 12222

[2]Department of Biometry and Statistics, School of Public Health, State University of New York, University at Albany, Rensselaer, New York 12144

[3]Department of Civil Engineering, University of Idaho, Idaho Falls, Idaho 83405

[4]Nicholas School of the Environment, Duke University, Durham, NC 27708

[5]EPRI, 3412 Hillview Avenue, Palo Alto, CA 94303

INTRODUCTION

Emissions-based three-dimensional photochemical modeling systems are the primary tools used in attainment demonstrations for the National Ambient Air Quality Standard (NAAQS) for ozone in the United States (U.S. EPA 1991, 1994 and 1999). These modeling systems take into account the complex effects of meteorology, emissions, transport, chemical transformation and removal processes on surface ozone concentrations. In the past, regulatory applications of these modeling systems entailed simulating a historic episode (typical duration of 2-5 days) of high ozone concentrations, evaluating the results of this "base case" simulation, and repeating the simulation with various "control cases" to determine emission reductions required to meet the ozone NAAQS (U.S. EPA 1994, 1999; Tesche et al., 1996).

This paper illustrates the shortcomings of the episodic-type modeling approach. To this end, we examine ozone predictions from different state-of-the-science modeling systems and those obtained with different configurations of the same modeling system for the ozone episode that occurred July 12 - 15, 1995. These simulations were performed in recent years, addressing different aspects of photochemical modeling (SMRAQ, 1997; Rao et al., 1998; Sistla et al., 1999; Biswas and Rao, 1999; Biswas et al., 2000). This paper discusses the implications of these studies to the regulatory use of photochemical models. After discussing the uncertainties associated with traditional episodic-type ozone modeling, we use spectrally decomposed time series of ozone observations to demonstrate the contribution of processes operating on time scales of 1 day to several weeks to high ozone concentrations, and illustrate that episodic-type modeling does not take properly into account key aspects of the ozone process.

The need for longer-term modeling is further emphasized by the recent results of seasonal photochemical simulations with two different modeling systems (SMRAQ, 1997; Hogrefe et al., 2000a; Hogrefe and Rao, 2000b). In particular, we demonstrate that the agreement between ozone observations and model predictions is best on time scales longer than one day (i.e., the synoptic and baseline time scales) for both modeling systems, and that model-to-model differences are least pronounced on these time scales. Also, we show that the ozone benefits estimated from an episodic-type modeling approach display a large amount of temporal variability (i.e., the estimates are highly dependent on the specific episode modeled) and that only longer-term modeling can reduce this variability. Hence, we conclude that photochemical modeling systems must be applied to time periods longer than the traditional episodic modeling to build confidence in the use of photochemical models for emissions management decisions.

DATABASE

Hourly ozone concentrations from various monitoring stations in the eastern United States were obtained from the U.S. Environmental Protection Agency's (U.S. EPA) AIRS data base.

Photochemical Modeling

Modeling Systems Used. Of the three main components of photochemical modeling systems (i.e., emissions modeling, meteorological modeling, photochemical modeling), this paper focuses on the differences in episodic ozone predictions stemming from differences in the meteorological and photochemical components. Emissions used in all modeling results presented here are largely updated versions of the 1995 emissions inventory compiled by the Ozone Transport Assessment Group (OTAG, 1997). Ozone predictions in this study are obtained from simulations with the RAMS3b/UAM-V modeling system (Walko et al., 1995; SAI, 1995), the SMRAQ modeling system (SMRAQ, 1997) in which MM5 (Grell et al., 1994) is used for meteorological modeling and MAQSIP (SMRAQ, 1997) for air quality modeling, and the MM5/SAQM modeling system (Grell et al., 1994; Chang et al., 1997). In addition, to examine the effects of different meteorological inputs on ozone predictions, UAM-V simulations were performed using meteorological fields derived from simulations with two different versions of the MM5 model and the results are compared with those of RAMS3b/UAM-V. Also, the effects of grid resolution on ozone predictions are assessed by performing three different RAMS3b/UAM-V simulations using the horizontal grid dimensions of 36 km, 12 km, and 4 km.

Model Setup and Simulation Length. For the comparison between the RAMS3b/UAM-V modeling system and the MM5/SAQM modeling system, both models were run with a 12-km inner grid covering most of the eastern United States (Figure 1). Details on the setup of the meteorological models can be found in Kallos et al. (1997) and Zhang and Rao (1999), and the setup of the photochemical modeling is described in Sistla et al. (1999). For the comparison between the RAMS3b/UAM-V modeling system and the SMRAQ modeling system, both modeling systems were run with the horizontal grid cell size of 36 km (Figure 1). Details on the setup of the SMRAQ modeling system can be found in SMRAQ (1997). For the simulations studying the effects of meteorology on ozone predictions, the results from the 12-km RAMS3b/UAM-V simulation were compared to ozone predictions obtained when using the MM5 meteorological fields as described in Zhang and Rao (1999) rather than RAMS3b to drive this simulation. The MM5/UAM-V modeling setup is described in Biswas and Rao (1999). An additional 12-km MM5/UAM-V simulation was carried out using meteorological fields from a different version of MM5; this simulation is described in Biswas et al. (2000). For the investigation of the effect of grid resolution on ozone predictions, a RAMS3b/UAM-V simulation with a 4-km inner grid was carried out and compared to the 12-km and 36-km RAMS3b/UAM-V simulations (Figure 1).

The simulation period for all of these modeling simulations covers the July 12-15, 1995 ozone episode. In addition, the RAMS3b/UAM-V simulations with horizontal grid dimensions of 36 km and 12 km were performed for the time period June 4 - August 31, 1995, and the SMRAQ simulation covered the time period May 15 - September 12. Only predictions for the July 13 -15, 1995 episode from all modeling systems are considered for the analysis of episodic ozone predictions. For the comparison of 36-km RAMS3b/UAM-V and SMRAQ predictions with each other and with ozone observations on different time scales, only observations and model predictions (interpolated to the observational sites using bilinear interpolation) within the 12 km grid of the RAMS3b/UAM-V simulation were analyzed for the time period June 4 - August 31, 1995. In addition to the base case for the seasonal 12-km RAMS3b/UAM-V simulation, the results of a NO_x/VOC emission reduction scenario reflecting a uniform reduction in emissions of magnitudes 50%/25% are discussed here.

Spectral Decomposition of Ozone Time Series

As discussed in Rao et al. (1997) and Hogrefe et al. (1999 and 2000a), seasonal time series of hourly ozone concentrations contain fluctuations on different time scales. Spectral analysis indicates that the single largest forcing in the hourly time series data is the diurnal forcing having a period of 24 hours. Additional frequency bands of interest are the intra-day range (periods less than 12 hours), the synoptic range (periods of 2-21 days), and longer-term (baseline) fluctuations (periods longer than 21 days). We use the Kolmogorov-Zurbenko (KZ) filter (Zurbenko, 1986) to estimate these components. A detailed discussion of the KZ filter along with a comparison to other separation techniques can be found in Eskridge et al. (1997). The ozone time series is log-transformed prior to analysis as in Rao et al. (1997) and Hogrefe et al. (2000a). Therefore, the sum of the components of interest in this study, namely, the intra-day (ID), diurnal (DU), synoptic (SY), and baseline (BL) components, is equal to the logarithm of the original time series:

$$\ln[O_3(t)] = ID(t) + DU(t) + SY(t) + BL(t) \qquad \text{Eq. (1)}$$

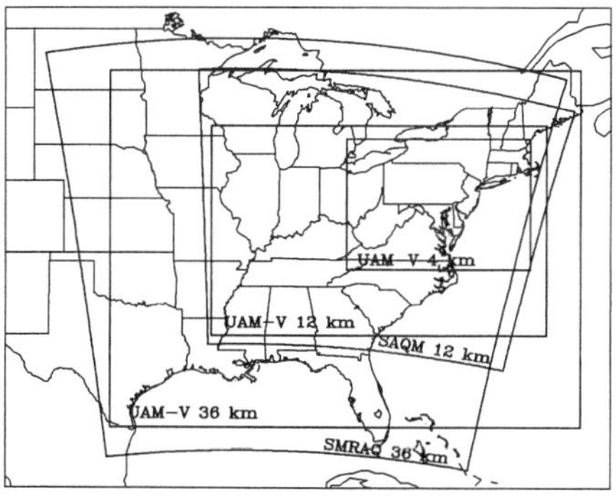

Figure 1: Modeling domains used in this study

RESULTS

Uncertainty in Ozone Predictions from Different Modeling Systems

Figures 2a-j present histograms of model-to-model differences of daily maximum ozone concentrations and distributions of differences of predicted and observed daily maximum ozone concentrations for the time period July 13 - 15, 1995 for the different model simulations described above. The model-to-model differences have the narrowest distribution for the case when the horizontal grid spacing was decreased from 12 km to 4 km for the RAMS3b/UAM-V system (Figure 2i). The distributions of differences between ozone predictions with the same photochemical model but different meteorological models (Figures 2e and 2g) are different, emphasizing the important impact of meteorology on ozone predictions. The distributions of the differences between two entirely different modeling systems (Figures 2a and 2c) again are different; these distributions have the

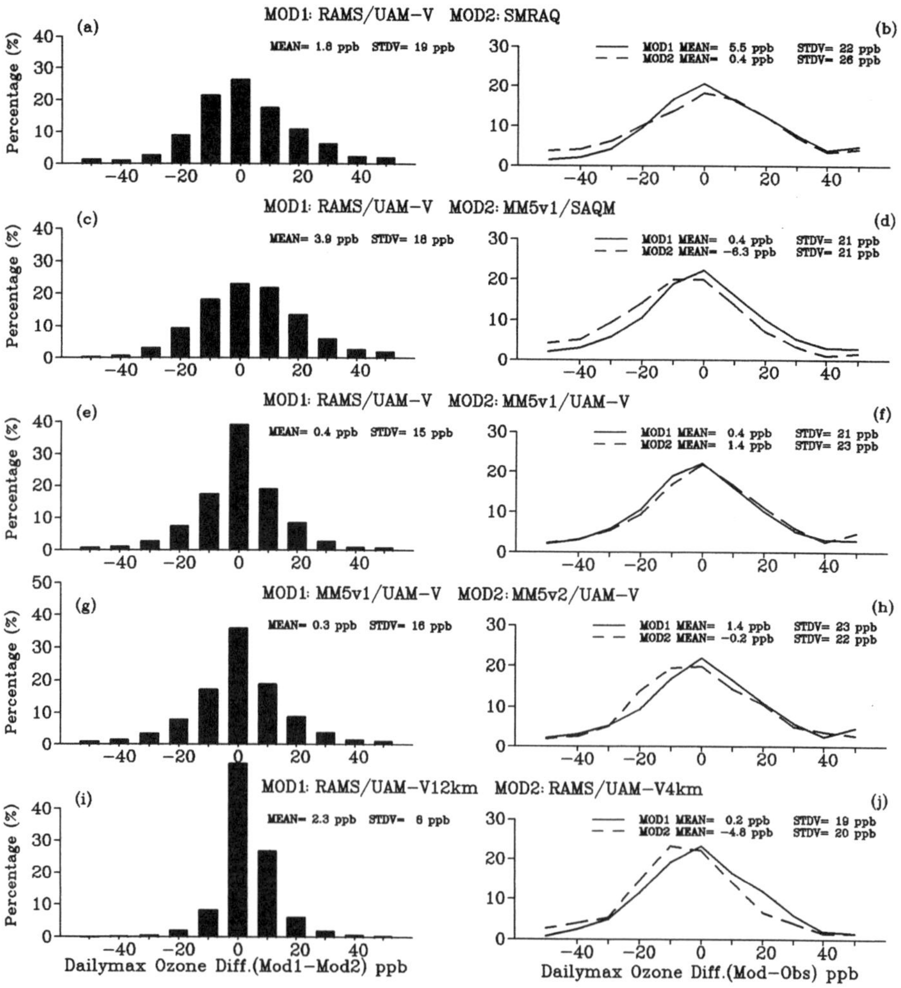

Figure 2: Distributions of model-to-model differences (left side) and model-to-observation differences (right side) for daily maximum ozone concentrations July 13 - 15, 1995 from different modeling simulations as described in the text.

largest standard deviations compared to the other simulations. Overall, these distributions of model-to-model differences illustrate that the choice of the modeling system, meteorological model, or horizontal grid spacing can cause differences in daily maximum ozone predictions of up to 40 ppb during an episode. When the distributions of the model-to-observation differences are compared (Figures 2b,d,f,h,j), it is evident that almost all of these distributions (with the exception of the distribution of differences between SMRAQ predictions and observations) have a standard deviation of about 20 ppb and that there are some differences in the means of these distributions. This is an indication that, while some simulations have a bias, none of these modeling systems performs significantly better than the others in terms of reducing the model-to-observation uncertainty of predictions of daily maximum ozone concentrations (see the standard deviation of these distributions). Therefore, it is clear that the predictions of daily maximum ozone concentrations with current-generation photochemical modeling systems during an individual ozone episode are highly uncertain, and that the choice of the modeling system itself introduces a large amount of uncertainty.

Time Scales Characteristic for High Observed Ozone Concentrations

To examine the differences in the dynamical forcings that distinguish episodic and nonepisodic conditions, sample distributions of each of the observed four temporal components at Queens, NY for days with high ozone concentrations only and for all days in the summer season (June, July and August) are examined (Figures 3a-d). The length of the data record used in this analysis is 18 years. High ozone days are defined as days having daily maximum 1-hr ozone concentrations greater than 100 ppb. It can be seen that the differences between episodic and nonepisodic conditions are most evident on time scales reflecting the synoptic and baseline forcings; similar results have been found at rural sites. The most striking feature is that the ID component is virtually the same for episodic and nonepisodic days. Both the positive mean value of the synoptic component and the upward shift of the mean baseline for the high-ozone-days distributions are evidence that high ozone concentrations tend to occur when the synoptic forcing is positive and the baseline forcing is stronger than average. This result implies that the longer-term processes (periods longer than one day) are essential forcing mechanisms for the observed high ozone levels.

Model Performance on Different Time Scales

Figures 4a-d show spatial images of the correlation coefficients between the seasonal time series of the observed and predicted synoptic and baseline components of ozone

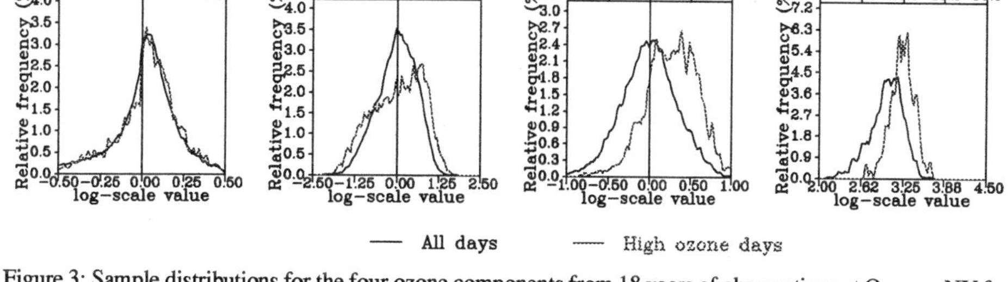

Figure 3: Sample distributions for the four ozone components from 18 years of observations at Queens, NY for both all summer days and summer days with ozone concentrations in excess of 100 ppb only.

(results for the RAMS3b/UAM-V and SMRAQ simulation are presented). Correlations between observations and model predictions are higher on these two time scales than for the ID component (spatial average correlation coefficient = 0.19) and for the amplitude of the diurnal component (spatial average correlation coefficient = 0.23) (Hogrefe et al.,1999). There are relatively small differences between the two modeling systems - for both modeling systems, correlations with observations are higher for the baseline than the synoptic component, and model performance is best along the coast and worst in the Midwest. This finding suggests that model-to-model differences, as presented above for a single episode, are most pronounced for individual days, but much smaller for fluctuations on the synoptic and baseline time scales. Most importantly, these longer time scales of best model performance and having the smallest model-to-model differences are also the essential forcing mechanisms behind days of high ozone concentrations as shown above.

Assessing the Variability of Model-Predicted Benefits of Emission Control Strategies

In its recent draft modeling guidance, the U.S. EPA defined the relative reduction factor (RRF) as the ratio of the mean daily ozone maxima for the emission control and base cases (U.S. EPA 1999). To determine if the simulated emission reductions would lead to compliance with the NAAQS at a given location, the observed design values would be

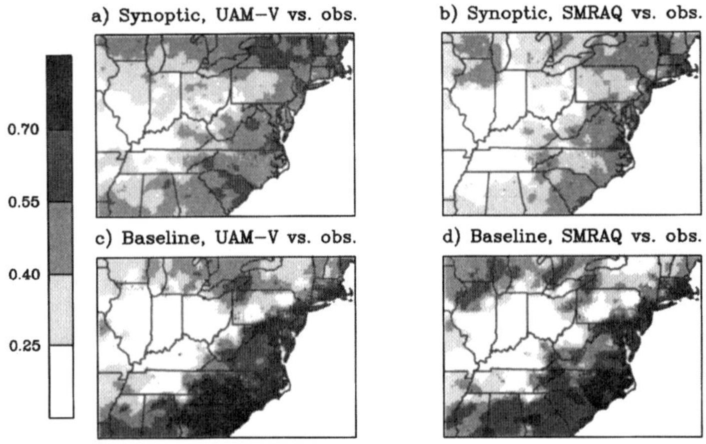

Figure 4: Spatial map of correlation coefficients between the observed and predicted time series of the synoptic and baseline components for ozone for the seasonal 36-km RAMS3b/UAM-V and the seasonal 36-km SMRAQ simulations.

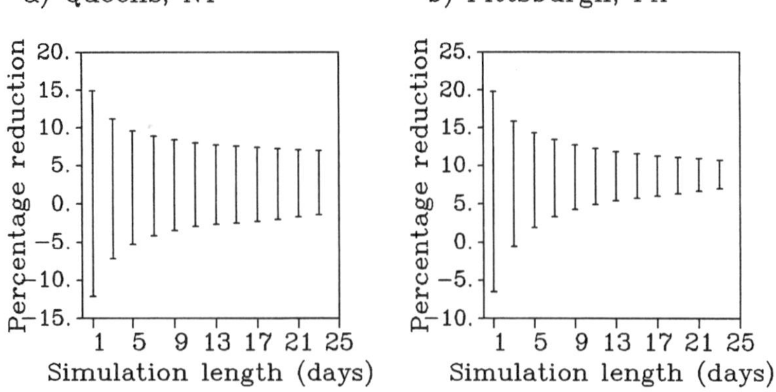

Figure 5: Mean percentage reduction factor ± 1 standard deviation as function of simulation length for Queens, NY, and Pittburgh, PA for the 50%NO_x/25%VOC emission reduction scenario

multiplied by the RRF. Hogrefe et al. (2000a) examined the uncertainties associated with the RRF approach and the relationship between the RRF and the simulation time period using the results from the 12-km RAMS3b/UAM-V base case and the seasonal RAMS3b/ UAM-V control case simulations.

In this study, we present percentage ozone reductions defined as (1-RRF)*100% rather than the RRF. Figures 5a-b illustrate the dependence of the variability of the reduction estimate on the length of the simulation used to calculate the estimate for two urban grid cells. These figures present the mean ± 1 standard deviation of the reduction estimate for a given simulation length. It can be seen that - while the mean reduction resulting from this control strategy varies from stations to station, the variability of this estimate is reduced at all stations as the simulation length is increased. Therefore, it is clear that the estimates of RRF are highly variable for short simulation lengths, and this variability decreases as the simulation length increases (which, by the definition of the RRF also implies that the RRF estimates are the results of base and control case concentrations that are averaged over a longer time period). In other words, it is important that the concept of RRF be applied to ozone predictions from longer modeling periods rather than from episodic modeling to reduce the dependence of the RRF estimates on the selected modeling period. If policy decisions were to be based upon one RRF for one episode only without accounting for the associated variability, there is a danger that hoped-for results can be achieved by selectively choosing the episodes or simulation lengths or "tuning" the model.

SUMMARY

This study demonstrated that several state-of-science photochemical modeling systems display significant model-to-model differences when applied to only episodic days, making their use in traditional ozone attainment demonstrations highly uncertain. It was shown that atmospheric processes operating on time scales ranging from several days to several weeks are essential contributors to the days of high ozone concentrations at both urban and rural locations. In addition, the results from two seasonal simulations with two different modeling systems showed that the highest model-to-observation as well as model-to-model correlations are present on the synoptic and baseline time scales. This strongly suggests that the current generation modeling systems should be applied to simulation periods longer than a single episode in the regulatory setting. This notion is further emphasized by the finding that the model-predicted efficacies of a particular emission control strategy are highly variable for short simulation periods. Moreover, the shift from 1-hr to the 8-hr ozone standard requires that we move away from modeling over small domains for shorter time periods to modeling over larger domains for longer time periods.

ACKNOWLEDGMENTS

This work is supported by the United States Environmental Protection Agency under STAR Grant R825260-01-0, NARSTO-North East through EPRI under Contract WO3189-12, and New York State Energy Research and Development Authority under Contracts 4914 and 6085. Thanks are extended to E. Zalewsky, S. Wu, W. Hao, G. Sistla, G. Kallos, K. Lagouvardos, V. Kotroni, K. Zhang, and the SMRAQ team for performing the modeling simulations.

REFERENCES

Biswas, J., and S. T. Rao, 1999: Uncertainties in episodic ozone modeling stemming from uncertainties in the meteorological fields. *J. Appl. Meteor.* Accepted

Biswas, J., S. T. Rao, K. Zhang, and N. Seaman, 2000: Sensitivity of the regional ozone modeling results to different physical processes within a meteorological model. Preprints, *11th Joint Conference of the American Meteorological Society and the Air & Waste Management Association on the Applications of Air Pollution Meteorology*, Long Beach, CA, Amer. Meteor. Soc.

Chang, J. S., S. Jin, Y. Li, M. Beauharnois, C. H. Lu, H. C. Huang, S. Tanrikulu, and J. DaMassa, 1997: The SARMAP Air Quality Model, final report. *Air Resources Board, California Environmental Protection Agency*, Sacramento, CA

Eskridge, R. E., J.-Y. Ku, S. T. Rao, P. S. Porter, and I. G. Zurbenko, 1997: Separating different scales of motion in time series of meteorological variables. *Bull. Amer. Meteor. Soc.*, 78, 1473 - 1483

Grell, G. A., J. Dudhia, and D. Stauffer, 1994: A description of the fifth-generation Penn State/NCAR Mesoscale Model (MM5). *NCAR Technical Note*, NCAR/TN-398 + STR

Hogrefe, C., S. T. Rao, and I. G. Zurbenko, 1999: Seasonal photochemical modeling over the eastern United States: Model performance on different space and time scales and implications to the use of models in a regulatory setting; Preprints, *Symposium on Interdisciplinary Issues in Atmospheric Chemistry*, Dallas, TX, Amer. Meteor. Soc, 42 - 49.

Hogrefe, C., S. T. Rao, I. G. Zurbenko and P. S. Porter, 2000a: Interpreting the information in ozone observations and model predictions relevant to regulatory policies in the Eastern United States. *Bull. Amer. Meteor. Soc.*, Accepted

Hogrefe, C., and S. T. Rao, 2000b: Evaluating meteorological input variables for seasonal photochemical modeling on different time scales. Preprints, *11th Joint Conference of the American Meteorological Society and the Air & Waste Management Association on the Applications of Air Pollution Meteorology*, Long Beach, CA, Amer. Meteor. Soc.

Kallos, G., K. Lagouvardos, and V. Kotroni, 1997: Atmospheric modeling simulations over the eastern United States with the RAMS3b model for the summer of 1995. Final report to *Electric Power Research Institute*, Palo Alto, CA.

Ozone Transport Assessment Group, 1997: *Technical reports of interest to the OTAG Air Quality Analysis Workgroup*. http://capita.wustl.edu/otag/Reports/reports.html

Rao, S. T., I. G. Zurbenko, R. Neagu, P. S. Porter, J. Y. Ku, and R. F. Henry, 1997: Space and time scales in ambient ozone data. *Bull. Amer. Meteor. Soc.*, 78, 2153 - 2166.

Rao, S. T., and Coauthors, 1998: An integrated modeling and observational approach for designing ozone control strategies for the eastern United States. Proceedings, *NATO/CCMS International Technical Meeting on Air Pollution Modeling and its Applications*, Bulgaria, September 1998.

Sistla, G., and Coauthors, 2000: An operational evaluation of two regional-scale ozone air quality modeling systems over the Eastern United States. *Bull. Amer. Meteor. Soc.*, In revision

SMRAQ, 1997: The SMRAQ project: development and application of a seasonal model for regional air quality. http://envpro.ncsc.org/SMRAQ/

Systems Applications International, 1995: Users Guide to the Variable Grid Urban Airshed Model (UAM-V). *Systems Applications International*, San Rafael, CA.

Tesche, T. W., D. E. McNally, J.G. Wilkinson, C. F. Loomis, and R. A. Emigh, 1996: Initial estimates of emissions reductions needed for ozone attainment in the Pittsburgh-Beaver Valley ozone nonattainment area. Report prepared for: *Southwestern Pennsylvania Clean Air Stakeholder Group*, Alpine Geophysics, LLC, Covington, KY.

United States Environmental Protection Agency, 1991: Guideline for regulatory application of the Urban Airshed Model. EPA-450/4-91-013, July 1991, *United States Environmental Protection Agency*, Research Triangle Park, NC 27711.

United States Environmental Protection Agency, 1994: Guidance on Urban Airshed Model (UAM) reporting requirements for attainment demonstrations. EPA-454/R-93-056, *United States Environmental Protection Agency*, Research Triangle Park, NC 27711.

United States Environmental Protection Agency, 1999: Draft report on the use of models and other analyses in attainment demonstrations for the 8-hour ozone NAAQS. EPA-44/R-99-0004, May 1999, *United States Environmental Protection Agency*, Research Triangle Park, NC 27711.

Walko, R. L., C. J. Tremback, and R. F. A. Hertenstein, 1995: RAMS - The Regional Atmospheric Modeling System, Version 3b, User's guide. *ASTER Division, Mission Research Corporation*, Fort Collins, CO

Zhang, K. and S. T. Rao, 1999: Mesoscale 4DDA analysis on meteorological characteristics of high ozone episodes in 1995. *New York State Department of Environmental Conservation Report*

Zurbenko, I. G., 1986: The spectral analysis of time series, *North-Holland*, Amsterdam, Chapter IV, 105-186

DISCUSSION

B. BORNSTEIN: Your first 'final conclusion' *[bullet 1 in the conclusion slide of the presentation: "The use of different modeling systems, different meteorological models, or different grid cell sizes can introduce large uncertainties to the predictions of daily maximum ozone concentrations on individual days"]* seems to disagree with your results, in that it seemed to say that the different model systems and configurations yielded different results.

C. HOGREFE: This point in my 'final conclusions' slide refers to the fact that different model systems and configurations do indeed yield different results since they are predicting different daily maximum ozone concentrations on individual days. However, this does not contradict the main result of the paper, namely, none of these modeling systems or configurations compare more favorably with the observations than any other (as measured by the standard deviations of the distributions of model-to-observations differences). In other words, the use of different modeling systems or configurations will affect the individual predicted values, but the overall modeling uncertainty with respect to observations remains unchanged. We, therefore, argue that – regardless of the modeling system or configuration used – modeling uncertainty can only be reduced by considering longer term averages, i.e. the daily maximum ozone predictions at a specific site averaged over the entire simulation period.

P. SEIBERT: Are the 'synoptic' and the 'baseline' contributions indeed that? I am wondering about the big variations in the baseline values.

C. HOGREFE: In our case, the baseline component contains fluctuations with periods longer than about 3 weeks. These fluctuations include seasonal changes of the solar flux, biogenic emissions or large-scale flow patterns among others. These fluctuations do play an important role in determining the levels of ambient ozone concentrations as illustrated in our paper. Please see our article in the September 2000 issue of the Bulletin of the American Meteorological Society (Hogrefe et al. 2000a in the reference section of the paper) for a detailed discussion on the scales.

F. MÜLLER: One of your results shows that all investigated model systems doing equally [poorly]. Is this a result of comparing 3-D model results, which are grid averages, with point measurements, or is this expressing the 'general' problem of the present day models by applying operator splitting? Is the methodology of operator splitting causing the presented differences?

C. HOGREFE: I do not believe that the method of operator splitting is causing these differences. But your question raises an important point: one always has to keep in mind the time and space scales that model predictions represent. Our time scale analysis showed that correlations between observations and model predictions on the intra-day time scale and for the amplitude of the diurnal time scale are very weak (see section "Model performance on different time scales" in the paper). This implies that the small spatial scales associated with these higher-frequency fluctuations cannot be resolved adequately by the model; in turn, the observation network is also not consistently dense enough to support higher resolution modeling. The presented result you are referring to was calculated for predictions of daily maximum ozone concentrations which are influenced by fluctuations on all time scales. The 'weak performance' can be thought of as the uncertainty of predictions of daily maximum ozone concentrations that is caused by the contribution from time and space scales that cannot be captured by any of the modeling systems presented here because of finite grid spacing and our inability to accurately specify precursor emissions in time and space.

INTEGRATED ASSESSMENT MODELLING OF ABATEMENT STRATEGIES: ROLE OF UNCERTAINTIES

Rachel F. Warren and Helen M. ApSimon

Air Pollution Group, TH Huxley School, RSM Building,
Imperial College of Science, Technology and Medicine,
Prince Consort Road, London SW7 2BP, UK

INTRODUCTION

In 1979 the Convention on Long-range Trans-boundary Air Pollution was set up under the auspices of the United Nations Economic Commission for Europe (UN ECE 1998, http:/www.unece.org/env), in recognition of the trans-boundary nature of air pollution problems such as acidification. In 1980 the first Sulfur protocol came into being in Europe, implying a 30% reduction of emissions of SO2 in European countries. This uniform emission cut took no account of the spatial non-uniformity of (a) the sensitivity of the land to acid deposition, subsequently mapped as critical loads (Posch et al. 1997) (b) the patterns of meteorological transport of air pollutants, subsequently studied by the Meteorological Synthesising Centre West of EMEP to attribute deposition to sources (Barret and Seland 1995) and (c) the regional variations in abatement costs, subsequently collated and respresented as cost curves by the Institutional Institute for Applied Systems Analysis (IIASA) (Amann, 1989; Amann et al. 1994) and by Imperial College (Cowell and ApSimon 1998). Integrated assessment models have been used as tools to investigate cost-effective strategies for further improvement across Europe as a whole. These indicated the pattern of emission reductions required to reduce deposition closer to critical loads by attaining intermediate 'target loads'. The Second Sulfur Protocol signed in Oslo in 1994 (UN ECE 1994) was the first international environmental agreement to be based on an integrated assessment approach, and in particular the use of critical loads was key to the whole process. Since 1994, IAM has extended to cover other pollutants, namely oxides of nitrogen (NOx) and reduced Nitrogen (NH3) and other effects, taking into account eutrophication. The RAINS model (Alcamo et al. 1990, Amann et al. 1998) now includes volatile organic compounds (VOC) in order to address the formation of tropospheric ozone as a result of emissions of NOx and VOC. The IAM approach was again used in the drawing up of the multi-pollutant multi-effect protocol signed in Gothenburg in December 1999. This paper describes uncertainty analysis which support other IAM work carried out by IIASA using the official RAINS model. Illustrative results are shown which demonstrate the robustness of results to uncertainties in atmospheric transport, in critical loads and in abatement costs, and to the influence of including additional effects.

Air Pollution Modeling and Its Application XIV, Edited by
Gryning and Schiermeier, Kluwer Academic/Plenum Publishers, New York, 2001

METHODS

Abatement Strategies Assessment Model (ASAM).

The ASAM model differs from the RAINS model in adopting a sequential approach to the identification of a cost-effective abatement strategy, thus allowing examination of strategies which make the best effort at attaining a particular environmental goal given the financial constraints even if the environmental goal is not attainable. The model has been described previously (ApSimon et al 1994, ApSimon and Warren 1996). Figure 1 illustrates the sequential approach used by the ASAM model. ASAM uses emission data and the meteorological data to calculate the deposition of Sulfur and or Nitrogen at the REFERENCE scenario, a set of emissions selected by policy makers as the starting point for optimization. This REFERENCE scenario typically incorporates existing commitments such as protocols and EU Directives, together with any additional country Current Reduction Plans (CRP). The meteorological data provides information concerning the long-range transport of a pollutant from each country to some 700 receptors. Each receptor is a 150x150 km sized grid cell defined by the 'EMEP grid'. ASAM compares the deposition at the REFERENCE scenario in each grid cell with the critical or target loads (see later) which are defined on the same scale as the deposition. The difference between the two is known as the 'exceedance' (Posch 1998). ASAM then draws on information concerning the (annualized) costs of abatement of different pollutants in different countries. It reviews the possible abatement measures that could be taken and selects that abatement step which is the most cost-effective. This is the step which has the highest ratio of benefit:cost, where the benefit is measured in terms of reduction in exceedance. ASAM then reassesses the deposition, recalculates the exceedance, and selects a new abatement step based on the same cost-effectiveness criterion. The process is repeated until critical/target loads are attained or no further abatement measures are possible. This is said to identify the 'Best Economic Environmental Pathway' from an initial set of country emissions to a final set of country emission ceilings satisfying environmental constraints. Examination of this pathway can be very instructive. Figure 2 shows how in ASAM, as expenditure increases in Europe, environmental improvement accrues rapidy at first, and then much more slowly as successively more expensive steps are required to protect the more 'difficult' areas from acidification. These 'difficult' areas are known as binding grid cells. Curves such as Figure 2 can be used to highlight the trade-off between abatement costs and environmental protection. Figure 2 uses ecosystem area unprotected from acidification as the measure of environmental benefit, but similar figures can be produced for other effects.

The use of Target Loads in IAM

Since critical loads cannot be attained even if the maximum feasible reductions (MFR) are implemented across Europe, intermediate target loads need to be set. Much effort has been devoted to the appropriate selection of these target loads. The UNECE Task Force on Integrated Assessment Modelling (TFIAM) has recommended the use of targets based on the 'gap closure' approach. This involves reducing the exceedance of critical loads occurring in 1990 by the same percentage in each EMEP grid cell. This choice of target load has been examined critically (Warren and ApSimon a, in press).

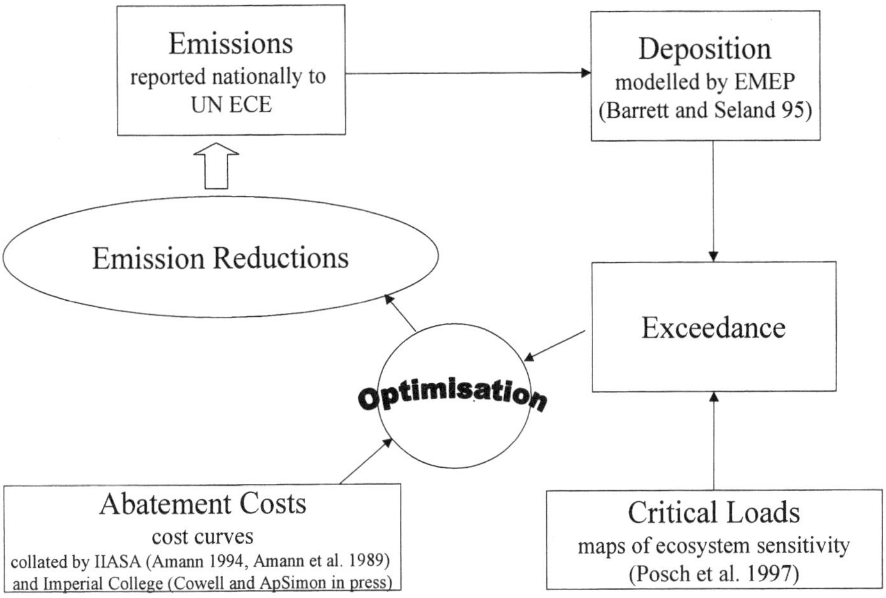

Figure 1. The Abatement Strategies Assessment Model, ASAM

Robustness Studies

If IAM model results are to be used for policy purposes, this demands that there should be a thorough understanding of the degree of robustness which models have to uncertainties. Extensive studies have been carried out with the ASAM model which help to back up the use of the IAM approach in the UN ECE negotiations (Warren and ApSimon 1999, Warren and ApSimon c in press.). Uncertainties may arise firstly from the data used in models (or the method in which the data in incorporated into the models) or secondly through model simplifications, omissions, or assumptions. This study addresses first the uncertainties that arise directly from the data or its use in the IAMs. Model robustness has been examined by comparing abatement strategies resulting from the use of alternative input data compared to the standard input data used in a base case strategy starting from the REFERENCE scenario mentioned above. 'Alternative' input data may differ from the standard data through either systematic or random variations. An example of the second type of uncertainty (in this case an important factor that is omitted from the modeling process) is addressed in the second part of the study.

It is easy to show that the 'end point' of the curve in Figure 2 where an environmental target is finally reached, is rather sensitive to small changes in some of the input data. A significantly different end-point to an ASAM derived abatement strategy can be obtained by 'tweaking' key input data such as critical loads in sensitive areas or source-receptor coefficients for countries that contribute deposition to sensitive areas. In the RAINS model, this effect has been greatly reduced by the application of the 'compensation mechanism' whereby exceedances at binding grid cells may be compensated for by 'equivalent' improvements in other areas within the same country (Amann et al. 1998). However, in ASAM it is not really the end point of the strategy that is the important result, but the nature of the curve shown in Figure 2 and the points along it. Thus robustness studies have concentrated on comparing the order in which countries cut back their emissions, i.e. comparing the emissions and abatement costs in strategies based upon alternative input data at the same overall cost to Europe.

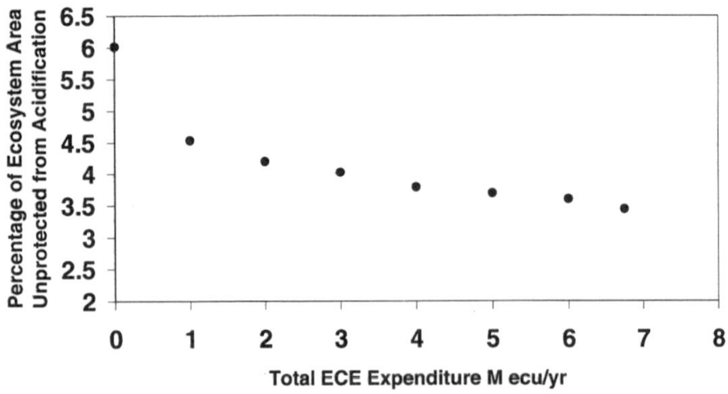

Figure 2. Ecosystem areas unprotected as a function of overall UNECE expenditure M ecu/yr

RESULTS

Uncertainties in long-range transport

The source-receptor matrices describing the atmospheric transport show sensitivity to the meteorology and to the representation of atmospheric processes and emissions. Any two dimensional Lagrangian model such as the EMEP model has limitations in representing three dimensional dispersion, the chemistry of the deposition, and the influence of local processes such as orographic enhancement (Fowler et al. 1955). The latter process, whereby deposition is greatly enhanced in upland areas, can be a very important issue particularly in the UK where many of the upland areas are those most sensitive to acid deposition. As an indication of the sensitivity Figure 3 illustrates the robustness of a typical ASAM-derived emission reduction pattern across countries (labeled 'base case' in the Figure) to 20% random variations in source-receptor matrices. The meteorological variability can be investigated by looking at inter-annual variation. Prior to use in the ASAM model the data are combined in order to represent an 'averaged' meteorological situation, obtained from individual matrices for each of 10 years. Thus Figure 3 shows also the sensitivity of the base case strategy to the use of alternative years' meteorology. An overall European investment level of 5 billion DM/yr has been selected for the robustness comparison.

It is clear that model shows great robustness to both the meteorological and the random variations in the source-receptor matrices when results are compared at the same overall European level of expenditure, robustness being greatest with resepect to random variations. EMEP are currently developing an Eulerian model to calculate source-receptor matrices (Jakobsen et al. 1997) and in future work we plan to compare the use of Eulerian and Lagrangian approach.

Uncertainties in critical/target loads

These may arise from (a) innate uncertainties in the assessment of critical loads in a particular area (b) whether the critical/target loads are exceeded by 'background' deposition. The latter is that deposition coming from natural sources, from sources outside the EMEP grid e.g. from North America. Apart from uncertainties in the critical loads, it would be undesirable if strategies were not robust to small changes in the target loads which policy makers might select as environmental constraints. As explained previously,

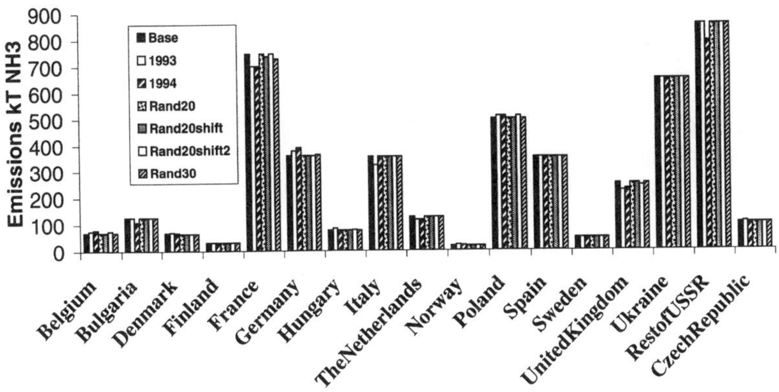

Figure 3. Sensitivity to variations in source-receptor matrices: NH3 emissions at UNECE expenditure 5 bn ecu/yr.

the overall cost of an abatement strategy is strongly dependent on the critical /target loads in the binding grid cells. However, the overall pattern of emission reductions that ASAM produces at a given expenditure level, is robust to uncertainties in critical/target loads even in the binding grid cells. For example, altering the % gap closure target in a binding grid cell from 88 to 90% can increase overall costs of a strategy from (for example) 6.4 to 8.3 billion e.c.u./yr; whilst examining country expenditures at an overall ECE expenditure of 5 billion e.c.u./yr, most country commitments remain identical whilst there are small changes in commitments in one or two countries (Warren and ApSimon 1999). In contrast, should the spatial variation in critical loads be ignored completely, technologies are then selected only on the basis of implementing the cheapest steps first, without regard to their location and consequent impacts. The result is a very different abatement strategy which produces far less environmental benefit (in terms of, for example, the area remaining unprotected from acidification) than the optimized strategies of the kind represented in Figure 2.

Uncertainties in abatement costs

ASAM's robustness to both random and systematic changes in abatement costs has also been examined. Random perturbations in marginal costs were introduced in order to simulate up to 10% random uncertainties. Remarkable robustness of strategies was observed when comparing at the same overall European abatement cost. Systematic changes, such as errors of a factor of 2 in all marginal costs in a single country, sometimes produced significant changes in the strategy for the country concerned, and occasionally also in a geographically close one (which might have to compensate for changes in the first country's role), delaying or hastening that country's participation in an abatement strategy. However, emission reductions and expenditures in other countries remained remarkably robust (Warren and ApSimon, 1999).

An overall representation of model robustness

Many uncertainty analyses have been carried out and are summarized in Figure 4, in order to show how environmental benefit, measured in terms of area protection from acidification, accrues as a result of increasing expenditure in Europe, taking into account the possible uncertainties. Figure 4 shows that despite the uncertainties there is still a consistent picture showing a clear trade off between expenditure and environmental protection. The outlying points on the curve show for contrast, a major change to which the abatement strategies are not robust, namely the non-implementation of any emission reductions for ammonia. This greatly reduces the cost-effectiveness of the possible abatement strategies. Unfortunately the recently signed Gothenburg Protocol contains little in the way of emission ceilings for NH3 below the REFERENCE scenario. This factor is clearly far more influential in undermining the cost-effectiveness of the strategy than any of the possible sources of uncertainty investigated.

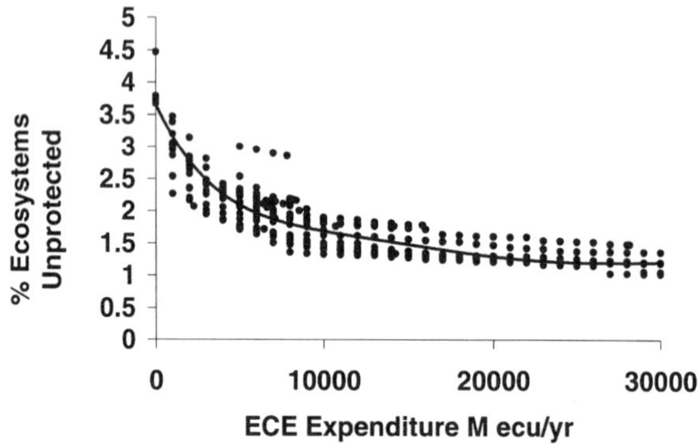

Figure 4. Ecosystem areas unprotected from acidification as a function of UN ECE expenditure

Uncertainties arising from omitting other atmospheric effects

Another form of 'uncertainty' is when a factor is missed out in the modeling exercise. In addition to the effects considered in the UN ECE program the health effects of human exposure to secondary particulates have become an increasing concern. These secondary aerosols (specifically, $(NH_4)_2SO_4$, and NH_4NO_3) form from oxides of sulfur, nitrogen and ammonia and are transported over long distances; they are thus trans-boundary pollutants in their own right. They contribute very substantially to the overall atmospheric particulate burden (ApSimon et al in press) and hence to the human exposure to PM10. In fact, economic assessments have indicated that the greatest benefits of reducing atmospheric emissions of sulfur and nitrogen compounds in Europe come from the reduction in human exposure to secondary PM10 (ApSimon et al. 1997). It is important, therefore, to investigate the strategies to reduce exposure to PM10 instead of other effects. To address this, a new version of ASAM has been constructed which seeks only to reduce human exposure to secondary particulate material. The model aims to minimize the annual average exposure to PM10 over the European population. Only the long-range contribution is included with no allowance for more local urban enhancement. The strategies derived using such a model (type A) have been directly compared with those aimed at reducing acid deposition (type B) (Warren and ApSimon b in press). It turns out that a strategy of type A also produces substantial reductions in acid deposition to sensitive ecosystems, delivering between 70 and 80% of the ecosystem benefits delivered by a strategy of type B; whilst a strategy of type B produces substantial reductions in human exposure to secondary particulates, delivering some 70% of the benefit of a strategy of type A.

Other Sources of Uncertainty

Since the cost curves used in IAM are mostly restricted to end-of-pipe or add-on technology alone, other options which are potentially cheaper and which may serve a dual purpose, such as energy-saving or reductions in CO2 emissions, are not considered. These options include fuel switching, energy efficiency and economic restructuring. Some of these omissions may be taken into account by sensitivity analysis. For example (Agren 1999) calculated that the application of measures to comply with the Kyoto protocol in Europe would cut the costs of attainment of the environmental goals of the proposed National Emissions Ceilings Directive (European Commission 2000) by about two thirds. Acid deposition varies spatially on a small scale and thus ecosystem exceedances may be wrongly estimated by using too coarse a scale. The approach omits the dynamic aspects of ecosystem resilience and recovery time, and ecosystem benefit is measured in terms of exceedance or areas where critical loads are exceeded, rather than by using sophisticated dose-response functions actually reflecting the damage to ecosystems at different levels of exceedance. Future work in IAM hopes to address some of these issues by using a reduction in the scale to 50x50 km , and also to use dynamic models such as MAGIC (de Vries 1989) in combination with the critical load approach to investigate the temporal aspects. Issues such as this are likely to have a larger effect on model robustness than the uncertainties in the data examined in this study.

CONCLUSIONS

IAM is a way of linking atmospheric modeling and other scientific aspects to the technological and economic considerations. It is a useful tool to explore cost-effective strategies to maximize the environmental benefits of emission abatement whilst minimising overall expenditure. The gain in cost-effectiveness achieved by carrying out such an optimization is very significant even taking into account the uncertainties in the data used by the models. The study has shown how the ASAM model can be used to produce a prioritised sequence of abatement steps which may be viewed as a plot of accruing environmental benefit versus increasing European expenditure. This approach has been used to demonstrate the robustness of the ASAM derived abatement strategies at given overall levels of European expenditure. Robustness has been shown with respect to systematic and random variations in meteorological information, in abatement costs, and in critical loads. Robustness was greatest with respect to random uncertainties. Inevitably the approach has major omissions such as the reduction of human exposure to secondary particulate material. ASAM has been used to illustrate the substantial reductions in secondary PM10 that occur as a result of a strategy derived to reduce acid deposition, and indeed how such a strategy produces 70 % of the reduction in exposure to PM10 that a strategy aimed specifically at reducing human exposure to secondary PM10. Other uncertainties need to be addressed by a change in methodology, such as the use of a Lagrangian as opposed to an Eulerian approach to modeling the long-range transport, or by incorporating a dynamically modelled response of ecosystems to changing levels of acid deposition, and these may influence derived strategies in a more fundamental way. In summary, the demonstration of robustness of IAM results to uncertainties is extremely important to policy makers and thus studies of this type have helped to underpin the work of the UN ECE Task Force on Integrated Assessment Modelling in its exploration of abatement strategies which could be used as a basis for new UN ECE protocols.

Acknowledgements

This work was made possible by the funding of the Air Quality Division at the Department of Environment, Transport and the Regions (DETR), UK.

REFERENCES

Agren, C. , 1999, *Getting more for less: An alternative assessment of the NEC directive.* European Environmental Bureau, Swedish NGO Secretariat on Acid Rain, European Federation for Transport for Environment, Goteborg.

Alcamo, J., Shaw, R., & Hordijk, L., (eds.), 1990, *The RAINS Model of Acidification: Science and Strategies in Europe*, Kluwer Academic Publishers, Dordrecht.

Amann, M., Bertok, I., Cofala, J., Gyarfas, F., Heyes, C., Klimont, Z., Makowski, M., Schopp, W., & Syri, S., 1998, *Fifth Interim Report: Cost-effective Control of Acidification and Ground-Level Ozone, Part A: Methodology and Databases*, International Institute for Applied Systems Analysis, Laxenburg, Austria

Amann, M., Cofala, J., and Klaassen, G. ,1994, *SO2 Control Cost Module in the RAINS 7.0 Model*, International Institute for Applied Systems Analysis, Laxenburg, Austria

Amann, M. ,1989, *Status Report: Potential and Costs for Control of NOx Emissions in Europe*, International Institute for Applied Systems Analysis, Laxenburg, Austria

ApSimon, H.M., Gonzalez del Campo, T., and Adams, H.S., Modelling Long-range Transport of Primary Particulate Material over Europe, *Atmospheric Environment, in press.*

ApSimon, H.M., Warren, R.F. & Wilson. J.J.N. , 1994, The Abatement Strategies Assessment Model – ASAM: Applications to Reductions of Sulfur Dioxide Emissions Across Europe, *Atmospheric Environment* 28:649

ApSimon, H.M. & Warren, R.F. , 1996, Transboundary Air Pollution in Europe, *Energy Policy*, **24**, 631-640

Bailey, P., Gough, C., Chadwick, M. and McGranahan, G., *Methods for Integrated Environmental Assessment: Research Directions for the European Union,* Stockholm Environment Institute, York

Barrett, K. & Seland, O. , 1995, *EMEP/MSC-W Report 1/95, European Transboundary Air Pollution: 10 years calculated fields and budgets to the end of the first Sulfur Protocol*, Oslo: EMEP MSC-W, Norwegian Meteorological Institute, P.O. Box 43-Blindern, N-0313 Oslo 3, Norway

Cowell, D., and ApSimon, H.M., 1998, Cost-effective Strategies for the Abatement of Ammonia Emission from European Agriculture, *Atmospheric Environment* 2:573

De Vries, W., Posch, M., and Kamari, J., 1989, Simulation of the long-t, ern soil response to acid deposition in various buffer stages. *Water Air and Soil Pollution* 48: 349

Fowler, D., Leith, I.D., Cape, J.N., Binnie, J., Crossley, A., Inglis, D.W.F., Choularton, T.W., Longhurst D.W.S., and Colan, D.E., 1955, Orographic Enhancement of Wet Deposition in the United Kingdom: Continuous monitoring. *Water, Air and Soil Pollution* 85: 2107

Jakobsen, H., Jonson, J. & Berge, E., 1997, *The multi-layer Eulerian model: Model description and evaluation of transboundary fluxes of sulfur and nitrogen for one year* EMEP MSC-W, Norwegian Meteorological Institute, P.O. Box 43-Blindern, N-0313 Oslo 3, Norway

Posch, M., Hettelingh, J.-P., de Smet, P.A.M. & Downing, R.J., (eds), 1997, *Calculation and Mapping of Critical Thresholds in Europe: Status Report 1997,* Coordinating Centre for Effects, Rijksinstitut Voor Volksgezondheid en Milieu, Bilthoven

Posch, M., 1998, *Defining an Exceedance Function:* Note to the Parties under the Convention on Long-Range Transboundary Air Pollution (unpublished)

United Nations Economic Commission for Europe, 1994, *Protocol to the 1979 Convention on Long Range Transboundary Air Pollution on Further Reduction of Sulfur Emissions,* UNECE, Oslo

United Nations Economic Commission for Europe, 1998, *Air Pollution: The Convention on Long-Range Transboundary Air Pollution,* UNECE, Geneva

Warren, R.F., and ApSimon, H.M., 1999, Uncertainties in Integrated Assessment Modelling of Abatement Strategies: Illustrations with the ASAM Model, *Environmental Science and Policy*, 2: 439

Warren, R.F., and ApSimon, H.M., a, Selection of Target Loads for Acidification in Emission Abatement Policy: the use of Gap Closure Approaches, in press, *Water, Air and Soil Pollution*

Warren, R.F., and ApSimon, H.M., b, The Role of Secondary Particulates in European Emission Abatement Strategies: Illustrations using the Abatement Strategies Assessment Model, ASAM, in press, *Integrated Assessment*

Warren, R.F. , and ApSimon, H.M., c, The Abatement Strategies Assessment Model, ASAM: Robustness Studies, Illustrative Results, and Investigation of the Role of European Countries in Sulfur Dioxide Emission Reduction Schemes in press, *Journal of Environmental Management*

DISCUSSION

S. T. RAO: Have you considered the issue of feasibility of implementing an emissions reductions control strategy in your optimisation module? Would it not be more useful to identify the relative ranking of different emissions reductions options, in terms of costs and environmental benefits, than a command and control type emissions reductions plan? Did you also consider an emissions-weighted type trading program to maximise environmental benefit from a given control strategy?

R. WARREN: To explain again how ASAM works, the model does exactly what you suggest, i.e. it examines all the possible emissions reductions options and ranks them in terms of costs and environmental benefits. It does this by ranking them according to the ratio of environmental benefit to cost. It then implements the emissions reductions control strategy so that the model itself can monitor the overall environmental benefits and costs that accrue from the emission abatement strategy. Owing to the sequential way in which ASAM works, it produces a picture of how environmental benefit accrues as expenditure increases, and thus a policy maker can identify the point at which there is an appropriate trade off between costs and benefits.

As well as watching how the environmental benefit accrues en route to a particular environmental target, the ASAM model can also produce information on the ease of attainment of strategies which satisfy different environmental constraints. These strategies may be compared in terms of their overall costs and benefits so that policy makers can select an appropriate choice.

The ASAM derived strategy will be, to the best of its knowledge, the most optimal way to reduce emissions towards the selected environmental constraints. The ASAM results are, of course, only as good as the data that has gone into the model. Thus, if the costs in the model do not reflect reality, or some countries find their emission reductions from a given strategy too expensive or politically unacceptable, then the cost curves need to be adjusted to reflect this. To cover the situation where some countries feel that their costs are unacceptably high I have carried out an extensive study of ways of placing limitations on marginal costs within ASAM so that the more expensive abatement options are excluded from the problem. This will shortly be published in the Journal of Environmental Policy and Management.

Since ASAM produces the most optimal strategy, it cannot be bettered in terms of cost-effectiveness by an emissions trading scheme. In particular an emissions-weighted trading scheme acts to maximise the ratio of emissions reductions: cost, rather than environmental benefit : cost. Thus the advantage of the effects-based approach, which is used in the technique of integrated assessment modelling to maximise the ratio of environmental benefit: cost, is lost.

ASAM has been used to show the great increase in cost-effectiveness seen when comparing optimised abatement strategies to strategies where emissions are reduced uniformly across Europe. However, trading could be made more effective by allotting permits weighted according to the environmental detriment caused by unit emission at each source. In order to thus combine the effects-based approach with a permit-trading scheme, I have carried preliminary simulations of a permit trading approach using weightings based on environmental detriment using ASAM, and have shown that it is more cost-effective than a straight emissions-weighted trading approach, but less cost-effective than the strategies derived from the standard ASAM model (unpublished).

INTEGRATED ASSESSMENT OF EUROPEAN AIR POLLUTION EMISSION CONTROL STRATEGIES AND THEIR IMPACT ON THE FLANDERS REGION

Clemens Mensink and Jan Duerinck

Vito (Flemish Institute for Technological Research)
Centre for Remote Sensing and Atmospheric Processes
Boeretang 200, 2400 Mol, Belgium

1. INTRODUCTION

Integrated environmental assessment techniques are applied to describe multiple effects of multiple pollutants. In particular, this study focuses on acidification, eutrophication and ozone formation (multiple effects) caused by SO_2, NO_x, NH_3 and VOC (multiple pollutants). The assessment includes a cost-benefit analysis to estimate both costs and benefits of the proposed emission reduction measures. Regional scale air quality models are essential tools in obtaining the relation between the pollutants and their effects. Another useful tool is the concept of critical loads (for acidification and eutrophication) and critical levels (for ozone formation). Both indicators estimate the impact on ecosystems and human health in terms of exceedances. Economical techniques, like the derivation and application of cost-curves and external cost evaluations of potential benefits, are required for an explicit monetarisation of both costs and benefits. Air quality models and economical cost functions are very sensitive to their input data and are therefore in some cases facing high uncertainties.

At the moment two integrated assessment studies have been carried out at a European level. First of all there is the work of the Convention on Long Range Transboundary Air Pollution (LRTAP) that has resulted in a multi-effect, multi-pollutant protocol signed in Göteborg, Sweden, on November 30, 1999. Secondly, the possibilities for further cost-effective emission reductions in Europe have been explored to support the European Commission's abatement strategies on acidification and ozone.

The preparatory studies for these tasks are based on calculations using the RAINS and EMEP models (Amman et al., 1999). The outcome of the policy related integrated assessment studies at a European level has important implications for policies on a national or regional level as well. The computations with the RAINS and EMEP models are carried out on a grid with a spatial resolution of 150 km x 150 km, providing emission reduction proposals on a country to country base. Therefore it is especially important for small countries and regions to verify *in more detail* the model results as well as the regional impact stemming from the proposed emission reductions.

This study focuses on three scenarios reflecting three different abatement strategies. A first scenario can be characterised as a reference scenario, based on the current national reduction plans in Europe and the current national and European legislative actions, including the national emission ceilings agreed on in the Göteborg protocol. In a second scenario the abatement policy as proposed in Flemish environmental policy strategy is embedded in the reference scenario. A third scenario aims at a reduction of the area of unprotected ecosystems in Europe by at least 50% in 2010 compared to 1990. The composition of the scenarios is discussed in section 2.

The impact of the three emission reduction scenarios on acidification and eutrophication was evaluated using the Operational Priority Substances (OPS) model. Results for yearly averaged SO_x -, NO_y -, NH_x- and total N-depositions have been obtained on a grid with a resolution of 5 km x 5 km covering Flanders. They are discussed in section 3. The impact of the emission reduction scenarios on ozone formation was evaluated on the basis of the model calculations carried out by the LOTOS model for Belgium for the summer period of 1994, in combination with a multiple linear regression technique. This is discussed in section 4. Section 5 shows some results of the cost-benefit analysis. On the benefit side, exceedances of critical loads for nitrogen and sulphur deposition were determined for 652 forest ecosystems in Flanders. On the cost side, the cost-curve for each of the pollutants was used to estimate the total cost per scenario. Section 6 concludes with a brief discussion of the results and includes an estimation of the uncertainties.

2. EMISSION REDUCTION SCENARIOS

In assessing the impact of current and future abatement strategies, three scenarios were defined, comparing the projected emissions towards 2010 with the reference situation in 1990. A first scenario reflects the impact of the current reduction plans and the current legislative actions and can be considered as a reference (REF) scenario. The current reduction plans are reflected in the officially adopted national emission ceilings (Göteborg protocol). The current legislative actions are those that are needed to comply with the adopted national and international regulations on emission control. These legislative actions include not only national emission control regulations in the individual European countries (e.g. Vlarem II in Flanders), but also the relevant European Directives. The implementation of the Auto/Oil-1 programme is included in the current legislative actions as well. Table 1 shows the emissions for the 15 EU countries as used in the REF scenario.

A second scenario is referred to as the Flemish Environmental Policy (FEP) scenario. It combines the REF scenario with the emission reduction policy for Flanders as formulated for different target groups in the environmental policy strategy for Flanders (MINA plan 2, 1997). Note that this regional policy is an independent policy that is not necessarily included in the REF scenario. Long term objectives of the Flemish environmental policy are to reduce depositions to levels causing no irreversible damage to ecosystems and to reduce concentrations of ozone and other photochemical oxidants to acceptable levels, avoiding harmful effects to human health and vegetation. For the reference year 1990 the OPS model calculated an average acid deposition over Flanders of 5110 mole H^+/ha/yr and an average nitrogen deposition of 44 kg/ha/yr. Intermediate and long term target values for acidification are an average acid deposition of 2900 mole H^+/ha/yr in 2002 and 1400 mole H^+/ha/yr (indicative) in 2010. The average N-deposition in Flanders should be reduced to 27 kg/ha/yr in 2002 and 5-20 kg N/ha/yr in 2010. Target values for ozone concentrations are not provided.

Table 1: Emissions of SO_2, NO_x, NH_3 and VOC for the REF and GAP scenario in 2010.

Country	SO_2 (kton)		NO_x (kton)		NH_3 (kton)		VOC (kton)	
	REF	GAP	REF	GAP	REF	GAP	REF	GAP
Austria	39	39	107	91	66	66	159	129
Belgium	106	76	181	127	74	57	144	102
Denmark	55	55	127	127	69	69	85	85
Finland	116	116	170	152	31	31	130	110
France	400	218	860	679	780	718	1100	932
Germany	550	463	1081	1051	550	413	995	924
Greece	546	546	344	264	73	73	261	173
Ireland	42	28	65	59	116	116	55	55
Italy	500	500	1000	869	419	419	1159	962
Luxembourg	4	3	11	8	7	7	9	6
Netherlands	50	50	266	238	128	104	191	156
Portugal	170	141	260	144	108	67	202	102
Spain	774	746	847	781	353	353	669	662
Sweden	67	67	148	148	57	48	241	219
UK	625	497	1181	1181	297	264	1200	964
EU-15 (total)	4044	3545	6648	5919	3128	2805	6600	5581

The third scenario is called the 50% gap closure (GAP) scenario. For acidification and eutrophication the GAP scenario aims at a reduction of the area of unprotected ecosystems in Europe by at least 50% in 2010 compared to 1990. An ecosystem is considered to be protected if the deposition to which the ecosystem is exposed does not exceed the critical load for that ecosystem. Below the critical load no significant harmful effects on sensitive elements of the environment occur according to the present knowledge (WHO, 1995). For ozone formation the AOT60 (Accumulative exposure Over a Threshold value of 60 ppb) is used as an indicator for the protection of human health. An eight-hour maximum value of 60 ppb is considered as a level at which acute adverse effects occur in the population (WHO, 1997). In the GAP scenario the AOT60 is treated as an hourly accumulative exposure exceeding a threshold value of 60 ppb. For damage on vegetation the AOT40 (Accumulative exposure Over a Threshold value of 40 ppb) is used as indicator. The GAP scenario wants to achieve a reduction of two thirds of the AOT60 in 2010 compared to the 1990 situation, and a reduction of one third of the AOT40 in 2010 compared to 1990. The GAP scenario selects the most stringent emission reductions from the H1 scenario in the IIASA study (Amman et al., 1999) and the national emission ceilings provided by the Göteborg protocol. Table 1 shows the emissions for the 15 EU countries as used in the GAP scenario.

3. IMPACT ON ACIDIFICATION AND EUTROPHICATION

The impact of the emission reduction scenarios on acidification and eutrophication was evaluated using the Operational Priority Substances (OPS) model, as developed by van Jaarsveld (1990). OPS is a Lagrangian trajectory model for long-term simulations, ranging from one month to several years. The OPS model computes concentrations and dry and wet depositions for primary (SO_2, NO_x, NH_3) and secondary (SO_4, NO_3, NH_4) acidifying components. It uses a climatological database in which statistical values for 18 meteorological parameters are provided. The climatological database is constructed by a meteorological pre-processor using hourly observations of wind direction and wind speed (measured at two heights), global radiation, temperature, precipitation intensity and precipitation amount. Other input data are related to the receptor characteristics (roughness length) and the emission sources (co-ordinates, dimensions, source strength). The model has been implemented and validated for the Flanders region (Mensink and Dumont, 1997). The scenario computations were performed on a grid containing 50 x 20 grid cells, covering the total area of Flanders with a mesh resolution of 5 km x 5 km. The climatological database for 1990 consists of a combination of hourly meteorological data from observatory stations in Gent, Antwerp and Mol. The meteorological data for 1990 were used for all computations, in order to allow a comparison that is not biased by meteorological conditions. The emission data for 1990 are derived from a detailed (1 km x 1 km) emission inventory for the Flanders region containing industrial point sources, domestic heating, road transport and agricultural emissions. Outside of Flanders, emission data from EMEP (Mylona et al., 1999), were used, in combination with the NUTS-3 distribution of the emissions as provided by CORINAIR. Table 2 shows the average acid deposition in 1990 and as expected in 2010 for the three scenarios, as well as the share of each of the three acidifying components.

For the impact of the three scenarios on eutrophication, the total nitrogen deposition (kg/ha/yr) was derived from the total depositions of the oxidised (NO_y) and reduced (NH_x) nitrogen, assuming that 1 mole H^+ of the acid deposition is equivalent to 1 mole of nitrogen. The last three columns in Table 2 show the average nitrogen deposition expected in 2010 and the individual contributions of oxidised and reduced nitrogen.

Table 2: Acid depositions and N-depositions obtained for the scenarios in 2010

Scenario	Acid deposition (mole H+/ha/yr)	% due to SO_2	% due to NO_x	% due to NH_3	N-deposition (kg/ha/yr)	% due to NO_x	% due to NH_3
1990	5110	38	21	41	44,0	34	66
REF	2538	22	21	57	27,6	27	73
FEP	2418	23	20	57	26,2	26	74
GAP	2126	19	20	61	24,2	25	75

4. IMPACT ON OZONE FORMATION

For the impact of the three scenarios on ozone formation, computed results from the LOTOS model were used. This model was used earlier to determine the influence of European emission reductions on ozone concentrations over Belgium for the period May 1 - August 31, 1994 (Builtjes and Boersen, 1996). The model computations were performed on a grid containing 70 x 70 grid cells covering a domain from 35°N to 70°N and from 10°W to 60°E. Transport phenomena on a regional scale as well as background concentrations on a continental scale are taken into account by the model. The photochemical processes in the model are driven by the Carbon Bound IV mechanism.

Using the hourly ozone concentrations computed by the LOTOS model, the impact of the three scenarios on AOT60 and AOT40 was calculated by means of a multiple linear regression technique as proposed by Amman et al. (1999):

$$AOT_{lj} = k_j + \sum_{i=1}^{M}(a_{ij}VOC_i + b_{ij}NOx_i + c_{ij}NOx_i^2) + \alpha_j \overline{en}_j^2 + \overline{en}_j \sum_{i=1}^{M} d_{ij}VOC_i \qquad (1)$$

Where:

AOT_{lj} = the calculated values for AOT60 (l=60) and AOT40 (l=40) at receptor j
k_j = back ground contribution to the AOT_{lj}
$a_{ij}VOS_i$ = linear contribution of the VOC emissions from country i to receptor j

$b_{ij}NOx_i$	= linear contribution of the NO_x emissions from country i to receptor j
$c_{ij}NOx_i^2$	= non-linear contribution of the NO_x emissions from country i to receptor j
$\alpha_j en_j^2$	= non-linear contribution of "effective" NO_x emissions ($\alpha_j = 0$)
$d_{ij}en_jVOS_i$	= non-linear contribution due to the interactions between NO_x and VOC

For three receptor points located in Flanders (j=1…3) the coefficients a_{ij}, b_{ij}, c_{ij}, d_{ij} and k_j are determined for the contribution of each emitting country (i=1…37). The 5 unknown coefficients in expression (1) are resolved by 5 equations related to 5 situations for which the AOT and the NO_x and VOC emissions are known (base case 1994 and 4 reduction scenarios: -30% NO_x, -30% VOC, -30% (NO_x + VOC) and –50% (NO_x + VOC)). The calculated coefficients are then used to calculate the AOT values for 1990 and the REF, FEP and GAP emission scenarios. The results for AOT60 and the AOT40 (crops and forests) are given in Table 3. Besides the average value for Flanders, the table gives the range of the values obtained in the three receptor points.

Table 3: Impact of REF, FEP and GAP scenarios in 2010 on ozone formation in Flanders.

	AOT60 (ppm.hour)	AOT40-crops (ppm.hour)	AOT40-forests (ppm.hour)
Base run 1994	1,8 (1,4 – 2,2)	7,8 (6,9 – 8,7)	8,5 (7,5 – 9,5)
1990	2,4 (1,9 – 2,9)	7,6 (6,3 – 9,0)	7,8 (6,4 – 9,4)
REF scenario	0,93 (0,83 – 1,0)	7,0 (6,9 – 7,1)	8,1 (7,9 – 8,2)
FEP scenario	0,93 (0,81 – 1,0)	7,0 (6,9 – 7,1)	8,1 (7,9 – 8,2)
GAP scenario	0,71 (0,49 – 0,85)	6,6 (6,5 – 6,7)	7,6 (7,5– 7,7)

5. COST-BENEFIT ANALYSIS

For acidification and eutrophication the benefit obtained from the three emission reduction scenarios could be estimated by evaluating the exceedances of critical loads for nitrogen and sulphur deposition as determined for 652 forest ecosystems in Flanders (Craenen et al., 1996). The critical load exceedance for acidification due to sulphur and nitrogen Ex(S+N) is given by:

$$Ex(S+N) = S_{dep} + N_{dep} - CL_{max}(S+N) \quad (2)$$

with S_{dep} and N_{dep} as actual deposition values, $CL_{max}(S)$ being the maximum critical load for sulphur and $CL_{max}(N)$ the maximum critical load for nitrogen (including uptake, immobilisation and denitrification). It is impossible to define unique critical loads of N and S and therefore impossible to define a unique exceedance in order to quantify the amount of S and N to be reduced (Posch et al., 1997). Many combinations of S and N deposition can be formulated which provide the same protection against acidification: only conditional critical loads of S or N can be evaluated, e.g. $CL(N|S_{dep})$. However, there is one exception, namely if:

$$S_{dep} > CL_{max}(S) \text{ OR } N_{dep} > CL_{max}(N) \quad (3)$$

i.e. if at least one of the maximum values is exceeded, then that deposition value has to be reduced to the related maximum value, *irrespective of the conditional critical loads.* Table 4 shows the percentage of forest ecosystems in Flanders for which the individual maximum values are exceeded. The fifth column in the table shows the percentage of ecosystems that satisfy condition (3), i.e. where the exceedance of the critical load for acidification is certain.

The critical load exceedance for eutrophication due to nitrogen as a nutrient $Ex(N_{nut})$ is given by:

$$Ex(N_{nut}) = N_{dep} - CL_{nut}(N) \quad (4)$$

The exceedance $Ex(N_{nut})$ has been determined assuming a dynamic denitrification, taking into account parameters like temperature, relative soil moisture saturation and pH of the soil (Craenen et al., 1996). The last column in Table 4 shows the percentage of ecosystems exceeding the critical load for nitrogen as a nutrient.

Table 4 : Number and percentage of forest ecosystems with critical load exceedances

	(a) $S_{dep} >$ $CL_{max}(S)$	(b) $N_{dep} >$ $CL_{max}(N)$	(a) OR (b)	%	$N_{dep} >$ $CL_{nut}(N)$	%
1990	518	650	650	100%	360	55%
REF	11	403	403	62%	252	39%
FEP	9	365	365	56%	249	38%
GAP	3	317	317	49%	240	37%

An estimation of the cost per scenario and per pollutant is based on cost curves as derived for Belgium (Duerinck et al., 1999). Cost curves are functions expressing the global cost of all measures to be taken in order to realise a certain emission reduction level, provided that the least expensive measure is taken first. The functions are based on selections of Best Available Technologies (BAT). Cost curves are facing large uncertainties in both the emission inventories and in estimating the cost of emission reductions. This uncertainty increases with increasing emission reduction levels, as illustrated for the VOC cost curve (Duerinck et al., 1999) in Figure 1. This figure shows the results of a Monte-Carlo simulation based on a log normal distribution with a standard deviation of 20% for the emission input data. Table 5 shows the marginal cost estimations per pollutant for the GAP scenario compared to the REF scenario. The last column of Table 5 gives an idea of the costs per reduced acid equivalent (1 Aeq. = 1 mole H^+) for SO_2, NO_x and NH_3 in a first attempt to make a cost-benefit comparison.

Table 5: Costs per pollutant and per reduced acid equivalent for the GAP scenario.

Pollutant	Emissions REF 2010 (kton)	Emissions GAP 2010 (kton)	Emission reduction (kton)	Costs (MEURO)	Benefit (Aeq./ha/jr)	Costs per reduced Aeq. (MEURO)
SO_2	106	76	30	96	164	0,59
NO_x	181	127	54	131	92	1,42
NH_3	74	57	17	363	90	4,03
VOC	144	102	32	81	-	-

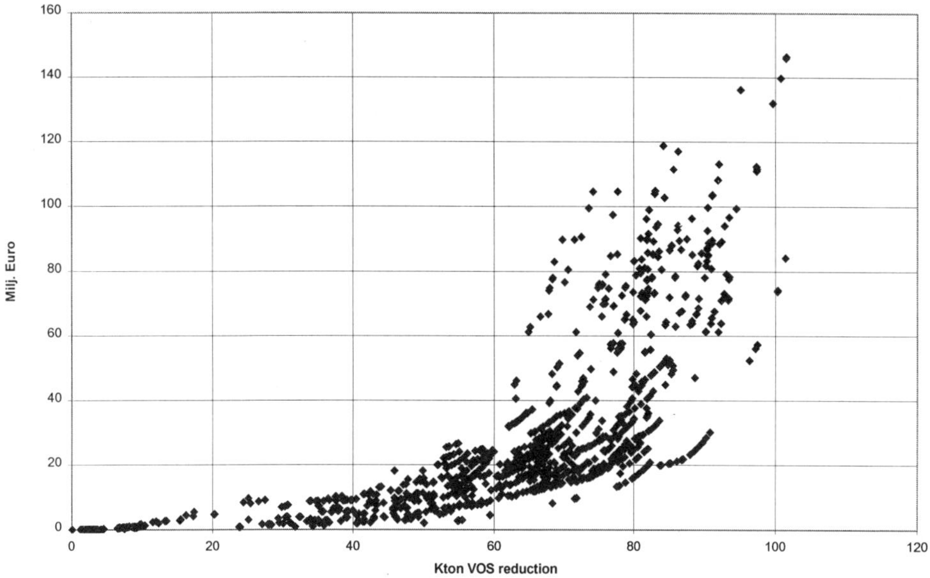

Figure 1: Uncertainties in VOC cost curves using a Monte-Carlo simulation

6. DISCUSSION

The uncertainties in the model calculations are estimated to be ± 20% for the OPS deposition calculations (van Jaarsveld, 1990) and 30% for the LOTOS calculation of maximum ozone concentrations (Builtjes, 1992). Geographical variations over Flanders with respect to AOT40 and AOT60 are also not negligible, as can be seen in Table 3. The AOT40 is reported to be very sensitive to the height at which it is modelled and to the ozone deposition velocity used. Both for the EMEP model and the LOTOS model uncertainties ranging from 200% to 300% were reported (Builtjes and Flossmann, 1998).

Besides the uncertainties in the model computations there are uncertainties in the emission data, the cost curves and in the determination of critical loads. The uncertainties in emission data are very difficult to estimate. A comparison between a number of emission inventories for VOC emissions in Flanders resulted in differences up to 200% (De Vlieger, 1993). For the other substances (SO_2, NO_x and NH_3) it is assumed that their emissions are better known. Figure 1 shows what an uncertainty of ± 20% means for the VOC cost curve. Finally, the uncertainty in determination of the critical loads is estimated to be 20% - 30% (Posch et al., 1997).

However, since the scenario computations are only focussing on relative comparisons between impacts and not on absolute differences between modelled and measured values, some observations are still valid, despite of the large uncertainties. From Table 2 it can be seen that the dominant role of ammonia in acidification and eutrophication will increase in the future for all scenarios. This is due to the fact that all scenarios focus on more cost effective SO_2-emission reductions rather than on expensive NH_3-emission reductions. The GAP scenario is not sufficient to achieve the long-term policy objectives in Flanders (1400 mole H+/ha/yr and 5-20 kg N/ha/yr in 2010). It seems that a further reduction of NH_3 emissions is needed to meet these objectives. However, Table 5 shows that they are relatively expensive and that there is still a potential for less expensive SO_2 and NO_x

reductions. From the results in Table 4 it can be concluded that the GAP scenario gives indeed a reduction in Flanders of 50% in the area of unprotected ecosystems, provided that the 652 forest ecosystems are representative in this case. The AOT60 was found to reduce with more than two thirds (70%) in the GAP scenario. The effect on the AOT40 is less pronounced: a reduction of 13% was found in the AOT40 for crops and only a reduction of 3% in the AOT40 for forests.

7. CONCLUSIONS

In 2010 ammonia is expected to be responsible for more than 50% of the acidification in Flanders and at least 70% of the nitrogen deposition in Flanders. The 50% gap closure scenario is not sufficient to achieve the long-term policy goals in 2010. On a first sight further reductions of ammonia inside (and outside) Flanders seem to be needed to achieve these objectives in 2010. However, they are relatively expensive and there seems to be still a potential for less expensive SO_2 and NO_x reductions.

The expected impact of the 50% gap closure scenario on acidification and AOT60 (reduction of two thirds in 2010 compared to 1990) could be confirmed. For the AOT40, a reduction of one third in 2010 compared to 1990 could not be confirmed. The uncertainties in determining the AOT40 seem to be rather high.

REFERENCES

Amman, M., Bertok, I., Cofala, J., Gyarfas, F., Heyes, C., Klimont, Z., Makowski, M. Schöpp, W. and Syri, S., 1999, *Cost-effective Control of Acidification and Ground-Level Ozone*, Seventh Interim Report to the European Commission, DG-XI, IIASA.

Mylona, S., Storen, E. and Grini, A., 1999, *EMEP Emission data*, Status report 1999, Norwegian Meteorological Institute, Research Report no 26, EMEP/MSC-W, Oslo.

Builtjes, P. and Flossmann, A., 1998, Model validation, *23rd NATO/CCMS International Technical Meeting on Air Pollution and its Application*, Preprints Volume II, September 28 – October 2, Varna, Bulgaria.

Builtjes P.J.H. and Boersen G., 1996, *Model calculations to determine the influence of European emission reductions on ozone concentrations over Belgium*, TNO-report TNO-MEP-R 96/274.

Builtjes, P.J. H., 1992, *The LOTOS Long Term Ozone Simulation-project*, Summary Report, TNO-rapport TNO-MW-R 92/240.

Craenen, H., Van Ranst, E., Groenemans, R., Tack, F. en Verloo, M., 1996, *Berekening en kartering van de kritische lasten voor Vlaanderen* (Calculation and mapping of the critical loads for Flanders), RUG Gent (in Dutch).

De Vlieger, I., 1993, Reliability of VOC-emissions from road transport in Belgium, in: *Proceedings of TNO/EURASAP workshop on the reliability of VOC emission databases*, IMW-TNO publictaie p93/040.

Duerinck, J., Van Rompaey, H. and Siebens, K., 1999, *Analyse van de reductiekosten voor NH_3, SO_2, NO_x en VOS* (Analysis of the reduction costs for NH_3, SO_2, NO_x and VOC), VITO report 1999/PPE/R/8 (in Dutch).

Jaarsveld J.A. van, 1990, *An Operational atmosferic transport model for Priority Substances; specification and instructions for use*, RIVM report nr. 222501002.

Mensink, C. and Dumont, G., 1997, Comparisons of model results with observations for acid depositions and acidifying air pollutants over Flanders, in: Power, H. ; Tirabassi, T., Brebbia, C.A. (eds): *Air Pollution V, Monitoring, Simulation and Control,* pp. 395-404, Computational Mechanics Publications, Southampton.

MINA-plan 2, 1997, *Het Vlaamse milieubeleidsplan 1997-2001* (Environmental Policy Strategy 1997-2001), Ministry of the Flemish Community, Dept. of Environment and Infrastructure, Brussels (in Dutch).

Posch, M., Hettelingh, J.-P., de Smet, P.A.M. and Downing, R.J., 1997, *Calculation and mapping of critical thresholds in Europe*, Status report 1997, Coordination Center for Effect, RIVM Report No. 259101007, Bilthoven.

WHO, 1997, *Air quality quidelines for Europe*, Second edition, Copenhagen

WHO, 1995, *Update and revision of the air quality guidelines for Europe*, Report on the WHO Working Group on Ecotoxic Effects, WHO, Copenhagen.

DISCUSSION

B. E. A. FISHER: Your work on Flanders shows the differences between an integrated assessment on a small length scale and a European assessment performed on a larger length scale (typically 150 km in the IIASA and ASAM calculations). Should one therefore conclude that every local region such as Flanders should perform its own integrated assessment? If so what is the value of the European scale assessment?

C. MENSINK: The regional assessment on a small length scale (typically 5 km) gives more detailed information on pollutant concentrations and depositions, which is very useful to support regional policies (e.g. for ammonia) and to verify the impact of European policies (e.g. the Göteborg Protocol) as shown in this paper. In my opinion both assessments are needed in order to tune regional policies and make them in compliance with the European context.

A HIGHLY PARAMETERIZED REGIONAL OXIDANT MODEL IMBEDDED IN AN INTEGRATED, GAME-LIKE MODEL DESIGNED TO SIMULATE INTERACTION BETWEEN ENVIRONMENTAL, SOCIAL AND ECONOMIC FACTORS

M. Rucker[1], D.G. Steyn[1], D. Biggs[2], M. Walsh[2], D. Rothman[3] and J.B. Robinson[4]

[1] Atmospheric Science Program, Department of Earth and Ocean Sciences The University of British Columbia, Vancouver, B.C., V6T 1Z2, Canada
[2] Envision Sustainability Tools Inc., 2386 East Mall, Vancouver, B.C., V6T 1Z3, Canada
[3] Biosphere 2, 32540 S. Biosphere Rd., P.O. Box 689, Oracle, AZ 85623
[4] Sustainable Development Research Institute, The University of British Columbia, Vancouver, B.C., V6T 1Z2, Canada

INTRODUCTION

QUEST is a game-like model that integrates environment, societal and economic variables in a single state-of-the-art computer model that has the look, user-friendliness and appeal of a computer game. The purpose of the model is to facilitate debate and discussion among a variety of stakeholders about regional sustainability. This is achieved by allowing users (who are in fact players) to actively explore different possible social, economic and environmental scenarios for the future of their region. QUEST is not intended to be a predictive model, but a tool for exploring and evaluating alternative futures. As a result, the modelling approach in QUEST is based on the principles of *backcasting* (Robinson, 1988) and the *design approach to modelling* (Gault, 1987). These methods emphasize the choice and consequences in decision making with the user at the heart of the modelling process. The user is given the freedom to simulate a wide range of choices and observe the consequences of those decisions.

QUEST was originally developed for the Lower Fraser Basin of British Columbia and is under development for a number of regions worldwide, including Manchester, England; Langat Basin, Malaysia; Mexico City, Mexico; Canterbury, New Zealand; and Bangalore, India. The modular design of QUEST allows it to emphasize issues which are important to the region in question, while keeping the modelling approach unchanged.

The version of QUEST presented in this paper is specific to the Lower Fraser Basin (LFB) of British Columbia. The Basin includes Greater Vancouver, a region experiencing rapid urban sprawl, and the Fraser Valley, an area of significant agricultural importance.

With high population and economic growth foreseeable in the future of this region, environmental concerns such as air and water quality, and loss of agricultural land have been the subject of much debate. LFB-QUEST has been used in the Vancouver region for public consultation, public education and outreach, policy debate, and at all levels of schooling. Although LFB-QUEST deals with a wide variety of issues surrounding sustainability, this paper concentrates on the air quality component of QUEST.

DESCRIPTION OF QUEST

Conceptual Framework

The conceptual framework of QUEST is shown in Figure 1. The user creates and explores 40-year future scenarios in four main steps. In *Invent-a-Future*, the user creates a scenario by choosing population and economic activity targets and by defining world views. The world view settings, as defined through ecological resilience, technological innovation and social adaptability, provide a way in which uncertainty surrounding sustainability is incorporated into QUEST. In addition to these parameters, the user specifies politics and land use preferences, and chooses how QUEST summarizes the scenario results at the end of each decade. These settings have no effect on the model calculations, but provide a forum in which the player discovers the complexities and tradeoffs of sustainability. In *Choose Policies*, the user continues to develop the scenario by making various policy and human activity decisions for each decade. These choices are then run through the various submodels and the consequences are presented to the user in *View Consequences.* In this step, the user faces the task of evaluating the desirability of the scenario before making new policy choices for the subsequent decades. This iterative process continues for each decade until a scenario has been generated for the entire 40-year period. In *Scenario Review*, the user reviews the completed scenario and compares his/her goals for politics and land use to the actual scenario outcomes.

Model Structure

QUEST consists of a set of extensively linked submodels, which are designed to run linearly with behavioral feedback occurring through the user. The hierarchy and major linkages of the submodels within LFB-QUEST are illustrated – in a simplified manner – in Figure 2. This figure shows the Demography submodel at the top of the hierarchy while the Economic Input/Output module feeds into all 9 human activity submodels. This dependency structure reflects the concept that human impacts on a region's environment are largely determined by scale (i.e. number of people) and intensity (i.e. level of economic activity). Of the human activity submodels, the Government, Land Use, Labour and Consumption submodels are derived from the Demography and Economic submodels. The Agriculture, Transportation, Housing and Industry submodels are primarily related to the allocation of space to accommodate people and economic activity, and are therefore dependent on the Land Use as well as the Demography and Economic submodels. The final submodel of the human activity group, the Energy submodel, calculates energy use in the Basin, and is dependent on virtually all of the other submodels. Of the environmental submodels, the Natural Habitat and Water Quality submodels flow from the Land Use submodel, while the Air Quality and Ecological Footprint produce outputs that are the result of all of the submodels and their interactions.

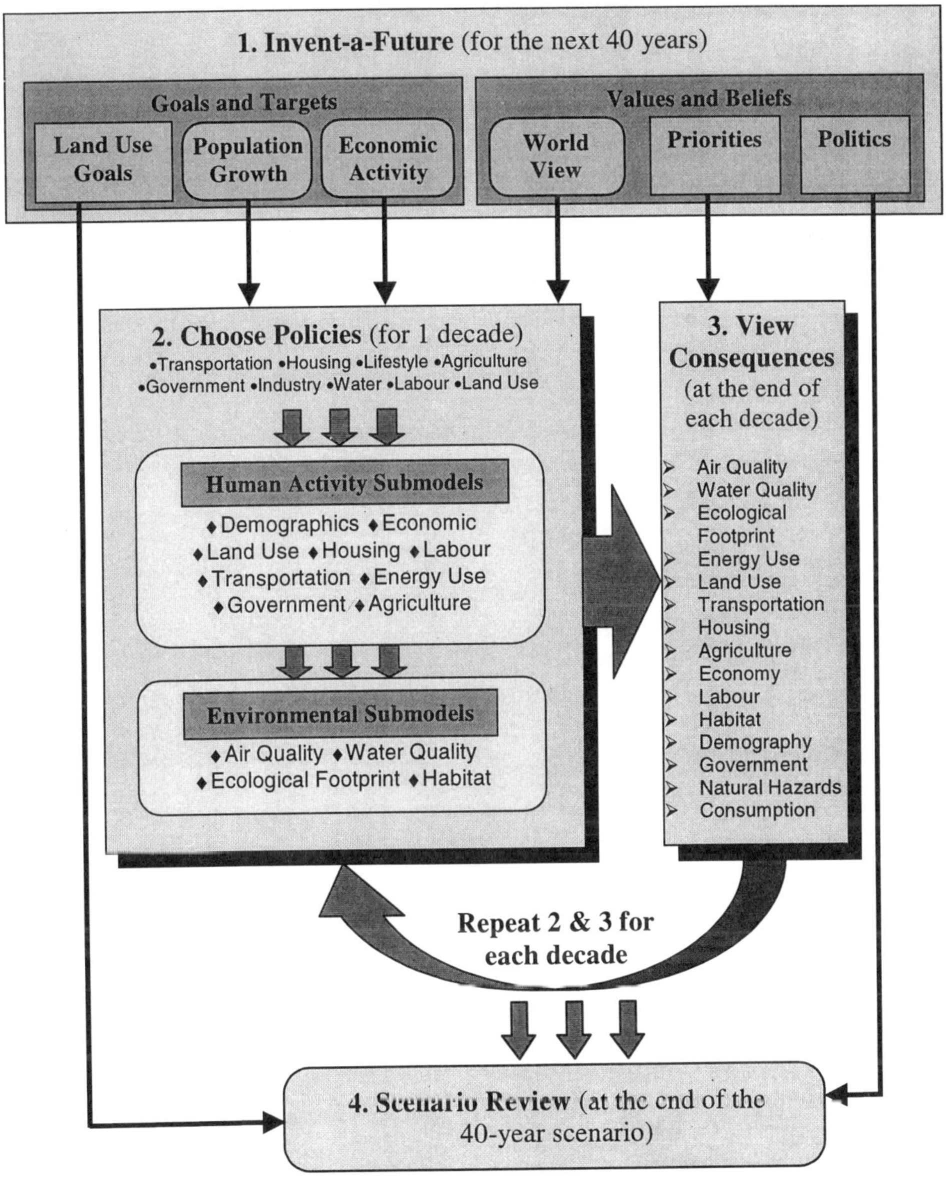

Figure 1. Conceptual framework of QUEST.

User Input

The user inputs into QUEST are listed in Table 1. In addition to setting population and economic activity targets, the user is able to customize his/her future scenario by setting the policy and human activity parameters in Table 1. These parameters are controlled through slider settings, whose actual values are based on observations of past trends, while allowing for a range of plausible future trends. The behaviours of the sliders are also controlled by the world view settings.

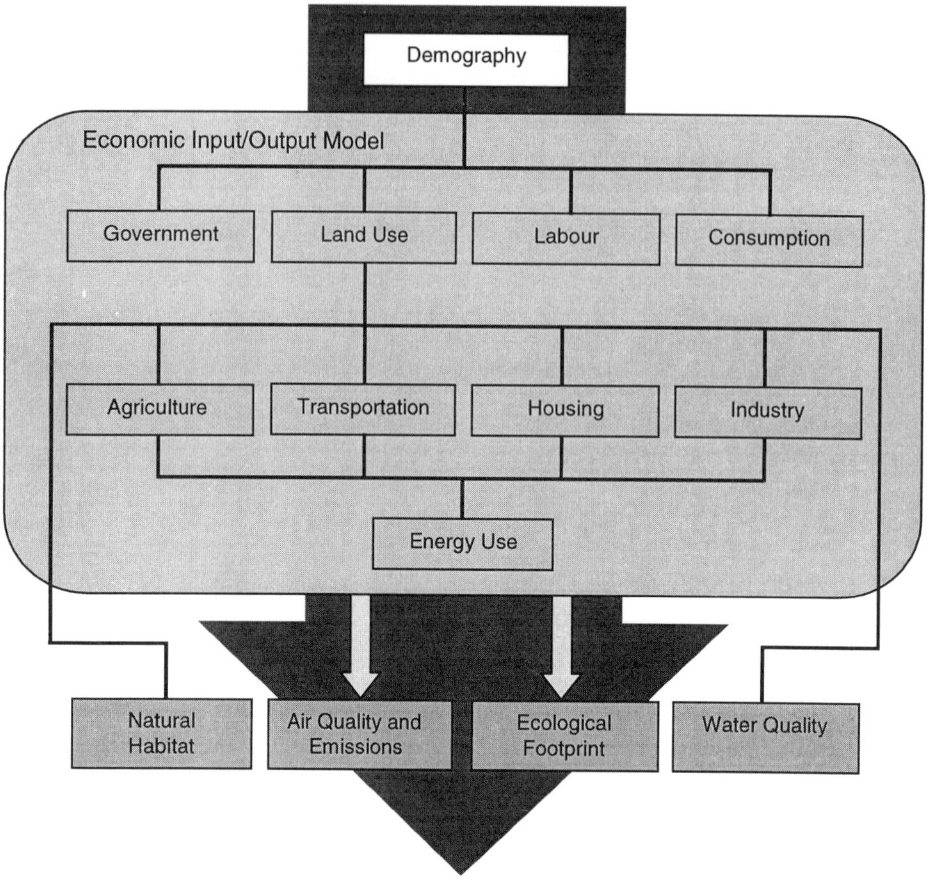

Figure 2. LFB-QUEST submodel dependency structure.

Table 1 further illustrates interactions between submodels in terms of direct and indirect effects of user input on the submodels. The air pollution submodel, for example, does not have direct user inputs, but is affected indirectly through user inputs in the Transportation, Housing, Industry, Labour and Land Use submodels.

AIR QUALITY SUBMODEL

The Air Quality submodel in LFB-QUEST performs two main functions. It aggregates emissions from mobile and stationary sources through outputs from other submodels and generates a spatial map of ozone pollution for the Basin.

Emissions

QUEST segregates emissions into common local air pollutants and greenhouse gases. Common local air pollutants include CO, NOx, SOx, VOCs, road dust and particulate matter, while greenhouse gases include CO_2, CH_4, N_2O and CFCs.
Emissions are determined through outputs from the Land Use, Transportation, Economic, Industry and Energy Use submodels. Aggregated totals of mobile emissions are provided through the Transportation submodel, which considers light- and heavy-duty vehicles.

Table 1. Effects of user inputs on the submodels. An 'o' indicates an input that *directly affects* a submodel. A '*' indicates an input that *indirectly affects* a submodel, by way of other submodels.

	Submodel													
Direct User Inputs	**Demographics**	**Economic I/O**	**Land Use**	**Transportation**	**Air Quality**	**Energy Use**	**Consumption**	**Labour**	**Government**	**Footprint**	**Housing**	**Habitat**	**Water Quality**	**Agriculture**
2030 Population Level	o	*	*	*	*	*	*	*	*	*	*	*	*	*
2030 Economic Level		o	*	*	*	*	*	*	*	*	*	*	*	*
Transportation:														
Transportation Priorities		*		o	*	*	*		o	*				
Vehicle Occupancy		*		o	*	*	*			*				
Fuel Efficiency		o		o	*	o	o			*				
Emission Levels				o	*	*				*				
Housing:														
Housing Density		*	o	*	*	*	*		o	*	*	*	*	*
Energy Efficiency		o				o	o							
Fuel Mix						o								
Lifestyle:														
Diet		o				*	o			o				
Energy & Water Use		*				*	*			o				
Shopping		o				*	o			o				
Waste Generation						*				o				
Agriculture:														
Agricultural Methods						o								
Agricultural Intensity		*				o								o
Government:														
Tax Rate Revenue		*							o					
Health Care Spending		*							o					
Education Spending		*							o					
Social Services Spending		*							o					
Industry:														
Energy Efficiency		o			*	o		*	*	*				
Pollution Control		o			*	*		*	*	*				
Research & Development Spending		o			*	*		o	*	*				
Structural Change		o			*	*		*	*	*				
Water:														
Sewage Treatment		*							o					
Drinking Water Treatment		*							o					
Storm Water Management		*							o					
Labour:														
Male Participation Rate								o						
Female Participation Rate								o						
Retirement Age								o						
Paid Workweek Length		o			*	*		o	*	*				
Land Use:														
Development Pattern			*		*							*	*	*
Source of New Development			*		*							*	*	*
Development in High Risk Areas			*		*							*	*	*

Light-duty vehicle emissions are a function of fuel efficiency and emission standard policies as well as total automobile distance travelled which in turn depends on land use type[1], modal split and automobile occupancy policy. Heavy-duty vehicle emissions are a function of the same factors as light-duty emissions, but GDP is used as a proxy measure for distance travelled.

Stationary source emissions are derived from the Economic and Air Quality submodels. The Economic submodel provides aggregated totals of industrial, agricultural and population related area and point emissions as a function of GDP levels. The Air Quality submodel further contains area emissions from natural lands and miscellaneous population sources, which are assumed constant per unit area and per capita, respectively. These constants are derived from the BC Emissions Inventory of Common Air Pollutants. Natural land area emissions include forest fires and natural emissions from forests, grasslands and shrubs, while miscellaneous area emissions from population sources includes open burning, incineration, landfills, dry cleaning and lawn equipment.

Spatially Allocated Air Pollution Module

Several models with varying degrees of complexity exist to model or forecast ozone concentrations. To thoroughly model the spatial ozone distribution, resulting from the non-linear interactions between emission sources, photochemistry and meteorology, requires the use of a coupled meteorological/photochemical model. Such a research level tool would require unmeetable input data demands and create lengthy processing delays with its high computational overhead costs. Long delays would prevent the user from exploring many alternative scenerios. Since QUEST teaches the user about the trade-offs associated with sustainability issues through the exploration of alternative futures, quick execution times are necessary.

The spatially allocated air pollution module in QUEST contains a highly parameterized regional oxidants model. The parameterization is based on the observed pattern of the urban ozone plume during summertime exceedance days. The levels and patterns of the historic observed plume are then scaled and stretched to respond to changes in emission levels, and shifts in population and transportation patterns. The dependence of the plume on emissions is based on detailed modelling studies using combined mesoscale and photochemical models (Hedley et al., 1997a,b) and observational models (Steyn et al., 1996), and various observational studies (e.g. Steyn et al., 1997). The population and transportation patterns are spatially distributed in the Air Quality submodel in order to modify the spatial pattern of the plume.

The historic observed plume is determined by averaging daily maxima of hourly averaged ozone values during exceedance days for each of the 22 monitoring stations in the LFB over a 5 year period centered around the year 1990 (all future scenarios in LFB-QUEST begin in the year 1990). The analysis is restricted seasonally to the months April through September and diurnally to the hours 1200 to 1900. By using only exceedance days (defined as days on which one or more monitoring station reports hourly average ozone concentrations in excess of 82 ppb), the model in effect mimics days on which meteorological conditions conducive to worst case ozone levels exist. These conditions correspond to limited vertical mixing, horizontal advection from the west and high temperatures. The averaged ozone values of the 22 monitoring stations are inter- and extrapolated to the 500x500m grid used in QUEST. To facilitate in the creation of the grid, ozone background levels of 15 ppb are assumed (Jiang et al., 1997). Outside the limits of the modelling domain as defined by the 500m contour level, ozone levels are forced to zero to avoid unrealistic increases in ozone concentrations over mountainous terrain.

[1] There are 7 land use types in LFB-QUEST: natural, agricultural, industrial, low density, medium density, high density and core.

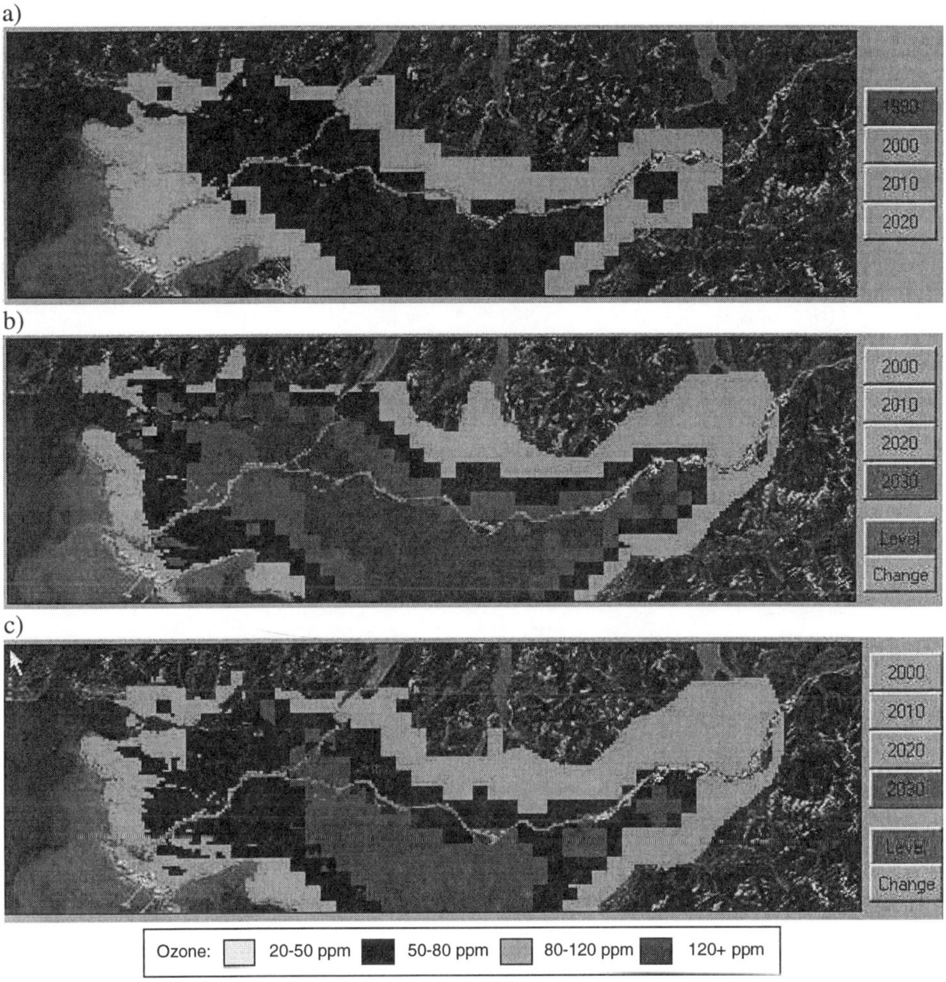

Figure 3. Ozone distribution for a) 1990 base case; b) do-nothing scenario; and c) take-action scenario. The scenarios are described in the text.

The ozone concentration for each grid point and for decade T is determined by multiplying the base case ozone concentration value of each grid cell by the ratio of the total upwind VOC emissions in decade T to those in 1990. It is assumed that the whole of the basin is VOC sensitive; hence any air pollution control strategy would involve controlling VOC emissions. The total upwind VOC emissions for a cell X are determined by adding up the VOC emissions per neighbourhood type for all cells which lie west upwind of cell X and within the angle of 53° centered at cell X[2]. The angle of 53° accounts for dispersion of the precursor emissions as well as change in wind direction due to large scale meandering and channeling of the winds by the topography in the LFV. The VOC emissions per neighbourhood type are determined by adding up the mobile, point and area sources per neighbourhood type for VOC. VOC mobile sources per neighbourhood type are determined by dividing the total VOC emissions by the total population and then multiplying by the population per neighbourhood type.

[2] The angle of 53°is used to accommodate the numerical coding of the model. The angle represents a 2:1 ratio of east-west to north-south grid cells.

EXAMPLE OUTPUTS

In this section, the air quality consequences for two scenarios, a 'do nothing' and a 'take action' future, are illustrated. Both assume the same growth in population and economic activity, and identical world views. In the 'do nothing' scenario, the future continues on the same trends as in the past. The 'take action' scenario assumes that the population growth occurs in denser urban nodes and that people become less car-oriented. In the 'do-nothing' scenario (Figure 3b), there is a significant increase in ozone levels over most of the Basin. In the 'take action' future (Figure 3c), there is still an increase in ozone levels, but the increase is significantly reduced.

SUMMARY

In this paper, we have presented the air quality submodel in LFB-QUEST. Although the air quality model is highly parameterized, as required by the relatively short execution times, the interactions between the air quality module and the other make it a powerful tool in exploring the links between human behaviour and air quality. It is hoped that through the exploration of these links, future air quality issues can be better understood by regional planners and the public.

REFERENCES

BC Environment – Air Resources Branch, December 1994, 1990 British Columbia Emissions Inventory of Common Air Pollutants.

Envision Sustainability Tools Inc., 1998, LFB QUEST Model Structure, 92pp.

Gault, F.E., K.E. Hamilton, R.B. Hoffman and B.C. McInnis, 1987, The design approach to Socio-Economic modelling. Futures, **19**, 3-25.

Hedley, M. and D.L. Singleton, 1997a, Evaluation of an air quality simulation of the Lower Fraser Valley – I. Meteorology. *Atmos. Env.*, **31**, 1605-1615.

Hedley, M., R. McLaren, J. Weimin and D.L. Singleton, 1997b, Evaluation of an air quality simulation of the Lower Fraser Valley – II. Photochemistry. *Atmos. Env.*, **31**, 1617-1630.

Jiang, W., D.L. Singleton, M. Hedley and R. McLaren, 1997, Sensitivity of ozone concentrations to VOC and NO_x emissions in the Canadian Lower Fraser Valley. *Atmos. Env.*, **31**, 627-638.

Robinson, John B., 1988, Unlearning and Backcasting: Rethinking some of the questions we ask about the future. Technological Forecasting and Social Change, **33**, 325-338.

Rucker, Magdalena, 1998, Spatially Allocated Air Pollution Module in QUEST. Report submitted to the Sustainable Development Research Institute, UBC. 15pp.

Steyn, D.G., M. Baldi, K. Stephens and D.R. Hastie, 1996, Observation based analysis of photochemical smog in the Lower Fraser Valley using two analytical schemes. UBC Department of Geography Occasional Paper No. 41. 44pp.

Steyn, D.G., J.W. Bottenheim and R.B. Thomson, 1997, Overview of tropospheric ozone in the Lower Fraser Valley, and the Pacific '93 field study, *Atmos. Env.*, **31**, 2025-2035.

DISCUSSION

R. WARREN: You mentioned a trade-off between environment and economy - isn't an important way forward for sustainability to develop air pollution scenarios such as energy saving schemes and are these included in the model?

D. G. STEYN: Underlying the entire QUEST model is the idea that sustainability can only be achieved by reconciling the competing requirements of environment, economy and society. A result of this reconciliation is that no trade-off exists. Because the model has full coupling between environment, economy (including energy) and society, all links such as the one you mention are built in.

C. MENSINK: One of the basic assumptions in the model is that the ozone parameterisation based on the results for 1990 is valid for four decades. Non-linear phenomena in the ozone formation processes are therefore not considered. Can you comment on this?

D. G. STEYN: Yes indeed, this is one of the major weaknesses of the model as it exists at the moment. We are working on a more complete version of this model, expanded to cover a larger domain, and to include a wider range of components of the environment. The air quality component will be modified to incorporate true responsiveness to emissions levels by multiple runs of a photochemical model. These model runs will be used to generate an ozone response surface (effectively an OZIPP) for the region, as well as more complete ozone spatial patterns.

EVALUATION OF AMMONIA EMISSION REDUCTIONS IN THE NETHERLANDS USING MEASUREMENTS AND THE OPS MODEL

J. (Hans) A. van Jaarsveld[1] and Albert Bleeker[2]

[1]National Institute of Public Health and Environmental Protection
P.O. Box 1
3720 BA Bilthoven, The Netherlands

[2]TNO-MEP
P.O. Box 342
7300 AH Apeldoorn, The Netherlands

INTRODUCTION

Atmospheric ammonia (NH_3) is a pollutant, being, along with sulphur dioxide (SO_2) and nitrogen oxides (NO_x), one of the three main primary pollutants leading to acidifying deposition. With the decreasing SO_2 emissions throughout Europe, and an increased appreciation of the role of NH_3 and NO_x in causing eutrophication of ecosystems, the scientific and political attention given to NH_3 has grown. Recently, this has led to activities within the UNECE to develop a second protocol to limit European NO_X emissions that would also consider controls on NH_3 emissions.

In the Netherlands the need for NH_3 emission control was already recognised in the early eighties (van Breemen et al. (1982); Heij and Schneider, 1991). Several abatement measures were taken since 1990 such as the coverage of manure storage basins, the development of low emission stable systems, and, when applying the manure in the field: the (obligatory) incorporation of manure into the soil. On the basis of statistical data and assumptions about the effectiveness of the regulations it was estimated that the ammonia emissions were decreased by 37% in 1996 relative to 1990.

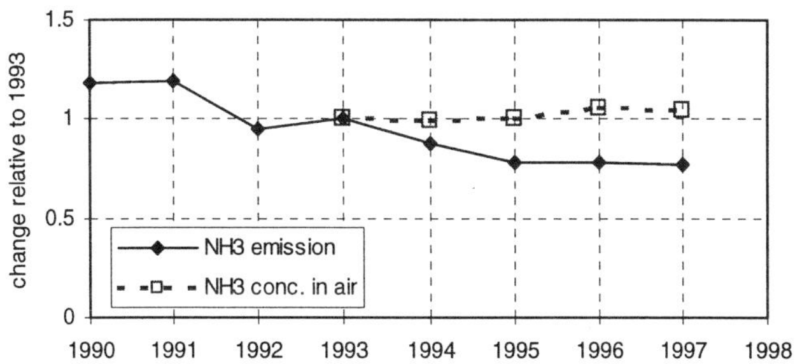

Figure 1. Comparison of trends in national NH_3 emissions and area averaged NH_3 concentrations

A national network was set up in order to monitor ammonia and ammonium concentrations in ambient air and in precipitation. The suggestion was that an emission decrease of this magnitude would have effected atmospheric concentrations significantly. At first sight, however, neither of the concentrations showed the expected decrease. The problem is made visual in Figure 1.
This paper describes an evaluation of the set of measurements and emissions with the help of the transport and deposition model OPS. The work is a quantitative extension to the analysis described by Erisman et al. (1998) as it comes up with an estimate of the achieved national emission reduction.

AMMONIA IN THE ATMOSHERE

The atmospheric cycle of ammonia is illustrated in Figure 2. Ammonia evaporates from manure and is carried by the wind. Because of the low height at which NH_3 enters the atmosphere (0-5m), relatively high concentrations will result, especially in low wind speed conditions. The atmospheric NH_3 concentration can be measured and used as the most direct indicator for NH_3 emissions. Some of the ammonia will be converted into ammonium salts. The most important reaction is with H_2SO_4 resulting in either $(NH_4)_2SO_4$ or $(NH_4)HSO_4$. Most of the atmospheric H_2SO_4 in NH_3 rich areas such as the Netherlands will disappear this way. The remaining NH_3 will be involved in an equilibrium reaction with HNO_3 giving $(NH_4)NO_3$. Because of the high NH_3 concentrations in the Netherlands there will be low HNO_3 concentrations on average.
The most important removal process is dry deposition, but it is also the least well quantified process. Several studies indicate the existence of so called canopy compensation points for plants (Farquhar et al., 1980; Sutton et al., 1993). When atmospheric concentrations are at or below this point then dry deposition stops or even changes sign (=emission from the plant). The compensation point concentration is plant species dependent and a function of temperature and probably also of the N-fertiliser condition of the soil Schøring et al. (1998). Another complication is the mutual influence of dry deposition of NH_3 and SO_2 through pH coupling (co-deposition, van Hove et al., 1989). Since the NH_3/SO_2 ratio is rapidly growing in the last ten years, an impact on the NH_3 concentration levels may be expected.
Wet deposition is also an important removal process. In the Netherlands most of the wet deposition is from direct uptake of gaseous NH_3 in cloud and precipitation droplets.

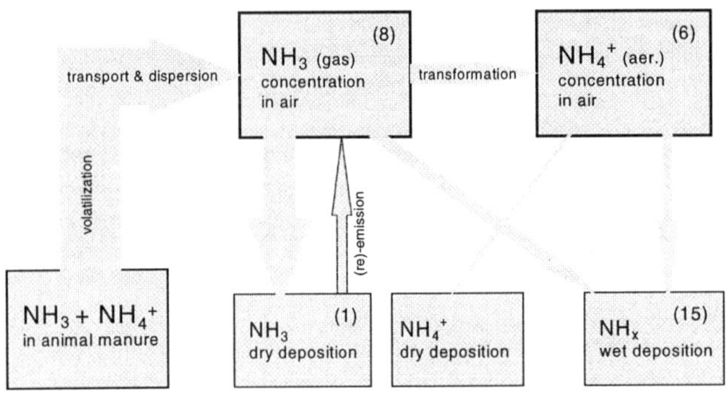

Figure 2. The emission, dispersion, transport, conversion and deposition chain for ammonia. The width of the arrows indicate the relative importance of the fluxes. The number of measuring sites in the Netherlands for the different species is given within brackets.

MEASUREMENTS

When measurements are to be used to derive time trends from it, then it is very important that the methodology and/or calibration do not change in time. Such a usage of the data should be known from the start otherwise one might end up with 'modifications' of which the effect is not sufficiently quantified. In the Netherlands a consistent set of ammonia concentrations is available since September

1992. These observations are part of the Dutch Air Quality Monitoring Network (LML) and consists of 8 stations where on an hourly basis ammonia is measured using automated wet denuder systems 'AMOR' developed by ECN (Wyers et al., 1993).In addition to these measurements also ammonium aerosol (NH_4^+) is measured at 7 locations (24h samples) and wet deposition at 14 locations(monthly basis).The location of the stations is given in Figure 2.

Analysing the hourly NH_3 results gives a lot of insight in both the temporal and the spatial differences in The Netherlands. It turns out that NH_3 concentrations in areas with high emissions behave completely different from those in more remote areas. For example, the highest concentrations in high emission areas occur during the night when the atmosphere is stable, while in low emission areas the highest concentrations are measured around noon when the atmosphere is unstable. Looking at the 5-year average annual cycle then the highest concentrations are found in spring (manure is spread out on the land) and in August (highest temperature, probably the lowest dry deposition velocity). All sites show a strong correlation of NH_3 concentrations with precipitation duration when this is analysed on a monthly basis (explained variance approx. 40%). This may point to a dependency of agricultural activities on weather conditions but also to a relation between dry deposition velocities and surface wetness.

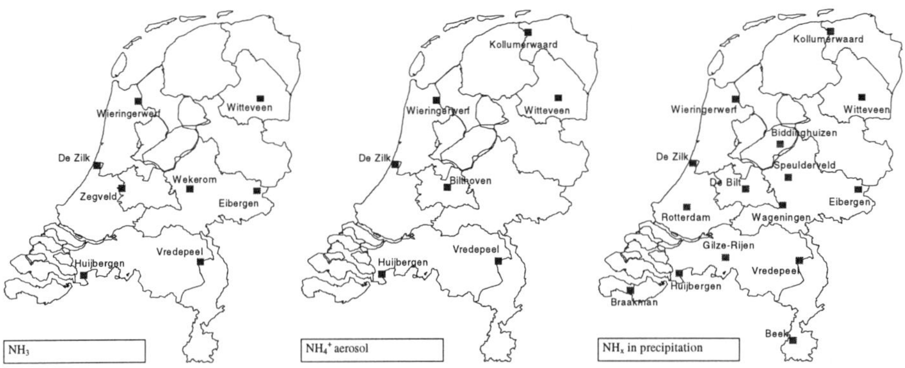

Figure 3. Locations of the monitoring stations in the Netherlands

Tabel 1. Ammonia emissions for the Netherlands for the relevant categories (kton $NH_3\ a^{-1}$).

	agricultural emissions					other	total
	grazing	housing	storage	land spreading	fertiliser use	emissions	emissions NL
1985	15.7	83.2	2.5	125.0	12.3	16.7	255.4
1988	15.0	83.7	4.6	106.8	11.1	11.7	232.9
1989	15.0	82.3	4.8	104.7	10.8	11.7	229.3
1990	15.8	83.7	5.2	104.9	10.0	11.7	231.2
1991	16.1	86.0	5.5	104.4	9.7	11.7	233.3
1992	15.4	85.2	5.8	57.8	9.5	11.7	185.4
1993	15.5	88.4	6.2	64.5	9.5	11.7	195.7
1994	14.2	85.3	5.9	45.3	9.0	11.7	171.5
1995	14.4	86.2	4.0	26.3	9.9	11.4	152.3
1996	14.9	86.5	4.1	25.6	9.5	11.4	152.0
1997	14.2	86.8	4.1	25.4	9.4	11.5	151.4

EMISSIONS

The Dutch State of the Environment publishes every year the official ammonia emission numbers for the Netherlands (RIVM, 1997). These numbers are calculated with models which include emission factors for the different agricultural sources and animal categories derived from research and generalised by the Working Group Agricultural Emission Factors, and statistical data on number of animals, housing systems, manure transportation and the implementation of measures for each municipality (Van der Hoek, 1994). Table 1 gives the emission numbers.
For modelling purposes the emissions per municipality are then distributed over a 5x5km grid covering the Netherlands. Foreign emissions are taken from data reported to UN-ECE (EMEP, 1998). The spatial distribution of the emissions in Belgium and Germany is based on Asman (1992). The resulting emission distribution map for the Netherlands and parts of Belgium and Germany is given in Figure 4.

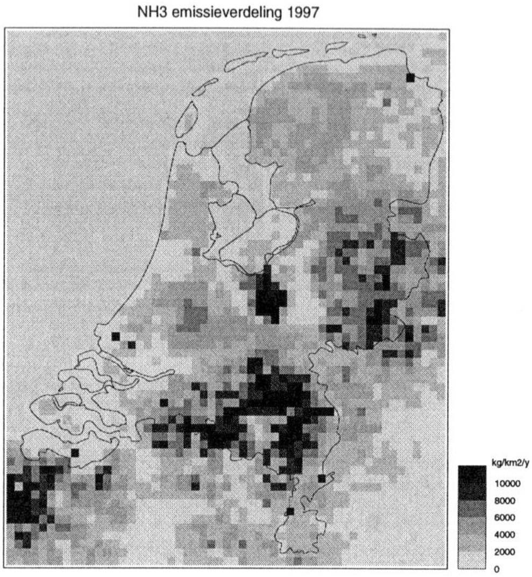

Figure 4. Distribution of ammonia emissions over the Netherlands

MODEL RESULTS

The model used to relate ammonia emissions with concentrations in air and precipitation is the OPS model (Operational Priority Substances, van Jaarsveld, 1995). This model is a combination of a Gaussian plume model for local scale application and a trajectory model for long range transport. It calculates dispersion and dry- and wet deposition for both NH_3 and the secondary formed NH_4^+ with a spatial resolution, which is mainly dependent on the resolution of the emission data. The model was successfully applied to deduce SO_2 and NO_2 trends from measurements of ambient concentrations. Asman and van Jaarsveld (1992) applied the model for the calculation of the NH_x distribution in Europe. For the current application it is important that the model has a high spatial resolution and even can calculate the contribution of individual stables to the NH_3 concentration at the monitoring stations. The model is recently extended with some features such as the calculation of time dependent NH_3 to NH_4^+ conversion rates on the basis of pre-described SO_2, NO_2 and NH_3 concentration fields over NW Europe. Another important extension is the inclusion of an ammonia volatilisation model (van Jaarsveld, 1996) which describes the emission rate of manure applied on the field as a function of temperature and atmospheric stability.

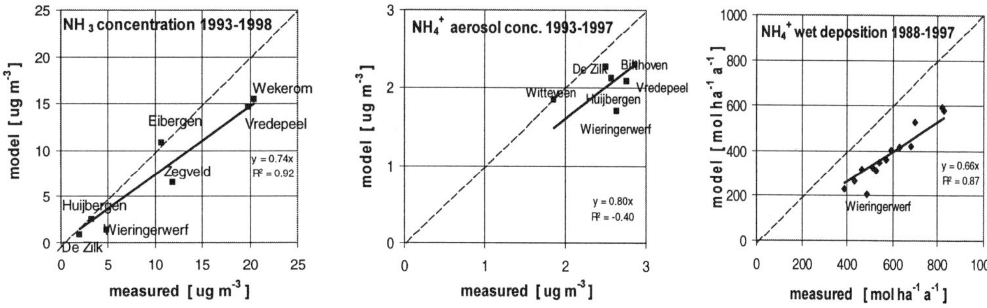

Figure 5. Comparison of time averaged model results with measurements of the LML network

In Figure 5 the capability of the model is showed to simulate the spatial distribution of NH_3, NH_4^+ and NH_x wet deposition. For this purpose temporal fluctuations have been eliminated as much as possible by taking both the modelled and measured results as averages over several years. It is shown in Figure 5. that the spatial agreement is satisfactory but all model outputs are systematically lower than the measurements. This points probably to a general underestimation of NH_3 emissions.

Time series of modelled and measured values are given in Figure 6.. This comparison is based on spatial averaged values as much as possible in an attempt to eliminate local influences. The modelled NH_3 concentration shows a slight decrease in the period 1993-1997, and the measured concentrations a slight increase. The discrepancy in trends is much less than in Figure 1 where the emission trend was compared with the measured concentrations indicating that the atmospheric conditions influenced the emission-concentration relations in this period. On the basis of modelled and measured NHx wet deposition one may come to similar conclusions as for the NH_3 concentration case. The measured NH4+ aerosol concentration decreases more rapidly than the other measured species. This is mainly due to the fact that these concentrations depend on available H_2SO_4 much more than on NH_3.

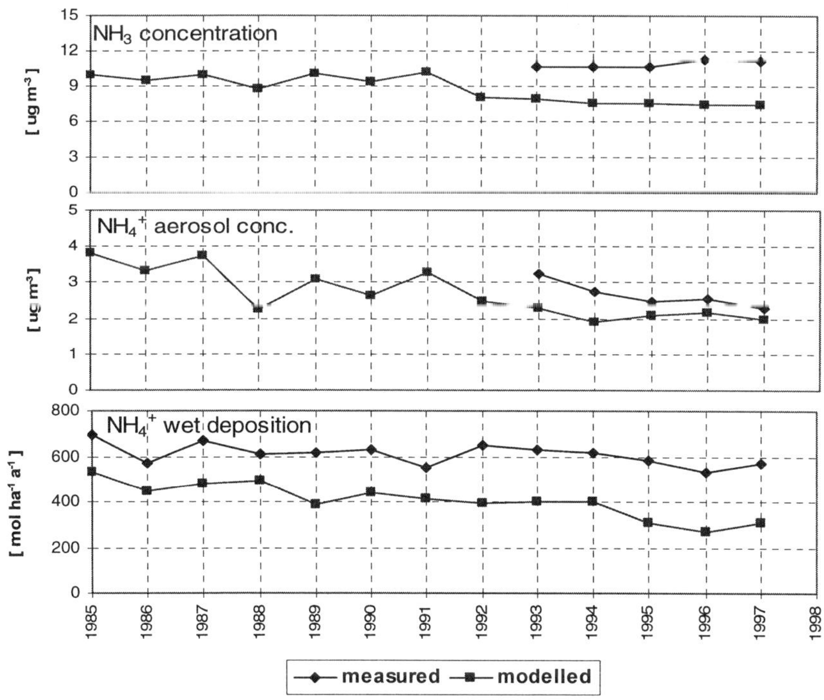

Figure 6. Comparison of time series of measured and modelled values.

Influence of chemistry and meteorology on NH_x time series

The influence of the changing composition of the atmosphere is determined by taking the ratio between results as calculated in Figure 6 and the results for the same years when pre-cursor concentrations SO_2 and NO_2 are fixed at their 1984 levels. The results are given in Figure 8 in terms of the impact on NH_3 concentrations and the wet deposition. For average NH_3 concentrations in the Netherlands there is a positive effect of almost 5% in the period 1993-1997. This means that if emissions have decreased by 5%, the effect on NH_3 concentrations is fully masked by the decreasing $NH_3 > NH_4^+$ conversion.

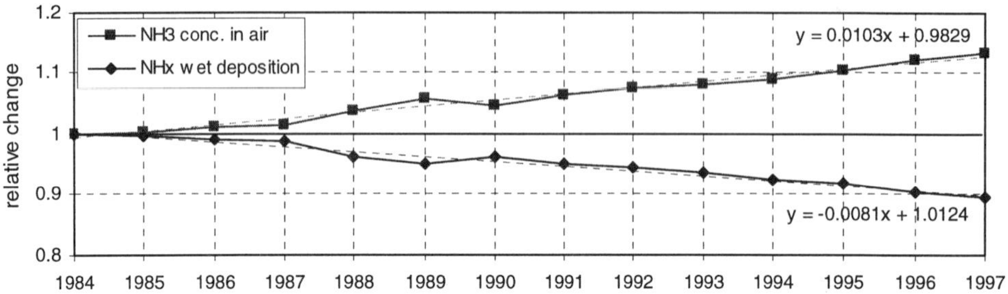

Figure 7. The influence of the changing chemical composition of the atmosphere on NH_3 concentrations and on wet deposition. The reference year is 1984

A similar procedure was applied to calculate the influence of meteorology + chemistry. Here model runs were carried out for all the years, using 1993 emissions but with actual meteorological and chemical data. The results are given in Figure 8. The largest change in time is in the NH_4^+ concentration. This is mainly due to the decreasing SO_2 emissions in this part of Europe. What is also clear from Figure 8 is that the meteorological conditions in 1995-1997 caused high NH_3 concentrations compared to 1993. The 1995-1997 period was relatively dry and the average wind speed was relatively low. The combined impact of all circumstances gives a 12% higher NH_3 concentration in 1997 compared to 1993.

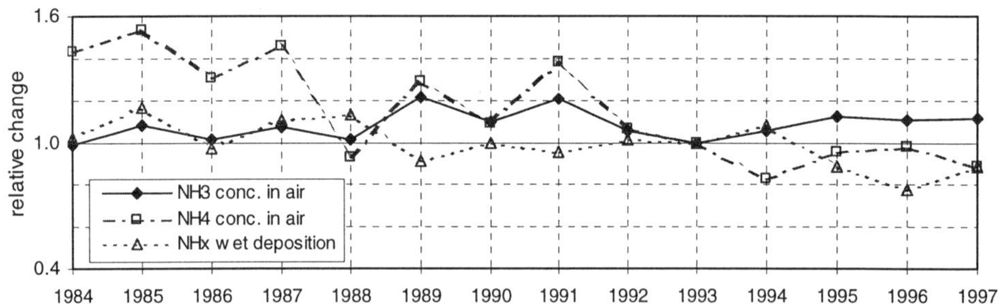

Figure 8 The influence of meteorological + chemical changes. The reference year is 1993

Estimation of the realised reduction of ammonia emission in the Netherlands

Although in the period 1993-1997 the difference between model calculated and measured NH_3 concentration trends (Figure 6.) is smaller than the difference between emission trends and NH_3 concentration trends (Figure 1.), the data still suggest that the real emission decrease is less than the official decrease. The net change in NH_3 concentration due to emission changes in the 1993-1997 period can be estimated as:

change in measured NH_3 concentration	+ 6.3	%
impact of meteorological changes	- 7	%
impact of changing chemistry	- 5	%
net change in NH_3 concentration due to emission changes	- 5.7	%

This net decrease cannot directly translated to an emission decrease in the Netherlands because the emission-concentration relation is not the same for all emission categories. Moreover, one has to take into account the influence of foreign emissions. In Table 2 the relative contributions of the most important emission categories are given. The striking thing is that emissions from housing systems have a much larger effect on NH_3 concentrations than emissions from the other categories (a factor 2.8 more effective!). This is mainly the result of the dependence of land spreading emissions on weather conditions (low emissions during stagnant conditions).

Tabel 2 Relative contributions of emission categories to concentrations and depositions in 1997. Values are calculated as averages for 7 selected NH_3 locations of the LML network.

	Foreign sources (in % of the total contribution)	*Dutch sources (in % of the Dutch contribution)*			
		Housing	Land spreading	Grazing + fertiliser	Other
emission		60.0*	16.8*	15.7*	7.6*
NH_3 concentration	6.4	80.8	9.7	7.7	1.9
NH_4^+ concentration	6.6	79.6	5.6	5.4	9.4
wet deposition	39.1	66.1	13.7	12.6	7.6
dry deposition	11.0	72.1	13.9	11.4	2.6
total deposition	16.8	71.2	13.9	11.6	3.4

* percentage of the total NH_3 emission in the Netherlands

Using the data in 1 and 2 the net decrease of NH_3 concentrations in the 1993-1997 period can be written as:

$$0.936 \times (E_{housing} \times EC_{housing} \times \Delta_{housing} + E_{oth} \times EC_{oth} \times \Delta_{oth})/(E_{housing} + E_{oth}) = 5.7\ \%$$

in which $E_{housing}$ represents the housing and storage emissions and E_{oth} the sum of the remaining emissions. The relative emission-concentration relations are denoted as $EC_{housing}$ and EC_{oth} while the real decreases (% between 1993 and 1997) for the two emission categories are denoted as $\Delta_{housing}$ and Δ_{oth}. The factor 0.936 is the fraction of the atmospheric NH_3 concentrations caused by foreign emissions. $E_{housing}$ and E_{oth} are given in Table 1 as 88.6 and 107.3 kton a^{-1} respectively for the reference year 1993. $EC_{housing}$ and EC_{oth} are given in Table 2 as 80.8/60.0 and 9.7/16.8. If one assumes that $\Delta_{housing}$ is very small (see Table 1) then Δ_{oth} can be calculated as 19.3 %, and the total emission reduction in between 1993 and 1997 as 20.7 kton. The claimed reduction was 44.3 kton so one can say that almost half the reduction is realised.

Determining the type and magnitude of the missing emissions

If the realised emission reduction is less than planned then the first suggestion is that the measures taken were less effective than foreseen. Since the introduction of land spreading techniques was a keystone in the emission reduction plans, it is interesting to see if there is any confirmation in the measurements for higher land spreading emissions than used in the calculations so far. There are number of cases where the typical characteristics of this source category may show up in the NH_3 concentration measurements:

1. The spatial pattern of housing and land spreading emissions is not the same. Land spreading emissions are more evenly distributed over the country then emissions from housings.
2. The relation with temperature and atmospheric stability is different. Land spreading emissions depend much more on these parameters.
3. The seasonal pattern of the emissions is different. Land spreading takes place, for example, mainly from February to May but only if the land can be accesses by heavy tractors.

On the basis of those 3 cases a number of model runs were performed to see which emission change would give the best agreement with the measurements. The correlation coefficient for the set model results and measurements was taken as the measure of agreement wherever possible.

1. The explained spatial variance for the set of 7 stations increased from 0.87 to 0.93 if the land spreading emissions were increased by a factor 7.
2. Dividing the individual hours over classes of atmospheric stability for both the measurements and model results showed a significant improvement in the agreement, both for a high emission area and a low emission area, when land spreading emissions were increased. The optimal agreement occurred at a factor 7.
3. When monthly values were compared then an optimal agreement was found when land spreading emissions were increased with a factor 4 in 1996. This case is shown in Figure 9.

Of this three cases the one on the basis of monthly variations is the most powerful to quantify the influence of land spreading emissions because these emissions are almost zero in winter months and high in spring. The multiplying factors ranged from 1.3-1.6 for the years 1994-1995 and 1.8-4 in 1996 and 1997. Considering all cases a multiplying factor of 3 for land spreading emissions may be considered as the most realistic for 1996 and 1997.

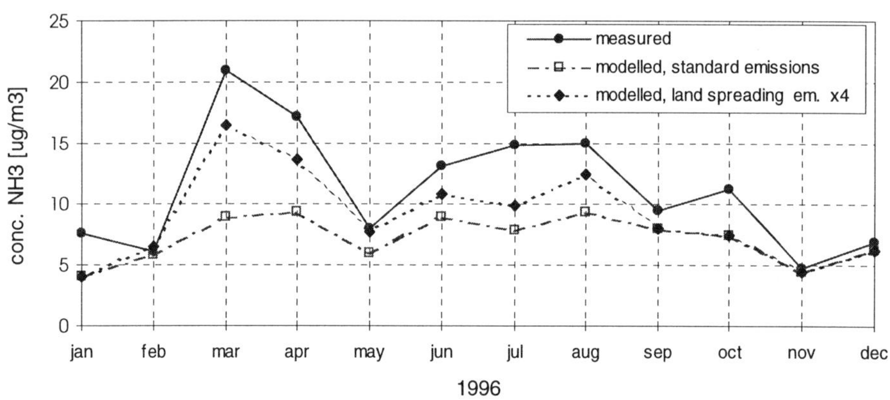

Figure 9. Comparison between measured and modelled NH_3 concentrations on a monthly basis, before and after optimising the land spreading emissions

SUMMARY AND CONCLUSIONS

Ammonia concentrations in ambient air are strongly related to local ammonia emissions and can act as the most direct indicators for spatial emission differences and trends in emissions. Nevertheless, the interpretation of measured time-series in terms of emission trends is not straightforward because year-to-year differences in meteorological factors and systematic changes in the chemical composition of the atmosphere have their impact on the emission-concentration relation. These impacts can be quantified by using a transport and deposition model.
In the Netherlands the official ammonia emissions decreased 23% between 1993 and 1997 mainly because of introducing manure application measures such as the obligatory surface injection of liquid manure. In the same period there was no visible change in measured ammonia concentrations.

Analysing the 1993-1997 period with the OPS model learned that:

- Emissions from land spreading of manure depend strongly on atmospheric stability, their impact on ambient ammonia concentrations is much less then emissions from stables. Since the 23% emission decrease is mainly due to the introduction of land spreading techniques such as sod-injection, only a 13% decrease in ammonia concentration is to be expected.
- Meteorological factors (e.g. low precipitation amounts in 1996 and 1997) masked a 7% decrease.
- The changing chemical composition of the atmosphere (mainly SO_2) masked another 5 % decrease

From the combination measurements and model calculations follows an emission decrease in 1997 of 20 kton NH_3 relative to 1993. This is roughly half the official emission decrease over the same period.

A comparison of the measured and modelled concentrations shows that the modelled diurnal variation, annual variation and the spatial distribution all improve significantly when emissions of the manure land spreading type are considerably increased (factor 2-8, best estimate: 3).

An explanation of the remaining differences between measurements and modelling results requires more research in the field of a) ammonia emissions from land spreading b) spatial distribution of ammonia concentrations over the Netherlands and c) the dry deposition process of ammonia particularly for dominant land cover types.

REFERENCES

Asman, W.A.H. (1992) Ammonia emission in Europe: updated emission and emission variations. Report No. 228471008, National Institute of Public Health and Environmental Protection, Bilthoven, The Netherlands.

Asman, W.A.H. and van Jaarsveld, J.A. (1992) A variable resolution transport model applied for NHx for Europe. *Atmospheric Environment* **26A**, 445-464.

EMEP (1998) Transboundary acidifying air pollution in Europe.MSC-W Status Report 1/98, Part 1; Calculation of acidifying and eutrophying compounds and comparison with observations. Norwegian Meteorological Institute, Oslo, Noorwegen.

Erisman, J.W., Bleeker, A., and van Jaarsveld, J.A. (1998) Evaluation of ammonia emission abatement on the basis of measurements and model calculations. *Environmental Pollution* **102**, 269-274.

Farquhar, G.D., Firth, P.M., Wetselaar, R. and Weir, B. (1980) On the gaseous exchange of ammonia between leaves and the environment: determination of the ammonia compensation point. *Plant Physiol.* **66**, 710-714.

van Breemen N., Burrough P.A., Velthorst E.J., van Dobben H.F., de Wit T., Ridder T.B. and Reinders H.F.R. (1982) Soil acidification from ammonium sulphate in forest canopy throughfall. *Nature* **299**, 548-550.

van der Hoek, K.W. (1994) Estimation of the ammonia emissions in the Netherlands in 1990, 1991 and 1992. Report No. 773004003, National Institute of Public Health and Environmental Protection, Bilthoven, the Netherlands (in Dutch).

Heij, G.J. and Schneider, T. (1991) Final report, Dutch Priority Programme on Acidification, Second Phase. Report No. 200-09, National Institute of Public Health and Environmental Protection, Bilthoven, The Netherlands.

Hove, L.W.A. van, Adema, E.H., Vredenburg, W.J. and Pieters, G.A. (1989) A study of the absorption of NH_3 and SO_2 on leaf surfaces. *Atmospheric Environment* **23**: 1479-1486.

van Jaarsveld, J.A. (1990) An operational atmospheric transport model for priority substances; specification and instruction for use. RIVM rapport no. 222501002, Bilthoven.

van Jaarsveld, J.A. (1995), Modelling the long-term atmospheric behaviour of pollutants on various spatial scales. PhD thesis, University of Utrecht.

van Jaarsveld, J.A. (1996) The dynamic exchange of pollutants at the air-soil interface and its impact on long range transport In: *Air Pollution Modeling and its application XI,* edited by Sven-Erik Gryning and Francis Schiermayer.

Schjørring, J.K., Husted S., and Mattson M. (1998) Physiological parameters controlling plant - atmosphere ammonia exchange. *Atmospheric Environment* **32**, 507-512.

Sutton, M.A., Pitcairn, C.E.R., Fowler, D. (1993) The exchange of ammonia between the atmosphere and plant communities. *Advances in Ecological Research* **24**: 301-393.

RIVM (1997) State of the Environment 1997. Samson-Tjeenk Willink, Alphen a/d Rijn, the Netherlands (in Dutch).

Wyers, G.P., Otjes, R.P. and Slanina J. (1993) A continuous-flow denuder for the measurement of ambient concentrations and surface exchange fluxes of ammonia. *Atmospheric Environment* **27A**:2085-2090.

CHEMICAL TRANSPORT MODEL ON-LINE COUPLED WITH RAMS FOR REGIONAL CHEMICAL CLIMATE

Itsushi Uno[1,2], Seita Emori[3], and Marina Baldi[4]

[1] Research Institute for Applied Mechanics/ Kyushu Univ
Kasuga Park 6-1, Kasuga 816-8580, Japan
[2] Frontier Research System for Global Change, Tokyo 105-6791, Japan
[3] National Institute for Environmental Studies, Tsukuba 305-0053, Japan
[4] Institute of Atmospheric Physics (IFA-CNR), Rome, Italy

INTRODUCTION

Fast growing East Asian countries with their rapid increase in population and economy and, consequently, with their augmented energy consumption and production, show a dramatic increase of the emissions of air pollutants like sulfur and nitrogen oxides and particulates. As a consequence, acid deposition and long range transboundary air pollution are becoming one of the major concerns of Environmental Protection Agencies in each country which have recently started to expand their monitoring activities.

Oceans such as the East China Sea, the Japan Sea between mainland China and Japan, as well as the Pacific Ocean cover a huge fraction of the region, and the interaction between Pacific maritime air masses and continental air masses strongly affect the meteorology and climatology of East Asia. In particular, as a result of this interaction, storm tracks and orographic precipitation during winter, synoptic scale persistent precipitation belt related to the summer monsoon, and the passage of fast moving typhoon represent some of the climate features characteristic of this region (Emori *et al.,* 2000). Therefore knowledge of both pollutant concentration and meteorological parameters over these oceans is critically necessary to understand the long-range transport and transformation processes. However there are quite few observation point over these oceans, therefore remote sensing observation (i.e., satellite) and/or simulations with a regional meteorological model become important tools to evaluate the spatial distribution of cloud and precipitation in this area.

To overcome these difficulties, an on-line regional scale chemical transport model fully coupled with CSU-RAMS (Regional Atmospheric Modeling System, Pielke *et al.*, 1992) was developed to study regional transboundary air pollution. In particular, the coupled model can be adopted to simulate regional meteorological climate and chemical composition climate over the Asian and West Pacific region.

This study shows the role of meteorological and climatological conditions in the transboundary air pollution transport in East Asia and indicates the climatological changes of chemical composition concentration and deposition by season.

NUMERICAL MODELS: RAMS AND ON-LINE TRANSPORT MODEL

The Mesoscale Model

The parallel version of the Regional Atmospheric Modeling System (RAMS version 4.28) has been used to simulate the regional scale 3-D meteorological fields including boundary layer turbulence, cloud and precipitation. In addition, an on-line tracer model fully coupled with RAMS was developed to study the sulfur long range transport and deposition. The Isentropic Analysis package (RAMS/ISAN) built into RAMS was used to model field initialization and four-dimensional data assimilation (FDDA) based on the ECMWF 2.5 degree global analysis data. FDDA involves the integration of effective time-dependent observational data into a predictive model. This can be done during a model run as in Newtonian relaxation (*nudging*) scheme based on the assimilation of time-dependent lateral boundary conditions provided by the ECMWF global data.

The on-line Long-Range Transport and Deposition Model

Sulfur transport and deposition is simulated using RAMS and an on-line long-range transport model. Chemical reaction and deposition processes are included to examine transport and deposition from East Asian mainland towards Japan and the Pacific Ocean. The tracer model is fully coupled with RAMS, and this is a unique approach because the regional meteorological conditions, which play significant role in the wet deposition and vertical diffusion of tracers, are continuously updated within the tracer model at the same time intervals, and transport and deposition processes can be directly handled by using real time meteorological parameters from the mesoscale model. The chemical tracer model can simulate the SO_2, sulfate and yellow sand transport and transformation processes, and aerosol concentration can be feedbacked to atmospheric radiation budget of RAMS (feedback is not included in this study).

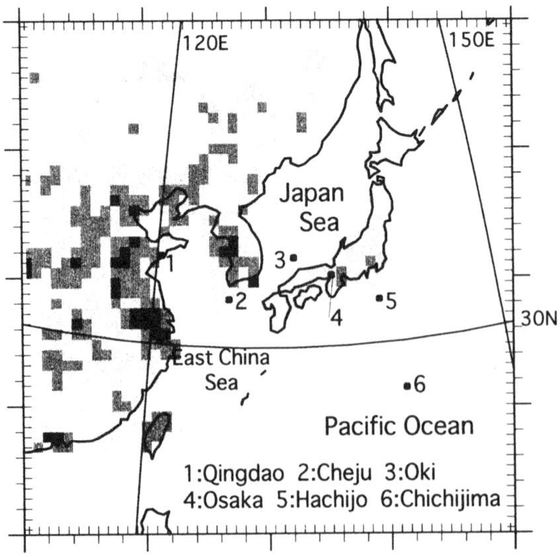

Figure 1. Model domain and site locations analyzed in the paper (#1-#6). Tone color darkness (3 levels) represents the SO_2 emission relative intensity of 1, 5, 20 strength.

A crucial point in transboundary air pollution and wet and dry deposition modeling are the removal rates and chemical conversion rates adopted in the model, and particular attention must be paid to SO_2 conversion rates to sulfate. Despite of these complex chemical conversion mechanisms, at present stage the on-line transport model includes linear chemical reaction from SO_2 to SO_4^{2-} in the gas phase by $SO_2 + OH \rightarrow SO_4^{2-}$. In aqua phase, transport model includes two major reactions of $S(IV) + O_3(aq) \rightarrow SO_4^{2-}$ and $S(IV) + H_2O_2(aq) \rightarrow SO_4^{2-} + H_2O$. The main gas and liquid phase chemical reactions between SO_2 and SO_4^{2-}, and their respective rate coefficients used in the on-line tracer model in their simplified form are taken from Pham *et al.* (1995) and Takemura *et al.*, (1999) Seasonal variation of O_3 and H_2O_2 concentration are assured based on the results reported by Takemura *et al.* (1999). It also includes dry deposition as well as wet scavenging of these species based on precipitation rate.

Numerical simulations setting

The simulation domain adopted is centered at 35N 130E, and it covers a large part of East Asia, including all of Japan, North and South Korea, most of China, and parts of Russia and Mongolia (Fig.1). The horizontal grid consists of 50 by 50 grid points, with a resolution of 80 km. In the vertical the domain is divided into 23 layers (top level is 20km) and a terrain following coordinate has been used. The mesoscale model has been initialized using ECMWF data at 2.5 degree resolution, and sea surface temperature (SST) from NCEP database (1 degree resolution). A strong nudging is prescribed every 12 hours for outer 5 lateral grid cells of the domain. Note that even though the ECMWF data has a resolution lower than the model, however it is still good enough to provide appropriate external forcing (i.e., the large-scale meteorological conditions) to the simulation region. To simulate transport and deposition of anthropogenic pollutants in East Asia, SO_2 emission dataset at 1 x 1 degree resolution is used (Akimoto and Narita, 1994). As shown in Fig. 1, a huge amount of SO_2 emission intensity is concentrated over east coast of China (approximately 20 million ton /year based on the year 1987 statistics).

The actual integration has been performed on a Pentium III Linux Cluster system of 16 processors, which gave the opportunity to test the entirely parallelized version of the code, including the on-line tracer module in several conditions. The use of a cluster of Pentium III processors not only reduces significantly the CPU time of the simulations, but also it abates the costs of the simulations to a minimal fraction compared to other parallel clusters.

The numerical integration covered a 12 months period: from April 01, 1994 0UTC until April 1, 1995 0UTC, and results from the simulations have been compared to observations from several monitoring stations (see location is shown Fig. 1). CPU elapsed time for 12 months simulation with 90 sec time step is 40.8 hours.

REGIONAL CLIMATE AND CONCENTRATION VARIATIONS

Previous studies pointed out that wind pattern variations associated with synoptic scale pressure system changes have a key role in the transboundary air pollution transport (Uno *et al.*, 1997) in East Asia. The dramatic variation of topography and land type of this region, and the alternance of large industrial/urban centers and agricultural/rural areas, together with the interactions between continental and marine air masses strongly influence the large scale flow patterns, and play a prominent role in determining the effects of pollutants production and transport in the region. In fact, East China sea, the Japan sea and the Pacific Ocean cover a huge fraction of east Asia, therefore the knowledge of pollutant concentration and meteorological parameters over these water masses are critically necessary to understand the long-range transport and transformation processes.

The important climate features of the region, schematically represented in Fig. 2, show, for each season, a typical pattern depending on the relative strength of Pacific high and continental high pressure systems.

A large scale traveling high/low pressure system, moving slowly eastward is characteristic of the spring/fall weather (Fig. 2a). The Baiu rainy season, characterized by the presence of a rain belt with heavy precipitation, strongly affects the wet deposition in the region during late spring - early summer (Fig. 2b). Distribution of this rainfall amount in the belt is non-uniform and the location of the associated rainbelt zone oscillates in the south-north direction due to the relative strength of the Pacific and continental high pressure systems. The air mass under the Pacific high pressure system is relatively clean while the air mass under the continental high is usually polluted due to the existence of the intense emission area.

When Pacific high pressure system becomes stronger, then rain-belt disappears and Japanese area is covered by Pacific high, which characterizes the typical summer monsoon (Fig. 2c). In this condition, relative clean and hot/moist air dominates the East Asia domain.

Finally typical winter monsoon is shown in Fig. 2d. From the middle of November to early March, strong pressure gradients between Siberia (Continental) High and Okhotsk Low

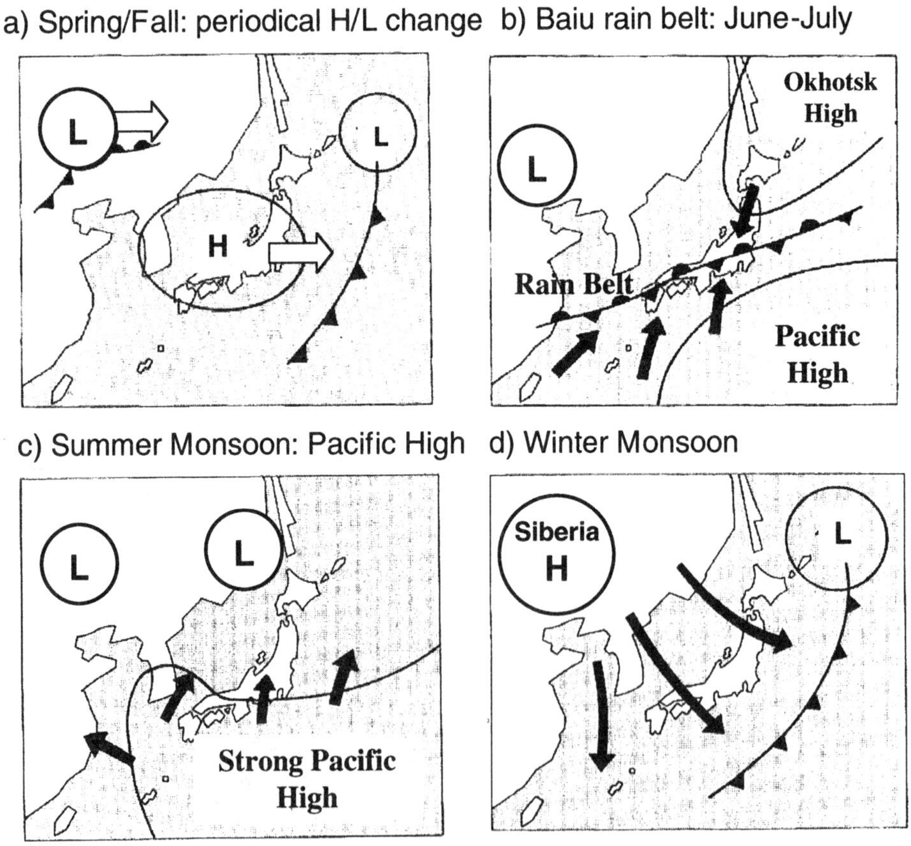

Figure 2. Important regional climate/weather pattern over East Asia. (a) Spring/Fall travelling high/low pattern, (b) Baiu rain season pattern, (c) Summer monsoon pattern, (d) Winter monsoon pattern.

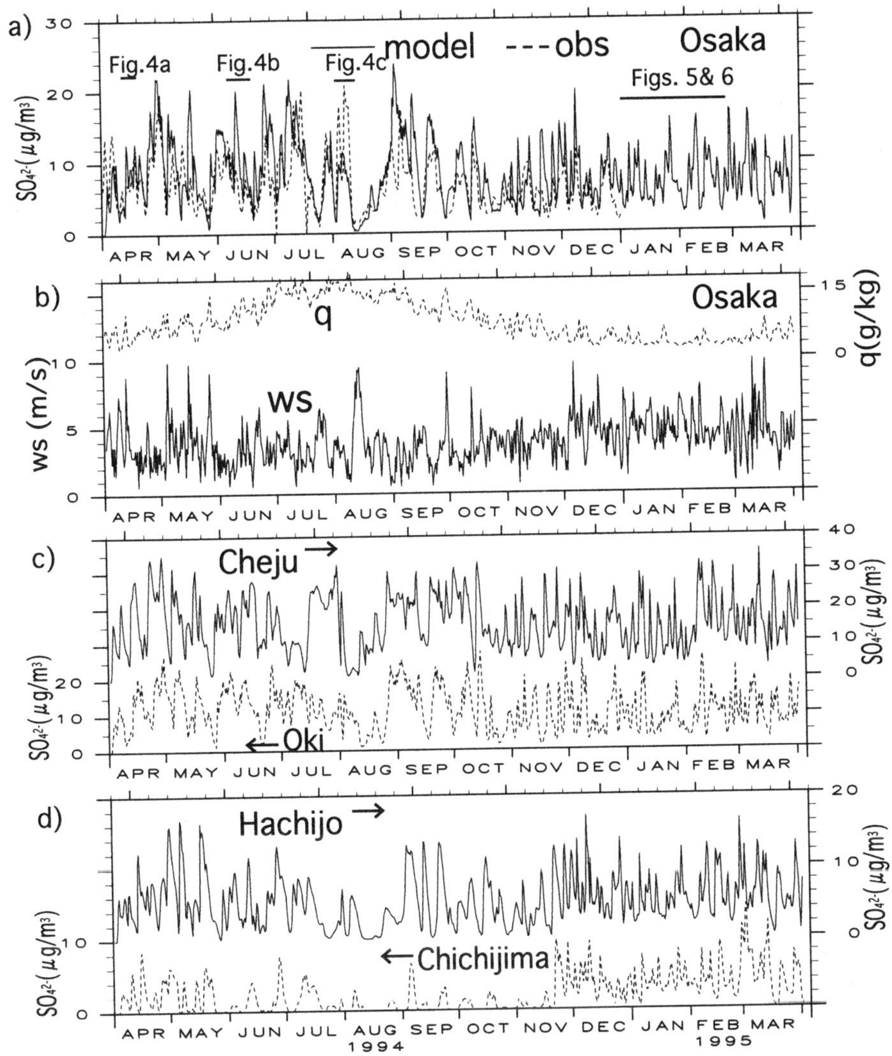

Figure 3. (a) SO_4^{2-} concentration variation at Osaka (location #4 in Fig 1). Models twelve hours averaged concentration (straight line) and daily averaged observation (dot line), (b) specific humidity and surface wind speed at Osaka, (c) SO_4^{2-} concentration variation at Cheju, Korea and Oki, (d) SO_4^{2-} concentration variation at Hachijo and Chichijima.

exists in average, which follows the strong winter monsoon in the region. This typical winter monsoon usually lasts a few days, and after this, low pressure system developes and moves from China continent to eastside of Japan as a transition of weather change. Such a transition pattern is usually observed 1 – 2 times per week.

These seasonal regional climate changes play a significant change in regional air quality. Twelve months time series of some key meteorological parameters such as wind speed, water vapor content at several sites in the region have been analyzed in order to show different climatological seasonal pattern (Fig. 3). Precipitation pattern prediction from RAMS plays an important role and the performance of precipitation fields have been analyzed in detail in previous studies (Emori *et al.*, 2000) and show a good agreement with observation data, and they will not be discussed here.

Fig. 3a shows the SO_4^{2-} concentration variation at Osaka (location #4 in Fig 1). Modeled twelve hours averaged concentration (straight line) and daily averaged observation (dot line) are shown in figure. Fig. 3b shows the specific humidity and surface wind speed at Osaka. Figs. 3c and 3d show SO_4^{2-} concentration variation at Cheju, Oki, Hachijo and Chichijima (Locations are shown in Fig. 1). Observed sulfate concentration at Osaka compared with model results show a good agreement, and the intermittency during the winter season and the periodicity typical of spring/fall rainy seasons, when the alternance of high/low pressure systems characterize the meteorology of the region, is nicely reproduced by the numerical model. One of the surprising facts is the time variation among these 5 stations in winter shows very similar spiky pattern that will be discussed later.

Some specific periods have been analyzed in order to have a better understanding of the seasonal differences in the typical regional climatology and the variations in the sulfate concentration (Figs. 4- 6).

At first, for each season, selected typical weather patterns (Fig. 2a, 2b and 2c patterns) have been chosen with their associated sulfate concentrations and accumulated precipitation amount (Fig. 4). Averaged flow field, SO_4^{2-} concentration and precipitation from April 14 to April 20 is shown in Fig. 4a, which represents typical spring time travelling high pressure pattern. Almost no precipitation observed and the clockwise outflow at the northern edge of high pressure system transports the pollutants toward the northern part of Japan. Fig 4b shows averaged field from June 6 to June 18, which represents the rainy season shown in Fig. 2b. It shows clearly that the high concentration is trapped north of the rain belt. Fig. 4c shows the typical summer monsoon pattern already shown in Fig 2c. Strong outflow from Pacific high and precipitation occur in the northern part of China and Korea, and pollutants are transported to these precipitation zones.

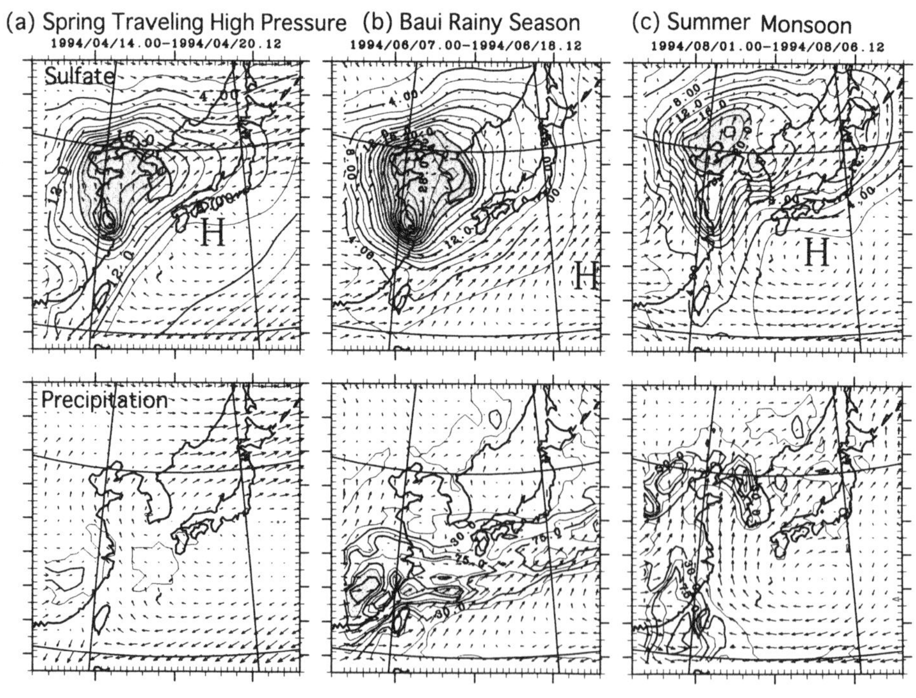

Figure 4. Upper raw is averaged surface SO_4^{2-} concentration (μg/m^3), wind fields and lower raw shows accumulated precipitation (mm) during (a) April 14 – April 20, (b) June 7 – June 18, (c) August 1 – August 6, 1994.

Secondly, some details analysis in winter monsoon is shown in Fig. 5. It shows the time variation of SO_4^{2-} concentration during winter monsoon in January – February 1995 at 6 locations shown in Fig. 1. Fig. 6 shows a snapshot of sulfate concentration outbreak occurred from Jan 25 to Jan 26, 1995. Winter season is characterized by the intermittent outbreak of cold/dry air masses, carrying air pollution from mainland Asia towards the Pacific Ocean and over Japan. Such an intermittence is clearly shown in the time series of SO_4^{2-} concentration. As shown in snapshot in Fig. 6, low pressure systems, moving southeasterly from China mainland to eastside of Japan, bring pollution as far as Chichijima, and a clear footprint of the motion of such cyclonic systems is given by the SO_4^{2-} concentration time series in several sites in the domain. Concentration peak time within 6 sites clearly shows time lag (it takes about 2-3 days from Qingdao to Chichijima). This time delay can be explained by 'ice-cream cone cup' like polluted air mass outbreak shown in Figure 6.

ANNUAL SULFATE CONCENTRATION AND DEPOSITION

Chemical climatic representation of sulfate field is important to understand the several environmental impacts for acidification. Figure 7 shows the surface level annual averaged sulfate concentration (Fig. 7a), DJF(Dec.-Jan.-Feb.) averaged sulfate concentration and wind field (Fig. 7b), model precipitation (Fig. 7c) and wet deposition field (Fig. 7d). Annual averaged sulfate concentration fields over the China mainland is strongly affected by the horizontal distribution of SO_2 emission intensity shown in Fig. 1. Contour lines of the annual averaged sulfate field show a distribution parallel to the ideal line connecting Taiwan and Japan Islands in the SW-NE direction, with a gradient more gentle than over the mainland. Sulfate concentration differences on the Japan mainland is within a factor of 2. During winter time (DJF), the southern part of China and Taiwan show concentration values higher than the annual averaged value, and the continental outflow pattern shown in Fig. 7b is responsible for these differences. In addition, Japan sea side, southern part of East China sea, north Pacific ocean are affected by large precipitation amounts. In particular, continental outflow passing over the warm Japan sea is the main reason of the large precipitation in Japan seaside. As shown in Fig. 7b, the sulfate concentration level on the Japan seaside is lower than on the mainland, however, the wet deposition amount in this region is higher than expected, because the wet scavenging is due to the combination of sulfate and precipitation rate.

Calculated wet deposition amount is strongly affected by the local precipitation rate (including in-cloud scavenging rate) and sulfate concentration field. Model predicted annual SO_4 wet deposition over Japanese area ranges between 100 – 500 SO_4-mg/m^2, which is still lower than the observed results. Further modeling study is necessary to reproduce the wet deposition distribution over east Asia, especially aqua phase SO_2 oxidation reaction model and in-cloud scavenging model are critically important to improve the current modeling performance.

CONCLUDING REMARKS

The present paper shows the role of meteorology and climatology in the transboundary air pollution transport in East Asia, and indicates the climatological changes of chemical composition concentration and deposition by season. The numerical experiments using the Pentium III - Linux 16 CPU cluster system successfully simulated the regional climate over a 12-months time period. Surface concentration, and dry/wet deposition amount of sulfate, modeled using a chemical model fully coupled with the mesoscale model, show

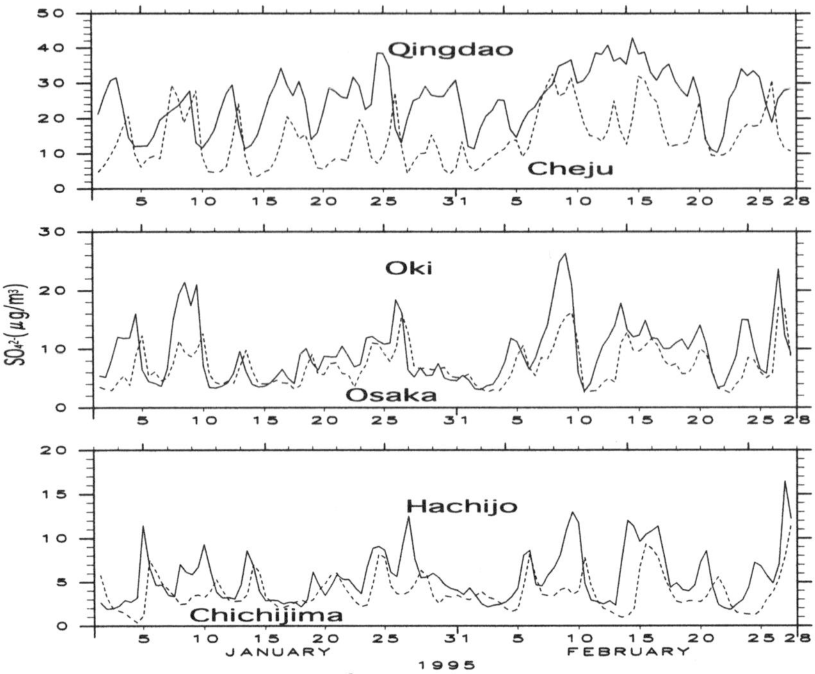

Figure 5. Time variation of SO_4^{2-} concentration during winter monsoon in Jannuary-February 1995 at Qingdao (China), Cheju (Korea), Oki, Osaka, Hachijo and Chichijima

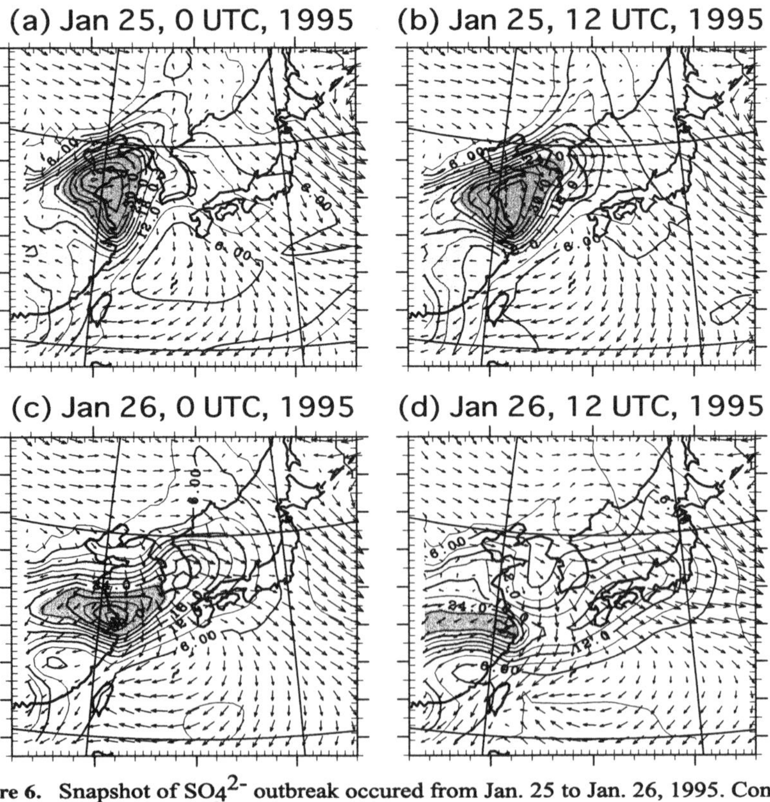

Figure 6. Snapshot of SO_4^{2-} outbreak occured from Jan. 25 to Jan. 26, 1995. Contour interval is every 3 (μ g/m^3)

patterns consistent with the climatology of the region both at annual and seasonal scale, explain the important meteorological condition for transboundary air pollutants transport in East Asia, and indicates the dramatic changes of chemical composition concentration by season. In addition the coupled model is able to correctly simulate specific episodes. However, absolute values of sulfate concentration and accumulated deposition, although comparable with observations, are not completely satisfactory and the chemical model requires more improvements. In particular, the wet scavenging and SO_2 oxidation reaction processes, not fully implemented in the present stage on-line tracer model, due to the complexity of those processes, must be parameterized more accurately.

Acknowledgments

This work is supported by Core Research for Evolution Science & Technology (CREST; PI Prof. Mitsuo Uematsu of Tokyo University) and Research and Development Applying Advanced Computational Science and Technology of Japan Science and Technology Corporation(ACT-JST). The authors are grateful to the Prof. Akira Mizohata of Osaka Prefectural University, for providing sulfate observational data. The authors want to express their thanks to Mr. Koji Ishihara and Dr. Craig Tremback for their support to run RAMS using Pentium Linux Cluster.

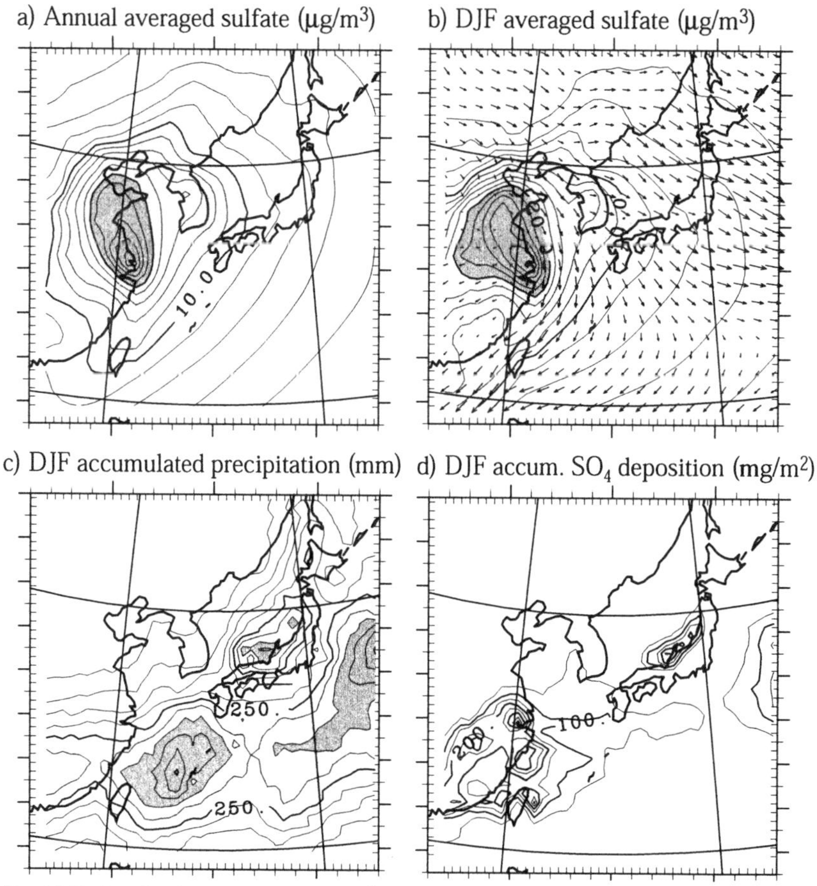

Figure 7. (a) Surface level annual averaged SO_4^{2-} concentration (μg/m^3), (b) DJF (Dec.-Jan.-Feb.) averaged SO_4^{2-} concentration (μg/m^3) and wind fields, (c) DJF accumulated model precipitation (mm) (DJF) and (d) DJF accumulated (μg/m^3) wet deposition field (mg/m^2) (DJF).

REFERENCES

Akimoto H. and Narita H., 1994: Distribution of SO_2, NO_x and CO_2 emissions from fuel combustion and industrial activities in Asia with 1 x 1 resolution, *Atmos. Environ.*, **28**, 213-225.

Emori S., T. Nozawa, A. Numaguti, I.Uno, 2000: A regional climate change projection over East Asia. *Proceedings of AMS 11th Symposium on Global Change Studies*, 15-18, January, Long Beach, US

Pham M., J.-F. Muller, G.P. Brasseur, C. Granier, G. Megie, 1995: A three-dimensional study of the tropospheric sulfur cycle. *J. of Geophys. Res.*, **100,** D12, 26061-26092.

Pielke,R.A., W.R.Cotton, R.L.Walko, C.J. Tremback, W.A. Lyons, L.D. Grasso, M.E. Nicholls, M.D. Moran, D.A. Wesley, T.J. Lee and J.H. Copeland, 1992: A comprehensive meteorological modeling system - RAMS, *Meteorol. Atmos. Phys.*, **49**, 69-91.

Takemura, T., H. Okamoto, Y. Maruyama, A. Numaguti, A. Higurashi, and T. Nakajima, 1999: Global three-dimensional simulation of aerosol optical thickness distribution of various origins. *J. Geophys. Res.*, submitted.

Uno, I., T. Ohara and K. Murano, 1997: Simulated acidic aerosol long-range transport and deposition over east Asia - role of synoptic scale weather systems, *22nd NATO/CCMS International Technical Meeting on Air Pollution Modeling and its Application*, 119-126, June, France.

DISCUSSION

G. SCHAYES: Can you show which emission inventory is used and what is it accuracy?

I. UNO: SO_2 emission inventory proposed by Akimoto and Narita (1992) is used in this calculation. Its base year is the year 1987. This SO_2 emission inventory is widely used in Asian study (such as GEIA emission inventory). Quite few additional data set is available to examine its accuracy, and we believe it is a best estimate for East Asian application.

A. BAKLANOV: Atmospheric pollution model on line coupled with a meso-scale meteorological model gives many additional possibilities to improve deposition parameterizations, for the wet deposition, first of all. Did you use 3-D fields of precipitation characteristics (e.g. cloud water, top and base of clouds.) for the wash out and rainout parameterization?

I. UNO: In the present on-line transport model, top and base of cloud layer are determined using the 3-D cloud water and pristine ice concentration distribution. Between cloud base and surface layer, surface level precipitation rate is used for wash out scheme. Of course, this wash out scheme must be improved based cloud microphysics information.

AN ASSESSMENT OF MODELLING OZONE CONTROL ABATEMENT STRATEGIES IN PORTUGAL: THE LISBON URBAN AREA

Carlos Borrego, Nelson Barros and Oxana Tchepel

Department of Environment and Planning
University of Aveiro
3810-193 Aveiro, Portugal

Summary

The Lisbon Region (LR), by its urban and industrial importance is the example of a coastal region subject to high emission levels and potentially favourable to the development of high levels of photochemical pollutants, particularly tropospheric ozone. The application of the photochemical modelling system, MAR IV, suggest that the current ozone formation is NO_x *- limited* and that the forest VOC emissions represent a negligible contribution toward the ozone formation in this region. On the other hand, small industrial sources and particularly road traffic emissions arise as being the principal precursors sources for the development of short-term ozone high concentration levels in the Lisbon Region. Finally, in accordance with the previous conclusions and keeping in mind the future evolution of precursor emissions, it was possible to modelling alternative scenarios so as to define an efficient abatement strategy for the formation of tropospheric ozone in the Lisbon Region.

Key words: Tropospheric Ozone; Lisbon Region; Emission control strategies

1. INTRODUCTION

Since 1985, Portugal is one of the participants of the CORINAIR program funded by the European Commission. This program has attempted the quantification of several pollutant emissions including the NO_x and non-methane volatile organic compounds (NMVOC) (the main ozone precursors) in the countries belonging to the European Union. According to the CORINAIR, between 1985 and 1990, in Portugal, the increase rate for the total NO_x emissions was about 190 % and for the total VOC emissions about 160 %. The large point sources are responsible for about 30% of the NO_x and a non-significant quantity of NMVOC. On the other hand, two activities are responsible for almost all the remaining precursors emissions: the road traffic, with almost 50% of the NO_x and the biogenic activity, in particular forest emissions, with 50% of the global NMVOC emissions.

It should also noted that, compared to its size, Portugal has a relatively extensive coastline (about 850 km) associated with significant terrain features and sea/land breeze circulations, which result in a complex wind field with strong implications on the formation and transport patterns of photochemical pollutants. Like other countries, the Portuguese coast is the region with the higher population density and economic development. On a narrow band of 50 km from coastline are where about 80% of the population live and almost all the main industries and power plants can be found. In fact, 85% of the total NO_x and 95% of NMVOC are emitted in this region.

The Lisbon Region (LR), by its urban and industrial importance is the example of a coastal region subject to high emission levels and potentially favourable to the development of high levels of photochemical pollutants, particularly tropospheric ozone (Figure 1).

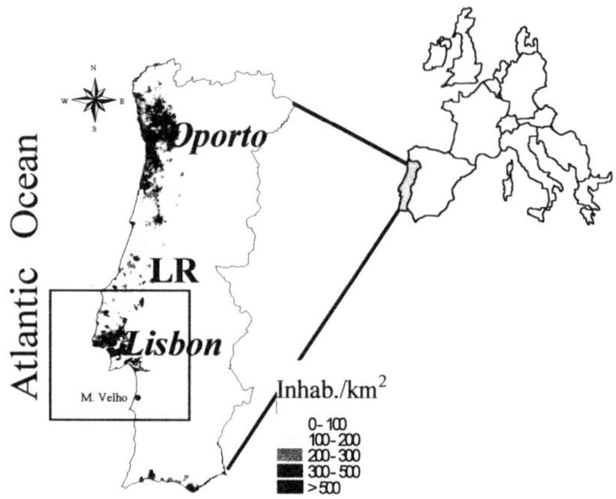

Figure 1. Population density of Portuguese littoral (50 km from the coastline).

However, the European Union policy concerning the protection of the environment and natural resources has steadily grown in importance since the 1980s and became one of the top priorities. The air quality is one of the areas in which Europe has been most active in recent years. In order to achieve environmental objectives set in the EU Fifth Environmental Action Programme, the European Commission has aimed to develop an overall strategy. Member States are required to transpose and implement new directives, which set long-term quality objectives (European Commission, 2000). The EU is also currently working on ozone abatement strategy, seeking cost-effective, effect-oriented and source-based solutions to achieve a set of ozone concentration standards. The standards will be established in a new Directive for ozone. The ozone abatement strategy is integrated in a more extensive approach, which final result will be a Directive proposal for setting national emission ceilings for SO_2, NO_x, VOC and NH_3 in order to reduce, for besides the atmospheric concentrations of ozone, reduce acidifying and eutrophying substances considerably. The European Commission has defined also a number of more specific measures to achieve the emission reductions required by the Environment Action Programme, which include reductions on atmospheric emissions from road transport in the context of Auto-Oil I programme and reduction of NMVOCs from industrial stationary sources, addressed in the solvent Directive and the Directive to Reduce Emissions from Storage and Distribution of Petrol (European Commission, 2000).

Portugal is a member of the European Community since 1986, but is still in a preliminary stage of emission reduction activities. Nevertheless, the adoption by Portugal of the EU Directives in the following years may change the actual relative distribution of the ozone precursors emissions in the Portugal and therefore the ozone formation in LR.

The object of the present work is closely related with the short-term production of photo-oxidisers in the lower troposphere of the Lisbon Region, ozone in particular, and the use of numerical tools adequate to its management as a disturbing element of the air quality at the sub-regional level (around several hundred kilometres).

2. TROPOSPHERIC OZONE IN THE LISBON REGION

To analyse the importance of ozone in the lower troposphere of the Lisbon Region, the following methodology was used: (i) the analysis of historical ozone records; (ii) the verification of its importance in light of the established threshold values from the current Ozone Directive (92/72/EEC) and the guideline established by the World Health Organisation (WHO) for protection of human health.

2.1 Analysis of historical ozone records

The first measurement of ozone made in Lisbon was reported in the beginning of the seventies. These were measurements made at the headquarters of the so-called National Meteorological Service using an 'ozonometer' (Figueira, 1976).

The current series of ozone observations in Lisbon, with ultra-violet absorption sensors and computer registering, initiated in the late eighties at the 'O Seculo' Street (Seculo Station) and in the early nineties at the Entrecampos Station.

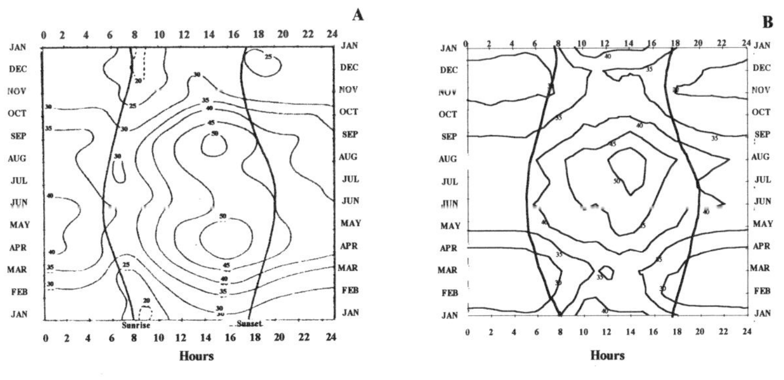

Figure 2. Ozone isopleths for the 1971-1975 (A) and for the 1989-1996 (B) series in μg.m^{-3}

In Figure 2, the ozone isopleths presented by Figueira (1976) for the 1971-1975 (A) series are represented as well as the registered series of the Seculo station, 1989-1996(B). Although the direct comparison of the observed levels in both series are questionable, due to the limitations associated with the methods and different measuring locations, the comparison of seasonal and daily distribution patterns is licit.

It is possible to verify that the evolution of ozone levels in the city of Lisbon currently has a more important local photochemical formation component than in the past. Namely, the areas of high Spring and end of Summer values found in the 1971-75 series were substituted by the current series for a single region of high values centred in the Summer, between July and August. The absence of high values during the Summer in the 1971-75 series indicates an incipient photochemical activity opposite to the current series where the Summer measurements (April through September) have always been higher than the Winter measurements.

2.2 Observed ozone levels and air quality targets

The analysis of the ozone values was made keeping in mind the guidelines of the current Community Directive (92/72/EEC) and the guidelines presented by the World Health Organisation (WHO) for the protection of human health and protection of the vegetation. To analyse the ozone concentration level observed in Lisbon, data from Seculo urban station has been processed. In order to evaluate the impact of Lisbon urban plume, data from the M. Velho station (100 km downwind Lisbon, see Figure 1.) were also considered in this study.

From the data observed since 1989 at the Seculo station, it is possible to conclude that the ozone threshold values established for the protection of human health by the current Directive (110 $\mu g.m^{-3}$, 8-hour period means) and of the WHO (120 $\mu g.m^{-3}$, floating 8-hour maximum) were exceeded several times during the year (Table I).

Table I Exceedances of the ozone concentration threshold values for protection of human health and exceedances to population information and warning threshold values at Seculo station (Sec) (Lisbon centre city) and M. Velho station (MV) (100 km downwind Lisbon).

Guidelines ($\mu g.m^{-3}$)		1989		1990		1991		1992		1993		1994		1995		1996		1997	
		Sec	MV	Sec	MV	Sec	MV	Sec	MV	Sec	MV	Sec	MV	Sec	MV	Sec	MV	Sec	MV
92/72/EEC	110	12	na	3	8	1	2	5	1	2	1	na	6	0	11	1	172	na	120
	180	72	38	5	12	5	0	17	0	0	0	na	0	0	0	0	1	na	56
	360	14	24	0	4	0	0	0	0	0	0	na	0	0	0	0	0	na	0
WHO	120	9	na	2	6	1	0	3	1	2	0	na	2	0	5	1	73	na	82

na: not available

To evaluate the impact on vegetation, the established guide values, were calculated for the series in analysis. As can be verified, at Seculo and M. Velho stations, the guide value for vegetation protection (65 $\mu g.m^{-3}$ daily average and 200 $\mu g.m^{-3}$ hourly average) were substantially exceeded (Table II).

Table II Exceedances of the ozone concentration threshold value for protection of vegetation (92/72/EEC) at the Seculo station (Sec) (Lisbon centre city) and M. Velho station (MV) (100 km downwind Lisbon).

Guidelines ($\mu g.m^{-3}$)	1989		1990		1991		1992		1993		1994		1995		1996		1997	
	Sec	MV	Sec	MV	Sec	MV	Sec	MV	Sec	MV	Sec	MV	Sec	MV	Sec	MV	Sec	MV
65 (24 h)	10	154	4	50	2	45	2	90	4	57	na	58	94	62	6	276	na	169
200 (1 h)	63	7	4	0	3	0	9	0	1	0	na	0	0	0	0	0	na	11

na: not available

Especially at Monte Velho in 1996, 276 exceedances of the established guideline of 65 $\mu g.m^{-3}$ were observed and in 1995, at the Seculo station, the same guideline was exceeded 94 times. On the other hand, the 200 $\mu g.m^{-3}$ guideline was exceeded at the Seculo station, in 1989, 63 times and at Monte Velho, in 1997, 11 times.

These results show the importance of the development of abatement ozone strategies in Lisbon Region either for protection of human health and for vegetation protection.

3. CONTROL STRATEGIES

The definition of short-term ozone production control strategies in the lower troposphere at the sub-regional level passes through two complementary steps: (i) an understanding of the target pollutant formation pathway in the area of interest (VOC-Limited or NO_x-Limited formation); (ii) evaluation of the most important precursor emission sources in the photochemical process.

Any one of these tasks, due to the complexity involved, can only be carried out by numerical modelling. In the meantime, the model or system of models used should have passed through an evaluation process of its performance so as to guarantee the quality of derived results.

In this work, the modelling of photochemical processes was carried out by the MAR IV system application on several emission scenarios (Barros, 1999). This is a modelling system especially developed for application in coastal zones, which are subject to mesoscale circulations, such as Lisbon. The system includes a 3D prognostic model for meso-meteorological fields and a photochemical model for production and transport of atmospheric pollutants both reactive and non-reactive. This system passed through an evaluation process of its performance which included: analysis of sensitivity for idealised applications, the comparison of results with other developed models in an independent way and a validation with real data collected during the LisbEx 96 and 97 campaigns made in the Lisbon Region and Alentejo Coast region. (Barros, 1999; Borrego et al.,1998). Also, a Quality Assurance / Quality Control system was initiated for model application.

3.1 Evaluation of Ozone formation

Ozone formation in polluted atmosphere is directly dependent on the concentration of organic compounds and nitrogen oxides. A widely used theoretical method to study ozone formation is the use of a maximum ozone level isopleth diagrams reached during the day in the conurbation plume for a given atmosphere characterised by a certain level of NO_x / NMVOC concentration.

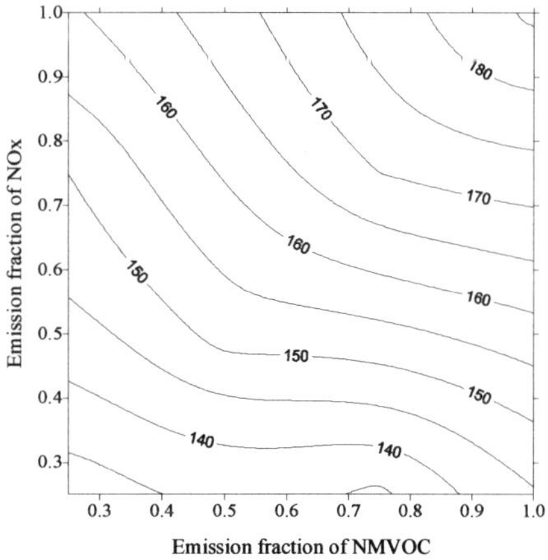

Figure 3. Isopleths of ozone maximum concentration as a function of different emission scenarios. NMVOC emissions fraction (%) and NO_x emissions fraction of (%).

Figure 3 shows a diagram of MAR IV predictions of ozone maximum concentration isopleths as a function of different emission scenarios for a typical Summer synoptical forcing. The analysis of Figure 3 reveals an ozone formation clearly controlled by the NO_x emissions. Alterations in the NO_x emissions almost always lead to significant variations in the ozone formation in contrast to the variations in the emissions values of MNVOC that practically do not change the ozone formation in the study area. The control of the ozone production will pass, therefore, to the control of the emissions in NO_x at the Lisbon Region.

The different emission sources that exist in the area affect in various ways the ozone formation according to the ratio NO_x/NMVOC emitted and the emission pattern (elevated point source, area source, etc.). For the control strategy, this step is fundamental for understanding the influence of the different types of emission sources in the ozone formation.

3.2 Contribution of the various emission sources

Applying the numerical model, it is possible to determine the relative contribution of each group of sources and define a specific and effective plan of action. In this way, it was sought by successive simulation for the typical Summer situation, to find out the relative importance of each group of emission sources for the ozone formation.

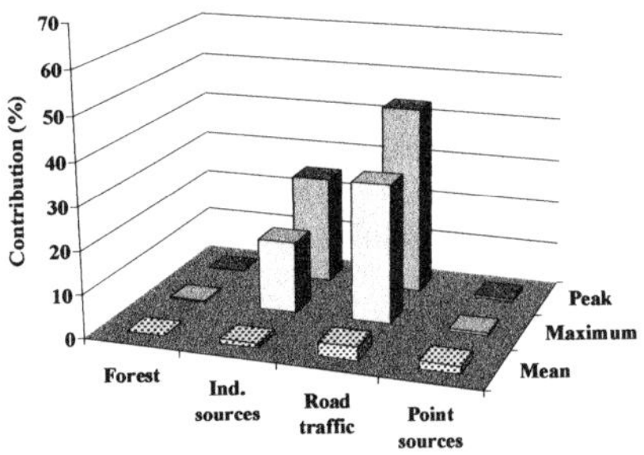

Figure 4. Relative contribution of various emissions sources for the ozone formation in the Region of Lisbon.

Figure 4 shows the evaluation results of relative contributions of the various emission sources for ozone formation in the study area. As can be verified, road traffic and the diffusive industrial sources (considered as area sources) contribute to about 70% of the peak ozone values verified in the area and about 50% of the mean maximum values. Contrary to what has been verified in other regions of Europe and the United States (Vogel et al., 1995; Scheffe et al, 1991) the forest emissions of the Lisbon Region do not significantly contribute to ozone production. This fact, already described by Builtjes et al. (1996) the LOTOS model for the Iberian Peninsula, show a strong NO_x - limited ozone formation. Furthermore, the elevated point sources, due to their emission conditions, do not significantly contribute to the short-term ozone production at this scale.

In this way, one can verify that for the simulation conditions small and medium-sized industry, considered in the MAR IV system as diffusion sources in area and especially the road traffic provide a very significant contribution to the verified ozone levels.

3.3 Emission Scenarios

The ozone production control strategies should always take into account the emission control of its precursors. As previously stated, the ozone production is essentially due to, among other factors, the NO_x emissions and the NMVOC being, therefore, dependent on its relative concentrations on a non-linear way. Depending on the scale of action, this control can be made from the global and/or regional level, through the international protocols and/or by the control within a specific region of action. Due to the context of the present work, the control strategies to be defined will be specifically oriented for the Lisbon Region with action only at the level of its emissions.

Reviewing the previous conclusions, the option was to test a control strategy of the tropospheric ozone levels in the Lisbon Region exclusively by the variation of the emissions attributed to the road traffic. Then, two idealised scenarios were created, if the evolution of the road traffic emission sources was related to the referred situation: (i) with half of the present emissions; (ii) double the present emissions.

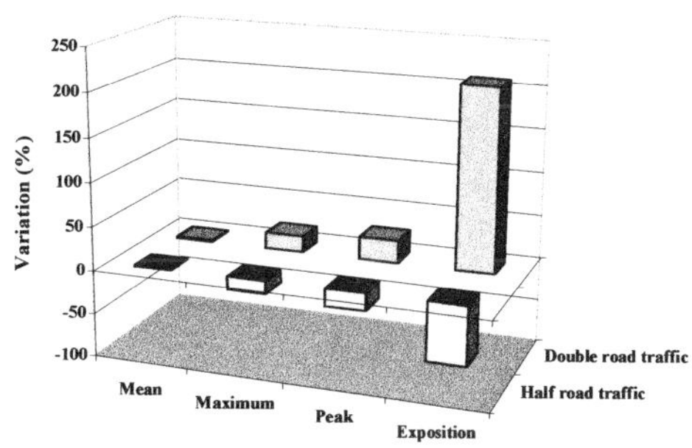

Figure 5. Comparison of the evolution of ozone production in Lisbon Region, when faced with the current reference situation, to a reduction scenario in half or a double increase scenario.

The first scenario seeks to verify the check of eventual policy on emission reduction resulting from road traffic, namely through the stabilising the private automobile traffic and or by transposition of the EU Directives to internal law. Half of the current emission of NO_x and NMVOV is also about the EU proposal (European Parliament and Council) for a Directive setting national emission ceilings for certain atmospheric pollutants and for a Daughter Directive relating to ozone in ambient air. The second scenario is a BAU (Business As Usual) type, which seeks to verify the long-term significance, for ozone production in the Lisbon Region, of an eventual non-controlled increase of road traffic in the coming years.

In Figure 5, it is possible to compare the two scenarios when faced with the current situation in the study area. As verified, the major differences can be seen for the exposition parameter, which represents the number of calculation cells with mean hourly concentration higher than 180 $\mu g.m^{-3}$ (92/72/EEC).

For the road traffic emission duplication scenario, an increase of more than 200% exists in an area attaining levels higher than those legally established. On the contrary, with a reduction in half, although areas continue to subsist with levels higher than180 $\mu g.m^{-3}$, there exists a reduction of 70% referred to base state. In relation to the maximum and medium values modelled, the differences are not very important. The ozone formation in the intervention area is clearly dominated by the emissions plume coming from the Lisbon urban zone by which the differences on average tend to dilute themselves. Only in the peak values are the differences significant.

4. CONCLUSIONS

The analysis of ozone data available for the Lisbon Region show a tendency for an increase in the photochemical activity which turn into an increment of the tropospheric ozone levels associated with the Summer season. The levels observed show that the guide values established, whether using the current Community Directive or WHO, were exceeded on various occasions, thereby leaving place for concerning not only at the level of public health but at the vegetation protection level as well. For this reason, a project based on the numerical simulation was developed with a view to establishing short-term ozone abatement strategies in the Lisbon Region (LR). In this way, and in keeping with the work of numerical modulation made through the MAR IV system, the LR presents an ozone formation clearly limited by the NO_x emissions. On the other hand, the ozone formation, in the study area as well as the Lisbon urban zone, is strongly conditioned by the road traffic emissions. The biogenic emissions, contrary to those that occur in other European countries and in the United States, particularly in the South, do not significantly affect the ozone formation. This aspect gives consistency to the idea of ozone formation limited by the NO_x emissions previously mentioned.

The study made by the application of the MAR IV system with idealised emission scenarios, has shown it is possible to significantly control the tropospheric ozone levels in the Lisbon Region by the management of emissions resulting from road traffic.

Acknowledgements

The authors would like to express their gratitude to the CZCM - Centre of Coastal Zone and Sea of the University of Aveiro for the Post-Doctorate grant to Nelson Barros, the PRAXIS XXI PhD grant to Oxana Tchepel and DGA - General Direction of Environment for the support to the Directives work.

REFERENCES

Barros N.. Poluição atmosférica por foto-oxidantes: O ozono troposférico na região de Lisboa. PhD thesis on Environmental Sciences, Universidade de Aveiro, Portugal, (1999).

Borrego, C., Barros, N., Miranda, A. I., Carvalho, A. C. and Valinhas, M. J. Validation of two photochemical numerical systems under complex mesoscale circulations. *Pré-Proceedings of the 23rd NATO-CCMS International Technical Meeting on Air Pollution Modelling and Its Application,* p. 411-418, (1998).

Builtjes, P., Boersen, G., Esser, P., Feitsma, T., Hout, D. Ozone over Spain - Results of the LOTOS model. TNO report, TNO-MEP - R96/351. (1996).

European commission. *EU Focus on Clean Air.* Directorate-General Environment, Nuclear Safety and Civil Protection (2000).

Figueira, M.. Near-ground ozone and natural radioactivity time-variation at Lisbon and their origin. In Joint Symposium on Atmospheric Ozone. Dresden, Germany, 1976 - *Proceedings of the Joint Symposium on Atmospheric Ozone.* WMO, Berlin, Vol.III, , p. 269-278 (1977).

Scheffe, R., Gipson, G. and Morris, R.. The influence of biogenic emissions estimates on ozone precursor control requirements for Atlanta. In *Air Pollution Modelling and its Aplication VIII.* New York: H. van Dop e D.G. Steyn, Plenum Press, p. 447-452 (1991).

Vogel, B., Fiedler, F. and Vogel, H.. Influence of topography and biogenic volatile organic compounds emission in the state of Baden-Württemberg on ozone concentrations during episodes of high air temperatures. *Journal of Geophysical Research,* Vol. 100, no. D11, p. 22907-22928, (1995).

DISCUSSION

G. TONNESEN: Was surprised that the O_3 isopleth diagram did not exhibit O_3 inhibition at high NO_x - low VOC conditions.

C. BORREGO: In fact, the work done over a smaller domain (only a few dozen of kilometres around the metropolitan area of Lisbon) shows an O_3 isopleth diagram with a classic ozone inhibition at high NO_x - low VOC conditions. Nevertheless, in the present work, the concern was about the short-term impact of the ozone Lisbon urban plume in a mesoscale domain (200 x 200 km^2 centred in Lisbon metropolitan area). At this scale, and for this particular domain, the amount of VOC emissions (in particular from forests) is so high that, keeping the NO_x and VOC emission scenarios in the same variation range, the NO_x emission scenario is never high enough to inhibit ozone production. Therefore, the ozone effective production in Lisbon region, based on the current emission pattern, is highly dependent of NO_x availability. This ozone formation pattern was also observed, at some Iberian Peninsula regions, by other research teams.

G. TONNESEN: Current emissions models in California significantly underestimate NO_x emissions from heavy duty diesel trucks. Is there uncertainty in the NO_x emissions modelling in Portugal?

C. BORREGO: Comparisons of emission data obtained by different methodologies demonstrate an uncertainty about 30%. An additional analysis of the data confirms the underestimation of the quantity of pollutants released to the atmosphere. Nevertheless, the high uncertainty is associated with the VOC and not the NO_x emissions.

THE DESIGNATION OF FUZZY AIR QUALITY MANAGEMENT AREAS

Bernard E A Fisher and Alexander G Newlands

University of Greenwich
School of Earth and Environmental Sciences
Medway Campus
Chatham, Kent ME4 4TB U.K.

MODELLING AIR QUALITY IN LONDON

A requirement of recent UK air quality legislation is to estimate the size of areas where air quality standards will be exceeded. Results of the modelling studies are dependent on a number of factors such as the emissions, meteorological conditions, assumptions regarding the conversion of NOx to NO_2 etc. The sensitivity of results to these factors could be determined by running the model many times and changing each of these factors in turn. This would be a rather elaborate exercise and certain conclusions can already be drawn from results already available.

Predictions based on the GRAM model (Fisher and Sokhi, 1998, Fisher and Newlands, 1998) taking account of the variation in emission density within London, have been compared with measurements at 8 sites in London for 1998. This is a small sample and so no attempt is made to separate the sites into roadside (5) and urban background sites (3). If measured and calculated annual mean NO_2 concentrations (assuming photochemical equilibrium) are compared the square root of the mean sum of the squared differences at each site can be calculated and equals 5.4 ppb. The mean measured concentration is 28 ppb corresponding almost exactly with the mean calculated concentration. An estimate of the error in predicted concentrations is $28\pm 1.65\times5.4=28\pm9$. It may be assumed that there is a 90% probability that the predicted concentration lies within this range.

The pattern of concentration over London may be understood by considering the shape of the basic equations underlying GRAM. There are two components the urban background and the concentration from roads. Considering a large urban area approximately circular in size, the annual average NO_2 concentration in ppb as a function of distance X in km from the centre is approximated by

$$C_A(X) = 3\,L^{0.413}\,10^{\left[0.104-0.574\left(\frac{2X}{L}\right)+0.071\left(\frac{2X}{L}\right)^2\right]} \tag{1}$$

where the average emission density of NOx (t/km²/y) in London extrapolated to the year 2005 is applied, the fraction of NOx in the form of NO_2 is assumed to be 0.3 and L is the width of the city taken to be 56km. An approximation of this form is applied in the dispersion model GRAM.

The corresponding formula for the annual average NO_2 concentration in ppb at a distance x (m) from a road carrying 60,000 vehicles per day is

$$C_B(x) = \sqrt{\frac{2}{\pi}} \frac{187}{2u\sigma_z(x)1.95} \tag{2}$$

with emissions extrapolated to the year 2005. u is the annual average wind speed and $\sigma_z(x)$ is the vertical dispersion coefficient taking account of the initial mixing due to vehicles travelling along the road. The fraction of NOx in the form of NO_2 is assumed to be 0.3

$C_A(X)$ and $C_B(x)$ are plotted in Figs (1) and (2) below.

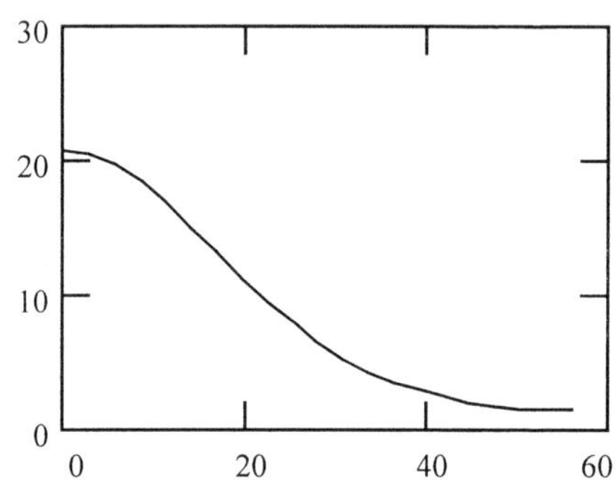

Fig 1 Estimated annual average NO_2 concentration $C_A(X)$ (ppb) in London in 2005 as a function of distance X from centre(km)

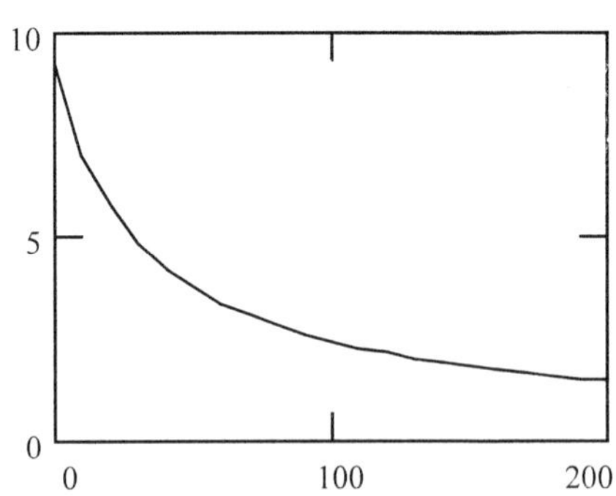

Fig 2 Estimated annual average NO_2 concentration $C_B(x)$ (ppb) in 2005 as a function of distance x from centre line of road (m) carrying 60,000 vehicles per day

Fig 1 shows a small exceedence of the NO_2 annual average air quality standard of 21ppb in the centre of the urban area. Fig 2 shows that the NO_2 standard is unlikely to be exceeded due to emissions from the road alone. In the centre of the urban region there will be areas subject to both contributions.

The predicted results for 2005 show a central area of radius a few km containing concentrations exceeding 21ppb. An estimate of the area of central London exceeding

21ppb can be made assuming there is a 95% probability that the calculated 16ppb contour line represents an actual concentration of less than 16(1+9/28) = 21ppb. An estimate by eye would suggest that the 16ppb contour line includes on average everywhere within about 10 km of central London. The choice of contour line depends on the decision maker. This is quite a precautionary approach. A less precautionary approach would consider one standard deviation below the standard, 18ppb, which would ensure with an 84% probability that the contour line encompassed all areas of exceedence.

An equivalent way of describing an air quality management area (AQMA) is to assume that concentrations are distributed about calculated concentration according to a log-normal distribution with the standard deviation given by the validation above. One can then consider the degree of membership of the AQMA depending on the probability of exceeding 21ppb, assuming an underlying log-normal distribution of concentration, and the degree of membership of the air quality management area can be drawn as a function of distance from the urban centre (see Fig 3).

Figure 3 shows the membership grade of an air quality management area. If one defines the air quality management area as all places with a membership grade greater than 0.5 one would include everywhere within a radius of about 5km from the centre of London. There is a substantial membership grade, equal to about 0.5, that nowhere is the air quality objective exceeded and no air quality management area need be declared.

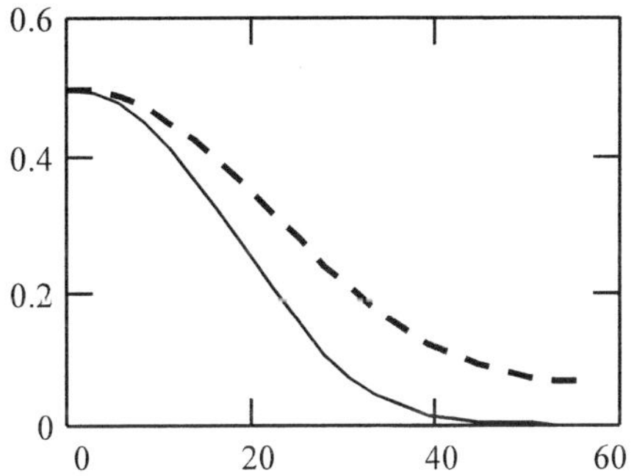

Fig 3 The dashed line shows the membership function of an AQMA as a function of distance(km) from the centre of London, and the solid line shows an alternative membership function

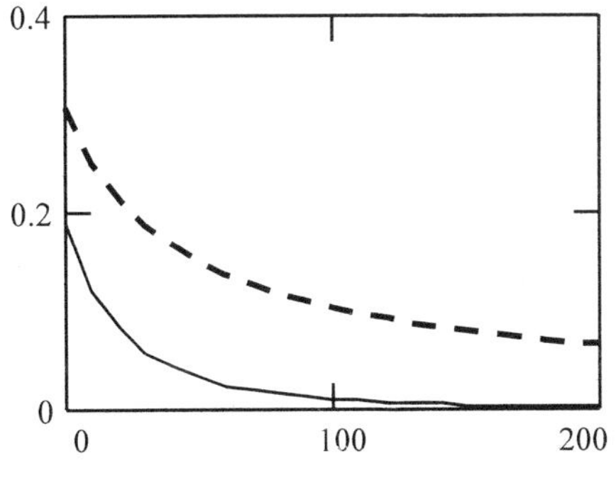

Fig 4 The dashed line shows the membership function of an AQMA as a function of distance(m) from the centre line of a road carrying 60,000 vehicles per day caused by emissions from the road alone. There appears to be low likelihood of an exceedence due to road traffic from the road in isolation. The solid line shows a different membership function for an AQMA around a road used later in this paper.

SENSITIVITY ANALYSIS USING MONTE CARLO SIMULATION

In many urban areas there will not be enough monitoring sites to perform a reliable validation exercise. Since the uncertainty in many of input parameters is large the only viable option is to perform a sensitivity analysis using Monte Carlo simulation. This was applied to a major road carrying 60,000 vehicles per day, within but not near the centre of London in 1998, using 100 simulations and the model GRAM, which runs almost instantaneously. In the Monte Carlo simulation the following input variables are varied: vehicle velocity, traffic flow, traffic growth, fraction of heavy duty vehicles and NOx emission rate per unit area of the city. The variation is done by considering rectangular distributions about a baseline value for each input parameter. The width of the distributions were taken somewhat arbitrarily to be 25%, 15%, 25%, 50% and 25% respectively, of the mean values of the vehicle velocity, traffic flow, traffic growth, fraction of heavy duty vehicles and NOx emission rate per unit area of the city. Half the width of the distribution represents the maximum deviation of possible values from the baseline. In each simulation values are sampled from the distributions. One is then able to generate the mean NO_2 concentration at a location and standard deviation around the mean (Table 1).

Table 1 Comparison of errors for three locations next to road within London

Location within London	Mean NO_2 (ppb)	Standard deviation Monte Carlo (ppb)	Standard deviation FOEA (ppb)
10m from road	29	1.4	1.3
50m from road	25	0.9	0.8
100m from road	23	0.7	0.6

It can be seen that the standard deviation from the mean concentrations are markedly lower at all locations, in relative terms, compared with those arising in the earlier validation study. Clearly the input parameters considered are those which would normally be supplied by the model user and neglect other variables, such as the function relating NOx to NO_2, which a model user would not be able to change. The function relating NOx to NO_2 has the effect of attenuating all the predicted values of NO_2 but the inherent uncertainty associated with the function is often neglected. Parameters associated with street canyons not discussed in this paper might also lead to large concentration fluctuations because of uncertainties in input. It is concluded that a partial sensitivity analysis, such as the one conducted here, can grossly underestimate the potential errors.

Vehicle velocity, traffic flow and fraction of heavy duty vehicle are inputs related to the road. The traffic growth is a parameter related to both the road and the urban background. The NOx emission rate per unit area of the city is directly related to the NO_2 background concentration. One limitation of the Monte Carlo analysis is that one does not know which factor is most important. In this case the traffic growth is not significant because the year 1998 is considered which is not much later than the base year of 1996.

A First Order Error Analysis (FOEA) was conducted to tells which of the variables is most important. If V is an input variable, and δV is the % variation about its mean value, the half width variation is $0.5V\delta V/100$ and the variance σ^2_V is $(0.5V\delta V/100)^2/3$. The sensitivity is the change in concentration δC brought about by a change in variable δV. This functional derivation $\delta C/\delta V$ can be estimated by taking small changes in V ($\delta V \pm 2.5\%$) and evaluating the change in concentration. The overall variance is

$$\Sigma \left(\frac{\delta C}{\delta V_i} \right)^2 \sigma^2_{V_i} \qquad (3)$$

where the summation is over the range of input variables being considered. The overall variance is a first order estimate of the error associated with all the variables. One can see from Table 1 that for this example the FOEA agrees well with the error estimated from the Monte Carlo simulation. From the FOEA one can determine which variables make the main contribution to the error. Somewhat surprisingly it turns out that the fraction of heavy duty vehicles is the most important factor of the five considered though of lesser importance at the more distant sites.

FUZZY AIR QUALITY MANAGEMENT AREAS

Considerations of uncertainty lead naturally to the notion of an Air Quality Management Area as a fuzzy set. If an Air Quality Management Area (AQMA) is described by an ordinary crisp set this would assign a value of either 1 or 0 to each location. A location would have a membership grade of 1 if it lay within the AQMA and would have a membership of 0 if it lay outside the AQMA. The concept of a fuzzy set generalises membership. Each location z is associated with a membership function $\mu(z)$ defined over the whole area under consideration. $\mu(z)$ takes values between 0 and 1. A membership grade of nearly 1 suggests that the point z lies within the AQMA; a membership grade of nearly 0 suggests that the point z lies outside the AQMA. Although the range of values between 0 and 1 is the most commonly used for representing membership grades, any arbitrary set with some ordering can be used. Although $\mu(z)$ may appear to indicate some form of probability this is not the case (Klir and Folger, 1988).

If an AQMA is defined as a fuzzy set, this does not prevent one defining an AQMA with sharp boundaries for administrative reasons. An α-cut of a fuzzy AQMA is the area containing all locations that have a membership grade in the AQMA greater than or equal to a specified value of α. One may decide to define the AQMA as all points at which the membership function $\mu(z) \geq 0.5$. Concepts that apply to ordinary crisp sets such as the union and intersection of two sets, A and B, may be generalised to fuzzy sets. Several classes of functions have been proposed which are functions of the membership grades of elements in A and B, $\mu_A(x)$ and $\mu_B(x)$, and also satisfy a series of axioms which ensure that the union or intersection have sensible and desirable properties. For example if one was considering two different fuzzy AQMAs defined for two different pollutants, such as nitrogen dioxide and PM_{10}, it might be desirable to consider the fuzzy union of these AQMAs before declaring an Air Quality Management Area.

For the present application we are concerned with a single pollutant, nitrogen dioxide, and the annual average air quality standard. However we are interested in defining an AQMA based on two kinds of sources: a line source, representing a road, and an area source, representing an urban area. For example $\mu_A(X)$ is taken to represent the membership grade of a fuzzy AQMA in an urban area A expressed as a function of distance X from the centre of the urban area, and $\mu_B(x)$ is taken to represent the membership grade of a fuzzy AQMA associated with a major road B. One wishes to know the membership function of the fuzzy set D consisting of the road B running through the urban area A.

Aggregation operations on fuzzy sets are operations by which several (in this case two) fuzzy sets are combined to produce a single fuzzy set. Fuzzy unions and intersections qualify as special aggregation operations on fuzzy sets. Aggregation operations must satisfy several axioms to ensure that the operation is sensible. The most interesting axiom is that arising when considering two pairs of points in A and B, so that the membership $\mu_A(X)$ of point X in A is greater than the membership $\mu_A(X')$ of point X' in A and the membership grade $\mu_B(x)$ of point x in B is greater than the membership grade $\mu_B(x')$ of point x' in B, then the aggregate membership of the pair X and x is greater than the aggregate membership of the pair X' and x'. The aggregation function is said to be *monotonically non-decreasing* in all its arguments. In other words if a point is closer to the centre of the urban area and also closer to the road, it should have a higher membership function in the aggregate fuzzy set D, consisting of the urban area and the road, than the membership function of a point that is further from the centre of the urban area and further from the road.

These concepts are clearer if a specific example is considered. The example requires membership functions of fuzzy air quality management areas to be defined. The membership function should express the uncertainty in the predicted air quality concentrations. If the annual average air quality standard for nitrogen dioxide C_0 is 21ppb, and C(z) is the concentration estimated at a point z using a dispersion model, the crisp definition of an AQMA would conclude that z lies within the AQMA if C(z) is greater than C_0. However if C has a multiplicative uncertainty associated with it, then even if C(z) is less than C_0, there is a probability $p(\lambda)$ of a multiplicative error $\lambda = C_0/C(z)$ which make the estimated concentration equal to C_0. Any multiplicative error greater than $C_0/C(z)$ would make the estimated concentration greater than C_0 and imply that z should be within the AQMA. Thus the membership grade of z can be written as

$$\mu(z) = \int_{C_0/C(z)}^{\infty} p(\lambda)d\lambda \tag{4}$$

For convenience μ is set equal to $1/(1+\lambda)$, so that $p(\lambda)=1/(1+\lambda)^2$.

We consider the aggregation of two AQMAs with these simple membership functions. $C_A(X)$ is the calculated concentration of NOx in a circular urban area with uniform emission density as a function of distance X from the centre, whose membership grade as a function of X is

$$\mu_A(X) = \frac{1}{1+\dfrac{C_0}{C_A(X)}} \tag{5}$$

$C_B(x)$ is the calculated concentration of NOx near to a uniform, straight road, expressed as a function of distance x from the road centre line, with the membership grade of the air quality management area given by

$$\mu_B(x) = \frac{1}{1+\dfrac{C_0}{C_B(x)}} \tag{6}$$

$C_A(X)$ and $C_B(x)$ may be calculated directly from appropriate area and line dispersion models. The aggregated concentration near to a road in an urban area is $aC_A(X)+bC_B(x)$, where a and b are the multiplicative errors on C_A and C_B with the joint probability $p_A(a)p_B(b)$, where $p_A(a)=1/(1+a)^2$ and $p_B(b)=1/(1+b)^2$. For a given value of a, the location (X,x) will lie just within the AQMA if $b=(C_0-aC_A(X))/C_B(x)$. The membership grade μ_D in the aggregated fuzzy set D of fuzzy sets A and B is given by

$$\mu_D = \int_0^{C_0/C_A} \int_{\frac{C_0-aC_A}{C_B}}^{\infty} p_A(a)p_B(b)da\,db + \int_{C_0/C_A}^{\infty} \int_0^{\infty} p_A(a)p_B(b)da\,db \tag{7}$$

This expression can be integrated because of the simple form of p_A and p_B and gives the following equation (8)

$$\mu_D(X,x) = \frac{C_A(X)+C_B(x)}{C_0+C_A(X)+C_B(x)} + \frac{C_A(X)C_B(x)}{(C_0+C_A(X)+C_B(x))^2} \ln \frac{(C_A(X)+C_0)(C_B(x)+C_0)}{C_A(X)C_B(x)}$$

This expression satisfies the *monotonically non-decreasing* requirement of aggregation and the other normal axioms of aggregation.

It may be seen that the first term in equation (8) for μ_D, the coherent term, would be the membership function if one just added together the concentrations C_A and C_B. The second term in equation (8) arises because high values of uncertainty in the estimate of concentrations in A in the urban area do not always coincide with high values of uncertainty in the estimate of concentrations in B near the road. It is clear that for this example and other cases the errors associated with urban scale dispersion models and roadside dispersion models should be considered separately. The coherent term ν_D is defined by

$$\nu_D(X,x) = \frac{C_A(X)+C_B(x)}{C_0+C_A(X)+C_B(x)} \tag{9}$$

To demonstrate the effect of aggregation these expressions for μ_D and ν_D have been evaluated using simple expression for C_A and C_B derived from dispersion models corresponding to idealised situations described earlier (equations (1) and (2)). It may be noted that the ratio of the annual average NO_2 to NOx are assumed to be different in the urban (0.3) and roadside models (0.15) for which there is some experimental evidence. The expression for the membership function for an urban area given by equations (1) and (5) has been plotted as the solid line in Fig 3. The expression for the membership function for a line source given by equations (2) and (6) has been plotted as the solid line in Fig 4.

The aggregate membership function of a busy road running through an urban area μ_D is plotted in Fig 5 below. The coherent term ν_D plotted in Fig 6 below is significantly different. This suggest that the errors associated with the urban area and the road need to considered separately and the use of a single combined error, such as that derived from in the validation study for London described in the introduction, could be misleading. Taking an α-cut at 0.5, so that a membership of 0.5 is the criteria for declaring an AQMA, one sees that distant from the road the AQMA would be at a distance of 10km from the centre, but near to the road it is 20km from the centre. The different scales on X and x should be noted. X is in units of km while x is in units of m. It would therefore be very difficult to designate the AQMA by plotting the AQMA on a map with a uniform scale.

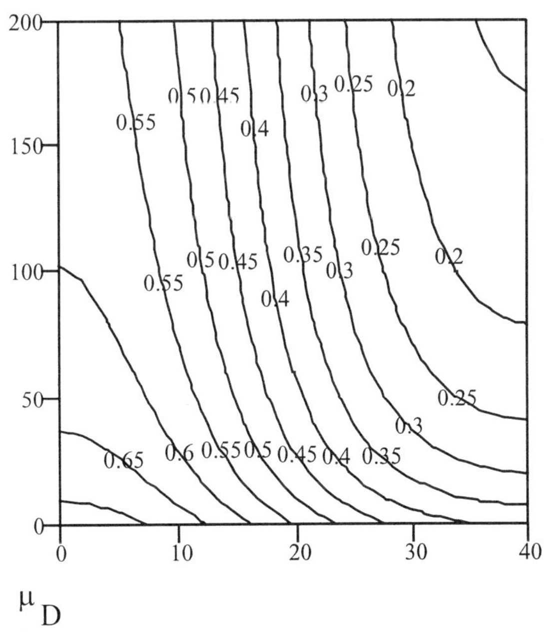

Fig 5 Aggregate membership function

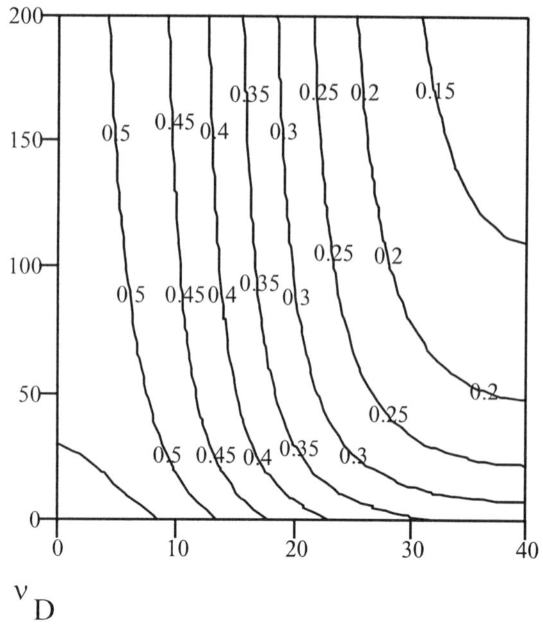

Fig 6 Coherent membership function

CONCLUSIONS

No dispersion model should be used without the associated errors in its predictions being presented. All model estimates should have ranges associated with them. The large number of parameters introduced into current dispersion models makes it difficult to estimate errors. This is a weakness of the models, not of the error analysis. Errors associated with different parts of a dispersion calculation should be distinguished. Errors associated with roads may be different to those for the urban background. The fuzzy approach has potential for use where one has separate well established, commonly used models e.g. CALINE4 for the road and OSPM for a street canyon, and one wishes to add the urban background using a well tested multi-source Gaussian model. The declaration of AQMAs in cities should be done in a way that takes account of the fuzziness in the setting of a AQMA. The AQMA can have sharp boundaries depending on choosing an α-cut of the underlying fuzzy set. The value of α chosen is a political decision depending on the degree of precaution selected. It is virtually impossible to display the declaration of an AQMA in a city on a map, because of the order of magnitude differences in the scale of the errors.

REFERENCES

Fisher B and Sokhi R, 1998, Investigation of roadside concentrations in busy streets using the model GRAM: conditions leading to high short-term concentrations, 5th Int Conference on harmonisation within Atmospheric Dispersion Modelling for Regulatory Purposes, 18-21 May 1998, Rhodes, Greece

Fisher B and Newlands A G, 1998, Clarifying the relationship between urban road structure and air quality exceedences using a training model, 23rd NATO/CCMS Int Technical Meeting on Air Pollution Modelling and its Application, Sept 28-2 Oct 1998, Varna , Bulgaria

Klir G J and Folger T A, 1988, *Fuzzy sets, uncertainty and information*, Prentice Hall, Englewood Cliffs

DISCUSSION

M. COPE: How may the fuzzy sets be interpreted in terms of a spatial distribution of air quality objective exceedences?

B E A FISHER: The region containing exceedences of an air quality objective is derived from a calculation based on some kind of air pollution model. The region is known as an air quality management area and can be thought of as a crisp set. A point either lies within or outside the region. If uncertainties in the air pollution model are considered, then it is no longer possible to state whether a point lies precisely inside or outside the region. The air quality management area may now be interpreted as a fuzzy set. The degree of fuzziness depends on the uncertainty in the air pollution model.

APPLICATION OF MICROMETEOROLOGICAL MODEL TO SIMULATE LONG-TERM DEPOSITION OF ATMOSPHERIC GASES AND AEROSOLS

Thomas F. Lavery
Randal S. Goss
Allan Fabrick
Christopher M. Rogers

Environmental Science & Engineering, Inc.
PO Box 1703
Gainesville, FL 32602-1703

INTRODUCTION

Atmospheric deposition takes place via two pathways: wet deposition and dry deposition. Wet deposition is the result of precipitation, which removes particles and gases from the atmosphere. Dry deposition is the transfer of particles and gases to the landscape through a number of atmospheric processes in the absence of precipitation. Wet deposition rates of acidic species across the United States have been well documented over the last 20 years; however, comparable information on dry deposition is just becoming available.

The direct measurement of dry deposition is not straightforward, but a number of investigations have shown that it can be reasonably inferred by coupling air concentration data with routine meteorological measurements (Shieh *et al.*, 1979; Hicks *et al.*, 1985; Meyers and Yuen, 1987; Wesely and Lesht, 1988, Meyers *et al.*, 1998). Shieh *et al.*, (1979) have shown that submicron particle and sulfur dioxide (SO_2) deposition rates for the eastern United States were strongly dependent on wind speed, solar radiation, and the condition and type of ground cover. For example, rapidly growing vegetation and forests were found to generally experience higher deposition rates than senescent vegetation, short grass, or snow. This approach has been expanded (Meyers *et al.*, 1998) to calculate deposition rates for various additional atmospheric species using site-specific meteorological data.

In 1986, the U.S. Environmental Protection Agency (EPA) established and began to operate the National Dry Deposition Network (NDDN). The objective of the NDDN was to obtain field data at approximately 50 sites throughout the United States to establish patterns and trends of dry deposition. The approach adopted by the NDDN was to estimate dry deposition using measured air pollutant concentrations and modeled deposition velocities (V_d) estimated from meteorological, land use, and site characteristic data.

Passage of the Clean Air Act Amendments (CAAA) in 1990 required implementation of a national network to monitor the status and trends of 1) air emissions, pollutant deposition, and air quality; 2) determine the effects of emissions on water quality, forests, and other sensitive ecosystems; and 3) assess the effectiveness of emission reduction requirements through operation of a long-term monitoring program. In response to these requirements of the CAAA, the EPA, in coordination with the National Oceanic and Atmospheric Administration (NOAA), created the Clean Air Status and Trends Network (CASTNet). CASTNet became operational in mid-1991, and the NDDN program was incorporated into CASTNet at that time. To increase spatial representation of CASTNet for the western United States, EPA and the National Park Service (NPS) agreed to share responsibilities in the operation of an additional 27 NPS monitoring sites for the measurement of dry deposition.

A micrometeorological model called the Multi-Layer Model (MLM) is used to simulate deposition velocities and dry deposition fluxes of sulfur and nitrogen gases and aerosols at 78 sites across the United States. The MLM is a version of the model described recently by Myers *et al.*, (1998). The model is based on the assumption that dry deposition processes can be modeled as resistances to deposition and that dry deposition flux can be estimated as the linear product of measured ambient concentrations and deposition velocities.

The MLM simulates deposition through a 20-layer canopy in which model parameters are modified by the redistribution of heat, momentum, and pollutants. The MLM requires hourly input data: wind speed and direction, sigma theta, temperature, relative humidity, solar radiation, surface wetness, leaf area index (LAI) vegetative species, and percent green leaf-out. The MLM was evaluated using extensive direct measurements of pollutant flux at nine different locations with different crops and land uses.

CASTNet includes measurements of weekly concentrations of sulfur dioxide (SO_2), particulate sulfate (SO_4^{2-}), nitrate (NO_3^-), ammonium (NH_4^+), and nitric acid (HNO_3) at 78 sites across the United States (Figure 1). The network also includes continuous measurements of meteorological parameters and ozone (O_3), which are archived as hourly averages. The CASTNet database currently covers the period from 1987 through the third quarter 1999. The database is updated quarterly.

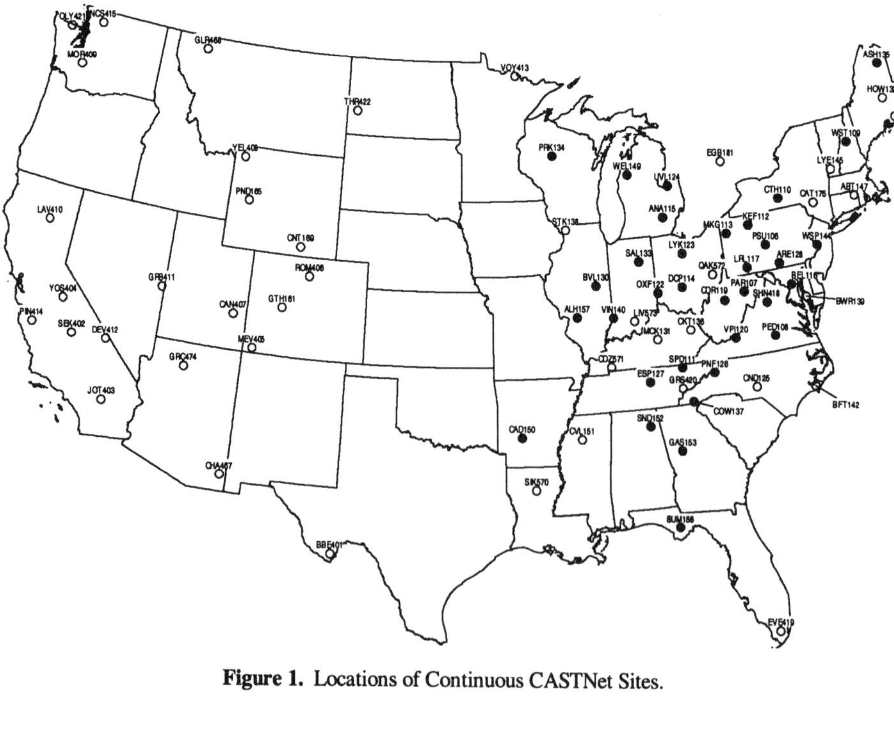

Figure 1. Locations of Continuous CASTNet Sites.

DEPOSITION MODEL

The network design was based on the assumption that dry deposition or flux could be estimated as the linear product of ambient concentration (C) and V_d:

$$Flux = \overline{C} * \overline{V}_d$$

where the overbars indicate an average over a suitable time period (Chamberlain and Chadwick, 1953).

The influence of meteorological conditions, vegetation, and chemistry is simulated by V_d. Dry deposition processes are modeled as resistances to deposition as shown in Figure 2.

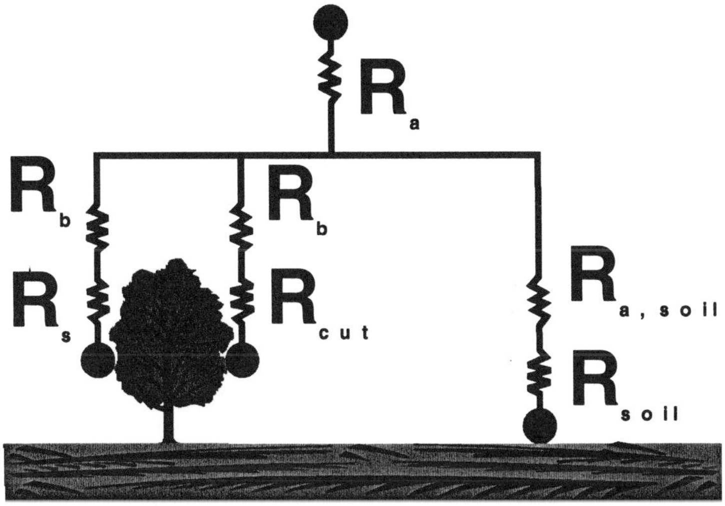

Figure 2. Resistance Model of Atmospheric Dry Deposition.

R_a, the aerodynamic resistance, is inversely proportional to the atmosphere's ability to transfer material downward from the planetary boundary layer to the surface layer by turbulent processes. R_b is the boundary layer resistance to vertical transport (molecular diffusion) through a shallow (approximately 1 millimeter) nonturbulent layer of air in direct contact with the surface. R_b depends on the aerodynamics of the surface and the diffusivity of the pollutant being deposited. The stomatal resistance is designated as R_s and is simulated using information on the type of vegetation, LAI, temperature, relative humidity, and solar radiation. The articular resistance (R_{cut}) is specified as constant for each pollutant and vegetation type. R_{soil} is the soil uptake resistance and is determined from information on soil type, surface wetness, and precipitation.

For pollutant species with low solubility or reactivity, such as O_3 and SO_2, the controlling component is the stomatal resistance, which has large diurnal and seasonal variability. For highly reactive species such as HNO_3, R_s is generally small, regardless of season or canopy type, and R_a and R_b control V_d. Deposition of the particle species SO_4^{2-} and NO_3 is primarily governed by turbulent processes and is represented in the model as a function of R_a.

Using this physical and mathematical framework, two dry deposition models (Big Leaf and MLM) have been used to calculate dry deposition for CASTNet. Both models were developed by NOAA Atmospheric Transport and Diffusion Division, Oak Ridge, TN. The Big Leaf model (Hicks *et al.,* 1985) treats the vegetation canopy as a one-dimensional surface. Big Leaf model results, aggregated to seasonal and annual averages for 1991, have been reported by Clarke and Edgerton (1993). The MLM accounts for water stress on the vegetation and deposition to snow surfaces. Additionally, several parameters (e.g., soil resistance) have been modified in the MLM from those used in the Big Leaf model. Dry deposition calculations for the CASTNet sites (and the results reported here) are currently made using a version of the MLM (Meyers *et al.,* 1998).

The meteorological variables used to determine V_d are obtained from the 10-m meteorological tower at each of the sites, normally located in a clearing over grass or another low vegetative surface. Data on vegetative species and percent green leafout are obtained from site surveys and observations by the site operator. LAI measurements were taken during 1991 and 1992 at times of summer maximum. LAI values that are used in the MLM are extrapolated from the 1991 and 1992 measurements using percent leafout observations. The resistance terms are calculated for each chemical species and major vegetation/surface type every hour. The V_d for a site is then calculated as the area-weighted V_d over vegetation types within 1.0 km of the site. Hourly V_d values are then averaged over a week and multiplied by the weekly integrated concentrations to produce weekly fluxes of HNO_3, SO_4^{2-}, NO_3^-, NH_4^+, and SO_2. O_3 flux is calculated using hourly O_3 measurements and hourly V_d values.

MODELED DEPOSITION VELOCITIES

The MLM was run using meteorological measurements and information on land use, vegetation, and surface conditions to calculate dry deposition velocities for SO_2, HNO_3, O_3, and particles. Weekly average deposition velocities were calculated for each CASTNet site for the period 1989 through 1998. Figure 3 shows weekly average deposition velocities for HNO_3 for four sites. The four sites represent a range of land use characteristics. The Arendtsville, PA site (ARE128) is in rolling terrain with nearby peach orchards. The Bondville, IL site (BVL130) is agricultural in a flat terrain setting. The Coweeta, NC site (COW137) is forested, complex terrain. The Sand Mountain, AL site (SND152) is rolling with agricultural land use. The MLM simulates the highest deposition velocities at Arendtsville with values about twice as high as the Bondville site. All sites show an annual cycle with significant weekly variability. Other examples of weekly deposition velocities can be found in the CASTNet Deposition Summary Report (EPA, 1998).

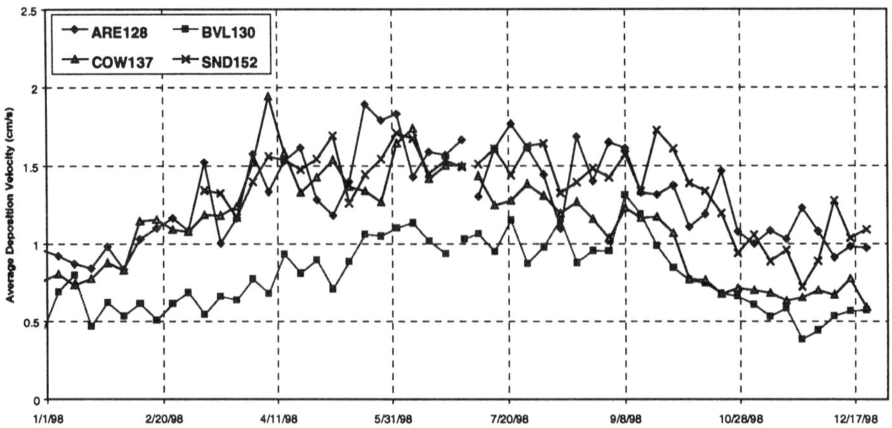

Figure 3. Weekly HNO_3 Deposition Velocities.

RESULTS OF MODEL SIMULATIONS

The CASTNet database provides an opportunity to examine trends in atmospheric concentrations and depositions of sulfur and nitrogen species as compared to trends in emissions of SO_2 and NO_x. The database was processed to calculate seasonal and annual average concentrations, dry depositions (from MLM), wet depositions, and total (dry plus wet) depositions at each CASTNet site. Concentrations and dry depositions include SO_2, SO_4^{2-}, SO_2 plus SO_4^{2-} (as S), HNO_3, NO_3^-, and HNO_3 plus NO_3^- (as N). The wet depositions include SO_4^{2-} (as S) and NO_3^- (as N). The total depositions represent the sum of the dry and wet sulfur and nitrogen species. The concentrations and depositions were also averaged over 34 eastern sites (see solid circles Figure 1) to obtain composite average concentrations and depositions. The 34 sites were selected based on complete dry deposition data for eight of the 10 years.

Annual average concentrations and depositions were analyzed for trends through simple linear regressions over the period 1989 through 1998. Figure 4 shows SO_2 concentrations versus year for the ten-year period. The figure also displays annual SO_2 emissions summed over all states east of and including the north-south line of states from Minnesota to Louisiana (EPA, 1998). Figure 5 shows annual dry, wet and total sulfur depositions for the period 1989 through 1998. The results depict the downward trend in sulfur depositions and show that dry deposition represents a significant contribution (35 to 40%) of total deposition. Figure 6 shows the trend in total (wet plus dry) deposition.

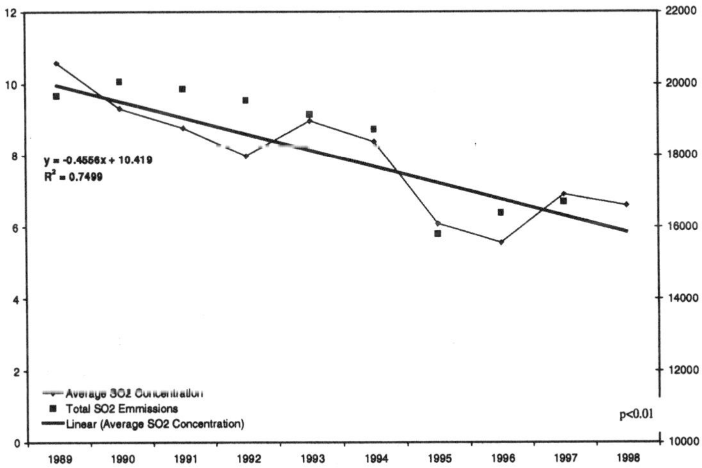

Figure 4. Trends in Annual SO_2 Concentrations.

The composite concentrations of SO_2 (Figure 4) and the total sulfur depositions (Figure 6) both decreased as the emissions decreased. SO_2 concentrations and total sulfur deposition averaged over all eastern sites have declined about the same amount as emissions during the ten-year period. Trends for individual monitoring stations vary from a few percent to almost 50 percent. Site-by-site variability in trends depends on the proximity of the monitoring station to major sources and how those sources have reduced their emissions. The linear regression analyses do not explain fully the factors (e.g., meteorology, atmospheric chemistry) that have contributed to the changes in the observed concentrations and depositions, but the relationship to emission changes is evident.

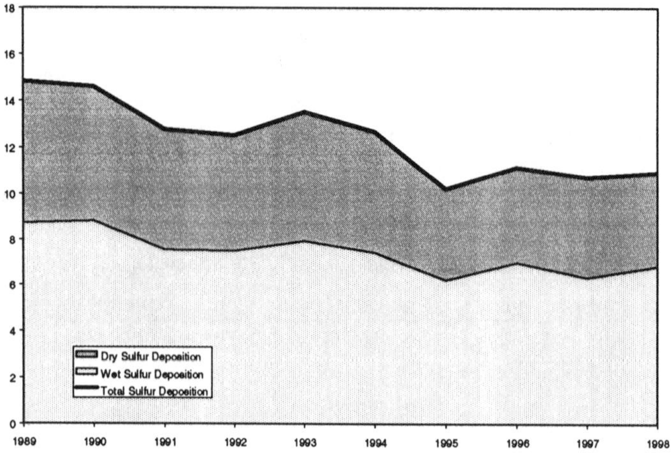

Figure 5. 1989 through 1998 Sulfur Deposition.

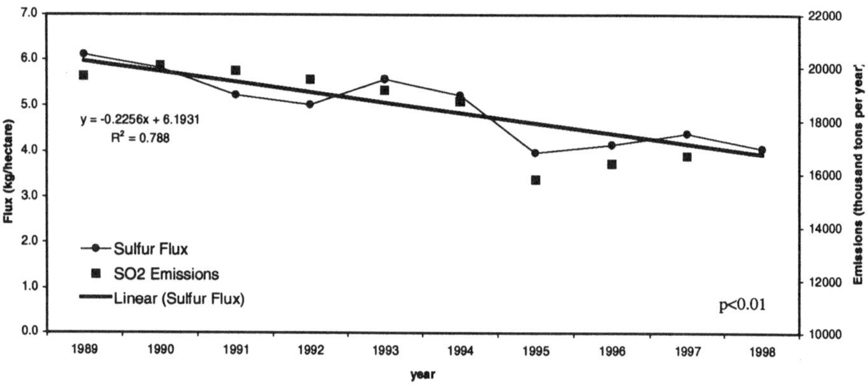

Figure 6. Trend in Average Annual Total Sulfur Depositions.

For nitrogen species, the evolution of concentrations and depositions is generally consistent with the annual distribution of NO_x emissions, i.e., no change over the last ten years.

REFERENCES

Chamberlain, A.C., and Chadwick, R.C. 1953. Deposition of airborne radioiodine vapor. *Nucleonics.* 8:22-25.

Clarke, J.F., and Edgerton, E.S. 1993. Dry Deposition Flux Calculations for the National Dry Deposition Network. Prepared for the U.S. Environmental Protection Agency, Atmospheric Research and Exposure Assessment Laboratory. Research Triangle Park, North Carolina.

Hicks, B.B., Baldocchi, D.D., Hosker, R.P., Jr., Hutchison, B.A., McMillen, R.T., and Satterfield, L.C. 1985. *On the Use of Monitored Air Concentrations to Infer Dry Deposition (1985).* NOAA Technical Memorandum ERL ARL-141.

Meyers, T.P., and Yuen, T.S. 1987. An assessment of averaging strategies associated with day/night sampling of dry-deposition fluxes of SO_2 and O_3. *J. Geophysic Res.* 92: 6705-6712.

Meyers, T.P., Finkeistein, P., Clarke, J., Ellestad, T.G., and Sims, P.F. 1998. A multilayer model for inferring dry deposition using standard meteorological measurements. *J. Geophysic Res.* Vol. 103, No D17, pp. 22,645-22,661.

Sheih, C.M., Wesely, M.L., and Hicks, B.B. 1979. Estimated Dry Deposition Velocities of Sulfur Over the Eastern United States and Surrounding Waters, *Atmos. Environ.* 13:1361-1368.

U.S. Environmental Protection Agency (EPA). 1998. *National Air Pollutant and Emissions Trends Report, 1997.* EPA454/R-98-016. OAQPS, Research Triangle Park, North Carolina.

U.S. Environmental Protection Agency (EPA). 1998. *Clean Air Status and Trends Network (CASTNet) Deposition Summary Report (1987-1995).* EPA/600/R-98/027. OAQPS, Research Triangle Park, North Carolina.

Wesely, M.L., and Lesht, B.M. 1988. Comparison of RADM dry deposition algorithm with a site-specific method for inferring dry deposition. *Water, Air, and Soil Pollution.* 44: 273-293.

DISCUSSION

J. A. van JAARSVELD: Did you observe any change in the dry/wet deposition ratio? In Europe, or at least in the Netherlands, it turns out that SO_2 concentrations decrease much faster than concentrations in precipitation.

T. F. LAVERY: No. Trends in sulphur and nitrogen depositions were analyzed for an aggregate of 34 CASTNet sites (Figure 1). The ratio of dry sulphur deposition to total (dry + wet) sulphur deposition stayed constant at approximately 40 % over the period 1989 through 1998. The ratio of dry nitrogen to total nitrogen deposition remained constant at approximately 35%.

S. T. RAO: While there might be no trends in nitrogen deposition when you consider whole eastern U. S., is it possible that trends might exist on regional or sub-regional scales?

T. F. LAVERY: The nitrogen depositions aggregated over the 34 stations show no trends. We have not investigated regional or sub-regional trends. Of course, it's possible there are trends on smaller scales or for individual monitoring sites. We will be investigating regional and sub-regional trends as part of the efforts in support of the CASTNet 1999 Annual Report, which will be published in the fall of 2000.

R. SAN JOSE: How are you planning to improve your LAI data for your deposition model. From our experience in Madrid we have compared Wesely (1989) model with the model from Nemarri and Running (1989) by using AVHRR data (from NOAA) which correlated NDVI data with LAI data to obtain canopy resistance.

T. F. LAVERY: We have investigated the direct use of satellite-derived LAI data. We feel this information is not sufficiently advanced to determine small scale estimates of LAI for use in the MLM. However, satellite data may be useful to determine the evolution of green leaves (i.e., green-out), which will help simulate the temporal distribution of LAI values for each CASTNet site. Currently, updated LAI values will be determined by field measurements during periods of maximum green-out at each site.

IMPROVING SHORT-TERM AIR QUALITY USING COMPUTER MODEL PREDICTIONS BASED ON METEOROLOGICAL FORECASTS TO MANAGE POWER STATION EMISSIONS

Tim A Hill, Gill C Hunter[1], David Acres[2], David N Futter and Alan Webb[1]

PowerGen UK plc
Power Technology Centre
Ratcliffe-on-Soar
Nottingham
NG11 0EE
[1] National Power PLC, [2] TXU Europe Power

INTRODUCTION

In 1996, upon the recommendation of the Expert Panel on Air Quality Standards, the UK Department of the Environment (DoE) adopted a new ambient air quality standard for SO_2 of 100 ppb as a 15 minute mean. An objective was subsequently set to reduce SO_2 ground level concentrations in the UK such that the 99.9th percentile of 15 minute means at all locations does not exceed 100 ppb by 2005.

At the request of the Environment Agency, the operators of coal-fired power stations undertook some preliminary modelling studies to establish the extent to which such power stations might cause exceedances of the 100 ppb SO_2 standard. From these studies it was concluded that these sources could cause exceedances of the standard at the 99.9th percentile level when operating at the high load factors typical at that time.

This precipitated a series of studies by the generating companies into circumstances under which high SO_2 ground level concentrations might result from plant operation and whether or not it might be possible to manage local air quality in an effective manner by predictive modelling using forecast meteorology and subsequent operational action: a management protocol for SO_2.

In this paper, the results of the studies are presented which culminated in three power companies running a year long trial of an SO_2 management system.

OBJECTIVES OF STUDIES

The first objective of the study was to identify, and if possible categorise, the meteorological conditions which would lead to exceedances of the SO_2 standard by power stations.

The second objective was an assessment of the accuracy of meteorological forecasts to predict the meteorological parameters relevant to dispersion and the limitations this would impose on a management protocol.

Finally, the conclusions of the desktop analyses were used to design a protocol for SO_2 air quality management and a 'real time' simulated trial undertaken at three coal-fired power stations.

OCCURRENCE OF GROUND LEVEL CONCENTRATIONS EXCEEDING 100 PPB

The dispersion model ADMS v2.1 (Atmospheric Dispersion Model System) was set up to simulate 2000 MW coal-fired power stations in different parts of the UK in order to assess time of day and seasonal dependence of the occurrence of ground level concentrations of SO_2 exceeding 100 ppb, and to establish the types of meteorological conditions which lead to such exceedances.

Using meteorological data and other input for a variety of locations around the UK, it was established that there were three broad categories of meteorological conditions that gave rise to ground level concentrations of SO_2 greater than 100 ppb:

- Low (convective) boundary layer (<500 m)
- Convective with boundary layer >500m and $h/L_{mo} < -2$ (h = boundary layer height and L_{mo} = Monin-Obhukov length)
- Windy (wind speed >8.2 ms^{-1} and $h/L_{mo} > -2$).

An examination of the distribution of meteorological conditions predicted to cause exceedances at full load with season and time of day showed that almost all the exceedances occurred during daylight hours. This is primarily because the boundary layer depth is much lower at night with the plume being emitted into the atmosphere above it. The exceedances can occur throughout the day and do not show any pronounced seasonal dependence except the obvious correlation between reduced convective activity in winter and the reduction in exceedances due to highly convective conditions.

Figure 1 shows the frequencies of occurrence of exceedances of the air quality standard as a function of meteorological category and distance downwind (power station on full load). Exceedances due to very windy conditions tend to occur within 8 km of the power station whereas the model predicts that those due to highly convective conditions or low boundary layer conditions tend to occur over much greater distances.

The next stage was to develop an understanding of the robustness of UK Meteorological Office predictions as a surrogate for actual measured meteorological parameters in modelling the dispersion of a plume.

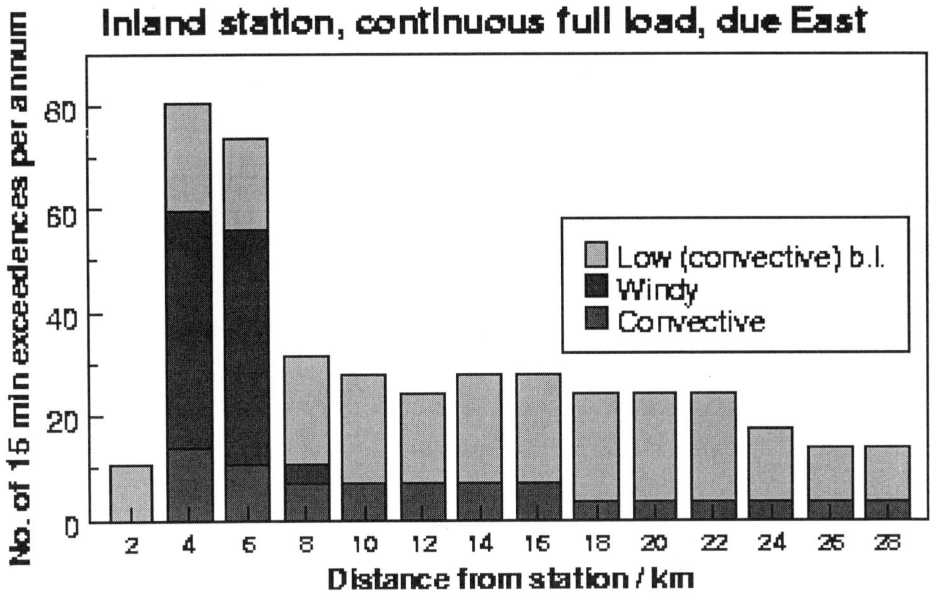

Figure 1. Number of 100 ppb exceedances predicted for a 2000MW power plant at full load as a function of weather category and distance dwnwind.

Predictions of the meteorological parameters required by ADMS were obtained from the UK Meteorological Office's Mesoscale model. This model is run from 6 am daily to provide hourly predictions of meteorological parameters until 12 noon the following day. In this study data were used from 12 noon on the day of the model run until 12 noon on the following day (i.e. predictions made between 6 and 30 hours in advance).

Predictions of hourly average values of wind speed, wind direction, cloud cover, temperature, Pasquill stability class and boundary layer depth were provided, together with their actual values at the UK Meteorological Office measurement site.

The correlation coefficients between the predicted and measured data were determined for each of the above parameters. The predicted and actual parameters were found to be significantly correlated at the 0.1% level in all cases. The predictions for wind direction and temperature were clearly better than those for other parameters. Those for cloud cover were found to be the poorest.

A comparison of ADMS modelled hourly average concentrations at ground level for unit emission concentration using predicted and actual input meteorological parameters is shown in figure 2. Again, the correlation is significant at the 1% level. However, the correlation coefficient, R^2 , is very low at 0.33, and even less, at 0.05, if the correlation at a specific point is determined. Therefore, it can be concluded that, on an hour-by-hour basis the prediction of exceedance is poor.

Figure 3 compares predictions for a typical 2000 MW power station based on predicted and actual meteorology for exceedances of the EPAQS equivalent 1 hour value of 80 ppb and WHO 1 hour guideline (122 ppb over 1 hour) for SO_2. The number of occasions when the forecast meteorological data and actual meteorological data both led to an exceedance is seen to be relatively small.

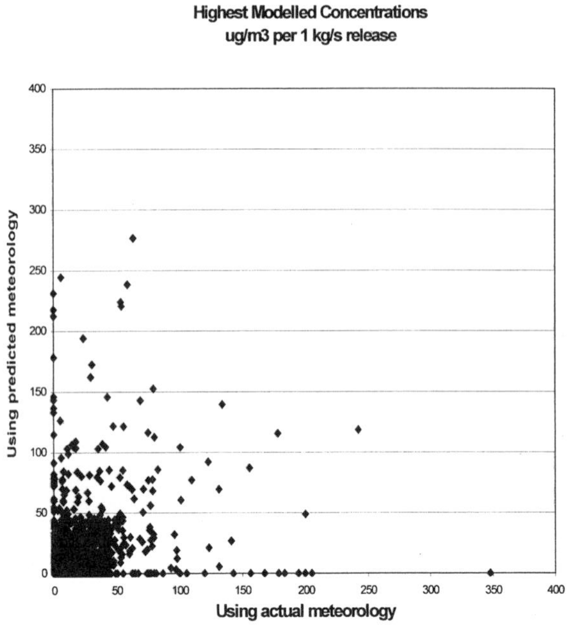

Figure 2. Scatter plot of predicted ground level concentration of SO_2 using forecasted meteorology versus predictions using actual meteorology.

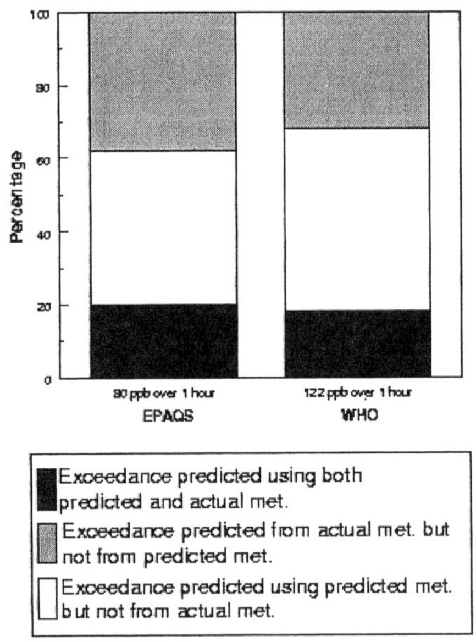

Figure 3. Percentage of time that exceedances of 100 ppb SO_2 standard using actual meteorology was also predicted using forecast meteorology.

Overall, it was concluded that prediction of key meteorological parameters, (particularly cloud cover which is an important input parameter to models such as ADMS) from 6 to 30 hours in advance was found to be poor. As a result, environmental air quality management based on concentration predictions from dispersion models on an hour-by-hour basis, is unsatisfactory, as on most occasions either unnecessary action will be taken to control emissions or pollution incidents will be missed.

A broader approach might involve taking action on a daily timescale rather than on an hour by hour timescale. A noticeable improvement in coincident predictions of days during which an exceedance occurred compared to those for the hour-by-hour comparison was found.

Therefore, it was concluded that, for the management of pollution incidents using model predictions based on meteorological forecast data to stand any chance of being effective, the timescale for action would have to be the whole day rather than for individual hours and to be irrespective of location. This timescale also corresponded more closely with the time taken to change sulphur emissions by switching coal supply.

Having determined the basic potential for an SO_2 protocol, a one year long trial was designed to study the feasibility in detail.

An in-depth trial of a method of air quality management for SO_2 was undertaken over a one year period (Jan-Dec 1998) at three major UK coal-fired power stations. The three power stations selected for the trial (and their owners at the time) were Ferrybridge C (PowerGen UK plc), West Burton (TXU Europe Power, formerly Eastern Generation Ltd) and Eggborough (National Power PLC).

The process involved using forecast meteorological data and a plume dispersion model to predict whether or not exceedances of air quality standards might occur the next day or later during the same day as the predictions.

Although the trial did not involve any actual changes to operation of any power station, it was designed to closely follow in real time what would happen were such a system operated in reality.

Key components of the system were:

- A dedicated computer (running the ADMS dispersion model and software to input data automatically) which predicts maximum SO_2 emissions or electricity generating capacity for each day and the following day.
- A daily Met Office Unified Model forecast of parameters for use in the ADMS dispersion model.
- Computer terminals in the Companies' Energy Management Centres to enable load/fuel change decisions to be made each day based on the predictions.

The exact way the trial was conducted varied in detail between companies. Figure 4 is a schematic diagram for the PowerGen arrangement.

The key part of the trial was the computer system which forecast the generation constraints that the environmental air quality considerations placed on the operation of the power station.

At 0600 GMT each day, the system dialled up the Meteorological Office and downloaded 60 hourly forecasts of the meteorological parameters required by the dispersion model.

Once the meteorological data had been down loaded the program ran the ADMS dispersion model for a series of power station loads for each hour of the meteorological dataset downloaded.

The hours for which predictions were made were 0600 to 1800 for the next day and 0800 to 1800 on the day. It had been established that ADMS did not predict any exceedances outside these hours from power stations.

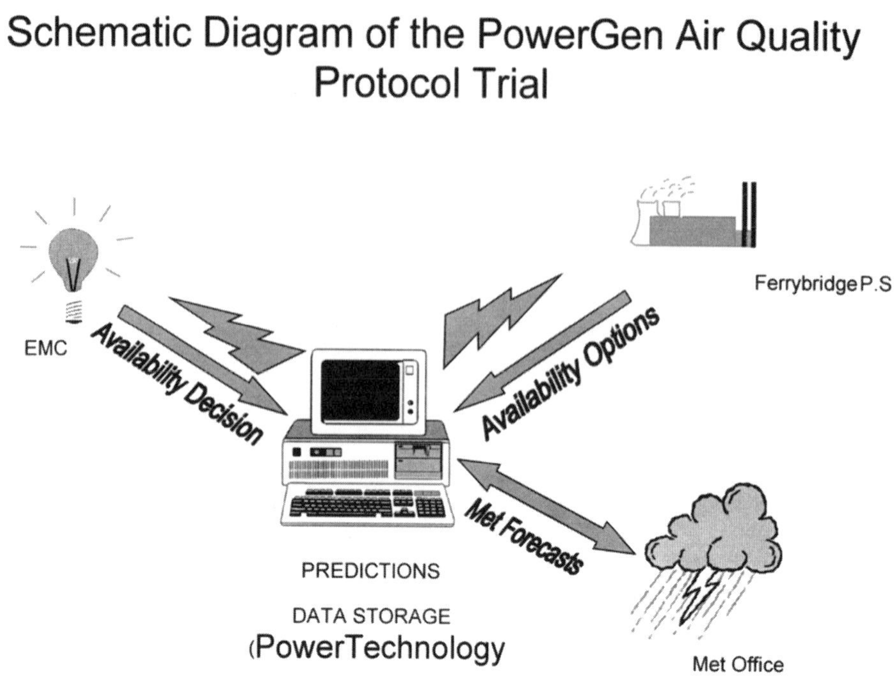

Figure 4. Schematic diagram of main components of SO_2 Trial Protocol.

The model predicted ground level concentrations of SO_2 on a 40x40 km square grid at a resolution of 1.5km, centred on the power station.

The maximum generation for the day and the day ahead were then determined based on the sulphur content of the coal stocks available.

The potential constraints to operation of the power station were then passed on to the company Energy Management Centre by intranet or e-mail.

The EMC were normally given 2 options – to burn standard sulphur content coal , with reduced generating capacity if necessary, or to switch to low sulphur coal, which might allow the generating capacity to be maintained.

This procedure continued each day for one year. The recorded data were then analysed and the ADMS predictions with the modified ('Trial') generation and SO_2 emissions compared with the predictions using the actual generation and actual SO_2 emissions (in both cases actual meteorological data recorded at the nearest UK Meteorological Office weather station were used).

By way of an example, the spatial distribution of ADMS predicted event hours in the vicinity of the one of the power stations is presented in figure 5. In figure 5 the predictions are for the actual power station generation and actual meteorological data. The result of actual implementation of the Protocol was simulated using the data from constrained generation brought about by the trial with the actual meteorological data measured at the weather station.

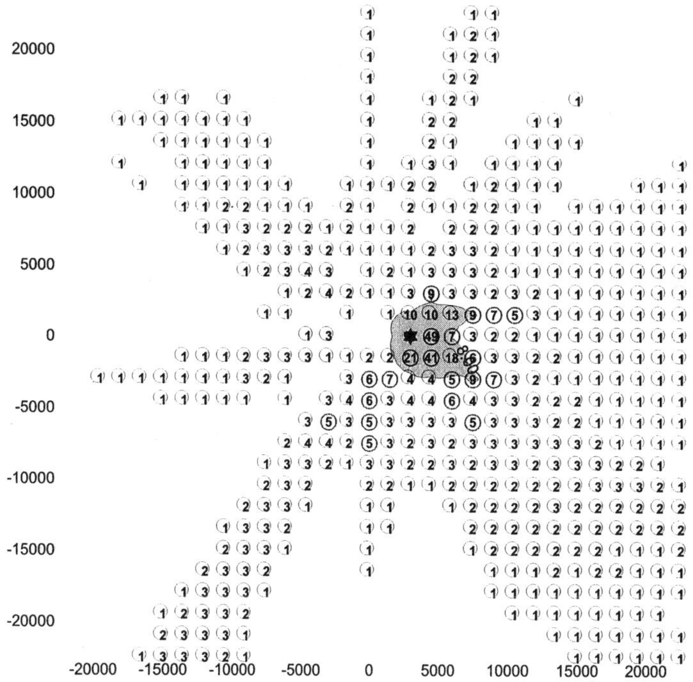

Figure 5. 40x40 km grid showing number of hours during year that an exceedance of the 100 ppb SO_2 standard was predicted around one of the power stations without the trial protocol in operation.

SO_2 monitoring sites were in operation at a number of locations around the power stations taking part in the trial and data from these sites were compared to the predicted ground level concentrations. As an example, there was a monitoring site 4.5km ENE of the power station whose impact on local air quality is predicted in figure 5. This is close to the position of maximum predicted impact.

An analysis of the data from this site showed that, when the wind was blowing from the direction of the power station (assumed to be 240-270° N) , the number of hours during which there was at least one 15 minute exceedance of 100 ppb was 32. These occurred on 23 different days during the trial. If the Protocol had been in operation, 11 of the event hours actually measured would have been predicted and 17 of the event days. This confirms the supposition that it is easier to predict that an exceedance will occur at some time during a day than to be more precise about the actual hour of occurrence.

(An 'event day' is defined as a day when ADMS predicted an exceedance of the 100 ppb SO_2 standard at any receptor point at any time during the day. An 'event hour' is defined as an hour during which ADMS predicted at least one 15 minute exceedance of 100 ppb).

Comparison of the predicted number of exceedances (10-46) with the actual measured number (32) suggests that the model predicts the number of event hours over the period of a year reasonably well. However, the timing of the events is not well predicted.

A series of tests were applied to the data from the trial to determine the degree to which it met the objectives, see table 1.

It would be hoped that the hit rate will be higher than the miss rate, and that the success rate will be higher than the failure rate for a successful trial.

It would be expected that the Random Chance Rate would be lower than the Trial Success Rate if the protocol is to have any hope of succeeding.

Table 4.2. Summary statistics of the ability of the Trial Protocol to manage air quality.

	PowerGen		National Power		Eastern Generation	
Test	**Day Result**	**Hour Result**	**Day Result**	**Hour Result**	**Day Result**	**Hour Result**
Hit Rate	0.68	0.75	0.53	0.67	0.73	0.71
Miss Rate	0.32	0.25	0.47	0.33	0.27	0.29
Success Rate	0.50	0.14	0.31	0.078	0.74	0.27
Failure Rate	0.47		0.57		0.28	
Action Rate	0.60		0.54		0.74	
Random Chance	0.25		0.17		0.55	

Footnote:

The Hit (Miss) Rate can be defined as: $\frac{\text{Event Days(hours) Saved(Missed)}}{\text{Event Days(hours)}}$

The Success Rate as: $\frac{\text{Event Days(hours) Saved}}{\text{Action Days(hours)}}$

The Action Rate as: $\frac{\text{Action Days}}{\text{Days in Trial}}$

The Failure Rate as: $\frac{\text{Unnecessary Action Days}}{\text{Action Days}}$

The Random Chance Rate as:

$\frac{\text{Number of Action Days}}{\text{Number of Days Trial Ran}}$ $\frac{\text{Number of Event Days}}{\text{Number of days in period}}$ or

Action Rate Event Rate

From the various tests it can be seen that the trial successfully prevents about 60% of all events (averaged over the three power stations), and misses about 35% completely (miss rate). The cost of this is a very high action rate, especially during the summer when most days required action.

For the trial as a whole the success rate was only marginally higher than the failure rate, so for every action which prevented an event, another was taken which was unnecessary.

The above tables show that the protocol is initiating action far too often to be an effective tool for use against local pollution incidents. The costs of implementing such a system due to loss in generation and low sulphur coal, would be as large as the costs of retro-fitting of flue gas desulphurisation plant.

CONCLUSIONS

- Predicting the occurrence of days during which a power station causes a ground level concentration of SO_2 to exceed the UK national air quality standard concentration for at least 15 minutes somewhere in the vicinity is subject to large uncertainties.
- Notwithstanding the inaccuracies of the physical dispersion algorithms implicit in any predictive model used, the accuracy of such a prediction also depended strongly on the accuracy and applicability of the meteorological data input. For example the best available forecast data (from the UK Meteorological Office) limits the correct day-in-advance prediction of exceedances of the air quality standard to about 50%.
- This makes a system of local air quality management based on forecast meteorological data and computer modelling very inefficient and expensive.

INTEGRATED REGIONAL MODELLING

chairpersons:	J. Kretzschmar S. T. Rao
rapporteurs:	C. Soriano P. Suppan

AIR QUALITY MODELLING FOR THE HONG KONG PATH PROJECT

Julie Noonan[1], William Physick[1], Martin Cope[1],
Manuela Burgers[2] and Marilyn Olliff[2]

[1] CSIRO Atmospheric Research, PB 1, Aspendale, Victoria 3195, Australia.
[2] Environment Protection Authority of Victoria, GPO Box 4395QQ
Melbourne Victoria 3001, Australia.

INTRODUCTION

The aim of the PATH (Pollutants in the Atmosphere and their Transport over Hongkong) project was the development of a multi-scale numerical air quality modelling system for the Hong Kong Environmental Protection Department (EPD). The system is based on eleven case-study periods that are representative of various meteorological and air quality situations that occur regularly in the Hong Kong region - generally conditions that are conducive to photochemical smog and high PM10 concentrations. The primary purpose of the system is to use the eleven case studies as an aid in the evaluation of various emission controls and strategic planning scenarios.

The PATH system consists of three coupled components: a meteorological model MM5, an air chemistry model SAQM and an emissions inventory model EMS-95. Emissions inventories were developed for each case-study period for Hong Kong and southern China. Simulations for each period were carried out using MM5 and SAQM on one-way and two-way nested grids, with grid spacings ranging from 40.5 down to 0.5 km.

The following sections briefly describe the major components of the study and present some meteorological and air quality results.

SELECTION OF CASE STUDY PERIODS

Seven case-study days were selected to calculate *annual-average concentrations* of PM10 and three days were chosen as *pollution episodes,* based on observations of higher than average concentrations of either PM10 (one day chosen) or photochemical smog (two days chosen). An additional period of two consecutive *special validation days* was also selected from the Aircraft Inventory Verifications Study (AIVS) undertaken in November 1997 as part of the PATH project. The AIVS was primarily for model and inventory validation purposes.

The annual-average calculations were based on the hypothesis that there are recurring weather patterns that can be characterised by the distribution of meteorological variables such as pressure, wind and precipitation, and that each pattern is associated with characteristic concentration levels of PM10. Data analysis showed this hypothesis to be valid, and allowed us to estimate long-term averages based on a manageable number of individual events, each of which is representative of a major meteorological category. The frequency with which each category occurs over the averaging period was calculated by examination of mean sea level pressure charts from 1990-1996.

For details of the case study selection process and the calculation of the annual average of PM10, see Physick *et al.* (1998) and Physick and Noonan (2000).

EMISSIONS

The emissions inventories in PATH were created using the Emissions Modelling System version 1995 (EMS-95; Emigh and Wilkinson 1995), which provides the primary emissions data for the air quality model. EMS-95 also uses data generated by the meteorological model (MM5) to determine emissions that are dependent on meteorological parameters such as temperature and wind speed.

EMS-95 calculates and processes emissions from point, area-based, mobile and biogenic sources. Large quantities of data were required for the compilation of the detailed emission inventory. The data relate to three main areas: the *scale* of the emissions; the *temporal* variation in the emissions over a given day, week, season or year; and the *spatial* allocation of the emissions to the modelling grid. EMS-95 allows the generation of alternative emission scenarios without affecting the base data, allowing investigations of the effects on air quality from various control strategy or development scenarios.

For the PATH inventory system EMS-95 provides emissions on an hourly timescale (the basic temporal resolution of the modelling components of PATH). The inventory includes all pollutants that contribute to the formation of photochemical smog and secondary particles, primary particulate emissions, and primary emissions of toxic organics. The organic compounds are speciated into chemical surrogates suitable for use in the chemical mechanism of SAQM, nitrogen oxides are speciated into nitrogen dioxide and nitric oxide, and particulate emissions are distributed between the three size ranges used by the model (less than 1 μm, between 1 and 2.5 μm, and between 2.5 and 10 μm). Emissions are allocated spatially to the SAQM grid structure.

Emissions from beyond Hong Kong's borders have the potential to impact on air quality within Hong Kong. In order to adequately represent and predict Hong Kong air quality it is therefore necessary to quantify the cross-border flux of emissions originating in the southern regions of mainland China. To enable this quantification, an emissions inventory has been produced for southern China, although in less detail than that of the Hong Kong inventory.

MM5

The multi-scale, non-hydrostatic model chosen for this study is the Pennsylvania State University - National Center for Atmospheric Research (NCAR) Mesoscale Model, usually referred to as MM5 (Dudhia, 1993). MM5 has a multiple nesting capability, with domains interacting through two-way or one-way nesting. It also includes a four-dimensional data assimilation (FDDA) scheme for introduction of observational data through analysis nudging and/or observational nudging.

Within PATH, MM5 was run for five grids with grid spacings of 40.5, 13.5, 4.5, 1.5 and 0.5 km (see Figure 1). The grid sizes are 115 x 75, 85 x 73, 61 x 55, 61 x 55 and 58 x 82 grid points, respectively. All modelling domains have the same vertical structure with 26 unequally spaced levels, 10 of which are below 1000 m. The 13.5-km grid is centred over southern China, the 1.5-km grid covers the Special Administrative Region of Hong Kong and the smallest urban domain covers Hong Kong Island, Kowloon and parts of the New Territories. This grid was used for issues involving primary pollutants, as pollutant residence time does not allow the full development of secondary products before they are transported beyond the boundaries of this domain.

The purpose of having several levels of resolution is to simulate the cross border fluxes of pollutants between southern China and Hong Kong, the transport of local and regional air contaminants within Hong Kong, and the intra-urban transport of local concentrations of emissions.

Simulations were initialised from the 0.5-degree latitude-longitude gridded European Centre for Medium-Range Weather Forecasts (ECMWF) analysis. Boundary conditions for the outer grid were obtained by linear interpolation between ECMWF analyses at 6-hourly intervals. Most runs were for two days, the first day a spin-up day for SAQM and the second day, the day of interest.

Observational data were assimilated from hourly values of surface wind from the Hong Kong Observatory anemometer network (28 sites within the SAR of Hong Kong) on the 1.5- and 0.5-km grids.

In the next two sections, results are presented from one case-study day. Statistics from the quantitative evaluation of all eleven case studies are also presented.

Case Study: 19-20 August 1996

This period was chosen as one of the two photochemically active episodes. On 19 August 1996, a tropical depression was situated about 500 km east of the Philippines. By the next day, it had intensified to a tropical storm and moved westward to be centred just

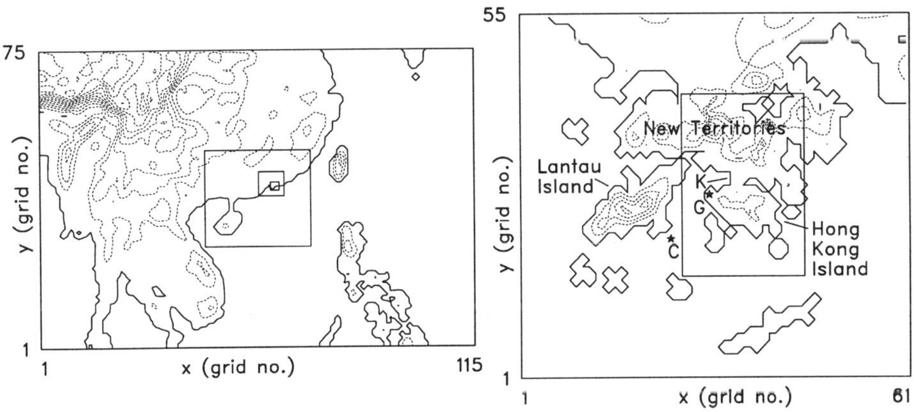

Figure 1. The 40.5-km spaced grid (left), showing the position of the nested 13.5-, 4.5- and 1.5- km spaced grids. The dashed lines are the terrain contours, contour interval is 500 m. The 1.5-km spaced grid (right), showing the position of the nested 0.5-km spaced grid. The terrain contour interval is 100 m. Place names mentioned in the text are shown: K – Kowloon, G – Green Island, C – Cheung Chan.

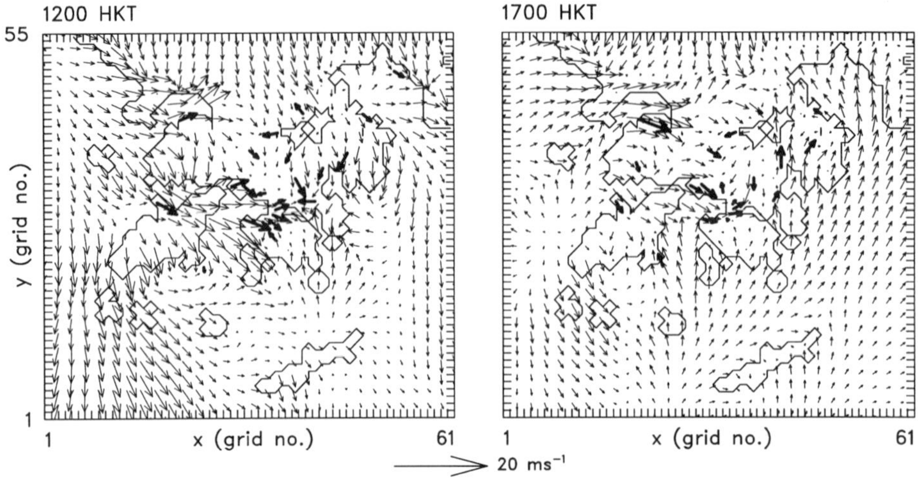

Figure 2. Modelled (1.5-km grid) and observed (darker arrows) surface winds at 1200 and 1700 HKT on 20 August 1996.

north of Manila. Low pressure generally prevailed over China, resulting in light northeasterly flow over the Hong Kong region.

Simulations were initialised at 0200 Hong Kong Time (HKT) on 19 August 1996 and continued until 2400 HKT on 20 August 1996. The 40.5-km grid was run independently of the other grids, with the results only being used by SAQM. Simulations on the 13.5-, 4.5-, 1.5- and 0.5-km spaced grids were run with two-way nesting between all grids. Observational data were assimilated on the 1.5- and 0.5-km grids.

Figure 2 shows modelled surface winds on the 1.5-km grid at 1200 and 1700 HKT on the 20 August 1996 with the observed winds overlain. At 1200 HKT a convergence line running northwest – southeast can be seen in both the observations and the model results. The convergence line separates west-northwesterlies from northeasterlies.

Sea breezes are also evident in both the observations and model results at various coastlines. At 1700 HKT, the model predicts that the southerly sea breeze, suggested by the observation at Cheung Chau just east of Lantau Island, lies just to the south of Green Island. By 1800 HKT (not shown), observations and model show that it has passed through Green Island and onto the mainland.

Overall, the important features of the observed surface wind patterns in this case study of complex winds are reproduced well, as are upper-air wind profiles (four daily sondes at King's Park).

Statistical MM5 Results

Evaluation of the MM5 simulations was performed through comparison of observed and predicted winds and temperatures, on both a qualitative and a quantitative basis. Quantitative measures used include Willmott's index of agreement *I*, lying between 0 and 1, where 1 indicates perfect agreement (Willmott *et al.*, 1985). It is a measure of how all the predictions' departures from the observed mean match the observations' departures from the observed mean, and in particular how well is the match in the sign of the departures. *SV*, the standard deviation of the predictions divided by the standard deviation of the observations (ideally equal to 1) and *SR*, the root mean square error divided by the standard deviation of the observations (ideally less than 1), are also used in the evaluation.

For each case-study period, Figure 3 shows the 24-hour mean statistics for simulations without FDDA of surface observations. Also included in the Figure are statistics from a number of other similar studies: the Vancouver sea-breeze simulation of Steyn and McKendry (1988) (SM1), simulations of two sea-breeze days in the Los Angeles basin by Ulrickson and Mass (1990) (UM1 and UM2), simulations in the same area by Seaman *et al.* (1995) (S95, S96) and a 5-case average value from simulations of the lake breeze around Lake Michigan by Lyons *et al.* (1995) (L95).

Figure 3 shows that SV for wind speed is less than or equal to 1 for all simulations, and greater than 0.8 for only two, indicating that the model is simulating less variability across the region than is observed, although the statistics still compare favourably with the studies from elsewhere. The inability of the model to match the variability in the wind observations is not surprising, considering the complex terrain and coastlines of Hong Kong. SR for wind speed is acceptable for all simulations, with six cases taking a value less than or equal to 1.03, probably a better performance than the North American simulations. More than half of the cases have an index of agreement value for wind speed, which is better than the best value from the North American studies, and all but one of the temperature values are better than the single value from the Steyn and McKendry study.

It should perhaps be mentioned again that the statistics in Figure 3 are for the simulations without any assimilation of observations. Assimilation of data into the final simulations for PATH considerably improved each case.

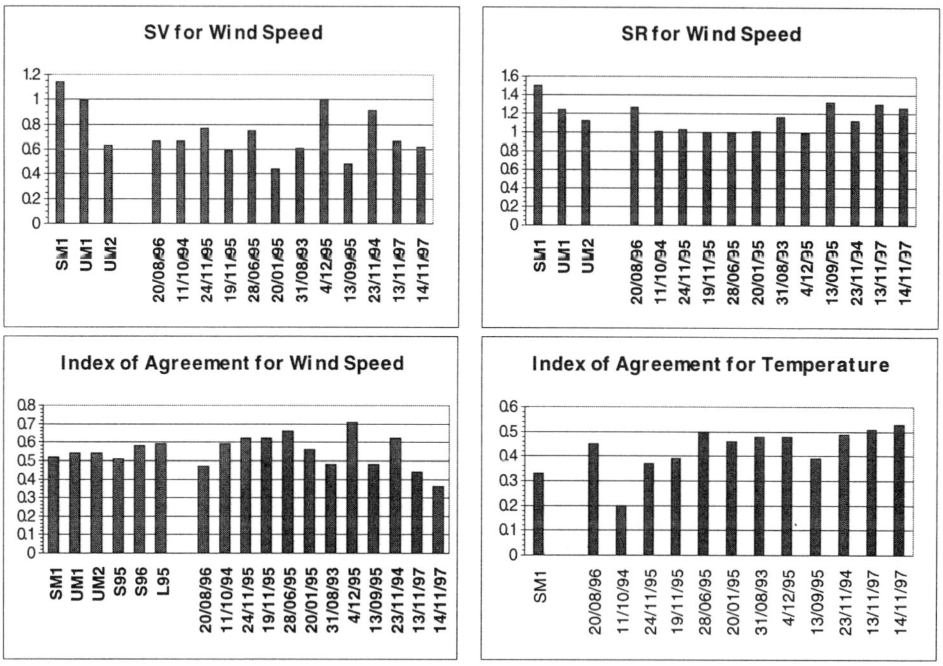

Figure 3. Values of various 24-hour mean statistics for the Case Studies and other similar studies.

SAQM

The air quality simulation component of PATH is the SARMAP Air Quality Model (SAQM) (Chang *et al.*, 1997). SAQM is a derivative of the Regional Acid Deposition Model (RADM), though its development was influenced by the ADOM, CALGRID and STEM-II models. The model is fully nestable within itself and non-hydrostatic compatible so that a mass balance is preserved in the presence of local pressure gradients. Modifications to SAQM for PATH are described in Cope *et al.* (2000).

In the Eulerian framework, SAQM solves a set of chemical species conservation equations. There are two major changes for a species in SAQM: transportation and chemistry; the chemical mechanism used in SAQM is Carbon Bond-IV (CB-IV).

The transport of species in SAQM utilises the meteorological data from MM5: three-dimensional wind, temperature and pressure. Additional data, such as eddy diffusivities, are also needed to calculate the sub-grid scale turbulent transport. The temperature and radiation fields are used in the calculation of chemical transformation rates. The SAQM meteorological pre-processor was modified after testing of SAQM found that there were mass inconsistencies in the MM5 two-way nested regions.

Since SAQM requires the input of meteorological fields generated by MM5, a complete run of the meteorological model must precede each case-study air quality simulation. However, SAQM can be re-run with modified emissions, but using the same meteorological input pre-computed by MM5.

SAQM operates within PATH using (basically) the same, five-level system of nested grids as the MM5 meteorological model. Similarly, the same terrain-following vertical coordinate and the same vertical structure are used. The PATH emissions inventory provides the second crucial part of SAQM input.

A reactive Particle Model (PM) has been incorporated into SAQM for use within the PATH system, and is described in detail in SAIC (1997). SAQM was modified to incorporate modules for modelling reactive and unreactive particles, as required for the simulation of PM10.

The following section presents some results from SAQM.

SAQM Annual Average PM10 Results

An important component of the PATH project was the development of a capability for modelling annual-average PM10 concentrations. Seven representative 24-hour PM10 events were selected for detailed modelling.

Observed and modelled annual-average PM10 concentrations for Central Western, Kwai Chung, Kwun Tong, Tsuen Wan and Sha Tin monitoring stations are shown in Figure 4. The observed annual-average concentrations are the mean of the annual-averages for the years 1993-95. 'STAT' corresponds to the observed annual-average concentration calculated at each station by weighting the daily mean concentration of the modelled day for each category by the annual frequency of that category, and summing (Physick and Noonan, 2000). The 1.5-km and 0.5-km SAQM results are calculated using the same methodology. The performance of the PATH system should be evaluated by comparing the results with 'STAT'.

From Figure 4 it can be seen that PATH has performed well in reproducing the statistical estimates of annual-average PM10 concentration for Central Western, Kwai Chung and Sha Tin monitoring stations. However, PATH has had more difficulty at Tsuen Wan and Kwun Tong monitoring stations, underestimating PM10 concentrations by 25 to 35%. One possible cause for this underestimation is the presence of local industrial

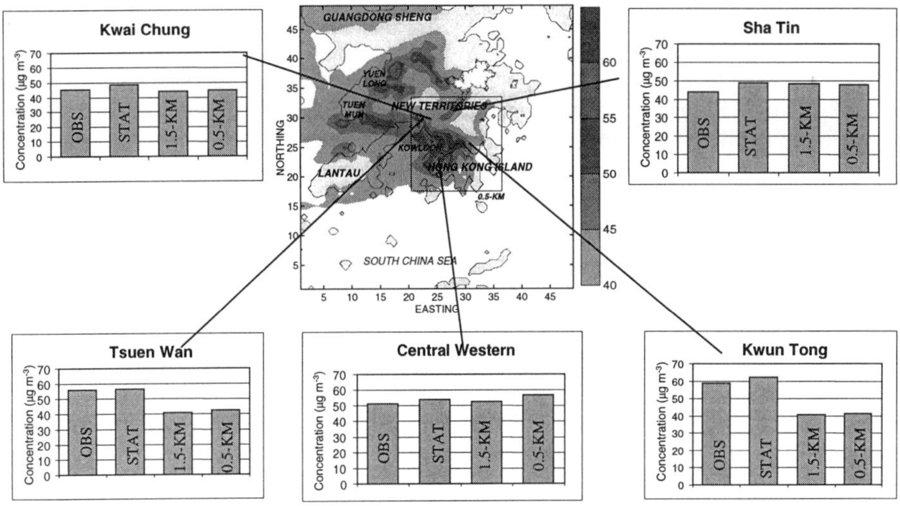

Figure 4. Observed and modelled annual-average PM10 concentrations for monitoring sites in the SAR. Also shown is the modelled spatial distribution of annual-average PM10 for the 1.5-km grid. Note STAT denotes the statistical model; 1.5-km and 0.5-km correspond to annual-average concentration estimates made using the 1.5-km and 0.5-km grid resolutions.

sources, which may influence the observed concentrations, but which fall below the resolvable scales of SAQM.

The spatial distribution of modelled annual-average PM10 concentration for the 1.5-km grid is also shown in Figure 4. It can be seen that the majority of Hong Kong is predicted to comply with the 55 $\mu g\ m^{-3}$ air quality objective. However, non-compliance is predicted for a number of small areas within the region. This outcome is in general accordance with the observed annual-average PM10 concentrations.

DISCUSSION

In summary, we found that all the MM5 simulations qualitatively compare well with observations. Statistics from a quantitative assessment are good, and compare favourably with those from studies elsewhere at the same scale in similar or less challenging geographic settings.

SAQM model outputs have also been qualitatively evaluated by comparing with observational data from a number of monitoring stations in Hong Kong and air quality data collected on instrumented aircraft flights for two of the case-study periods. In general, there was good agreement between modelled and observed data when the meteorological conditions favoured southerly clean airflows over Hong Kong. The air quality in Hong Kong under such conditions is dominated by local emission sources, which are well quantified. However, when the meteorological conditions are such that the air quality in Hong Kong is influenced significantly by emissions from southern China, the disparity between modelled and observed data is greater. This disparity may be attributed largely to the much greater uncertainty in emission estimates for southern China.

ACKNOWLEDGMENTS

The authors wish to thank the Hong Kong Environmental Protection Department (EPD) and Hong Kong Observatory for provision of data. We are also grateful to the EPD for permission to publish these results.

REFERENCES

Chang J. S., S. Jin, Y. Li, M. Beauharnois, C. –H. Lu, H. –C. Huang, S. Tanrikulu and J. DaMassa, 1997: The SARMAP Air Quality Model Final Report. California Air Resources Board, CA.

Cope, M. E., M. Burgers, and M. Olliff, 2000: Application of the SARMAP Air Quality Model (SAQM) to the modelling of air pollution in Hong Kong for the PATH study. *11th Joint Conf. On the Applications of Air Pollution Meteorology with the A&WMA,* Long Beach, California, USA, 96-101

Dudhia, J., 1993: A nonhydrostatic version of the Penn State-NCAR Mesoscale Model: Validation tests and simulation of an Atlantic cyclone and cold front. *Mon. Wea. Rev.,* **121**, 1493-1513

Emigh R.A. and J.G. Wilkinson, 1995: The Emissions Modelling System (EMS-95) User's Guide. Alpine Geophysics.

Lyons, W. A., C. J. Tremback, and R. A. Pielke, 1995: Applications of the Regional Atmospheric Modeling System (RAMS) to provide input to photochemical grid models for the Lake Michigan Ozone Study (LMOS). *J. Appl. Meteorol.*, **34**, 1762-1786.

Physick W. L., M. Cope, C. Fung, D. Hearn, A. Jackson, K. Tilly, S. Bentley, J. Carras, W. Farrell, M. Finn, Galbally, R. Goudey, B. Hellyer, P. Hurley, M. Meyer, J. Noonan, Y-L. Ng, M. Olliff, L. Bonadio, R. Serebryanikova, F. Vukovich, Z. Wang, I. Weeks, D. Williams, C. Wilson, N. Wong and R. Ye. 1998: The Territory-Wide Air Quality Modelling System (TWAQMS) for Hong Kong. *Proc. 14th Int. Clean Air and Environment Conf.*, Melbourne, Australia, 125-131.

Physick, W. L. and J. A. Noonan, 2000: Mesoscale modelling with MM5 for the PATH study (Pollutants in the Atmosphere and their Transport over Hong Kong). *11th Joint Conf. On the Applications of Air Pollution Meteorology with the A&WMA,* Long Beach, California, USA, 90-95.

SAIC, 1997, The Particle Model (PM) in SAQM for the Territory-Wide Air Quality Modelling System (PATH Model). *Science Applications International Corporation, Technical Note*, August 1997.

Seaman, N. L., D. R. Stauffer, and A. M. Lario-Gibbs, 1995: A multi-scale four dimensional data assimilation system applied in the San Joaquin valley during SARMAP. Part I: Modeling design and basic performance. *J. Appl. Meteorol.*, **34**, 1739-1761.

Steyn, D. G. and I. G. McKendry, 1988: Quantitative and qualitative evaluation of a three-dimensional mesoscale numerical model simulation of a sea breeze in complex terrain. *Mon. Wea. Rev.*, **116**, 1914-1926.

Willmott, C. J., S. G. Ackleson, R. E. Davis, J. J. Feddema, K. M. Klink, D. R. Legates, J. O'Donnell, and C. M. Rowe, 1985: Statistics for the evaluation and comparison of models. *J. Geophys. Res.*, **90**, 8995-9005.

Ulrickson, B. L. and C. F. Mass, 1990: Numerical investigation of mesoscale circulations over the Los Angeles Basin. Part I: A verification study. *Mon. Wea. Rev.*, **118**, 2138-2161.

DISCUSSION

R. BORNSTEIN: Could you define "non-hydrostatic coordinates"?

M. COPE: In the context of the SAQM model this merely indicates that the model's coordinate system is the same as that of the non-hydrostatic MM5 model. This is advantageous because it helps to maintain the consistency between the mass and momentum fields when MM5 fields are used in SAQM to transport air pollutants.

R. BORNSTEIN: Is a z_0-formulation sufficient for such tall urban buildings, or should you use a displacement height and/or urban building topographic heights?

M. COPE: In fact the urban canopy was modelled through the use of a displacement height and through adjustments to the topographic height fields within the vicinity of the urban areas. We were also aided by the fact that the majority of the air quality measurements were made on the tops of the buildings i.e. at or above the displacement height.

STUDY OF THE TRANSPORT AND DIFFUSION PROCESSES IN THE PBL USING THE RAMS AND SPRAY MODELS: APPLICATION TO THE TRACT EXPERIMENT

J. C. Carvalho,[1] G. A. Degrazia,[1] D. Anfossi,[2] and S. Trini Castelli [2]

[1] Universidade Federal de Santa Maria, Departamento de Física - 97119-900, Santa Maria, RS, Brazil
[2] Consiglio Nazionale delle Ricerche, Istituto di Cosmogeofisica - 10133, Turin, Italy

INTRODUCTION

In this work, the meteorological model RAMS and the Lagrangian particle model SPRAY were coupled to simulate the dispersion of a tracer emitted during the TRACT experiment, performed in the Rhine valley region, in Central Europe.

The Lagrangian particle model SPRAY is a three-dimensional model to simulate the pollutants dispersion over complex terrain. The model is based on the Langevin equation and the wind field and the turbulent parameters are the input quantities. The wind field is obtained from the meteorological model RAMS and the turbulent field is obtained through an interface program that connects the models RAMS and SPRAY. This interface program named MIRS (Method for Interfacing RAMS and SPRAY) obtains the meteorological data from the RAMS and supplies the turbulence parameterisations for the particle model SPRAY.

The results of the flow simulations are evaluated by means of statistical indexes calculated from quantities simulated by RAMS and observed during the TRACT experiment. Concentrations data measured in the surface level are used to evaluate the concentrations simulated by SPRAY. The analysis of the results shows that the model system reproduces well the general behaviour of the tracer plume, the temporal and spatial concentration distribution and the location of the concentration maxima.

MODEL SYSTEM

The model system RAMS-MIRS-SPRAY is based on a combination of the meteorological model RAMS, the interface code MIRS and the Lagrangian particle model SPRAY. RAMS (Regional Atmospheric Modeling System) is a prognostic model, that was developed at the Colorado State University (Pielke et. al, 1992). RAMS has been used to simulate any flow situation from microscale to synoptic scale. The model includes a large number of options that can be selected by the user. The main options comprehend hydrostatic and non-hydrostatic versions, two-way interactive grid nesting, terrain-following coordinates, stretched vertical coordinates, nudging system, different options of numerical schemes, several top and lateral boundary conditions and a set of physical parameterisations. RAMS includes also a model for soil and vegetation temperature and moisture.

The interface code MIRS (Trini Castelli and Anfossi, 1997), uses the RAMS outputs (wind speed, turbulent kinetic energy - TKE, diffusion coefficients, potential temperature, surface fluxes, the superficial layer parameters - friction velocity u_*, temperature scale θ_* and Obukhov length L - and topography) and calculates the parameters for the Lagrangian model SPRAY, not directly given by the RAMS, i.e.: the convective velocity scale w_*, the variances of the velocity fluctuation σ_i^2, the local velocity decorrelation timescale τ_{L_i}, the third and fourth moments of the vertical wind velocity fluctuations and PBL height, where $i = 1,2,3$. Besides reading the RAMS outputs and calculating the above-described parameters, MIRS prepares a single file having the correct format, with the temporal sequence of interest, to be used by SPRAY as an input information. Three options are available in MIRS for the velocity variances and the decorrelation timescales: TKE values from RAMS (Mellor and Yamada, 1982), Hanna (1982) and Degrazia et al. (2000). For the third moment of the vertical velocity, MIRS uses the Chiba (1978) expressions, and for the fourth moment it uses the following relationship: $\overline{w^4(z)} = 3.5[\overline{w^2(z)}]^2$ (Ferrero and Anfossi, 1998). For the PBL height, three methods are considered: McNider and Pielke (1981), Maryon and Buckland (1994) and Kalthoff et al. (1998). In the first two methods, the profile of gradient Richardson number Ri is considered. The third method is based on work of Kalthoff et al. (1998) carried out by using data collected during the TRACT experiment. They showed that, during clear sky and weak subsidence conditions, the height of the convective boundary layer (CBL), approximately follows the height of the underlying orography in the TRACT area.

SPRAY (Tinarelli et al., 2000) is a Lagrangian stochastic particle model designed to study the pollutants dispersion in complex terrain. It is based on a three-dimensional form of the Langevin equation for the random velocity (Thomson, 1987). The velocity and the displacement of each particle are given by the following equations (Rodean, 1996):

$$du_i = a_i(\boldsymbol{x},\boldsymbol{u},t)dt + b_{ij}(\boldsymbol{x},\boldsymbol{u},t)dW_j(t) \tag{1a}$$

and

$$d\boldsymbol{x} = (\boldsymbol{U} + \boldsymbol{u})dt\,, \tag{1b}$$

where $i, j = 1,2,3$, $\boldsymbol{x}$ is the displacement vector, $\boldsymbol{U}$ is the mean wind velocity vector, $\boldsymbol{u}$ is the Lagrangian velocity vector, $a_i(\boldsymbol{x},\boldsymbol{u},t)dt$ is a deterministic term, $b_{ij}(\boldsymbol{x},\boldsymbol{u},t)dW_i(t)$ is a stochastic term and the quantity $dW_j(t)$ is the incremental Wiener process. The deterministic coefficient depends on the Eulerian probability density function (PDF), $P_E(\boldsymbol{x},\boldsymbol{u},t)$, of the turbulent velocity and is determined from the Fokker-Planck equation. In the two horizontal directions the PDF is assumed to be Gaussian. In the vertical direction the PDF is assumed to be non-Gaussian (to deal with non-uniform turbulent conditions and/or convection). In this case, two different approaches can be adopted in order to calculate the Fokker-Planck equation: a bi-Gaussian one and a Gram-Charlier one (Anfossi et al., 1997), truncated to the third and fourth order. The diffusion coefficient $b_{ij}(\boldsymbol{x},\boldsymbol{u},t)$ is obtained from the Lagrangian structure function and is related to the Kolmogorov constant,

C_0, for the inertial subrange. In SPRAY it is possible to choose between constant and variable time step. The model can account for the plume rise if necessary.

TRACT EXPERIMENT

The TRACT (TRAnsport of Air Pollutants over Complex Terrain) experiment was part of the sub-project of EUROTRAC that aimed at investigating the transport and chemical transformation of trace substances and pollutants in the troposphere over Europe. TRACT (Zimmermann, 1995) took place from 7 to 23 September 1992 and represented one of the main field programs of EUROTRAC. The main objectives of this experiment were the study of pollutants transport and diffusion atmospheric processes, which are forced by thermal and dynamic influences due to the non-homogeneous terrain.

The TRACT investigation area covered an area of about 300 km^2 and included the south-eastern part of Germany, eastern part of France and the northern part of Switzerland. The terrain elevations in the TRACT region varies from about 100 m to 1500 m above sea level (asl). The most important orographic structures that influence the flow are many valleys (Rhine, Swiss Midland, Kraichgau, Neckar) and mountains (Vosges, Black Forest, Schwabische Alb, Swiss Jura).

During the campaign three intensive measurement periods (IMPs) were organized, in which the available equipments collected data during at least 24 consecutive hours. A very large number of different kinds of measurements were made. In particular, tracer release and concentration measurements were performed during the first day of the second IMP (16 – 17 September) by a ground network of 20 sampling stations with 28 samplers. The tracer, perfluordimethylcyclohexano (PDCH) – C_8F_{16}, was released at the ground level in Sasbach (48.65N, 08.088E), in the Rhine valley, during about 3 hours (from 05:02 to 07:58 UTC) with an e emission rate of 5 g/s.

The periods during which the TRACT area was influenced by high pressure systems were selected as IMPs. The weather situation during the 2nd IMP was characterized by an eastward moving high-pressure system. On 16 September, the winds within the PBL and above blew from westerly directions during the day. In the evening, the winds turned north-easterly when the high-pressure centre moved northward of the TRACT area. On 17 September, as a consequence of the eastward moving of the high-pressure system, the flow at the surface blew from easterly. Above of PBL, the winds blew from westerly-northwesterly.

RESULTS

RAMS configuration included three nested grids. The outer grid had 20 x 20 points and 16 km horizontal resolution, the intermediate grid had 38 x 38 points and 4 km horizontal resolution, and the inner grid had 62 x 62 points and 1 km horizontal resolution. The domain top was set equal to 15 km. A vertical stretched grid with forty levels was utilized on each grid. The resolution was of 50 m close to the surface and increased to a maximum of 500 m and then kept constant above. The simulation time step was set to 30 s for the outer grid, 10 s for the intermediate grid and 3.3 s for the inner grid. RAMS was initialised using the European Centre for Medium-Range Weather Forecasts (ECMWF) gridded (0.5° lat./lon.) analyses field. Data assimilation was performed on the input data interpolating them on the model grids from isentropic surfaces. New analyses data were used to nudge the lateral boundaries of the outer grid every 6 hours. The 2.5 level Mellor and Yamada scheme (1982) was utilized for the vertical diffusion parameterisation. Data of land use (15" x 15") and topography (30" x 30") of the TRACT area were supplied to

RAMS. For the soil model, the initial values of moisture and temperature in the ground were obtained from ECMWF. The moisture processes were treated as passive water vapour only.

The RAMS model evaluation was performed using the following statistical indexes: fractional bias (*fb*) (for scalars and for the wind speed module), root-mean-square error (*rmse*) (for scalars) and root-mean-square vector error (*rmsve*) (for the wind speed components; Cox et al., 1998). Data observed from 15 surface and 15 radiosounding stations were available. For sake of conciseness, the results from a sample of some stations will be presented only.

Table 1 presents the statistical indexes of observed and simulated values at the Colmar-Meyenheim, Guettingeen, Karlsruhe, Muehlacker, Sinsheim, Strassburg and Stoetten surface stations. Colmar-Meyenheim, Karlsruhe, Strassburg, Muehlacker and Sinsheim are located into the valleys, whereas Guettingeen and Stoetten are mountain stations. Indexes are calculated for wind speed components (u,v), wind speed (V), air temperature (T) and specific humidity (q). Figure 1 illustrates the time evolution of the same statistical indexes at the Bruchsal (lat./lon.: 49.13/8.56; height: 110 m asl), Musbach (lat./lon.: 48.50/8.48; height: 695 m asl) and Oberbronn (lat./lon.: 48.94/7.61; height: 274 m asl) radiosounding stations. Bruchsal and Oberbronn are situated in the Rhine valley and Musbach is situated over the Black Forest mountain. In the radiosounding stations, the indexes are computed for u,v, V, potential temperature (θ) and q. As a general consideration, the statistical results (surface and profile data) reveal that RAMS simulations can be considered satisfactory.

Table 1. Statistical indexes results for the ground stations.

station	lat. (degree)	lon. (degree)	height (m asl)	*rmsve* (u,v)	*fb* (V)	*rmse* (T)	*fb* (T)	*rmse* (q)	*fb* (q)
Colmar-Meyenheim	47.92	7.40	209	1.48	0.30	4.90	-0.007	1.02	0.04
Guettingeen	47.60	9.28	440	3.25	0.80	3.71	-0.14	1.81	0.01
Karlsruhe	49.03	8.36	112	1.27	-0.02	2.03	-0.004	1.02	-0.08
Muehlacker	48.96	8.86	244	1.46	0.15	4.42	-0.14	2.03	-0.15
Sinsheim	49.25	8.88	169	1.47	-0.16	2.95	-0.087	1.28	-0.08
Strassburg	48.55	7.63	150	1.61	0.26	4.55	-0.011	1.88	-0.14
Stoetten	48.66	9.86	734	2.12	0.64	2.35	-0.006	0.89	-0.02

Figure 2 shows a comparison between observed and simulated time evolutions at the Sinsheim station. The variables V, *dd*, T and q are considered. The wind speed and temperature results are satisfactory. The result for the specific humidity is not good. It can be related to the choice of humidity initial values in the soil. Figure 3 shows a comparison between observed and simulated profiles at the Musbach station. The variables considered are V, *dd*, θ and q. From the graphs it is possible to notice that the simulation quality obtained over complex terrain is quit good. The comparisons relative to the others stations present similar results.

In the Lagrangian dispersion model SPRAY, the concentration is determined by counting the number of particles in the sampling volume (cell). The domains for the concentration calculation coincide with the RAMS simulation grids. The grid 3 was used to calculate the dispersion close to the source, while the grids 1 and 2 were used to simulate the plume splitting in several tracer puffs. The cell size for the calculations of ground level concentrations was 500 m x 500 m x 50 m. The integration time step was set equal to 5 s. The PDF Gram-Charlier truncated to the fourth order was used. One hundred particles were emitted at each time step. The simulations started at 05:02 UTC and ended at 09:00 UTC for the grid 3, at 10:00 UTC for the grid 2 and at 18:00 UTC for the grid 1.

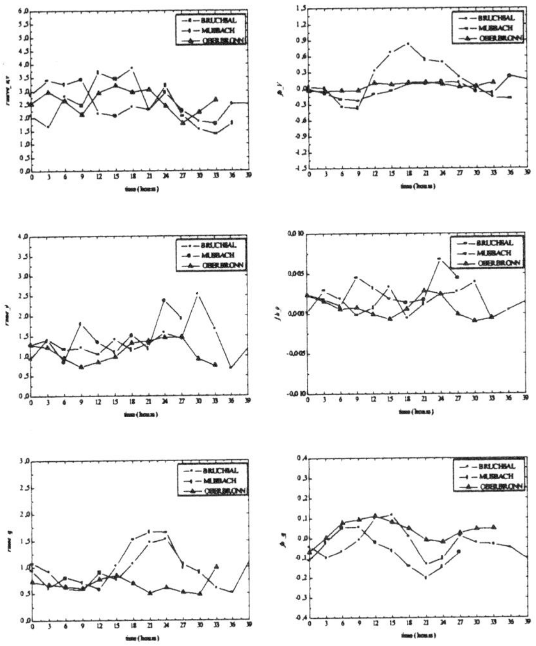

Figure 1. Time evolution of statistical indexes at stations Bruchsal, Musbach and Oberbronn. Hour 0 indicates 00 UTC, day 16 September.

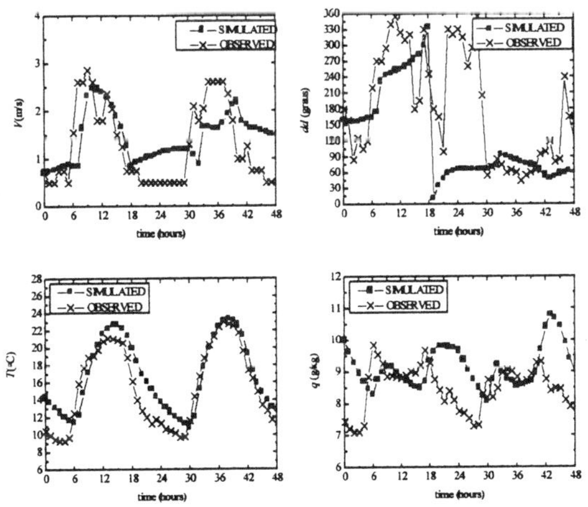

Figure 2. Simulated and observed time evolution at ground level at station Sinsheim. Wind directions 0° and 360° represent the same direction. Hour 0 indicates 00 UTC, day 16 September.

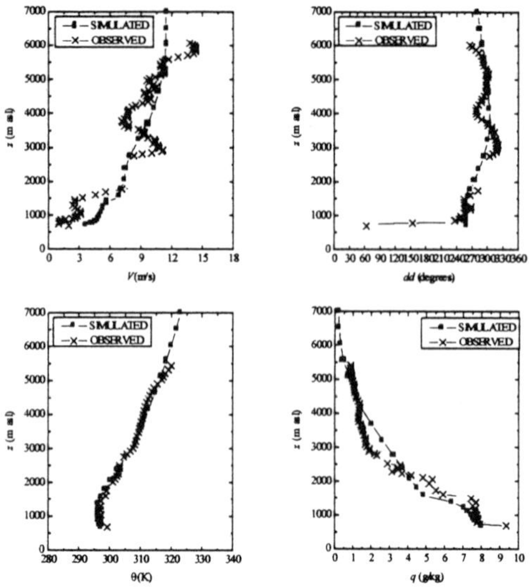

Figure 3. Simulated and observed profiles at station Musbach at 12 UTC on 16 September. Wind directions 0° and 360° represent the same direction.

Table 2 shows the comparison between observed and simulated ground level concentration values (in μ g/m^3) at the sampler stations for the grid 1 . All the available parameterisations in MIRS were tested. The best results were obtained by the combination between Degrazia et al. and Maryon and Buckland, and Hanna and Degrazia et al. in combination with PBL height of Kalthoff et al. The order of magnitude of the maximum (06:00 – 06:30 UTC) is well simulated, confirming that the model system reproduces well the observations. The model is able to calculate significant concentrations during the 12 hours where there are observed values, simulating the puffs distribution of tracer.

Table 2. Comparison of simulated and observed concentration values for the grid 1.

			McNider and Pielke			Maryon and Buckland			Kalthoff et al.	
hour (UTC)	sampler	obs.	TKE - RAMS	Hanna	Degrazia et al.	TKE - RAMS	Hanna	Degrazia et al.	Hanna	Degrazia et al.
5:30 – 6:00	32	0.0321	0.0026	0.0046	0.0008	0.0068	0.0080	0.0002	0.0405	0.0011
6:00 – 6.30	1	3.0276	8.7116	2.1669	1.6005	8.6340	2.1687	1.6356	2.2660	1.5473
6:30 – 7:00	32	0.0241	0.0020	0.0077	0.0070	0.0033	0.0171	0.0103	0.0701	0.0200
7:00 – 7:30	1	1.3202	3.0485	0.1341	0.1047	3.1284	0.1605	0.1094	0.2304	0.2545
7:00 – 7:30	4	0.0325	0.0949	0	0.0010	0.0901	0	0.0012	0.0026	0.0046
7:30 – 8:00	6	0.0034	0.0097	0	0.0008	0.0089	0	$4x10^{-5}$	0.0114	0.0121
8:00 – 8:30	1	0.0196	0.0011	0	0.0003	0.0003	0.0090	0.0013	0.0088	0.0051
9:30-10:00	32	0.0012	0	0	0	0	0	0	0.0005	0
9:30-10:00	37	0.0019	0	0	0	0	0	$5x10^{-6}$	0.0023	0.0002
10:30-11:00	34	0.0017	0	0	0	0	0	0	0	0.0015
11:30-12:00	32	0.4586	0	0.0015	0	0	0	0	0.0015	0.0015
11:30-12:00	34	0.0014	0	0	0	0	0	0.0005	0.0001	0.0015
12:30-13:00	23	0.0023	0	0	0	0	0	0.0002	0	0
12:30-13:00	32	0.0012	0	0	0.0006	0	0	0.0003	0.0005	0.0018
13:30-14:00	21	0.0028	0	0	0.0016	0	0	0.0004	0	0
13:30-14:00	22	0.0085	0	0	0.0021	0	0	0.0003	0	0
13:30-14:00	38	0.0028	0	0	0	0	0	0	0.0005	0.0017
15:30-16:00	23	0.0023	0	0	0	0	0	0.0011	0	0
16:30-17:00	21	0.0019	0	0	0	0	0	0	0	0

Figures 4 a-d present a time evolution of particle positions, using the Degrazia et al. parameterisation and the Kalthoff et al. method for the PBL height. In the figures the wind field simulated by RAMS in 10 m is also shown . The distances in the axis x and y are in km, referred the absolute coordinates $x=0$ (Greenwich meridian) and $y=0$ (equator). The graphs refer to 06, 08, 12 and 16 UTC. The location of the main valleys and mountains in the simulation domain is shown in figure 4 c. At 06 UTC, the plume is very narrow and follows the wind direction (northward) within the Rhine valley. A small error in the wind direction reproduction can produce consistent errors in the concentration calculation in the first hours of emission. At 08 UTC, some puffs separate from the main plume and start to travel to the west and north of the domain; the displacement towards west is caused by the horizontal wind shear in the valley. Some particles reach a height of 500 m above ground level (agl). At 12 UTC the plume is divided in two parts. A part remains almost stationary within the Rhine valley, and the other part moves from the Rhine to Kraichgau and Neckar valleys. At 12 UTC the particles reach a height of 800 m agl. At 16 UTC a cloud of particles covers the Kraichgau and Neckar valleys and few particles are found close to the emission region.

From the sequence of figures 4 a-d, it is possible to conclude that a tracer transport from Rhine valley to Kraichgau and Neckar valleys occurred. Löffler-Mang et al. (1998) observed that the displacement of air between regions with high densities of pollution and industrialization, due to the geostrophic wind direction and orographic effects, will also strongly influence the transport of air pollutants in the TRACT area. This observation seems to be proved by the present numeric simulation results.

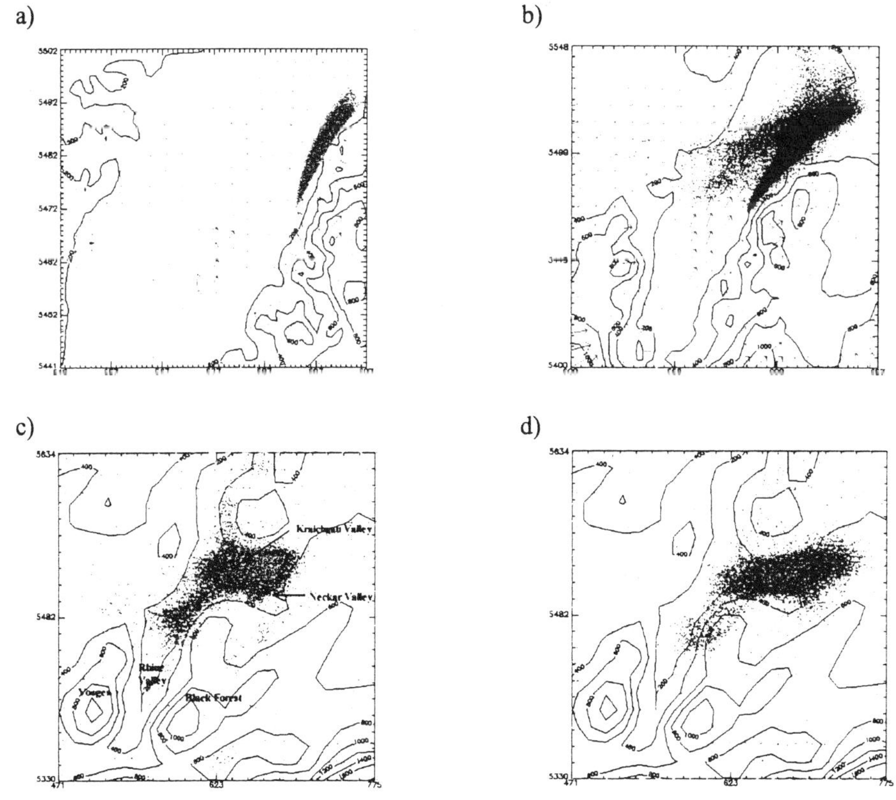

Figure 4. Particle positions and wind speed at 10m level: (a) grid 3, at 06 UTC; (b) grid 2, at 08 UTC; (c) grid 1, at 12 UTC and (d) grid 1, at 16 UTC. Contour labels indicate topography levels in meters.

CONCLUSIONS

The main objectives of this study was to carry out the link between the meteorological model RAMS and the Lagrangian particles model SPRAY in order to simulate the transport and diffusion processes and to verify its ability to describe the dispersion of a tracer emitted during the TRACT experiment. According to the results here presented, it is possible to state that such goals were reached. In particular, the comparisons of modelled and observed meteorological variables show that RAMS simulates very well the TRACT experiment. The model system correctly reproduces the general behaviour of the contaminant plume, the temporal and spatial distribution of the concentration and the location of the concentration maximum. It is also shown that the connection between RAMS and SPRAY, through MIRS, is fully operative.

ACKNOWLEDGEMENTS

The TRACT field campaign (leadership: Franz Fiedler, Karlsruhe, Germany) was performed by many teams including one of the authors' teams. We wish to acknowledge the efforts and contributions of all TRACT participants and, in particular, Dr. H. Zimmermann for his friendly assistance in the collection and transmission of the field data. The model SPRAY has been used in this work in the frame of a Research Contract with ENEL/Area Ambiente, Milano, Italy. The financial support provided by CAPES, CNPq and FAPERGS is also acknowledged.

REFERENCES

Anfossi, D., Ferrero, E., Sacchetti, D., and Trini Castelli, S., 1997, Comparison among empirical probability density functions of the vertical velocity in the surface layer based on higher order correlations, *Bound.-Layer Meteor.* 82:193.

Chiba, O., 1978, *J. Meteor. Jpn.* 56:149.

Degrazia, G.A., Anfossi, D., Carvalho, J.C., Mangia, C., Tirabassi, T., and Campos Velho, H.F., 2000, Turbulence parameterisation for PBL dispersion models in all stability conditions, *Atm. Env.* (in press).

Ferrero E., and Anfossi D., 1998, Comparison of PDFs, closures schemes and turbulence parameterizations in Lagrangian Stochastic Models, Int. J. Environment and Pollution, 9, 384-410.

Hanna S.R., 1982, *Applications in air pollution modeling, Atmospheric Turbulence and Air Pollution Modeling*, F.T.M. Nieuwstadt and H. Van Dop eds., Reidel-Dordrecht, Cap. 7.

Kalthoff, N., Binder, H.J., Kossmann, M., Vögtlin, R., Corsmeier, U., Fiedler, F., and Schlager, H., 1998, The temporal evolution and spatial variation of the boundary layer over complex terrain, *Atm. Env.* 32:1179.

Löffler-Mang, M., Zimmermann, H., and Fiedler, F., 1998, Analysis of ground based operational network data acquired during the september 1992 TRACT campaingn, *Atm. Env.* 32:1229.

Maryon, R.H., and Buckland, A.T., 1994, Diffusion in a Lagrangian multiple particle model: a sensivity study, *Atm. Env.* 28:2019

McNider, R.T., and Pielke, R.A., 1994, Diurnal boundary-layer development over sloping terrain, *J. Atmos. Sci.* 38:2198.

Mellor, G.L., and Yamada, T., 1982, Development of a turbulence closure model for geophysical fluid problems, *Rev. of Geophys. and Space Phys.* 20:851.

Pielke R.A., Cotton W.R., Walko R.L., Tremback C.J., Lyons W.A., Grasso L.D., Nicholls M.E., Moran M.D., Wesley D.A., Lee T.J., Copeland J.H. (1992) "A comprehensive meteorological modeling system - RAMS". Meteorology and Atmospheric Physics, 49, 69

Rodean, H.C., 1996, *Stochastic Lagrangian models of turbulent diffusion*, Amer. Meteor. Soc., Boston.

Thomson, D.J., 1987, Criteria for the selection of stochastic models of particle trajectories in turbulent flows, *J. Fluid Mech.* 180:529.

Tinarelli, G., Anfossi, D., Bider, M., Ferrero, E., and Trini Castelli, S., 2000, A new high performance version of the Lagrangian particle dispersion model SPRAY, some case studies. *Air Pollution Modelling and its Applications XIII*, S.E. Gryning and E. Batchvarova eds., Plenum Press, New York, in press

Trini Castelli, S., and Anfossi, D., 1997, Intercomparison of 3-D turbulence parameterisations for dispersion models in complex terrain derived from a circulation model, *Il Nuovo Cimento* 20C:287.

Zimmermann, H., 1995, *Field phase report of the TRACT field measurement campaign EUROTRACT report*, Garmisch-Partenkirchen.

DISCUSSION

A. BAKLANOV: You use the RAMS model with the 2.5 level turbulence scheme. Did you try to estimate the PBL height directly from the eddy or TKE profiles and to compare with other methods? Other comments. You mentioned the bad correspondence between the measured and modelled PBL height for the nocturnal BL. It is quite expected for the bulk Richardson number method in the formulation from your transparency. e.g. due to the overestimation of the shear production. To avoid this problem, you can use, for example, the modified Vogelezang & Holtslag (1996) Ri-number method, which combines the effects of shear in the outer region of BL with surface friction.

D. ANFOSSI: Yes we tried, in previous RAMS-MIRS-SPRAY applications to estimate the PBL height directly from the eddy or TKE profiles. The results compared satisfactorily well in some cases but only during day-time. This method completely failed during night-time.

R. BORNSTEIN: You must be careful in determining h_i at night from TKE values from a mesomodel. How did you make this determination?

D. ANFOSSI: We agree that one must be careful. Usually, to fix a value for h_i, we examine the TKE vertical profiles from the bottom upward and we look at the level at which a significant decrease in the TKE value appears. Since during night-time these values are always very low it was nearly always impossible to find any value for h_i.

F. MÜLLER: The bifurcation shown in your last slides could be caused also by the passage of a convergence during the noon hours.

D. ANFOSSI: It could be. However the splitting of the main plume in different puffs, following different pathways, started well before noon.

SIMULATION OF PHOTOCHEMICAL SMOG EPISODES IN EUROPE USING NESTING TECHNIQUES AND DIFFERENT MODEL EVALUATION APPROACHES

Adolf Ebel, Michael Memmesheimer, Hermann–Josef Jakobs, Christoph Kessler, Georg Piekorz, Michael Weber*

University of Cologne, Institute for Geophysics and Meteorology
and Rhenish Institute for Environmental Research
Aachener Str. 201–209, 50931 Cologne, Germany

* now: MPI for Meteorology, Bundesstr. 55, 20146 Hamburg, Germany

ABSTRACT A regional chemical transport model system (EURAD: European Air Pollution Dispersion model system) has been used to analyse peculiar features and improvements of air quality simulations using the method of subsequent model nesting. An episode with enhanced production of photo–oxidants in July 1994 in Central Europe is mainly studied. Effects of increasing resolution on model accuracy, treatment of boundary conditions, efficiency of chemical transformation and tracer flux divergence are discussed. Results of diagnostic model evaluation are presented.

INTRODUCTION

Transport and chemical transformation of air pollutants are controlled by processes which act on a wide spectrum of temporal and spatial scales. Their resolution in chemical transport models poses particular problems regarding numerical methods, computer resources, subgrid–scale parameterizations and adequate treatment of input data, e.g. for emissions and land use. The aim of this study is to contribute to the exploration of nesting techniques applied in Eulerian air quality models regarding their ability to handle these problems.

The advantage of the method of subsequent nesting as applied in this paper is that it allows consistent simulations of air quality from regional down to local scales by the use of the same modelling system. Using the method of nesting for smaller scale and local simulations will usually lead to an improvement of calculated pollution fields and yield more reliable results than stand-alone local models since a more realistic treatment of initial and boundary conditions becomes available. On the other hand, high resolution calculations may help to check and improve coarse grid simulations in various ways. For instance, till now no systematic approach exists dealing with the role of smaller scale tracer fluctuations of reactive tracer concentrations for larger scale ave-

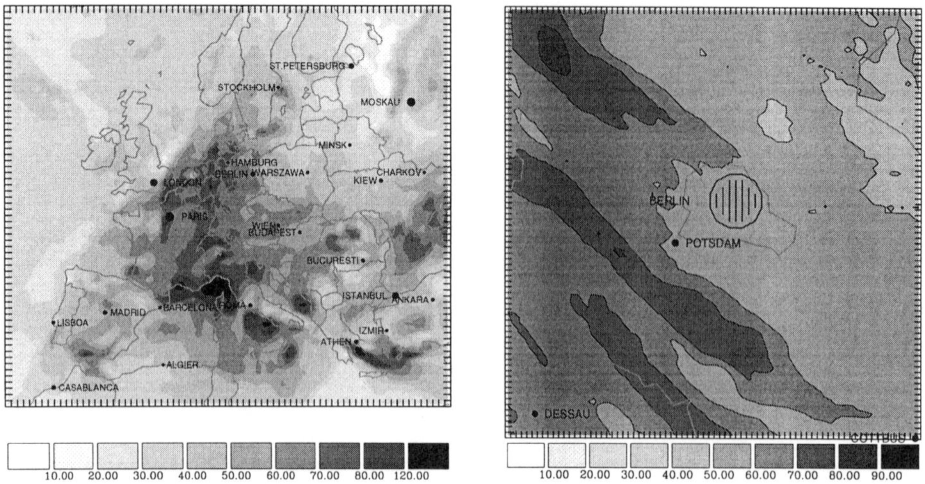

Figure 1: Simulation of ozone mixing ratios (ppbV) in the surface layer (0 – 75 m) with the EURAD model system for Europe (left panel, resolution 54 km) and the area around Berlin (right panel, resolution 2 km). July 22, 1994, 14 UTC. Note different scales of shading.

rage chemical transport calculations. Experiments with nested models could be used for systematic studies of perturbations of reactive tracer concentrations, particularly regarding the interdependence of large and small scale processes.

Though measurements of a larger spectrum of pollutants gradually become available it is still ozone which is usually and advantageously chosen to characterize the state of pollution of ambient air by photo-oxidants. In this paper we will adjust to this situation for convenience and the sake of brevity exploiting mainly simulation results for ozone. Attempts to include NO and NO_2 have been made and a few results are shown. Yet the simulations would require more comprehensive analyses for these species which are out of the scope of this study. The principal results for NO and NO_2 show clear disagreements between local and regional observations and chemical transport model (CTM) calculations (Ebel et al., 1995) as it has also been stated by Thakur et al. (1999) for global scales. The solution to the problem of local and regional discrepancies between measurements and simulations of nitrogen compounds has to await more reliable field experiments than available till now.

For the discussion of model application and simulation results a few examples have been selected out of a larger number of episodic studies with the model system described in Section 2. Details of model runs and analyses of specific features are presented in Section 3 and 4, respectively. Questions of evaluation are addressed in Section 5. The final Section 6 contains conclusions of this study.

THE MODEL SYSTEM

The European Air Pollution Dispersion (EURAD) model system has been employed for this study. Its main components are the Eulerian Mesoscale Meteorological Model, Version 5 (MM5; Grell et al., 1993), the EURAD Chemistry Transport Model, Version 2 (EURAD-CTM2; Hass, 1991) and the EURAD Emission Model (EEM; Memmesheimer et al., 1995). Recently, the range of application of the system has been extended through the adaptation of a γ-meso-scale meteorological driver (the system "Chemistry and Air Transport on the Regional and Local Scale", CARLOS, Brücher et al., 1999, Kessler et al., 1999). The meteorological model is initialized with ECMWF analyses. Emission input data can either be derived from EMEP emission inventories or

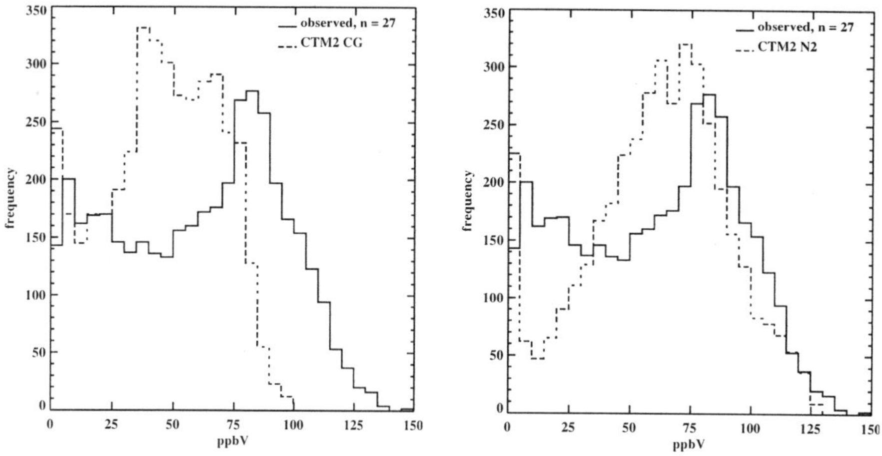

Figure 2: Frequency distribution of ozone mixing ratios (ppbV) as observed in the ABL at 27 locations (continuous curve) and simulated for these locations (dotted curve). Left panel: low resolution, 27 km. Right panel: high resolution, 3 km, emission and land use data adjusted. Absolute number of counts (vertical axis) in 5 ppbV bins (horizontal axis). Episode July 21 – 28, 1994, North-Rhine Westphalia.

are provided by IER, University of Stuttgart, using the GENEMIS data base (Friedrich, 1997). Biogenic emissions are calculated within EEM employing land type data which can also be obtained from GENEMIS. Using coarse resolution the model domain can be chosen to cover most of Europe. This may be scaled down to subsequently nested domains of the order of 100 x 100 km^2. The vertical resolution may be adjusted to the problem under consideration. It is larger in the boundary layer than in the free troposphere. In this study a vertical resolution of about 75 and 38 m has been used for the lowest model layer in different simulations. More detailed descriptions of the EURAD system may be found in various publications (e.g. Hass et al., 1995; Jakobs et al., 1995; Ebel et al., 1997). The EURAD system has been applied to the simulation of a larger number of episodes, studying air pollution processes and phenomena, the effects of emission reductions and the impact of air traffic exhaust on ozone in the tropopause region (Petry et al., 1994). Transport, microphysical and chemical transformation of aerosols (Ackermann et al., 1998) as well as the use of four-dimensional variational data assimilation for the EURAD-CTM (Elbern and Schmidt, 1999) are major foci of present model development.

SIMULATIONS, EPISODES

In this study we are mainly dealing with results from simulations of a photochemical smog episode which occurred in Central Europe July 19 – 30, 1994. We also refer to simulations of the so-called BERLIOZ episode (Berlin Ozone Experiment, an intensive photo-oxidant field campaign, conducted in July and August 1998). BERLIOZ provided a comprehensive set of measurements carried out with ground based and airborne methods which are presently employed for the evaluation of a hierarchy of CTMs including the EURAD system. Airborne and extended ground based observations are also available for a part of the photochemical smog episode in July 1994 for Berlin (FluMOB experiment, July 23 – 27, 1994; Stark and Lutz, 1995). Nested high resolution areas selected for this study are, therefore, Berlin for both episodes and, in addition, the Rhine-Ruhr region for July 1994 where a peculiar response of the simulations to large- and small-scale emission scenarios has been found.

DISCUSSION: HIGH VERSUS LOW RESOLUTION

In this section we emphasize some selected features which are demonstrating improvements of air pollution simulations resulting from model nesting. As shown in Fig. 1b it is obvious that small scale structures which are important for the assessment of local structures like plumes and their chemical characteristics can be analyzed with much more detail and accuracy than in the case of coarse grid resolution (Fig. 1a). Source-receptor relationships can more clearly be identified. In this example the air pollution plume of Berlin is an easily identifiable structure. Yet Fig. 1b also shows two additional plumes which partly originate from sources outside the nested domain. Such features could not be simulated adequately if the larger model domain would not provide suitable boundary conditions. Obviously, without generating realistic initial and boundary conditions through the nesting technique the reliability and accuracy of small to local scale simulations would considerably decrease. For instance, there are situations where the plume of a well identifiable source like the city of Berlin can only be extracted from the simulated tracer distributions if a careful analysis of background pollutant fields is carried out. This has, for instance, been found during the time of the BERLIOZ experiment in July 1998.

Comparing observed time series of ozone with those simulated for nested domains with increasing resolution one usually finds increasing accuracy of calculated maxima and minima during both episodes studied. The tendency of overpredicting nighttime values of ozone and underpredicting daily maxima is significantly decreasing with the use of finer grids (and reliable initial and boundary conditions). The growth of the quality of calculations is demonstrated by comparing the frequency distributions of ozone concentrations for a nested simulation as exhibited in Fig. 2. The improvements are most significant in areas with highly structured sources, in this case in densely populated and industrialized areas of the Rhine-Ruhr region. Nitrogen always behaves less friendly than ozone in this respect. High resolution sometimes helps to calculate the impact of strong NO_x sources with higher precision. But it still fails in quite a large number of cases. This is again an indication of the NO_x problem from which most, if not all, air quality models still suffer.

Nested simulations can be used to find preliminary answers to the question as to how strongly perturbations of chemical tracer concentrations which result from irregular emission source distributions and small-scale land type differences can affect coarse grid calculations. The role of perturbations of reactive species concentrations still lacks systematic investigations. Comparing the volume estimate of the surface layer obtained for the single box covering Berlin with low horizontal resolution (54 x 54 km^2, average surface layer height 75 m) with the average for this volume calculated with higher resolution (down to 2 x 2 km^2, i.e. a fine mesh with 729 small boxes) one generally finds the expected decrease of high resolution estimates for evening nighttime hours (up to 30% of the coarse grid estimates). Surprisingly, one also obtains an overall decrease of noon and afternoon values (up to 15% in individual hours) during four days of high ozone concentrations (July 24 – 27, 1994) though rather strong local maxima may originate. The average reduction was 7% in the Berlin area. The finding that high resolution leads to a lower area average than coarse resolution and the fact that high resolution calculations usually meet observed local maxima more accurately while low resolution infers a strong tendency to underprediction can only be explained by a peculiar stochastic behaviour of the specific time dependent ozone field under consideration. It is a clear indication of the relevance of concentration perturbations and the processes causing them for the formation of photo-oxidants. It also shows the difficulty to establish monitoring networks which are fully representative under all conditions of photochemical smog formation.

The effect of increased resolution of concentration fields, processes and precursor

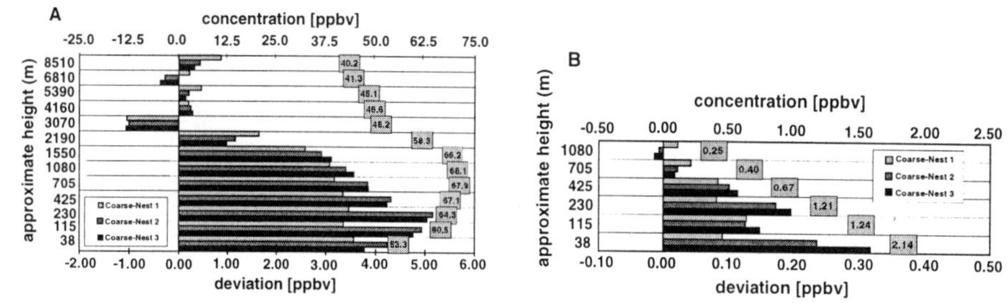

Figure 3: Deviations of high resolution averages of ozone (A, left) and NO_2 mixing ratios (B, right) from the temporal mean grid box value for Berlin obtained with coarse grid (54 x 54 km^2) simulation for the photochemical smog episode from July 24 – 27, 1994. Coarse minus fine grid (nest 1, light grey, 18 km; nest 2, grey, 6 km; nest 3, dark, 2 km) differences are plotted. Concentrations obtained from coarse grid calculations are also shown by labelled boxes. Horizontal scales: concentrations and deviations in ppbV. Vertical scale: Average altitude of model layers in meters.

sources on the simulations is strongest at lower levels as shown in Fig. 3 for ozone and NO_2. NO_2 shows a more pronounced dependence on grid size than ozone in the lower and middle atmospheric boundary layer (ABL) which may be ascribed to smaller spatial scales of the NO_2 concentration distribution and processes causing them (e.g. spatially irregular emission of precursors).

Estimating the contributions of the most relevant processes (chemical transformation, turbulent diffusion, deposition, horizontal and vertical advection) to the ozone budget over Berlin below 6 km altitude during the episode from July 21 - 27 with increasing ozone concentrations one finds a significant decrease in modelled net chemical production with increasing resolution (22% for nest 3 with a grid size of 2 x 2 km^2) and an increase of deposition losses (24%). Yet changes of the advection terms compensate these reductions to a great deal so that the net difference of the estimated budget amounts to a drecrease of only 7% if a resolution of 2 km is applied. This compensating behaviour of modelled processes controlling ozone budgets has also been found for other conditions and episodes (Memmesheimer et al., 1997).

EVALUATION STUDIES

A larger number of evaluation studies has been carried out with EURAD applying various methods and measures of performance (Dennis et al., 1991). In this study additional work regarding reliability and performance of the model is presented. A diagnostic approach is used. Process-oriented evaluation generally lacks from the availability of experimental process data applicable to regional scales. The comparison of model performance employing varying resolution together with process analysis may be regarded as a step in this direction. Yet, of course, observational results about processes would be needed for reliable process evaluation.

A comparison between simulations and observations has already been shown in Fig. 2 using frequency distributions of both types of data. Using scatter diagrams one can easily identify areas with differing correlations between observations (in this case obtained from ground level monitoring networks) as it is evident from Fig. 4. A suburban station (panel A) and a site in a predominantly rural area (panel B) have been chosen to demonstrate typical differences of model performance. Results from a simulation employing a horizontal resolution of 27 km are shown in order to enable the comparison of two rather distant locations with sufficiently different surroundings. A strong tendency of the model for overpredicting low (nighttime) ozone mixing ratios and underpredicting the high (daytime) ones is found for rural conditions. In contrast, the selected suburban station only shows weak dependence of the negative bias of the

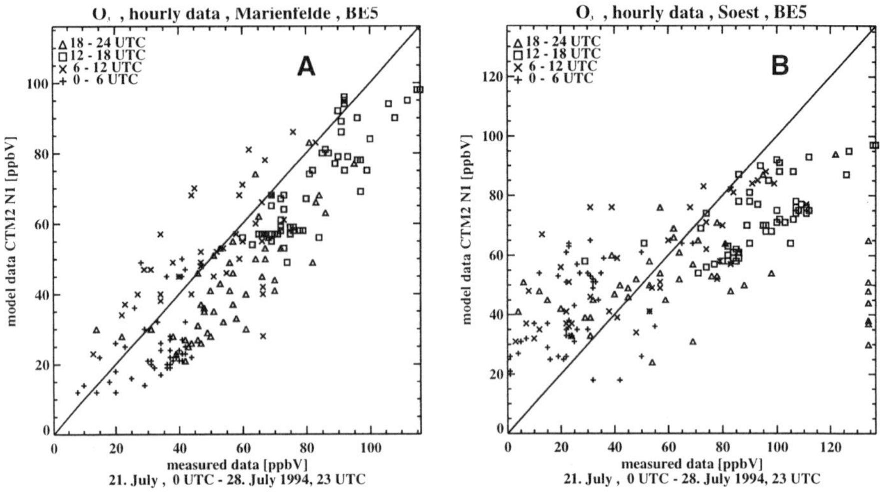

Figure 4: Scatter diagrams of observed versus calculated mixing rations (ppbV) of ozone. Episode July 21 – 28, 1994. Stations Berlin–Marienfelde (A, suburban surroundings) and Soest (B, rural surroundings). Horizontal resolution of the model is 27 km.

calculations on time and concentrations. This behaviour of the model is more frequently found in densely populated areas than in other places. The reasons for this could not yet be identified. What is evident from the NO_2 data and their scatter diagrams (not shown) are clear differences of their statistical characteristics which could at least explain the reduction of overprediction of nighttime O_3 due to increased concentrations of nitrogen compounds in areas with stronger NO emissions. Yet in general the agreement between measured and modelled NO_x is by no means satisfactory. This may be due to problems of emission inventories used or transport estimates based on erroneous meteorological fields. Yet it could also be caused by observational errors. This means that comprehensive model evaluation should not only focus on the simulation results but also include the evaluation of model input (emissions, meteorology) and, of course, measurements which are chosen to diagnose the performance of the model.

Instead of using temporal correlations it is also possible to study spatial relationships with scatter diagrams as done in Fig. 5. Though rather reasonable simulated time series could be obtained with regularly employed emission inventories for the Rhine-Ruhr region during the July '94 episode for low (27 km) and medium (9 km) resolution, the spatial scatter of measured versus calculated ozone maxima (Fig. 5, left panel) exhibits a negative correlation and a large bias. I.e., higher ozone mixing ratios are produced in areas with lower observed values by the model. This is quite unusual for most simulations. In this case it was possible to use an independent set of emission and land use data for higher resolution (3 km) runs leading to a more plausible scatter (Fig. 5, right panel) with considerably reduced bias. This is a strong indication that the emission data employed for larger scale simulations of the episode and, possibly, the land use data, too, need to be revised for the Rhine-Ruhr area.

Preliminary results from comparisons of airborne measurements over Berlin with simulated ozone mixing ratios for a few days in July 1994 confirm the tendency of the model to underestimate daylight ozone concentrations for higher levels in the ABL, too, though the structure of the city plume seems to be reproduced rather well.

CONCLUSIONS

It is obvious that the technique of model nesting can lead to considerable impro-

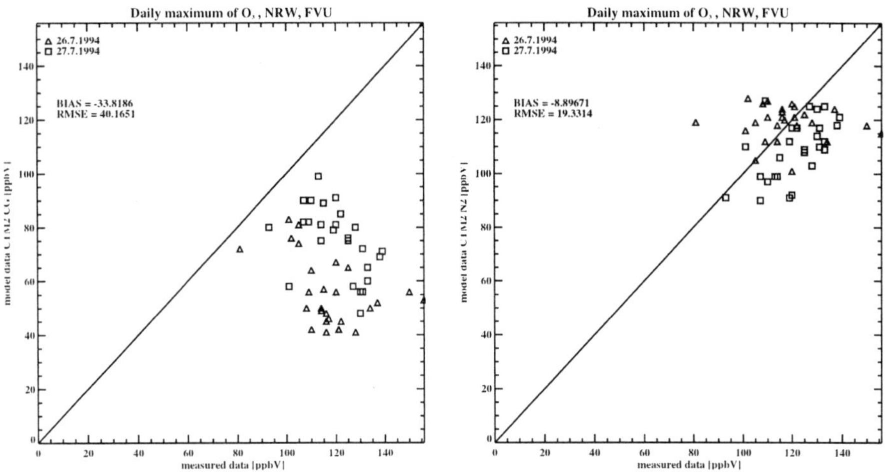

Figure 5: Ozone maxima on July 26 and July 27, 1994 as observed and simulated for the Rhine–Ruhr area. Left panel: horizontal resolution of 27 km, emissions modelled with EMEP inventory. Right panel: horizontal resolution 3 km, LUA emission and land use data.

vements of air quality simulations regarding the reproduction of realistic temporal and spatial structures of the distribution of photo–oxidants and other trace substances in the ABL. But there is no guarantee that the improvements are always such as intuitively exptected. Nesting, i.e. the application of higher resolution, should not just be taken as simple grid refinement for selected areas. It has to be accompanied by the use of appropriately resolved and structured input data sets for emissions, land type and topography. Otherwise, the gain of accuracy by nesting may remain low.

Putting emphasis on the analysis of an episode with enhanced photochemical activity it could be shown that an integral response of the chemistry-transport system to spatial resolution changes exists. This may cause problems for specific applications of the CTMs to questions regarding air quality and impact assessment. For instance, the use of models for critical load studies may require a specific simulation strategy to explore the dependence of model estimates on varying resolution including structural changes of model input fields.

An interesting finding of this study is the dependence of simulated volume averages of ozone and NO_2 (and other chemical species) on the resolution and thus fine structure of concentration fields and on processes causing structural differences. A single case could only be analyzed till now resulting in a surprising decrease of the volume average of the ozone mixing ratio when high instead of coarse resolution was used. There seem to exist other cases where the opposite change is encountered. Further investigations of this intriguing phenomenon are planned.

ACKNOWLEDGEMENTS

We gratefully acknowledge the permission of the German Weather Service to use meteorological data provided by the ECMWF. Emission data were obtained from the IER University of Stuttgart, EMEP (NILU) and the Environmental Agency (LUA) of the state North-Rhine Westphalia. The Research Center Jülich supported the study by giving access to its computer facilities (NIC). Financial support came from the German Federal Ministry of Education and Research (BMBF) under grants 07TFS10/LT1-A1 and C1.

REFERENCES

Ackermann, I.J., Hass, H., Memmesheimer, M., Ebel, A., Binkowski, F.B., and Shankar, U., 1998, Modal aerosol dynamics model for Europe: development and first applications, *Atmos. Environm.* 32:2891–2999

Brücher, W., Kerschgens, M.J., Kessler, Ch., and Ebel, A., 1999, Modelling of regional and local air pollution based on dynamical simulation of traffic, in: *Traffic and Mobility, Simulation – Economics – Environment*, W. Brilon, F. Huber, M. Schreckenberg, H. Wallentowitz, eds., pp. 225–240, Springer, Berlin

Dennis, R.L., Barchet, W.R., Clark, T.L., Seilkop, S.T., and Roth, P.M., 1991, Evaluation of regional acid deposition model (part 1), in: *Acid Deposition: State of Science and Technology*, Vol. 1, P.M. Irving et al., eds., pp. 5-1 – 5-E8, U.S. National Acid Precipitation Program, Washington, D.C.

Ebel, A., Elbern, H., Feldmann, H., Jakobs, H.J., Kessler, Ch., Memmesheimer, M., Oberreuter, A., and Piekorz, G., 1997, *Air Pollution Studies with the EURAD Model System (3): EURAD – European Air Pollution Dispersion Model System*, Mitteilungen aus dem Institut für Geophysik und Meteorologie, A. Ebel, M. Kerschgens, F.M. Neubauer, P. Speth, eds., Nr. 120, Cologne

Ebel, A., Feldmann, H., Fiedler, F., Hass, H., Jakobs, H.J., Klemm, O., Nester, K., Schaller, E., Schwartz, A., and Werhahn, J., 1995, Contributions to the evaluation of chemical transport models within the SANA project, in: *Air Pollution III*, Vol. 4, A. Ebel, N. Moussiopoulos, eds., pp. 103–110, Computational Mechanics Publications, Southampton

Elbern, H., and Schmidt, H., 1999, A four-dimensional variational chemistry data assimilation scheme for Eulerian chemistry transport modeling, *J. Geophys. Res.* 104:18583–18598.

Friedrich, R., 1997, GENEMIS: Assessment, improvement, and temporal and spatial disaggregation of European emission data, in: *Tropospheric Modelling and Emission Estimation*, (Series on Transport and Chemical Transformation of Pollutants in the Troposphere, Vol. 7), A. Ebel et al., eds., pp. 179-214, Springer, Berlin

Grell, G.A., Dudhia, J., and Stauffer, D.R., 1993, *A Description of the Fifth-Generation PennState/NCAR Mesoscale Model (MM5)*, NCAR Technical Note, NCAR/TN-398+IA

Hass, H., Jakobs, H.J., and Memmesheimer, M., 1995, Analysis of a regional model (EURAD) near surface gas concentration predictions using observations from networks, *Meteorol. Atmos. Phys.* 57:173–200

Hass, H., 1991, *Description of the EURAD Chemistry-Transport-Model Version 2 (CTM2)*, Mitteilungen aus dem Institut für Geophysik und Meteorologie, A. Ebel, F.M. Neubauer, P. Speth, eds., Nr. 83, Cologne

Jakobs, H.J., Feldmann, H., Hass, H., and Memmesheimer, M., 1995, The use of nested models for air pollution studies: an application to a SANA episode, *J. Appl. Meteor.* 34:1301–1319

Kessler, Ch., Brücher, W., Memmesheimer, M., Kerschgens, M.J., and Ebel, A., 1999, Interaction of traffic and other anthropogenic emissions in polluted regions and their environment, in: *Traffic and Mobility, Simulation – Economics – Environment*, W. Brilon, F. Huber, M. Schreckenberg, H. Wallentowitz, eds., pp. 241–257, Springer, Berlin

Memmesheimer, M., Ebel, A., and Roemer, M., 1997, Budget calculations for ozone and its precursors: seasonal and episodic features based on model simulations, *J. Atmos. Chem.* 28:283–317

Memmesheimer, M., Hass, H., Tippke, J., and Ebel, A., 1995, Modeling of episodic emission data for Europe with the EURAD Emission Model (EEM), in: *Regional Photochemical Measurement and Modeling Studies*, A.J. Ranzierei, P.A. Solomon, eds., Vol. 2, pp.495–499, Air & Waste Management Associaton, Pittsburgh

Petry, H., Elbern, H., Lippert, E., and Meyer, R., 1994, Three dimensional mesoscale simulations of airplane exhaust impact in a flight corridor, in: *Impact of Emissions from Aircraft and Spacecraft Upon the Atmosphere*, U. Schumann and D. Wurzel, eds., pp. 329–335, DLR–Mitt. 94–06, Cologne

Stark, B., and Lutz, M., 1995, *Flugzeug- und Bodenmessungen von Ozon und Vorläuferstoffen zur Abschätzung der Wirksamkeit emissionsmindernder Maßnahmen im Großraum Berlin–Brandenburg (FluMOB-Project)*, Senatsverwaltung Berlin, Abt. Immissionsschutz, Berlin

Thakur, A.N., Singh, H.B., Mariani, P., Chen, Y., Wang, Y., Jacob, D.J., Brasseur, G., Müller, J.-F., and Lawrence, M., 1999, Distribution of reactive nitrogen species in the remote free troposphere: data and model comparisons, *Atmos. Environm.* 33:1403–1422

DISCUSSION

D. W. BYUN: It was an interesting presentation. We at EPA also are in the process of studying multi-level nesting of air quality models. However, unlike to your case, the meteorological descriptions are changing from one resolution to another. How could you isolate the effects of grid resolution on air quality model output in such cases?

A. EBEL: It is also our experience that processes controlled by meteorology may change considerably when the horizontal and vertical resolution is increased through multiple nesting. A possible way to check and analyse resulting effects on the concentration distribution of modelled chemical species is the calculation of tendencies or budget terms for the domain under consideration. The following example of budget estimates for ozone shows that significant changes may be found for individual processes (in particular for those controlled by meteorology) when high resolution calculations are carried out though the total budget only shows a small deviation from the coarse resolution result. The example is related to Fig. 3 (Berlin box, 54 x 54 km^2). Yet it only covers the period of July 21 – 24, 1994.

Change of ozone in tons per period calculated with coarse (54 km x 54 km, first column) and high (2 km x 2 km, second column) horizontal resolution in a layer between 80 and 1800 m resulting from

gas phase chemistry	1,461	1,075
cloud processes	0	0
vertical diffusion	-257	-356
vertical advection	424	114
horizontal advection	-1,484	-696
total ozone mass change	*144*	*137*

It is evident and can be shown by more detailed analysis of the model results that the fluxes of ozone precursors and their divergence are also considerably changing with higher resolved meteorological fields leading to a high degree of compensation of the different ozone budget terms in this case. This points to the complexity of the interrelationship between the meteorological and chemical processes in the atmospheric boundary layer as discussed in this paper and in the free troposphere.

MODELLING AND DATA ASSIMILATION OF OZONE

M. van Loon[1*], P.J.H. Builtjes[1], and A.J. Segers[2]

[1]TNO Institute of Environmental Sciences, Energy and Process Innovation, P.O. Box 342, 7300 AH Apeldoorn, The Netherlands
[2]Delft University of Technology, Dep. of Applied Mathematics, P.O. Box 5031, 2600 GA, Delft, The Netherlands

INTRODUCTION

Results of simulations of air pollutants with atmospheric models sometimes show large differences with measured concentrations, despite the fact that the underlying physical and chemical processes are quite well-known in the case of ozone. Even if these processes were exactly known, it is still impossible to match model results with measurements, because of several "uncertainties". Roughly, three categories of uncertainties can be distinguised:

- uncertainties in the input of the model (meteorological fields, emission input etc.)
- approximations made in the model, for example by considering average concentrations over a certain volume (typically a grid cell in Eulerian type of models)
- measurement errors

The idea behind the data assimilation technique as applied in this work, a special implementation of the well-known Kalman Filter, is that if the uncertainties as mentioned above can be quantified in statistical terms, model simulations can be improved by adding information that has the potential of reducing (part of) the uncertainties: measurements.

For this reason, the application of data assimilation, which has already proved to be useful in various fields of application, see e.g. Verlaan and Heemink (1996), received increasing attention among scientists in the field of atmospheric chemistry and transport modelling in recent years, see e.g. Elbern (1997, 2000), KNMI (1999) and Van Loon et. al (2000).

In this paper, two data assimilation simulations with the atmospheric transport chemistry model LOTOS for a part of July and August 1997 are presented. First, a short description of LOTOS and the principles of the Kalman Filter technique is given, followed by specifications of the simulations and their results. The paper ends with conclusions.

*corresponding author, e-mail: loon@mep.tno.nl

Air Pollution Modeling and Its Application XIV, Edited by
Gryning and Schiermeier, Kluwer Academic/Plenum Publishers, New York, 2001

SHORT DESCRIPTION OF LOTOS

The Long Term Ozone Simulation model is an Eulerian grid model that uses a 70x70 equidistant grid, covering the domain [10W-60E]×[35N-70N]. This domain is large enough to be able to simulate ozone concentrations in Europe for both episodic and long term calculations. In the vertical direction there are 3 layers: the mixing layer and two layers above it. The height of the mixing layer is varying in space and time. The two layers above the mixing layer have equal height of at least 100m, with the model top at 1800m, unless the (local) mixing height is greater than 1600m. The concentrations close to the ground (e.g. at measurement height) are computed diagnostically, by apllication of a vertical profile using the mixing layer concentrations and "subgrid-scale" considerations. This procedure is necessary, since concentrations of ozone (and in general of species which are deposited) close to the ground can be up to, say, 30 percent lower than the average concentration in the mixing layer, especially during night. The model includes the processes: horizontal and vertical advection and diffusion, dry deposition and chemistry. The chemical mechanism used in LOTOS is the CBM-IV mechanism, see Gery et al. (1989). Wet deposition is neglected, since the necessary meteorological input is missing. Fortunately, wet deposition plays no significant role in ozone formation. The meteorological input consists of 3-hourly fields with the horizontal winds, water vapour concentration, temperature, layer depth for the three model layers and of fields with the surface wind and temperature and cloud fraction.

For more details on the model, the interested reader is referred to Builtjes (1992) and Roemer (1996).

KALMAN FILTERING

In order to apply Kalman Filtering both the model and the measurments need to be described in statistical terms. For the model, this means that instead of the deterministic formulation,

$$x^{k+1} = f(x^k),$$

in which x^k denotes the state vector x of the model at time level k and f the model operator, a stochastic extension of the form

$$x^{k+1} = f(x^k, \gamma^k),$$

is needed, where γ^k is the so-called noise input vector. Each element of this vector represents an uncertainty in the model. The state vector x is not necessarily restricted to the concentrations of the species in all grid cells, but may contain other variables as well. In the work reported here the vector γ^k is part of the state vector as well. In that way the filter explicitly estimates these variables.

A measurement vector y^k, representing a set of measurements at time level k, is written as

$$y^k = C^k x^k + v^k,$$

where C^k represents the operator that computes the modelled value from the model state at the measurement location. The vector v^k is the measurement error, assumed to be Gaussian. It this study it is assumed that the measurement errors are uncorrelated in time and space.

The Kalman Filter uses the error description of the model to create an approximation of the covariance matrix for the entries of the state vector, simply by evaluating a model time step with different values of elements of the vector γ. Hence a model time step is repeated several times in order to build the covariance matrix and propagate it in time. Based on this matrix and the observations including their uncertainties the Kalman Filter computes a best fit of the model with the available observations at each time level.

Since the original Kalman Filter was designed for linear applications, many variants extists for non-linear applications, all using different ways of constructing the (approximate) covariance matrix for the entries of the model state. The processing of the measurements remains the same as in the original Kalman Filter. For reasons of convenience, the specific implementation used in this study is the Ensemble Kalman Filter. In future applications we intend to use the Reduced Rank Square Root implementation (see Verlaan and Heemink (1996), Van Loon and Heemink (1997) and Segers et. al (1998a, 1998b)), since this implementation has the potential of reducing the necessary number of model evaluations per time step significantly compared to the Ensemble Kalman Filter at the cost of some extra linear algebra.

SPECIFICATION OF THE SIMULATIONS

Uncertainties ("noise")

In both data assimilation simulations, the NO_x and VOC emissions emissions in a number of areas, formed by (groups of) countries, are considered uncertain. These areas are listed in Table 1 together with the index number of the noise parameters used in the simulations.

Table 1. Definition of the areas to which an individual noise parameter acts with the index number of these noise parameters.

	noise parameter index number		
area number + definition	Nox emission	VOC emission	deposition
1. Netherlands, Belgium, Luxembourg	1	2	11
2. Germany	3	4	12
3. France, Spain	5	6	13
4. UK, Ireland	7	8	14
5. Eastern Europe, Yougoslavia, Albania	9	10	15

Hence in both runs 10 noise parameters are used to describe the uncertainty in the emissions. In the second simulation also noise has been added to some stability parameters that influence the NO_x and O_x depostion rates and hence also the assumed vertical concentration profile. The domain is divided in 5 somewhat different areas as for the emissions.The first two areas remain the same. The third area is formed by France only, the fourth area consists of all sea areas and the fifth of the rest of the domain that is not included in one of the first four areas. For each of these areas separate noise parameter is defined resulting into 15 noise parameters in total in the second simulation.

If the Kalman Filter wants to do a time step in which one of the elements of the vector γ is nonzero, the emissions of the species concerned (i.e. NO_x or VOC) or the stability parameter in the corresponding area is multiplied by $\exp(\gamma)$. The reason to take the exponent of γ is to avoid a negative multiplication factor.

Colored noise is used for the elements of the vector γ. The standard deviation σ is set to 0.22 for entries corresponding to the emissions and 0.1 for entries corresponding to the deposition. The correlation length τ of all elements of γ is set to 6 hours. Each time step γ (as part of the state) needs also to be updated and estimated. This is done by

$$\gamma^k = \gamma^{k-1} \exp(-1/\tau) + w \cdot \sigma \cdot \mathrm{sqrt}\{1 - \exp(-2/\tau)\},$$

The vector w is a white noise vector with zero mean and standard deviation one. In fact w is the variable that is estimated by the filter. If at a certain stage during the simulation, the Kalman Filter sees no reason to take a nonzero value for w, the corresponding value of γ will not become zero immediately as would be the case if uncolored noise is used. Instead, the estimated absolute value of γ will become somewhat closer to zero. This means that there is some persistency in the model. Only if at a number of consecutive times w is estimated zero, γ will become zero as well. The white noise parameter w is only estimated zero by the filter if either the measurements are 'insensitive' for the corresponding noise process (a nonzero values of w will not improve the fit) or no measurements are taken into account.

Measurements

Ozon time series with resolution of 1 hour were collected from 42 measurement sites. For use in the assimilation time series of ozone in 21 stations are used:

Netherlands:	Braakman, Eibergen and Wieringerwerf
Germany:	Deuselbach, Herleshausen, Kyritz, Lueckendorf, Melpitz,
Rottenburg,	Schleiz, Waldhof and Wiesenburg
Austria:	Achenkirch, Hochburg and Illmitz
Denmark:	Lille Valby and Keldsnor
Great Brittan:	Eskdalemuir, Glazebury, Lullington and Bottesford

The other 21 stations are used for diagnostic purposes: comparion of the modelled and measured values of ozone at these sites provides information on the consistency of the system. If the modelled concentrations improve at the measurement locations that are taken into account, but at the same time become worse at other locations, it is an indication that the specified noise is not or only to a very limited extent responsible for the differences between modelled and measured concentrations.

The standard deviation of the measurement errors is assumed to be 15% with a minimum of 0.5 ppb and a maximum of 3 ppb. Measurements of other species than ozone are not used.

RESULTS OF THE SIMULATIONS

In order to get an overall impression of the performance of the assimilation, averages residues for all stations are computed. The residue for a station is defined as the average of the absolute differences between the measured and modelled concentration at each hour for which a valid measurement is available.

In the scatter diagrams (Figure 1) the residues are plotted of the the time series in the stations that are used in the first assimilation similation in which only noise was added to the emissions as decribed in the previous section. Residues from the simulation with assimilation are plotted against residues from a simulation without assimilation ("the model as it is").

Figure 1 shows clearly that data assimilation improves the modelled concentrations not only at the measurement locations used but also at other locations.

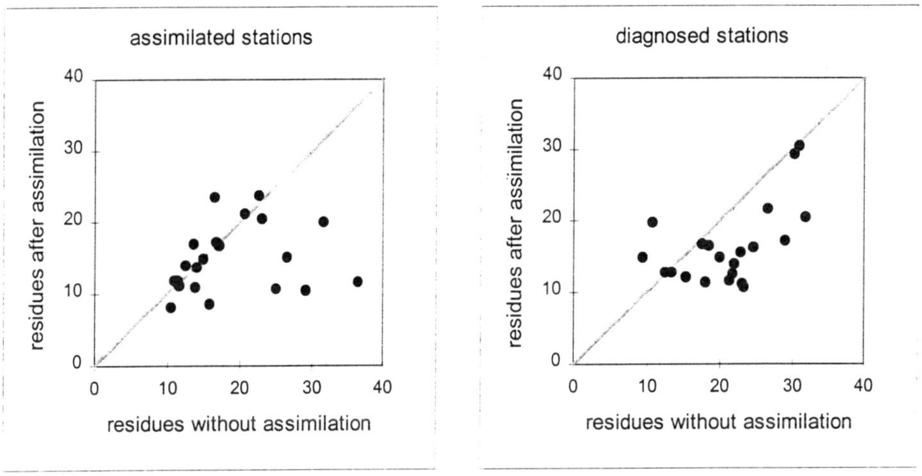

Figure 1. Scatter diagrams of the residues resulting from a model simulation without assimilation and an assimilation simulation with only noise on the emissions.

In the second simulation also noise was added to the deposition. In Figure 2 the residues of the three different simulation in the assimilated stations are plotted. The term 'model' in the figure caption denotes a model simulation in which no data assimilation is applied ("the model as it is").

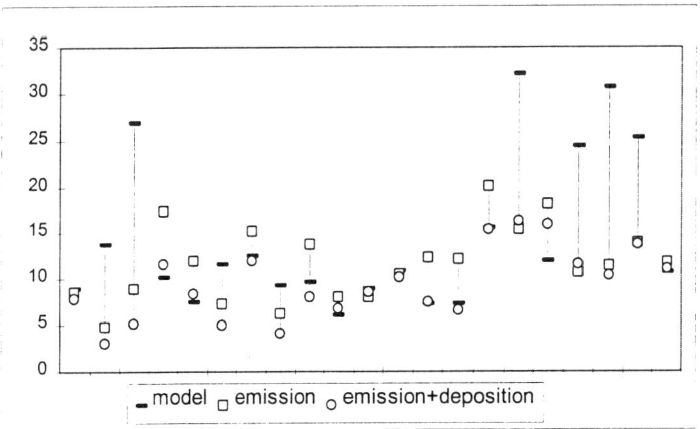

Figure 2. Residues in the assimilated stations computed the three different simulations.

From Figure 2 it is clearly seen that also adding noise to the deposition gives improved results (smaller residues on average) compared to the simulation in which only noise was added to the emissions. At some locations the additional decrease of the residues is quite substantial. Figure 3 shows a time series plot at the location Eibergen, The Netherlands.

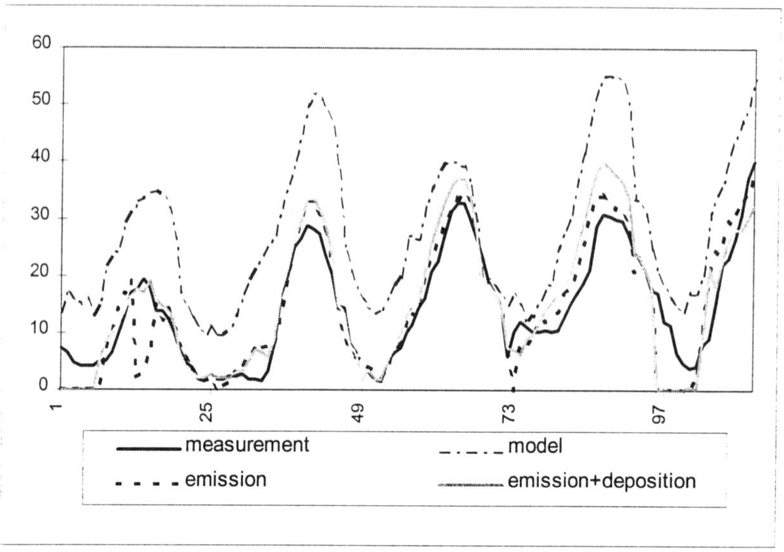

Figure 3. Simulated and measured time series at station Eibergen, 2 - 6 August 1997.

From Figure 3 the improvement by the data assimilation is clearly visible. It is also seen that at this particular location the second simulation (emission+deposition) is not always the one closest to the measurements, allthough on average it is.

It is also interesting to see whether temporal patterns and the absolute values of the estimates of the noise parameters (the γ) in the emissions made by the filter in both assimilation simulations are of the same order or that they differ significantly. In Figure 4 the values of exponent of the noise parameters corresponding to the NO_x emission in the Netherlands for the first days of the simulations are plotted. The exponent is shown because according the its definition (see above) the emissions are multiplied by the exponents of the noise parameters. Hence the numbers in the figure can be viewed as estimated emission correction factors.

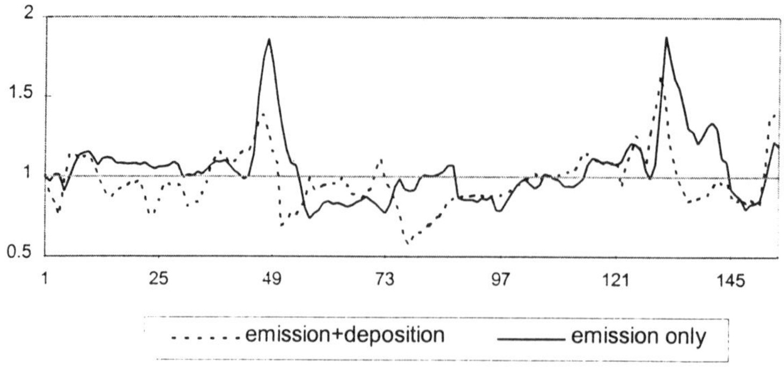

Figure 4. Estimated NO_x emission correction factors in The Netherlands.

As can be seen from Figure 4 there is some correlation in time between the two time series of correction factors, but the deviations from one are generally smaller in case also noise is added to the deposition. This means that part of the differences between modelled results and measurements which was attributed to the NO_x emissions in The Netherlands in the first simulation, is according to the second simulation more likely to be caused by the deposition.

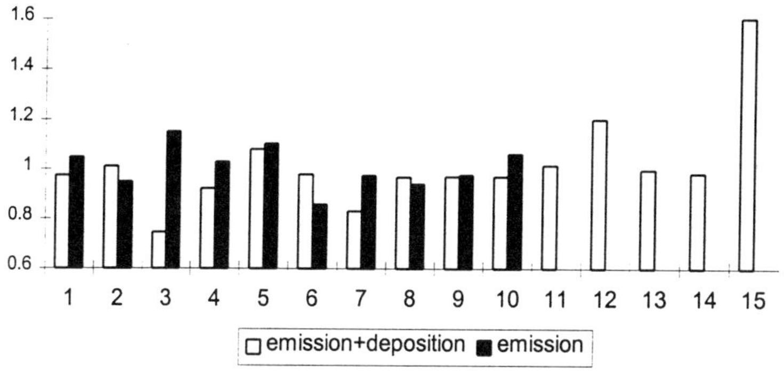

Figure 5. Average values of the exponent of the noise parameters; the category number refer to the index number of the noise parameters defined in Table 1 and the text below.

Average values of the correction factors (i.e. the exponents of the γ) are given in Figure 5. This figure reveals that there are clear differences between the emission correction factors estimated by the first and the second simulation, although in most cases the differences are small. It also shows that from the results of the simulations no conclusions can be drawn about the accuracy of the emission data base, because apparently not all uncertainty that is present in the model has fully be quantified. Additional noise processes need to be defined that together with the noise on the emissions and deposition fully explain the differences between model and measurements, before the estimated emission correction factors can be considered realistic.

CONCLUSIONS

From the results reported in this paper the following conclusions are drawn:

- successful application of the Kalman Filter in combination with an atmospheric transport chemistry model is possible
- the uncertainties taken into account in this study (emission, deposition) are responsible for the differences between modelled and measured concentrations only to a limited extent, since the specified uncertainties do not lead to a ‘perfect’ match between modelled and measured concentrations.
- the multiplication factors for the emissions and the deposition are indicative, not only because no other (noise) processes are taken into account yet, but also because of conceptual limitations of the model itself and the fact that the filter assumes the impact of a change in the emission on the ozone concentration to be a linear.

The following recommendations are made:

- other noise processes need to be taken into account. Possible processes are: vertical diffusion, mixing layer formulation, meteorological input.
- measurements of other species should be taken into account, keeping in mind the model restrictions.
- measurements of emitted species at remote locations can be useful in order to get a better estimate for the emissions, if used in combination with other noise processes.
- modifications in the filter need to be implemented in order to take non-linear effects into account.

A paper on results of data assimilation simulations using additional noise input is planned for the near future.

REFERENCES

P.J.H. Builtjes, 1992, The LOTOS - Long Term Ozone Simulation - project. Summary report. *TNO report TNO-MW-R92/240*, Delft, The Netherlands.

H. Elbern et al., 1997, Variational data assimilation for tropospheric chemistry modelling. *J. Geophys. Res.*, 102:15967.

H. Elbern and H. Schmidt, 2000, A Four-Dimensional Variational Chemistry Data Assimilation Scheme for Eulerian Chemistry Transport Modelling. *To be published in J. Geophys. Res.*

M.W. Gery et al., 1989, A photochemical kinetics mechanism for urban and regional scale computer modelling. *J. Geophys. Res.*, 94: 925

KNMI, 1999, Proceedings of the SODA Workshop on Chemical Data Assimilation. *KNMI Publication 188*, KNMI, De Bilt, The Netherlands.

M. van Loon and A.W. Heemink, 1997, Kalman Filtering for non-linear atmospheric chemistry models: first experiences. *Technical Report, MAS-R9711*, CWI, Amsterdam, The Netherlands.

M. van Loon et al., 2000, Data Assimilation of Ozone in the Atmospheric Transport Chemistry Model LOTOS. *To appear in a special issue of Environmental Modelling and Software.*

M.G.M. Roemer. Trends of Tropospheric Ozone over Europe. Ph.D. Thesis, University of Utrecht, 1996.

A.J. Segers et al., 1998a, Kalman Filtering for non-linear atmospheric chemistry models: first experiences, in: *Proceedings of the first GLOREAM workshop* pp. 111-118.

A.J. Segers et al., 1998b, Large scale data assimilation based on RRSQRT-filters; application on atmospheric chemistry models, In: *Proceedings Air Pollution IV*, C.A. Brebbia et al. (ed.), pp. 25-34, Wessex Institute of Technology, Computational Mechanics Publications, (1998).

M. Verlaan and A.W. Heemink, 1996, Tidal Flow Forecasting using Reduced Rank Square Root Filters. *Technical Report, Delft University of Technology, Delft, The Netherlands.*

DISCUSSION

D. DAESCU: Please present some details about the computational costs of the Kalman filter. Among different methods to perform data assimilation please explain your choice for the Kalman filter.

P. BUILTJES: The Kalman filter increases the computational costs roughly by a factor 50, depending of the number of modes required. Next to Kalman filter the other possibility would be 4-D var/adjoint. I have to admit that my choice was primarily intuition. I have the feeling that I still understand the basic procedure in Kalman filtering, and I have much more difficulty in understanding what is actually happening in the case of adjoint modelling.

J. BARTNICKI: Can you estimate contributions of different emission sources to concentrations and depositions modified in data assimilation approach?

P. BUILTJES: Yes, you can do that in principle. The fundamental question is in which way to perform emission scenario runs. I would be in favor of performing them on the base run, and on the assimilated run in which the parameters, including for example the emissions themselves, have been adjusted. In this way an indication of the accuracy of the scenario results would be obtained.

G. KALLOS: Do you perform assimilation only for ozone (or other pollutants) or for meteorology as well? Do you use 3-D Kalman filtering or 1-D (timeseries). If you do not do the meteorology, the system is incomplete in the sense of the errors (uncertainties) introduced.

P. BUILTJES: We use 3-D Kalman-filter in space and also in time. Concerning the meteorology, we use for LOTOS diagnosed meteorological input, which is by definition (very) close to the observations. This means that data-assimilation is not useful in this case, although assimilation of derived quantities like the mixing height might still be possible. In case prognostic meteorology is used as input, I agree with you that before assimilation is performed on chemical species, first the meteorology should be assimilated.

ATMOSPHERIC TRANSPORT MODEL STUDIES FOR SWEDEN – COMPARISON TO EMEP MODEL RESULTS AND EVALUATION OF PRECIPITATION CHEMISTRY STATION NETWORKS

Christer Persson and Robert Bergström

Swedish Meteorological and Hydrological Institute (SMHI)
SE-601 76 Norrköping, Sweden

INTRODUCTION

Deposition of air pollutants has been studied in Sweden since the 1950's when a precipitation chemistry network was established. Since then considerable efforts have been made on both national and European scale. The work on the European scale has been co-ordinated within EMEP (see e.g. Tarrason and Schaug, 1999). Since the mid 1990's the MATCH modelling system (Multi scale Atmospheric Transport and Chemistry model), developed at SMHI, has been used, combined with results from national atmospheric chemistry measurements for air pollution assessment studies in Sweden. The purpose of the national studies is to obtain estimates of air pollution concentrations and depositions over Sweden with higher geographical and temporal resolutions compared to the EMEP results, and to perform national scenario studies.

Since economic resources for environmental monitoring are limited, the existing station network needs to be optimised. This could release resources for investigations of new important problems and pollutants. In this study MATCH model system results using an existing network are compared to results with reduced monitoring networks. The results indicate that there is a possibility to shrink the existing station network without serious loss of information.

MODELLING SYSTEM

The MATCH modelling *system* consists of three parts: 1) a regional atmospheric dispersion model, 2) a system for data assimilation of concentrations in air and precipitation, and 3) an objective analysis system for meteorological data.

Part 1 is a three-dimensional Eulerian atmospheric dispersion model, the MATCH *model*, which is used for calculations of the Swedish contributions to concentrations and depositions of air pollutants (Robertson et al., 1999). This model has been used for 1997. For earlier data a somewhat simpler version of the model (Persson et al., 1996) has been applied. The horizontal resolution for applications to data from 1997 is 11x11 km^2; for earlier data the resolution is 20x20 km^2.

The long-range transport (LRT) contributions are estimated in part 2. To derive distributions of the contribution from sources outside Sweden the following method has been applied. Model calculated daily contributions from Swedish sources (part 1) are deducted from observed daily values of concentration in air and precipitation at background atmospheric chemistry stations. The residuals are termed LRT contributions. These are analysed using an optimum interpolation (OI) method to give distributions of LRT contributions of concentrations in air and precipitation *over the whole model domain*. The idea behind the method is that the LRT contributions to the concentrations should vary more smoothly in space than the total concentrations, which are affected to some extent by local sources. LRT concentrations should therefore be more suitable for interpolation. LRT of wet deposition is calculated by multiplying the concentration in precipitation with the observed precipitation field. LRT of dry deposition is computed by running the LRT air concentrations through the dry deposition module of the dispersion model.

The OI technique has been widely used in meteorological applications. For this study we have used essentially the same system as the one applied to meso-scale meteorological analyses at SMHI (Häggmark et al., 2000). In OI, observations are normally used together with a background field, often called the first guess. In this study a linear interpolation of measured values is used as the first guess. Structure functions have been assumed to be isotropic, i.e. the first guess error correlation is only dependent on distance and not on direction. Since the concentrations can vary over some orders of magnitude, we determine the logarithms, perform the analyses and then recalculate the values.

This technique to determine the LRT contribution has been tested against independent data with good results (Langner et al., 1996). A successful verification of the MATCH model on the European scale (Graziani et al., 1998) also gives some support for the Sweden-scale model version.

The meteorological input data are based on operational runs at SMHI of the numerical weather prediction model HIRLAM, and the meso-scale analysis system MESAN (Häggmark et al., 2000). Some additional improvements to obtain detailed precipitation fields, with corrections for wind loss effects in precipitation samplers and topographical effects for the precipitation amounts have also been used (Vedin and Raab, 1995).

COMPARISONS TO EMEP MODEL RESULTS

At present, five years of comparable data for deposition over Sweden are available. We have compared the results obtained with the MATCH system with EMEP results. The total emissions from Swedish sources are roughly the same in both models. The EMEP models have a coarser horizontal resolution (50x50 km^2 for 1997 data and 150x150 km^2 for earlier data) than the MATCH model. Moreover, the MATCH model includes data assimilation of concentrations in air and precipitation, which the EMEP models do not. Results for the total deposition of sulphur, oxidised nitrogen and reduced nitrogen in Sweden, for the years 1991 and 1994-1997, are given in Figure 1 together with a comparison of the contributions from Swedish sources to the deposition during 1997.

The EMEP results for 1997 are obtained from Tarrason and Schaug (1999) (EMEP *Eulerian* Acid Deposition model), while earlier data are based on the EMEP *Lagrangian* model (EMEP, 1998). The MATCH estimates for total deposition over Sweden are larger than the EMEP results, for all compounds and years, but the differences decrease for 1997 with using the EMEP Eulerian model. The largest differences occur for reduced nitrogen, for which the EMEP results are about 40% lower than the MATCH results.

The EMEP and MATCH results referring to *contributions from Swedish sources* agree very well for sulphur and reduced nitrogen, while for oxidised nitrogen the values obtained from EMEP are substantially larger. The reasons for this are not yet understood. In Table 1 the MATCH results for deposition over Sweden are separated into dry and wet deposition.

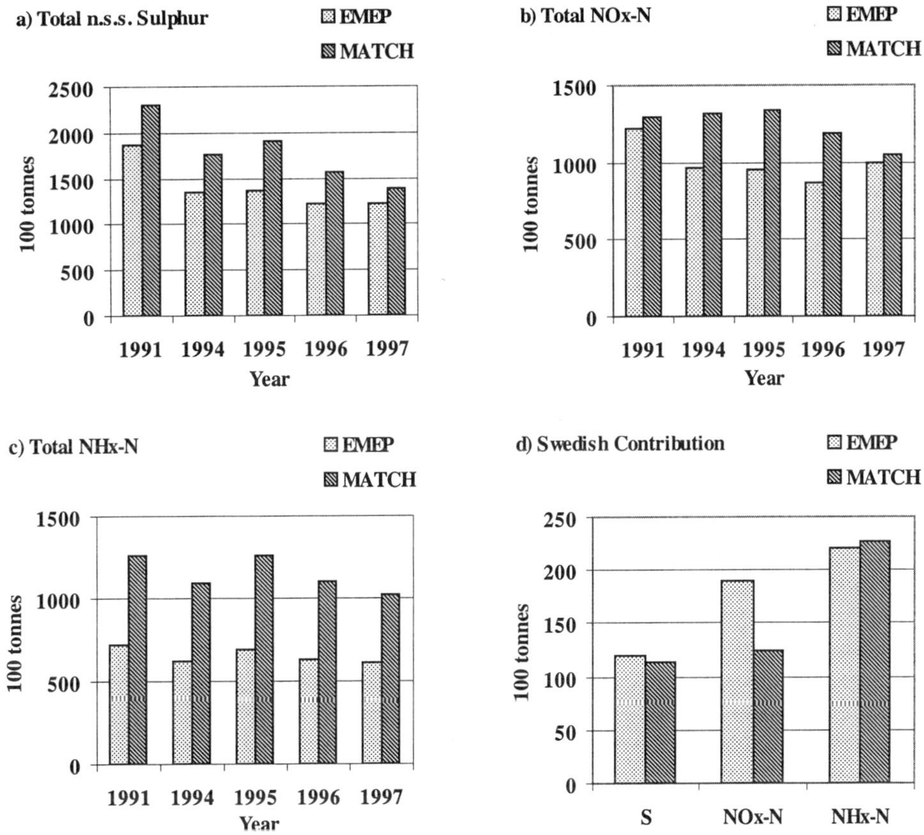

Figure 1. Comparisons between EMEP and MATCH estimates of deposition (100 tonnes of S or N) to Sweden. a) Total deposition of non-sea salt sulphur 1991, 1994-1997, b) total deposition of oxidised nitrogen 1991, 1994-1997, c) total deposition of reduced nitrogen 1991, 1994-1997 and d) Swedish contributions to the deposition of sulphur, oxidised nitrogen and reduced nitrogen during 1997.

Table 1. MATCH estimates of dry and wet deposition of sulphur, oxidised nitrogen and reduced nitrogen (100 tonnes of S or N) to Sweden during 1997. The dry deposition percentage of the total deposition is also indicated.

Compound	Dry deposition (100 tonnes of S or N)	Wet deposition (100 tonnes of S or N)
Non-sea salt sulphur	461 (33%)	926
Sulphur	-	1 042
Oxidised nitrogen	341 (33%)	708
Reduced nitrogen	396 (39%)	622

EVALUATION OF PRECIPITATION CHEMISTRY NETWORKS

The MATCH modelling system has been used as a tool to evaluate the importance of different atmospheric station networks for the national and regional air pollution assessments in Sweden, concerning sulphur, oxidised nitrogen and reduced nitrogen.

In the present study we have used data from several different precipitation chemistry monitoring networks in Sweden. Data for the years 1994-1997 have been available from: a) 4 Swedish EMEP stations, b) 27 stations in the National Precipitation Chemistry Network (of which 3 sites coincide with Swedish EMEP sites), c) 2 stations in the Swedish Integrated Monitoring Programme, and d) 8 EMEP or EMEP-equivalent stations in neighbouring countries.

The EMEP stations took daily precipitation samples until 1995 and changed to weekly samples 1996. The National Network and the Integrated Monitoring stations take monthly precipitation samples. The monitoring stations are shown in Figure 3.

At three sites, Vavihill (southern Sweden, with high deposition), Aspvreten (south central Sweden, medium deposition) and Bredkälen (northern Sweden, low deposition), measurements for both the EMEP Network and the National Network are available. It is of interest to compare these measurements internally, and to compare measurements from the National Network with MATCH estimates. At these three sites only EMEP measurements have been used in the MATCH data assimilation for the operational calculations of the deposition. In Figure 2 scatter plots are given for comparisons between the two independent monitoring networks, as well as between the MATCH estimates and the measurements from the National Network for 1994-1997.

The obtained correlation for sulphur, oxidised nitrogen and reduced nitrogen is somewhat higher for the MATCH estimates than for the EMEP-measurements. This is due to the OI technique used in the MATCH system, where strongly diverging measured values are damped. The wet deposition values from the MATCH estimates are probably also more representative than the point measurements, due to a better analysis of the "true" precipitation amounts in the MATCH system.

For the purpose of evaluation and as a basis for optimising the precipitation chemistry networks in Sweden, we have made some model experiments using the MATCH system and observed data from 1997. The LRT analysis and the deposition parts of the system have been re-run for several cases based on different assumed precipitation chemistry station networks. This was done in order to test the sensitivity of the estimated wet deposition of sulphur and nitrogen to the selected network. Wet deposition and precipitation concentration calculations were based on the following six assumed station networks: Case 1: All available EMEP stations + 100% of the stations in the National Precipitation Chemistry Network. (Reference case, which has also been used for the operational Swedish assessment studies.) Cases 2-5: All available EMEP stations + 75% of all stations in the National Network. Case 6: All available EMEP stations.

Thus, in all cases we have included all available EMEP stations. Cases 2-5 were used as a kind of cross validation. 75% of the National Network data is used for analysis and the verification is done on the remaining 25%. This procedure is then permuted for the whole data set. Case 6 is chosen just as a theoretical example with only 4 Swedish stations.

Scatter plots of the cross correlation study are shown in Figure 3. The explained variance is 0.7 - 0.8. The pollution gradient over Sweden, with high concentrations in the south and low in the north increases the correlation somewhat. The geographical spread of the stations in the National Network is however comparable to the spread of the three stations in Figure 2. Figures 2 and 3 therefore should be roughly comparable, which means that the explained variance in the cross correlation study is 5-10% lower than when comparing measurements from two different station networks at the same sites.

In Table 2 the annual integrated wet deposition values to Sweden during 1997, for the six different network cases, are presented. Cases 2-5, with a 25% reduction of the number of National Network stations, only deviate marginally from case 1 with all stations included. For case 6 (only EMEP stations) there is a clear tendency towards larger integrated deposition to Sweden. For reduced nitrogen, case 6 gives about 23 % larger values than case 1.

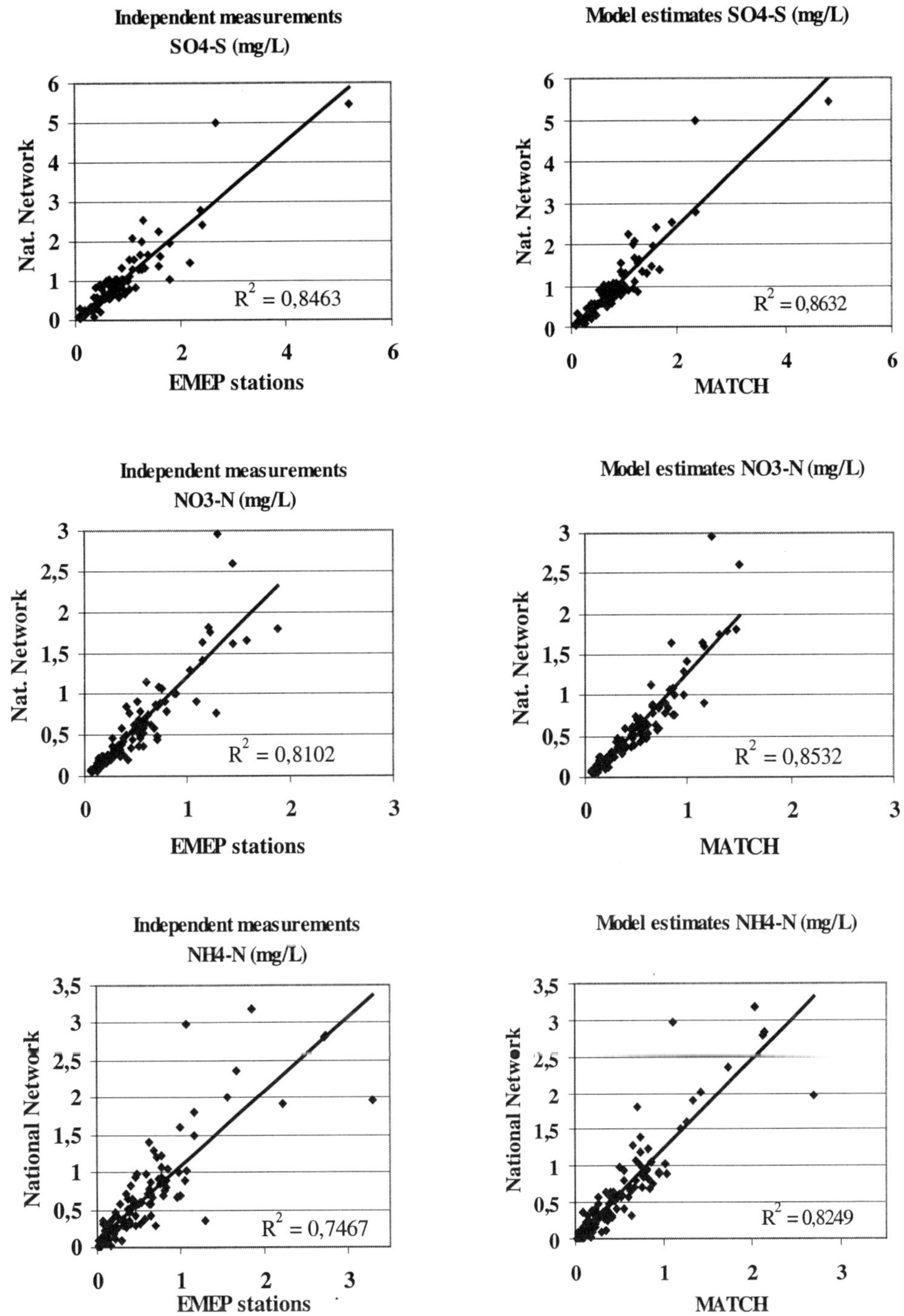

Figure 2. Scatter plots showing monthly precipitation concentration data (sulphur, oxidised nitrogen and reduced nitrogen) from 1994-1997 for three sites: Vavihill (southern Sweden), Aspvreten (south central Sweden) and Bredkälen (northern Sweden), where EMEP and National Network are located together. In the left column independent measurements for EMEP and National Network are compared. In the right column corresponding MATCH model estimates are compared to the independent National Network data.

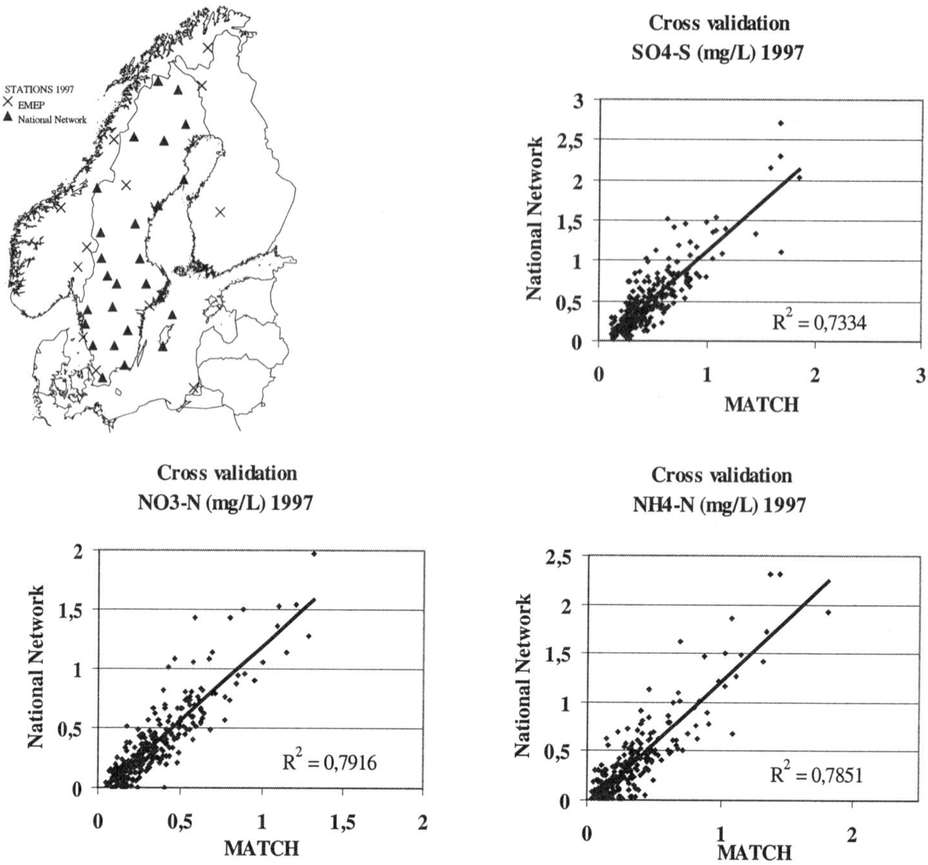

Figure 3. Upper left: Precipitation chemistry stations, which have been used in this study. × = EMEP and black triangles = National Network. Rest of Figure: Scatter plots for cross validation of monthly data, 1997, for sulphur, oxidised nitrogen and reduced nitrogen.

Table 2. Integrated wet deposition over Sweden during 1997 for non-sea salt sulphur, total sulphur, oxidised nitrogen and reduced nitrogen (100 tonnes S/year and 100 tonnes N/year) for the different station network cases considered.

		Wet Deposition (100 tonnes of S/year or N/year)			
Case no	**Stations**	**Non-sea-salt $SO_4 - S$**	**Total SO_4 - S**	**$NO_3 - N$**	**NH_4 - N**
1	EMEP + 100% of stations in National Network	926	1 042	708	623
2	EMEP + 75% of National Network stations (alt. A)	915 (-1.2%)	1 030 (-1.2%)	708 (0.0%)	615 (-1.3%)
3	EMEP + 75% of National Network stations (alt. B)	933 (+0.8%)	1 055 (+1.2%)	718 (+1.4%)	651 (+4.5%)
4	EMEP + 75% of National Network stations (alt. C)	927 (+0.1%)	1 040 (-0.2%)	712 (+0.6%)	637 (+2.2%)
5	EMEP + 75% of National Network stations (alt. D)	942 (+1.7%)	1 063 (+2.0%)	722 (+1.9%)	632 (+1.4%)
6	EMEP stations only	947 (+2.3%)	1 117 (+7.2%)	800 (+13.0%)	764 (+22.6%)

In Figures 4 and 5, maps are shown for the MATCH estimates of the annual wet depositions. The general patterns for the depositions are very similar and to a large degree dependent on the detailed information about precipitation amounts, which has been included in the calculations. However, over the northern half of Sweden the deposition values are larger in case 6, especially for reduced nitrogen.

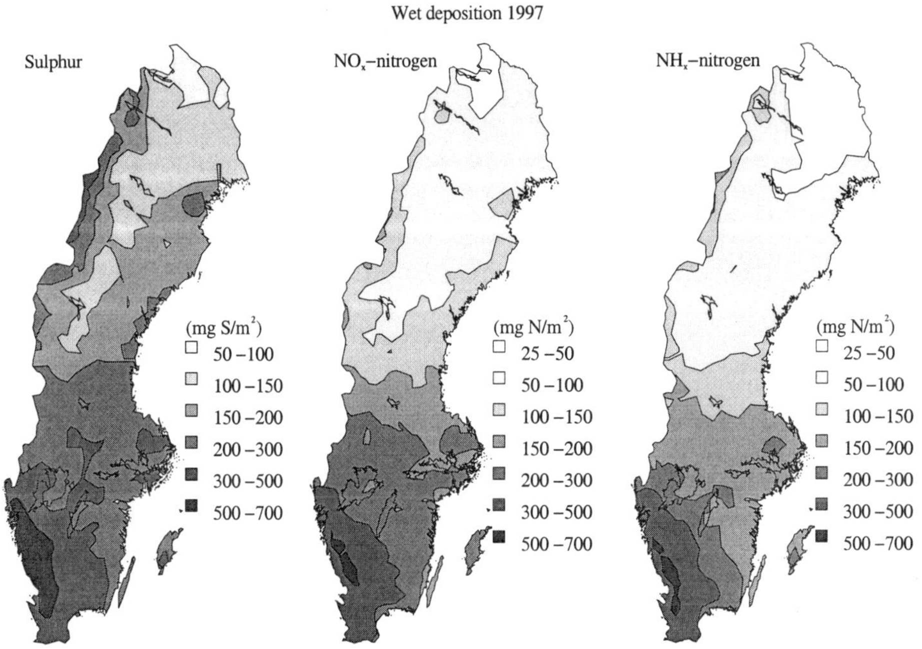

Figure 4. Wet deposition to Sweden of sulphur, oxidised nitrogen and reduced nitrogen, 1997, as obtained by MATCH calculations based on station network case 1 (all stations).

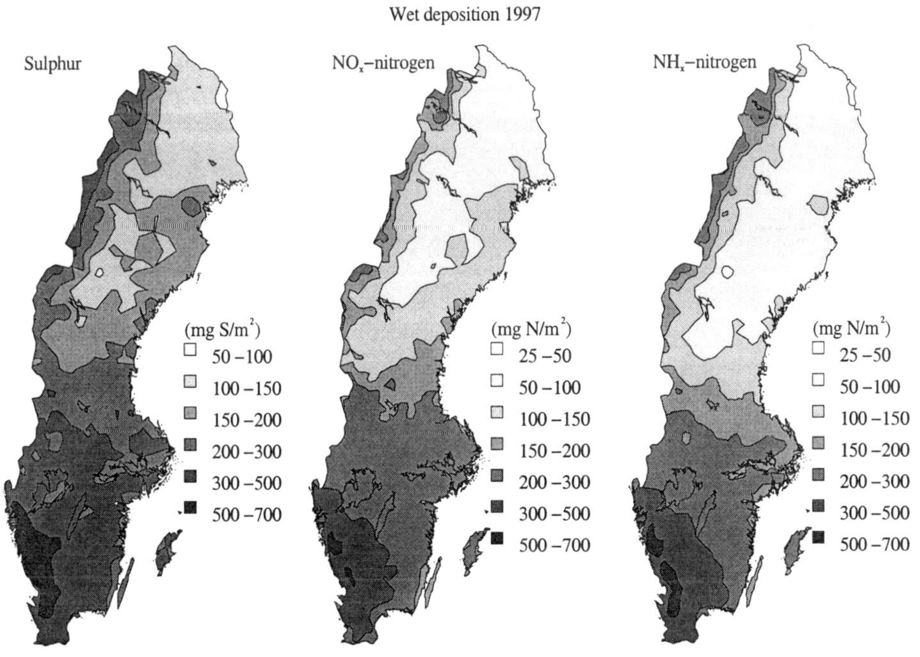

Figure 5. Wet deposition to Sweden of sulphur, oxidised nitrogen and reduced nitrogen, 1997, as obtained by MATCH calculations based on station network case 6 (only EMEP stations).

CONCLUSIONS

Comparisons between MATCH and EMEP results for total deposition over Sweden show larger values for the MATCH estimates, especially for reduced nitrogen. The differences are however smaller for the last year of the study (1997).

Annual wet deposition patterns over Sweden and the total integrated deposition to Sweden of sulphur and nitrogen are not very sensitive to a reduction of about 25% of the present stations in the National Precipitation Chemistry Network. The results indicate that about 30 stations of good quality can give satisfactory information concerning annual deposition on a national scale. The additional regional station network in Sweden, with about 100 stations, has not been studied so far, but the present results imply that there might be a possibility to reduce the number of samplers in that network.

Our conclusion, that it is possible to reduce the National Precipitation Chemistry Network with only small changes in the calculated annual wet deposition pattern for sulphur and nitrogen, agrees well with what was presented by Oehlert (1996) for the US.

ACKNOWLEDGEMENTS

The Swedish Environmental Research Institute Ltd (IVL) has provided air and precipitation chemistry data from Swedish stations. Data from Norway, Finland and Lithuania were provided by the Norwegian Institute for Air Research (NILU), the Finnish Meteorological Institute (FMI) and the Institute of Applied Environmental Research at Stockholm University (ITM), respectively. This work has received financial support from the Swedish Environmental Protection Agency.

REFERENCES

EMEP, 1998, EMEP/MSC-W Status Report 1/98. Transboundary Acidifying Air Pollution in Europe. Part 2: Numerical addendum. Norwegian Meteorological Institute, Oslo, Norway.

Graziani, G., Klug, W. and Mosca, S., 1998, *Real-time long-range dispersion model evaluation of the ETEX first release.* Office for Official Publications of the European Communities. Luxembourg.

Häggmark, L., Ivarsson, K.-I., Gollvik, S. and Olofsson, P.O., 2000, MESAN, an operational mesoscale analysis system, *Tellus A* 52:2.

Langner, J., Persson, C., Robertson, L. and Ullerstig, A., 1996, Air pollution assessment study using the MATCH modelling system. SMHI RMK No 69. SMHI, Norrköping, Sweden.

Oehlert, G. W. 1996, Shrinking a wet deposition network. *Atm. Environment.* Vol 30, No 8, 1347-1357.

Persson, C., Langner, J. and Robertson, L., 1996, Air pollution assessment studies for Sweden based on the MATCH model and air pollution measurements. *Air Pollution Modeling and Its Application,* Vol **XI**. Eds Gryning S-E and Schiermeier F.A., Plenum Press, New York and London.

Robertson, L., Langner J. and Engardt M., 1999, An Eulerian limited area atmospheric transport model. *J. Applied Meteorology* 38:190-210.

Tarrason and Schaug, 1999, Transboundary acid deposition in Europe. EMEP summary report 1999. Report 1/99. Norwegian Meteorological Institute, Oslo, Norway.

Vedin, H. and Raab, B. eds., 1995, *Climate, Lakes and Rivers.* Swedish National Atlas, Stockholm Sweden.

ESTABLISHMENT OF A MODEL NETWORK AND ITS APPLICATION FOR THE PREDICTION OF THE AIR POLLUTION IN A MESOSCALE AREA

Klaus Nester, Franz Fiedler, Walburga Wilms-Grabe, and Tianliang Zhao

Institut für Meteorologie und Klimaforschung
Forschungszentrum Karlsruhe / Universität Karlsruhe
Postfach 3640, D-76021 Karlsruhe

INTRODUCTION

In the frame of the German Tropospheric Research Programme (TFS) a model network has been established. The main purpose of the close co-operation of four groups in this network is the prediction of the air pollution, especially the ozone concentration, from the European scale down to the smallest mesoscale in different regions. The emission data will be provided by the „Institut für Energiewirtschaft und Rationelle Energieanwendung" (IER) of the University Stuttgart. The „Deutsche Wetterdienst" (DWD) in Offenbach provides the routine meteorological data and uses additionally a chemical transport model based on that of the EURAD group. The „Institut für Geophysik und Meteorologie" (IGM) at the University of Cologne contributes with the EURAD model, which runs on the European scale with nesting options. At the „Institut für Meteorologie und Klimaforschung" (IMK) in Karlsruhe the model system KAMM/DRAIS is applied in the mesoscale α range.

Aim of the activities at the node IMK was the disposition of a network version of the mesoscale model system KAMM/DRAIS in order to perform real-time predictions of the relevant air pollutants in mesoscale areas. This new model version will be especially applied to predict the ozone concentration distribution during summer smog episodes. The main purpose of the predictions will be to answer the following questions:

How reliable are the predictions and how they can be improved?
Can short-term measures avoid an excess of the ozone concentration limits, if such an excess is predicted?

In the following a preliminary answer can be given to the first question but not yet to the second, because a greater number of prediction simulations with modified emissions have not yet been carried out .

THE MODEL SYSTEM KAMM/DRAIS IN THE MODEL NETWORK

The model system KAMM/DRAIS is a complex 3-d Chemical-Transport Model (CTM), which is especially suited for mesoscale applications. It consists of the non hydrostatic model KAMM (Adrian and Fiedler, 1992) and the dispersion model DRAIS (Nester et al., 1998) including chemical reactions (Stockwell et al., 1990). The model system KAMM/DRAIS needs input data from the results of larger scale model simulations as well as emission data. Its inclusion in the network is demonstrated in fig. 1.

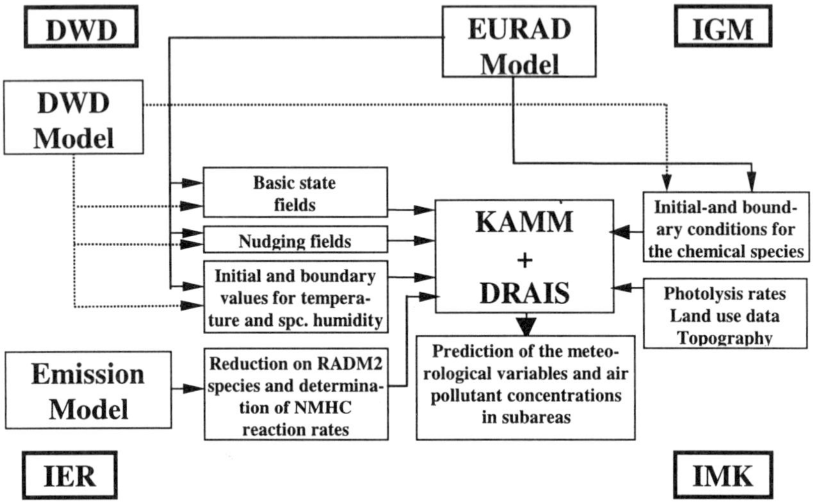

Figure 1. Inclusion of the model system KAMM/DRAIS in the model network

The emission data are provided by IER, Stuttgart (Friedrich et al.,1999). The emission preprocessing at IMK reduces the data for the VOC's to the classes of the model and calculates the reaction rates for all classes depending on the averaged emissions of the original species. Point sources belonging to the same grid point are also combined provided that the source height and the other emission conditions are the same.

The meteorological data from the EURAD model (Ebel et al.,1997) or the DWD model are used to determine the basic state variables (pressure, temperature and geostrophic wind) of the KAMM model. Additionally, the initial and boundary conditions for the humidity and the temperature are calculated. Usually, a nudging procedure improves the model results. The nudging flow fields are derived from the wind data of the EURAD or the DWD model. The initial and boundary conditions for the chemical species are also interpolated from the corresponding large scale model results.

In order to use the model system KAMM/DRAIS as real-time model it is necessary to manage the running of the whole model system automatically from the beginning of the data transfer to the IMK node. This is realised in a special program, which has the following tasks:

- Check of the data from the network partners
- Control of the data transfer between the different computers
- Start of all preprocessing programs on different processors to get the model input data
- Check of availability of the input data files

- Start of the model run and all consecutive runs

The interactions of the model system KAMM/DRAIS with its main subprograms and input data are presented in fig. 2.

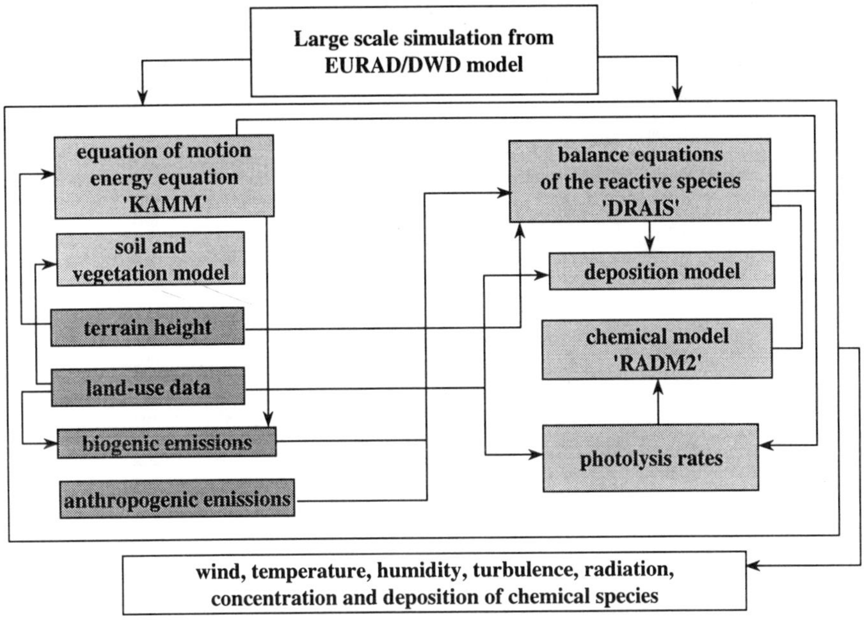

Figure 2. Schematic representation of the model system KAMM/DRAIS

PERFORMANCE OF THE PREDICTIONS

In summer 1999 the model network was applied to carry out predictions of the air pollution in Europe and Southwest Germany and Alsace in France. On the European scale the simulations are performed operationally at the DWD with a grid size of about 21 km (Rißmann et al., 1999). For episodes with nice weather conditions, when high ozone concentrations could be expected, the simulations for the smaller scale region are carried out with the KAMM/DRAIS model using a grid size of 5 km. Usually, the emission data are prepared already the day before the forecast is simulated. After the first results from DWD arrived at IMK, the managing program, which handles all procedures at IMK, is started. The calculated results are stored on an hourly basis, that they can be evaluated later on.

Two examples of the ozone concentration distribution near ground level in the afternoon are shown in the figure 3. The upper figure shows a typical distribution with only minor spatial variations between 50 ppb and 60 ppb, whereas the distribution in the lower figure is characterised by a strong increase of the ozone concentration from west to east. In both cases this general behaviour agrees well with the measured data. Especially, in the latter case the larger scale transport of ozone causes the strong spatial differences.

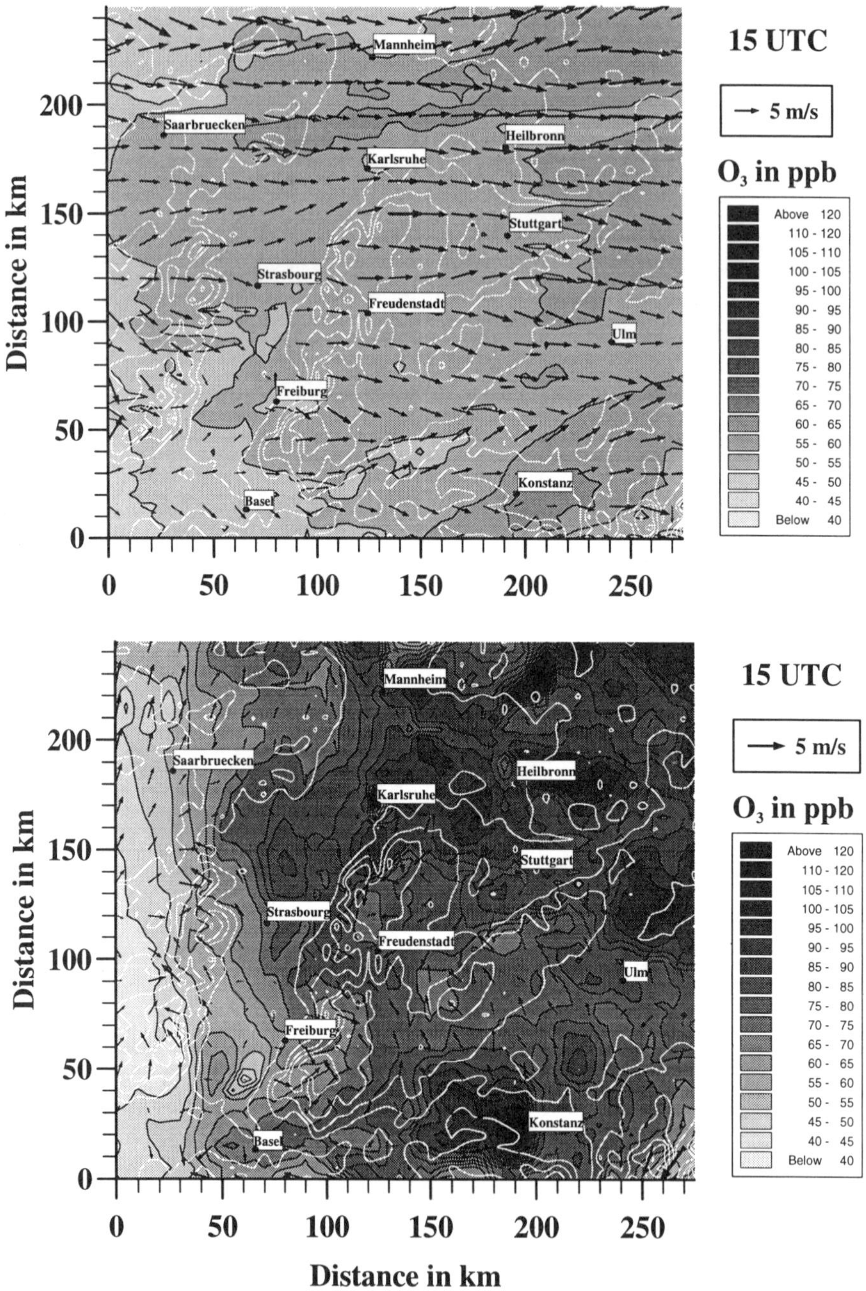

Figure 3. Ground level ozone concentration distribution at 15 UTC
Upper figure : 20 August 1999
Lower figure : 14 September 1999

COMPARISON WITH MEASURED OZONE CONCENTRATIONS

For 11 days the simulated and measured ozone concentrations at about 60 stations in Baden-Württemberg have been compared and analysed. In order to check, whether the diurnal cycles of the ozone concentrations are realistically simulated the diurnal cycle of the measured and simulated ozone concentrations averaged over all stations and all days are evaluated and compared. The result is shown in fig. 4. From this figure it can be stated that the mean diurnal cycle of the measured ozone concentration is well predicted by the simulation. Only the simulated increase of the ozone concentration in the morning begins too early. Different reasons for this discrepancy are possible:

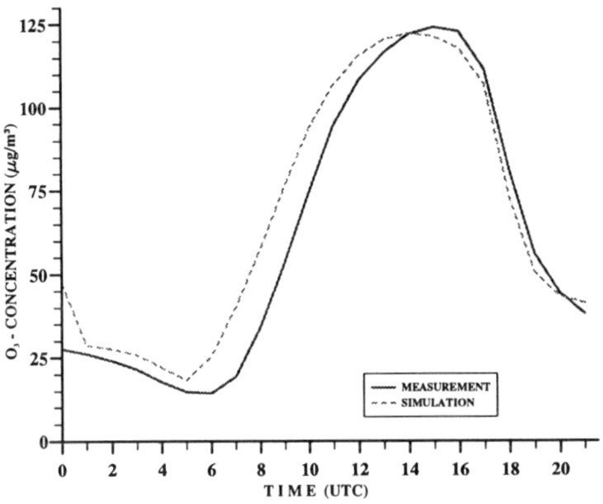

Figure 4. Average diurnal cycle of the ozone concentration

- The increase of the mixing height begins in the simulation too early in the morning.
- The peak emissions occur later in the morning.
- The local concentrations of the emitted species at the stations are much higher than the grid values in the model.

Further studies are necessary to find out, which is the dominant effect.

The direct comparison of the measured and simulated ozone concentrations for all evaluated days is shown in figure 5 (upper scatter plot). The comparison is restricted to the time period between 3 UTC and 21 UTC. The slope of the regression line has a value of 0.71, which demonstrates that the lower ozone concentrations are overestimated and the higher values are underestimated by the model simulations. Despite the great scatter of the data, the correlation coefficient of 0.8 is rather high. The statistical analysis provides that in 50% of the compared data the difference between the measured and simulated ozone concentrations is lower than 10 ppb (see fig. 6).

Because the ozone concentration usually reaches its maximum in the time period between 11 UTC and 16 UTC, the data for this period have also been analysed. The lower scatter plot of figure 5 shows the comparison between the measured and simulated ozone concentrations for this period. The slope of the regression line (0.44) and the correlation coefficient (0.57) are remarkably lower than for the longer period. But the percentage of the points which are close to the line of agreement is greater. The statistical analysis provides, that in 60 % of the cases the predicted and measured ozone concentrations differ less than 10 ppb from each other (fig. 6). This reliability confirms the result of the model evaluation study carried out with other data (Fiedler et al., 1999). The fact, that on an average the higher ozone concentrations are underestimated and the lower ozone concentrations are overestimated by the simulation has not changed. This effect is still more pronounced if individual days are considered. figure 7 shows a typical example. The simulated ozone concentrations at the measuring stations for the 11 days agree better with the measured data as can be seen in figure 8, which is a typical example for such a comparison. This means, that the spatial variation of the ozone measurements in the whole area is greater than the

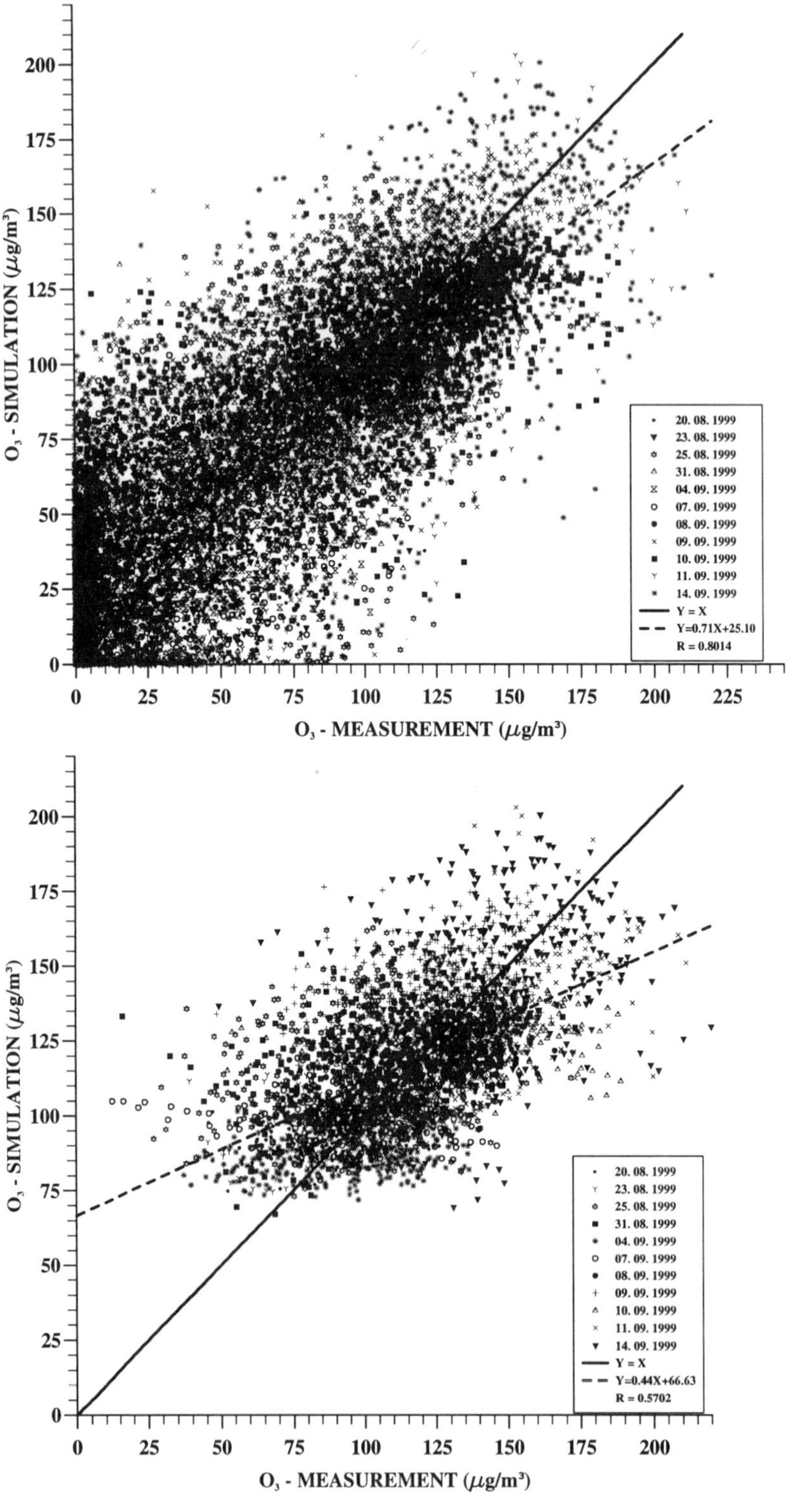

Figure 5. Scatter diagram of the ground level ozone concentrations
Upper figure: time period 3 UTC - 21 UTC; Lower figure: time period 11UTC - 16UTC

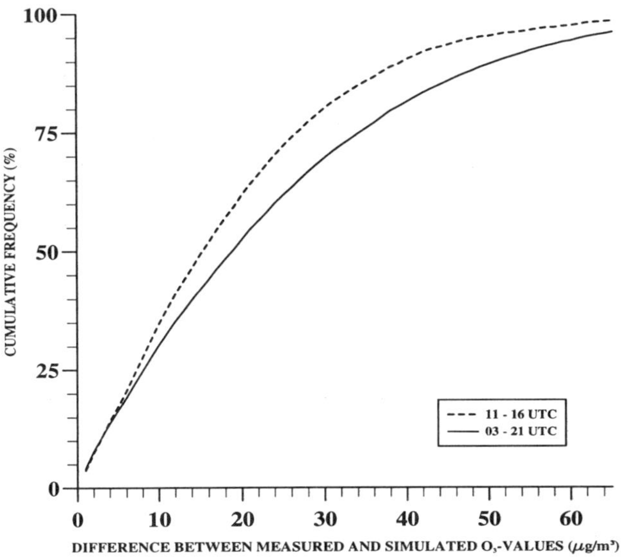

corresponding variation in the simulations. The results seem to indicate that the horizontal grid resolution of the model simulations is still too coarse. Because most of the measuring stations are located in cities, influences at these stations caused by local meteorological and emission conditions, which cannot be resolved by mesoscale models, may also contribute to the observed discrepancies.

Figure 6. Cumulative frequency distribution of the difference between measured and simulated ozone concentrations

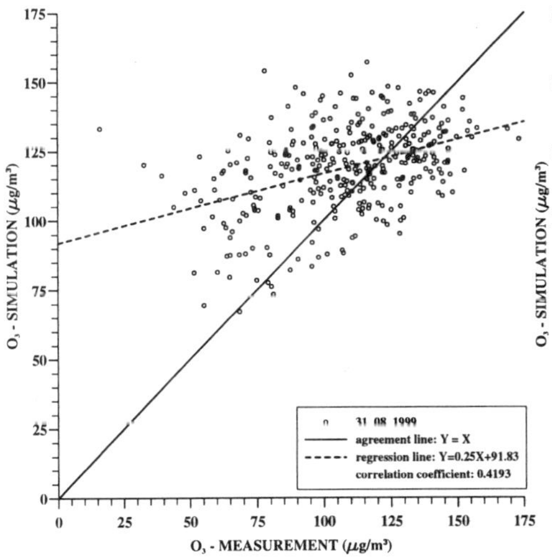

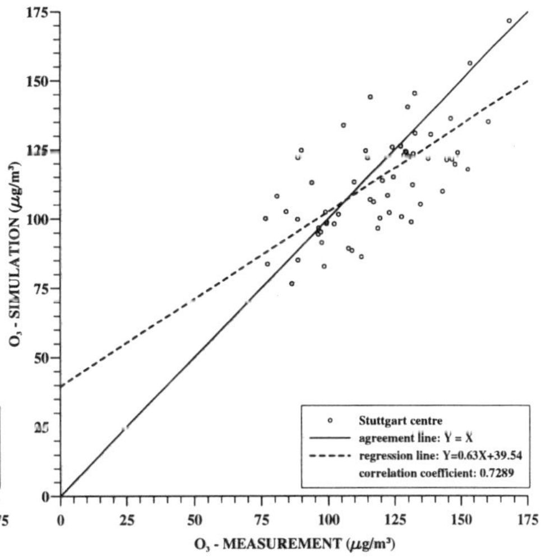

Figure 7. Scatter diagram of the ground level ozone concentrations for 31 August 1999, 11 UTC - 16 UTC

Figure 8. Scatter diagram of the ground level ozone concentrations for Stuttgart centre, 11 UTC - 16 UTC

CONCLUSIONS

First applications of the model network to predict the air pollution on the European scale and in a subregion have been successfully carried out. In general, the average diurnal cycle of the ozone concentration is quite well predicted. Only the increase of the mean ozone concentration in the morning begins in the simulation earlier than in the measurements. The statistical evaluations show that the model simulations in the subregion Southwest Germany predict the ozone concentrations rather well. In the time period between 3 UTC and 21 UTC the deviation between the simulated and measured ozone values was less than 10 ppb in 50% of the cases. This percentage increases to 60% for the time period between 11 UTC and 16 UTC, where the highest ozone concentrations occur. But the slope of the regression line decreases from 0.71 to 0.44 and the correlation coefficient from 0.80 to 0.57, respectively. These data prove that the lower ozone concentrations are overestimated and the higher values are underestimated by the simulations. Mainly, the spatial variability of the ozone concentration for the individual days in the period between 11 UTC and 16 UTC is not sufficiently simulated. On the contrary, the temporal variability of the simulated ozone concentrations at individual stations agrees better with the measured one . In order to improve the model simulations, the reasons for the observed discrepancies have to be analysed by corresponding sensitivity studies.

FUTURE ACTIVITIES

The predictions will be continued, if the partners in the network provide the necessary data. A parallel version of the model system KAMM/DRAIS will also be applied in order to reduce the run-time of the model. Additionally, realistic short-term emission reduction scenarios will be taken into account if an excess of the ozone limits is predicted. From the results of such a simulation it can be found out, whether the intended measures are successful or not.

ACKNOWLEDGEMENTS

We thank the ministry for science and technology for the financial support of the project. All colleagues involved in the network activities are acknowledged for the excellent co-operation. We also thank the UMEG for providing the ozone concentrations at the network stations in Baden-Württemberg.

REFERENCES

Adrian G and Fiedler F, 1991, Simulation of unstationary wind and temperature fields over complex terrain and comparison with observations, *Contr.Atmos. Phys.*, 64, pp 27-48

Ebel A, Elbern H, Feldmann H, Jakobs H J, Kessler C, Memmesheimer M, Oberreuter A, and Piekorz G, 1997, Air pollution studies with the EURAD Model System(3): EURAD-European air pollution dispersion model System. Mitteilungen aus dem Institut für Geophysik und Meteorologie der Universität Köln, Heft 120, p.172

Fiedler F., Nester K., Wilms-Grabe W., and Zhao T., 1999, Evaluation of the model system KAMM/DRAIS, to be published in: *Proceedings of the III GLOREAM Workshop*, Ischia

Friedrich R., Heidegger A., and Kudermann F., 1999, Development of an emission calculation module as a part of a model network for regional atmospheric modelling, *Proceedings of EUROTRAC Symposium '98*, Garmisch-Partenkirchen, Vol. 2, pp 247-250

Nester K and Fiedler F, and Panitz H.-J., 1998, Simulation of mesoscale air pollution with the model system KAMM/DRAIS, *Proceedings of the 11th World Clean Air and Environment Congress*, Durban, South Africa, Vol.4, Paper-No. 10D2

Rißmann J., Jakobs H., Heidegger A., and Nester K., 1999, First results of operational ozone forecasts with the TFS Network Model System at the DWD, to be published in: *Proceedings of the III GLOREAM Workshop*, Ischia

Stockwell W R, Middelton P, and Chang J S, 1990, The second generation Regional Acid Deposition Model, chemical mechanism for regional air quality modelling. *J. Geoph. Res.* Vol. 95, No. D10, pp 16343-16367

DISCUSSION

J. KRETZSCHMAR: What is the time required to produce one ozone map for the whole area on an half an hour basis?

K. NESTER: The model needs 12 hours computing time for a 24 hours forecast. The postprocessing of the results in order to produce ozone maps is negligible compared to this time. In summer we will have a faster computer and we can use the parallel version of our model. Then the computing time can be reduced by about a factor of 10.

J. KRETZSCHMAR: Will the model go on-line to predict/forecast ozone levels in real time?

K. NESTER: It is intended to publish the model results on our Web site, if the predictions will be continued this year.

J. KRETZSCHMAR: Is there any link between this approach and "the system for local forecast of surface ozone in Germany" of the University of Berlin?

K. NESTER: There is no link to the system applied by the University of Berlin. We use more complex models in our model network.

R. SAN JOSE: Have you tried to make some comparisons following the last EU Directive (April,99); although this Directive is not for ozone?

K. NESTER: We did not apply this EU Directive. Our comparison is only based on half an hour mean values of the ozone concentrations. The aim of our simple statistical evaluation was only to see how realistic the forecasts of the ozone concentrations are.

QUANTIFYING THE EXPORT OF POLLUTANTS FROM THE BOUNDARY LAYER

Elisabeth Donnell[1], Deb Fish[1], and Alan Thorpe[2]

[1]Department of Meteorology, University of Reading, RG6 6BB, UK
[2]Hadley Centre, UK Met. Office, Bracknell, RG12 2SZ, UK

INTRODUCTION

Pollutants are longer lived in the free troposphere. This means that the transport of pollutants from the boundary layer to the free troposphere has implications for long range transport and global warming. It is therefore important to understand the mechanisms which transport such pollutants from the boundary layer to the free troposphere and to be able to quantify this transport. Ozone is a particular secondary pollutant which is harmful to health and a green house gas. Measurements have shown that ozone in the free troposphere has more than doubled over Europe since pre–industrial times (*e.g.* Staehelin *et al.*, 1994).

The aims of this work are therefore to firstly estimate the proportion of a passive tracer initially well mixed in the boundary layer, that is transported to the free troposphere. The second aim is to quantify the contribution of the different transport mechanisms, namely advection, convection and turbulent mixing, to this exchange.

It is difficult to quantify the exchange of ozone into the free troposphere from the boundary layer by measurements alone, as there is no unique chemical signature for the boundary layer (Bethan *et al.*, 1998), unlike the stratosphere. We have therefore chosen to undertake a number of idealised passive tracer studies in a mesoscale model to try and quantify this transport under different meteorological scenarios. As most previous studies have concentrated on the transport due to convection (*e.g.* Gimson, 1997), we concentrate more on the role of fronts.

METHODOLOGY

To answer the questions put forward we have chosen to incorporate a passive tracer into the mesoscale version of the UK Meteorological Office's Unified Model (Cullen, 1993). We have chosen an Eulerian approach as we want to represent the boundary layer and convective processes which are not easily parameterised in a Lagrangian model. The model is run at a horizontal resolution of approximately 13 km so that the fronts can be resolved.

In the model the tracer is transported in three ways; by turbulent mixing, by convection and by advection. The advection of the tracer is carried on in the model using the Roe scheme (1985), which is necessarily conservative and positive definite. The convection is carried out using a mass flux scheme and the turbulent mixing is carried out using a local mixing scheme.

The different transport mechanisms are isolated so that the transport by each mechanism can be quantified. That is, the model is run with a four tracers, which are transported by either advection only, convection only, turbulent mixing only or all three mechanisms.

The model is initialised with the tracer filling the boundary layer over the whole model domain. This initial field was chosen because it gives a clear comparison between the different mechanisms. If the tracer was initialised only at the first model level, then for the case where the tracer is only advected, very little of the tracer would be transported to the free troposphere due to the fact that vertical velocities are very small near the ground. In this case turbulent mixing is required to mix the tracer to a height where the velocities are stronger and can be advected.

Each simulation is run for 24 hours and the amount of tracer in the free troposphere is calculated at the end of each hour. The definition of the free troposphere is now described.

Definiton of the Boundary Layer and the Free Troposphere

Our definitions of the boundary layer top and the free troposphere are based on values of the moist Richardson number (Durren and Klemp, 1982). The free troposphere is defined to be any grid location, except in the first model level, where the moist Richardson number is greater than one. The boundary layer is defined to be any region which is vertically connected to the surface where the Richardson number is less than or equal to one. This therefore leaves regions of air which are not vertically connected to the surface but have low Richardson number. It is likely that tracer in these regions would eventually be defined to be in the free troposphere.

RESULTS

The Case Study Days

Results will be shown from three case study days. As our initial motivation was to investigate the role of fronts we look at two frontal days and compare it with a high pressure day. Figure 1 shows the synoptic situation for each of the case study days at the end of the simulation. Each simulation was run for 24 hours. The three case study days are summarised as:

(A) May case study day (17/05/95): There was a low pressure system located in the southern half of the model domain. It slowly moved towards the north east (Fig 1a). This case study day was chosen because of the relatively slow moving meteorology.

(B) January case study day (15/01/99): This case study day has a much more active front. The low pressure system deepened quite rapidly over the time period of the model simulation. By the end of the simulation a cold front is situated crossing the UK (Fig 1b).

(C) July case study day (09/07/99): At the start of the day the high pressure system is located over the UK, and it slowly moves towards the east during the simulation (Fig 1c).

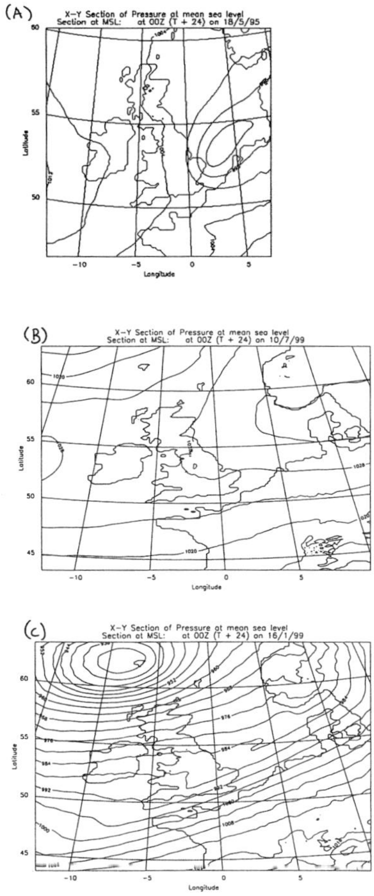

Figure 1. Synoptic situation at T+24 for the three case study days, (A) May, (B) January and (C) July.

Table 1. Percentage of tracer in the free troposphere after 24 hours for each case study day and for each tracer.

Case Study Day	Mix only	Advection Only	Convection Only	All
May 17/05/95	29.1	31.5	46.7	51.4
Jan. 15/01/99	31.3	30.3	26.8	54.2
July 09/07/99	34.2	25.5	29.5	37.2

We now discuss the results in terms of the different mechanisms.

Turbulent Mixing

For all the case study days with only the turbulent mixing transporting the tracer a perhaps surprisingly large proportion of the tracer is in the free troposphere at the end of 24 hours, with more than 29% of the tracer in the free troposphere at the end of the simulation. Table 1 lists the amount of tracer in the free troposphere after 24 hours for each of the simulations. Why is so much tracer being transported to the free troposphere by mixing alone when it would expected that the top of the boundary layer would cap the tracer?

Figure 2 shows horizontal cross–sections of the vertically integrated tracer amount in the free troposphere after 24 hours. The first column shows the simulations where tracer was transported by turbulent mixing only for the different case study days. The pattern of the tracer in the free troposphere is rather patchy. For the two frontal cases the mixing is generally in the relatively cooler air behind the front, and for the high pressure day it is mostly over the land.

Figure 3 shows vertical cross section of tracer concentration through the front for the May case study day. Where the tracer is transported by turbulent mixing only, it appears that the tracer does remain in a relatively confined layer adjacent to the surface (Fig 3). It is also shown that the tracer is transported slightly higher in the region of the front most likely due to the lifting of the boundary layer in that region. This figure appears to show that the tracer is capped at the top of the boundary layer. It should be remembered however that this is a particular snap shot in time, the Richarson number can be rather noisy, and the boundary layer top changes rapidly in the region of the front. This means that tracer which was in the boundary layer in one timestep can be defined to be in the free troposphere in the next.

To further investigate the relatively large transport of tracer into the free troposphere by turbulent mixing alone, we calculated the percentage of tracer in the free troposphere depending on whether it is located over land or over sea (Fig. 4). It is clearly seen that for the January day (frontal day) a lot more tracer is mixed to the free troposphere over the sea than over the land (Fig 4). It is over the sea where the boundary layer is not very well defined. There is a clear diurnal cycle in the tracer transported to the free troposphere over land on the July day. For the case of turbulent mixing only we see that the tracer is mixed to the free troposphere slowly after the start of the simulation. After sunrise the boundary layer deepens, and hence some of the tracer which was in the free troposphere is now defined to be in the boundary layer. This is why it appears that the amount of tracer in the free troposphere is decreasing between 4 and 9 am. Very little is mixed out of the boundary layer during the day, but as soon as the boundary layer collapses at sunset all the tracer which was defined to be in the boundary layer now suddenly finds itself in the free troposphere (Fig 4). This is also why we see more tracer over the land after 24 hours for the july case (Fig 2).

The reason why more than 29% of the tracer is transported to the free troposphere by mixing alone is twofold. Firstly it is due to the diurnal cycle in boundary layer depth, such that when the boundary layer collapses at the end of a convective day tracer that was in the boundary layer is suddenly defined to be in the free troposphere. The second reason is due to a relatively 'leaky' boundary layer top in stable regions due to the top of the boundary layer being ill defined, and in the region of the front.

Advection

Advection is never the dominant transport mechanism for any of the 3 case study days, as seen in Table 1 where between 25.5 and 31.5 % of the tracer is transported to the free troposphere by advection alone. This is perhaps surprising as our motivation was to look at the transport by frontal advection. However when examining Figures 2 and 3 the role of frontal advection does seem to be important. The horizontal cross–sections showing the vertically integrated tracer amount clearly show tracer being transported in the region of the front and low pressure system (Fig 2). The advection in the region of the front also appears to be the dominant process when we look at the vertical cross sections (Fig 3). However the transport by the front is in a relatively narrow region, so averaged over the whole model domain it does not appear to be significant. It is significant that the tracer is transported to more than 8 km and with the highest concentrations by advection alone.

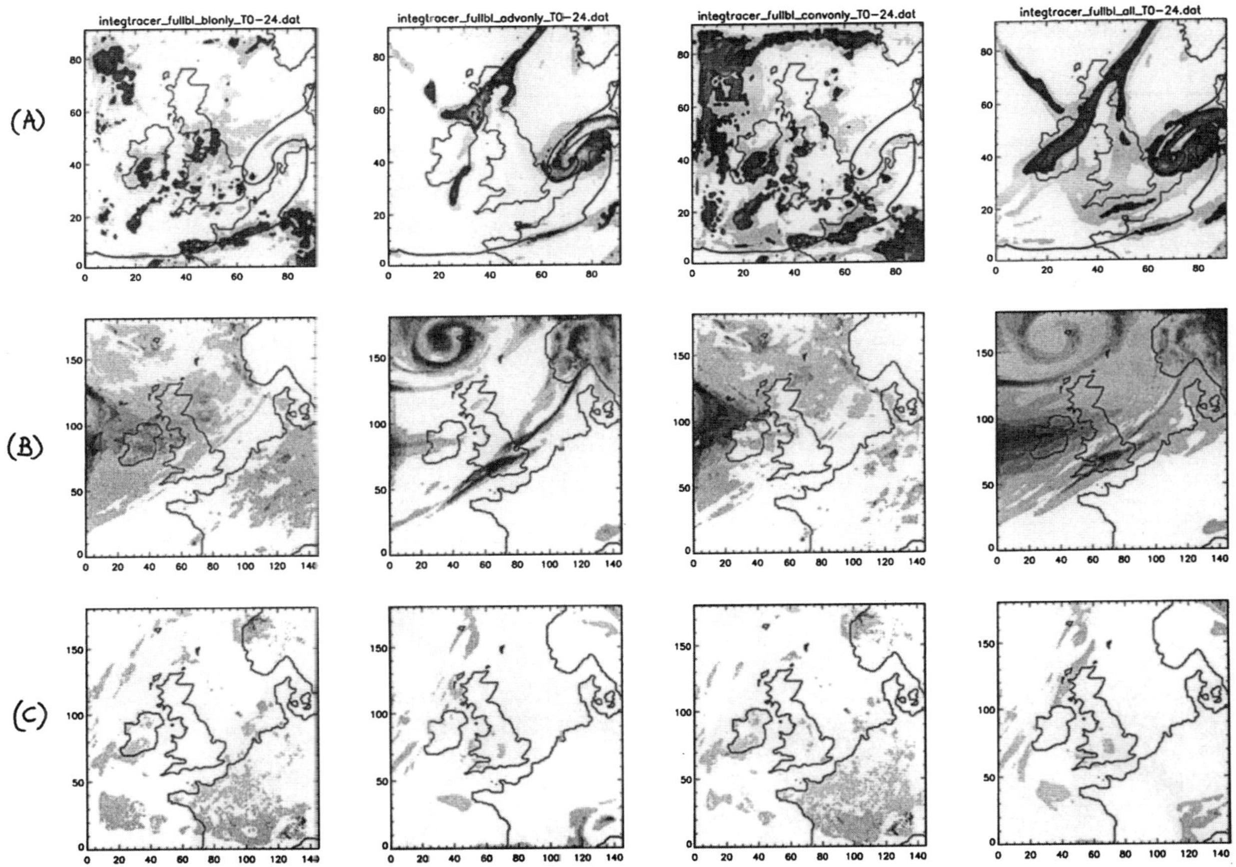

Figure 2. Horizontal cross sections showing vertically integrated tracer amount for each of the case study days (rows) and each of the transport mechanisms (columns from left to rights; turbulent mixing only, advection only, convection only and all mechanisms).

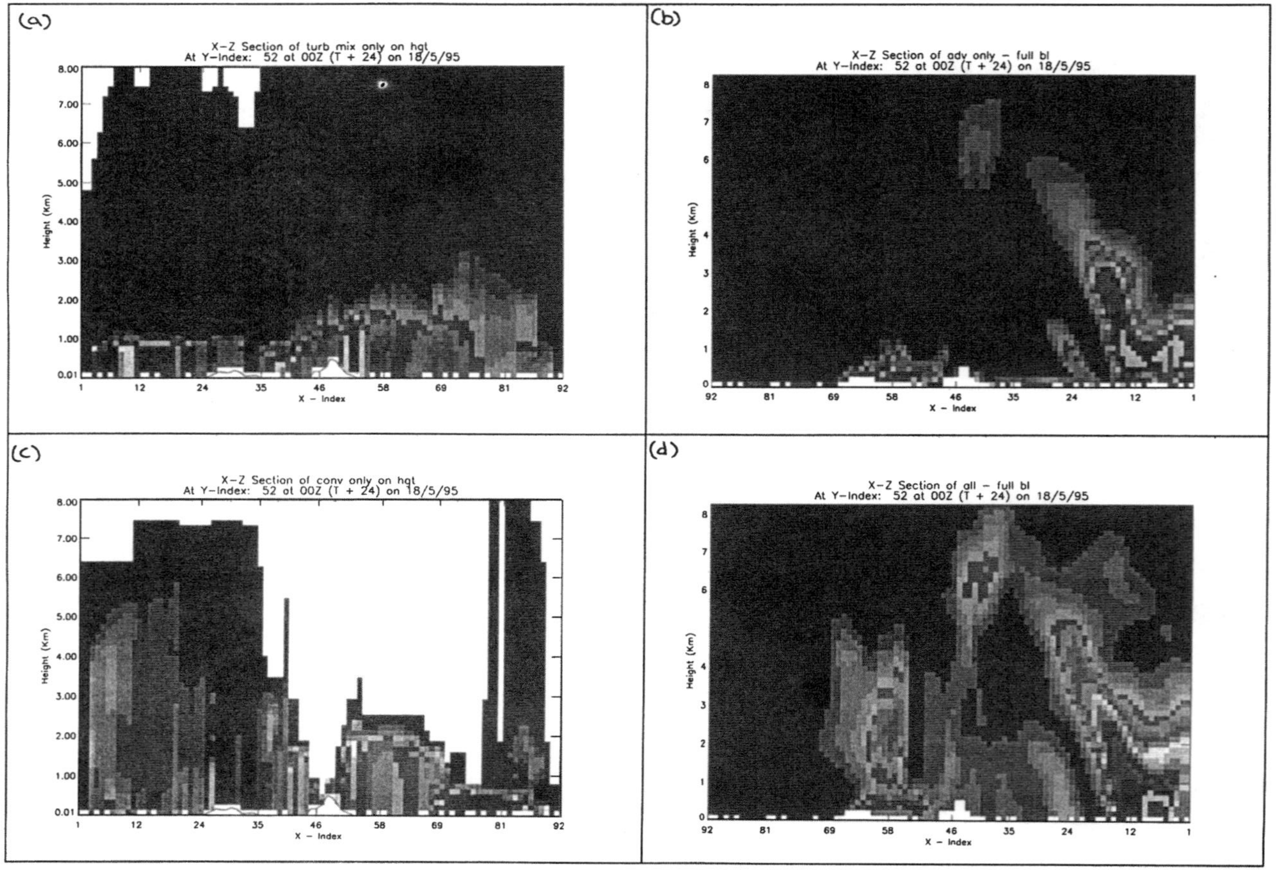

Figure 3. Vertical cross sections showing tracer concentration through the front on the 17/05/95, for the different mechanisms (a) turbulent mixing only, (b) advection only, (c) convection only and (d) all mechanisms.

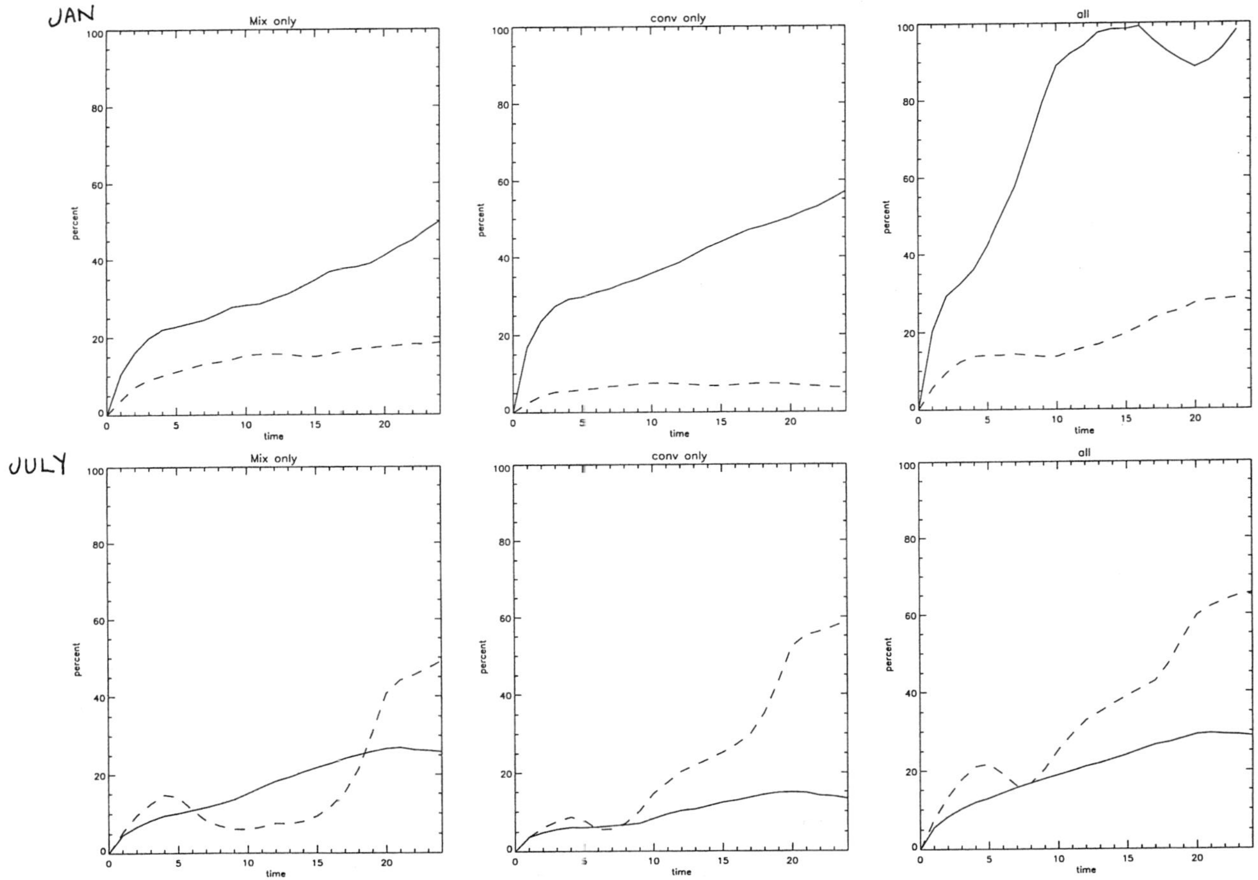

Figure 4 Percentage of tracer mixed to the free troposphere versus time for the January and July case study days.

Convection

Convection appears to be the dominant transport mechanism on the May case study day with a large value of 46.7% of the tracer transported to the free troposphere after 24 hours (Table 1). Similar to the transport by turbulent mixing only, the horizontal cross section of vertically integrated tracer shows the tracer in the free troposphere is over a wide spatial region (Fig 2). The effect of the diurnal cycle is seen in the July case study day where at the end of the simulation most of the tracer is located over the land, due to the collapse of the boundary layer. Convection also takes the tracer to a significant height in the free troposphere at more than five kilometers (Fig 3.). Again similar to the turbulent mixing there is a large difference in the amount of tracer transported over sea compared to the amount transported over the land for the different case study days (Fig 4.) For the frontal cases the convection is mostly over the warmer sea behind the front, and for the high pressure day the convection is dominantly over the warmer land.

All Mechanisms

The greatest amount of tracer is transported to the free troposphere when by all transport mechanisms are employed. More than 50% is transported to the free tropsphere for the two frontal cases (Table 1). It is also clear from Table 1 and Figures 2 and 3 that the different transport mechanisms are not additive. This is because the tracer may be advected away from a region which is convective. Even with all mechanisms the role of the front and convection is clear in the cross–sections (Fig. 3 and Fig. 4).

CONCLUSIONS

In this idealised study of trancer transport is was seen that more of the tracer was transported to the free troposphere in the frontal cases compared with the high pressure case. More than 50% of the tracer was in the free troposphere at the end of the simulation for the frontal cases. The tracer was transported to heights greater than 8 km by the front and more than 5 kilometers due to convection behind the front. It was also seen that the tracer was mostly transported over the sea in the frontal cases and mostly over the land in the high pressure case.

Future work will include the addition of a photochemistry and depositions schemes.

Acknowledgments

Thanks to Ed Dicks for assistance with modifying the Unified Model code.

REFERENCES

Bethan, S., G. Vaughan, C. Gebig, A. Volz–Thomas, H. Richer and D.A. Tidderman, 1998, Chemical air mass differences near fronts, *J. Geophys. Res.*, 103:13413.

Cullen, M.J.P., 1993, A unified forecast climate model, *Met. Mag.* 122:81.

Durren, D.R. and J.B. Klemp, 1982, The effects of Moisture on trapped mountain lee waves, *J. Atmos. Sci.*, 39:2490.

Gimson, N.R., 1997, Pollution transport by convective clouds in a mesoscale model, *Q.J.R. Meteorol. Soc.*, 123:1805.

Roe, P.L., 1985, Large scale computations in fluid Mechanics, *Lectures in Applied Maths*, PUBLISHER

Staehelin, J., J. Thudium, R. Buehler, A. Volz–Thomas, and W. Graber, 1994, Trends in surface ozone concentrations at Arosa (Switzerland), *Atmos. Environ.*, 28:75.

DISCUSSION

B. E. A. FISHER: The calculated percentages of tracer in the free troposphere are based on assuming that the tracer is initially distributed throughout the domain. If one considered tracer released in a specific region or location e.g. close to a front, the calculation would presumably be very different? The fraction advected into the troposphere could be much higher.

E. DONNELL: This is true. Our study was however looking at the relative importance of different mechanisms, and the amount of air transported between the boundary layer and the free troposphere. We have carried out studies where instead of initialising the model with tracer filling the boundary layer, the tracer has been emitted at the surface following emissions of CO. In this case we found that after 24 hours 30% of the tracer was in the free troposphere. This is less than the 50% of the previous case due to the two main reasons. Firstly because the tracer is emitted at the surface, where the vertical velocities are lower, therefore the tracer needs to be mixed to heights in the boundary layer where the vertical velocities are stronger before it can be exported out of the boundary layer. Secondly the emissions are emitted as a function of time, which is not directly comparable to the case discussed in the paper.

R. YAMARTINO: You indicated clouds as part of "future work", but I'd imagine that many of the vertical exchange mechanisms are built into the sub-grid scale (SGS) parameterisations of the delta x=13km model. Could you comment on what SGS are already in the model (e.g. clouds)?

E. DONNELL: When I mentioned clouds as future work it was in relation to the chemistry. In the unified model there is SGS parameterisation of convection. The model parameterises convection by a mass-flux scheme, and the non-advective transport of tracer by this convection is also represented using the same mass-fluxes. Convection is an important exchange mechanism, where in our study, for the frontal cases, of convection contributed approximately 20% of the total tracer that was transported to the free troposphere.

D. STEYN: Can you comment on your ability to resolve Richardson's number near the top of the boundary layer. I ask this because your model vertical resolution at that height (approximately 100m) is in the range 100 to 200m, a resolution so coarse that Richardson number must be very difficult to determine with any certainty.

E. DONNELL: Calculating the boundary layer top using Richardson number is consistent with the mixing scheme in the model, as the mixing coefficients are calculated as a function of Richardson number. Often the boundary layer top is not well defined, especially in the frontal cases, so it is not possible to determine the boundary layer top with greater accuracy than 100m anyway. Using Richardson number to calculate the top of the boundary layer is a common method.

NONLINEARITIES IN THE SULFATE SECONDARY FINE PARTICULATE RESPONSE TO NO_X EMISSIONS REDUCTIONS AS MODELED BY THE REGIONAL ACID DEPOSITION MODEL

Robin L. Dennis[1*], Gail S. Tonnesen[2], and Rohit Mathur[3]

[1]Atmospheric Sciences Modeling Division, MD-80, Air Resources Laboratory, NOAA, Research Triangle Park, NC 27711
[2]CE-CERT, University of California-Riverside, Riverside, CA 92507
[3]Environmental Programs, MCNC, Research Triangle Park, NC 27709

INTRODUCTION

The production of many pollutants in the atmosphere is driven by the oxidizing capacity of the atmosphere. Therefore, pollutants like ozone, acid rain and fine particles are connected because the same oxidant fields that determine the oxidizing capacity are involved in the production of these pollutants (Seinfeld and Pandis, 1998). A one-atmosphere perspective of air pollution conceptually recognizes that there is a connection (Meng, et al., 1997; Dennis, 1998).

Regional reductions in NO_X emissions, principally from utility plants, are a focus for reducing urban and regional ozone (O_3) concentrations in the United States (U.S. EPA, 1998). In current regulatory analyses of benefits and costs of proposed NO_X emissions reductions for O_3 control, the change in fine particulate concentrations is also taken into account, because of major health concerns regarding fine particulates. Fine particulates are defined as particles with a diameter less than 2.5 microns. Regional oxidant production is generally NO_X-limited. Therefore, significant reductions in NO_X emissions and thereby NO_X concentrations would be expected to change the regional oxidizing capacity of the atmosphere. From the one-atmosphere perspective, it is then natural to ask the question whether proposed NO_X reductions will have any feedback on other pollutants. Fine particulate sulfate, henceforth sulfate, is a potential candidate because its production is associated with three different oxidants: hydroxyl radicals (OH), hydrogen peroxide (H_2O_2), and ozone (O_3). In this paper we examine the feedback of NO_X emission reductions on the production of sulfate.

---

*On assignment to the National Exposure Research Laboratory, U.S. Environmental Protection Agency, Research Triangle Park, NC 27711.

CHEMISTRY OF SULFATE PRODUCTION AND THE OH CYCLE

Sulfate is formed by the oxidation of SO_2 through two main pathways: gas-phase oxidation of SO_2 by OH and aqueous-phase oxidation (in water droplets) of SO_2 by H_2O_2 and O_3 with subsequent evaporation of the droplets. While aqueous-phase oxidation of SO_2 can be limited by the amount of available oxidant, modeling analysis performed by the National Acid Precipitation Program (NAPAP, 1991) found that only an area along the upper Ohio River Valley is expected to have some oxidant limitation. Furthermore, results from the Regional Acid Deposition Model (RADM) showed that H_2O_2 is the dominant aqueous-phase oxidizer and that aqueous-phase oxidation is much more important than gas-phase oxidation in producing sulfate in the eastern U.S. (McHenry and Dennis, 1994).

A major oxidizing specie in the atmosphere is the hydroxyl radical, OH. The largest single source of OH for regional domains is the photolysis of ozone, O_3, by ultraviolet light in the presence of water. As shown in Figure 1, OH attacks hydrocarbons to produce organic peroxy radicals, RO_2. Organic peroxy radicals can react with an NO molecule and convert it to NO_2, and in the process, a peroxy radical, HO_2, will be formed. The peroxy radical can react with another NO molecule and convert it to NO_2, and in the process, the OH radical is regenerated or recreated, keeping the cycle going.

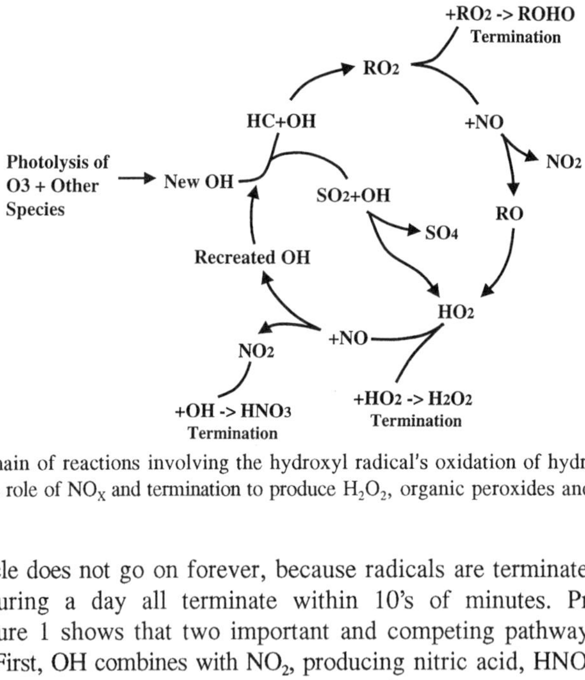

Figure 1. Cyclic chain of reactions involving the hydroxyl radical's oxidation of hydrocarbons and sulfur dioxide, showing the role of NO_X and termination to produce H_2O_2, organic peroxides and nitric acid, HNO_3.

The OH cycle does not go on forever, because radicals are terminated. Radicals that are produced during a day all terminate within 10's of minutes. Production equals termination. Figure 1 shows that two important and competing pathways of termination break the cycle. First, OH combines with NO_2, producing nitric acid, HNO_3, thus removing one radical. With high concentrations of NO_X, this pathway out-competes and inhibits the availability of radicals by removing many OH's very quickly. If NO_X is scarce, then this pathway competes poorly against the second termination pathway. In this second main termination pathway, the peroxy radical, HO_2, combines with itself, to form H_2O_2; and/or, organic peroxy radicals combine with peroxy radicals to form organic peroxides. In this instance two radicals are removed, yet, an important oxidant (H_2O_2) for the aqueous-phase formation of sulfate is produced.

Hence, NO_X emissions and sulfate production are linked through the photochemical cycle. In urban areas with relatively high NO_X emissions, radical formation and propagation is inhibited. If NO_X emissions are reduced by small to moderate amounts, O_3 and OH will increase. If SO_2 is available, then where the OH increases, an increased amount of SO_2 will be oxidized to SO_4 in the gas phase. This produces one type of "nonlinear" response: SO_4 increases to a reduction in NO_X emissions even when SO_2 emissions are unchanged.

In rural areas with relatively modest NO_X levels, O_3 and OH will decrease when NO_X is reduced. Percentage-wise, O_3 and OH will reduce less than NO_X and NO_2. The ratio of OH to NO_2 will increase and more of the OH budget will terminate as H_2O_2. This can lead to an increase in H_2O_2. If there is excess SO_2, then where H_2O_2 increases, an increased amount of SO_2 will be converted to SO_4 in the aqueous phase. This produces the most efficient "nonlinear" response in SO_4 to a reduction in NO_X emissions when there is no change in SO_2 emissions. Where OH decreases, then less SO_2 will be oxidized to sulfate in the gas-phase, producing a proportionate reduction in sulfate for a reduction in NO_X emissions. The nonlinear increases and proportionate reductions compete for influence on the overall change in sulfate concentrations levels.

MODEL DESCRIPTION AND EMISSIONS REDUCTION STUDIED

The model used in the analysis is the Regional Acid Deposition Model, RADM, which was developed for NAPAP and designed to treat both gas and aqueous phase oxidation of sulfur and nitrogen compounds and their wet and dry deposition with a high degree of mechanistic detail (Chang, et al., 1987, 1990). A full gas-phase photochemical mechanism (Stockwell, 1990) and a modestly detailed aqueous-phase mechanism (Walcek and Taylor, 1986) are employed in the model. RADM has also been extended to incorporate secondary inorganic fine particulate production based on the algorithms in the Regional Particulate Model (Binkowski and Shankar, 1995). The RADM was "instrumented" with process analysis (Jeffries and Tonnesen, 1994) to provide information on chemical rates in addition to species concentrations. The aggregation methodology that uses a statistical sample of synoptic transport categories (Brook, et al., 1995a and 1995b) is employed to calculate seasonal and annual averages of ambient air concentrations and wet and dry deposition associated with representative, long-term meteorology. The RADM employs a region-wide 80-km grid over eastern North America from the southern tip of Florida to James Bay and the Texas Gulf Coast to Nova Scotia. A nested 20-km grid to provide better detail of photochemical processes is also available, but the runs have not yet been analyzed.

Figure 2 shows the 80-km RADM prediction of average daily sulfate concentrations for the ozone season (May - September) using representative meteorology and 1990 emissions. Compared to CASTNet data averaged over 1988-1992, the RADM has a tendency to overpredict the seasonal average. Over the Ohio River Valley and northward, predictions are within 50% of measured concentrations. For two sites in the southern part of the domain the overprediction reaches a factor of two. When compared directly to a two-month average in the spring of 1990, RADM predictions (not using aggregation) of daily sulfate are basically unbiased and within +/- 20% of observed data. Thus, it is suspected that the long-term statistics on non-precipitating clouds used in the aggregation is not representative of the southern U.S. around the 1990's.

The U.S. Environmental Protection Agency has been assessing the effectiveness of reducing NO_X emissions from electric power plants (utilities) during the ozone season to reduce regional and urban levels of O_3 in the eastern United States (U.S. EPA, 1998). The Agency proposal to reduce utility emissions in 22 states is termed the NO_X SIP Call. In

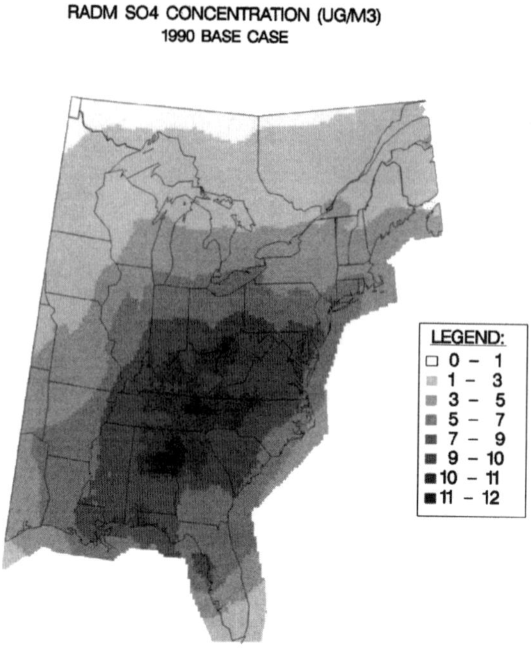

Figure 2. Ozone season average concentration of daily sulfate concentration predicted by RADM for representative meteorology (using the aggregation method) and 1990 emissions. Note the nonuniform scale.

the base case modeled here, the utility NO_X emissions during the ozone season account for 27% of the total NO_X emissions eastern U.S. In the control case modeled here, the utility emissions are reduced by 62% and the total eastern U.S. NO_X emissions are reduced by 17%. In local areas of high utility emissions of both NO_X and SO_2, such as the Ohio River Valley, the local NO_X concentration reductions are greater than 35%. Thus, the NO_X emissions reductions in this study are not spatially uniform, but rather are focused spatially. Because SO_2 emissions do not remain constant due to the effects of trading and control changes, even when only NO_X controls are modeled, the SO_2 emissions were forced to be identical between the Base and Control Cases to avoid unnecessarily complicating the analysis and masking the potential signal.

RESULTS OF THE MODELING ANALYSIS

The 80-km RADM results are analyzed first to provide an overall perspective about the nonlinear system. Since the NO_X emissions reductions are planned only for the ozone season, or warm season, we examine the sulfate changes only for that period.

Figure 3 shows the location of the major reductions in seasonal NO_X from a termination perspective, examining changes in HNO_3. The reductions are localized regionally with the largest change, greater than 35%, occurring in the upper Ohio River Valley. This region also has high SO_2 emissions. There is also a broad reduction of HNO_3 greater than 25% surrounding the Ohio River Valley and extending south across Tennessee, Alabama, Georgia and South Carolina with small areas of embedded reductions greater than 35%.

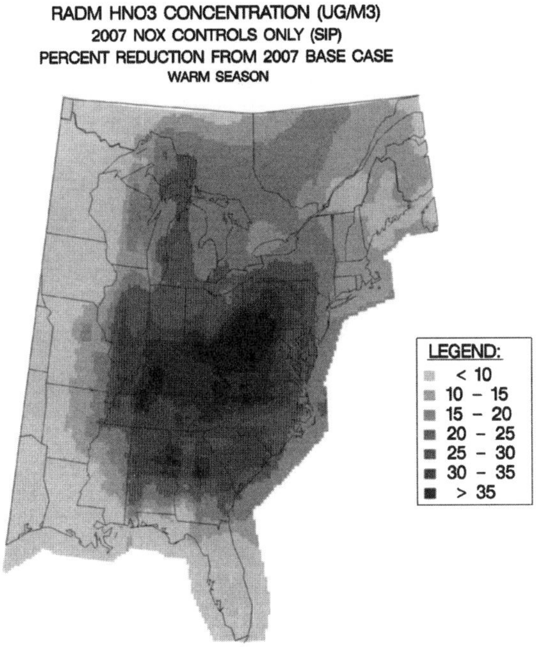

Figure 3. Percent change in average ozone-season daily HNO_3 concentrations due to the power plant NO_X emissions reductions, as predicted by RADM.

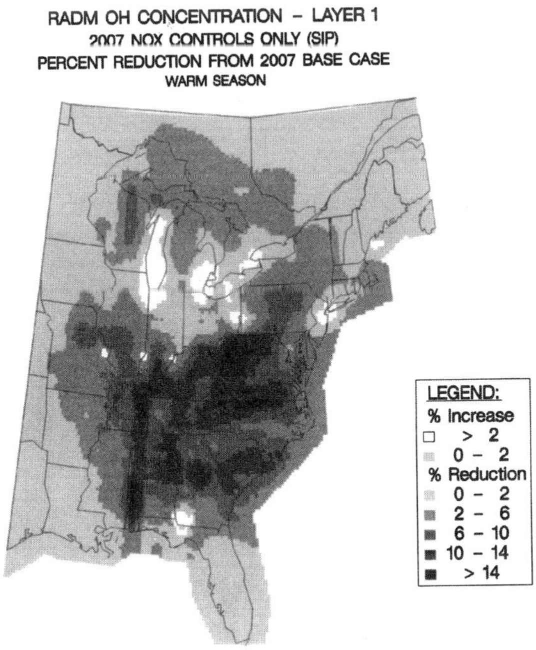

Figure 4. Percent change in average ozone-season daily hydroxyl radical (OH) concentrations due to the power plant NO_X emissions reductions, as predicted by RADM.

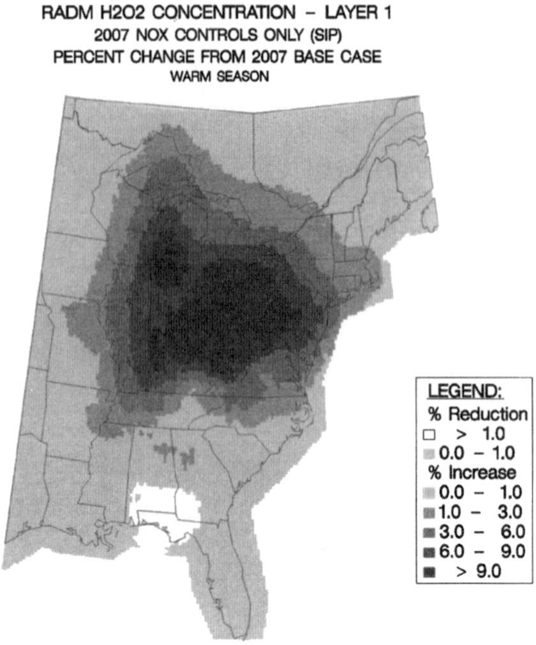

Figure 5. Percent change in average ozone-season daily hydrogen peroxide (H_2O_2) concentrations due to the power plant NO_X emissions reductions, as predicted by RADM.

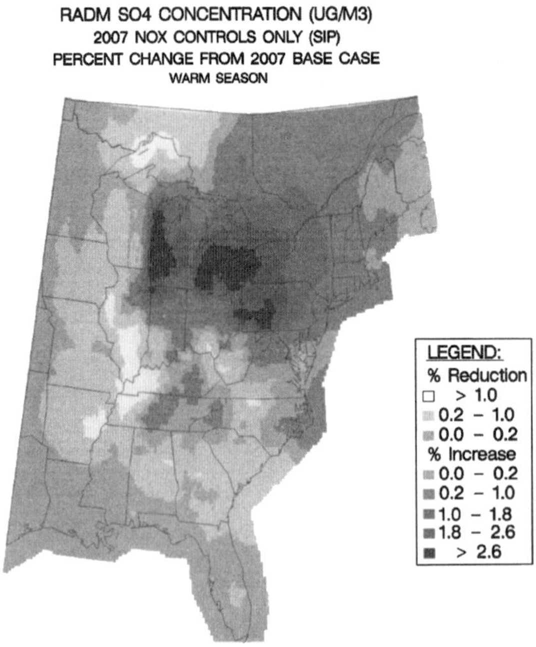

Figure 6. Percent change in average ozone-season 24-hour average sulfate concentrations predicted by RADM for the power plant NO_X emissions reductions.

Figure 4 shows the percent change in mixed layer seasonal OH concentrations. The largest reductions are south of the Ohio River Valley across West Virginia and western Virginia and in the south along a line between Birmingham, Alabama and Atlanta, Georgia. There are increases in OH around Chicago, Detroit Toronto, New York and west of Pittsburgh. The pattern of percent change in ozone is similar. The main differences are that the reductions are smaller and the increases are comparable or smaller. The reductions in seasonal ozone are about one-third those of OH, i.e., where there is a decrease in OH of greater than 14%, the ozone decrease is greater than 5%. The increase in daily average ozone for Chicago is comparable to and for Detroit less than their respective OH increases, but over New York average ozone decreases slightly. The change in the seasonal oxidant fields is smaller than the change in NO_X termination or nitric acid.

Figure 5 shows the percent change in seasonal H_2O_2 concentrations. The largest increases are over much of Ohio and western Pennsylvania, covering much of the area of HNO_3 reduction along the upper Ohio River Valley. Across the Florida panhandle there is a reduction in H_2O_2 because the entire system is becoming less reactive. There is very little change in H_2O_2 in the Birmingham and Atlanta areas.

Figure 6 shows the percent change in seasonal sulfate. The largest increase, greater than 2.6% is around the Pittsburgh area, over Lake Michigan (Chicago) and over the Detroit area and Lake Ontario. The region of sulfate increase is downwind of the area with the large SO_2 emissions and increase in H_2O_2 concentrations. In lower West Virginia and and the Atlanta area the reduction in OH has the larger influence and sulfate concentrations decrease slightly.

DISCUSSION

There are several pathways in which changes in NO_X emissions can feed back and affect the production of sulfate. Changes in OH (increases and decreases) and H_2O_2 were both important. In some areas (e.g., Detroit) both oxidants increased and sulfate increased. In other areas the oxidants went in opposite directions: H_2O_2 prevailed in western Pennsylvania, increasing sulfate, and OH prevailed over West Virginia, decreasing sulfate. We can explain the feed backs based on our understanding of the photochemical system, but we need the model to put it together in detail. In this analysis the change in particulate nitrate is very small, so the change in sulfate dominates any change in inorganic fine particulate mass. While modeled changes in sulfate were not large, these changes can be important because in the U.S. fine particulate health effects are calculated in terms of human mortality, whereas ozone effects are in terms of human morbidity. Very roughly, a 0.1 $\mu gm/m^3$ reduction in annual median fine particulate produces approximately the same human health monetary benefit as a 10-12 ppb reduction in 8-hour average ozone (U.S. EPA, 1999). Hence, it is useful to understand and account for these feedbacks.

The 80-km model is not expected to resolve adequately the feed backs for strong sources like large urban areas or major utility emissions. We plan to repeat this analysis at 20-km resolution over a domain containing the major changes. We find from analysis of modestly different spatial patterns of similar regional utility NO_X emissions reductions that the feed backs can be sensitive to these differences. In future work we believe it is important to examine the impact of greatly different spatial patterns of the NO_X emissions changes on the feed backs.

ACKNOWLEDGMENT

This paper has been reviewed in accordance with the U.S. EPA's peer review policies and approved for publication. Mention of trade names or commercial products does not constitute endorsement or recommendation for use.

REFERENCES

Binkowski F. S. and U. Shankar, 1995. The regional particulate model 1. Model description and preliminary results, *J. of Geophysical Research*, **100**, 26191-26209.

Brook, J. R., P. J. Samson, and S. Sillman, 1995a. Aggregation of selected three-day periods to estimate annual and seasonal wet deposition totals for sulfate, nitrate, and acidity. Part I: A synoptic and chemical climatology for eastern North America, *J. of Applied Meteorology*, **34**, 297-325.

Brook, J. R., P. J. Samson, and S. Sillman, 1995b. Aggregation of selected three-day periods to estimate annual and seasonal wet deposition totals for sulfate, nitrate, and acidity. Part II: Selection of events, deposition totals, and source-receptor relationships, *J. of Applied Meteorology*, **34**, 326-339.

Chang, J. S., R. A. Brost, I. S. A. Isaksen, S. Madronich, P. Middleton, W. R. Stockwell, and C. J. Walcek, 1987. A three-dimensional Eulerian acid deposition model: Physical concepts and formation, *J. of Geophysical Research*, **92**, 14681-14700.

Chang, J. S., F. S. Binkowski, N. L. Seaman, J. N. McHenry, D. W. Byun, P. J. Samson, W. R. Stockwell, C. J. Walcek, S. Madronich, P. Middleton, J. E. Pleim, and H. H. Lansford, 1990. *The Regional Acid Deposition Model and engineering model*, State of Science & Technology Report No. 4, National Acid Precipitation Assessment Program, Washington, D.C.

Dennis, R. L, 1998. The Environmental Protection Agency's third generation air quality modeling system: An overall perspective, in *Preprints of the 10th Joint conference on the Applications of Air Pollution Meteorology with the A&WMA*, 11-16 January 1998, Phoenix, Arizona, American Meteorology Society, Boston, Massachusetts, 255-258.

Jeffries, H. E. and S. Tonnesen, 1994. A comparison of two photochemical reaction mechanisms using mass balance and process analysis, *Atmospheric Environment*, **28**, 2991-3003.

McHenry, J. N. and R. L. Dennis, 1994. The relative importance of oxidation pathways and clouds to atmospheric ambient sulfate production as predicted by the Regional Acid Deposition Model, *J. of Applied Meteorology*, **33**, 890-905.

Meng, Z., D. Dabdub and J. Seinfeld, 1997. Chemical coupling between atmospheric ozone and particulate matter, *Science*, **277**, 116-119.

NAPAP, 1991. *Nation Acid Precipitation Assessment Program: 1990 integrated assessment report.* National Acid Precipitation Assessment Program, Washington, D.C.

Seinfeld, J. and Pandis, 1998. *Atmospheric chemistry and physics: From air pollution to climate change*, John Wiley & Sons, Inc. New York.

Stockwell, W. R., Middleton, P., Chang, J. S., and Tang, X., 1990. The second generation Regional Acid Deposition Model chemical mechanism for regional air quality modeling, *J. Geophysical Research*, **95**, 16,343-16,367.

U.S. Environmental Protection Agency (U.S. EPA), 1998. *Regulatory impact analysis for the regional NOx SIP Call*, U.S. Environmental Protection Agency, Office of Air Quality Planning and Standards, Research Triangle Park, NC.

U.S. Environmental Protection Agency (U.S. EPA), 1999. *The benefits and costs of the Clean Air Act 1990 to 2010, EPA report to Congress*, U.S. EPA Report No. EPA 410-R-99-001, U.S. Environmental Protection Agency, Office of Air and Radiation and Office of Policy, Washington, D.C.

Walcek C. J., and G. R. Taylor, 1986. A theoretical method for computing vertical distributions of acidity and sulfate production within cumulus clouds, *J. of Atmospheric Science*, **43**, 339-355.

DISCUSSION

S.T. RAO: There is strong seasonality in the PM concentrations. It appears that you are presenting results for the warm (ozone) season only. Have you looked at the impacts on annual PM concentrations?

R. L. DENNIS: Yes. The health impact dose response curves are in terms of changes in the median concentration of fine particulates. The original analysis was performed for annual median PM. The warm season concentrations basically determine the median concentrations. It turns out that the change in warm season average inorganic PM is very close to the change in annual median inorganic PM. Thus, the warm season average changes that were presented are good surrogates for the change in annual median concentrations and it makes sense to focus the explanation for the changes on what is happening in the warm season.

S. T. RAO: Also, these is a phase lag between ozone maximum and PM maximum in the time series of these pollutant concentrations. Have you looked at these differences regarding the efficacy of emissions reduction options?

R. L. DENNIS: The seasonal average changes in PM are more associated with a variety of conditions, including the presence of clouds. Sulfate is produced by cloud process and clouds can get in the way of ozone production and SO_2 must compete for the OH. So one does not necessarily expect high sulfate to be tightly correlated with high ozone. The aggregation method samples this variety across the warm season. The non-linear response of sulfate appears to be more associated with moderate ozone than with maximum ozone, according to the model. That is because the NO_x disbenefit is stronger for moderate ozone days than it is for maximum ozone days. In the latter, the photochemical processing is faster, taking the system farther into the NO_x-limited regime and thus mitigating the NO_x disbenefit.

R. WARREN: Considering the disbenefits of reducing NO_x in terms of the cases you presented where SO_4 increases, to what extent might this be compensated for by the reduction in NO_3 aerosol?

R. L. DENNIS: The amount of nitrate is determined by thermodynamic relations and competition for ammonia. In the eastern US the sulfate is not at all close to being neutralized so the nitrate levels are quite low. While the pattern of the percent change in nitrate is broadly similar to the percent change in HNO_3, the absolute levels of change are very small. Also, the ammonia associated with the inorganic system is more influenced by the increase in sulfate than by the decrease in nitrate and ammonium concentrations increase overall in the areas where the sulfate increases, further reducing the influence of the reductions of nitrate. The net result is that the change in nitrate only compensates about 5%, possibly up to 10%, of the change in sulfate. Thus, the non-linear response of sulfate production comes through the inorganic fine particulate system with very little attenuation for conditions representative of the eastern US.

SENSITIVITY OF OZONE AND AEROSOL PREDICTIONS TO THE TRANSPORT ALGORITHMS IN THE MODELS-3 COMMUNITY MULTI-SCALE AIR QUALITY (CMAQ) MODELING SYSTEM

Daewon W. Byun and Jonathan E. Pleim

Atmospheric Sciences Modeling Division, NOAA, Research Triangle Park, NC 27711, USA (on assignment to the National Exposure Research Laboratory, U.S. Environmental Protection Agency)

INTRODUCTION

Recently, the Office of Research and Development (ORD) of the U. S. Environmental Protection Agency (EPA) developed and released to the public the Models-3 Community Multi-scale Air Quality (CMAQ) modeling system. Models-3 represents the computational framework and CMAQ refers to the multi-scale (urban and regional) and multi-pollutants (ozone, aerosol, and acid/nutrient depositions) air quality model. CMAQ's science subsystems are composed of the MM5 meteorological model, Models-3 Emissions Processing and Projection System (MEPPS), the CMAQ chemistry transport model (CCTM), and several interface processors. The interface processors include a meteorology-chemistry interface processor, an initial condition and boundary condition processor, a photolytic rate processor, an emissions-chemistry interface processor, and a chemical mechanism reader. Theoretical formulations and numerical algorithms of CMAQ can be found in Byun and Ching (1999).

One of the goals of the Models-3 CMAQ system is to develop a community modeling paradigm that allows continuous improvement of the one-atmosphere modeling capability in a unified fashion. The system is designed to be flexible so that different levels of model configuration can be achieved. Key input components such as the meteorological model and the emissions processor can be replaced, or different science process modules in the CMAQ CTM (CCTM) can be selected to build appropriate air quality models according to the user's needs. CMAQ's modular design promotes incorporation of several sets of science process modules representing different algorithms and parameterizations of physical and chemical processes. For example, there are several different atmospheric transport algorithms available with CCTM. In principle, the atmospheric transport processes, except for the sub-grid scale cloud mixing, are divided into advection and diffusion processes. They are further separated both in the horizontal and vertical directions, respectively. One objective of the present study is to demonstrate benefit of the modularity of the CMAQ system. Another is to assess the effects of different transport algorithms on air quality predictions.

CONFIGURATIONS OF CMAQ TESTED

The meteorological model used is the MM5 version 2.10 with the Kain-Fritsch cloud parameterization, Dudhia surface radiation, Blackadar high-resolution PBL options, and four-dimensional data assimilation with ECMWF analysis data. Emissions data are from the EPA's 1995 summer weekday inventory, with continuous emissions monitoring (CEM) data. CMAQ can be configured with various degrees of complexity and choices of optional science modules. Many transport modules are available or in the process of being incorporated in CCTM.

Here we configured CCTMs with combinations of the following choices:(1) two flux-based advection algorithms–Colella and Woodward (1987) Piecewise Parabolic Method (PPM) and Bott (1988) scheme (BOTT); (2) two vertical diffusion methods–a local scheme based on the semi-implicit K-theory (EDDY) and a nonlocal scheme asymmetric convective model (ACM) by Pleim and Chang (1992); and (3) two different science processes where emissions are injected–in vertical diffusion and chemistry modules, respectively. Table 1 summarizes the different CCTM configurations tested. Other key specifications of the CCTM versions used in the simulations were:

- Mass conservation adjustment (Byun 1999) .
- Horizontal diffusion with scale dependent diffusivity.
- QSSA gas-phase reaction solver (Young et al. 1993) .
- Convective cloud mixing and explicit and sub-grid aqueous-phase reactions.
- Modal approach aerosol size distribution and dynamics .
- Carbon Bond 4 (CB-4) chemistry mechanism

Refer to Byun and Ching (1999) for the detailed descriptions of these modules.

In the present study, plume-in grid module was not activated in the chemical transport model. In the following, we summarize results of preliminary analyses of CMAQ runs for an ozone episode period. CCTMs with different configurations were run for the 11-day period, July 6-July 16, 1995 at 36 km resolution 75x69 cell domain covering the eastern half of USA. The initial and boundary conditions used are constant values representing tropospheric background conditions. As with other Eulerian modeling exercises, the simulation results for the first few days are not expected to be realistic. However, to compare the impacts of using different model configurations, all days were analyzed to demonstrate differences in the predicted time series. Comparisons of model predictions with observations were also performed, but only for the July 11-15 time period. The period 11 - 15 July 1995 is known for a high ozone episode occurred in the Northeast Ozone Transport Region, which covers mid Atlantic states and New England. An overview of synoptic meteorological conditions and local flow characteristics during the episode was presented by Byun et al. (1998).

COMPARISON AMONG CCTMS WITH DIFFERENT CONFIGURATIONS

Figures 1 and 2 provide evolution of ozone and NO_y (the sum of all the nitrogen species) concentrations for different CMAQ configurations relative to the base runs. Some of the photochemical characteristics can be studied by comparing the ratio of ozone produced to the reacted nitrogen species, represented as NO_z (equal to NO_y minus NO_x). Ozone concentrations in cells with higher $\Delta O_3 / \Delta NO_z$ ratios are more likely to be sensitive to changes in NO_x emissions, whereas lower ratios imply a greater sensitivity to emissions of volatile organic compounds (VOC). Figure 3 summarizes the O_3-NO_z scatter diagram at 21 UTC, July 13, 1995. We use these figures to highlight differences in CMAQ simulations among the different model configurations.

Table 1. Summary of tested CMAQ chemistry transport model configurations at 36-km grid resolution.

CMAQ CTM	Advection Algorithm	Vertical Diffusion Algorithms	Emissions Injected
Base (S36)	Piecewise Parabolic Method (PPM)	Eddy Diffusion	In Vertical Diffusion
A36 (without aerosol)	Bott Scheme	Asymmetric Convective Mixing (ACM)	In Vertical Diffusion
B36	Bott Scheme	Eddy Diffusion	In Vertical Diffusion
C36	PPM	ACM	In Vertical Diffusion
D36	PPM	Eddy Diffusion	In Chemistry
E36	PPM	ACM	In Chemistry
F36	Bott Scheme	Eddy Diffusion	In Chemistry

First, we discuss the differences between the CMAQ runs with the two diffusion algorithms. In meteorological models, a non-local scheme may be used to simulate the over-turning of large scale turbulence eddies more realistically than the local scheme (EDDY) for the convective atmosphere. Here, we tested if the nonlocal modeling can effectively be used for simulating mixing of precursor and secondary pollutant species in photochemical air quality models. For this test, CMAQ was configured with a simple nonlocal mixing scheme ACM for the convective conditions and EDDY for stable conditions. The results in Figures 1 and 2 show that CCTMs with ACM produce higher ozone and NO_y concentrations. ACM mixes NO_x and VOC emissions much more rapidly, thus resulting in more vigorous photochemical reaction. VOC emissions are mostly associated with surface sources and NO_x emissions are from both the surface (mobile and urban sources) and elevated sources. Therefore VOC emissions injected into the lowest model layer are moved to the upper boundary layer very quickly by the nonlocal scheme, thus changing the relative composition of the nitrogen and VOC compounds in the simulated air. Figure 3 also demonstrates this. The regimes controlled by the higher O_3/NO_z ratio (say, 9:1 slope) are similar to each other for all the configurations. However, regimes covered by the lower slope (say, 3:1 slope) are wider for the CCTMs with ACM (Figures 3 A, C and E) than those with EDDY (Figures 3 B, D, F and S). Compared with surface ozone measurements, the concentrations with ACM are generally too high (see Figure 5). We suspect that this is due to the incoherent representation of the ACM mixing algorithm and the emissions in the system. Note that in meteorological models, heat and moisture exchanges between the surface and atmosphere are represented as fluxes, not as direct injections from heat and moisture sources. In CCTM, pollutants are directly injected into the cell as the molar fluxes. Treatment of isoprene emission, for example a species that may be affected by the turbulence fluxes, in such a way may not be adequate. Further research is needed to reconcile these differences.

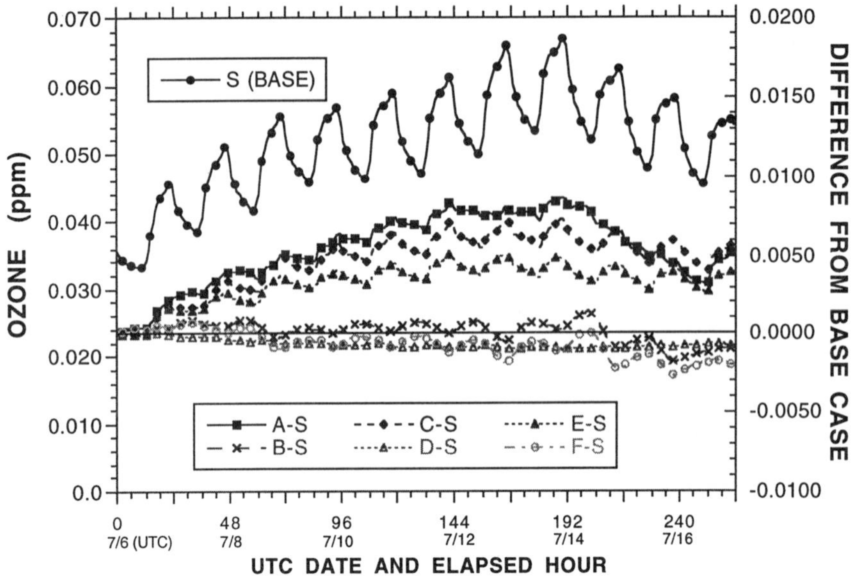

Figure 1. Time series of mixed-layer domain average ozone concentrations for the base case (S) and the difference between the each sensitivity run and the base case for CCTMs A to F.

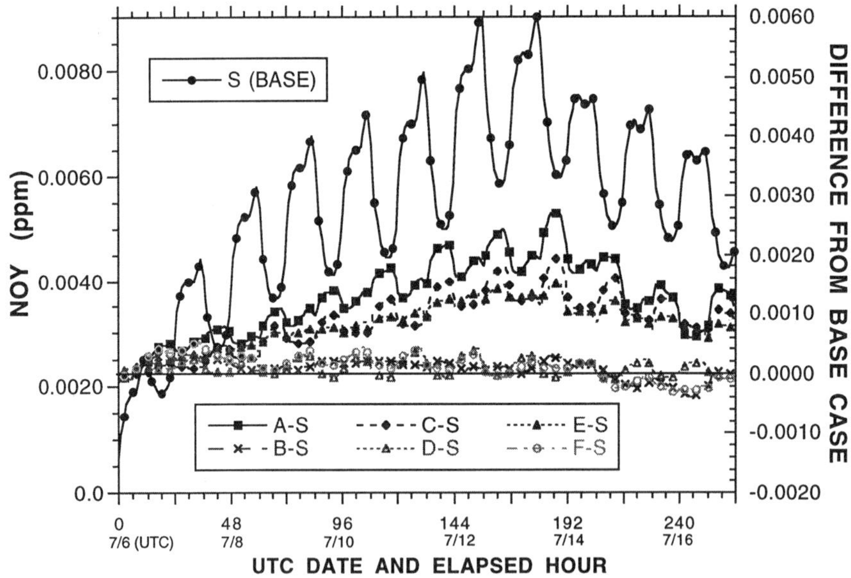

Figure 2. The same as Figure 1, but for NO_y concentrations.

Second, we discuss the differences between the CMAQ runs with the two advection algorithms, PPM and BOTT. PPM is absolutely positive definite and monotonic while the

Bott scheme is non-monotonic and could generate unwanted local extreme values for trace species such as aerosol number density that have large concentration gradients. Numerical tests with CMAQ have revealed that PPM is much more robust for multi-pollutant simulations under various meteorological and chemical conditions. Although simulations with gas-phase species alone were satisfactory, the runs with aerosol dynamics with BOTT and the ACM modules were not successful. Apparently, a minor problem with CMAQ's BOTT module that replaces very small negative values with zero causes inadvertent re-initialization of aerosol number concentrations in certain cells. In comprehensive AQMs, the choice of advection algorithm can affect the simulated efficiency of photochemical ozone production by modulating pollutant mixtures. Analysis results (not shown here) show that the Bott scheme is less diffusive than PPM, and thus maintains peaks of concentrations better. Even for the domain average concentrations, lines A-S vs. C-S in Figure 1 show generally higher ozone concentration for BOTT except for last two days. However, differences in the ranges of the two O_3/NO_z slope regimes between PPM and BOTT are small (see Figures 3-A vs. 3-C, 3-D vs. 3-F, and 3-B vs. 3-S pairs).

Third, we compare CMAQ results for the cases where emissions were injected either in the vertical diffusion module or in the gas-phase chemical reaction module. This modeling construct affects CMAQ results due to the different titration rates of ozone with fresh NO emissions, thus changing the propagation of radicals at different stages of photochemical evolutions. The ozone concentrations were lower for the runs with emissions in chemistry than those with emissions in the vertical diffusion routines. For the species that have short chemical time scales competing with the transport time scales, the concentrations differences are substantial depending on where we inject emissions. Again, further studies are warranted to establish the proper configuration of process modules in Eulerian air quality models with the time-splitting approach and to determine optimal inter-process time steps.

Next, we present effects of different configurations on aerosol predictions for PM2.5 (particular matter with diameter smaller than 2.5 μm). In CMAQ PM2.5 concentration is estimated as the sum of all the Atiken mode and accumulation mode aerosol species. In the current implementation of CMAQ, aerosol precursor emissions are injected in the aerosol module. There were little differences in SO_2 and aerosol concentrations depending on which module the sulfur emissions were injected (results not shown here). However, the different transport modules produced some differences in aerosol concentrations, as demonstrated in Figure 4. At concentrations higher than 1.0 $\mu g/m^3$, aerosol concentrations corresponding to B36 (with BOTT advection scheme) are about 10% higher than the base case S36. For the C36 configuration (with ACM vertical diffusion), they are about 15% higher.

CONCLUSIONS

We have demonstrated that the choice of modules in transport processes interacts with other model configurations. Comparison with observations, especially with secondary species such as O_3, may not be sufficient to allow selection of the best modules. For example, we have compared first layer ozone concentrations with those from EPA's AIRS database. Figure 5 shows that the configuration F36 has least bias compared with observations. However, this alone is not sufficient to determine which transport algorithms are superior. Factors such as the representation of emissions inputs, the treatment of plume-in-grid, the use of different chemical mechanisms, the selection of different chemical solvers, and the model grid structure (i.e., vertical and horizontal resolutions and domain size), all contribute to different model results. Establishment of the best configuration of science process modules in a comprehensive AQM requires balanced representations of transport algorithms with other physical and chemical processes.

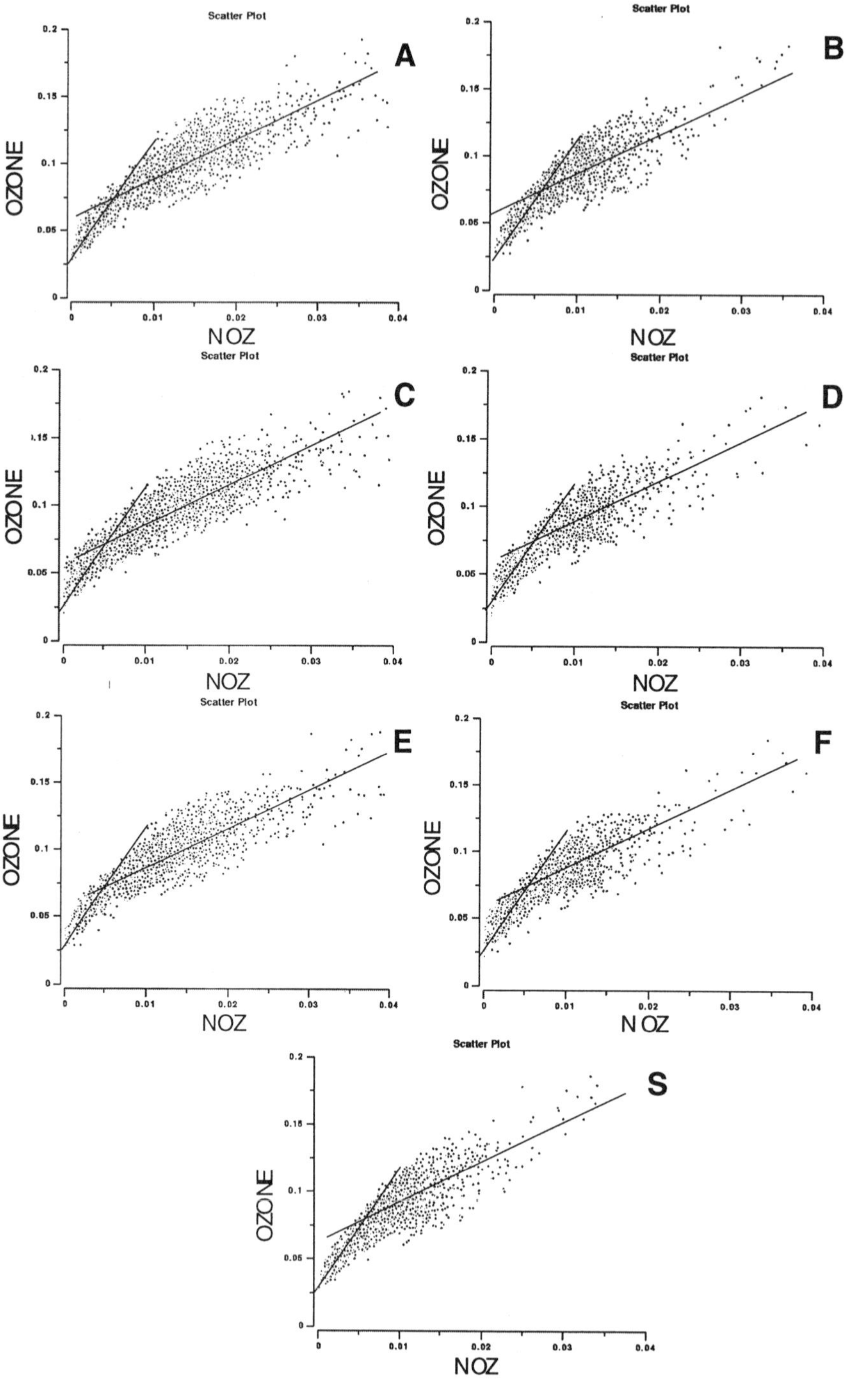

Figure 3. Ozone-NO_z relationships at layer 1 (surface) for different CMAQ configurations (A to F) at UTC hour 21, July 13, 1995. Reference lines in each plate represent 9:1 and 3:1 ozone to NO_z slopes, respectively.

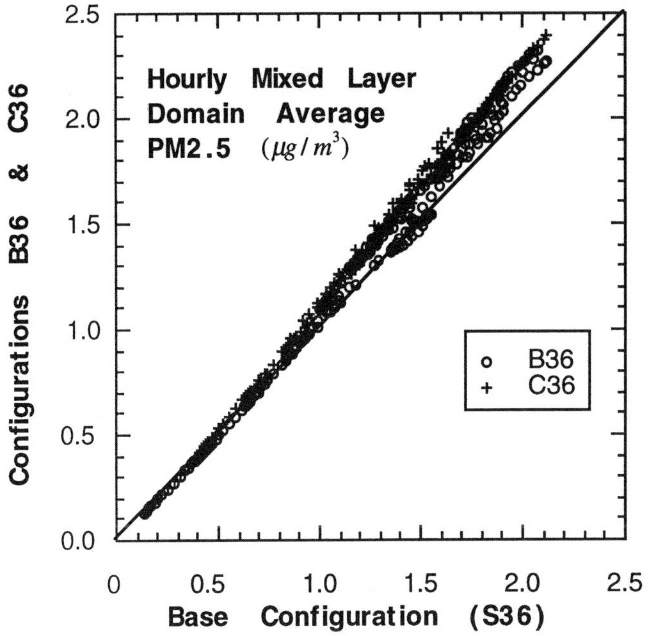

Figure 4. Comparison of hourly mixed layer PM2.5 concentrations (averaged over the simulation domain).

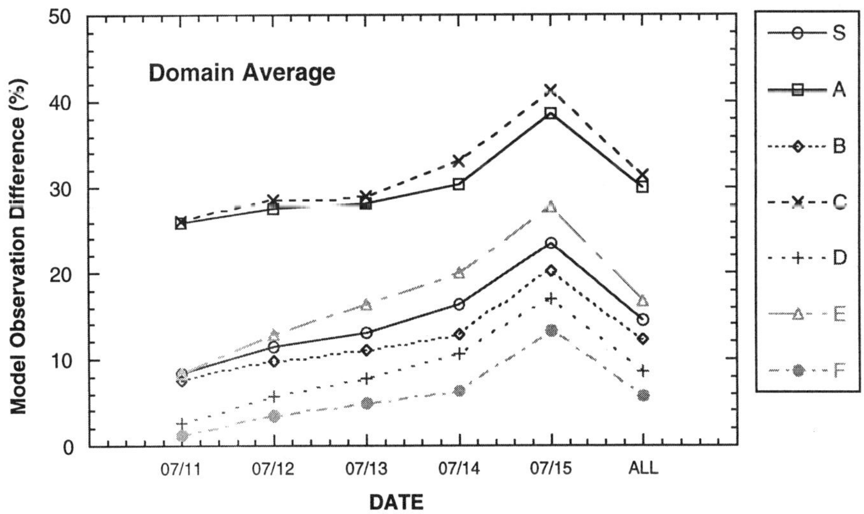

Figure 5. Comparison of surface ozone concentrations with AIRS database. All the configurations tested produced higher domain average than the observations.

The authors express their appreciation to U.S. EPA's Models-3 CMAQ science team members. Also, thanks are extended to Mr. Jerry Gipson and Dr. Avi Lacser for their thorough reviews and U.S. EPA's National Environmental Supercomputing Center for providing necessary computation resources for the study. This paper has been reviewed in accordance with the US Environmental Protection Agency's peer and administrative review policies and approved for presentation and publication. Mention of trade names or commercial products does not constitute endorsements or recommendation for use.

REFERENCES

Bott, A., 1989, A positive definite advection scheme obtained by nonlinear renormalization of the advection fluxes. *Mon. Wea. Rev.* **117**, 1006-1015.

Byun, D.W., J. Young., G. Gipson., K. Schere, J. Godowitch., J. Pleim, F. Binkowsk., S. Roselle, B. Benjey, J. Ching., J. Novak, S. LeDuc, 1998, An urban air quality simulation with the community multi-scale air quality modeling system. The Second Urban Environment Symposium, Albuquerque, NM, Nov. 2-6, 1998. Amer. Meteor. Soc.

Byun, D.W., 1999, Dynamically consistent formulations in meteorological and air quality models for multi-scale atmospheric applications: Part II. Mass conservation issues. *J. Atmos. Sci.*, Vol 56, 3808-3820.

Byun, D.W. and J.K.S. Ching, ed., 1999, *Science Algorithms of the EPA Models-3 Community Multi-scale Air Quality (CMAQ) Modeling System*, NERL, Research Triangle Park, NC. [Available from National Exposure Research Laboratory, U.S. Environmental Protection Agency, Research Triangle park, NC 27711]

Colella, P. and P. R. Woodward, 1984, The piecewise parabolic method (PPM) for gas-dynamical simulations. *J. Comp. Phys.*, **54**, 174-201.

Pleim, J.E. and J. Chang, 1992, A non-local closure model for vertical mixing in the convective boundary layer. *Atmos. Env.*, **26A**, 965-981.

Young, J. O., E. Sills, and D. Jorge, 1993, *Optimization of the Regional Oxidant Model for the Cray Y-MP*. EPA/600/R-94-065. Research Triangle Park, NC. [Available from National Exposure Research Laboratory, U.S. Environmental Protection Agency, Research Triangle park, NC 27711]

DISCUSSION

A. ELIASSEN: We have seen that the nonlinearities of the advection schemes influence the calculated concentrations of the chemical species. Would you say that this would influence the model's response to emission changes? In other words, can we be reasonably certain that the chemical nonlinearities of the model is not adversely affected by the nonlinearities of the advection scheme?

D. W. BYUN: Thank you for the question. Model simulation results with the piecewise parabolic method (PPM), for example, show much less nonlinearity than the previous schemes, such as the Smolarkiewicz iterative upwind scheme. Considering other uncertainties in the system, it may not cause serious effects on nonlinearity in chemistry. On the other hand, the differences in the diffusive characteristics of advection schemes may influence the chemical nonlinearity because the relative composition of NO_x, and VOC mixture in a cell can be modified.

S. T. RAO: Have you examined the relative contributions to the model to observation uncertainty of advection versus vertical mixing schemes? Is the uncertainty introduced by vertical mixing larger than that introduced by the advection scheme used in the model? If that, in fact, is the case, do you expect that these would be a significant uncertainty in the relative efficiency VOC and NO_x emission reduction options? What are the implications of these modeling uncertainties to emission management decisions? In other words, how confident can we be in using these models in a regulatory setting?

D. W. BYUN: Although I have not analyzed the issue in depth, present modeling results show that ozone and other pollutant concentrations are affected significantly by the vertical diffusion schemes used. As long as wind flow is well characterized by the meteorological model, uncertainty caused by advection is much less than the one by vertical mixing process. Your question on the implication of these modeling uncertainties to emissions management decision is a very serious one. There are situations that we may have consistent model responses to emission control strategies regardless of which diffusion algorithm was used. On the other hand, when model responds differently to different vertical algorithms, we must be very careful to draw emissions management decisions.

P. SEIBERT: Also the Bott scheme cannot be considered to give true solutions. Did you think about creating a real reference run with considerably higher resolutions, especially in the vertical, to find out how much you are off the true solution?

D. W. BYUN: I have compared numerical results with analytical solutions for advection and diffusion processes independently as a module verification exercise. However, in the complex modeling form, I have only tested the numerical convergence at 12 and 4 km horizontal grid resolutions both with direct MM5 simulation at 4 km and with interpolated meteorology from 12 km MM5 results. I could extend the test for higher vertical resolution as well. But it is really difficult to generate a reference meteorological input when we change the vertical resolution of MM5. Also, there are many problems in interpolating meteorological parameters and redistributing area emissions vertically in the chemistry transport model.

INTEGRATED REGIONAL MODELING OF OZONE, PARTICULATE MATTER, AND DEPOSITION OVER THE EASTERN UNITED STATES: COMPARISONS OVER WET AND DRY EPISODES

James Wilkinson, James Boylan, Talat Odman, and Armistead Russell

School of Civil and Environmental Engineering
Georgia Institute of Technology
Atlanta, GA 30332-0512

INTRODUCTION

As part of the Southern Appalachians Mountains Initiative (SAMI), the effects of emissions controls on air quality are being assessed using an integrated, "one-atmosphere" modeling approach. The modeling system consists of RAMS for meteorology, EMS-95 for emissions and the Urban-to-Regional Multiscale (URM) model for transport and chemistry. With this system, the evolution of primary and secondary gas-phase and aerosol-phase pollutants, can be followed and both PM_{10} and $PM_{2.5}$ can be simulated. EMS-95 has the ability to inventory size- and species-resolved aerosol emissions. URM, as applied to this project, includes a wet deposition module, an aerosol module, and a direct sensitivity analysis module, which simultaneously yields three-dimensional sensitivities of pollutant concentrations and deposition fluxes to emissions. Nine episodes (each seven to ten days in duration) will be simulated over the eastern United States using a fine grid of 12 km in and around the SAMI states, and successively coarser grids, up to 192 km near the boundaries of the domain. Here, four episodes are studied, including two periods that that experienced significant precipitation levels and high wet deposition rates and two periods that tended to have less rain. The results are used to develop seasonal and annual air quality metrics and their response to emission controls, for an assessment of the visibility and acid deposition in the region. In this paper, URM is evaluated for its ability to accurately simulate ozone and $PM_{2.5}$ concentrations, as well as, the deposition of acids in four episodes. When compared to data from IMPROVE sites, the model predicted concentrations for most fine aerosol species with less than 50 percent normalized error. The wet deposition amounts are in good agreement with the NADP observations when the high spatial and temporal variability of precipitation is factored into the performance metrics. While not always the case, the model results tend to be better for dry periods rather than wet periods. Part of the reason is believed to be the over-scavenging of aerosols when there is rain. This could be due to either the parameterizations of the scavenging (e.g., wash-out ratio), the assumed fraction of the grid where rain exists, or the prediction of more rain than was observed. However, the all-around results found good comparison between observed and simulated values. These results suggest that the modeling system can be confidently used for assessing control strategy impacts. In this study, the domain has been divided into eight source regions, and using the direct sensitivity analysis feature of URM, the relationships between the aerosol levels in the southern Appalachians and the emissions from various regions have been studied. It was observed that SO_2 emission reductions from different regions displayed very different levels of impact at various sites within the SAMI region.

Studies that have been conducted in national parks and national forest wilderness areas of the southern Appalachian Mountains have documented adverse air pollution effects on visibility, streams, soil, and vegetation. Although it is known that air pollution levels which currently affect park and wilderness resources come from existing sources of pollution - large and small, mobile and stationary, near and distant, the relative contribution of each source type to the regional air pollution problem is not well quantified. The 1990 Clean Air Act Amendments (CAAA) required major reductions in airborne pollutants. Although the reductions are expected to produce air quality improvements, there is uncertainty whether the results will be enough to protect and preserve the ecosystems and natural resources of the southern Appalachians, especially in Class I areas.

The Urban-to-Regional Multiscale (URM) model[5] has been applied to the SAMI modeling domain to characterize the air pollution formation processes that affect air quality in the southern Appalachian Mountains. The URM model results will be used to assess emission control strategies designed to reduce atmospheric pollutants in the Class I areas.

The URM model is a three-dimensional Eulerian photochemical model that accounts for the transport and chemical transformation of pollutants in the atmosphere. The URM model uses a finite element, variable mesh transport scheme[6]. The gas-phase reaction kinetics are calculated using the SAPRC chemical mechanism[2]. This mechanism has been updated to account for a more accurate isoprene chemistry among other reactions. URM uses variable size grids in its horizontal domain to effectively capture the details of pollution dynamics without being computationally intensive. The URM model has been expanded to include aerosol dynamics and wet deposition scavenging processes. The ISORROPIA aerosol module[4], the Reactive Scavenging Module[1] for acid deposition, and the Direct Decoupled Method[8] for gas and aerosol sensitivities have been incorporated into the URM framework to produce an integrated, "one atmosphere" airshed model.

AEROSOL MODULE

The aerosol module is capable of simulating concentrations of all major primary and secondary components of atmospheric PM. There are three groups of aerosol species that are considered in the aerosol routine. These include inert species, inorganic equilibrium species, and organic species. The inert species include magnesium, potassium, calcium, elemental carbon, and an "other PM" group which includes all other inert PM species. The inorganic equilibrium species include sulfate, nitrate, ammonium, sodium, chloride, and hydrogen ion. The organic aerosols are represented by a lumped species which is the sum of numerous condensable organics resulting from the oxidation of organic gases.

A sectional approach is used for characterization of the continuous aerosol size distribution by using four size bins: smaller than 0.156 μm, 0.156 - 0.625 μm, 0.625 - 2.5 μm, and 2.5 - 10.0 μm. The module simulates mass transfer and particle growth occurring between the gaseous and aerosol species during condensation and evaporation. The effects of nucleation and coagulation are ignored. The ISORROPIA algorithm is used to model inorganic atmospheric aerosols. ISORROPIA is a computationally efficient and rigorous thermodynamic model that predicts the physical state and composition of the sodium-ammonium-chloride-sulfate-nitrate-water aerosol system. The aerosol particles are assumed to be internally mixed, meaning that all particles of the same size have the same composition. The possible species for each phase are:

Gas Phase: NH_3, HNO_3, HCl, H_2O
Liquid Phase: NH_4^+, Na^+, H^+, Cl^-, NO_3^-, SO_4^{2-}, HSO_4^-, OH^-, H_2O
Solid Phase: $(NH_4)_2SO_4$, NH_4HSO_4, $(NH_4)_3H(SO_4)_2$, NH_4NO_3, NH_4Cl, $NaCl$, $NaNO_3$, $NaHSO_4$, Na_2SO_4, H_2SO_4

The ISORROPIA mechanism contains fifteen equilibrium reactions that are solved for in conjunction with their equilibrium constants[4].

The production of condensable organic species from oxidation of gaseous organic compounds is based on the organic yields reported by Pandis *et al.*[7] The formation of condensable organic aerosol species is done in the chemistry module, followed by the distribution of condensed organic aerosols to the four size bins in the aerosol routine. Also, an algorithm to simulate particle deposition and gravitational settling for particles of various sizes has been added. Inputs to the aerosol module include temperature, relative humidity, air density, and gas and aerosol concentrations. Outputs from the module are the updated equilibrium concentrations for the gas (HNO_3, NH_3, and HCl) and the aerosol species.

REACTIVE SCAVENGING MODULE

The Reactive Scavenging Module (RSM) uses synoptic scale temperature and precipitation rates to simulate a field of representative clouds which are defined by scavenging rates, water profiles, and wind fields. The module simulates the time dependent chemical kinetic interaction of these clouds with the gas and aerosol species, as well as, simulating vertical convective transport within a column of air[1]. Scavenging processes within the module include gas, aerosol, and microphysical scavenging. Gas scavenging can occur in cloud water (equilibrium process based on species solubilities), rain water (mass transfer rates are calculated for most species), and snow (limited to nitric acid). Aerosol scavenging is treated by nucleation and by inertial impaction processes. Microphysical scavenging refers to processes that not only transfer water from one water category to another, but also transfer the water-bound pollutants.

There are a number of input parameters that are required to be passed from URM to the RSM module, including temperatures and pressures for all vertical layers at each computational node. There are five additional meteorological parameters that are required to be passed: (1) total area averaged precipitation rate of convective and stratiform clouds, (2) fraction of precipitation associated with convective rain, (3) fraction of grid area covered by convective storms, (4) cloud top height of convective storm, and (5) cloud top height of stratiform storm.

The other required inputs are the gas and aerosol concentration profiles for the species that will be scavenged and/or transported via the convective clouds. The scavenged species are sulfur dioxide, aerosol sulfate, ozone, nitric acid, aerosol nitrate, hydrogen peroxide, ammonia, ammonium aerosol, and soluble crustals (Mg and Ca). Other gas and aerosol species are passed into the RSM module where vertical convective transport is simulated. Output from the module includes updated concentration profiles for all the species affected by scavenging and convective cloud transport, in addition to wet deposition mass fluxes for SO_2, SO_4, NO_3, H_2O_2, NH_4, Mg, Ca, K, and H-ion.

SAMI MODELING DOMAIN

The URM modeling domain covers the eastern half of the United States. The multiscale grid dimensions correspond to 192, 96, 48, 24, and 12 km. We have placed the fine 12 km grid over the southern Appalachian Mountains and the adjacent areas that are expected to most directly influence the air quality in the region of interest. The coarse grid is placed over the boundary cells and in areas that are not expected to significantly contribute to the air quality in the southern Appalachian Mountains. The domain height is 12,867 m and is divided into seven vertical grid layers. The thickness of each layer from the ground to the top of the domain are: 19 m, 43 m, 432 m, 999 m, 1779 m, 3588 m, and 6007 m. The use of finer resolution near the surface of the domain, as compared to the coarser resolution aloft, allows the steeper concentration gradients that typically exist in the near-surface troposphere and the evolution of the mixing depths during the day to be captured. Also, the SAMI modeling domain has been divided up into sub-domains in order to perform sensitivity analysis. Figure 1 shows the URM modeling grid along with the eight sub-domains that will be considered in assessing emission reductions.

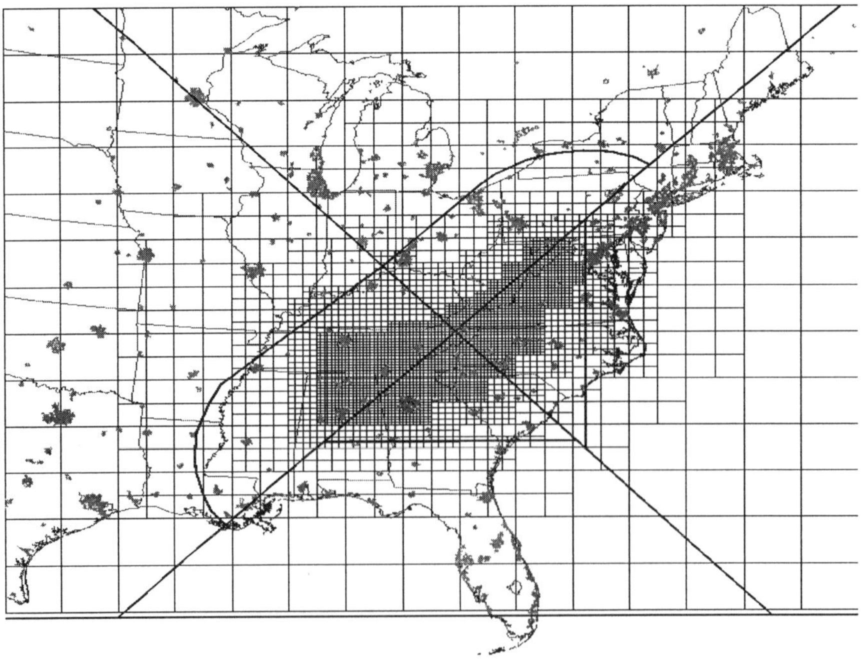

Figure 1. SAMI modeling domain and sensitivity sub-domains.

MODEL PERFORMANCE

A total of nine episodes will be used to develop seasonal and annual air quality metrics to assess visibility and acid deposition problems in the region. This paper focuses on four episodes: July 11 - 19, 1995; May 24 - 29, 1995; May 11 - 17, 1993; and March 23 - 31, 1993. The July and May 1995 episodes are considered "dry", and the May and March 1993 episodes are considered "wet". A comprehensive set of statistical calculations has been performed to determine the ability of the model to accurately estimate ozone and ambient aerosol concentrations and acid deposition mass fluxes. Among the statistical measures examined are mean bias, normalized bias, mean error, and normalized error. The statistical calculations were done using the Modeling Analysis and Plotting System (MAPS) package[3]. The results will be summarized below.

Ozone Performance

Ozone performance was evaluated by comparing modeling results to observations taken from the routine ozone monitors located throughout the 12 km modeling region. The results are shown below (Fig 2). Ozone monitoring is typically done in seasons of higher ozone; therefore, there are no observations for the March 1993 episode to compare our model results against. The model performance in terms of bias and error were generally within the typical EPA guidelines, although there tended to be a high bias during the May 1993 episode.

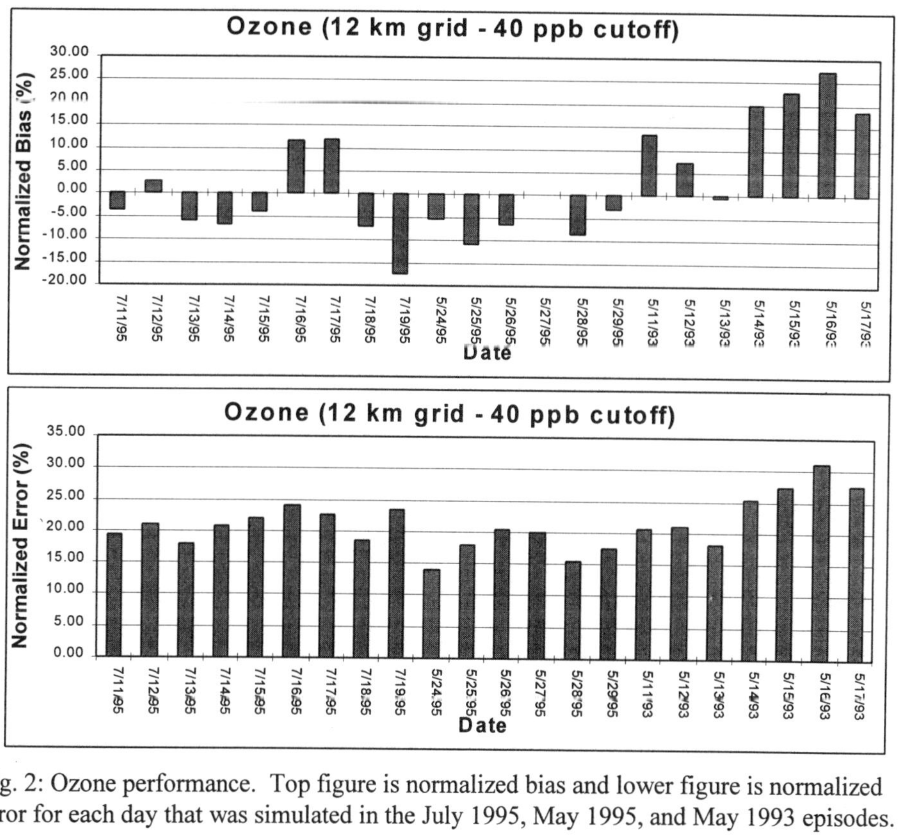

Fig. 2: Ozone performance. Top figure is normalized bias and lower figure is normalized error for each day that was simulated in the July 1995, May 1995, and May 1993 episodes.

Aerosol Performance

Aerosol performance was evaluated by comparing modeling results to observations taken from the Interagency Monitoring of Protected Visual Environments (IMPROVE) monitoring network. The species that were compared include fine (< 2.5 μm) sulfate, fine nitrate, fine ammonium, fine elemental carbon, fine organic carbon, fine soils (crustals), total $PM_{2.5}$, and total PM_{10}. There are eighteen IMPROVE monitoring sites in the modeling domain. However, five of those sites are located near or on the boundary of the modeling domain and are easily influenced by the boundary conditions. Therefore, only observations from the remaining thirteen stations are used to determine the aerosol model performance. Table 1 lists the resolution of the URM grid cells containing these stations. IMPROVE measurements are taken twice each week (Wednesday and Saturday) and are reported as a twenty-four hour average concentration. The four nearest nodes to each IMPROVE station are distance weighted and used to determine the aerosol concentration at each monitoring site.

Table 2 shows a summary of the normalized mean error for all the stations in the SAMI states for each day that IMPROVE measurements were available throughout the four modeled episodes. The normalized mean error (*NME*) is calculated as:

$$NME = \frac{1}{N}\sum_{i=1}^{N}\frac{\left|c_i^e - c_i^o\right|}{c_i^o} \times 100\% \qquad (1)$$

where c_i^e is the model-estimated 24-hour aerosol concentration at station *i*, c_i^o is the observed 24-hour aerosol concentration at station *i*, and *N* equals the number of estimate-observation pairs drawn from all valid monitoring station data on the simulation day of interest.

Table 1. URM grid resolution for IMPROVE stations.

IMPROVE STATION	Grid cell size (km)
Brigantine (NJ)	48
Dolly Sods/Otter Creek (WV)	12
Great Smoky Mountains (TN)	12
Jefferson/James River Face (VA)	12
Lye Brook (VT)	96
Mammoth Cave (KY)	24
Okefenokee (GA)	96
Cape Romain (SC)	48
Shenandoah (VA)	12
Shining Rock (NC)	12
Sipsy (AL)	12
Upper Buffalo (AR)	96
Washington, D.C.	24

Typically, the largest portion of fine PM consists of sulfate. Sulfate is produced by the gas phase reaction of SO_2 with OH or heterogeneously by reacting SO_2 with H_2O_2 and/or ozone when a rain, cloud, or fog droplet is present. The mean normalized mean error is calculated by averaging the normalized mean errors for each of the 10 IMPROVE days that measurements were taken. These values are reported for each species in Table 2. The mean normalized mean error for sulfate is less than 45%. Figure 3 shows a scatter plot of observed versus modeled sulfate concentrations for all the IMPROVE stations in the domain for each episode. Aerosol nitrate is formed by the condensation of nitric acid into the aqueous or salt complex form. Its concentration depends on the amount of gas-phase nitric acid, ammonia, and sulfate that is available. The nitrate concentrations are usually low (less than 1 $\mu g/m^3$); therefore, the normalized error can be very high. Ammonium usually presents itself in the form of ammonium sulfate ($(NH_4)_2SO_4$), ammonium bisulfate (NH_4HSO_4), ammonium nitrate (NH_4NO_3), and/or mixed salts. The ammonium concentration primarily depends on the amount of sulfate and gas-phase ammonia that is available. There are no direct measurements of ammonium at the IMPROVE stations. Therefore, for the purpose of ammonium performance evaluation, it has been assumed that the sulfate and nitrate are completely neutralized with NH_4. The mean normalized mean error for the ammonium aerosol is similar to that of sulfate (i.e. less than 45%).

Table 2: Aerosol mean normalized errors for July '95, May '95, May '93, and March '93 episodes.

Sulfate

Date	Norm. Error (%)
7/12/95	16.60
7/15/95	14.54
7/19/95	41.33
5/24/95	51.34
5/27/95	31.74
5/12/93	46.10
5/15/93	32.24
3/24/93	27.94
3/27/93	67.55
3/31/93	107.60
MEAN	43.70

Nitrate

Date	Norm. Error (%)
7/12/95	331.12
7/15/95	133.81
7/19/95	73.05
5/24/95	238.15
5/27/95	128.08
5/12/93	193.39
5/15/93	1092.96
3/24/93	412.42
3/27/93	88.12
3/31/93	261.20
MEAN	295.23

Ammonium

Date	Norm. Error (%)
7/12/95	23.97
7/15/95	33.14
7/19/95	42.80
5/24/95	31.51
5/27/95	17.28
5/12/93	42.51
5/15/93	31.82
3/24/93	33.83
3/27/93	58.39
3/31/93	125.88
MEAN	44.11

Organics

Date	Norm. Error (%)
7/12/95	61.71
7/15/95	67.64
7/19/95	40.09
5/24/95	51.90
5/27/95	38.92
5/12/93	39.50
5/15/93	57.83
3/24/93	24.38
3/27/93	59.36
3/31/93	48.13
MEAN	49.63

Elem. Carbon

Date	Norm. Error (%)
7/12/95	37.22
7/15/95	42.28
7/19/95	50.20
5/24/95	28.17
5/27/95	26.43
5/12/93	73.95
5/15/93	38.17
3/24/93	39.89
3/27/93	74.19
3/31/93	49.13
MEAN	46.76

Soils

Date	Norm. Error (%)
7/12/95	212.36
7/15/95	104.91
7/19/95	321.54
5/24/95	89.42
5/27/95	229.66
5/12/93	176.63
5/15/93	114.63
3/24/93	703.68
3/27/93	744.38
3/31/93	239.58
MEAN	293.68

PM_2.5

Date	Norm. Error (%)
7/12/95	35.34
7/15/95	32.50
7/19/95	28.51
5/24/95	41.01
5/27/95	10.67
5/12/93	35.02
5/15/93	22.99
3/24/93	22.52
3/27/93	66.20
3/31/93	109.75
MEAN	40.45

PM_10

Date	Norm. Error (%)
7/12/95	26.88
7/15/95	26.80
7/19/95	34.30
5/24/95	27.26
5/27/95	37.53
5/12/93	37.76
5/15/93	21.84
3/24/93	62.50
3/27/93	60.89
3/31/93	85.68
MEAN	42.14

PM_Coarse

Date	Norm. Error (%)
7/12/95	122.11
7/15/95	60.17
7/19/95	104.70
5/24/95	43.44
5/27/95	202.34
5/12/93	85.58
5/15/93	54.42
3/24/93	283.30
3/27/93	128.55
3/31/93	131.28
MEAN	121.59

The amount of organic aerosols formed in the atmosphere is determined by using measured organic aerosol yields. Almost all of the model predictions are biased low. This can possibly be attributed to an under prediction in the organic aerosol yields that were used or a deficiency in the emission inventory. The mean normalized mean error is still less than 50%. Elemental carbon (EC) is an inert primary emission species. The observations for EC are typically very low (around 0.5 $\mu g/m^3$). The model results match well with observations and have an average mean error of less than 0.25 $\mu g/m^3$. The aerosol species that are lumped into "soils" consist of calcium and "other" PM. The predictions typically are biased high. There are high uncertainties in the emission inventories for these crustal species, which may lead to the overprediction.

In order to determine $PM_{2.5}$, all the aerosol species for the first 3 size bins were summed together. To determine the PM_{10} concentrations, the coarse aerosol fractions (2.5 – 10.0 μm) were added to the $PM_{2.5}$ concentrations. There is good agreement between model predictions and observation (typically less than 40% normalized mean error) for most of the episode days, except for the March 1993 episode. This larger discrepancy in the March 1993 episode is mainly due to the high errors associated with the sulfate, ammonium, and soils.

Acid Deposition Performance

Wet acid deposition performance was evaluated by comparing modeling results to observations taken from the National Atmospheric Deposition Program (NADP) monitoring network. The species that were compared include sulfate, nitrate, ammonium, hydrogen ion, and crustal cations (Mg and Ca). NADP measurements are taken once each week (Tuesday) and the concentrations and precipitation are reported as a 7-day cumulative. Dividing the measured concentration by the precipitation results in depositions with units of mass flux. Since there is a large spatial variation in precipitation and the RAMS model results do not match observed precipitation at exact locations, the NADP observations are compared to the best (i.e., closest to the observation) model result within a 30 km radius from the monitoring site. There are eighty-three NADP monitoring sites in the modeling domain. However, since wet deposition is very localized, only data from the fourteen stations in the 12 km grids are used to determine model performance. These stations are listed in Table 3.

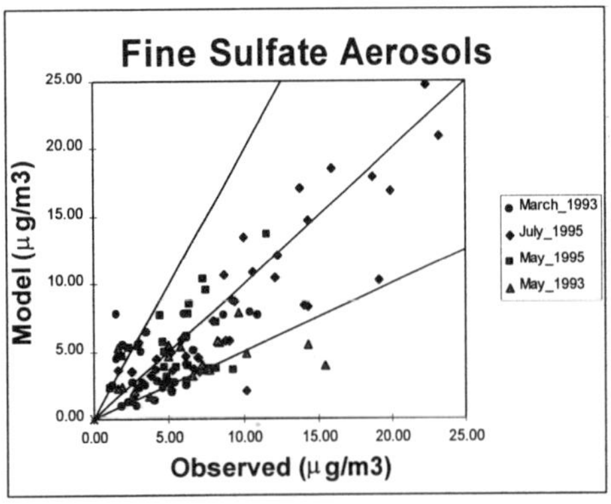

Figure 3: Modeled vs. observed sulfate aerosol concentrations

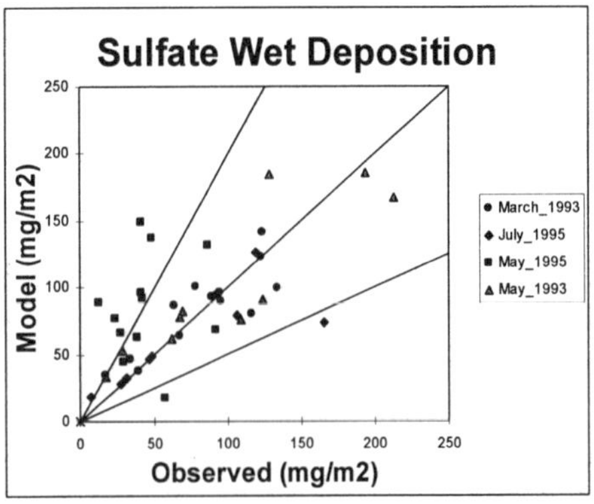

Figure 4: Modeled vs. observed sulfate wet deposition mass fluxes

Sulfate and nitrate are two of the most important wet deposition species. They both show low normalized mean error. The average mass fluxes and normalized mean errors for the four wet deposition episodes are located in Table 4. The mean normalized mean errors for sulfate and nitrate are 35.4% and 31.8%, respectively. Figure 4 shows a scatter plot of observed versus modeled wet sulfate deposition mass fluxes for all the NADP stations in the 12 km grid for each episode. The correlation between modeled and observed sulfate mass fluxes are very good, except for the May 1995 episode which shows an overprediction.

Table 3: NADP stations falling into the 12-km URM grid.

STATION	COUNTY	STATE
Sand Mountain Exp. Station	Dekalb	AL
Georgia Station	Pike	GA
Lilley Cornett Woods	Letcher	KY
Coweeta	Macon	NC
Piedmont Research Station	Rowan	NC
Mt. Mitchell	Yancey	NC
Walker Branch Watershed	Anderson	TN
Wilburn Chapel	Giles	TN
Great Smokey Mountains National Park – Elkmont	Sevier	TN
Charlottesville	Albemarle	VA
Horton's Station	Giles	VA
Shenandoah National Park	Madison	VA
Babcock State Park	Fayette	WV
Parsons	Tucker	WV

Table 4. Wet deposition performance for July '95, May '95, May '93, and March '93 episodes.

Sulfate

	Mass Flux (mg/m2)	Norm. Error (%)
July 11-18, 1995	64.49	26.97
May 23-30, 1995	42.56	79.34
May 11-18, 1993	113.81	14.73
March 23-30, 1993	83.40	20.57
MEAN	76.07	35.40

Nitrate

	Mass Flux (mg/m2)	Norm. Error (%)
July 11-18, 1995	34.03	27.52
May 23-30, 1995	29.29	42.80
May 11-18, 1993	55.66	44.65
March 23-30, 1993	56.21	12.26
MEAN	43.80	31.81

Ammonium

	Mass Flux (mg/m2)	Norm. Error (%)
July 11-18, 1995	7.78	160.72
May 23-30, 1995	6.79	264.14
May 11-18, 1993	15.01	83.56
March 23-30, 1993	7.70	484.68
MEAN	9.32	248.28

H-ion

	Mass Flux (mg/m2)	Norm. Error (%)
July 11-18, 1995	1.36	46.92
May 23-30, 1995	0.85	13.87
May 11-18, 1993	1.93	16.55
March 23-30, 1993	1.97	64.37
MEAN	1.53	35.43

Calcium

	Mass Flux (mg/m2)	Norm. Error (%)
July 11-18, 1995	2.45	31.48
May 23-30, 1995	2.30	81.74
May 11-18, 1993	4.94	10.96
March 23-30, 1993	2.13	191.81
MEAN	2.95	79.00

Magnesium

	Mass Flux (mg/m2)	Norm. Error (%)
July 11-18, 1995	0.30	30.66
May 23-30, 1995	0.42	144.67
May 11-18, 1993	0.79	59.05
March 23-30, 1993	0.46	419.17
MEAN	0.49	163.39

The ammonium wet deposition is biased high in all four episodes. The mean normalized mean error is approximately 250%. The hydrogen ion deposition is biased low in all four episodes with a mean normalized mean error of approximately 35%. Both the calcium and magnesium deposition fluxes are biased low in the July 1995 episode and biased high in the other three episodes. The mean normalized mean errors for calcium and magnesium are 79% and 163%, respectively. Some of the discrepancies between modeled and measured fluxes can be attributed to differences in the modeled and observed precipitation rates.

SENSITIVITY ANALYSIS

Using air quality models to calculate the sensitivity of gas and aerosol phase concentrations and wet deposition fluxes to input and system parameters is important in determining the most effective control strategies. In order to most efficiently and accurately determine sensitivities, the Decoupled Direct Method (DDM) has been integrated into the model. Using direct derivatives of the equations governing the evolution of species concentrations, the local sensitivities to a variety of model parameters and inputs are computed simultaneously with the species concentrations.

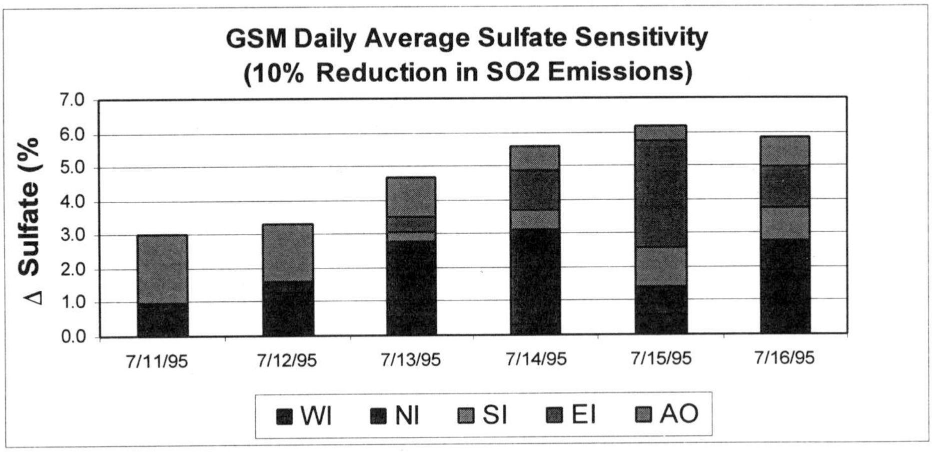

Figure 5. Fine sulfate sensitivity to a 10 percent SO_2 reduction.

Preliminary sensitivity runs were made for the July 2010 episode using the 1995 meteorological data with a 2010 emission inventory. Sensitivity coefficients were computed each hour for all the grid cells in the domain. By examining sensitivities at a specific station, we can determine the sub-domain from which emission reductions would have the greatest effect. Figure 5 shows a 6-day stacked bar chart for daily averaged sulfate sensitivities at the Great Smokey Mountains National Park. The sulfate sensitivities represent the percent change in sulfate concentrations due to a 10% reduction in the west-inner sub-domain (WI), north-inner sub-domain (NI), south-inner sub-domain (SI), east-inner sub-domain (WI), all outer sub-domains (AO), and domain wide (sum of WI, NI, SI, EI, and AO). In the Great Smokey Mountains National Park, it can be seen that different sub-domains can have varying contributions to the overall reduction of sulfate from day-to-day depending on the specific meteorology. For example, on July 11, a 10% reduction in domain-wide SO_2 emissions will result in a 3% reduction in the sulfate concentration at GSM, with 2% due to SO_2 reductions in AO and 1% due to SO_2 reductions in WI. It should be noted that the sensitivity results reported above are just an example of the type of results that are available. Since this is a work in progress and the data presented above is a partial data set, any inferences that might be drawn could be significantly altered when the rest of the data becomes available.

CONCLUSIONS

A "One-Atmosphere" modeling approach has been taken to help assess the impact of control strategies on air quality in the Southern Appalachians. The modeling system, consisting of RAMS, EMS-95 and URM, simulates gaseous and condensed phase pollutants, and uses a sectional approach to provide size distributions of the aerosol. It is also used to simulate dry and wet deposition. In this paper, the model's results were compared against a suite of observations for ozone and PM concentrations, as well as wet deposition, for four episodes. Two of those episodes were relatively wet, and two drier. While the results tended to be good for both sets of episodes, there was a tendency to overestimate ozone and underestimate aerosols during the wet episodes.

ACKNOWLEDGMENTS

This project is funded by the Southern Appalachian Mountains Initiative. The authors thank the members of the SAMI Atmospheric Modeling Subcommittee for their constructive criticism, innovative ideas and invaluable help throughout this project.

REFERENCES

1. Berkowitz, C.E.; Easter, R.C; Scott, B.C. "Theory and results from a quasi-steady-state precipitation-scavenging model", *Atmos. Environ.* 23:1555-1571 (1989).
2. Carter, W.P.L. "A detailed mechanism for the gas-phase atmospheric reaction of organic compounds", *Atmos. Environ.* 24:481-518 (1990).
3. McNally, D.E.; Tesche, T.W. *Modeling Analysis and Plotting System User Manual.* Alpine Geophysics (1991).
4. Nenes, A.; Pilinis, C; Pandis, S.N. "ISORROPIA: A new thermodynamic equilibrium model for multicomponent inorganic aerosols", *Aquatic Geochem.* 4:123-152 (1998).
5. Odman, M.T.; Russell, A.G. "Multiscale modeling of pollutant transport and chemistry", *J. Geophys. Res.* 96:7363-7370 (1991).
6. Odman, M.T.; Russell A.G. "A multiscale finite element pollutant transport scheme for urban and regional modeling", *Atmos. Environ.* 25A:2385-2394 (1991).
7. Pandis, S.N.; Harley, R.A.; Cass, G.R; Seinfeld, J.H. "Secondary organic aerosol formation and transport", *Atmos. Environ.* 26A:2269-2282 (1992).
8. Yang Y.J., Wilkinson J.G., and Russell A.G. "Fast, Direct Sensitivity Analysis of Multidimensional Photochemical Models", *Environ. Sci. Technol.* 31:2859-2868 (1997).

DISCUSSION

R. YAMARTINO: To what do you attribute the poor O_3 model performance in wet periods (e.g., to the chemistry)?

T. ODMAN: I have presented normalized bias and normalized error as the primary measures of the O_3 performance of the model. For May 1993, the only wet period in our O_3 performance evaluation, the mean O_3 concentrations are much lower than the mean concentrations for the dry episodes. Therefore, for the same amount of absolute error, the normalized error would be larger for this wet episode and the same is true for the bias. However, the positive bias on all the days of this period is indicative of a systematic error that leads to overestimation of O_3. This may be due to underestimating the cloud cover and consequent reduction of photolysis rates. To be certain, we have to perform additional analysis.

C. MENSINK: Classical finite element schemes are known to be very diffusive. How do you deal with that in your model?

T. ODMAN: The finite element transport scheme used in URM is based on the Streamline Upwind Petrov-Galerkin (SUPG) method. Being an upwind method, SUPG is considerably less diffusive than classical finite element schemes. A comparison of SUPG to classical finite element schemes as well as other popular advection schemes in terms of numerical diffusion can be found in Odman and Russell (1991).

GLOBAL AND LONG-RANGE TRANSPORT

chairpersons:	R. Bornstein E. Genikhovich
rapporteurs:	C. Soriano J. H. Sørensen

LONG-TERM CALCULATION OF HG POLLUTION OVER SOUTH-EAST EUROPE

Dimiter Syrakov

National Institute of Meteorology and Hydrology
1784 Sofia, Bulgaria

INTRODUCTION

Mercury is one of the several chemicals emitted by sources such as coal-fired power plants and incinerators. It has received special attention in the recent years as certain forms of Hg exhibit elevated toxicity and bio-concentrate in vegetation and fish. In the past, mercury was viewed as a local pollutant near industrial and municipal discharges of trace metals. The recent discovery of fish with high level of mercury concentration in lakes far from industrial sources shows that the long-range transport of atmospheric mercury and its subsequent deposition in soils and watersheds is the primary cause of mercury in remote territories. The complexity of Hg anthropogenic and natural sources accompanied by the complex chemical behavior of Hg in the air, water (clouds, precipitation, lakes and seas) and soil make the problem of Hg pollution very difficult. A great number of works are devoted to investigation and modeling of different sides of Hg atmospheric and biospheric cycles. Here, a simple parameterization scheme for Hg chemical transformation, deposition and re-emission is presented, conjugated to the 3D Eulerian dispersion model EMAP (BC-EMEP, Syrakov, 1995), developed in the frame of the Bulgarian contribution to EMEP (1994-1997).

SHORT DESCRIPTION OF EMAP MODEL

EMAP is a simulation model, which allows describing the dispersion of multiple air pollutants. The processes as horizontal and vertical advection, horizontal and vertical diffusion, dry deposition, wet removal, gravitational settling (aerosol version) and the simplest chemical transformation (sulfur version) are accounted for in the model. Within EMAP, the semi-empirical diffusion-advection equation for scalar quantities is treated. The governing equations are solved in terrain-following co-ordinates. Non-equidistant grid spacing is settled in the vertical direction. The numerical solution is based on discretization applied to a staggered grid. Conservative properties are fully preserved within the discrete

Air Pollution Modeling and Its Application XIV, Edited by
Gryning and Schiermeier, Kluwer Academic/Plenum Publishers, New York, 2001

model equations. The advective terms are treated with the TRAP (Syrakov, 1996, Syrakov and Galperin, 1997b) scheme, which is a Bott type one. Displaying the same simulation properties as the Bott scheme (explicit, conservative, positively definite, transportive, limited numerical dispersion), the TRAP scheme occurs to be several times faster. The advective boundary conditions are zero at income flows and "open boundary" at outcome ones. The turbulent diffusion equations are digitized by means of the simplest schemes. The bottom boundary condition for the vertical diffusion equation is the dry deposition flux; the top boundary condition is optionally of the "open boundary" or "hard lid" type. The lateral boundary conditions for diffusion are of the "open boundary" type. In the surface layer, a parameterization is applied, permitting to have the first computational level at the top of SL. It provides a good estimate for the roughness level concentration and also accounts for the action of continuous sources on the earth surface (Syrakov and Yordanov, 1996, 1997).

As to the numerical method, time splitting is applied to transform the complex problem into a number of simple tasks. The discretized equations are solved numerically on an Arakawa C-type staggered grid. In one time step, the one-dimensional schemes are applied sequentially for every dimension, for advection, diffusion, etc.; in the next time step their order is reversed. One time step source produced concentrations are added at each time step at the respective grid points.

Only 850 hPa U-, V- and θ-fields, as well as surface θ-field, are necessary as meteorological input, θ being potential temperature. A simple PBL model (Yordanov et al., 1983) is built in EMAP producing U-, V-, W- and K_z-profiles at each grid point. It provides also u_* and SL universal profiles necessary in SL parameterization. The roughness and Coriolis parameter fields are a pre-set additional input to the PBL model.

DESCRIPTION OF THE HG PARAMETERIZATION SCHEME

The aim of this modeling is to estimate Hg pollution in the region of south-east Europe, taking a territory of 8x9 EMEP 150x150 km^2 grid cells, with Bulgaria in the center (see Figures later on). Each cell is divided into 36 25x25 km^2 cells. The chosen territory includes entirely Albania, Bulgaria and Moldova, parts of Greece, Romania, Turkey, Yugoslavia, Ukraine, the Black, Adriatic and Aegean seas. In the created versions of EMAP, a 5-layer structure is used. The first four layers have representative levels at 50, 200, 650 and 1450 m with layer boundaries 20-100, 100-375, 375-995, 995-1930 m. The 5th layer accounts parametrically for the free atmosphere.

Sources of atmospheric Hg

There are two types of sources of Hg in the atmosphere - anthropogenic and natural ones. The anthropogenic sources are divided into large point sources (LPS) and area sources (AS). LPS are the electric power plants with fossil fuel, notably coal, incinerators, chemical industry, smelting of non-ferrous metals. There is a wide variety of diffuse anthropogenic area sources, such as discarded batteries, discarded fluorescent lights, latex paints, fungicides used in agriculture and paper production etc.

In this study, the Bulgarian anthropogenic Hg source inventory for 1995 (Ivancheva et al., 1999) is used. All LPSs are given with their model co-ordinates and release rate. A common height of h=200 m is prescribed to these sources. The area sources are distributed over the whole territory of the country; each cell considered as an area source. These sources are supposed to release in the first computational level.

Similarly, there are many point and diffuse natural sources. The primary natural sources of Hg are the earth crust and mantle. The main sources are volcanoes, faults, and other opening in the crust, especially in mobile belts along plate boundaries. Hg is also emitted from various rocks, soils and the ocean surface; there are countless diffuse secondary sources all over the earth surface. When the impact of the Hg pollution is assessed, one important complication must be taken into consideration - a good deal of airborne mercury (both anthropogenic and natural) deposited in aquatic and terrestrial environment is re-emitted back to the atmosphere through natural processes, such as microbial activities. It is difficult, often impossible, to separate the natural diffuse emissions from the re-emissions. A great number of estimates for the natural and re-emission fluxes exist, depending on the type of surface, geographical position, season, even time of the day. Sometimes these estimates differ by orders of magnitude. In this work, primary natural sources are not taken into account, because such sources are missing in the chosen territory. The emission from diffuse natural sources and the re-emission are dealt with together; using the possibility of the SL parameterization built in EMAP to account for surface source action. The re-emission parameterization will be described later on.

Forms, transformation and emission of atmospheric Hg

There are two dominant states of mercury in the atmosphere: Hg^0 and Hg^{II}. Elemental mercury Hg^0 is vapor and is not very soluble. It could be oxidized in gaseous and aqueous phase. In gaseous phase (dry oxidation) it can be oxidized by oxidants such as chlorine (Cl_2), hydrogen peroxide (H_2O_2) and ozone (O_3). However, significant uncertainties still remain in the kinetics of these reactions. Much faster is the oxidation of Hg^0 by O_3 in the aqueous phase; however, its low solubility limits the amount of Hg^0 per volume of air that can be converted to Hg^{II}. Divalent mercury Hg^{II} exists in various chemical forms and phases. Part of the species are gaseous, another part - weightless aerosols. The mercury chlorides, sulfates, oxides etc. can then be reduced back to Hg^0. The kinetics of these processes is also uncertain. Here, dividing of Hg^{II} into gas and aerosol phase will not be made. As common oxidation rate $Hg^0 \rightarrow Hg^{II}$, a value of

$$\alpha_{tr} = 0.01\ h^{-1}$$

will be used which is the lower limit of the estimate by Iverfeld and Lindqist (1986) for the liquid phase. At that stage of parameterization, it is assumed to be a constant.

The anthropogenic sources emit both Hg forms. The variable *Fract* is introduced to determine the ratio between them. If Q is the total emission, $Q.Fract$ is emitted as Hg^0 and $Q(1\text{-}Fract)$ is emitted as Hg^{II}. Here, a common value of $Frac$=0.9 is supposed.

The natural emission and the re-emission of mercury are mainly in the form of Hg^0. A very small amount is in the form of organic mercury compounds, which are reduced very soon to metal vapor condition in the atmosphere. In this work, it is adopted that the earth surface emits Hg^0 only.

Dry deposition and wet scavenging

Dry deposition in EMEP is introduced as bottom boundary condition for the vertical diffusion equation. The built in SL parameterization produces these boundary conditions, together with a surface level concentration estimate, on the base of SL dynamics and stability parameters, as well as roughness level deposition velocity V_d. Here, the linear approach is used, although the resistant approach can also be used (in this case, the aerodynamic resistance must be excluded because it is included in the SL parameterization). Both main atmospheric mercury forms differ drastically in their dry deposition velocities. V_d of Hg^{II} is some orders of magnitude bigger than this one of Hg^0. The dry deposition velocity is assumed to depend on the type of the surface. Taking into account only the main earth surface inhomogeneities, the next values for dry deposition velocities are adopted here (after Galperin et al., 1994):

for Hg^0: $V_d = 0.0001$ m/s for land and
$V_d = 0.0003$ m/s for sea;
for Hg^{II}: $V_d = 0.005$ m/s for land and
$V_d = 0.015$ m/s for sea.

Similar estimates for Hg^{II} dry deposition velocities are calculated by Pai et al.(1997) on the base of the resistance approach after aggregating the individual estimates for big territory and long period of time. These authors even neglect the Hg^0 dry deposition.

The masses scavenged by precipitation are determined according to the first order low

$$\Delta C = C[1\text{-}\exp(-\gamma R\Delta t)],$$

where C is the species' concentration, R - precipitation intensity in [mm/h], γ - wet removal constant [mm^{-1}], Δt - time step [h]. In this modeling, the following γ's are adopted:

for Hg^0 γ=0.000003 mm^{-1} and
for Hg^{II} γ=0.3 mm^{-1}.

These values reflect the low solubility of Hg^0 and the good solubility of Hg^{II}. It can be supposed that in clouds, where the droplets are in suspended condition, they are able to capture the mercury molecules by their whole surface ($4\pi r^2$). So, the efficiency of removal must be four times better than the removal efficiency in the falling droplets under clouds, which capture the molecules by their cross-section (πr^2). The forth layer in EMAP is assumed to be a cloudy one and wet removal constants are multiplied by 4, there.

Re-emission

Here, the scheme developed in MSC-E (Galperin and Maslyaev, 1996) for describing the re-emission and fixation of POPs is applied to mercury. In general form, it can be presented by the following system of equations:

$$\begin{aligned} &\partial Q_{av}/\partial t = D + W - \alpha_{reemis}Q_{av} - \alpha_{fix}Q_{av}\,, \\ &\partial Q_{fix}/\partial t = \alpha_{fix}Q_{av}\,, \\ &RE = \alpha_{reemis}Q_{av}\,, \end{aligned} \tag{1}$$

where $Q_{av}(t)$ is that part of the deposited substance which is capable to be re-emitted (called re-emission mass); $D(t)$ is the dry deposition flux; $W(t)$ - the wet deposition flux; α_{reemis} and α_{fix} – the re-emission and fixation factors, depending on temperature, the underlying surface type and state (sea, soil, vegetation, glacier etc.). $Q_{fx}(t)$ is the substance quantity fixed in the soil (note, that the introduced Q_{fx} characterizes not only the fixed part of the substance but also its accumulation in biota); $RE(t)$ - the re-emission flux.

It is shown in Syrakov (1997) that this simple re-emission and fixation scheme can be combined easily with the SL parameterization built in EMAP. Scheme (1) was chosen to account for the re-emission when simulating the long-range transport of mercury and estimating its deposition on regional and continental scale.

In this Hg parameterization, it is assumed that the re-emission acts as a continuous surface source of Hg^0. The following expressions for the re-emission and fixation factors are used:

for sea: $\alpha_{reemis} = 0.005\delta_t$, $\alpha_{fix} = 0.000002\delta_t$;
for land: $\alpha_{reemis} = 0.0002t\delta_t$, $\alpha_{fix} = 0.00002t\delta_t$,

where t is the temperature in [°C], δ_t being a cut-off function (δ_t=0 at $t \le 0$, δ_t=1 at t>0).

One can notice that, in this parameterization, the re-emission and fixation factors depend only on the main surface characteristics - the sea/land division. The sea parameters do not depend on temperature keeping in mind the high temperature inertia of sea waters. In case of land, linear temperature dependence is supposed. The cut-off function δ_t is introduced to reflect the fact that the re-emission is a function of microbial activity. When τ is below zero this activity is suspended; it grows up with the temperature increase.

This simple parameterization of the extremely complicated re-emission processes introduces a new accumulative variable (the re-emission mass Q_{av}) which has to be initialized. In the work, the problem is solved using the evidence data for the natural and re-emission mercury fluxes. With each grid point of the model area, some mean re-emission mass is associated which produces at τ=10°C the Hg flux typical for that region. The flux estimates of Münch et al. (1992) and of Moisseev (1998), where lots of measurement and modeling results are aggregated and generalized, are used. The Münch et al. data is mainly exploited. Moisseev's estimates are used only for some newly created states. On **Table 1**, the typical fluxes and the respective calculated re-emission masses are displayed.

Table 1. Total (natural + re-emission) fluxes of mercury in south-east Europe

Territory	Adriatic sea	Aegean sea	Albania *	Black sea	Bulgaria *	Greece *	Moldova *	Marble sea*	Romania	Turkey	Ukraine	Yugoslavia
abbreviation	ADS	AES	ALB	BLS	BUL	GRE	MOL	MRS	ROM	TUR	UKR	YUG
Flux [ng/m^{2}h]	1.0	1.0	7.16	1.0	6.38	7.78	3.39	1.0	5.19	7.00	3.63	7.65
Re-em.factor at τ=10°C [h^{-1}]	0.005	0.005	0.002	0.005	0.002	0.002	0.002	0.005	0.002	0.002	0.002	0.002
Re-emission mass [ng/m^2]	200	200	2080	200	3190	3890	1695	200	2595	3500	1815	3825

Note: Stars mark territories, included entirely in the model area.

The simple parameterization of mercury emission, transformation and deposition in the atmosphere, combined with the dynamical schemes of the 3D Eulerian dispersion model EMAP, produce an easy-to-use tool for study of mercury transport on regional scale. The re-emission and fixation scheme (1) is effectively combined with the SL parameterization built in EMAP. The sensitivity tests performed in Syrakov (1999) show that the behavior of the examined charecteristics is quite reasonable from physical point of view. At the same time they point out that even slight re-emission causes considerable redistribution of the pollutant.

PROVISIONAL ESTIMATE OF THE IMPACT OF BULGARIAN SOURCES OF ATMOSPHERIC HG FOR 1995

In this part, a long term integration is performed on the base of the Hg emission inventory of Bulgarian sources presented in Ivancheva et al.(1999) and meteorology data for 1995 produced by the "Europa-Model" of Deutscher Wetterdienst. As only Bulgarian sources are used, the results can be considered as an estimate of the Bulgarian impact in the region's pollution.

In **Fig. 1**, the Bulgarian mercury emissions for 1995 are shown, divided into large point sources and area ones. The total Hg emission in Bulgaria for 1995 is estimated to 6.877 tones.

The calculations are carried out on 120 MHz Pentium PC. An year integration lasts about 5 hours on such a platform. The output at the end of each month is some fields and some accumulated quantities called **totals**.

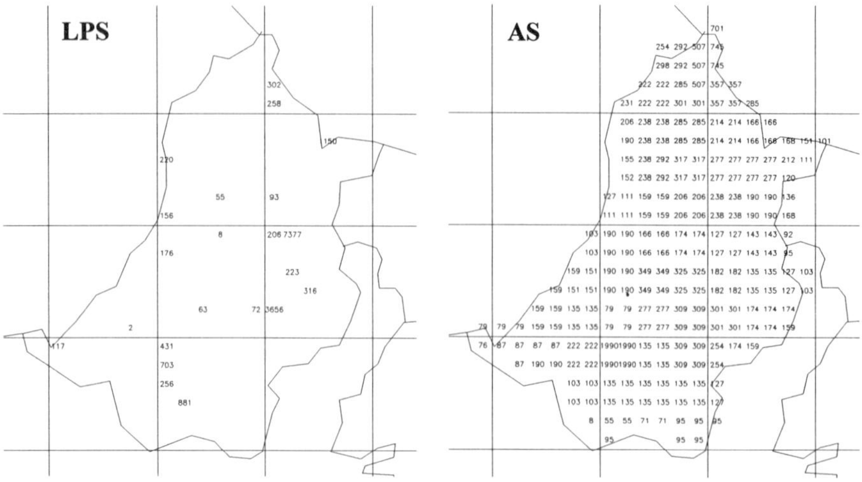

Figure 1. 1995 Bulgarian Hg inventory: large point sources in 10 μg/s
area sources in μg/s

The output fields are: **CA** - mean surface concentration; **DD** - dry deposition; **WD** - wet deposition; **TD** - total deposition; **SP** - sum of precipitation; **RM** - re-emission mass; **FM** - fixed mercury mass, lost for atmospheric mercury cycle.

The totals are periodically calculated during the integration and the mass balance error is then displayed for conservativity control. It is worth to be mentioned that this error is less than 0.1% during the whole period of integration. After finishing the 12-month integration, annual averaging of surface concentration fields, as well as accumulation of the other fields (except RM whose transformations are transmitted from month to month via the initial RM) is made. Finally, the annual concentration in precipitation (CP) is calculated, dividing WD by SP.

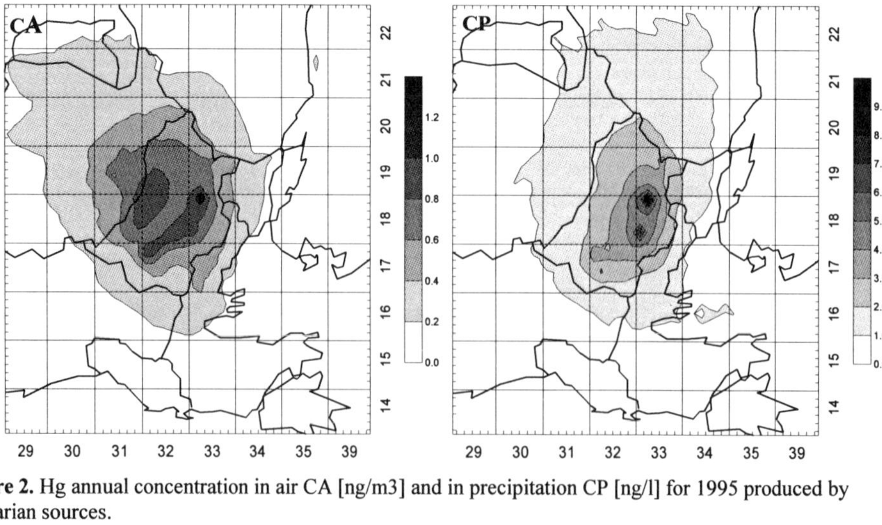

Figure 2. Hg annual concentration in air CA [ng/m3] and in precipitation CP [ng/l] for 1995 produced by Bulgarian sources.

In **Fig. 2**, the annual fields of air concentration and concentration in precipitation are shown. It can be seen that the highest levels of the pollutant characteristics are over the Bulgarian territory. Values between 0.5 and 1.2 ng/m^3 and 3 and 10 ng/l (for CA and CP, respectively) prevail for the country and for some neighboring areas. In the rest part of south-east Europe, CA does not exceed 0.4 ng/m^3 while CP varies between 0.3 and 2 ng/l. The maximal values (1.27 ng/m^3 for CA and 9.82 ng/l for CP) are observed in the neighborhood of the powerful sources (especially for the region of the "Maritza Iztok 1,2,3" power plants). The presented concentration fields look quite good, at least as an order of magnitude in comparison with measurements and other model calculations, in spite that no tuning of the model parameters as to better fit experimental data has been made.

The distribution of the total Hg loads for the different territories listed on **Table 1** is presented on **Table 2** where the monthly variations can be seen, too. The last row and column show the percentage of the respective quantity from the total anthropogenic emission, which for 1995 is estimated to 6.877 Mg. As can be seen, only 11% of the Bulgarian Hg pollution are deposited in the country. A relatively big receiver is Romania - 8.3%. Only 43 % of Bulgarian lead emission is located in the neighboring territories, the other quantity leaves the domain.

Table 2. Distribution of total Hg deposition produced by Bulgaria for 1995.

	Jan	Feb	Mar	Apr	May	Jun	Jul	Aug	Sep	Oct	Nov	Dec	Year	%emit
ADS	13.0	2.5	7.6	2.3	6.8	1.6	3.5	5.1	1.0	6.1	1.3	2.4	53.0	.8
AES	12.9	24.3	24.0	25.6	12.8	11.4	18.7	14.9	7.8	28.2	20.9	13.3	214.8	3.1
ALB	4.0	3.5	2.6	3.4	6.3	1.1	4.0	2.8	.6	2.2	.4	2.5	33.3	.5
BLS	79.0	64.2	49.3	47.0	29.1	35.2	12.6	15.4	72.4	33.4	58.9	50.0	546.5	7.9
BUL	44.2	75.8	74.4	74.0	63.6	101.8	75.6	71.1	48.3	29.0	49.9	44.4	752.2	10.9
GRE	18.7	20.5	20.2	18.9	15.3	12.6	31.2	18.4	5.9	18.4	12.2	10.8	203.0	2.9
MOL	2.7	8.2	6.8	5.7	7.6	11.8	2.2	3.7	8.9	1.3	2.8	1.5	63.1	.9
MRS	1.8	2.2	2.0	2.7	1.2	1.7	1.0	.6	1.7	1.8	1.9	1.4	20.0	.3
ROM	27.5	63.1	66.5	72.1	71.5	101.8	27.8	34.3	38.8	12.8	20.8	32.5	569.6	8.3
TUR	19.2	29.0	21.4	29.2	14.7	16.9	7.0	5.0	13.3	13.6	15.5	10.5	195.2	2.8
UKR	3.8	9.0	7.2	6.3	6.5	8.9	1.8	3.1	7.9	1.1	4.4	2.1	62.2	.9
YUG	16.5	23.4	21.4	28.5	27.7	12.3	34.1	30.5	5.0	11.7	3.3	20.6	235.1	3.4
All	243.3	325.7	303.5	315.7	263.2	317.1	219.3	204.9	211.5	159.6	192.2	192.0	2948.1	42.7
%emit	3.5	4.7	4.4	4.6	3.8	4.6	3.2	3.0	3.1	2.3	2.8	2.8	42.7	

If this percentage be compared with the respective percentage for other pollutants as sulfur, lead and benzo(a)pyrene (BC-EMEP, 1994, 1995, 1996, 1997), one can notice that these figures are very small for Hg. This fact can explain by two reasons. The first one is the re-emission. The emitted mercury is transported by atmospheric circulation, deposited, re-emitted, transported far on etc. It is natural that when re-emission exists, the transport abilities of the pollutant are greater. The other reason is that the main part of airborne mercury exists in the state of Hg^0. This compound has very small dry deposition velocity and is almost impossible to be removed by precipitation. The most effective way of depositing mercury is the transformation of Hg^0 to Hg^{II}. The transformation rate $Hg^0 \rightarrow Hg^{II}$ is limited and as a result mercury is transported to long distances.

CONCLUSIONS

The mercury version of the dispersion model EMAP, in spite of the simple parameterization of mercury specific processes in air, soils and seas, as well as the first glance setting of the ruling parameters, and rough initialization of the re-emission mass, shows quite good results when applied to real data. Much work has to be done further for adjusting the model to the measurement data, both in making parameters more precise and improving and complicating the parameterization by including additional processes and variables.

REFERENCES

BC-EMEP (1994, 1995, 1996, 1997): Bulgarian contribution to EMEP, Annual reports for 1994, 1995, 1996, 1997, NIMH, EMEP/MSC-E, Sofia-Moscow.

Galperin M. and A. Maslyaev (1996): A tentative version of airborne transport POP model; Description and preliminary results of sensitivity studies, EMEP/MSC-E report 3/96, June 1996, Moscow, Russia.

Galperin M., A.Gusev, S.Davidova, V.Koropalov, E.Nesterova, M.Sofiev (1994): An approach to model evaluation of the airborne mercury transport, EMEP/MSC-E Technical report 7/94, April 1994, Moscow, Russia.

Iverfeldt A. and O.Lindqvist (1986): Atmospheric oxidation of elemental mercury by ozone in the aqueous phase, *Atmos. Environ.*, **20**, 553-564.

Ivancheva J., P. Videnov and S. Bogdanov: (1999) Emissions of cadmium and mercury in Bulgaria for 1990 and 1995, in "Bulgarian contribution to EMEP, Annual reports for 1998, NIMH, EMEP/MSC-E, Sofia-Moscow, January 1999, 3-17.

Moisseev B.N. (1997): Simple model for assessment of mercury emission fluxes from soils and their mapping for Europe, EMEP/MSC-E Technical Note 8/97, June 1997, Moscow, Russia.

Münch J., J.Pacyna, F.Axenfeld (1992); European Emission Database on Mercury Compounds for Modelling Purposes, Belastung von Nord- und Ostsee durch okologisch gefahrliche Stoffe am Beispiel atmospharisher Quecksilberverbindungen. - GKSS 92/E/I 11, Berlin.

Pai P., P.Karamchandani, C.Seigneur (1997): Simulation of the regional atmospheric transport and fate of mercury using a comprehensive Eulerian model, *Atmospheric Environment*, **31**, No. 17, 2717-2732.

Syrakov D. (1995), On a PC-oriented Eulerian Multi-Level Model for Long-Term Calculations of the Regional Sulphur Deposition, in Gryning S.E. et al. (eds), Air Pollution Modelling and its Application XI, NATO - Challenges of Modern Society, Vol. 21, Plenum Press, N.Y. and London, pp. 645-646.

Syrakov D. (1996), On the TRAP advection scheme - Description, tests and applications, in Geernaert G., A.Walloe-Hansen and Z.Zlatev <Eds.>, Regional Modelling of Air Pollution in Europe. *Proceedings of the first REMAPE Workshop, Copenhagen, Denmark, September 1996*, National Environmental Research Institute, Denmark, pp. 141-152.

Syrakov D. (1997): Influence of re-emission on pollution distribution: one dimensional multi-level model, in V.Andreev (eds), Bulgarian contribution to EMEP, Annual reports for 1997, NIMH, EMEP/MSC-E, Sofia-Moscow, 21-25.

Syrakov D. (1999): Multi-layer parameterization scheme for Hg (first version). An analysis of re-emission influence on deposition distribution, in "Bulgarian contribution to EMEP, Annual reports for 1998", NIMH, EMEP/MSC-E, Sofia-Moscow, January 1999, 19-27.

Syrakov D. and M. Galperin (1997b) On a new Bott-type advection scheme and its further improvement, in H. Hass and I.J. Ackermann <Eds.>, *Proc. of the first GLOREAM Workshop, Aachen, Germany, September 1997*, Ford Forschungszentrum Aachen, pp. 103-109.

Syrakov D. and D.Yordanov (1997), Parameterization of SL Diffusion Processes Accounting for Surface Source Action, , *Proc. of 22nd NATO/CCMS International Technical Meeting on Air Pollution Modelling and its Application, 2-6 June 1997, Clermont-Ferrand, France*, 111-118.

Yordanov D., D.Syrakov D. and G.Djolov (1983) A barotropic planetary boundary layer, *Boundary Layer Meteorology*, **25**, 363-373.

DISCUSSION

T. IVERSEN: When the dispersion of Hg is global in nature, how can you defend choosing such a small domain?

D. SYRAKOV: As far as this is an estimate of the Bulgarian impact in Hg pollution of the region and only Bulgarian sources are dealt with such calculations are quite informative. If estimate of all Hg pollution is necessary to be made, non-zero values at the inflow boundaries must be set up or zooming procedure must be applied.

A COMPREHENSIVE EULERIAN MODELING FRAMEWORK FOR AIRBORNE MERCURY SPECIES: MODEL DEVELOPMENT AND APPLICATIONS

Gerhard Petersen,[1] Robert Bloxam,[2] Sunny Wong,[2] Olaf Krüger,[1] and Stefan Schmolke[1]

[1]GKSS Research Center
Institute of Hydrophysics
Max-Planck-Strasse
D-21502 Geesthacht, GERMANY
[2]Ontario Ministry of Environment
Environmental Monitoring and Reporting Branch
125 Resources Road
Etobicoke ON M9P 3V6, CANADA

INTRODUCTION

Mercury has been recognized as one of the most important toxic contaminants worldwide. The major features of mercury impacts on the environment and human health are

(i) slow or not at all degradation in the environment ('persistent'),

(ii) occurrence to a significant extent in both the gas phase and condensed states at environmental temperatures ('semi-volatile')

(iii) the tendency to accumulate in the lipophilic tissues of organisms ('bioaccumulative')

(iv) the potential to harm wildlife and human populations exposed to trace amounts ('toxic')

Unlike other heavy metals that are associated with atmospheric aerosols, mercury exists in ambient air predominantly in gaseous elemental form, which is estimated to have a global atmospheric residence time of about one year making it subject to long-range atmospheric transport over spatial scales from about 100 km to continental and global. Hence, mercury is a pollutant of concern in remote areas far away from anthropogenic sources, such as the polar regions, European regional seas, and inland lakes in the Northern U. S., Canada and Scandinavia.

The 1990 U. S. Clean Air Act Amendments, the major European marine environment protection conventions (OSPAR, HELCOM, MEDPOL), the Arctic Monitoring and Assessment Program (AMAP), the U. S./Canada Great Lakes Water Quality Agreement (GLWQA) and the recently signed UN-ECE protocol on reducing the atmospheric transboundary transport of mercury in Europe have intensified the scientific interest to relate the spatial and temporal information on the release of mercury into the atmosphere with the pattern of atmospheric deposition fluxes to various ecosystems by means of numerical modeling on continental geographical scales. In this context, efforts have been made to simulate the atmospheric transport and fate of mercury and to derive estimates of ambient concentrations and dry and wet deposition fluxes of mercury over North America (Shannon and Voldner, 1995; Bullock et al., 1997; Pai et al., 1997) and Europe (Petersen et al., 1995; Ryaboshapko, 1998) through either relative simple Lagrangian formulations or Eulerian approaches employing extensive gas- and aqueous phase chemical mechanisms and tracking explicitly numerous species concentrations.

This paper describes the development, testing and evaluation of a comprehensive Eulerian mercury simulation model as a joint activity within the Canada-Germany Science & Technology Cooperation agreement.

DEVELOPMENT OF THE ADOM MERCURY MODEL

Recent progress in understanding the atmospheric mercury cycle (Schroeder and Munthe, 1998) has allowed for direct modeling of the complex non-linear mercury chemistry by fully three-dimensional Eulerian models. As a first step in this direction, the cloud mixing, scavenging, chemistry and wet deposition modules of the Acid Deposition and Oxidants Model (ADOM), originally designed for regional-scale acid precipitation and photochemical oxidants studies (Venkatram et al., 1988; Misra et al., 1988) have been restructured to accommodate recent developments in atmospheric mercury chemistry. A stand-alone version of these modules referred to as the Tropospheric Chemistry Module (TCM) was designed to simulate the meteorology and chemistry of the entire depth of the troposphere to study cloud mixing, scavenging and chemical reactions associated with precipitation systems that generate wet deposition fluxes (Petersen et al., 1998). The TCM chemistry scheme was developed by systematic simplification of the detailed Chemistry of Atmospheric Mercury (CAM) process model, which is based on current knowledge of physico-chemical forms and transformation reactions of atmospheric mercury species (Pleijel and Munthe, 1995).

After comprehensive testing under different environmental conditions the TCM has been implemented into both a North American and a European version of the full ADOM model. Within the constraints of the available computer resources and input data, these models incorporate an up-to-date understanding of the detailed physical and chemical processes in the atmosphere. In both models, the vertical grid consists of 12 unequally spaced levels between the surface and the top of the model domain at 10 km. The North American and the European version are run for a grid cell size of 127 by 127 km (fine mesh Canadian Meteorological Center (CMC) grid) over a 33 by 33 grid domain and of 55 by 55 km (High Resolution Limited Area Model (HIRLAM) grid) over a 76 by 76 domain, respectively.

The major modules making up the mercury version of ADOM together with the model input data sets are schematically depicted in Figure 1. The transport and diffusion module uses a sophisticated cell-centered flux formulation solver for the 3-dimensional advection-diffusion equation. Dry deposition is modeled in terms of a deposition velocity for gaseous and particle associated mercury species, which is calculated as the inverse of the sum the aerodynamic, deposition layer and surface canopy resistance. Gas phase chemistry is currently restricted to the oxidation of elemental mercury (Hg^0) by ozone. Chemical transformations of mercury species mainly occur in the aqueous phase, i. e. in cloud water droplets and in precipitation. Clouds are classified as stratus (layer clouds) or cumulus (convective clouds) according to the diagnostic output from the weather prediction models.

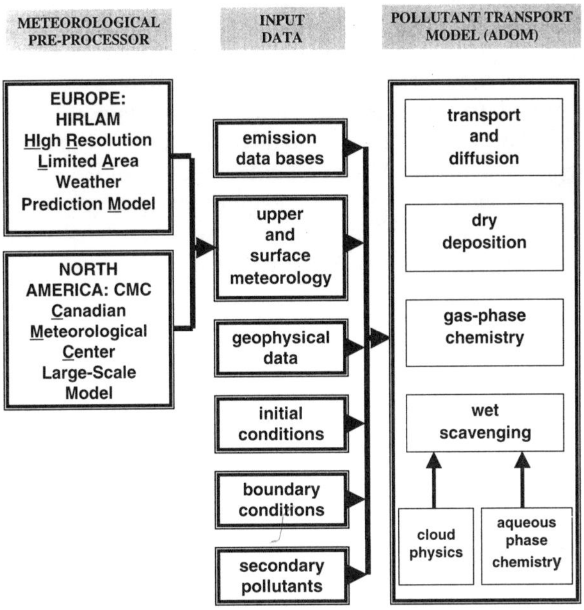

Figure 1. The ADOM model system.

The details of each module comprising the original ADOM version for acid rain studies are given in ERT (1984). The development and testing of the mercury wet scavenging module consisting of cloud physics and mercury gas and aqueous phase chemistry sub-modules is described in detail in Petersen et al., 1998.

PREPARATION OF MODEL INPUTS

Emissions

The databases for anthropogenic mercury emissions in Europe employed in the model calculations have been compiled for 1990 (Umweltbundesamt, 1994). The emission rates and their spatial distribution in the model grid are based on location and capacity of their dominating source categories such as combustion of fossil fuels in power plants, non-ferrous metal smelters, waste incinerators, chlor-alkaline factories and other industrial installations. The emission rates in each grid square are speciated with respect to gaseous elemental mercury (Hg^0), gaseous divalent mercury ($HgCl_2$), and particulate mercury (Hg(part.)), using estimated sector speciation percentages (Lindqvist et al., 1991). For modeling purposes, this speciation provides a possibility of separate treatment of mercury emitted in different physico-chemical forms and hence an assessment of the potential importance of these species for the mercury deposition pattern in Europe.

Prior to the political changes of the 1990s the total European anthropogenic mercury emissions into the air were estimated to be about 700 tons per year. Due to the sharp decline of industrial activities in Eastern Europe mercury emission have been reduced to about 450 tons in 1990. Compared to SO_2 and NO_x, mercury emissions in Europe still show a distinct geographical distribution characterized by a single very pronounced emission peak in Central Europe, which is clearly reflected in almost all of the model predicted mercury concentration and deposition patterns.

Meteorological Fields

The meteorological input data needed by ADOM are three-dimensional fields of wind speed, wind direction, pressure, temperature, relative humidity, vertical velocity and vertical diffusivity , and two-dimensional fields of surface winds, surface pressure, surface air temperature, friction velocity, Monin-Obukhov length, mixing height, cloud base and top height, amount of cloud cover and the amount of precipitation at every one hour model time step. These data sets are derived diagnostically using the weather prediction model HIRLAM for Europe and the Canadian Meteorological Center's model for North America .

Geophysical Data

The geophysical data include files for 8 land use categories (i. e. deciduous forest, coniferous forest, grassland, cropland, urban, desert, water and swamp) ans 12 soil categories. The database also includes information on terrain height and the growing season. This geophysical data affects meteorology, dry deposition processes and air-surface exchange of gaseous mercury species.

Initial and Boundary Conditions

Initial and boundary conditions are needed for all advected species in the model. This includes the emitted compounds and mercury oxide (HgO) formed by gas phase oxidation of Hg^0. In the atmospheric boundary layer (layers 1-4), constant initial and boundary concentrations of 1.5 ng m^{-3}, 0.2 pg m^{-3}, 20 pg m^{-3} and $0.7*10^{-6}$ pg m^{-3} are used for Hg^0, $HgCl_2$, Hg(part.) and HgO, respectively. Above layer 4, concentrations of $HgCl_2$ and Hg(part.) are allowed to decrease with height to a value of about 10 % of the boundary value at the model top, whereas the very low value for HgO is constant with height. Due to its relative long residence time in the atmosphere, the vertical decrease of Hg^0 is more gradual, i. e. approximately 80% of the boundary value at the top of the modeling domain.

Chemical Concentrations

The mercury chemistry in ADOM, described in Petersen et al. 1998, requires the specification of O_3, SO_2, and soot carbon concentrations in ambient air as well as Cl^- concentrations in cloud water and cloud water pH. For the results with the European version of ADOM, the concentrations of O_3, SO_2, and soot carbon were fixed at 35 ppb, 1 ppb and 1 microgram per m^3. The cloudwater concentration of Cl^- and the cloudwater pH were specified as $2*10^{-6}$ mol l^{-1} and 4.5, respectively.

RESULTS OF MODEL APPLICATIONS

The model has been applied to various mercury deposition episodes in North America and Europe. As an example for model testing and application in Europe, a winter 1998 simulation period is described subsequently. This episode has been studied as a part of the European Union-Marine Science and Technology (EU MAST III) Baltic Sea System Study (BASYS). The BASYS data base, which includes extensive mercury measurements in the atmosphere, provides a unique opportunity to test the ADOM mercury model system.

Concentrations in ambient air

Figures 2a, 2b and 2c show calculated Hg^0 , $HgCl_2$ and Hg(part.) air concentration patterns obtained from the BASYS study averaged over the entire month of February 1998. The concentrations refer to a vertical averaging over the first model layer (1-56 m). As

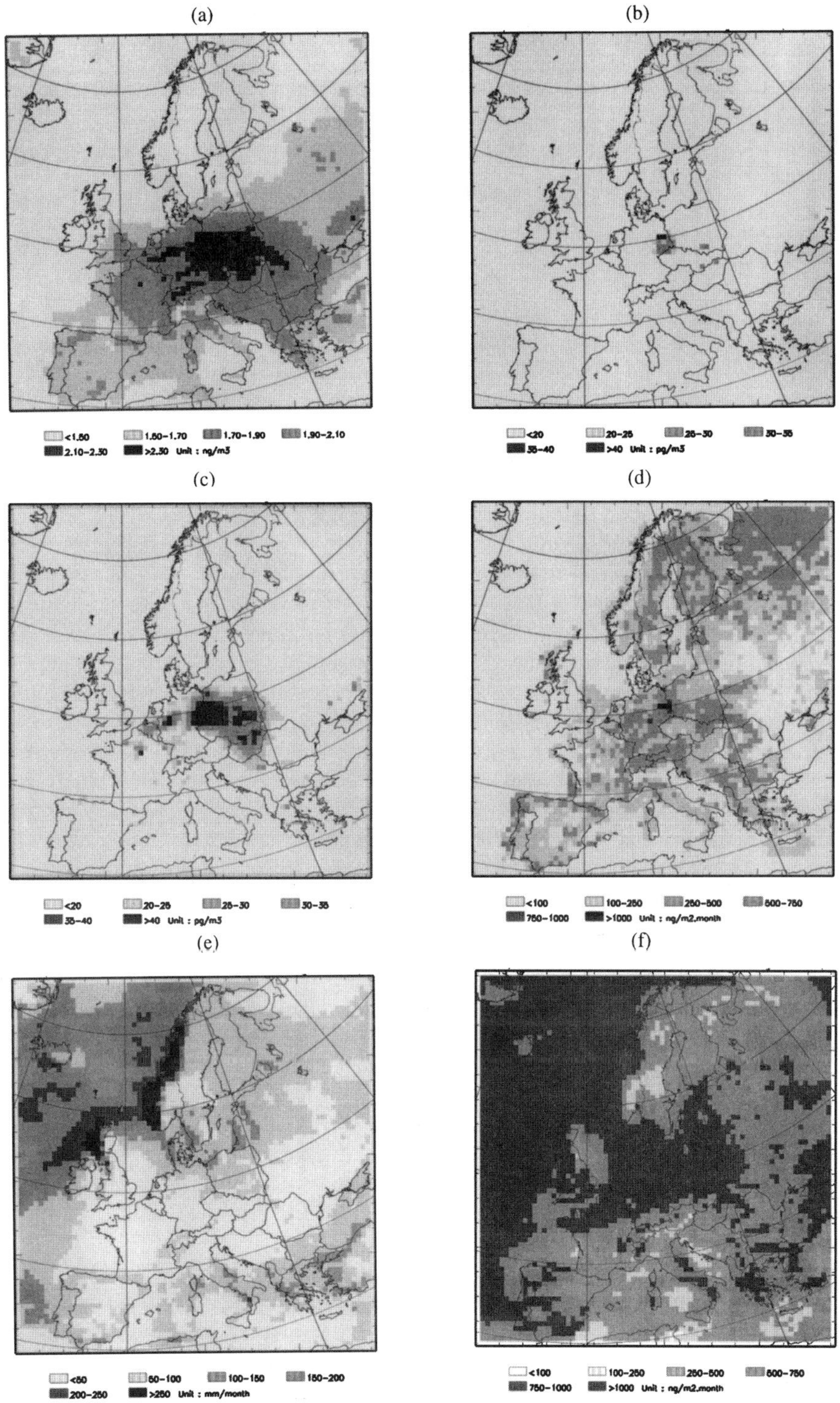

Figure 2. Concentration and deposition pattern of mercury species over Europe for February 1998 predicted by ADOM.

(a) elemental mercury (Hg^0) (b) mercury chloride ($HgCl_2$) (c) particulate mercury (Hg(part.))
(d) total dry deposition flux (e) precipitation amount (f) total wet deposition flux

(a) (b)

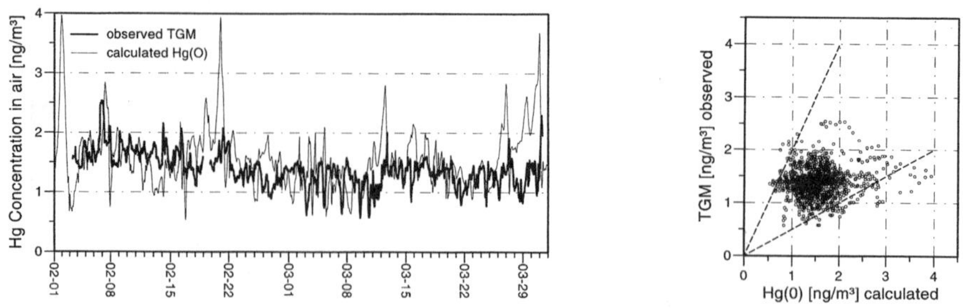

Figure 3. Comparison of observed and model predicted hourly mercury concentrations for February and March 1998 at a location at the German coast of the Baltic Sea.(a) time series (b) scatter diagram

expected the maximum concentrations of all three species are closely related to the source areas. Concentration values of Hg^0 generally range from about 1.1 ng m^{-3} in remote locations to 3.0 ng m^{-3} in the main emission area in Central Europe. The concentration pattern of Hg^0 is substantially elongated towards Eastern Europe, i. e. in the direction of the mean wind during that month. As can be seen from figures 2b and 2c, the concentration levels of $HgCl_2$ and Hg(part.) are about two orders of magnitude lower in the main emission areas due to lower emission rates of these species. The elongation of the concentration patterns towards Eastern Europe is less pronounced for $HgCl_2$ and Hg(part.), since these species have a significantly shorter atmospheric life time mainly due to their effective dry and wet deposition.

The observations and model predictions in Figure 3a are in reasonable agreement. Observed peaks are generally lower than the peak values of model predictions, probably due to underestimated vertical diffusion by the model at nighttime. The scatter plot of these data (Figure 3b) indicates an agreement within a factor of two in more than 90% of the cases.

Dry and wet deposition

The dry deposition fluxes shown in Figure 2d denote the sum of Hg^0, $HgCl_2$ and Hg(part.) deposition across the entire model domain. As can be seen, dry mercury deposition fluxes mainly occur over land surfaces with elevated levels in the major emission areas. The reason for that is twofold: First, due to its very low solubility the dry deposition rate of Hg^0 to all surfaces except forests was set to zero in the model simulations. Second, $HgCl_2$ and Hg(part.) are readily dry deposited in the vicinity of sources resulting in very minor dry deposition over sea areas.

For precipitating clouds, dissolved and particle associated mercury species present in cloud water are deposited at the end of the model time step. The wet deposition flux depicted in Figure 2f is derived by summing up the product of the hourly average cloudwater concentration of the aqueous species and the hourly precipitation amount over the entire month. As expected, the wet deposition pattern is determined by the precipitation distribution (Figure 2e) and to a certain extend to the concentrations of mercury species in ambient air (Figures 2a-2c).

In Figure 4 average levels of observed and model predicted concentrations of mercury in precipitation at 4 monitoring stations in Baltic Sea coastal areas are compared. The observed data are based on 64 precipitation events during the monitoring network study

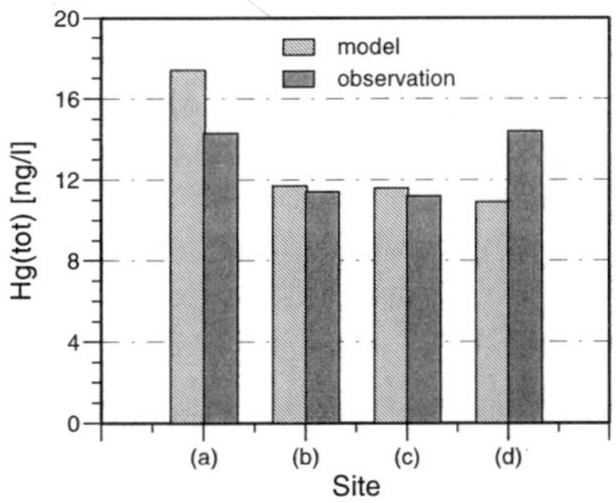

Figure 4. Observed and model predicted mercury concentrations in precipitation BASYS winter network study 1998 at
(a) Kap Arkona, Germany (b) Hel,Poland (c) Preila, Lithuania (d) Hoburg, Sweden

from February 3 to March 30 1998, whereas the calculated numbers are obtained from hourly values during that time period. The overall agreement is good with a tendency to overpredict concentrations relative close to sources (Kap Arkona at the German coast) and underpredict concentrations at more remote locations (Hoburg on the island of Gotland in the Central Baltic).

SUMMARY AND CONCLUSIONS

Comprehensive mercury model systems using the Eulerian reference frame of the Acid Deposition and Oxidants Model (ADOM) for Europe and North America have been developed to calculate the atmospheric long-range transport, chemical transformations and deposition fluxes of mercury species. The evaluation of the model performance led to four main issues:

(1) Model predictions of hourly Hg^0 concentrations in ambient air compare with observations at a location in the Baltic Sea coastal area within a factor of two in more than 90 % of the cases, suggesting that Hg^0 emissions and transport calculations are simulated satisfactorily by the European version of the model. However, the available data material is scarce, and more simultaneous measurements from representative locations with an hourly time resolution are necessary to evaluate the ultimate model performance with respect to Hg^0.emission and transport.

(2) To improve model derived estimates of mercury deposition it is necessary to compile emission inventories based on measurements of mercury species in flue gases of the most important source categories. At present the quality of the emission data for $HgCl_2$ and Hg(part.) limits the confidence with which the effects of these species on the deposition pattern can be predicted.

(3) Air/surface exchange fluxes of gaseous mercury species as a function of meteorological and geophysical parameters need to be measured in order implement these processes into the model system.

(4) Time dependent vertical Hg^0 concentration profiles at the model inflow boundies calculated by hemispheric or global scale models should be introduced. There is evidence now from long-term measurements at a location close to the western boundary of the European modeling domain (Ebinghaus and Schmolke; 2000), that Hg^0 concentrations show a pronounced seasonal variability which may have a significant effect on the temporal deposition pattern in the model domain at least in areas far from the major anthropogenic sources.

Overall, the development level of the model is such, that its application for an assessment of mercury concentration and deposition patterns over Europe and North America can be justified. The model incorporates the current state of knowledge of mercury chemistry and represents a significant progress in predicting long-range transport induced mercury deposition fluxes to aquatic and terrestrial ecosystems. The model system will be developed further when the information above mentioned becomes available.

ACKNOWLEDGMENTS

The authors wish to acknowledge the financial support for this work which was provided by the European Union within the MArine Science and Technology – BAltic Sea SYstem Study (MAST-III-BASYS), Subproject 5 'Atmospheric Load', and by the International Bureaus of the GKSS Forschungszentrum Geesthacht (GKSS) and the Deutsche Forschungsanstalt für Luft-und Raumfahrt (DLR) in the framework of the Germany/Canada Science and Technology Co-operation Agreement (Project No. CAN 99/002).

REFERENCES

Bullock, R. O., Benjey, W. G., and Keating, M. H., 1997, Modeling of regional scale atmospheric mercury transport and deposition using RELMAP, in: *Atmospheric Deposition of Contaminants to the Great Lakes and Coastal Waters,* J. E. Baker, ed., SETAC Press, Pensacola, 1997, pp.323-347.

Ebinghaus, R., and Schmolke, S. R., 2000, Spatial and temporal variability of atmospheric mercury concentrations in northwestern and central Europe, in: *Proceedings of the NIMD 99 Forum*, Minamata, Japan, October 11-13, 1999, in press.

ERT, 1984, ADOM/TADAP Model Development Program, Vols. 1-8, ERT No. P-B980-535, July 1984, Environmental Research and Technology, Inc., Newbury Park, California 91320, USA.

Lindqvist, O., Johansson, K., Aastrup, M., Andersson, A., Bringmark, L., Hovsenius, G., Hakonson, L., Iverfeldt, A., Meili, M., and Timm, B., 1991, Mercury in the Swedish environment – recent research on causes, consequences and corrective methods, *Water, Air, & Soil Pollution,* Vol. 55-1991.

Misra, P. K., Bloxam, R., Fung, C., and Wong S., 1989, Non-linear response of wet deposition to emission reductions: A case study, *Atmospheric Environment*, Vol. 23, No. 3, pp. 671–687.

Pai, P., Karamchandani, P., and Seigneur, C., 1997, Simulation of the regional atmospheric transport and fate of mercury using a comprehensive Eulerian model, *Atmospheric Environment*, Vol. 31, No.17, pp. 2717 – 2732.

Pai, P., Karamchandani, P., and Seigneur, C., 1999, Sensitivity of simulated atmospheric mercury concentrations and deposition to model input parameters, *Journal of Geophysical Research*, Vol. 104., pp. 13,855-13,868.

Petersen, G., Iverfeldt, A., and Munthe, J., 1995, Atmospheric mercury species over Central and Northern Europe. Model calculations and comparison with measurements from the Nordic air and precipitation network for 1987 and 1988, *Atmospheric Environment* Vol. 29, pp. 47-67.

Petersen, G., Munthe, J., Pleijel, K., Bloxam, R., and Kumar, A. V., 1998, A comprehensive Eulerian modeling framework for airborne mercury species: Development and testing of the tropospheric chemistry module (TCM), *Atmospheric Environment*, Vol. 32, No. 5, pp. 829-843.

Pleijel, K., and Munthe, J., 1995,. Modeling the atmospheric mercury cycle - chemistry in fog droplets, *Atmospheric Environment*, Vol. 29, No. 12, pp. 1441-1457.

Ryaboshapko, A., Ilyin I., Gusev A., and Afinogenova O., 1998, Mercury in the atmosphere of Europe: Concentrations, deposition patterns, transboundary fluxes, EMEP Meteorological Synthesizing Center-East, EMEP/MSC-E Report 7/98, June 1998, Moscow, 55 p.

Schroeder, W. H., and Munthe J., 1998, Atmospheric mercury – an overview, *Atmospheric Environment* Vol. 32, No 5, pp. 809-822.

Shannon, J. D., and Voldner, E. C., 1995, Modeling atmospheric concentrations of mercury and depositions to the Great Lakes, *Atmospheric Environment* Vol. 29, No 14, pp. 1649-1661.

Umweltbundesamt, 1994, The European atmospheric emission inventory of heavy metals and persistent organic pollutants for 1990, Umweltbundesamt (Federal Environmental Agency), P. O. Box 33 00 22, D-14191 Berlin, Germany.

Venkatram, A., Karamchandani, P., and Misra, P. K., 1988, Testing a comprehensive acid deposition model, *Atmospheric Environment*, Vol. 22,pp. 2717 – 2732.

DISCUSSION

J. A. van JAARSVELD: You mentioned air-soil exchange of mercury species. Does this mean that air and soil concentrations are in equilibrium? If so, where does the mercury finally end up? Is this the ocean sediment, the Arctic region (cold-condensation effect)? Can you comment on this?

G. PETERSEN: Indeed, air/soil as well as air/water exchange processes are playing a major role in the global geochemical cycling of mercury. Environmental mercury behaves like other persistent chemicals, not only with respect to the interaction with biota (strong tendency to bioaccumulate), but also considering its multihopp properties, as described in the literature for persistent organic pollutants. That means, the air/soil exchange of mercury is characterized by highly dynamic processes with alternating flux direction, depending on the surrounding conditions.

MODELING OF THE MERCURY CYCLE IN THE ATMOSPHERE

G. Kallos[1], O. Kakaliagou[1], A. Voudouri[1], I. Pytharoulis[1],
N. Pirrone[2], L. Forlano[2], and J. Pachyna[3]

[1] University of Athens, Department of Physics, Atmospheric Modeling and Weather Forecasting Group, University Campus, PHYS-5, 15784 Athens, Greece.
E-mail: kallos@mg.uoa.gr
[2] National Research Council, Institute for Atmospheric Pollution, Rende, Italy
[3] Norwegian Institute for Air Research, (NILU), Norway

INTRODUCTION

The mercury cycle in the atmosphere is considered as very complicated because of the various physicochemical processes involved. In the aquatic environment, important processes like biomethylation occur. With these processes, the highly toxic methylmercury compounds are entered in the aquatic nutrition chain and therefore, in the food chain. During the last years, a considerable effort has been devoted to reduce the mercury emissions. These control efforts are of limited efficiency for several reasons. The most important reasons are

(i) the increase of emissions from sources like coal burning, waste incinerators, cement production, mining etc,
(ii) the lack of understanding of important physicochemical processes like fluxes, transport, transformation and deposition,
(iii) lack of accurate emission inventories, and
(iv) lack of appropriate models.

The physical and chemical processes involved in the mercury cycle in the atmosphere are very complicated and need special treatment. At the framework of the EU/DG-XII project MAMCS a significant effort has been devoted for the development of appropriate models for studying the mercury cycle in the atmosphere. In addition, an improved emission inventory is created while monitoring data in various locations in Europe are selected and used for model calibration and inter-comparison. Our model development includes the incorporation of almost any type of source (point or area), gas and aqueous phase chemistry, gas-to-particle conversion, wet and dry deposition, air-water exchange processes etc. The development was performed within two well-known atmospheric modeling systems: the Regional Atmospheric Modeling System (RAMS) and the SKIRON/Eta. There are several reasons for performing the development of the mercury cycle modeling within these two models: The main reason for using RAMS is its unique capability of two-way interactive nesting of any number of grids which is considered as absolutely necessary for studying near-source dispersion of mercury species. Additional capabilities are the full microphysical parameterization for wet processes, the detailed parameterization of surface processes and the non-hydrostatic formulation. The main reason for using the SKIRON/Eta model for development is its unique capability of describing the dust cycle (uptake, transport, deposition) and the existence of a viscous sub-layer formulation which is necessary for description of mercury fluxes from the sea surface. In both models the mercury cycle formulation is called simultaneously at each time-step in order to eliminate unnecessary effects related to transport and diffusion processes. The configuration of the models is very flexible and can be used in any place on the earth.

In this work, the first results from both models for the Mediterranean Region are discussed.

THE MERCURY MODELING SYSTEMS

The model development was performed within the framework of two well-known atmospheric modeling systems having the desired capabilities. For the mercury processes in the atmosphere (emissions, transport, chemical-physical transformations, deposition) existed knowledge for various processes was integrated in these two systems. A quick description of the two modeling systems is provided below.

The RAMS model: RAMS is a highly versatile numerical code, developed at Colorado State University and Mission Research Inc/ASTeR Division. It is considered as one of the most advanced modeling systems available today. It is a merger of a non-hydrostatic cloud model and a hydrostatic mesoscale model. Its most important capabilities are the two-way interactive nesting of any number of grids, the incorporation of one of the most advanced cloud microphysical process algorithms, a surface parameterization scheme able to utilize information on land-use and soil texture at subgrid scale, an advanced radiative transfer scheme able to describe radiative processes at cloudy environment, a full soil temperature and moisture model and a hydrological model providing partitioning of rain water. It can include any number of passive scalars. A general description of the model and its capacities is given in Pielke *et al.* (1992).

The SKIRON/Eta system: This modeling system was developed recently at the University of Athens from the Atmospheric Modeling & Weather Forecasting Group (Kallos et al., 1997a). It is based on the Eta model, which was originally developed by Mesinger (1984) and Janjic (1984). It uses either the "step-mountain" vertical co-ordinate or the customary pressure or sigma (or hybrid) one. Major development of the Eta model has been at the National Center for Environmental Predictions (NCEP) in Washington. The most important features of this modeling system are the use of 2.5 order turbulence closure scheme, the incorporation of a viscous sublayer scheme for better parameterization of the surface fluxes over water and full physics for surface and cloud processes. Another practical advantage of the SKIRON system is that it provides all the necessary parameterization for precipitation on an efficient way and therefore it does not require expensive computer installations. In addition, this version of the Eta model is easily configurable for any place on earth.

The mercury modules

In both modeling systems (RAMS and SKIRON) the developed modules for the physico-chemical processes of mercury have been incorporated. On each model, basic processes like advection and diffusion are the ones already existing for passive tracers and have been modified accordingly. The mercury processes are constructed in modular form. The modules for the various atmospheric and surface processes of mercury species are briefly described below:

Emissions processor: This module deals with the preparation of emissions from anthropogenic and natural sources. It utilizes the data stored in the Mercury Emissions Database (MEI), allocates the various sources within the model domain (point sources), defines emissions according to the land use (area sources), initial and lateral boundary conditions according to the type of simulation initial run (coldstart) or continuation run (hotstart). The entire module is very flexible because the MEI should be updated easily and the source allocation is automatic according to the geographic co-ordinates and the type of sources. Stack characteristics can be added easily in case they are available.

Chemical-kinetics: Chemical and physical transformations of mercury in the atmosphere with changing meteorological conditions play an important role in the cycle of this contaminant in the environment. RAMS includes explicit treatment of cumulus convection while in SKIRON cloud and rain processes are parameterized. The chemical and physical transformations of mercury and its compounds in the atmosphere are described in the **Chemical-Physical (C-P) module**. The C-P module is a merger of a **Gas-Particle Partitioning (G-P) module** (Pirrone and Keeler (1995) and a **Chemical-Kinetics (C-K) model** (Pleijel and Munthe, 1995; Petersen et al., 1996).

G-P and C-K modules: The G-P and C-K modules are coupled and describe all the mechanisms involved in the dynamics of gaseous and particulate phase mercury in the atmosphere. Gas phase reactions include oxidation of elemental mercury (Hg^0) to divalent mercury (Hg^{2+}) by ozone, hydrogen peroxide, perchlorate and other oxidants. Particulate phase reactions involve the formation of HgO, HgS, $HgCl_2$, $HgSO_4$ and $Hg(NO_3)_2$. The influence of relative humidity on the aqueous oxidation of Hg^0 by ozone and reduction of Hg^{2+} by sulfite is also considered.

The G-P module describes the diffusive uptake or release of a gaseous Hg (Hg^0 and Hg species) by atmospheric particles. Many assumptions are inherent in the formulation of the G-P model. Mass transport in the air is assumed to be sufficiently rapid that the gas-phase concentration of Hg^0 and Hg species is uniform up to the external surface of the particle. The particles are considered only fractionally porous. In spherical co-ordinates, the diffusion of mercury and mercury species in atmospheric particles is governed by the following partial differential equation $\partial S(r) / \partial t = n\, D_m [\partial^2 B(r) / \partial r^2 + 2/r\, \partial B(r) / \partial r]$, where S($r$) is the total volumetric concentration of Hg at a radial distance r from the center of a particle, B(r) is the concentration of Hg in the micropores of a particle, D_m is the molecular diffusion coefficient of elemental mercury and mercury species in the air, and n is the intraparticle porosity of the porous shell and t is the time.

The C-K module accounts for all the major chemical reactions of atmospheric Hg in the gas phase, particulate phase and aqueous phase (i.e., cloud droplets, aqueous phase adsorbed in the total suspended particulate) during the transport of air masses on local and regional scale. In addition, two types of heterogeneous processes are considered in the C-K module: gas-solid processes and liquid-solid processes. As showed in recent studies of atmospheric deposition and transport of Hg on local-urban scales, the contribution of airborne mercury to the overall budget of mercury released from the atmosphere to water and terrestrial receptors may be substantial, although the concentration of atmospheric mercury in particle phase is low compared to that in the gas phase.

Dry deposition module: This module consists of two sub-modules in order to account for dry deposition over water surface and over land.

Dry deposition over water surface: The estimation of dry deposition fluxes and gas exchange rates of particulate and gaseous atmospheric Hg to and/or from the water surface is based on the Hybrid Receptor-Deposition (HRD) model developed by Pirrone et al. (1995). The HRD model accounts for the effect of water wave dynamics and spray formation by considering a separate formulation for smooth and rough areas of the water wave surface, changing meteorological conditions, change of the particle size distribution due to particle growth in the humidity gradient above the water surface, and roughness of the water surface on the transfer processes of contaminants to and/or from natural waters.

Dry deposition over land: Several models exist for dry deposition of particles and gases to soil and vegetation. To predict dry deposition fluxes of particles there is a number of physical-based models able to predict deposition velocities to land surfaces (Sehmel and Hodgson, 1978; Slinn and Slinn, 1981). This module takes into account super micron particles eddy diffusivity, gravitational settling and particle inertia as the main mechanisms influencing the deposition to a terrestrial receptor. Finally, the model combines this term with terminal settling velocity and Brownian diffusion to predict deposition velocities. At the framework of MAMCS project the suggestions of Hicks et al. (1985) have been adopted in order to reduce the uncertainty associate with deposition fluxes of atmospheric mercury to terrestrial. Other formulation (e.g. Giorgi, 1986) have been coded and are incorporated in the models for alternative testing.

Wet deposition module: At the framework of MAMCS project a state-of-the-art wet deposition model that has been developed and linked with the other modules and the atmospheric part of the models. It is based on the existing knowledge (Williams, 1982) and calculates the amount of gaseous and particulate mercury from the atmosphere by precipitation. The wet deposition module has been validated and calibrated by using a long-term record of mercury in rainfall precipitation collected in Europe during the last decade. The wet deposition module has been integrated in the atmospheric models in order to calculate the amount of mercury deposited by precipitation scavenging.

Both modeling systems include a number of pre-processing modules in order to handle all the necessary input data like topography, land cover, soil textures, SST, mercury emissions from point and area sources, gridded ECMWF or NCEP meteorological analysis fields etc. The post-processing modules are mainly for graphics and statistical processing and evaluation. Both systems have been developed for Europe and the Mediterranean but they are easily configurable and transferable to any place, domain size and grid spacing. The block diagrams for both systems are shown in **Fig. 1**.

MODEL SIMULATIONS FOR THE MEDITERRANEAN REGION AND EUROPE

As it has been described in previous work (e.g. Kallos et al., 1997b and references therein) the physiographic characteristics of the Mediterranean Region define a trend in general circulation from North to South. This is mainly due to differential heating between the land of North Africa, South Europe and the Mediterranean waters. This trend dominates during all seasons but it is more pronounced during the warm period of the year. Polluted air masses from Europe are transported towards Mediterranean and North Africa through well defined paths. There are mainly two paths followed: one across the Aegean and the other from

West Mediterranean to the East (e.g. Kallos et al., 1997b). The characteristic time scales for such transport towards North Africa and Middle East are between two and four days.

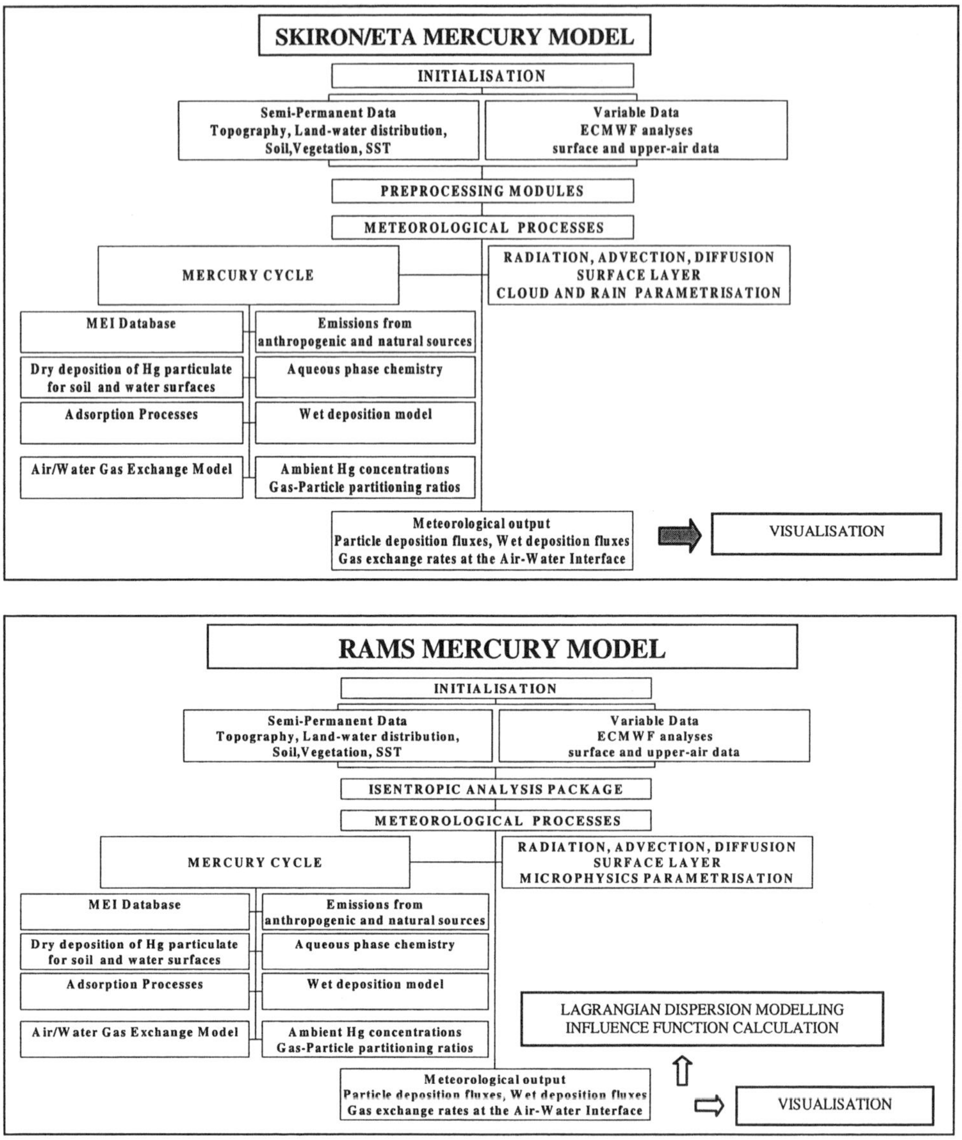

Fig.1: Schematic diagrams of the atmospheric and mercury processes for the two modelling systems

All this knowledge is directly applicable to the present study of the mercury budget in the Mediterranean Region since the long-range transport is considered as important. The main sources of mercury are power and cement plants, refineries, chemical plants (e.g. alkal-cloral plants) and in general from combustion. In addition, in the Mediterranean Region there are two mercury-mining activities in Spain and Slovenia while the volcanoes have a significant contribution too. The processes involved in mercury transport and transformation are rather complicated and require special treatment. Because of the small concentrations of some mercury species and the processes involved, especially the gas to particle conversion, stiff differential equations solvers were used. This requires significant computer resources which makes the simulations for long periods and high resolution very difficult. In addition, the aqueous phase processes are very important and the atmospheric models must include detailed cloud microphysical algorithms, which require also significant computer power. The two atmospheric models used for the development (RAMS and SKIRON/Eta) have such capabilities through different approaches. RAMS has a detailed cloud microphysical scheme and the two-way interactive nesting capabilities which make it appropriate for simulations near the sources and simultaneously over larger areas. The computer power required for such simulations is beyond the limits of the conventional workstations and servers available and require parallel computations. For this reason, most of the simulations performed so far are in a rather coarse grid covering the entire Mediterranean Region and Europe since most of the sources are in this broad area. The SKIRON/Eta system has a microphysical scheme which is less demanding in computer resources but accurate enough for precipitation calculations. Therefore, it is preferable for several sensitivity calculations for several days. The intercomparison of the results between the two models is an absolutely necessary process in order to avoid systematic errors since there are no systematic measurements available for the mercury species in several locations for intercomparison. At the framework of the MAMCS project there are five stations available in the Mediterranean Region (in Mallorca, Sicily, Calabria, Antalya, and Haifa). Measurements were made during four experimental campaigns, for 10-12 days each. In addition, five other stations operated in Northern Europe at the framework of another project funded by EU the Mercury Over Europe (MOE). Since the main target of the MAMCS project is to provide at least an indication of the mercury transport, transformation and deposition over the Mediterranean Region, the results presented here are related to deposition of mercury species through various processes. The simulation period is 1-18 May 1999. Both models were used with similar configurations (horizontal grid increments of 48 km and similar distribution of the same number -32- of vertical layers). The models were initialized on a horizontally-homogeneous manner with representative values of Hg^0, Hg^2 and Hg^P. Several tests were performed with the various processes triggered on or off in order to better understand the response of the entire systems. In **Fig. 2** the Hg^0 concentrations at the first model layer are shown after six days of simulation (RAMS simulation). The pattern follows the well-known long-range, one defined by the synoptic scale flow regime.

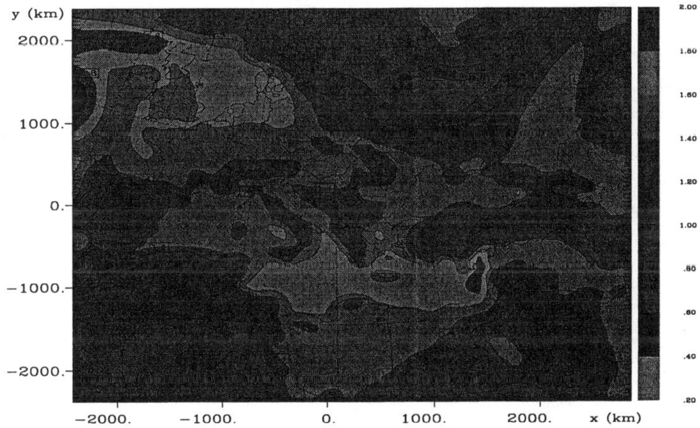

		Grid 1			
z = 61.5 m	1999−05−06−1200 UTC 0 s	min	max	inc	lab*
contours	Hg0 (ng/m3)	.4000	2.000	.2000	1e 0

Fig.2: Hg^0 concentrations at the first model layer after six days of simulation (RAMS simulation)

In **Fig. 3** the dry and wet total depositions of Hg^0, Hg^2 and Hg^P are presented after 17 days of simulation (SKIRON/Eta simulation). The Hg^0 wet deposition pattern is controlled by the areas affected from rain during the days of simulation. The opposite is true for dry deposition, something expected. Hg^2 is deposited rapidly in the vicinity of the sources, mainly due to its reactivity and solubility. The deposition patterns of Hg^P need to be examined carefully. Dry deposition was found to be higher over the sea rather than over land despite the fact that all anthropogenic sources are over land. This is due to the mechanism of particle growth in the wet environment induced by the model. Finally, as it was expected the washout mechanism of the Hg^P follows the rain pattern of these days of simulation.

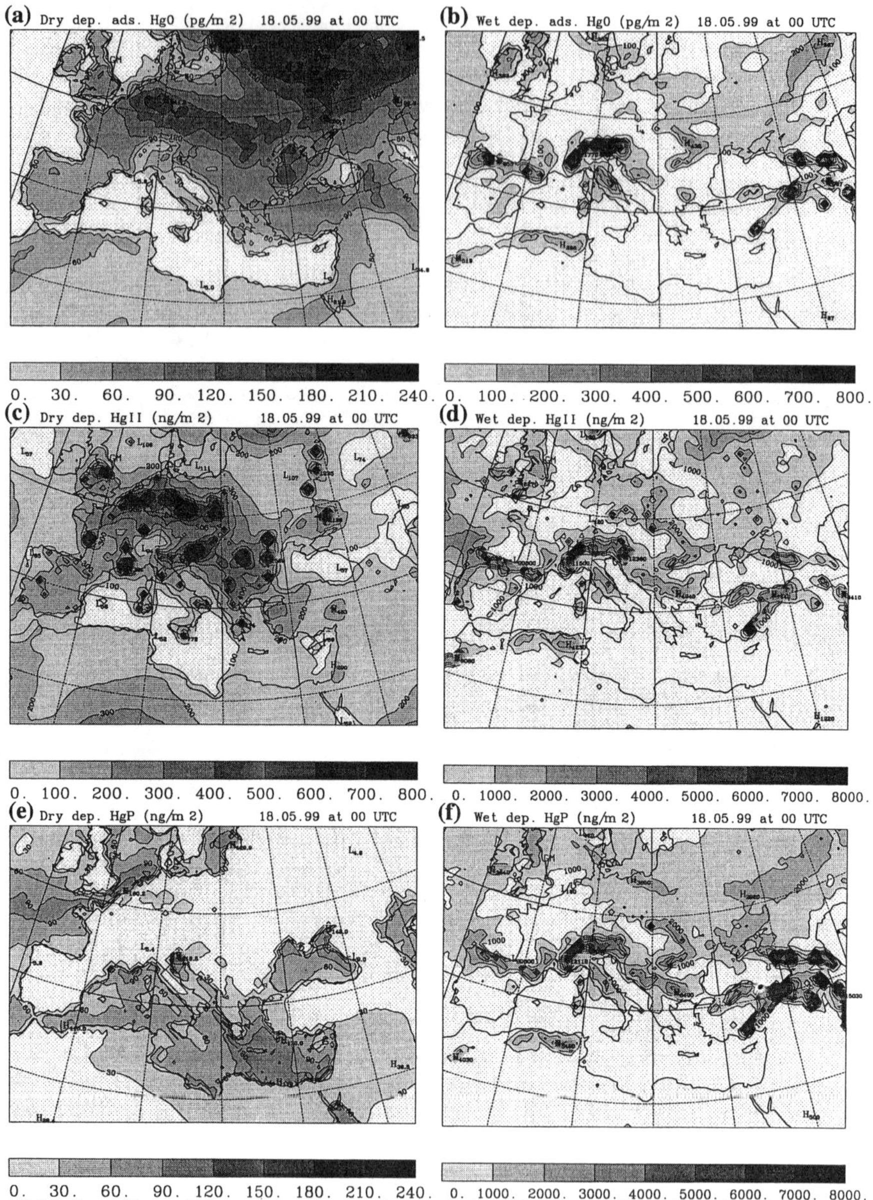

Fig.3: Total deposition of Hg species for the time period 1-18 May 1999. a) Dry deposition of adsorbed Hg^0 (pg/m^2), b) wet deposition of adsorbed Hg^0 (pg/m^2), c) dry deposition of Hg^2 (ng/m^2), d) wet deposition of Hg^2 (ng/m^2), e) dry deposition of Hg^P (ng/m^2), and f) wet deposition of Hg^P (ng/m^2). From the SKIRON/Eta model.

Despite the fact that a one to one intercomparison of the model results with the observations during this experimental period is extremely difficult an attempt was made and the results from two locations, Sicily and Calabria, are presented (**Fig. 4**). The bold lines represent the 6-hours moving averaged (panel a) and the 21-

hours averaged (panel b) concentrations measured at these locations. The thin lines indicate the modelled concentrations (in 6-hourly increments) at the lowest model level (~62 m. above ground) in two runs in which the *background* and lateral boundary concentration of Hg^0 (in the lowest 2 km) was equal to 1.6 and 1.4 ng/m^3. The model follows the trends of Hg^0 in Sicily and Hg^P in Calabria quite satisfactorily. Nevertheless, significant differences between the model forecasts and the observations appear in some relatively short periods of the simulation. Sensitivity tests showed that the predefined *background* and lateral boundary conditions play an important role in the simulations. By lowering the background and lateral boundary concentration of Hg^0, the concentrations of both Hg^0 and Hg^P become smaller. Figure 4a shows that changes in the background and lateral boundary concentration of Hg^0 impose a constant bias in its concentration throughout the run. Several other uncertainties are still not clarified for such intercomparisons. Sensitivity tests also showed that the predicted concentrations of Hg^0 and Hg^P are sensitive to the predefined value of lower tropospheric O_3; its concentration was set equal to 35 ppbv in the control run. Higher values of O_3 result in decrease of Hg^0 and increase of Hg^P, as expected; the opposite occurs when the background value of O_3 is reduced. According to the analysis, the significant deviation between the modelled and observed concentrations of Hg^P in Calabria after 12 May (Fig. 4b) are due to the misrepresentation of the nearby sources of Hg^0 and Hg^P, and to errors in the predicted model precipitation because of the coarse model grid used (~48 km). Higher resolution simulations indicated significant improvement of the predicted mercury concentrations in certain periods. Any time lags between the model forecasts and the observations are mainly due to the way the experimental groups averaged their observations. As far as the quality of the measurements is concerned, it is worth noting that sometimes different measurement methods exhibited significant differences, with magnitude higher than 100%, from each other. All these results are based on the emission inventory prepared at the framework of MAMCS project but it is difficult to provide even a crude estimation of the order of magnitude of the approximation. This is mainly due to difficulties in collecting such data and estimating the emission factors, which may vary to a large degree.

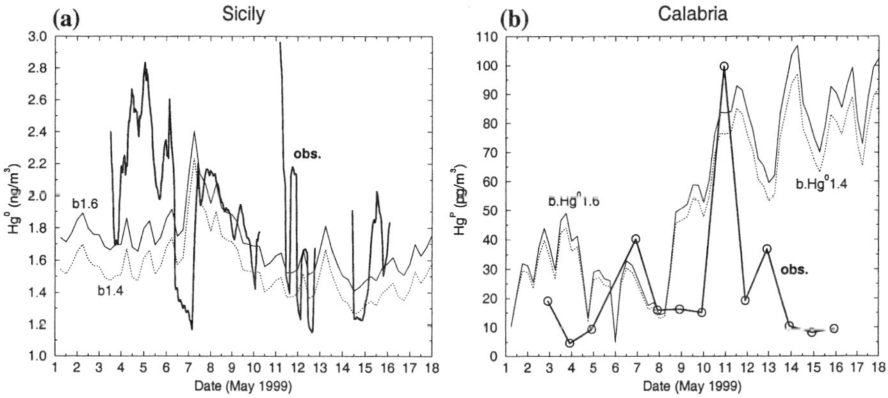

Fig.4: Concentration of a) Hg^0 (ng/m^3) in Sicily and b) Hg^P (pg/m^3) in Calabria at the lowest model level in May 1999. In both panels the thin lines are the SKIRON/Eta model forecasts, in which the lowest 2 km background and lateral boundary Hg^0 concentration was equal to 1.6 ng/m^3 (solid line) and 1.4 ng/m^3 (dotted line). Initial time at 0000 UTC, 1 May 1999. The bold line corresponds to a) a 6-hours moving average and b) a 21-hours average of the observations.

SOME CONCLUDING REMARKS

The modeling of the mercury processes has similarities with conventional air pollution modeling to a high degree. The processes involved and especially the aqueous phase chemistry and Gas-to-Particle conversion are very complicated and need special treatment. The model development performed so far gave encouraging results, which need further analysis. The processes included so far seem to work quite satisfactorily but a systematic model evaluation is difficult unless some other controlling factors, like emission inventories and observation quality, are not improved significantly. The present status of the emission inventories is not considered as satisfactory (the order of magnitude of the approximation is still unknown). Also, the accuracy provided by the various measurement techniques certainly needs significant improvement. Several other processes like fluxes of mercury species from/to water bodies or soil are poorly understood so far but play a very important role in the system. In addition, re-emission processes are considered as very important but there is lack of knowledge to a certain degree. Modeling attempts are not only constrained by adequate knowledge but also from available computer resources, despite the fact that the computer power available today is cheap. From our modeling experience with both models, the mercury processes require at least an order of magnitude more the computer power needed for the atmospheric processes, even with the cloud microphysical processes on. Although significant modeling effort is devoted so far, there is still need for further development. There is a need for a better representation of the gas-phase chemistry, which is poorly represented in the current version of the models. A more systematic way for measurements of the various mercury species and a better understanding of the air-water interactions is absolutely necessary.

Acknowledgements: This research was supported by the research project MAMCS (ENV4-CT97-0593) of the DG-XII of EU. Acknowledgement is also made to NCAR, which is sponsored by the National Science Foundation, and ECMWF for the gridded analysis data used for the model simulations.

References

Giorgi, F. (1986) Development of an Atmospheric Aerosol Model for Studies of Global Budgets and Effects of Airborne Particulate Materiel. PhD Thesis. Georgia Institute of Technology.

Hicks, B.B., D.D. Baldocchi, R.P. Hosker, Jr., B.A. Hutchison, D.R. Matt, R.T. McMillen and L.C. Satterfield (1985) On the use of monitored air concentrations to infer Dry Deposition, NOAA Technical Memorandum ERL ARL-141, Silver Spring, MD.

Janjic, Z.I. (1984) Nonlinear advection schemes and energy cascade on semi-staggered grids. Mon. Wea. Rev., 112, 1234-1245.

Kallos, G., S. Nickovic, A. Papadopoulos, D. Jovic, O. Kakaliagou, N. Misirlis, L. Boukas, N. Mimikou, G. Sakellaridis, J. Papageorgiou, E. Anadranistakis and M. Manousakis (1997a) The Regional weather forecasting system SKIRON. Proceedings of the Symposium on Regional Weather Prediction on Parallel Computer Environments, 15-17 October 1997, Athens, Greece. 109-122.

Kallos, G., V. Kotroni, K. Lagouvardos, A. Papadopoulos, M. Varinou, O. Kakaliagou, M. Luria, M. Peleg, A. Wanger and M. Uliasz (1997b) Temporal and spatial scales for transport and transformation processes in the Mediterranean. Proc. of the 22nd NATO/CCMS ITM on Air Pollution Modelling and Its Application, 2-6 June 1997, Cl. Ferrand, France. Edited by S-E. Gryning and N. Chaumerliac, Plenum Press, New York, Vol 20, pp.8.

Mesinger, F. (1984) A blocking technique for representation of mountains in atmospheric models. Riv. Meteor. Aeronaut., 44, 195-202.

Petersen, G., Munthe, J., Bloxam, R. and A.V. Kumar (1996) A comprehensive Eulerian modelling framework for airborne mercury species: Development and application of a tropospheric chemistry module. In the book of abstracts of the 4th Intern. Conf. on Mercury as a Global Pollutant. August 4-8, 1996, Hamburg, Germany.

Pielke, R.A., W.R. Cotton, R.L. Walko, C.J. Tremback, W.A. Lyons, L.D. Grasso, M.E. Nicholls, M.D. Moran, D.A.Wesley, T.J. Lee and J.H. Copeland (1992) A comprehensive meteorological modeling system - RAMS. *Meteorol. Atmos. Phys.*, 49, 69-91.

Pirrone, N. and G.J. Keeler (1995) Numerical Modeling of Gas-Particle Partitioning of Atmospheric Mercury in Urban Areas. In Proceedings of the *1995 Annual Meeting of the American Association for Aerosol Research (AAAR)*, October 9-13, 1995, Pittsburgh, Pennsylvania, U.S.A.

Pirrone, N., G.J. Keeler and T.M. Holsen (1995) Dry Deposition of Trace Elements over Lake Michigan: A Hybrid-Receptor Deposition Modeling Approach. *Environmental Science and Technology*, 29, 2112-2122.

Pleijel, K. and J. Munthe (1995) Modeling the atmospheric chemistry of mercury - The importance of a detailed description of the chemistry of cloud water. *Water, Air & Soil Pollut.,* 80, 317-324.

Sehmel, G.A. and W.H. Hodgson (1978) Model for Predicting Dry Deposition of Particles and Gases to Environmental Surfaces, PNL-SA-6721, Battelle Pacific Northwest Laboratory, Richland, Washington.

Slinn, S.A. and W.G.N. Slinn. (1981) Modeling of Atmospheric Particulate Deposition to Natural Waters. In: *Atmospheric Pollutants in Natural Waters*, S.J. Eisenreich, Ed., Ann Arbor Science, Ann Arbor, MI. pp.22-53.

Williams, R.M. (1982) A model for the dry deposition of particles to natural water surfaces. Atm.Environ., 16, 1933-1938

DISCUSSION

R. BORNSTEIN: Can any generalizations be made as to the fraction of Hg pollutants that are locally removed versus these involved in Long Range Transport?

G KALLOS: Some generalizations can be made but it is very difficult. For example, most of Hg(II) is removed within "a mesoscale area" around the source while the Hg(0) and Hg(P) are more "regional or global". But, do not forget that we have also chemical transformations on all scales.

P. SEIBERT: Both this talk and the previous one show practically zero correlation between measurements and modelled values. What would you blame for that?

G. KALLOS: I can speak only for my presentation. First, it is your conclusion for poor correlation with which I strongly disagree. According to my best knowledge on this subject, I can say that the single point intercomparison and therefore correlation according to your interpretation is not the best method. We need spatiotemporal correlations between several observational points and the model results. This is especially true for Hg modeling where there are no systematic monitoring stations and the difficulties on performing such measurements are many. In addition, there are no accurate emission inventories for Hg emissions while the emission factors are rather heuristic. The model algorithms should be blamed only for the accuracy of the transformations and the meteorology. Since both models are well tested for the meteorological processes (forecasting models) one very important factor is guaranteed. For the other, we still need to continue the work. Most of the practices have been transferred from the conventional air pollution modeling (e.g. advection, diffusion, source treatment, deposition) which means the model uncertainties are more bounded. To my opinion, we need more work on emissions, measurements of various Hg species and of course better understanding the air-sea interaction and scavenging processes.

LONG-RANGE TRANSPORT OF OZONE FROM THE NORTH AMERICAN BOUNDARY LAYER TO EUROPE: OBSERVATIONS AND MODEL RESULTS

Andreas Stohl[1] and Thomas Trickl[2]

[1]Lehrstuhl für Bioklimatologie und Immissionsforschung
Technische Universität München
Am Hochanger 13, D-85354 Freising-Weihenstephan, Germany

[2]Fraunhofer-Institut für Atmosphärische Umweltforschung (IFU)
Kreuzeckbahnstr. 19, D-82467 Garmisch-Partenkirchen, Germany

INTRODUCTION

Because of the long photochemical lifetime of ozone of the order of 1-2 months in the free troposphere (Liu et al., 1987), ozone transport over long distances is likely to occur quite frequently above the atmospheric boundary layer. Indeed, outflow from the North American continent with high ozone concentrations has been found over the middle of the Atlantic ocean (Parrish et al., 1993; Parrish et al., 1998). However, although photochemical model calculations (e.g. Wild et al., 1996) suggest that ozone transport also occurs over intercontinental distances, there is still little direct evidence from observations.

Stohl and Trickl (1999) presented one of the first clear examples of this kind: Ozone, formed photochemically in the boundary layer over eastern North America, was transported offshore. There the airmass, containing nearly 100 ppb of ozone, got entrained into a warm conveyor belt (WCB), an airstream concomitant with a cold front, which lifted it up to 11000 m above sea level. Finally, it was transported with the jetstream towards Europe, where it started to descend again. The chain of evidence for this case is especially good: Ozone concentrations of nearly 100 ppb were observed in the North American boundary layer, by three MOZAIC aircraft traversing the WCB, and by continuous lidar measurements over Europe. The WCB deposited its outflow above an airmass intruding from the stratosphere, turning the usual vertical layering of the atmosphere upside-down. Stratospheric air (dry, high potential vorticity) was found at low levels, and air from the North American boundary layer (moist, low potential vorticity) was present in the uppermost troposphere.

Here we present another example of intercontinental ozone transport from North

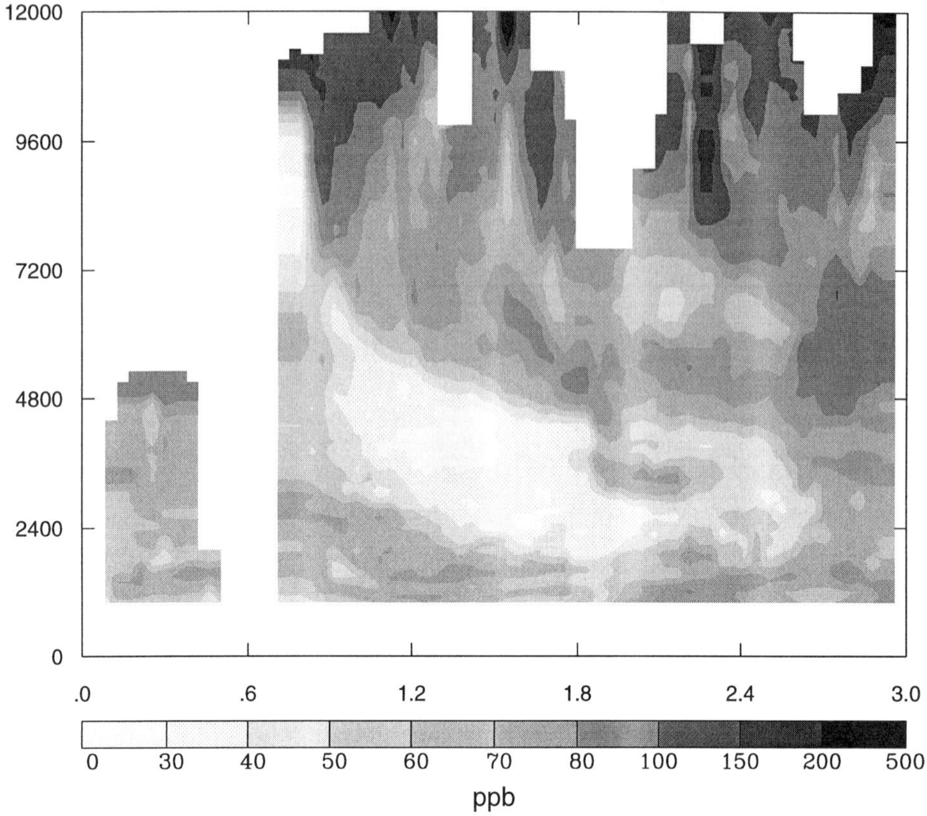

Figure 1: Time-height section of ozone as captured by lidar measurements in Garmisch-Partenkirchen from 28 May 1996 11 UTC to 31 May 11 UTC. Time is given in days, height in meters. The white areas correspond to missing data or cloud layers, for which no useful data have been obtained.

America to Europe, and we show that the transport patterns are the same as those described by Stohl and Trickl (1999). In addition, we establish a one-year "climatology" of WCBs which are responsible for the ozone transport.

A CASE STUDY

Figure 1 shows a time-height section of ozone lidar measurements in Garmisch-Partenkirchen (11.06 °E, 47.5 °N, 740 m asl). The range of these measurements was from 150 m above the ground to approximately 5 km above the tropopause and their vertical resolution was between 50 m at lower levels and about 500 m near the tropopause. The accuracy was of the order of 3 ppb in most of the troposphere and in the absence of clouds. A more complete description of these measurements was given by Eisele et al. (1999), where also a color plate (their Figure 4) of the lidar data can be found.

To elucidate the origin of the different airmasses seen in the lidar profiles (Figure 1), 168-h backward trajectories were initialized every 250 m asl along the lidar axis every three hours. The trajectory calculations were carried out with the FLEXTRA model (Stohl et al., 1995), driven with model-level wind fields from the European Centre for Medium-Range Weather Forecasts (ECMWF) with a resolution of 1 ° and a time interval of three hours.

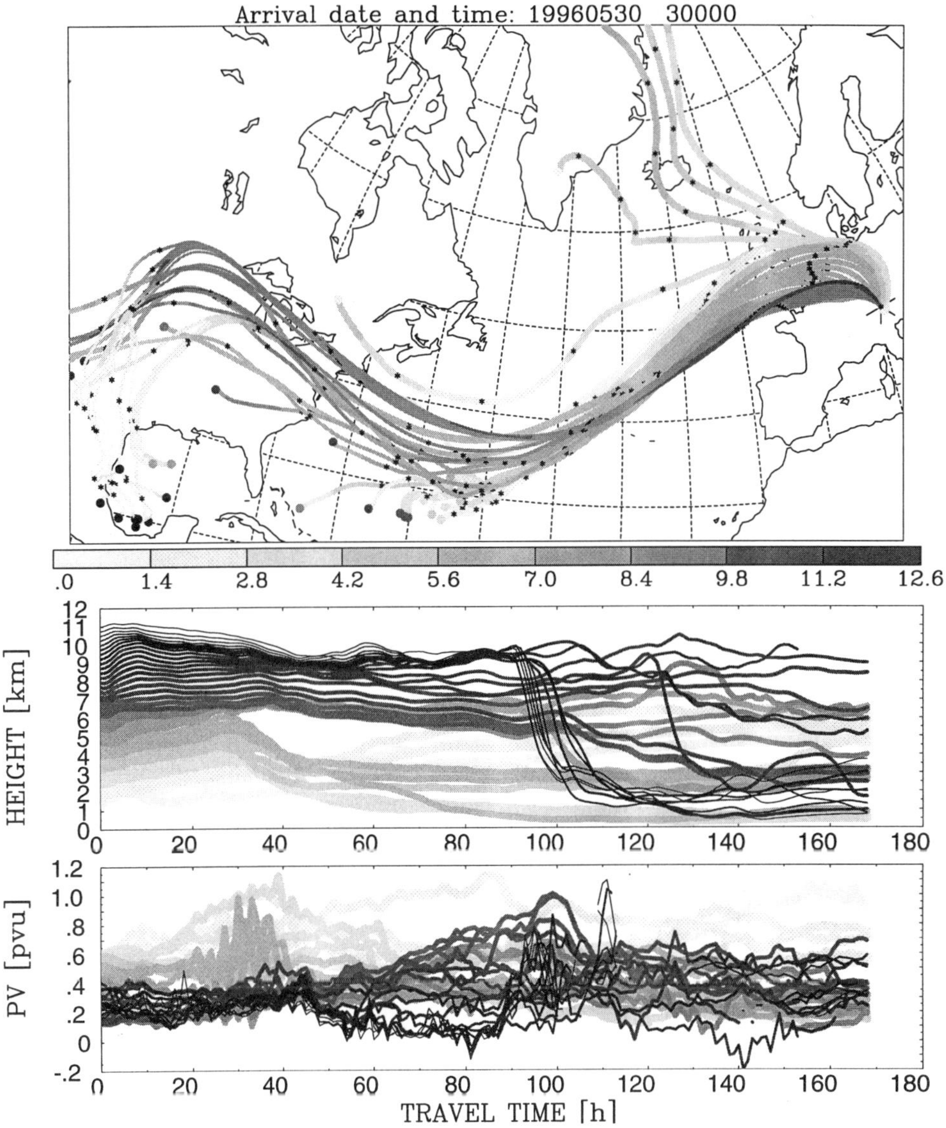

Figure 2: 178 h three-dimensional backward trajectories terminating at the lidar measurement site every 250 m in height on 30 May at 3 UTC (corresponding to day 1.6 in Figure 1): the top figure shows a horizontal projection of the trajectories, with the grayscale according to the label bar referring to the actual heights (in km asl); positions are marked with asterisks every 24 h. The dots at the trajectory starting points indicate the heights where the trajectories terminated along the lidar profile, with the grayscale of the dots corresponding to the middle figure. The same information is also provided by the line widths (thinner lines representing trajectories terminating at higher levels). The middle figure shows time-height profiles of the trajectories with both grayscale and line width indicating their ending height at T=0. The bottom figure represents PV along the different trajectories with both the line grayscale and line width corresponding to the middle figure. PV data is sometimes missing in the boundary layer in unstable situations. Note that time is backward along the trajectories.

Figure 2 shows the results for 30 May at 3 UTC (day 1.7 in Figure 1). The trajectories arriving at the lowest levels, where high (60-80 ppb) ozone concentrations are present in the lidar data, come from the north-north-west and descend strongly. Potential vorticity (PV) along these trajectories fluctuates between 0.6 and 1.1 potential vorticity units (pvu). Actually, this airmass is the wing of a strong intrusion of stratospheric air which led to elevated ozone concentrations at mountain stations in the Alps and Northern Apennines and has been studied in detail by Stohl et al. (2000). This structure of the intrusion is no longer fully resolved by the trajectory calculations, but calculations for earlier times clearly show the stratospheric origin of this airmass (see Stohl et al., 2000).

Very low (20-40 ppb) ozone concentrations are present between 2000 and 5000 m asl. The trajectories ending at these heights ascend from the subtropical maritime boundary layer (position seven days before arrival approximately 60 °W, 28 °N). Ozone destruction in the sunlit moist maritime boundary layer is very fast at these latitudes, explaining the low concentrations.

Above 5000 m, ozone concentrations are again much higher. The highest concentrations (up to 110 ppb) are found at approximately 10000 m. Trajectories starting at these levels are over the Caribbean Sea seven days before arrival, enter North America at low altitudes, i.e. within the boundary layer (below 1000 to 2000 m), one day later and then quickly ascend over the United States to the upper troposphere within approximately one day. Subsequently, the airmass is transported with the jetstream towards Europe. Since the ascent, albeit at a slower rate, continues on transit until shortly before the arrival at the lidar site, relative humidity in this airmass is still higher than that above and below. The radiosonde in Munich (90 km north-north-east of the lidar site) shows a layer between 10000 and 13000 m of 30 to 40% relative humidity with respect to water vapor (the values with respect to ice would be higher) already at 29 March 23 UTC – 4 hours before the main arrival of the WCB airmass started. 12 hours later this relatively moist layer is found again in the Munich sounding. Potential vorticity at these altitudes is lower than in the entire potential vorticity profile at the lidar site.

The transport mechanism for this episode is the same as in Stohl and Trickl (1999). Again, boundary layer air from North America is lifted within a warm conveyor belt, is injected into the uppermost troposphere and is subsequently transported towards Europe. As in the case of Stohl and Trickl (1999), this boundary layer air in fact overruns an airstream intruding from the stratosphere, causing a reversed profile – as compared to more typical conditions – of relative humidity and potential vorticity: high potential vorticity and low relative humidity at low levels, low potential vorticity and high relative humidity at upper levels. In contrast to the episode of Stohl and Trickl (1999), an airmass from the subtropical boundary layer was inserted in between, leading to very low ozone concentrations in the middle troposphere.

WARM CONVEYOR BELTS: A ONE-YEAR "CLIMATOLOGY"

Since it was shown above that ozone transport from North America to Europe appears to occur often with warm conveyor belts, it was interesting to study where on Earth WCBs are found most frequently and how the frequency and the locations change over the course of a year. We used the method of Wernli and Davies (1997), slightly modified, to identify WCBs using trajectory calculations carried out with the FLEXTRA model (Stohl et al., 1995). 48-hour forward trajectories were started daily at

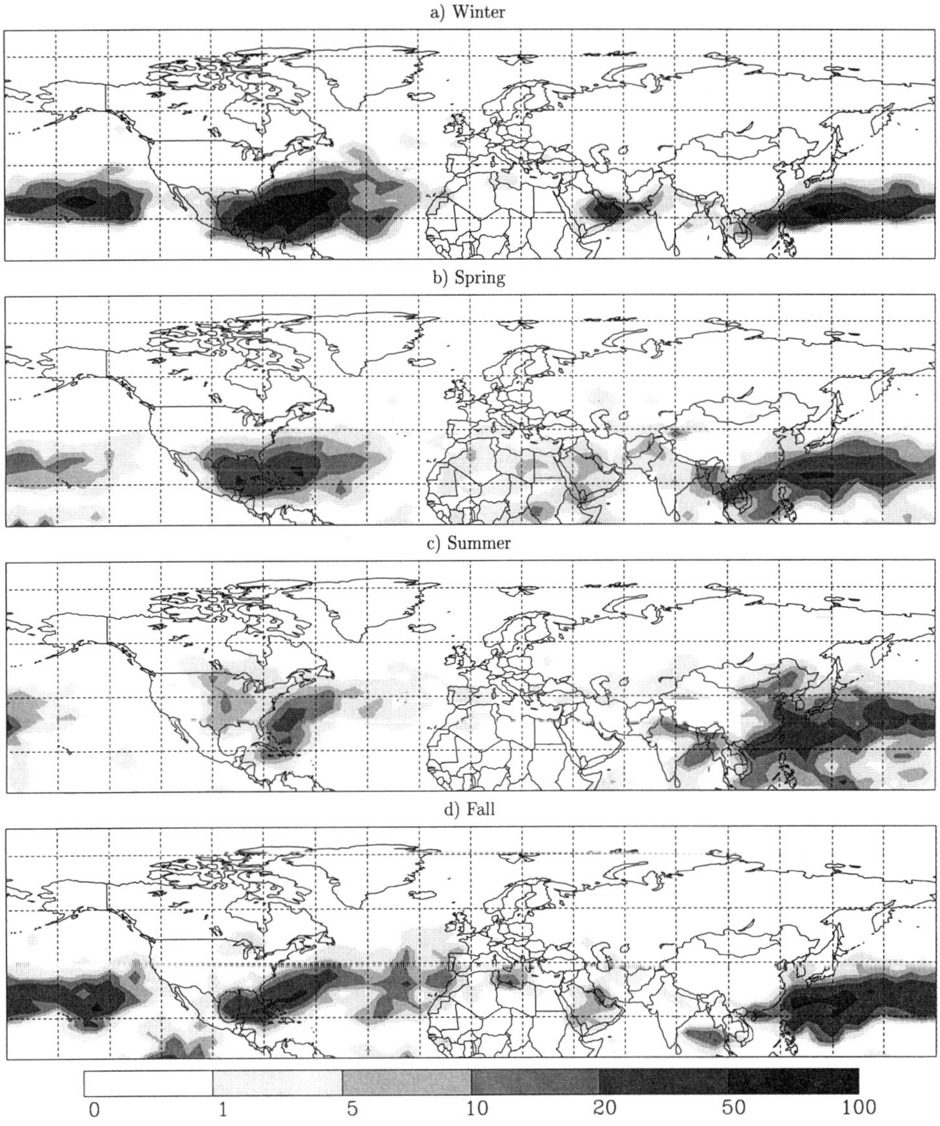

Figure 3: Seasonal and spatial distribution of WCB starting points. Depicted is the fraction (per mille) of trajectories starting from 500 m and 1500 m that fulfill the WCB criterion. Results are averaged over 5 °×3 °grid cells and are shown separately for the four seasons. In general, the frequency of WCBs is higher for the 500 m starting level than for the 1500 m starting level, but the geographical patterns are the same.

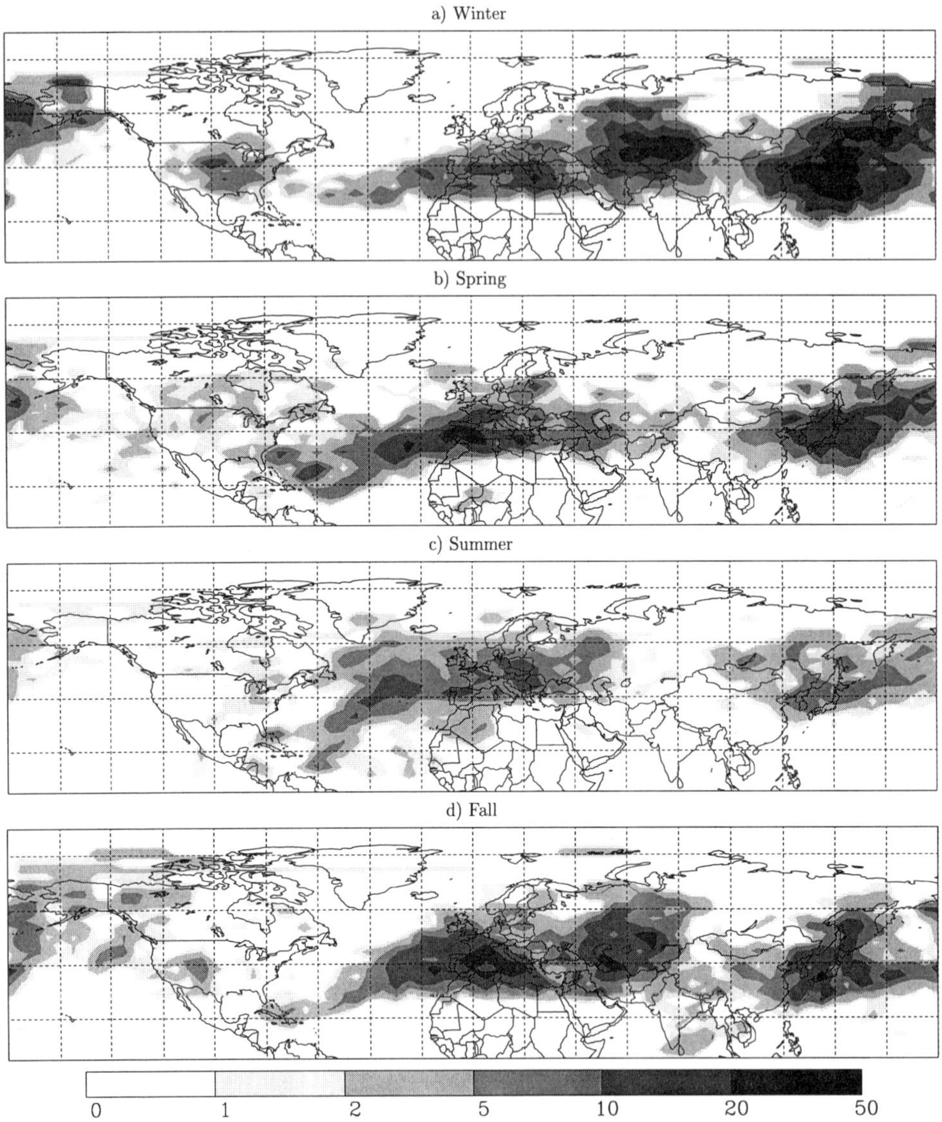

Figure 4: Seasonal and spatial distribution of WCB ending points. Depicted is the fraction (per mille) of trajectories ending between 8500 m and 10500 m that fulfill the WCB criterion.

500 m and 1500 m above sea level on a uniform $1° \times 1°$ latitude-longitude grid during the period 1 April 1997 to 3 April 1998. A trajectory was defined to represent a WCB if it ascended by more than 8000 m. To separate WCBs from similarly fast ascending airstreams occurring with tropical convection, it was furthermore required that the trajectories travel to the northeast (at least 5 °N and 10 °E), as is typical for WCBs.

Figure 3 shows the frequency of WCB trajectories relative to all airstreams (i.e. both those fulfilling and those not fulfilling the WCB criterion). WCB inflow occurs only at latitudes below 50 °N, in winter and spring below 40 °N, with maxima over the warm water pools at the southeastern seaboards of North America and Asia. The proximity to the continental east coasts is significant, since the worldwide highest anthropogenic emissions of sulfur, nitrogen oxides and other substances are located there. Extending the trajectories back in time (not shown) confirms that many airmasses entering WCBs indeed traverse the high-emission regions, while the remainder originate over the Gulf of Mexico and the Caribbean for the North American WCBs and the tropical Pacific for the Asian ones. Thus, the chemical characteristics of WCB airmasses likely range from very clean to highly polluted, depending on their exact path.

WCBs deposit ABL air parcels in the upper troposphere in the middle latitudes. Their outflow, as characterized by the ending points of the 48-hour trajectories, occurs most frequently over Europe, northern Africa and western Asia on the one hand, and eastern Asia on the other, throughout the year (Figure 4). Most of the WCBs originating at the eastern seaboard of North America deposit their airmass at the end of the North Atlantic stormtrack over Europe. In contrast, Asian WCBs run more parallel to the coast and reach the upper troposphere while being still close to Asia. Only a small number directly affects North America, and those few originate relatively far away from the Asian coast.

CONCLUSIONS

It was shown that the "textbook example" of intercontinental ozone transport from North America to Europe given by Stohl and Trickl (1999) indeed seems to be a very typical transport episode. All other episodes observed so far (two more were found in the years 1998 and 1999) followed the same pattern, namely lifting within a WCB, followed by transport towards Europe. WCBs frequently originate close to the North American east coast and have a high probability to draw polluted air from the continental rim into the upper troposphere. Emissions from Europe, which is in the main corridor of the "North American" WCBs, on the other hand have little chances of reaching the upper troposphere directly.

A closer examination of the maximum WCB inflow region at the North American seaboard shows that it shifts closer to the main emission regions in late spring. Export of ozone from North America to Europe may, thus, strongly increase from a winter minimum during the spring. In fact, all of our observations of intercontinental ozone transport episodes so far took place in spring. It is, therefore, a tempting hypothesis that transport from North America is a major ingredient to the often-observed springtime maximum in European background ozone concentrations.

Acknowledgments: ECMWF and DWD kindly provided access to ECMWF data. This study was part of the EU projects VOTALP II, funded by the European Commission under Framework Programme IV, contract number ENV4-CT97-0413, and STAC-

CATO, funded by the European Commission under Framework Programme V, EVK2-CT-1999-00050. The VOTALP II funding for AS was provided via a subcontract with the Institute of Meteorology and Physics in Vienna.

REFERENCES

Eisele, H., H.E. Scheel, R. Sladkovic, and T. Trickl, 1999, High-resolution lidar measurements of stratosphere-troposphere exchange, *J. Atmos. Sci.* 56: 319-330.

Liu, S.C., M. Trainer, F.C. Fehsenfeld, D.D. Parrish, E.J. Williams, D.W. Fahey, G. Hubler, and P.C. Murphy, 1987, Ozone production in the rural troposphere and the implications for regional and global ozone distributions, *J. Geophys. Res.* 92: 4191-4207.

Parrish, D.D., J.S. Holloway, M. Trainer, P.C. Murphy, G.L. Forbes, and F.C. Fehsenfeld, 1993, Export of North American ozone pollution to the North Atlantic Ocean, *Science* 259: 1436-1439.

Parrish, D.D., M. Trainer, J.S. Holloway, J.E. Yee, M.S. Warshawsky, F.C. Fehsenfeld, G.L. Forbes, and J.L. Moody, 1998, Relationships between ozone and carbon monoxide at surface sites in the North Atlantic region, *J. Geophys. Res.* 103: 13357-13376.

Stohl, A., N. Spichtinger-Rakowsky, P. Bonasoni, H. Feldmann, M. Memmesheimer, H.E. Scheel, T. Trickl, S. Hübener, W. Ringer, and M. Mandl, 2000, The influence of stratospheric intrusions on alpine ozone concentrations, *Atmos. Environ.* 34: 1323-1354.

Stohl, A. and T. Trickl, 1999, A textbook example of long-range transport: Simultaneous observation of ozone maxima of stratospheric and North American origin in the free troposphere over Europe. *J. Geophys. Res.* 104: 30445-30462.

Stohl, A., G. Wotawa, P. Seibert, and H. Kromp-Kolb, 1995, Interpolation errors in wind fields as a function of spatial and temporal resolution and their impact on different types of kinematic trajectories, *J. Appl. Meteor.* 34: 2149-2165.

Wernli, H., and H.C. Davies, 1997, A Lagrangian-based analysis of extratropical cyclones. I: The method and some applications, *Q. J. R. Meteorol. Soc.* 123: 467-489.

Wild, O., K. S. Law, KS. McKenna, B J. Bandy, S.A. Penkett, and J. Pyle, 1996, Photochemical trajectory modeling studies of the North Atlantic region during August 1993. *J. Geophys. Res.* 101: 29269-29288.

DISCUSSION

B. YAMARTINO: Did you use periodic boundary conditions in your modeling, so that you could carry trajectories much longer than you presently do (i.e. six days)?

A. STOHL: For this study, FLEXTRA and FLEXPART were not driven with global fields, but this is possible. If global fields are used, no boundary conditions are necessary. Then the length of the trajectories is virtually unlimited.

R. SAN JOSE: A few weeks ago we had a nocturnal ozone episode in Madrid city and regional area with concentrations up to 500 ppb, does your paper explain such an unique ozone episode?

A. STOHL: No, I can't imagine how intercontinental transport could be responsible for such high ozone concentrations. If these values are real, my best guess would be that it was a stratospheric intrusion.

D. W. BYUN: Lagrangian model is good to provide source/receptor relationship for horizontal movement of air. However, the "warm conveyor belt" phenomenon associated with the frontal system: considering the movement of air in the cloud and precipitation associated with the frontal system, do you think the Lagrangian trajectory model is still valid?

A. STOHL: Yes. I use (diabatic) three-dimensional trajectories. The vertical wind thus includes the effect from condensational heat release. Since the frontal system is rather well resolved by the ECMWF model, the vertical wind is realistic. It may be that additional vertical motions induced by especially active convective cells embedded within the warm conveyor belt are not fully resolved. That may lead to an underestimation of the vertical movement, but not for the whole conveyor belt.

H. VAN DOP: Are you suggesting that large amounts of polluted US boundary layer air is affecting the European atmosphere (also at ground level)?

A. STOHL: So far, we've only done case studies of the pollution transport. But the basic airflows that are responsible for the transport are very frequent, and we already have several cases which we believe to be due to intercontinental transport. I think that intercontinental transport of ozone, carbon monoxide and other substances is very important and should receive more attention in the future.

SIMULATION OF SULFATE AEROSOL IN EAST ASIA USING MODELS-3/CMAQ WITH RAMS METEOROLOGICAL DATA

Seiji Sugata[1], Daewon W. Byun[2], and Itsushi Uno[3]

[1]National Institute for Environmental Studies
Tsukuba, Ibaraki, Japan 305-0053
[2]National Exposure Research Lab, EPA
Research Triangle Park, NC 27711
[3]Kyushu University
Kasuga, Fukuoka, Japan 816-8580

INTRODUCTION

Interest has grown to understand the transport and chemistry of pollutants originating from eastern Asia, which is undergoing rapid industrialization. Acid deposition has become one of key environmental issues due to the increased use of high-sulfur fossil fuels in the area. During the last decade, comprehensive air quality models have been used to assess the severity of acid deposition problems and to develop effective emissions control strategies for North America and Europe. Recently, U.S. EPA has developed and publicly released the Models-3 Community Multiscale Air Quality (CMAQ) modeling system (Byun and Ching, 1999). It is a comprehensive modeling system that consists of a meteorological model, an emissions processing and projection system, and several interface processors such as Meteorology-Chemistry Interface Processor (MCIP), as well as the CMAQ Chemical Transport Model (CCTM). One of key design objectives of Models-3 CMAQ was to achieve flexibility that enables linkage of different science processors and modules to build appropriate air quality models to meet user's needs. It adapts a generalized coordinate system with governing equations for the fully compressible atmosphere to allow linkage of different description of atmospheric dynamics for multi-scale applications. It uses efficient modular structure with minimal data dependency and a set of a generalized chemistry solver module and chemical mechanism reader to handle multi-pollutant problems. The Models-3 CMAQ modeling system utilizes the Mesoscale Model Generation 5 (MM5) as the default meteorological driver. Until now, it has been tested only with a few applications in USA.

The present study attempts to address a few challenges in utilizing the flexibility of the system. We apply the CMAQ system with the meteorological data provided by the Regional Atmospheric Modeling System (RAMS) and to a different geographical area - East Asia covering the eastern half of China, Korean peninsula, and the Islands of Japan.

To demonstrate the model performance, we compare the results with observed time series of non-sea salt sulfate that are available at several sites in the southern part of Japan during January 1997.

ADAPTATION OF MODELS-3/CMAQ TO RAMS

There are some significant differences between RAMS and MM5, such as the vertical coordinate, horizontal grid system, and available meteorological parameters in the output. Although the CMAQ CTM uses a generalized coordinate system, the interface processors released were designed to work with MM5 output exclusively. Some of the interface programs both in the RAMS and CMAQ systems were modified to allow this linkage. They are a postprocessor routine of RAMS to output data in I/O API (input/output application programming interface) format as required by the Models-3/CMAQ system and a reader subroutine in the MCIP of CMAQ.

Except for a few minor changes in the CTM code to output wet-deposition data at the interval defined by the user and to incorporate daily photolytic rates for long-term simulations, we used the default CMAQ CTM version as released in 1999. The default options are:

- Advection with piece-wise parabolic method (PPM)
- Vertical diffusion with K-theory parameterization
- Deposition flux as bottom boundary condition for the vertical diffusion
- Mass conservation adjustment scheme
- Horizontal diffusion with scale dependent diffusivity
- Carbon Bond 4 (CB-4) chemistry mechanism with isoprene chemistry
- QSSA gas-phase reaction solver
- Emissions injected in the vertical diffusion module
- Aqueous-phase reactions and convective cloud mixing
- Modal approach aerosol size distribution and dynamics

For the detailed description of the science algorithms, refer to Byun and Ching (1999).

APPLICATION OF RAMS/CMAQ FOR EAST ASIA

To provide meteorological data for CMAQ simulations, RAMS was run from 27 December 1996 to 31 January 1997. The modeling domain is 4,000 x 4,000 km^2 on the rotated polar-streographic map projection centered at the (35 N, 130 E) with 80 km mesh. The model extends vertically up to 18 km, which is represented with the 23 layers in the sigma-z coordinates. For example, lowest four vertical layers are placed at elevations 47.7, 157.3, 288.7, and 446.5 m. RAMS options used for the run were; non-hydrostatic dynamics, simplified Kuo for cloud parameterization, Mellor-Yamada 2.5 for vertical diffusion, and Louis (1979) surface flux parameterizations. The RAMS runs used large-scale meteorological data provided by the ECMWF analysis. Then, the RAMS output is fed into MCIP to generate all the necessary meteorological parameters for the CMAQ CTM simulations.

There is no detailed emissions inventory yet for the East Asia region. For the simulation, we modified the emissions database from the Global Emissions Inventory Activity (GEIA) and SO_2 emissions data from Akimoto and Narita (1994). The data obtained from the GEIA web site were in latitude and longitude coordinates and thus were further processed for the RAMS grid. It was necessary to adapt the simple volatile organic compound (VOC) emissions from the GEIA database and Piccot et al. (1992) into multiple CB-4 hydrocarbon species. In addition, VOC emissions were modified by factors from 4 to 7.5 through the repeated simulations of the model and comparisons with ozone observations in the region. Ammonia emissions are obtained from Murano et al. (1995) and Zhao and Wang (1994). Descriptions of emissions processing methods can be found in Carmichael et al. (1998) and Uno et al. (1997).

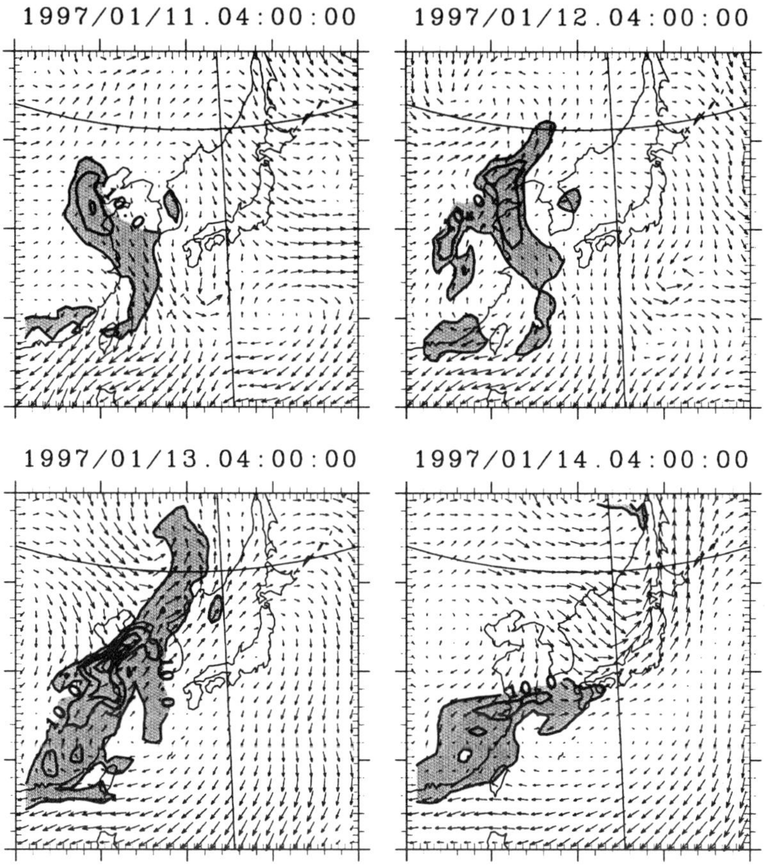

Figure 1. Sulfate aerosol (ASO_4, Aitken mode and accumulation mode combined) concentration in ($\mu g / m^3$) and horizontal wind at the surface from 11-14 January 1997. Time is in GMT.

Figure 1 illustrates one of the typical evolution patterns of sulfate aerosol (ASO_4) distributions during winter in the East Asia. The center of high concentration area was near Qingdao on 11 January. Then the area started to deform in an elongated shape with major axis from southwest to northeast direction on 13th. Although the peaks became smaller on 14th, the edge of high concentration area reached southern Japan. The eastward movement of a high-pressure system, with strong southwesterly winds behind it, caused the elongated pattern. After the system passed through on 13th, wind became northerly around the Korean peninsula and southern Japan. The northerly wind brought high sulfate concentrations to the southern Japan. This sulfate aerosol distribution is associated with a typical winter pressure pattern often referred to as, "high in the west and low in the east" illustrated in Figure 2.

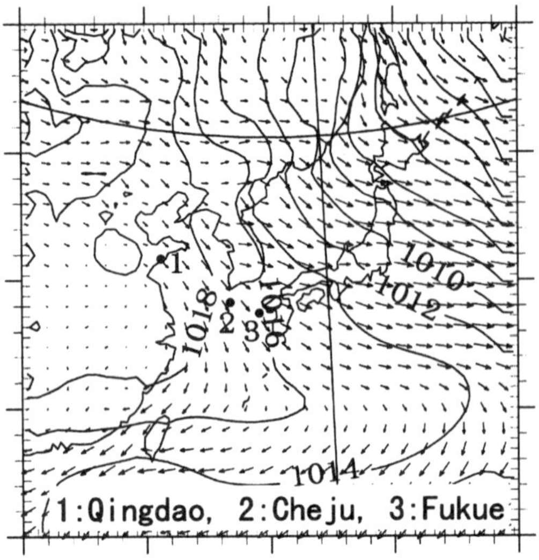

Figure 2. Typical weather pattern for January. Monthly averaged sea level pressure and horizontal wind at the surface calculated by RAMS in January 1997. Three observation sites (Qingdao, Cheju, and Fukue) are identified.

Figure 3 (a) presents observed monthly precipitation and Figure 3 (b) shows the same simulated by RAMS. RAMS successfully reproduced three large precipitation areas: area near the center of Japan facing the Japan Sea; the sea east of Japan; and the area between Taiwan and Kyushu island. Figures 3 (c) and (d) show simulated monthly mean SO_2 and sulfate aerosol (ASO_4) concentrations. Regions of high sulfur concentrations are directly related with the high anthropogenic emissions sources while sulfate aerosol concentrations are distributed with much smoother gradient fanning out from source regions toward downwind areas. Figures 3 (e) and (f) present dry and wet monthly deposition amounts of ASO_4, respectively. The total amount of wet deposition is much larger than the dry deposition, with the ratio of 20:1.

Observed time series of sulfate aerosol concentrations are available at several sites in China, Korea, and Japan. These observations were obtained from the ion-chromatography chemical analysis of Teflon filter measurements with 6- to 8-hour sampling periods. Figure 4 compares the model results with observations at Qingdao, Cheju, and Fukue. Model data show delay of peaks in the time series from upwind to downwind sites, i.e., from Qingdao to Cheju to Fukue.

COMPARISON OF SULFUR BUDGET WITH OTHER STUDIES

Because there have been only a few comparable regional modeling studies on the sulfur budget in this region, we compare the present modeling results with previous studies on the global sulfur budget (Chin et al. 1996, Langner and Rodhe 1991, Pham et al. 1995, and Takemura 2000). Using the process analysis feature available with CMAQ, we have estimated sulfur cycle budget for the eastern Asia simulation. Figure 5 summarizes the monthly sulfur budget obtained from the CMAQ simulation for January 1997. The CMAQ budget analysis indicates that one-third of the total sulfate aerosol in the model domain has resulted from gas-phase oxidation of SO_2. This fraction is much larger than that in the previous global studies while I note that this is approximately what we found with RADM over the eastern U.S. (McHenry and Dennis, 1994). One of possible reasons for the large fraction is that the regional model has much smaller area over the oceans with little or no fresh sulfur emissions, which gives clouds less time to find SO_2 and do the aqueous-phase conversion. The other possible reasons is that winter is the dry season in East Asia, which gives sulfur more opportunity for gas-phase conversion relative to the annual average. Therefore, we expect to predict a higher fraction of sulfate from gas-phase conversion than that for an annual global analyses for both reasons.

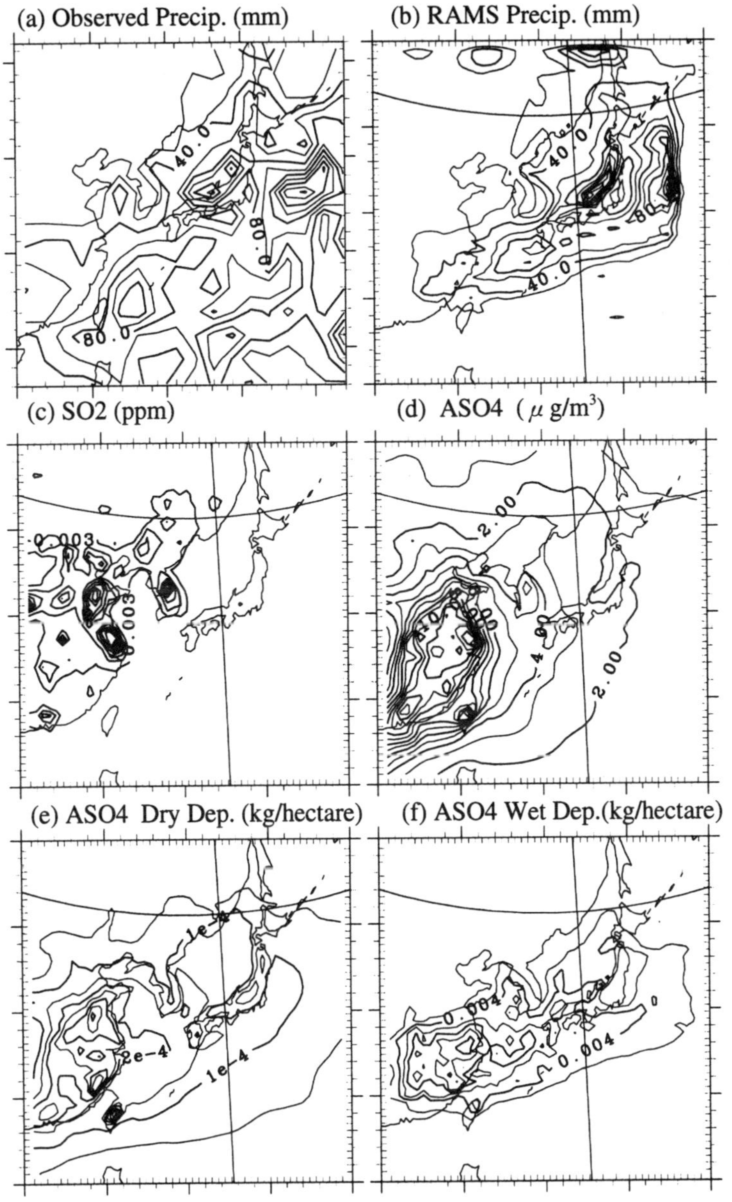

Figure 3. (a) Observed monthly total precipitation, (b) RAMS simulated monthly precipitation; and CMAQ simulated (c) monthly mean SO_2 concentration, (d) monthly mean sulfate aerosol (ASO_4), (e) monthly total sulfate dry deposition amount, and (d) monthly total sulfate wet deposition amount for January, 1997.

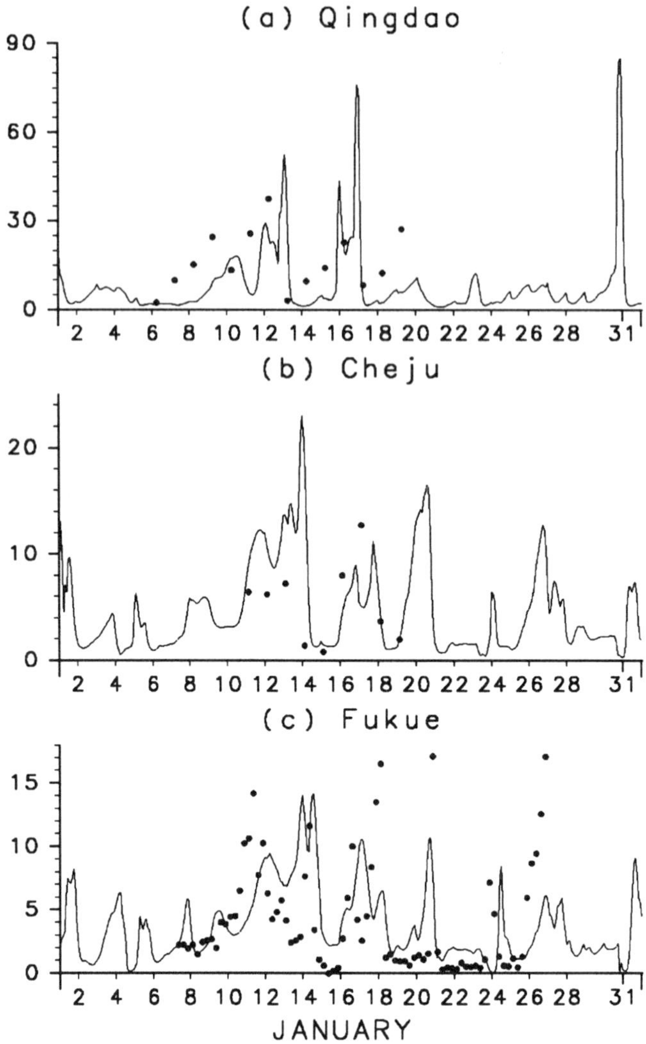

Figure 4. Observed and simulated time series of non-seasalt sulfate aerosol concentrations in $(\mu g / m^3)$ in January 1997. Dot (•) points represent measurements.

We have found that 40% of the total sulfur is removed from the simulation domain by horizontal advection and diffusion. This loss or gain (if there are more fluxes coming in then going out) due to the horizontal transport is a natural feature of the regional models (budget for this component is zero for the global models). From the wind field analysis, the loss due to the transport is mostly at the western boundary of the domain. Vertical transport and in-cloud mixing result in zero net budget as expected. Simulated lifetimes (turnover times) of SO_2 and sulfate were 1.4 day and 3.1 day, respectively. These numbers were close to those with the previous studies. Table 1 summarizes the ratio of gas- and aqueous-phase conversion of sulfur species and time scales of SO_2 and sulfate aerosol of the present study and the previous global studies in the literature.

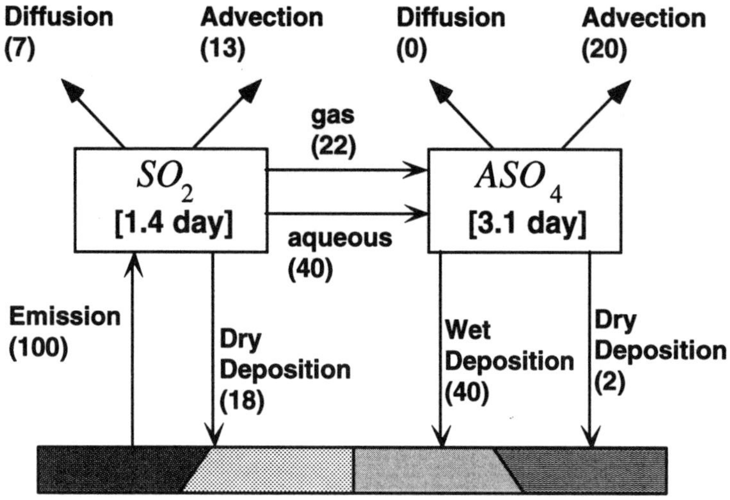

Figure 5. Process analysis illustration of sulfur circulation in the East Asia.

Table 1. Comparison of sulfur budget and life time scales of SO_2 and sulfate aerosol among global scale studies and the present study.

	Ratio between gas- and aqueous-phase conversion of SO_2	Lifetime of SO_2	Lifetime of sulfate aerosol
Present Study	1: 1.8	1.4 day	3.1 day
Chin et al. (1996)	1: 5.5	1.2 day	3.9 day
Pham et al. (1995)	1: 9.0	0.6 day	4.7 day
Takemura (2000)	1: 9.5	1.2 day	2.6 day
Langner and Rodhe (1991)	1: 5.4 or 1: 1.8	1.2 or 1.8 day	3.2 or 6.1 day

CONCLUSIVE REMARKS

We have presented simulation results of RAMS/CMAQ system applied to East Asia covering eastern half of China, Korean peninsula, and Japan Islands for January 1997. The simulation results were reasonable when compared with sulfate observations available at a few sites in the region. The estimated sulfur budget is comparable with other studies and it provides regional insights on the importance of different physical and chemical processes in determining source-receptor relations of the acid deposition problem in East Asia.

ACKNOWLEDGMENTS AND DISCLAIMER

The authors express their appreciation to Drs. Shiro Hatakeyama of NIES and S.-G. Shim of KAIST for providing the observation data. This paper has been reviewed in accordance with the U.S. Environmental Protection Agency's peer and administrative review policies and approved for presentation and publication. Mention of trade names or commercial products does not constitute endorsement or recommendation for use.

REFERENCE

Akimoto, H., and H. Narita, 1994, Distribution of SO_2, NO_x and CO_2 emissions from fuel combustion and industrial activities in Asia with 1° x 1° resolution. *Atmos. Environ.*, 28, 213-225.

Byun, D.W. and J.K.S. Ching, ed., 1999, *Science Algorithms of the EPA Models-3 Community Multi-scale Air Quality (CMAQ) Modeling System,* NERL, Research Triangle Park, NC. [Available from National Exposure Research Laboratory, U.S. Environmental Protection Agency, Research Triangle park, NC 27711]

Carmichael, G.R., I. Uno, M.J. Phandis, Y. Zhang, and Y. Sunwoo, 1998, Tropospheric ozone prediction and transport in the springtime in East Asia. *J. Geophys. Res.*, 103, 10,649-10,671.

Chin, M., D.J. Jacob, G.M. Gardner, M.S. Foreman Fowler, and P.A. Spiro, 1996, A global three-dimensional model of tropospheric sulfate. *J. Geophys. Res.*, 101, 18,776-18,690.

Langner J. and I. Rodhe, 1991, A global three-dimensional model of the tropospheric sulfur cycle. *J. Atmos. Chem.*, 13, 225-263.

Louis, J.-F., 1979, A parameteric model of vertical eddy fluxes in the atmosphere. *Boundary-Layer Meteor.*, 17, 187-202.

McHenry, J. N. and Dennis, R. L., 1994, The relative importance of oxidation pathway and clouds to atmospheric ambient sulfate production as predicted by the Regional Acid Deposition Model, *J. Applied Meteor.,* 33, 890-905.

Murano, J., S. Hatakeyama, T. Mizuguchi, and N. Kuba, 1995, Gridded ammonia fluxes in Japan, *Water Air and Soil Pollution,* 85, 1,915-1,920.

Pham, M., J. F. Muller, G. P. Brasseur, C. Granier, and G. Megie, 1995, A three-dimensional study of the tropospheric sulfur cycle. *J. Geophys. Res.*, 100, 26,061-26,092.

Piccot,S., S. D. Watson, J.W. Jones, 1992, A global inventory of volatile organic compound emissions from anthropogenic sources, *J.Geophy.Res.,* 97, 9,897-9,912.

Takemura, T., H. Okamoto, Y. Maruyama, A. Numaguti, A. Higurashi, and T. Nakajima, 2000, Global three-dimensional simulation of aerosol optical thickness distribution of various origins, *submitted to J. Geophys. Res.*

Uno, I., T. Ohara and K. Murano, 1998, Simulated acidic aerosol long-range transport and deposition over East Asia - Role of synoptic scale weather systems, *Air Pollution Modeling and its Application* Vol., 22, S.E. Gryning and N. Chaumerliac ed., 185-193, Plenum Pub. Co.

Zhao, D. and A. Wang, 1994, Estimation of anthropogenic ammonia emissions in Asia, *Atmos. Environ.,* 28, 689-694.

DISCUSSION

P. SEIBERT: I'm wondering why you employ a prognostic meteorological model at a resolution of 80km which is resolved by operational global weather forecast models. So taking the analyses from these models you may save a lot of work and maybe even get better analyses.

S. SUGATA: I would say this is just a first stage. My target will be finer grid scales. The operational models are no longer working in the scales.

MODELING OF MUDDY RAIN DUE TO THE LONG-RANGE TRANSPORT OF YELLOW-SAND IN EAST ASIA

Zifa WANG[1], Takahisa MAEDA[2] and Horimasa UEDA[3]

1 Frontier Research System for Global Change, Institute for Global Change Research, 1-18-16, Hamamatu-Cho, Minato-ku, Tokyo 105-0013 Japan
2 National Institute for Resources and Environment, AIST, 16-3 Onagawa, Tsukuba, Ibaraki 305-8569, Japan
3 Disaster Prevention Research Institute, Kyoto University, Kyoto 611, Japan

1. INTRODUCTION

The Asian dust (also referred to as yellow-sand or Kosa) originates mainly from the arid regions of central and eastern Asia: the Takla Makan, the Gobi, the Ordos deserts and Loess plateau located at 1500m over the sea level. Evidence of the long-range transport of Asian dust from the Asian continent to the central Pacific has been obtained at Hawaii, at the Eniwetok Atoll and in Japan for these two decades (*Duce et al.*, 1980; *Iwasaka et al.*, 1983; *Uematsu et al.*, 1983; *Parrington et al.*, 1983; *Murayama,* 1988; *Gao et al.*, 1992). The investigations display a clear seasonal cycle, with the major events of Chinese dust storms, the Kosa events, and the long range transport episodes occurring most frequently in Spring (between March and May) (*Merrill,* 1989). The transport flux of dust particles from the Asian continent to Pacific Ocean is considered to be much larger than that from Sahara to the Atlantic (*Uematsu et al.,* 1983; *Kotamarthi and Carmichael,* 1993). Numerical models have been developed to investigate the summer Sahara dust outbreak over Atlantic (*Welstphal et al.*, 1988; *Toon et al.*, 1988; *Nickovic, S. and S. Dobricic*, 1996). Modeling studies of the Asian dust are also presented in these years (*Kotamarthi and Carmichael*, 1993; *Sheng and Qin*, 1994; *Xiao et al.*, 1997). However, large uncertainties still exist in the above estimates for Asian dust. For example, blow-up process of yellow-sand, the weather responsible for the dust storms, the amount of the dust reaching the ocean in one transport event, changes of the size distribution and the chemical composition during the long range transport are still unknown.

Muddy rain has much deeper influence on the citizen life than expected. Dust storms always occur in dry season when the chance is very low to produce the muddy rain in Northern China. A heavy muddy rain was recorded in Beijing during the April 15-16,1998.

The ground surface was precipitated with a layer of dirty yellow-sand muddy mixed with polluted particles. Only to consider cleaning the cars of Beijing would it be more than two million US dollars. The episodes of dust storm in April 1998 are very serious in these ten years characterized by its large scale and duration and to be found by satellite to transport long distance over the Pacific Ocean even arriving to North American. In the study, based on the analysis of the weather conditions resulted in the muddy rain, a three-dimensional Eulerian long-range transport model for yellow-sand was carried out to simulate this muddy rain occurred at Beijing city and surrounding areas.

2. DESCRIPTION OF THE LONG-RANGE TRANSPORT MODEL

The model includes all the major processes in the atmospheric yellow-sand life cycle: deflation, horizontal and vertical advection, diffusion, dry deposition, wet deposition and micro-physical processes. The yellow-sand mass conservation equation using a spherical and terrain-following coordinate can be described mathematically by

$$\frac{\partial}{\partial t}(\Delta H \cdot C_l)+\frac{\partial}{R\cos\theta\partial\varphi}(u\cdot\Delta H\cdot C_l)+\frac{\partial}{R\cos\theta\partial\theta}(v\cos\theta\cdot\Delta H\cdot C_l)+\frac{\partial}{\partial\sigma}(W\cdot C_l)=$$
$$\frac{K_\varphi\partial}{R^2\cos^2\theta\partial\varphi}\left(\Delta H\cdot\frac{\partial C_l}{\partial\varphi}\right)+\frac{K_\theta\partial}{R^2\cos\theta\partial\theta}\left(\Delta H\cdot\cos\theta\frac{\partial C_l}{\partial\theta}\right)+\frac{\partial}{\partial\sigma}\left(\frac{K_\sigma\partial C_l}{\Delta H\partial\sigma}\right)+Q_l-D_l, \quad (1)$$

with $C_l(l=1,9)$ the concentration of yellow-sand for l-th particle size range(size bin), t the time, θ and φ are the latitude and longitude, R the Earth radius, K_θ, K_φ, and K_σ the diffusion coefficients in different directions, u, v the horizontal wind velocities, W the equivalent vertical velocity, Q_l the emission intensity, and D_l the removal term for particles. The terrain-following height coordinate σ reads

$$\sigma(\theta,\varphi)=\frac{z-h(\theta,\varphi)}{H(\theta,\varphi)-h(\theta,\varphi)}=\frac{z-h}{\Delta H} \quad (2)$$

with $H(\theta,\varphi)$ the height of the tropopause layer, $h(\theta,\varphi)$ the height of relief, and z the geopotential height.

The removal term D_l is usually separated into dry removal D_{dl} due to sedimentation and deposition at surface and wet removal D_{wl} due to precipitation. The dry removal mechanism reconstructed by Zhang et al. (1998) is used here to estimate D_{dl}.

The efficiency of wet removal (D_{wl}) by rain is usually described by the scavenging ratio Λ, which is calculated by Pruppacher and Klett (1978),

$$\Lambda=10.8\cdot E\cdot R^{0.16}/Dp \quad (3)$$

where E is the coagulation kernel (set as 0.83), R is the rainfall intensity (mm/hour), and Dp is the depth of cloud or the model layer.

Yellow-sand particles are divided into nine bins (Table 1) from 0.5μm to 90μm in the model. In each of the size bins we distinguish three types of aerosols. The first type is not allowed to grow by depleting vapors, but can grow by coagulation. Examples of this type include yellow-sand, sea salt and carbon particles. The second type undergoes condensation growth such as water droplets and sulfuric acid aerosols. The final type is the mixture of the first two kinds. The size spectrum of yellow-sand depends on the deflation, gravitational settling, micro-physical processes in clouds and washout by precipitation. The role of micro-physical processes can be ignored under dry condition. However, most of long-range transport events of yellow-sand are accompanied with a frontal cyclone and precipitation. A sub-model to deal with the micro-physical processes of yellow-sand is developed for the muddy rain simulation and can be switched on in the transport model under suitable conditions. It considers coagulation, condensation, evaporation and washout by rain and can be written as (Prupacher and Klett 1978),

$$\frac{\partial C(v,t)}{\partial t} = \int_0^v K_c(u,v-u)C(u,t)C(v-u,t)du - C(v,t)\int_0^\infty K_c(u,v)C(u,v)du$$
$$-\frac{\partial C(v,t)g(v,t)}{\partial r} + S(v,t) + R(v,t)C(v,t), \tag{4}$$

where u,v represent particle volumes, Kc is the coagulation kernel, g(v,t) the condensation growth rate, r the particle radius, S the source term and R the removal rate coefficient.

Table 1 Classification of particles size and its meridian radius and representative gravitational settling velocity Vg (m/s) in the transport model of yellow-sand (unit of radius is μm)

bin	1	2	3	4	5	6	7	8	9
radiu	0.5	0.96	1.83	3.5	6.7	12.8	24.5	46.7	90.0
Vg	1.9×10^{-5}	4.01×10^{-5}	1.46×10^{-4}	5.34×10^{-4}	1.96×10^{-3}	7.15×10^{-3}	2.61×10^{-2}	9.57×10^{-2}	0.35

Deficiencies in source parameterization will result in a large part of discrepancies in simulating the dust concentration in its transport. A new parameterization for the dust uptake accepted here is designed by detailed analyzing the meteorological condition, landforms and climate background with the daily weather reports at about 300 local weather stations in North China (Wang et al., 2000). Most of dust deflation occurs in the area from 35° to 50°N and 90° to 110°E, where is coincided with the arid region and the Loess Plateau. The potential source areas are divided into several distinctive types including deserts, the Loess plateau, grassland, cultivated land and deciduous forest. The size distribution and weight factor of the dust load are considered to change with latitude and season for each land type. Three major predictors, characterizing the mobilization process, are used in the module: (I) the friction velocity threshold, (II) the surface humidity threshold and (III) the predominant weather condition. The comparison of module results, using these 3 parameters altogether or separately, with observed data of dust deflation, showed that the best estimated can be obtained considering the three predictors altogether when we accept the criteria as the minimum of the total error ratio. A good performance is obtained by the numerical model coupled with the deflation module for the main features of the yellow-sand life cycle (Wang et al., 2000). The emission intensity $Q_{i,j,l}$, which depends on kinematics condition at the lowest model level, is calculated by

$$Q_{i,j,l} = C_1 C_2 u^{*2}_{i,j,l}(1 - u^*_{0i,j,l}/u^*_{i,j,l}) \cdot W_{i,j,l} \cdot R_{i,j,l}, \tag{5}$$

where the unit of $Q_{i,j,l}$ is $kg.m^{-2}.s^{-1}$), C_1 is the weight factor for different landtype(Wang, et al. 2000), C_2 is a constant depending on experiments and set as 2.9×10^{-11}, $u^*_{i,j,l}$ is the kinematic friction velocity, $u_{*0i,j,l}$ is the threshold value of kinematics friction velocity $u_{*0i,j,l}$, $R_{i,j,l}$ is the fraction of l-th bin of deflating yellow-sand and is obtained from the spectrum distribution of dust in source areas. The humidity factor $W_{i,j,l}$ is defined as

$$W_{i,j,l} = \begin{cases} (1 - RH/RH_0) & for \quad RH < RH_0 \\ 0 & for \quad RH \geq RH_0 \end{cases} \tag{6}$$

where RH, RH_0 are the surface relative humidity and its critical value.

A simplified but very accurate, mass conservative advection algorithm is used to solve the three-dimensional mass conservation equation (1) with a time-splitting technique (*Walcek and Aleksic,* 1998). Time step was chosen as 10 minutes by the CFL condition i.e., the Courant number ($u\Delta t/\Delta x$) less than 1.

The NCAR/Penn State Fifth-Generation Mesoscale Model (MM5) has been used as a meteorological driver to the model system to afford the comprehensive meteorological conditions such as winds, precipitation, cloud microphysics, boundary turbulence, which play significant roles for the transport and deflation of yellow-sand. MM5 includes a multiple-nest capability, nonhydrostatic dynamics and a four-dimensional data assimilation capability as well as more physics options and is used for a broad spectrum of theoretical and real-time studies, including applications of both predictive simulation and four-dimensional data assimilation to monsoons, hurricanes, cyclones, mesoscale convective systems, fronts, land-sea breezes, mountain-valley circulation, and urban heat islands. The boundary conditions used to solve equation (1) are set as these: yellow-sand can transport out of but not into the boundary. The bottom surface is assumed as an absorbing boundary while the top boundary as a closed one. The current model domain covers the East Asia from 16°N to 60°N, and from 72°E to 146°E with horizontal grids at 1°×1° resolution. The vertical extending consists of eighteen layers from the surface to tropopause along a terrain-following height coordinate.

3 SIMULATION RESULTS

Figure 1 shows the MM5 output of sea level pressure and precipitation distribution on April 16 and 17. The strong low-pressure system coupled with light rain located at Beijing area. The light rain is one of the important parameter of forming of muddy rain. The eastward traveling strong low-pressure system and the cold-frontal line is responsible for the strong episode. High winds were blown up over Xinjiang province and Loess Plateau and invaded into the northern China by the two eastward traveling strong low-pressure system and the cold-frontal line. As a result, more than three dust storm episodes occurred and the floating dust covered the most regions of China, and reached to Japan even to the North America. Satellite image clearly retrieved the dust front and dust cloud in Xinjiang province on April 14, in the Loess Plateau on April 15 (Fig.3a), which results in the heavy muddy rain in Beijing in the mid night of April 15. The second heavy dust storms occurred on April 19 and 20, which covered from the Takla Makan desert, Qinghai province and the Loess Plateau and transported southward to Shanghai and eastward to Japan. Figure 2 shows the distribution of the column content of yellow-sand on April 15 and 16. The column content represents the total amount of yellow-sand per square meter with unit of mg/m^2. The yellow-sand is mainly distributed in Xinjiang province, the Loess Plateau and north China on April 15, when the light rain is precipitated in Beijing area, just in front of the flowing dust. The micro-physics processes among the yellow-sand particles and cloud and rain drops caused the muddy rain.

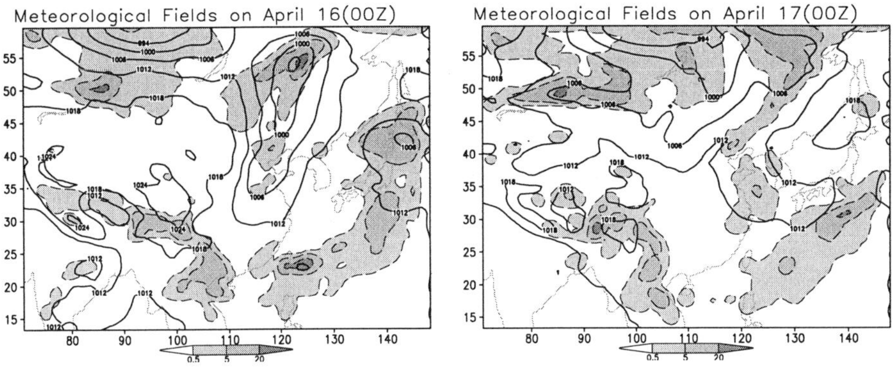

Fig.1 Sea level pressure and precipitation distribution simulated by MM5 for April 16 and 17. The shaded areas represent the precipitation amount more than 0.5mm for the last 24 hours.

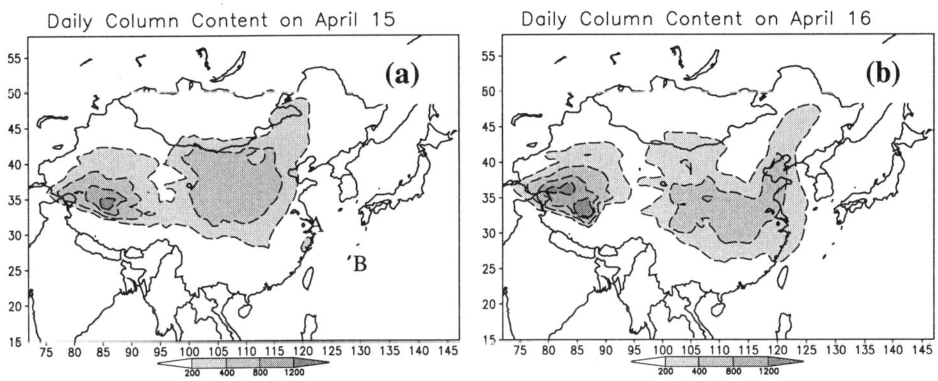

Fig. 2 Distribution of daily averaged column content (mg/m^2) of yellow-sand for April 15 and 16. The shaded contours are 200, 400, 800 and 1200 mg/m^2.

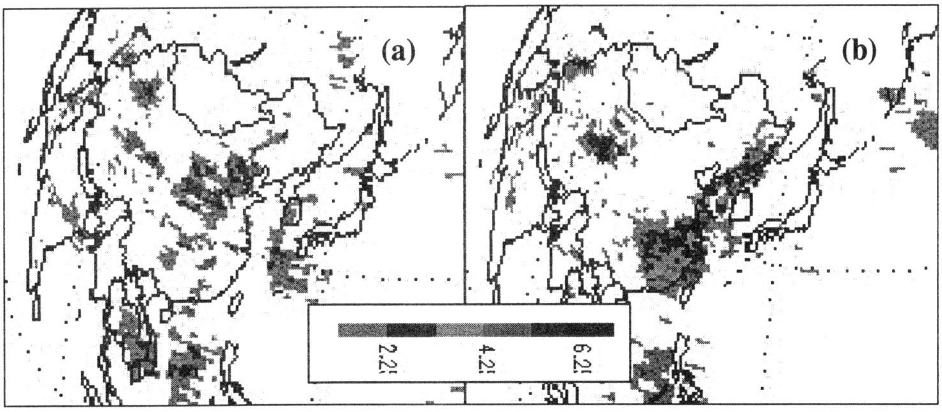

Fig. 3 Aerosol index distribution obtained from satellite data for April 15 (Fig. a) and April 16 (Fig. b).

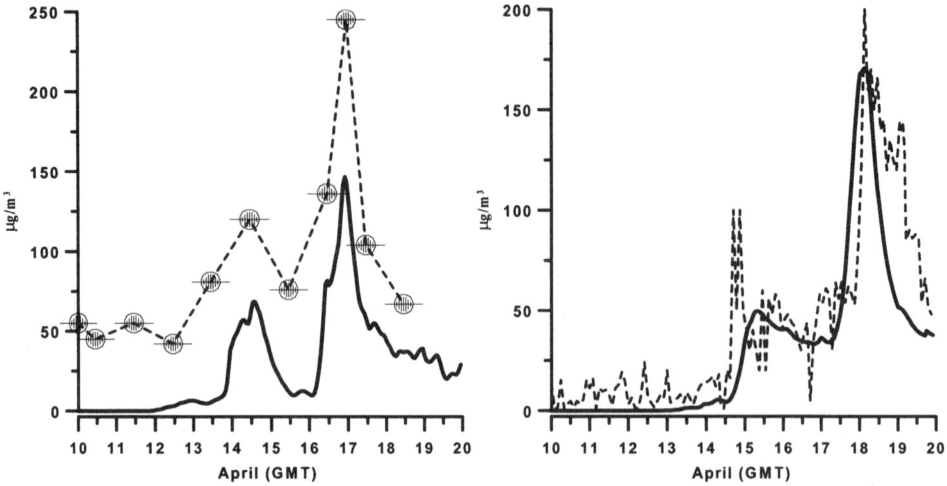

Fig. 4 Comparison of observed SPM concentration and simulated yellow-sand levels at Shanghai (Fig. a) and at Okinawa (Fig. b) on April 10 to 20, 1998. The dashed lines represent the observed values and the other black line is model result. The locations for two stations are shown in Fig. 2a. with A for Shanghai and B for Okinawa.

Figure 3. shows the aerosol index obtained from satellite data. It clearly shows the high concentrations of aerosols are distributed in Xinjiang province, the Loess Plateau and north China on April 15. Then the high levels are located at the Southeast China and Korea peninsular on April 16, showing the long-range transport of yellow-sand. Comparing with the model results (Figure 2), they have reasonable agreement. Figure 4 compared the observed SPM concentration with the simulated yellow-sand levels at Shanghai and at Okinawa. The tendency predicted by the model fits very well with observations. The observed values are larger than simulated ones partly because observed SPM concentration also include other anthropogenic particles and sea-salt particles.

The mass particle-size distributions of dust vary depending on the locations, which also are simulated by the model. The dominant particle-size range for dust over deserts is 6-20µm in mass median diameter. The concentrations of atmospheric dust in Northern China are dominated by dust particles of 2-6µm, which account for 40-70% of total dust mass. Over the off-shore areas of the East China, the dominant mass median diameter ranged from 1-3µm, indicating that the size distributions of dust shifted toward smaller sizes as the dust was transported further away from the continent. The cloud physics and precipitation process mixed with plenteous of yellow-sand with the size less than 2µm, led to the heavy muddy rain in Beijing.

Fig.5 shows the predicted dry and wet deposition amount of yellow-sand on April 15, 16 and 17, 1998. The distributions of dry deposition fit very well are mainly located in Xinjinag and North China on April 15. The wet deposition of yellow-sand are only located at limited areas in conjunction with the precipitation. The contents in the precipitation in many areas such as Taiwan, Korea and Japan are deeply influenced by this transport episode both in pH value and ion concentrations.

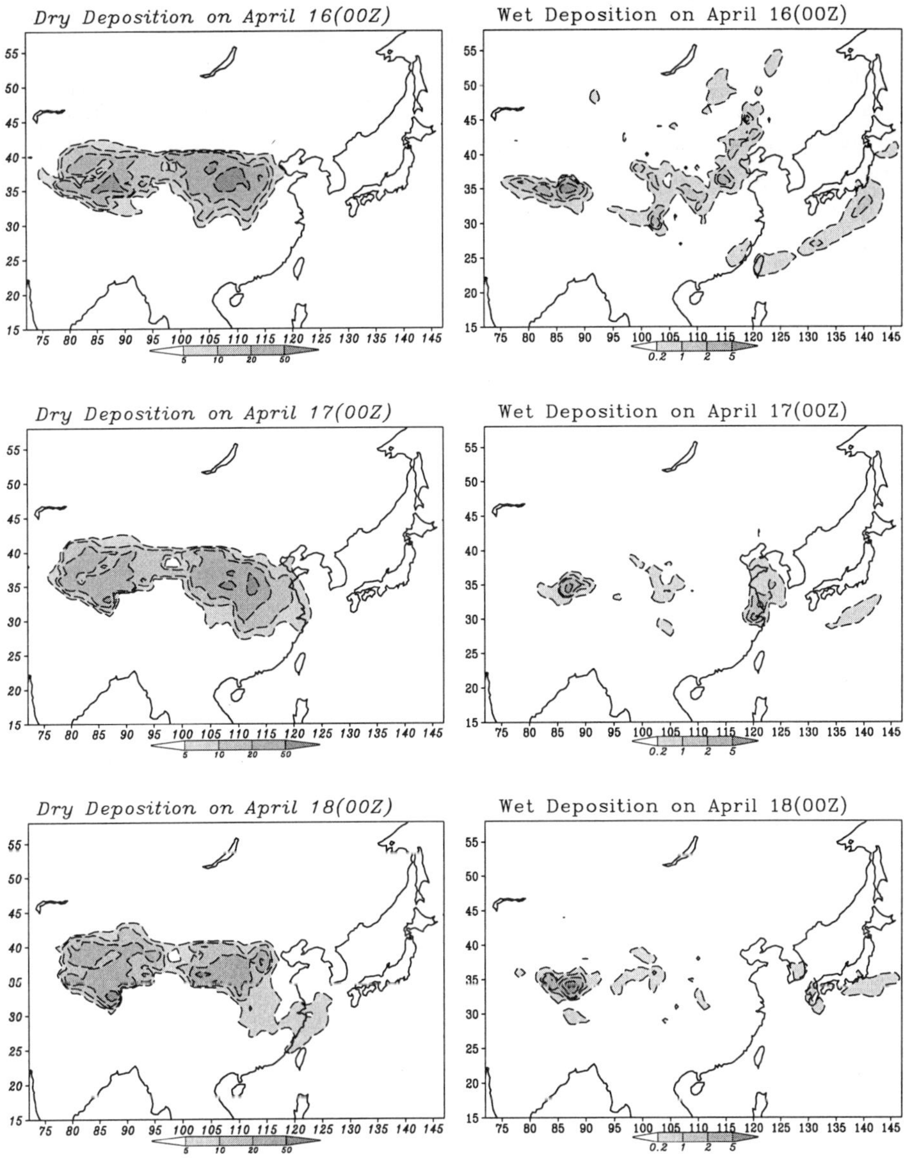

Fig. 5 Distribution of dry and wet deposition of yellow-sand on April 16, 17 and 18,1998. The shaded contours are 5, 10, 20, and 50mg/m^2/day for dry deposition and 0.2, 1, 2, and 5mg/m^2/day for wet deposition for the past 24 hours.

5 CONCLUSIONS

The 3-D regional Eulerian long-range transport model coupled with a new deflation module for yellow-sand in East Asia has been applied to simulate the heavy muddy rain recorded in Beijing on April 1998. Nine size bins for yellow-san particles and their micro-physical processes are considered in the model. The comprehensive meteorological conditions are obtained from MM5 (the NCAR/Penn State Fifth-Generation Mesoscale Model) results.

Results show the model, in conjunction with the deflation module, is reliable to predict the yellow-sand concentration and size distribution due to the up-lift, transport, diffusion and removal processes. Generally, there is a good agreement between the observed and predicted results. The vertical distribution of one transport process shows clearly that the small particles are transported as two levels, one is near the ground, the other in the middle troposphere. The long-range transport processes of yellow-sand and the light rain are responsible for the formation of muddy rain. The cloud physics and precipitation process mixed with plenteous of yellow-sand from loess plateau, especially the size less than 2μm, led to the heavy muddy rain in North China.

REFERENCES

Duce, R. A., Unni, C. K., Ray, B. J., Prospero, J. M., and Merrill, J. T., 1980, Long-range atmospheric transport of soil dust from Asia to the tropical North Pacific: Temporal variability, *Science,* 209: 1522-1524.

Gao, Y., Arimoto, R., Merril, J. T., and Duce, R. A., 1992, Relationship between the dust concentration over eastern Asia and the remote North Pacific, *J. Geophys. Res.***,** 97: 9867-9872.

Iwasaka, U., Minouru, H., and Nagaya, K., 1983, The transport and spatial scale of Asian dust-storm clouds: A case study of the dust-storm event of April 1979, *Tellus*, 35B:189-196.

Kotamarthi, V.R. and Carmichael, G.R., 1993, A modeling study of the long range transport of Kosa using particle trajectory methods, *Tellus*, 45B:426-441.

Merrill, J., 1989, Modeling long-range transport using trajectory techniques, in *Paleoclimatology and Paleometeorology: Modern and Past Patterns of Global Atmospheric Transport, NATO Adv. Study Inst. Ser. C,* vol. 282, edited by M. Leinen and M. Sarnthein, pp. 207-225, Kluwer Academic, Dordecht, Netherlands.

Nickovic, S., and Dobricic, S., 1996, A model for long-range transport of desert dust, *Mon. Wea. Rev.*, 124:2337-2345.

Parrington, J. R., Zoller, W. H., and Aras, N. K., 1983, Asian dust: Seasonal transport to Hawaiian Islands, *Science,* 220:195-197.

Pruppacher, H.R. and Klett, J.D., 1978, *Microphysics of Cloud and precipitation*, 714-724, Reidel.

Toon, O.B., Turco, R.P., Westphal, D., Malone, R., and Liu, M.S., 1988, A multidimensional model for aerosols, description of computational analogs, *J. Atmos. Sci*, 45:2123-2143.

Uematsu, M., Duce, R.A., Prospero, J.M., Chen, L., Merrill, J.T., and McDonald, R.L., 1983, Transport of Mineral Aerosol from Asia over the North Pacific Ocean, *J. Geophys. Res.*, 88: 5343-5352.

Wang, Z., Ueda, H., Huang, M., 2000, A deflation module for the long-range transport of yellow-sand over East Asia *J. Geophys. Res. (revised).*

Welcek. C. J. and Aleksic, N. M., 1998, A simple but accurate mass conservative peak-preserving, mixing ratio bounded advection algorithm with Fortran code, *Amtos. Environ.*, 32: 3863-3880.

Westphal, D., Toon, O.B., and Carlson, T. N., 1987, A two-dimensional numerical investigation of the dynamics and microphysics of Saharan dust storms, *J. Geophys. Res*, 92: 3027-3049.

Xiao, H., Carmichael, G. R., and Durchenwald, J., 1997, Long-range transport of SOx and dust in East Asia during the PEM B Experiment, *J. Geophys. Res.,* 102: 28,589-28612.

Zhang X. Y., Akimoto, R., Zhu, G.H., Chen, T., and Zhang, G.Y., 1998, Concentration, size-distribution and deposition of mineral aerosol over Chinese desert regions, *Tellus*, 50B: 317-330.

DISCUSSION

D.W. BYUN: Your transport model is formulated with spheric coordinates. However, you used MM5 which used a conformal map projection. These two are fundamentally different horizontal coordinates. When you interpolated MM5 data to the spherical coordinate grid, what kind of errors do you think you have introduced?

Z. WANG: Thanks for your question. MM5 have three kinds of map projection to select: Lamcon, Polstr and Mercat. I choose the third one, which would make little error of interpolation into spheric coodinate for the model domain we selected. I think this kind of error is not so serious compared with others such as the emission set.

P. BUILTJES: You showed AOD from satellites. Which instrument did you use? Was it ATSR?

Z. WANG: Thanks for your question. Yes. It is ATSR. I download the data from NCAR web page.

EFFECT OF BIOMASS BURNING ON TROPOSPHERIC CHEMISTRY IN SOUTHEAST ASIA - A NUMERICAL SIMULATION FOR SEPTEMBER AND OCTOBER, 1994 -

Toshihiro Kitada[1], Masato Nishizawa[1], Gakuji Kurata[1], and Yutaka Kondo[2]

[1]Department of Ecological Engineering, Toyohashi University of Technology, Tempaku-cho, Toyohashi 441-8580, Japan
[2] RCAST, University of Tokyo, Meguro-ku, Tokyo 153-8904, Japan

INTRODUCTION

In the dry season of 1993, 1994 and 1997, severe biomass burning occurred in tropical Southeast Asia, i.e. the islands of Sumatra and Kalimantan, Indonesia, and it resulted in various types of air pollution over large area by emitting aerosol particles and photochemical precursors (Matsueda, et al. 1998; Tsutsumi, et al. 1999). That event is caused by mixed reasons of human activity and natural condition, and occurs periodically every two or three years in association with "El Niño and Southern Oscillation (ENSO)". Thus we have evaluated effects of the biomass burning on the concentration fields of trace chemical species and on the tropospheric chemistry in East and Southeast Asia in September and October, 1994 by using our transport/chemistry model described in Kitada and Nishizawa (1998) and Kitada et al.(1998) and trajectory analysis. The calculation included anthropogenic emissions of various chemical species, and NOx emissions of lightning and soil microbial activity. For biomass burning, EDGAR annual emission was modified by considering fire activity map archived by IGBP-DIS Global Fire Product. The biomass burning in northern Australia, which was not included in EDGAR, was estimated in this study.

Obtained results were compared with observed vertical profile of ozone at Kuala Lumpur (KL) in Malaysia and Watukosek (WK) in Java, Indonesia as well as surface ozone at Singapore. Transport characteristics in tropical Southeast Asia was also discussed in conjunction with the ozone profiles and trajectory calculation. Judging from calculated column density of ozone produced by human-induced emissions, the importance of the area of both active deep cumulus convection and strong emission source, in the global scale mass transport/chemistry, was stressed; in this context, southern coastal China and Southeast Asia should be an interesting place. Furthermore, contribution of the biomass burning to several key chemical species was evaluated; for example, the biomass burning typically increased surface ozone by 10% at KL and 20% at WK, and CO by 13% at KL and 15% at WK at one time during the simulated time.

TRANSPORT/CHEMISTRY/DEPOSITION MODEL FOR TRACE CHEMICAL SPECIES AND EMISSION SOURCES

Equations for transport/chemistry/deposition (Kitada and Nishizawa, 1998; Kitada et al. 1998) were numerically integrated from 00GMT on 15 Sept. to 00GMT on 5 Oct., 1994. The domain was 80°E - 160°E, 19°S - 60°N, and the earth surface - 10hPa. Grid size was 1° in horizontal directions, and variable 17 layers in vertical direction. The model includes the processes of advection, diffusion, sub-grid scale cumulus convection, chemical reactions, and wet and dry depositions. Chemical species treated in the model were 38, 25 of which were advected. These species were tied with a system of 90 chemical reactions; the reaction system was an adaptation from Lurmann et al. (1986). ECMWF operational analysis data were used for wind, temperature, and water vapor etc. and renewed every 6 hours. Precipitation was provided every 24 hours.

Table 1. NOx emissions in calculation domain,
(a) Whole area:80E～160E and 19S～60N(AUSMIN2)

Source type		Strength(kgN/mon)	Relative Strength	Reference
Fossil fuel		3.78×10^{8}	0.554	EDGAR(1990)
Biomass Burning	Biofuel	6.48×10^{7}	0.095	EDGAR(1990)
	Biomass	$(1.76+0.57^{*})\times10^{7}$	0.034	EDGAR & this study
	Waste	5.80×10^{7}	0.085	EDGAR(1990)
Soil		5.95×10^{7}	0.087	Levy and Yienger(1995)
Lightning		4.19×10^{7}	0.061	Based on Kumar et al.(1995)
Aircraft		5.40×10^{6}	0.008	NASA(1990)
Total		6.82×10^{8}	1	

*Biomass burning source in Australia (this study).

(b) Limited tropical area: 90E～120E and 10S～10N

Source type		Strength(kgN/mon)	Relative
Fossil fuel		2.05×10^{7}	0.313
Biomass Burning	Biofuel	0.68×10^{7}	0.104
	Biomass	1.16×10^{7}	0.177
	Waste	0.55×10^{7}	0.084
Soil		1.17×10^{7}	0.179
Lightning		0.88×10^{7}	0.134
Aircraft		0.06×10^{7}	0.009
Total		6.55×10^{7}	1

Various types of emission sources were considered in the simulations. Those of NO_X, for example, are listed in Table 1a,b. EDGAR database for annual emission of biomass burning was modified so that it is matched with those for September and October; fire activities archived in IGBP-DIS Global Fire Product were used for the modification. Furthermore, since the biomass burning emission in Australia was not compiled in EDGAR, we estimated it with the Global Fire Product, vegetation map by NASA Goddard DAAC, and emission factors.

The burned biomass, M (kg/grid cell), was estimated by $M=\sum_{(i)}(A*B*CF)i$, where A denotes burned area in ha/grid cell, B the biomass density in kg/ha, CF the combustion factor (ratio of the burned biomass to existing biomass), and "i" the vegetation type. The burned area A was estimated using IGBP-DIS Global Fire Product. The factor B was also evaluated with vegetation map by NASA Goddard DAAC and biomass density information in Hao et al. (1990). We have assumed the following relation between combustion factor CF and monthly precipitation P (mm): $CF_{forest} = 0.4 - 4.2^{-4}*P$, and $CF_{grass} = 1 - 7.7^{-4}*P$ for forest and grass land respectively. In these equations information on CF in Hao et al. (1990), Ward et al. (1992), Cautenet et al. (1999) etc. was taken into account. For example, when using this relation in the western part of the Middle Sumatra where monthly precipitation reached 500 mm in September, 1994, $CF_{forest} = 0.2$ can be obtained. Finally, emissions of chemical species from the burned biomass M were calculated with an equation $M_j = \sum_{(i)}(EF_j MC)_i$ where M_j denotes emission strength of the j-th chemical species in kg/kg-fuel-C, MC the carbon content in the burned biomass M (the ratio in weight was assumed 0.45 as in Ferek et al. 1998), and EF_j the emission factor for the "j" th species; the values in Ferek et al.(1998) were employed for EF_j.

Table 1a,b shows relative importance of each emission source. In the whole calculation domain of East Asia (Table 1a), for example, fossil fuel most contributes to the NOx emission by 55%, and biomass burning (fire) accounts for only 3.4%. However, if we focus on Southeast Asia, the biomass burning (fire) can be a dominant emission source of NOx (Table 1b) and others; the magnitude of biomass burning emission can exceed 50% of the emission by fossil fuel.

SIMULATION CASES

Table 2 lists simulation cases. BASE2 includes all the emission sources, but its biomass burning source uses EDGAR only and does not cover Australia. AUSMIN includes Australian biomass burning. Although biomass burning is expressed with a flux form at earth surface in most of the cases, only AUSMIN2 adopts a "volume" source for the emission from biomass burning in Australia. The "volume" source was used to see the effect of initial plume rise by sub-grid scale fires on the global scale chemical transport; it was applied for northern Australia because the area is under strong influence of high pressure system with persistent clear weather and synoptic scale stable stratification. NOBB does not include biomass burning, and NOEMI no emission sources.

Table 2. Simulation cases.

	All the sources except biomass burning	Biomass burning (EDGAR)	Biomass burning (Australia)
BASE2	○	○	
AUSMIN	○	○	○ (Flux type*)
AUSMIN2 (Standard)	○	○	○ (Volume#)
NOBB	○		
NOEMI			

* The emission source was expressed with a flux form at the earth's surface.
# The emission was given by uniform volume source from the surface to about 2.5km. Initial plume rise associated with biomass burning was assumed.

MASS TRANSPORT CHARACTERISTICS IN TROPICAL AREA AS SUGGESTED BY BACKWARD TRAJECTORY ANALYSIS

By using vertical profiles of ozone observation and backward trajectory, general feature of air mass movement over the area was examined. During the simulated time from 15 September to 4 October in 1994, vertical sounding of ozone was performed at Kuala Lumpur (KL) on 30 Sept. (Tsuruta et al. 1997) and Watukosek (WK), Indonesia on 27 Sept. (Fujiwara et al. 1998 and 1999); these are shown in Fig.1a and b respectively. The profiles in Fig. 1 show interesting features that reflect transport characteristics, such as thermal stratification, in air mass over the sites. The ozone profile at KL (Fig.1a) shows rather uniform mixing ratio below about 12km, suggesting a well mixed air mass by strong deep cumulus convection. On the other hand, the ozone profile at WK

(Fig.1b) has its complex structure with several pronounced local minima and maxima, indicating suppressed vertical mixing and thus a stable stratification in the air mass. This difference can be observed also in the frequency map of sub-grid scale cumulus convection (Fig.2); i.e. strong activity of cumulus convection (KL) at Sumatra and near the Malay Peninsula, and very few activities of cumulus convection (WK) over East Timor Sea and northern Australia.

Another interesting feature in Fig.1a,b is clear minimum ozone concentration at 15 km high around tropopause height. ECMWF data suggests the low ozone mixing ratios accompany local high humidity. Thus it is indicated that these air masses of the low ozone mixing ratio and high water vapor have their origin near the surface level over remote ocean, i.e. Pacific Ocean.

To estimate origin of the air mass showing the characteristic ozone profiles at KL and WK (Fig.1a,b), 7-day backward trajectories, started at heights of 1.5, 3, 4.5, 7, 10, and 15 km high over KL and WK, were calculated (Fig.3a,b). (In the calculation vertical wind velocity in resolvable scale was used.) Figure 3a suggests at Kuala Lumpur the air mass reaching at 15 km has its origin far in Pacific Ocean and all other air masses below 10 km high came from lower troposphere over near KL area. The first finding well correlates with the low ozone and high humidity at 15 km, and the second one suggests the air mass below around 10 km near KL rather stagnates and strong cumulus convection likely occurs in it, resulting in the ozone profile in Fig.1a.

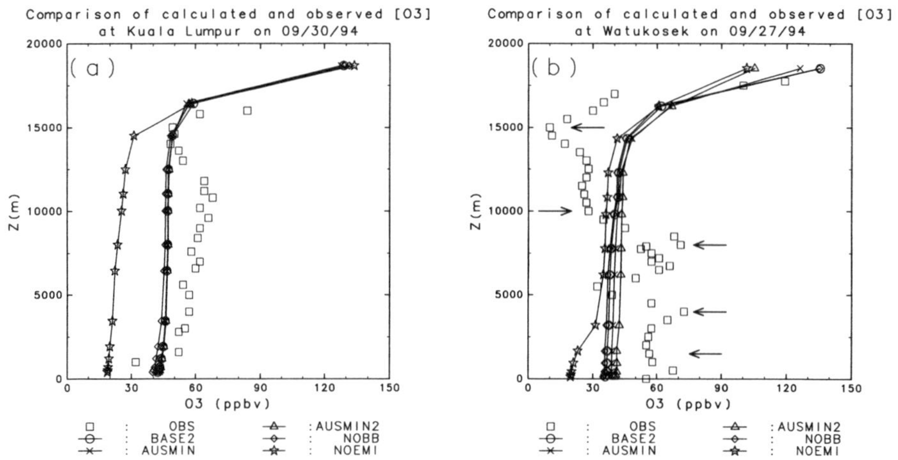

Figure 1. Observed and calculated vertical profiles of ozone: (a) Kuala Lumpur, Malaysia at 06GMT on 30 Sept 1994, (OBS: Tsuruta et al., 1997), and (b) Watukosek, Indonesia at 06GMT on 27 Sept 1994, (OBS: Fujiwara et al. 1999). See Table 2 for the simulation cases. Arrows in Fig.1b indicate characteristic local minima and maxima. Locations of Kuala Lumpur and Watukosek are shown in Fig.3.

Backward trajectories of WK (Fig.3b) show completely different features from those of KL (Fig.3a). Below the height of 7 km all the trajectories show very weak vertical motion in resolvable scale and hence suggest stable stratification over the area. From Figs.1b and 3b it can be observed that high ozone concentration below 5 km (Fig.1b) should be strongly affected by emission sources in northern Australia, and another ozone maximum at 7 km was formed by the transport of ozone rich air from upper layer, i.e. 9 km high, over Indian Ocean to the west of Australia. This ozone maximum at 7 km may suggest long range transport from high latitudes in southern hemisphere or Africa. Furthermore extremely low ozone at 15 km (Fig.1b) may be explained by the migration of the air mass with low ozone and high humidity air, from western Pacific Ocean; the air mass was brought into the upper troposphere three days ago by the flow associated with the typhoon T26.

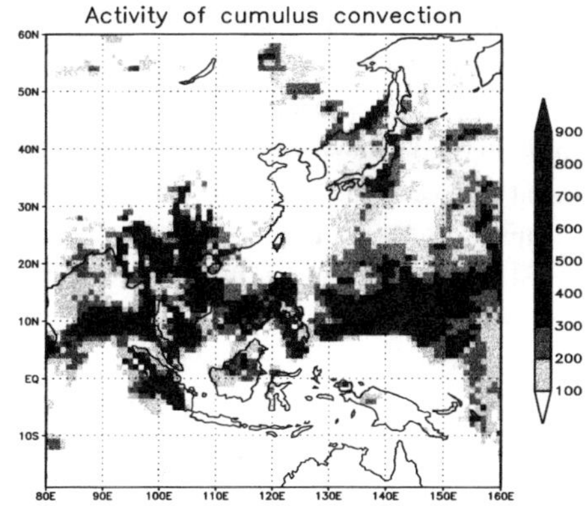

Figure 2. Activity of cumulus convection; spatial distribution of number of subgrid-scale cumulus convection applied during three weeks from 15 Sept. to 4 Oct., 1994. The application of the cumulus convection was judged every time step, i.e. 30 minutes.

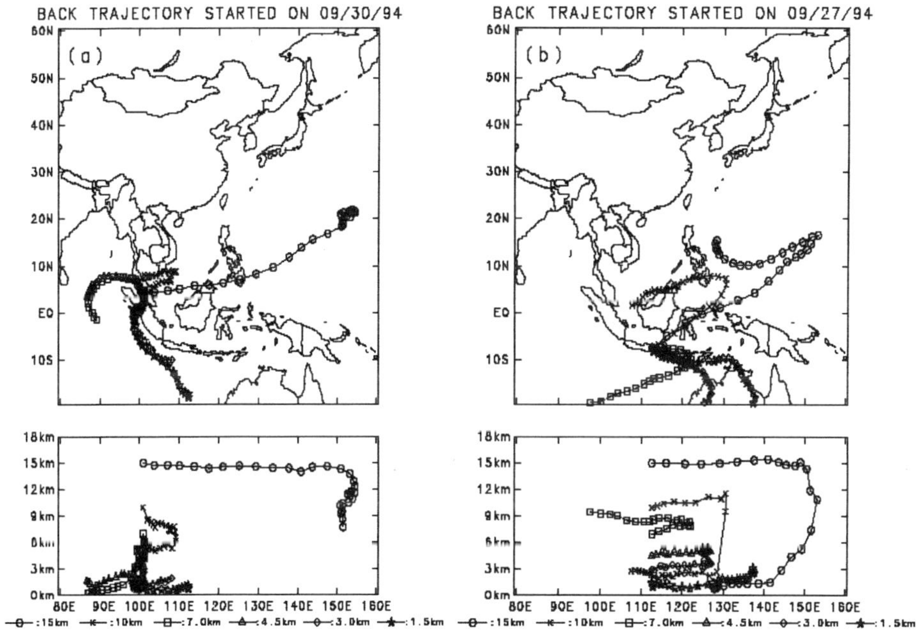

Figure 3. Seven day backward trajectory released at 1.5, 3, 4.5, 7, 10, and 15 km high: started from (a) Kuala Lumpur at 06GMT on 30 Sept 1994, and (b) Watukosek at 06GMT on 27 Sept 1994. Marked every 6 hours. In the calculations, vertical velocity in resolvable scale was used. KM(Kalimantan), SM(Sumatra), JV(Java), and TM(Timor).

COMPUTED VERTICAL PROFILES OF OZONE AND CO AT KUALA LUMPUR AND WATUKOSEK, AND SURFACE OZONE AT SINGAPORE

Computed ozone (reaction product) and CO (primary pollutant) profiles at KL and WK are shown in Figs.1 and 4 respectively. At KL (Fig.1a) all the calculated ozone profiles capture the feature of observed ozone, i.e. rather uniform mixing ratio, although NOEMI predicts much smaller value. Since NOEMI uses no emission sources, difference between AUSMIN2 (a standard case) and NOEMI shows the ozone produced with the precursors released from all the emission sources in the domain; hence it is suggested ozone at KL is produced largely from local sources. Surface ozone at Singapore (Fig. 5) shows similar tendency; without emission sources (i.e. NOEMI), temporal variation of surface ozone can not be explained.

At Watukosek, in contrast to KL, the simulations failed in reproducing some of the characteristics of the observed ozone (Fig.1b), i.e. the local maximum at 7 km in particular. As discussed in the backward trajectory (Fig.3b), the air mass at 7 km at 06GMT on 27 Sept. probably came directly from the south boundary of the domain. Hence this observed ozone peak was due to transport of ozone rich air mass from outside of the domain. Therefore that the boundary condition in the simulation could not express it may be the reason of this failure. In fact, the ozone and CO mixing ratios between 5 and 10 km in NOEMI are close to those in AUSMIN2 (Figs.1b and 4b), indicating the ozone and CO at these heights are not affected by emission sources in the calculation domain. The simulations are partly successful in reproducing increase in ozone (Fig.1b) and CO (see Fig.4b, though no observation is available) mixing ratios below the heights of 3-4 km.

Effects of biomass burning on calculated concentrations can be examined by comparing NOBB and AUSMIN2. From Figs.1,4, it is estimated that biomass burning increased (1) surface ozone by 3-4 ppb (i.e. about 10% of ozone mixing ratio of AUSMIN2) and surface CO by 20 ppb (i.e. 13% of total CO in AUSMIN2) at KL (06GMT, 30 Sept), and (2) surface ozone by 8- 9 ppb (i.e. about 20% of total ozone, AUSMIN2) and surface CO by 25 ppb (i.e. 15% of total CO, AUSMIN2) at WK (06GMT, 4 Oct). Ozone at Singapore (Fig.5) also suggests that increase of ozone by biomass burning is about 4-5 ppb.

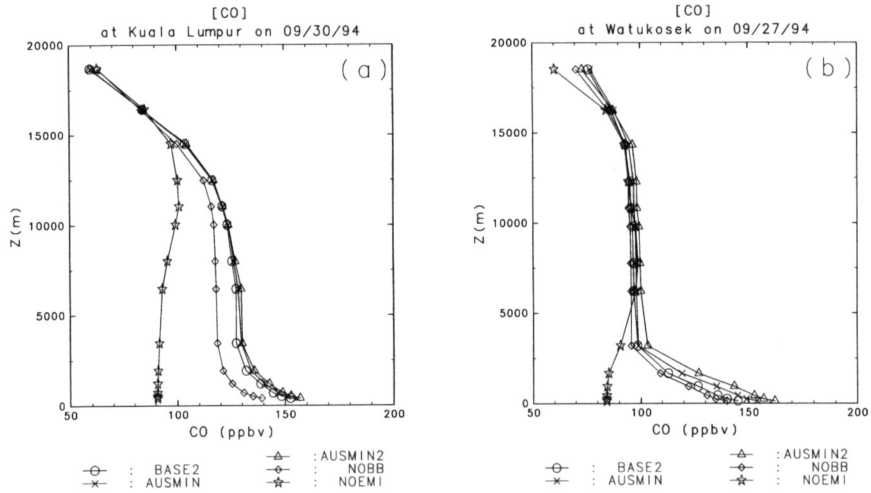

Figure 4. Same as in Fig. 1 except for CO

SPATIAL DISTRIBUTION OF OZONE MIXING RATIO AND COLUMN DENSITY AFFECTED BY BIOMASS BURNING

Effects of biomass burning on the calculated spatial distribution of ozone were evaluated. In stably stratified clear weather area such as northern Australia and eastern Java, initial plume rise by the biomass burning may affect subsequent horizontal transport and lifetime of the released pollutants, while, as we have already seen in Figs. 1a and 4a, in the area of active cumulus convection such as western Sumatra, northern Kalimantan, and the Malay Peninsula pollutants from the biomass burning could be quickly transported over the depth of troposphere and thus may affect environment in global scale (e.g., Matsueda et al. 1998).

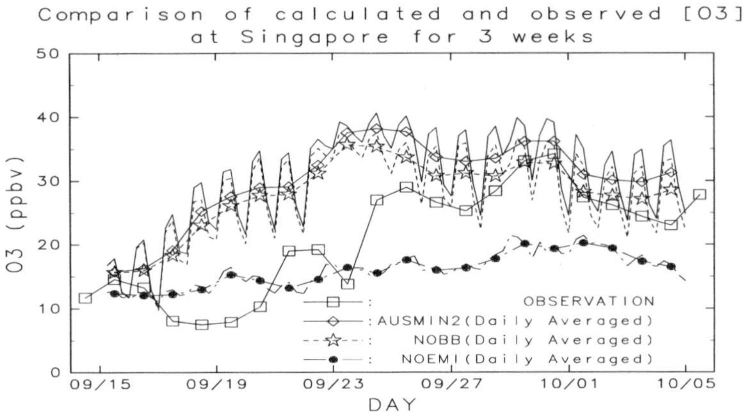

Figure 5. Temporal variation of ozone mixing ratio from 15 Sept. to 5 Oct. 1994 at Singapore: observation and calculations.

Figure 6a-d shows ozone related concentrations at 2 km high in a tropical area, at 06GMT on 27 Sept.: (a) ozone mixing ratio in AUSMIN2, (b) increase in ozone mixing ratio by all the biomass burning sources (i.e., AUSMIN2-NOBB), (c) ozone mixing ratio increased by Australian biomass burning source (i.e., AUSMIN2-BASE2), and (d) increase in ozone mixing ratio by the assumption of the elevated (volume) source for Australian biomass burning (AUSMIN2-AUSMIN). Figure 6e,f expresses (e) increase in ozone column density below 12 km high above mean sea level, by all the biomass burning sources (i.e., AUSMIN2-NOBB), and (f) the same as (e) but for by all the emission sources (i.e., AUSMIN2-NOEMI); both (e) and (f) at 06GMT on 27 Sept.
Figure 6a, which corresponds to a part of Fig.7a, indicates higher ozone mixing ratios appear over active emission source area extending from northern Australia to the Malay Peninsula, although it was suggested in Figs.1a and 5 that the biomass burning source may play rather minor role on the ozone compared with other sources. Figure 6b shows that the effect of biomass burning source on ozone mixing ratio spreads downwind-wards over tropical Asia, but at this height of 2 km it is rather confined in narrow band between 15°S and 8°N, indicating ITCZ (see Fig. 2) prohibits further northward transport. The maximum contribution of the biomass burning source to ozone seems to be about 10-15 ppb at 2 km. Figure 6c demonstrates that Australian biomass burning source explains most of the increase in ozone shown in Fig.6b in western Java and east Timor sea. Figure 6d shows effect of the initial plume rise by the biomass burning (fires) over northern Australia on increase in ozone mixing ratio. Comparison between ozone distributions in Figs. 6c and 6d suggests that to consider the initial plume rise of Australian biomass burning source in the simulation can enhance, in western Java area, increase in ozone mixing ratio by biomass burning by more than 50% at 2 km high. This means that fast transport of ozone precursors such as NOx and NMHC into upper layer strengthens photochemical reaction rate and also increases lifetime of the produced ozone in stably stratified situation.

Figure 6e shows column density of ozone (x10^{-5} kg/m^2) below 12 km high, produced by biomass burning source, at 06GMT on 27 Sept 1994, and Fig. 6f is the same as Fig. 6e except that the effect of all the emission sources in the domain is included. Commonly used Dobson Unit can be interpreted as 1 DU = 2.14x10^{-5} kg/m^2. Hence from Fig.6e the ozone column density produced by the biomass burning can be estimated as about 2 DU over western Java and 1 DU over the southern Malay Peninsula. Figure 6f suggests the other sources than biomass burning such as fossil fuel combustion produce more ozone in the southern Malay Peninsula, and it reaches almost 7 DU.

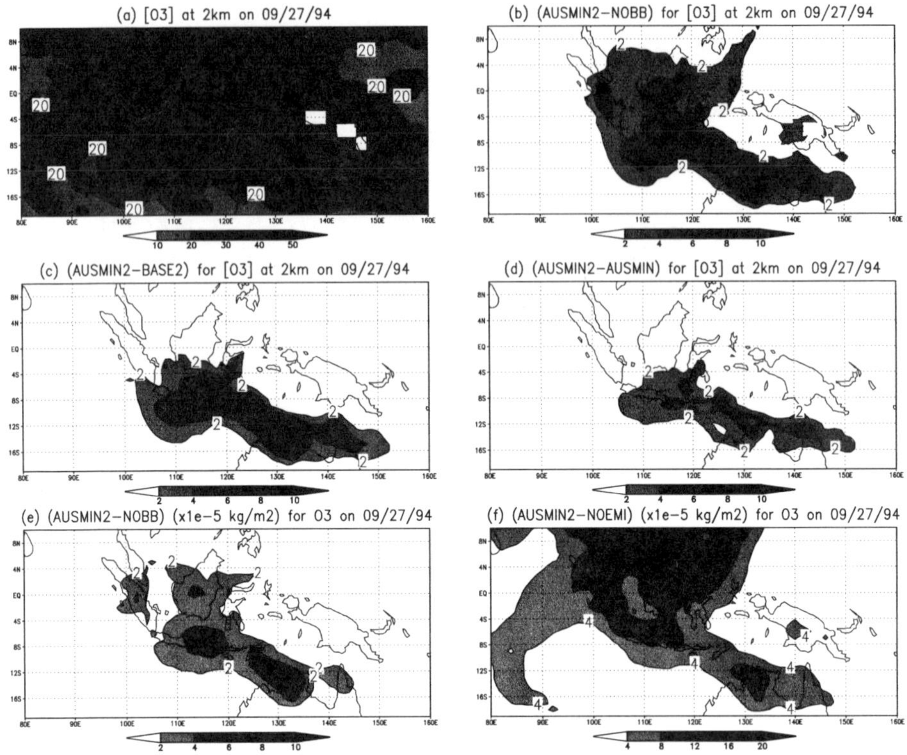

Figure 6. Calculated ozone at 06GMT on 27 Sept 1994: (a) mixing ratio at 2 km (AUSMIN2), (b) increase in mixing ratio by all the biomass burning sources at 2 km (AUSMIN2-NOBB), (c) increase in mixing ratio by the Australian biomass burning (AUSMIN2-BASE2), (d) increase in mixing ratio by applying elevated emission height to the Australian biomass burning (AUSMIN2-AUSMIN), (e) ozone column density, below 12 km high produced by all the biomass burning sources (AUSMIN2-NOBB), and (f) ozone column density below 12 km high, produced by all the emission sources (AUSMIN2-NOEMI). Mixing ratio (ppb) in (a)-(d); column density (x10^{-5} kg-ozone/m^2) in (e) and (f). See Table 2 for simulation cases.

Figure 7a shows computed ozone mixing ratio at 2 km high (AUSMIN2) and Fig. 7b is the column density of ozone produced by all the emission sources (AUSMIN2-NOEMI); both are plotted in the full calculation domain at 06GMT on 27 Sept 1994. (Figure 6f is a part of Fig. 7b). Figure 7a depicts the typhoon T26, named by Japanese Meteorological Agency, incorporates clean marine air into its eye region and meanwhile brings polluted continental air mass over Pacific ocean to the south of the typhoon. Figure 7b, compared with Fig.7a, suggests the area of large ozone column density generated by the emission sources can somewhat differ from the area of high ozone mixing ratio at 2 km high; i.e., the large column density can appear in the area where both large emission source and strong cumulus convection activity coexist, indicating fast upward transport of pollutants by cumulus convection enhances ozone production in upper layer and contributes to its longer lifetime. In this sense, southern coastal China should be an interesting place since there are both strong anthropogenic emission sources and cumulus convection. In fact, the ozone column density in Fig. 7b shows its enhancements of about 15 DU in coastal southern China and about 8 DU over the Malay Peninsula. This ozone column density agrees well with the

15-day averaged Meteor3/TOMS total ozone map for Sept 16-30, 1994, which is shown in Fujiwara et al. (1998). Additionally, from both Figs.7a and b, in the typhoon eye area, not only ozone mixing ratio at 2 km but also the ozone column density is small.

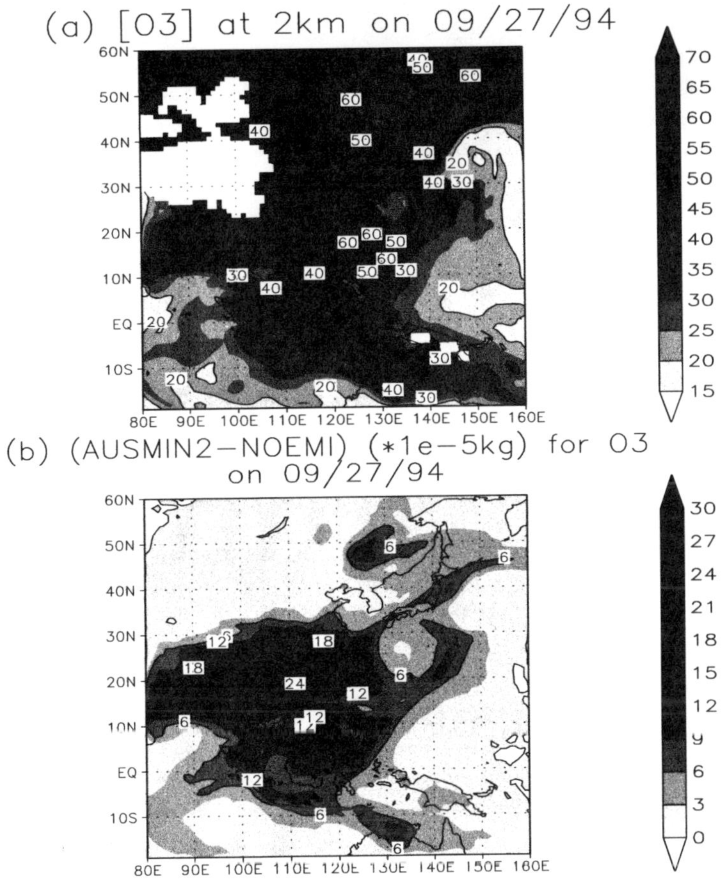

Figure 7. (a) Calculated ozone mixing ratio (ppb) at 2 km (AUSMIN2), and (b) ozone column density below 12 km (x10^{-5} kg/m^2) produced by all the emission sources (AUSMIN2-NOEMI) at 06GMT on 27 Sept 1994.

SUMMARY AND CONCLUSIONS

Numerical simulation of the transport/chemistry/deposition of trace chemical species was performed to see the influence of biomass burning in tropical Asia on concentrations of ozone and CO and tropospheric chemistry for September and October in 1994. Some of the obtained results are as follows: (1) In tropical Asia, two different meteorological situations in synoptic scale coexisted; dry and stably stratified atmosphere over Java and southern Kalimantan, Indonesia, and wet and unstable atmosphere with active cumulus convection over Sumatra, northern Kalimantan, and the Malay Peninsula. Effect of these different atmospheric conditions on ozone and CO were evaluated. (2) Increase in ozone and CO mixing ratios by biomass burning source was quantified: for example, ozone increased roughly by 3-4 ppb at Kuala Lumpur (which was much smaller than the increase of 20 ppb by other anthropogenic emission sources), and by 8-9 ppb at Watukosek, Java. Surface ozone at Singapore (Fig. 5) showed similar tendency to that at KL; ozone increased 3-4 ppb by the biomass burning and 10-15 ppb by the other sources. (3) Predicted ozone column density (Fig. 7) suggested importance of emission sources in southern coastal China and Southeast

Asia because of coexistence of strong source and cumulus convection. (4) Inclusion of initial plume rise for biomass fire emission can be important because of different chemical reaction rate and longer lifetime in free troposphere.

REFERENCES

Cautenet, S., Poulet, D., Delon, C., Delmas, R., Grègoire, J.M., Pereira, J.M., Cherchali, S., Amram, O., and Flouzat, G., 1999, Simulation of carbon monoxide redistribution over Central Africa during biomass burning events (EXPRESSO experiment); *J. Geophys. Res.*, 104:30641-30657.

Emission Database for Global Atmospheric Research (EDGAR) Version 2.0, 1997, ftp://info.rivm.nl/pub/lae/EDGARV20/

Ferek, R.J., Reid, J.S., Hobbs, P.V., Blake, D.R., and Liousse, C., 1998, Emission factors of hydrocarbons, halocarbons, trace gases and particles from biomass burning in Brasil, *J. Geophys. Res.*, 103:32,107-32,118.

Fujiwara, M., Kita, K., Kawakami, S., Ogawa, T., Komala, N., Saraspriya, S., and Suripto, A., 1999, Tropospheric ozone enhancements during the Indonesian forest fire events in 1994 and in 1997 as revealed by ground-based observation, *Geophys. Res. Let.*, 26:2417-2420.

Fujiwara, M., Kita, K., S., Ogawa, T., Komala, N., Saraspriya, S., and Suripto, A., 1998, Total ozone enhancement in September and October 1994 in Indonesia, *Proc. XVIII Quadrennial Ozone Symposium*, L'Aquila, Italy, 12-21 Sept., 1996, Vol.1, 363-366.

Hao, W.M., Liu, M.H. and Crutzen, P.J., 1990, Estimates of annual and regional releases of CO2 and other trace gases to the atmosphere from fires in the tropics, based on the FAO statistics for the period 1975-1980, in *Fire in the Tropical Biota*, Ecological Studies, Vol.84, ed. J.G. Glodammer, 440-462, Springer-Verlag.

IGBP-DIS, 1999, http://www.mtv.sai.jrc.it/projects/fire/gfp/

Kitada, T., and Nishizawa, M.., 1998, Modeling study of the long range transport of acidic pollutants over East Asia and the west Pacific Ocean –Sensitivity of acid deposition to scavenging model parameters and emission source distribution, *J. Global Environ. Eng.*, Japan Society of Civil Eng., 4:1-29.

Kitada, T., Nishizawa, M., and Kondo, Y., 1998, Long range transport of NOx, NOy, O_3, and Sox over East Asia and the western Pacific Ocean in winter season, in *Air Pollution Modelling and Its Application XII*, Plenum Pub. Co., 205-215.

Kumar, P. P., Manohar, G. K., and Kandalgaonkar, S. S., 1995, Global distribution of nitric oxide produced by lightning and its seasonal variation. *J. Geophys. Res.*, 100: 11,203-11,208.

Matsueda, H., Inoue, H., Sawa, Y., Tsutsumi, Y., and Ishii, M., 1998, Carbon monoxide in the troposphere over the western Pacific between 1993 and 1996, *J. Geophys. Res.*, 103:19093-19110.

NASA Goddard DAAC, 1999, http://daac.gsfc.nasa.gov/

NASA Langley RC, 1999, http://hyperion.gsfc.nasa.gov/AEAP/AEAP.html

Tsuruta, H., Yonemura, S., Peng, L.C., Fook, L.S., 1997, The increase of tropospheric ozone over Peninsular Malaysia by the 1994 forest fires in tropical east Asia, Paper presented at *International Symposium on Atmospheric Chemistry and Future Global Environment*, Int. Global Atmos. Chem., Nagoya , Japan, Nov.11-13, 1997.

Tsutsumi, Y., Sawa, Y., Makino, Y., Jensen, J.B., Gras, J.L., Ryan, B.F., Diharto S., and Harjanto, H., 1999, Aircraft measurements of ozone, Nox, CO, and aerosol concentratios in biomass burning smoke over Indonesia and Australia in October 1997: Depleted ozone layer at low altitude over Indonesia, *Geophy. Res. Let.*, 26:595-598.

Ward, D.E., and Hardy, C.C., 1991, Smoke emissions from wildland fires, *Environment International*, 17:117-134.

Yienger, J. J., and Levy II, H., 1995, Empirical model of global soil-biogenic NOx emissions. *J. Geophys. Res.*, 100: 11,447-11,464.

APPLICATION OF A NEW LAND-SURFACE, DRY DEPOSITION, AND PBL MODEL IN THE MODELS-3 COMMUNITY MULTI-SCALE AIR QUALITY (CMAQ) MODEL SYSTEM

Jonathan E. Pleim and Daewon W. Byun

Atmospheric Sciences Modeling Division, NOAA, Research Triangle Park, NC 27711 (on assignment to the National Exposure Research Laboratory, U.S. Environmental Protection Agency)

INTRODUCTION

The U.S. EPA has developed a new comprehensive air quality modeling system, known as the Models-3 Community Multi-scale Air Quality (CMAQ) model (Byun and Ching, 1999). The CMAQ system includes a comprehensive emissions processor, a Chemical Transport Model (CTM), and a meteorology model. The community modeling approach aims to provide a focal point for diverse model development research which will lead to many alternative algorithms and model components that can be inter-compared and evaluated. An early example of this process is the development of a more advanced surface exchange, planetary boundary layer (PBL), and dry deposition model and its incorporation into the CMAQ system.

Accurate simulation of air quality depends on realistic modeling of land surface and PBL processes. Quantities that exert first-order control on trace chemical concentrations and photochemistry include PBL height, temperature, wind speed, and dry deposition velocity. These quantities are all directly related to air-surface exchange processes of heat, moisture, momentum, and trace chemical species. In addition, cloud cover, which is important for photolysis rates, is greatly influenced by surface flux and PBL processes. For these reasons it is especially critical to apply realistic techniques for land-surface and PBL modeling within an air quality system.

Like most air quality modeling systems, CMAQ divides the treatment of meteorological and chemical/transport processes into separate models run sequentially. A potential drawback to this approach is that it creates the illusion that these processes are minimally interdependent and that any meteorology model with a good reputation is adequate for air quality work. However, most mesoscale meteorology models are developed for operational weather forecasting and meteorological research. These foci do not emphasize all the same critical capabilities as air quality applications. Conversely, CTMs are often developed to accept basic meteorological inputs from a variety of sources with little regard to its quality and even less regard to consistency between physical parameterizations in the meteorology model and the CTM. The work reported here attempts to address some of these weak links in the system, particularly where improvements in land-surface modeling in the meteorology model and consistency with similar components in the CTM can have significant effects on the air quality simulation. Therefore, this development cuts across several system components. A new land-surface model (LSM), which features explicit simulation of soil moisture and vegetative evapotranspiration, has been coupled with the Fifth Generation Penn State/NCAR Mesoscale Model (MM5). An attendant dry deposition model has been developed to take advantage of the more sophisticated treatment of surface fluxes, stomatal conductance, and surface layer diffusion in the new LSM. The Meteorology Chemistry Interface Processor (MCIP) has been modified to include the new dry deposition model as well as to make the additional information resulting from the new LSM available to the CTM. Also, a new non-local closure PBL scheme that is compatible with the modifications made to the MM5 has been added to the list of vertical diffusion module options of the CMAQ CTM.

This paper is not meant to be an evaluation of the models described here, since many of the components have been evaluated against field experiment data and reported elsewhere, but rather a demonstration that such modeling developments are important to air quality modeling objectives. Comprehensive evaluation of the CMAQ system, including these components, is ongoing.

MODEL DESCRIPTION

There has recently been heightened appreciation for the importance of more sophisticated land surface models (LSMs) in mesoscale meteorology models. Currently, there are at least 3 new LSMs being added to the MM5 system. Pleim and Xiu (1995) describe the prototype development and testing of an LSM (hereafter referred to as PX) for use in the MM5 and the CMAQ system. Pleim et al. (1996) describe testing and evaluation of the LSM as implemented in the MM4 with the new attendant dry deposition model through comparison to surface flux measurements of heat, moisture, and ozone dry deposition at two field sites. Xiu and Pleim (2000) describe implementation into the MM5 and evaluation compared to field measurements of surface fluxes, temperature, and PBL height. The PX LSM includes explicit simulation of soil moisture and temperature in two soil layers and three pathways for surface evaporation including soil evaporation, canopy evaporation, and evapotranspiration. The soil moisture model is based on the Interaction Soil Biosphere Atmosphere (ISBA) model (Noilhan and Planton, 1989; Jacquemin and Noilhan, 1990), which was specifically designed for mesoscale modeling and has an extensive record of evaluation and comparison. Stomatal conductance is parameterized according to root zone soil moisture, air temperature and air humidity, photosynthetically active radiation (PAR), and several vegetation parameters such as leaf area index (LAI) and minimum stomatal resistance. Although originally based on the ISBA model the stomatal and canopy parameterizations are almost entirely new. New features include a canopy shelter factor to account for shading within denser canopies, new stomatal functions with respect to environmental parameters, and inclusion of a data assimilation scheme similar to the technique described by Bouttier et al. (1993). A simple parameterization for describing seasonal growth of vegetation, including leaf-out of deciduous trees, has also been developed and tested.

Dry deposition is a good example of the close interaction of chemistry and meteorology. However, dry deposition models are usually part of the CTM and therefore inconsistent with the LSM with regard to land-use and vegetation characterization as well as aerodynamic, canopy, and stomatal resistance parameterizations. Clearly, the treatment of common parameterizations and datasets should be as consistent as possible between the LSM used in the meteorology model and the dry deposition scheme used in the CTM. Therefore, the new dry deposition module for CMAQ (known as M3Dry) directly uses the same bulk stomatal resistance, aerodynamic resistance, and vegetation parameters such as LAI, roughness length and vegetation coverage as in the PX LSM. Hence, the stomatal uptake of gaseous trace chemicals is simulated in exactly the same way (adjusting for molecular diffusivity) as evapotranspiration in the LSM. Other dry deposition pathways, such as deposition to leaf cuticles and soil, are parameterized according relative reactivity and solubility of the chemical species (see Byun and Ching, 1999 for more details).

The M3Dry dry deposition model also utilizes PBL and surface layer parameters directly from the MM5-PX. These codependencies not only ensure greater consistency between meteorology and chemistry models but also benefit the dry deposition calculation by using more responsive stomatal conductance simulations. In particular, the stomatal conductance, and therefore the dry deposition, can respond to soil moisture conditions, which are rarely considered in air quality models. Also, the indirect soil moisture data assimilation scheme provides realistic constraints on the stomatal conductance.

PBL processes are another critical modeling component for air quality systems. The PX LSM includes a simple non-local closure scheme known as the Asymmetric Convective Model (ACM) (Pleim and Chang, 1992). The ACM is quite similar to the Blackadar non-local scheme (Blackadar, 1978) which has long been the most widely used PBL option in the MM5 system. Both schemes use non-local transport in the convective boundary layer to simulate transport by rapidly rising buoyant plumes. The ACM differs from the Blackadar scheme in its treatment of downward transport, which is prescribed to be local, one layer at a time, to simulate gradual compensatory subsidence, hence asymmetrical.

MODEL COMPARISON

The modified CMAQ model system using the MM5 with the PX LSM (MM5-PX) and the new dry deposition model was run for a study period of July 6-16, 1995 at 36 km grid resolution and July 11-15, 1995 on a 12 km nested grid. The 36 km domain covered eastern United States and southeastern Canada while the 12 km nest covered most of the Northeast US and southern Ontario. Two sets of CMAQ-CTM (CCTM) runs were made using the MM5-PX and new dry deposition model; one with the standard eddy diffusion model for vertical mixing and the other using the ACM in the CCTM. These two sets of runs were compared to a set of base case runs which used the base case MM5 simulations, with the Blackadar PBL scheme and static moisture availability factors for surface moisture, and the RADM dry deposition model (Wesely, 1989). Other than the model modifications outlined above the three sets of runs used the same model options and input data (emissions, meteorology, and initial/boundary conditions) for both MM5 and the CCTM as described by Byun and Pleim (2000).

Table 1. Model experiment configurations.

Experiment	Meteorology	Dry Deposition	CCTM Vertical Mixing
Base	Base MM5	RADM	Eddy Diffusion
PX	MM5-PX	M3Dry	Eddy Diffusion
PX-ACM	MM5-PX	M3Dry	ACM

The three sets of model runs, referred to by experiment name as Base, PX, and PX-ACM as shown in Table 1, were compared to all of the ozone measurements from the EPA's AIRS monitoring network within the 12 km nested domain (208 sites). Figure 1 shows time series of ground level hourly ozone concentration averaged over the 208 measurements and the collocated 12 km grid cells from the three experiments. Note that this is not a spatial average since the measurement sites are not evenly distributed. Also, the AIRS sites tend to be located near urban or suburban areas. Thus, inclusion of the AIRS data is meant as a rough reference for the models rather than an evaluation goal.

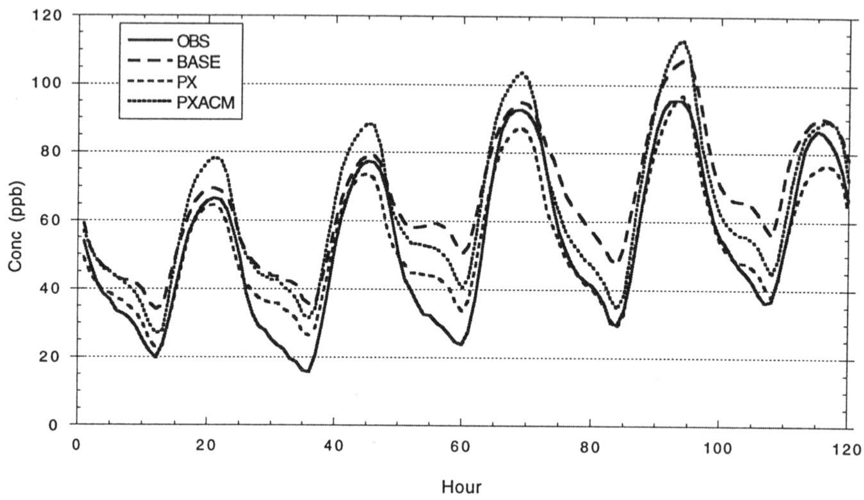

Figure 1. Modeled and measured ozone concentrations averaged over 208 AIRS monitoring sites.

With these disclaimers in mind, Figure 1 shows that the 3 models usually bracket the observations during the peak ozone periods but over-estimate low concentrations. The order of model ozone concentration magnitude is very consistent throughout the simulation with PX, Base, PX-ACM from low to high during the afternoon peaks but the order changes to PX, PX-ACM, Base at night and early morning. Compared to the observations, the peaks are best simulated by either the Base or PX while the PX is closest to the observation in the troughs. Note that over-prediction of nocturnal ozone minima is an endemic artifact of Eulerian grid models with coarse vertical resolution (the first layer thickness is about 35 m). The PX-ACM is generally the highest at the daytime peaks. This is because of much more rapid upward transport from the surface layer throughout the convective boundary layer (CBL). Therefore, in the lowest model layer the ACM simulates lower concentrations of surface emitted precursors, such as NO_x and Isoprene, but higher concentrations in the mixed layer resulting in faster ozone production aloft. The higher concentration ozone quickly mixes throughout the CBL. As noted by Byun and Pleim (2000) this apparent over prediction and over-vigorous upward transport may be an artifact of the simplistic treatment of surface emissions. Since ACM is based on similarity with sensible heat fluxes (as are most other PBL schemes) the lack of gradient diffusion surface layer fluxes for chemical emission may result in significant over-estimation of both surface flux rate and layer 1 concentrations.

To help understand the differences between the Base and PX runs Figure 2 shows PBL height, surface solar radiation, temperature at 1.5 m, and ozone dry deposition velocity averaged over the same observation site locations as for Figure 1. On the average, surface solar radiation is less for the PX case than the Base case. Other studies (Xiu and Pleim, 2000) have shown that solar radiation is essentially identical between these two versions of MM5 under clear sky conditions. Therefore, differences in the surface radiation are caused by differences in cloud cover. Averaged over the entire 12-km domain the cloud cover from the MM5-PX ranged from 0-20% (usually 5-10%) greater than from the Base MM5 during this 5-day period. It is difficult to say which cloud cover simulation is better because they both differed significantly from satellite photos. Hence, the simulation of cloud cover is a critical weak link in current modeling systems. Techniques for assimilation of satellite data are under development (McNider et al., 1998) that may lead to substantial improvements in meteorology and air quality modeling.

PBL heights are generally comparable but with significant differences on some days in either direction. The 1.5 m temperature simulations are similar but with higher temperatures from the PX runs on the second and third days even though the average surface radiation is less. It is interesting that the lesser average solar radiation in the PX experiment does not result in generally lower temperatures or lower PBL height, which demonstrates that the relationships between surface energy and PBL processes are somewhat different for the PX LSM.

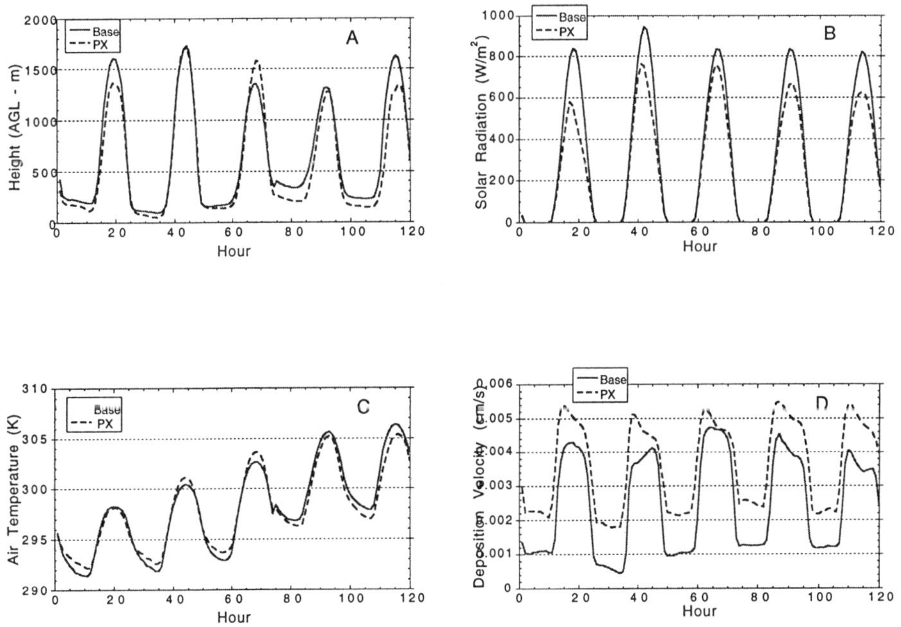

Figure 2. Modeled meteorology parameters averaged over AIRS site locations: A) PBL height, B) surface solar radiation, C) temperature at 1.5 m, D) ozone dry deposition velocity.

The ozone dry deposition velocity is considerably higher from the PX runs during both day and night, with differences from the RADM model often up to about 0.1 cm/s. The PX results also show a more consistent morning peak in ozone deposition velocity (usually about 9 LST) which is more realistic in vegetated areas (Finkelstein et al., 2000). The much higher nighttime deposition velocities simulated by the PX runs are partially attributable to the minimum friction velocity of 0.1 m/s in the MM5.

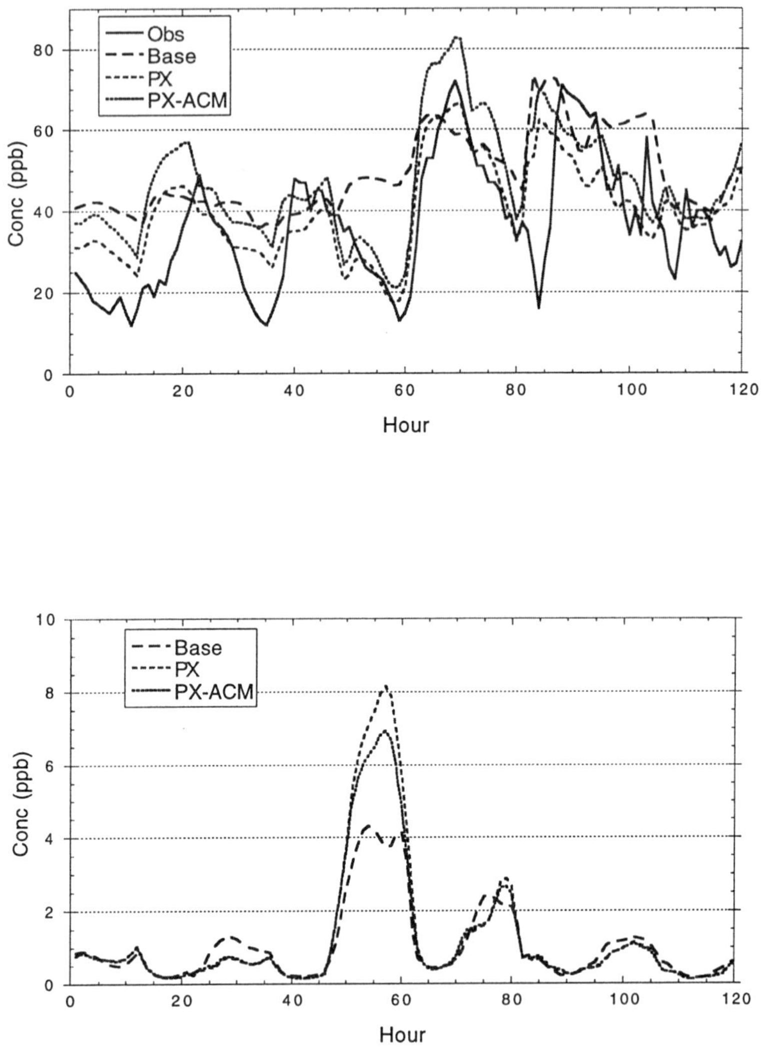

Figure 3. Modeled and measured ozone concentration (top) and modeled NO_x concentration (bottom) at the AIRS site near Plymouth, NH.

Cause and effect relationships are hard to understand from analysis of averaged quantities. Therefore, we have also selected a single site near Plymouth, NH for further study. This particular site was chosen because of relatively large differences among the models. Figure 3 shows the ozone and NO_x concentrations from the 3 modeling experiments along with the observations of ozone and Figure 4 shows the modeled meteorological parameters at this site. The solar radiation plot shows that both models simulated the second and third days as essentially clear while the other three days had various amounts of partial cloudiness with the PX simulation generally cloudier. The concentration plots show considerable difference among the experiments particularly on the third day. It is interesting that the base run shows very little diurnal change in ozone compared to the other runs and the observations. The lesser peak NO_x concentration on the morning of the third day in the Base run explains why the Base run shows no ozone trough. The lower NO_x peak is probably due to difference in the nocturnal PBL between the PX and Base versions and the higher nocturnal ozone dry deposition velocities from the M3dry model (Figure 4D). On the afternoon of the third day the Base ozone declines while the observations and other two runs increase to a late afternoon peak. A similar relative dip in temperature and PBL height is also evident in the Base MM5 output that is not reflected in the radiation and therefore not caused by clouds. This is an interesting example of how differences in meteorological simulations propagate through the chemical simulations.

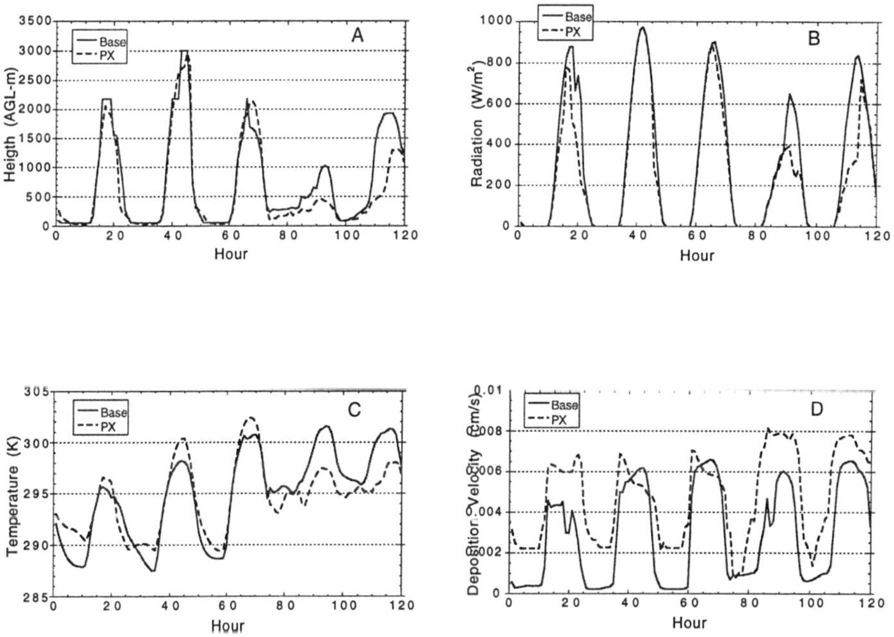

Figure 4. Modeled meteorology parameters at the AIRS site near Plymouth, NH: A) PBL height, B) surface solar radiation, C) temperature at 1.5 m, D) ozone dry deposition velocity.

CONCLUSIONS

Preliminary analysis of the sensitivity of the CMAQ system to inclusion of a more sophisticated land surface model with consistent treatment of dry deposition shows substantial effects on one of the most important model outputs, namely ozone concentration. The PX runs generally produce lower ozone concentrations than the base runs. This difference is probably mainly due to the tendency of the MM5-PX to produce greater amounts of cloudiness, which reduces photolysis rates and the greater ozone dry deposition velocities from the M3Dry deposition model. It is interesting that the greater cloudiness simulated by the MM5-PX does not similarly reduce temperature and PBL height compared to the Base MM5. Therefore, the PX LSM not only changes many of the surface, PBL, and cloud parameters but also alters the relationships between them.

The Plymouth, NH case study does not show such systematic differences in ozone concentration between the PX and Base runs as seen in the averaged results, particularly during the clear period on days 2 and 3. This case does show considerable differences caused by the different meteorological simulations such as the huge difference in concentration during the early morning of the third day and the difference in the late afternoon peak. The different dry deposition models may play a significant role in the nighttime differences but probably not during the daytime when the dry deposition velocities from the two models were very similar (see Figure 4D). It will take further analysis to differentiate the effects of the model changes in meteorology and dry deposition.

This preliminary study demonstrates the value of a flexible, comprehensive, air quality modeling system for developing and evaluating new modeling techniques. Evaluations of modeling advancements need to be performed at both the component level and the integrated level to assess relevance to the ultimate products. In this way the greatest attention can be paid to the weakest links in the system.

This paper has been reviewed in accordance with the US Environmental Protection Agency's peer and administrative review policies and approved for presentation and publication. Mention of trade names or commercial products does not constitute endorsements or recommendation for use.

REFERENCES

Blackadar, A.K., 1978, Modeling pollutant transfer during daytime convection, in: *Preprints, Fourth Symp. on Atmospheric Turbulence, Diffusion, and Air Quality*, Reno, NV, Amer. Meteor. Soc. 443-447.

Bouttier, F., Mahfouf, J.F., and Noilhan, J., 1993, Sequential assimilation of soil moisture from atmospheric low-level parameters. Part I: Sensitivity and calibration studies, *J. Appl. Meteor.* 32:1335-1351.

Byun D.W., and Ching, J.K.S., ed., 1999, Science Algorithms of the EPA Models-3 Community Multiscale Air Quality (CMAQ) Modeling System, NERL, Research Triangle Park, NC, EPA/600/R-99/030.

Byun D.W., and Pleim, J.E., 2000, Sensitivity of ozone and aerosol predictions to the transport algorthms in the Models-3 Community Multi-scale Air Quality (CMAQ) Model, *this volume.*

Finkelstein P.L., Ellestad, T.G., Clarke, J.F., Meyers, T.P., Schwede, D.B., Hebert, E.O., and Neal, J.F., 2000, Ozone and sulfur dioxide dry deposition to forests: Observations and model evaluation, *Accepted by J. Geophys. Res.*

Jacquemin, B., and Noilhan, J., 1990, Sensitivity study and validation of a land surface parameterization using the HAPEX-MOBILHY data set, *Bound.-Layer Meteor.* 52:93-134.

McNider, R. T., W. B. Norris, D. Casey, J. E. Pleim, S. J. Roselle, W. M. Lapenta, 1998, Assimilation of satellite data in regional air quality models. In: *Air Pollution Modeling and Its Application XII*, Gryning and Chaumerliac, Eds, Plenum Press, New York.

Noilhan, J., and Planton, S., 1989, A simple parameterization of land surface processes for meteorological models. *Mon. Wea. Rev.* 117:536-549.

Pleim, J.E., and Chang, J.S., 1992, A non-local closure model for vertical mixing in the convective boundary layer, *Atmos. Environ.*, 26A:965-981.

Pleim, J.E., Clarke, J.F., Finkelstein, P.L., Cooter, E.J., Ellestad, T.G., Xiu, A., and Angevine, W.M., 1996: Comparison of measured and modeled surface fluxes of heat, moisture and chemical dry deposition, In: *Air Pollution Modeling and Its Application XI*, Gryning and Schiermeier, Eds, Plenum Press, New York.

Pleim, J.E., and Xiu, A., 1995, Development and testing of a surface flux and planetary boundary layer model for application in mesoscale models. *J. Appl. Meteor.*, 34:16-32.

Wesely, M.L., 1989, Parameterization of surface resistances to gaseous dry deposition in regional-scale numerical models, *Atmospheric Environment*, 23:1293-1304.

Xiu, A., and Pleim, J.E., 2000, Development of a land surface model part I: Application in a mesoscale meteorology model, Accepted by *J. Appl. Meteor.*

DISCUSSION

B. YAMARTINO: A very nice feature of the ACM is its ability to describe counter-gradient transfers; however, in the examples you show, the more rapid vertical mixing could be achieved by simply using a larger K_{zz}. Have you seen some cases where the unique features/ advantages of the ACM are exhibited?

J. PLEIM: One of the unique features of the Asymmetric Convective Model (ACM) is that the upward and downward mixing is different (hence the name Asymmetric). The upward mixing is non-local so it transports material from the surface throughout the PBL very quickly, while the downward mixing is local so downward transport is more gradual. You are correct that this is not much different from eddy diffusion with large K_{zz} for surface sources but it is very different for elevated sources. Using the ACM, elevated plumes, for example, disperse slowly while the centerline descends until the plume hits ground then it mixes quickly throughout the PBL. Observations and LES suggest this behavior is realistic.

R. SAN JOSE: Are the USGS maps you referred in your talk from the NDVI data from AVHRR/NOAA satellite data? Is your dry deposition model capable to respond to relative humidity changes? If not the nocturnal ozone peaks can be catched by incorporating these features (see Erisman, 1994).

J. PLEIM: The USGS land use data was derived from NDVI from the AVHRR satellite data. The stomatal conductance model that is used for both evapotranspiration and dry deposition does respond to relative humidity. The dry deposition model also accounts for nocturnal dew formation such that the surface resistance is different (lower for ozone) when surfaces are wet.

TRANSPORT OF AIR POLLUTION FROM ASIA TO NORTH AMERICA

J.J. Yienger[1], G.R. Carmichael[1], M.J. Phadnis[1], S.K. Guttikunda[1], T.A. Holloway[2], M.K. Galanter[2], W.J. Moxim[2], and H. Levy II[2]

[1]Center for Global and Regional Environmental Research (CGRER)
204 Iowa Advanced Technology Laboratories
University of Iowa
Iowa City, Iowa 52242
[2]Geophysical Fluid Dynamics Laboratory (GFDL)
Princeton Forestall Research Campus, Rt 1
Princeton, NJ 08542

INTRODUCTION

Recognizing a potential for trans-Pacific transport, scientists are beginning to look for an Asian pollution signal over North America (NA). Kritz et al. [1990] previously found that Asian boundary layer air can be transported to the upper troposphere over California in two to four days. More recently, Jaffe et al., [1999] published evidence of Asian air pollution transported to the surface of North America during spring. The year following Jaffe's 1997 observations, a strong April Asian dust storm carried a visible plume of dust to North America [Husar, 1998]. There is growing awareness of the need to include an intercontinental perspective when analyzing North American chemical and aerosol data. Indeed, evidence of trans Pacific transport continues to mount; Jaffe's group has once again detected in the spring of 1999 significant plumes of Asian pollution off of Washington state [Dan Jaffe, private communication].

In this study we employ the Geophysical Fluid Dynamics Laboratory (GFDL) Global Chemistry Transport Model (GCTM) for CO and O_3 to specifically address the episodic nature of trans-Pacific pollution. The model's resolution (265 km x 265 km) is sufficient to produce synoptic-scale tracer fluctuations similar in magnitude and nature to those observed in the real atmosphere [see Levy and Moxim, 1989; Moxim, 1990; Moxim et al., 1996]. We address questions such as how well is the Asian pollution signal over North America represented by a mean value? How frequent are strong transport events (i.e., like those observed by Jaffe et al. [1999] at CPO) expected during and outside of spring, and how does this frequency differ with location and altitude?

MODEL DESCRIPTION

GCTM

The Geophysical Fluid Dynamics Laboratory (GFDL) Global Chemical Transport Model (GCTM) has a horizontal resolution of ~265km [2.4°x2.4° in the tropics and 3°-3.5°x2.4 in midlatitudes], 11 sigma levels in the vertical at standard pressures of 10, 38, 65, 110, 190, 315, 500, 685, 835, 940, and 990 mb and is driven by 12 months of 6-hour time-averaged winds, temperatures, and precipitation fields from a general circulation model [Manabe et al., 1974]. The transport portion of the GCTM employees a numerical scheme which is 2nd order in the horizontal and 4th order in the vertical for resolved (grid-scale) advection [see Section 3. of Mahlman and Moxim, 1978] and includes diffusion-based parameterizations for horizontal sub-grid scale transport and for vertical sub-grid scale transport due to dry and moist convection throughout the troposphere and shear dependent mixing in the boundary layer [Kasibhatla et al., 1996; Levy et al., 1999 and references therein for details].

CO and O_3 Simulations

This work relies on the analysis of global CO and O_3 simulations that have been described and extensively evaluated previously [Levy et al., 1997; Klonecki and Levy, 1997; Yienger et al., 1999; Holloway et al., 1999]. The GCTM simulation of CO is designed for 1990 conditions that include emissions from fossil fuel [300 Tg CO/yr], biomass burning [748 Tg CO/yr], biogenic hydrocarbon oxidation [683 Tg CO/yr] and CH_4 oxidation [758 Tg CO/yr]. The only CO destruction pathway is OH oxidation based on pre-calculated monthly-mean 3D OH fields [Spivakovsky et al., 1990] that have been scaled by 1.15 to give a CH_3CCl_3 global lifetime of 4.8 years..

The GCTM simulation of ozone is a one tracer experiment with four component sources and sinks. At the tropospheric boundaries, ozone is carried down from the stratosphere by Stratosphere-Troposphere Exchange [STE] and destroyed at the surface via dry deposition. In the polluted BL, net ozone production is computed with a NO_x-dependent proportionality constant that relates NO2 oxidation to HNO_3 to net ozone production. In the free troposphere and clean BL, instantaneous production and destruction are based on diurnal average steady-state solutions to a CH_4-CO/Acetone-H_2O-NO_x photochemical box model applied over land and sea for the global range of latitudes and model levels and for every month. This battery of solutions for the tropospheric ranges of CO/Acetone, NO_x, O_3 and H_2O comprise an off-line table that is accessed every time-step in every grid-box with the appropriate values of NO_x and CO from previous simulations (Levy et al., 1999; Holloway et al., 1999), O_3 from the current simulation, and monthly averaged water vapor (Oort, 1983; Soden and Bretherton, 1996). Both a detailed discussion of this method and an extensive comparison to observations can be found in Levy et al. [1997], Klonecki, [1998], and Yienger et al. [1999].

Definition of Asian and North American CO and O_3 Tracers

To isolate pollution from Asia, we made three separate 2-year CO simulations: one with full global emissions, one without surface emissions from Asia (defined to include South, Southeast, and East Asia), and one without surface emissions from North and Central America. In the last two cases, biogenic hydrocarbon oxidation is considered an instantaneous surface source of CO and is removed along with fossil fuel and biomass burning emissions of CO. The first year of the run is used to equilibrate the model and the second year used in analysis. We define the contribution of either Asia or North America to total CO (tracers we refer to as "Asian CO" and "North American CO") as the difference between the full emission simulation and the respective simulations without the local emissions. For ozone, analogous pairs of simulations were first run and differenced for NO_x and then O_3 simulations were run with the appropriate CO and NO_x fields to produce both 1990 and future "Asian O_3" tracers. See Levy et al. [1999] for a description of the NO_x sources used in the 1990 base-year simulations, and Yienger et al. [1999] for the methodology used to scale these global sources to 2020. Both 1990 and future-projected NO_x emissions for South, Southeast and East Asia are taken from Van Aardenne et al. [1998].

ASIAN CO TRANSPORT TO NORTH AMERICA

Signal Variability Across North America

In Figure 1 we show the time series of Asian CO for each of the three points considered in this study (California, CPO, and Central Canada as shown in Figure 1) for both the BL (940 mbar) and the mid troposphere (500 mbar). In order to illustrate the number of events analogous in strength to those observed at CPO, we include a line at 60 ppb that defines the average amount of Asian CO we estimate to have been present during the real events observed by Jaffe et al., [1999]. In arriving at 60 ppb as the indicator of Asian CO events, we noted that the actual episodes at CPO had a mean difference of ~20 ppb between trajectories segregated as Asian and non-Asian (168 ppb vs. 150 ppb).

Figures 1a-f reveals that the frequency and magnitude of Asian CO episodes typically increase with height and towards the mid-latitudes. In the middle troposphere (500 mb) over CPO, Figure 1b, the springtime Asian signal is stronger than in the BL (940 mb). There are numerous spikes over 60 ppb implying that many significant Asian pollution events pass overhead without being seen at the ground. Throughout the whole spring, such a CO signal occurs 20% of the time, as opposed to only 4% of the time in the BL. Trajectories from the GCTM reveal that flow from Asia is much more common aloft than in the BL. At the ground, flow reaching CPO was more likely to have been diverted from higher latitudes around the Aleutian Low than to have come straight from East Asia. Moving southward to central California the Asian CO signal in the BL is similar to CPO (Figure 1c). There are four distinct 60+ ppb episodes and a few extended periods in March and April that are close. The individual events are not as strong, and as a whole the Asian signal is a little less noisy than at CPO. This may be because the CO was more likely to have been transported to the surface via subsidence than at CPO where events episodically arrived via low level flow (< 700 mbar) all the way across the Pacific (a more detailed description of transport will be presented in the Section 5). Aloft at 500 mbar (Figure 1d) there is a large increase in the Asian CO signal that is even stronger than the relative increase over CPO. Here, Asian CO exceeds 60 ppb in spring nearly 50% of the time, and exceeds 100 ppb in six separate episodes. In contrast to California, the signal in north central Canada is much weaker. The BL signal (Figure 1e) is muted and nearly always representative of the mean enhancement, even in spring. Aloft (Figure 1f) it is much weaker than in the lower latitudes but still significantly enhanced over the BL. About eight separate spring events equal or exceed 60 ppb Asian CO, and their respective back trajectories originate from Asia. Most come from northern China and Japan, although the large spike in early February originates from South East Asia. In this particular transport event, air containing more than 60 ppb Asian CO was lifted from the south east Asian BL and transported, in 9 days, all the way to north central Canada.

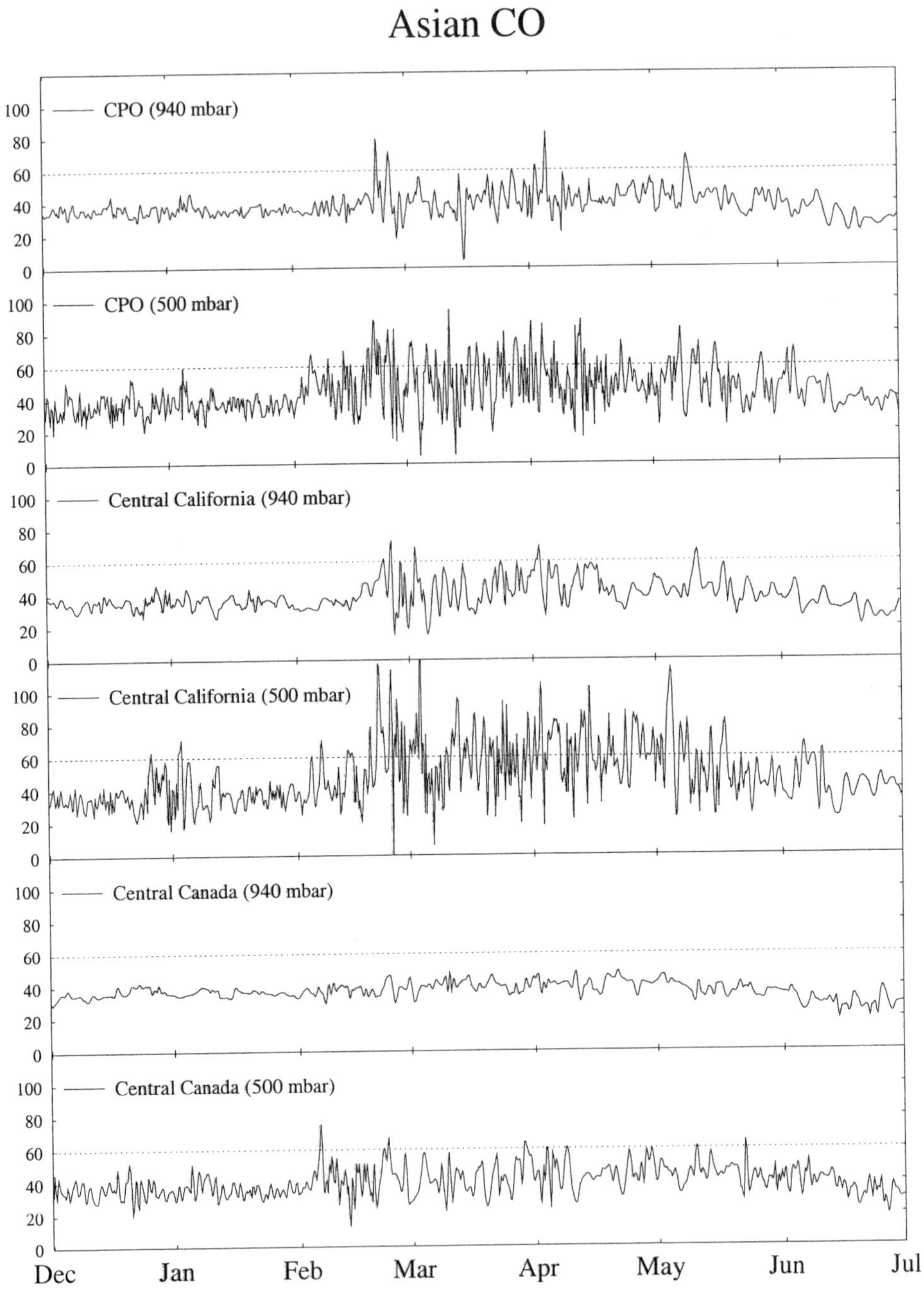

Figure 1. Time series at 6-hour intervals of Asian contribution to CO at all three points at both 940 mbar (BL) and 500 mbar (mid-troposphere). A line at 60 ppb illustrates the amount of Asian CO we estimate to have been present in the pollution parcels observed by Jaffe et al. [1999] at CPO.

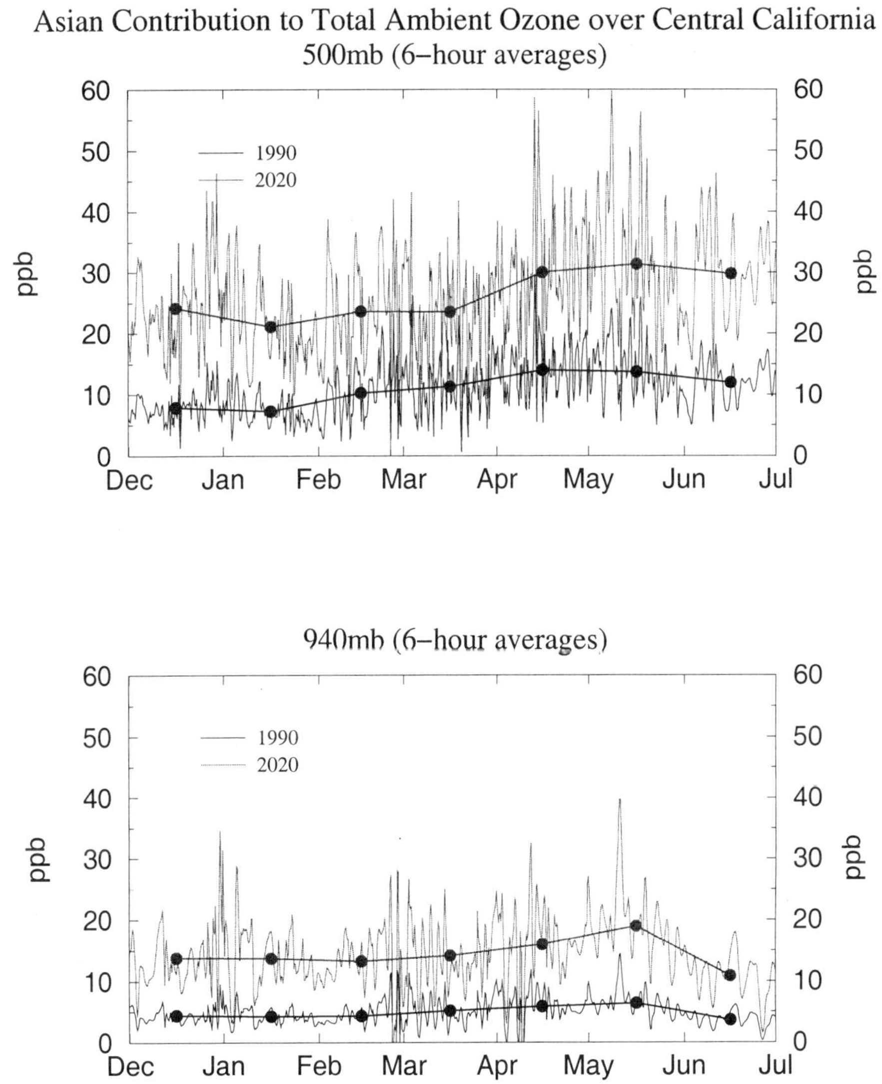

Figure 2. Asian contribution to total ozone offshore of central California at (a) 500 mbar and (b) 990 mbar. In both figures we present the time series for both 1990 emissions and the worst-case-scenario 2020 emission rates.

IMPACT OF ASIA ON OZONE ACROSS NORTH AMERICA

Episodic Enhancements

If there is a significant impact from Asia on US ozone levels, whether it be on surface air quality or on ozone variability in the free troposphere, we believe it probably occurs during episodic events when Asian contributions are unusually high. In Figure 2 we present the time series of Asian O_3 for 1990 in the grid box used to analyze CO directly off the coast of California. On the same figure we also plot the corresponding contributions when Asian emissions are increased under the future scenario to demonstrate the sensitivity of the episodes to increasing emissions. The Asian signals are highly episodic. In each case the strongest events occur in April-May, as opposed to late February-early May for CO, probably due to a lag between the maxima in physical outflow from Asia and net photochemical ozone production. At 500 mbar with 1990 emission levels (Figure 2a), 4 events with greater than 20 ppb of Asian ozone occur in May and two each in March, April and late February. Under the future scenario, extreme springtime events at 500 mbar now approach 60 ppb, and most events greater than 40 ppb have shifted to April and May, periods which have more active photochemistry. This behavior is consistent with previous global 3-D simulations that show that increasing emissions shift the timing of the mid-latitude springtime ozone maximum to later into spring (e.g. Yienger et al., 1999).

In the BL (Figure 2b), at 1990 emission levels a few Asian episodes exceed 10 ppb in spring but never exceed 5 ppb in summer. This suggests that even on an episodic basis, the impact of Asia on ozone this past decade has been small. On the other hand, as Asian NO_x emissions grow to 25 Tg N/yr or more, short-term spikes of Asian ozone in the BL could reach 30-40 ppb in May, and perhaps 10-20 ppb in summer. The 40 ppb May episode seen in Figure 2b accounts for roughly half of the total ozone in that parcel (~80 ppb). From the perspective of air quality accounting, if only three 8-hour 80 ppb episodes will be needed for exceedance, then a few strong trans-Pacific events may aggravate local pollution enough to tip the balance in some areas. This is particularly true in higher altitude urban or suburban areas more exposed to free-tropospheric air. Even if actual future emissions are not so strong, smaller episodes of 20-30 ppb Asian ozone (out of perhaps 60-70 ppb total) would still be significant in the US BL.

CONCLUSIONS

There are a number of implications of these results. For example, those measuring atmospheric chemistry over NA typically have very little information about potential significance of trans-Pacific transport, particularly on short time scales when most measurements are taken. Our results suggests that NA aerosol and chemical data may need to be reconciled with intercontinental transport to explain synoptic-scale variability, particularly for enigmatic high episodes that appear to have no local origin. For regional scale chemical modeling that relies on static or monthly averaged boundary conditions for either Pacific air or descending free tropospheric air over NA, this study provides insight into the potential synoptic-scale consequences of these boundary condition assumptions. Results presented in this paper suggest that regional scale models that pose static chemical boundary conditions for Pacific "background" air may suffer from unexplained variability in tracer time series when compared to observations. Finally, from an ozone policy perspective there could be significantly different consequences of an Asian ozone signal that arrives as a smooth constant background increase, versus one that arrives episodically as a mixture of weak and strong events. Episodes that contribution 40 ppb of ozone to surface ozone levels in NA due to Asian emissions are certain to exacerbate local pollution events and contribute significantly to the frequency of elevated (80 - 100 ppb) ozone exposures. This is especially of concern in elevated areas more frequently exposed to free tropospheric air.

We invite the reader to the following web site *[http://www.cgrer.uiowa.edu/asiaimpact]* that contains a number of related animations of trans-Pacific transport.

Acknowledgments - This work was supported in part by NASA grants NAG5-3855 and NAGW-2428. We also acknowledge Dan Jaffe for discussions related to PHOBEA.

REFERENCES

Galanter M., H. Levy, II, and G.R. Carmichael, 1999, Impacts of biomass burning on tropospheric CO, NO_x, and O_3, Submitted to *J. Geophys Res.*

Holloway, T.A., H. Levy, II., and P.S. Kasibhatla, 1999, The global distribution of carbon monoxide, Sumbitted to *J. Geophys. Res.*

Husar,R.B., 1998, The Asian dust event of April 1998, http://capita.wustl.edu/Asia-FarEast/.

Jaffe, D., A. Mahura, J. Kelley, J. Atkins, P.C. Novelli, and J.T. Merrill, 1997, Impact of Asian emissions on the remote North Pacific atmosphere: Interpretation of CO data from Shemya, Guam, Midway, and Mauna Loa, *J. Geophys. Res.*, 102, 28,627-28,635.

Jaffe, D, et al., 1999, Transport of Asian air pollution to North America, *Geophys.Res. Lett.*, 26, 711-714.

Klonecki., A.A., 1998, Model study of the tropospheric chemistry of ozone, Ph.D.thesis, Princeton University, Princeton, NJ.

Klonecki., A.A., and H. Levy, II, Tropospheric chemical ozone tendencies in CO-CH_4-NO_y-H_2O system: Their sensitivity to variations in environmental parameters and their application to a global chemistry transport model study, *J. Geophys. Res.*, 102, 21,221-21,237.

Kritz, M., 1990, The China Clipper-fast advective transport of radon rich air from the Asian boundary layer to the upper troposphere near California, *Tellus*, 42B, 46-61.

Levy, H. II, and W.J. Moxim, 1989, Influence of long-range transport of combustion emissions on the chemical variability of the background atmosphere, *Nature*, 338, 326-328.

Levy H., II, P.S. Kasibhatla, W.J. Moxim, A.A. Klonecki, A.I. Hirsch, S.J. Oltmans, and W.L. Chameides, 1997, The human impact on global tropospheric ozone, *Geophys. Res. Lett.*, 24, 791-794.

Levy II, H., W. J. Moxim, A. A. Klonecki, and P.S. Kasibatla, 1999, Simulated Tropospheric NOx: Its Evaluation, Global Distribution and Individual Source Contributions, *J. Geophys. Res.,* 104, 26,279-26,306.

Mahlman, J.D., and W.J. Moxim, 1978, Tracer simulation using a global general circulation model: Results from a midlatitude instantaneous source experiment, *J. Atmos. Sci.*, 35, 1340 - 1374.

Manabe S., Hahn, D.G., and J.L. Holloway Jr., 1974, The seasonal variation of the tropical circulation as simulated by a global model of the atmosphere, *J. Atmos. Sci.*, 43-83

Moxim, W.J., 1990, Simulated transport of NOy to Hawaii during August: A synoptic study, *J. Geophys. Res.*, 95, 5717-5729.

Moxim, W. J., H. Levy II, and P. S. Kasibatla, 1996, Simulated Global Tropospheric PAN: Its transport and Impact on NOx, *J. Geophys. Res.*, 101, 12,621-12,638.

Oort, A.H., 1983, Global atmospheric circulation statistics, 1958-1973, NOAA Professional Paper No. 14, U.S. Government Printing Office, Washington, D.C., 180p.

Soden, B. J., and F. P. Bretherton, 1996 Interpretation of TOVS Water Vapor Radiances in Terms of Layer-average Relative Humidities: Methods and Climatology for the Upper, Middle, and Lower Tropospere, *J. Geophys, Res.*, 101, 9,333-9,343.,

Spikovsky, C. M., R. Yevich, J. A. Logan, S. C. Wofsy. and M. B. McElsoy, 1990, Tropospheric OH in a Three-dimensional Chemical Tracer Model: As Assessment Based on Observations on CH3CCl3, *J. Geophys, Res.*, 95, 18,441-18,471.

van Aardenne J.A., G.R. Carmichael, H. Levy II, D. Streets, and L. Hordijk, 1999, Anthropogenic NOx emissions in Asia in the period 1990 to 2020, submitted to *Atmos. Environ.*, 33, 633-646.

Yienger, J.J., A.A. Klonecki, H.Levy, II, W.J. Moxim, and, G.R. Carmichael, 1999, An evaluation of chemistry's role in the winter-spring ozone maximum found in the northern midlatitude free troposphere, *J. Geophys. Res.*, 104, 3655-3667.

REGIONAL AIR POLLUTION AND CLIMATE

chairperson: E. Batchvarova

rapporteur: P.-E. Johansson

A COMPARATIVE STUDY OF TWO PHOTO-OXIDANT DISPERSION MODELS AND THEIR APPLICABILITY FOR REGIONAL AIR QUALITY FORECASTING

Allan Gross,[1] William Ross Stockwell,[2] and Jens Havskov Sørensen[1]

[1]Meteorological Research Division, Danish Meteorological Institute (DMI), Lyngbyvej 100, Copenhagen DK-2100, Denmark
[2] Department of Atmospheric Sciences, Desert Research Institute (DRI), 2215 Raggio Parkway, Reno, NV 89512-1095, USA.

INTRODUCTION

Anthropogenic and biogenic sources affects the air quality especially in conurbations but also at remote areas. In order to understand the effects of these sources, Regional Air Quality Models (RAQMs) are used. There are several examples showing that one of the central parts of RAQMs is the chemical mechanism (e.g. Stockwell, 1986; Müller and Brasseur, 1995). If RAQMs are to be able to predict realistic concentrations of surface ozone and other air pollutants, it is important that the gas-phase mechanism includes all the important atmospheric chemical reactions, and that the utilized rate constants and product yields for the gas-phase reactions, as well as the quantum yields and absorption cross sections for the photolysis reactions, are of high accuracy. This is a rather problematic task because the chemical reactivity of the organic compounds in the polluted troposphere is extremely complicated. Another problem is related to the numerical integration of the chemical mechanism. The system of Ordinary Differential Equations (ODEs) of atmospheric chemical mechanisms that describes the chemical species' concentrations as a function of time is extremely stiff. Therefore, it is not an easy task to solve these ODEs - special methods are required.

Due to the above described problems of chemical mechanisms used in RAQMs we will in this paper show results from two 3-D Lagrangian RAQMs, one handling the atmospheric chemistry in a detailed manner, the other in a more traditional manner. The differences between the models are as follows:

The traditional model uses

- the EMEP mechanism (Simpson et al., 1993),
- the Quasi-Steady-State Approximation (QSSA) solver (Hesstvedt et al., 1978), and
- parameterized photolysis (Simpson et al., 1993).

Table 1: Locations of the measurement sites used for the 3-D simulations.

STATION NAME	COUNTRY	LATITUDE	LONGITUDE	HEIGHT ABOVE (M)
Kosetice	Czech Rep.	49.58 N	15.08 E	633
Ulborg	Denmark	56.29 N	8.43 E	10
Frederiksborg	Denmark	55.97 N	12.33 E	10
Jægersborg	Denmark	55.76 N	12.53 E	39
Virolahti	Finland	60.52 N	27.68 E	8
Uto	Finland	59.78 N	21.38 E	2
Westerland	Germany	54.93 N	8.31 E	12
Deuselbach	Germany	49.76 N	7.05 E	480
Neuglobsow	Germany	53.17 N	13.03 E	65
Zingst	Germany	54.43 N	12.73 E	1
Meinerzhagen	Germany	51.12 N	7.63 E	510
Birkenes	Norway	58.38 N	8.25 E	190
Osen	Norway	61.25 N	11.78 E	440
Stara Lesna	Slovakia	49.15 N	20.28 E	808
Rörvik	Sweden	57.42 N	11.93 E	10
Vavihill	Sweden	56.02 N	13.15 E	175
Norra-Kvill	Sweden	57.82 N	15.57 E	261
Strath Vaich	United Kingdom	57.73 N	4.78 W	270
Harwell	United Kingdom	51.57 N	1.32 W	137

This Lagrangian model is presently used in the Danish Atmospheric Chemistry FOrecasting System (DACFOS), in the following we will call it L_DACFOS (Jensen et al., 1996; Kiilsholm et al., 2000).

The model that handle the chemistry in a more detailed manner uses

- the Regional Atmospheric Chemistry Mechanism (RACM) (Stockwell, 1997),
- the Gear solver (Jacobson and Turco, 1994; Jacobson, 1995 and 1998), and
- modeled photolysis (see (Gross, 2000; Gross et al., 2000a)).

The advantages of using a Gear algorithm are that a complex chemical scheme can be integrated without special ad-hoc adjustments to the rate equations to remove stiffness, and that the order and the time steps are chosen by an adaptive method ensuring high accuracy for a minimum use of computer time. Thus, the Gear algorithm is considered to be a benchmark solver for solving stiff chemical mechanisms. The disadvantage of using the Gear algorithm is that it can be computationally very expensive if the chemical compounds appearing in the chemical reactions are influenced by external sources, e.g. from transport, emissions or depositions. Therefore, certain improvements of the Gear algorithm used in this paper have been made in order to enhance the computational speed. These improvements are based on the Sparse-Matrix Vectorized Gear (SMVGEAR) solver developed by Jacobson and Turco (1994) and Jacobson (1995, 1998). We have modified the SMVGEAR solver such that it vectorizes over trajectories which is highly preferable compared with codes that vectorize over the chemical species (Jacobson and Turco, 1994; Jacobson, 1995 and 1998). The model used herein is therefore a highly vectorizable Multi-trajectOry Ordinary-differential-equation Numerical-box (MOON) model (Gross, 2000; Gross et al., 2000a) which vectorizes over trajectories.

Table 2: Simulated 3-D scenarios for "Comparison 1". All scenarios are simulated at: longitude 0°and latitude 45° N, under clear skies with ground albedo of 0.10, solar declination 23°, and altitude 0 km. The temperature is chosen as 288.15 K and the pressure 1013.25 mbar. The VOC distribution keys used are described in (Gross, 2000) and (Gross et al., 2000a).

COMPOUND	LAND CASE	URBAN CASE
N (molecules/cm^3)	2.55 X 10^{19}	2.55 X 10^{19}
H_2O (%)	1.00	1.00
O_3 (ppbV)	30.0	50.0
HNO_3 (ppbV)	0.10	0.10
CO (ppbV)	100.	294.5
CH_4 (ppbV)	1700.	1700.
H_2 (ppbV)	500.	500.
H_2O_2 (ppbV)	2.00	2.00
HCHO (ppbV)	1.00	1.00
O_2 (%)	20.9	20.9
N_2 (%)	78.1	78.1
NO (ppbV)	0.10	37.5
NO_2 (ppbV)	0.10	12.5
SO_2 (ppbV)	0.00	8.66
VOC (ppmC)	0.00	0.50

Since we wish to focus on how the different treatment of the atmospheric chemistry will affect the model output, the two models have the following similarities: Both models

- are Lagrangian,
- use meteorological data from DMI-HIRLAM,
- use the same physical parameterization of dry and wet depositions, and
- uses the emission inventory of EMEP.

More detailed descriptions of the L_DACFOS and MOON models are given by Jensen et al. (1996), Kiilsholm et al. (2000), Gross (2000) and Gross et al. (2000a).

In this paper the performance of the two models in simulating surface ozone is discussed. This task is accomplished as follows:

- a verification of the L_DACFOS and MOON models against measurement data (19 sites in Europe for the period August 11 to August 24, 1995, see Table 1).
- a comparison of the L_DACFOS and MOON models under two artificial scenarios (a rather clean and a rather polluted case, see Table 2).

In Section "Model Simulations" the European sites used and the model simulations performed are presented. The results of these simulations are presented in Section "Results". Finally, in Section "Conclusion" we conclude on the utility of the MOON model compared with the L_DACFOS model.

MODEL SIMULATIONS

The 3-D simulations presented herein are carried out for every third hour over a two week period from August 11 to August 25, 1995. At each arrival time and location, trajectories are run arriving at five equidistant heights between the ground and the top of the atmospheric boundary layer.

Table 3: 1000 LAND and 1000 URBAN case simulations of the L_DACFOS and MOON models for August 11, 1995, grouped according to their final ozone concentrations. The simulations are performed without sources and sinks. Initial concentrations are according to the LAND and URBAN case scenarios described in Table 2.

LAND CASE			URBAN CASE		
Final Ozone Conc. Interval	Number of Scenarios		Final Ozone Conc. Interval	Number of Scenarios	
(ppbV)	L_DACFOS	MOON	(ppbV)	L_DACFOS	MOON
[15; 20[	31	0	[50; 100[	7	0
[20; 25[	75	31	[100; 150[	16	3
[25; 30[	677	299	[150; 200[	34	51
[30; 35[	215	590	[200; 250[	97	543
[35; 40[	2	80	[250; 300[	253	390
			[300; 350[	179	13
			[350; 400[	114	0
			[400; 500[	97	0
			[500; 600[	54	0
			[600; 700[	53	0
			[700; 800[	27	0
			[800; 900[	37	0
			[900; 1000[	13	0
			[1000; 1400[	19	0

All of the arriving trajectories are used to calculate the surface ozone concentration at a receptor point since trajectories arriving at different heights can vary considerably due to the dependence of the wind on the height above ground. These trajectories correspond to at least four days.

Based on the calculated trajectories two types of comparisons between the models have been carried out. In Section "Comparison 1", *2.1*, 2×1000 artificial case simulations by the L_DACFOS and MOON models without sources and sinks using trajectories for 25 locations in Europe are compared. In addition to the 19 locations described in Table 1, this comparison also involves Keldsnor (Denmark), Lille Valby (Denmark), Brotjacklriegel (Germany), Witteven (Netherlands), Bilthoven (Netherlands) and Ladybower (United Kingdom). The artificial case simulations are described in Table 2.

In Section "Comparison 2", 3-D realistic simulations of the L_DACFOS and MOON models from August 11 to August 24, 1995, for the locations presented in Table 1 are performed and compared with measurement data.

RESULTS

Comparison 1

In Table 3 we have grouped the final ozone concentrations from the simulations based on the artificial cases given in Table 2. For the LAND case simulations the MOON model gives approximately 5 ppbV more ozone than the L_DACFOS model. The highest frequency of final ozone concentrations is obtained in the 25-30 ppbV interval for the L_DACFOS model and in the 30-35 ppbV interval for the MOON model. In a recent study of the EMEP and RACM mechanisms, it was observed that under rural conditions the differences between the mechanisms are very limited

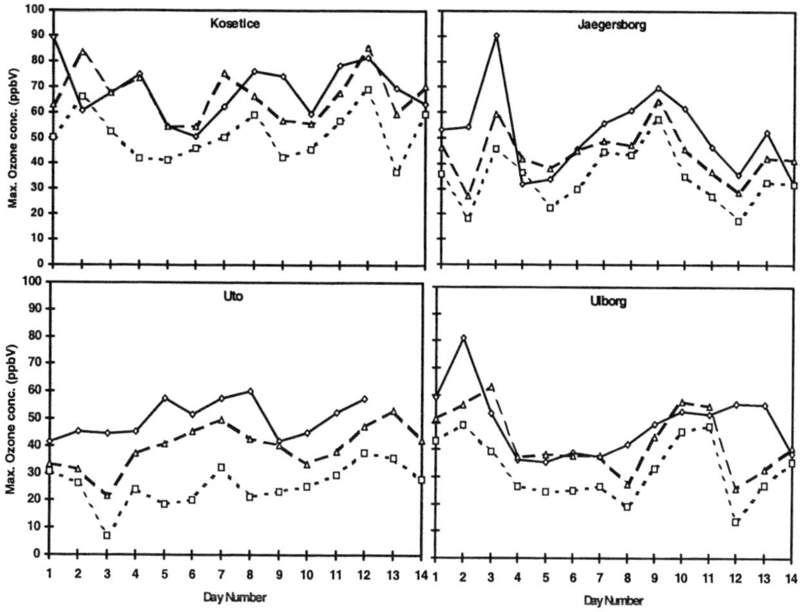

Figure 1: Modeled and measured daily maximum surface ozone concentration peaks for the period August 11 to August 24, 1995, for the four class representatives. Note that if measurement data are missing for a day of interest, then no maximum ozone concentration peak is plotted. Measured data are depicted by the solid line, the L_DACFOS model the dotted line, and the MOON model the dashed line.

(Gross, 2000; Gross and Stockwell, 2000). However, by using the parameterization of the photolysis, as done in the L_DACFOS model, the photolysis rates are overpredicted during the daytime (Gross, 2000: Gross et al., 2000a). This is the reason for the differences between the L_DACFOS and MOON models for the LAND case (Gross, 2000; Gross et al., 2000a).

Similar groupings as that done for the LAND case simulations are shown in Table 3 for the URBAN case simulations. It is a well known fact that for those cases (polluted scenarios) ozone is produced throughout the simulation (Kuhn et al., 1998; Gross, 2000; Gross and Stockwell, 2000). The URBAN case simulation grouping of the ozone concentrations shows that 41% of the chemical boxes from the L_DACFOS model gives too high ozone concentrations, while the MOON model gives expected results. These unrealistically high final ozone concentrations calculated by the L_DACFOS model has to do with the QSSA solver forced by the variation of the photolysis rates during the day.

Comparison 2

Based on an analysis of the measurement data between August 11 to August 25, 1995, for the 19 European locations given in Table 1 (Gross, 2000), we found it within reason to group these into the following four classes:

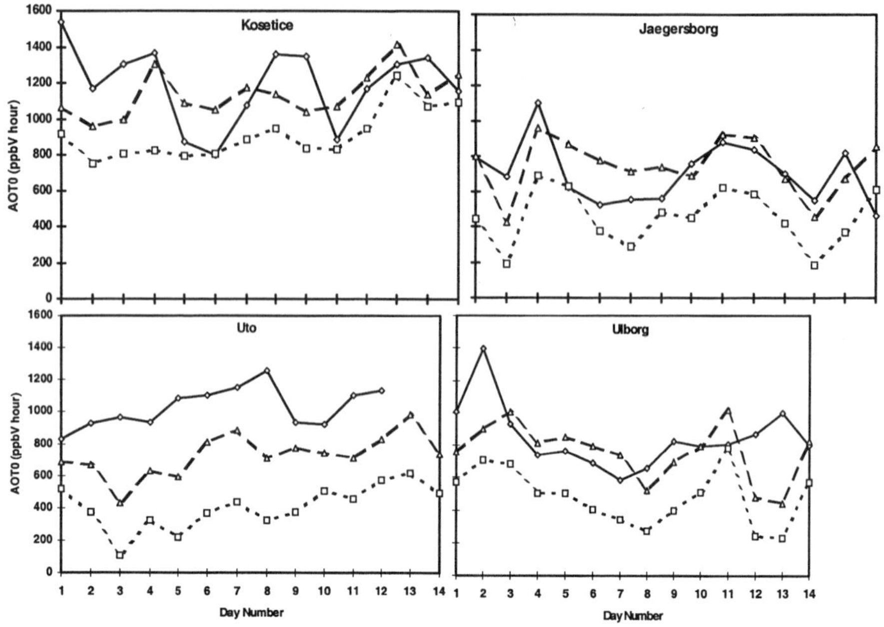

Figure 2: Modeled and measured AOT0 values for the period August 11 to August 24, 1995, for the four class representatives. Note that if measurement data are missing for a day of interest then no AOT0 value is calculated. Measured data are depicted by the solid line, the L_DACFOS model the dotted line, and the MOON model the dashed line.

1. "High" photochemical activity and "high" ozone concentration (Kosetice, Deuselbach, Neuglobsow, Meinerzhagen, Stara Lasna and Harwell). Kosetice is selected as representative of this class.
2. "High" photochemical activity and "low" ozone concentration (Jægersborg, Frederiksborg, Zingst, Birkenes, Osen and Rørvik). Jægersborg is selected as representative of this class.
1. "Low" photochemical activity and "low" ozone concentration (Virolahti and Uto). Uto is selected as representative of this class.
2. Both "high" and "low" photochemical activity and "high" and "low" ozone concentration (Ulborg, Westerland, Vavihill, Norra-Kvill and Strath Vaich). Ulborg is selected as representative of this class.

Due to the limited space we will not show time series of the 3-D simulations. However, such plots are shown for the four class representatives by Gross (2000) and Gross et al. (2000b), and for all 19 locations by Gross (2000).

In Figures 1, 2 and 3 the four class representatives' abilities to model the daily maximum ozone peak, AOT0 (Accumulated Ozone explusure above Threshold) and AOT40 values, are shown, respectively. AOT60 values have also been calculated for all 19 locations but these quantities are very sparse.

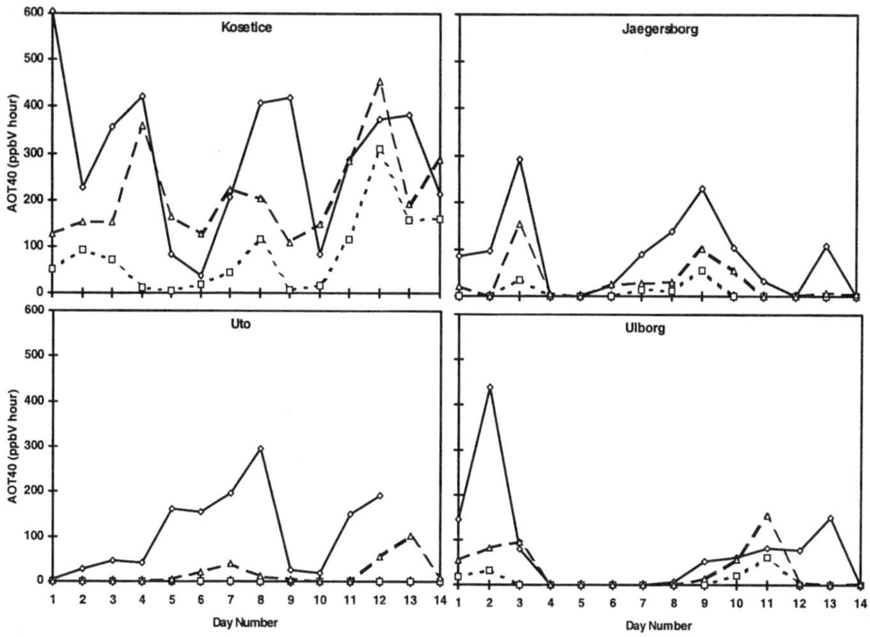

Figure 3: Modeled and measured AOT40 values for the period August 11 to August 24, 1995, for the four class representatives. Note that if measurement data are missing for a day of interest then no AOT40 value is calculated. Measured data are depicted by the solid line, the L_DACFOS model the dotted line, and the MOON model the dashed line.

In Figure 1 we observe that there is a good agreement between measurement data and the results of the MOON model while the L_DACFOS model in general gives a too low daily maximum ozone peaks during the entire two week period for all four stations.

It is highly important that a complete data set is available for the calculation of AOT values since missing data points can introduce large errors. The AOT values presented herein are therefore only calculated when complete data sets for the day of interest are available. Furthermore, AOT40 measurements are very sensitive to the height above ground of the station (Builtjes, 1998). Thus, it is required that AOT40 values from model simulations are calculated at the same height above the surface as the position of the measurement site. However, in the L_DACFOS and MOON models complete vertical mixing is assumed within the atmospheric boundary layer. This makes those models robust with no sensitivity to steep gradients near the surface.

The plots in Figure 2 shows that the MOON model's AOT0 values are in good agreement with the measurement data while the L_DACFOS model gives too low values. Furthermore, the agreement between model results and measurement data become less admirable with increasing x in AOTx (compare Figure 2 with Figure 3), but the MOON model still gives the best results. This confirms the fact that the photochemical activity is higher for the measurements than for the model results.

Finally, we have in Figure 4 calculated the ForeCast-skill (FC-skill) for all the locations split in the four classes. The FC-skill has been calculated within two forecasting intervals [0; 10] ppbV and

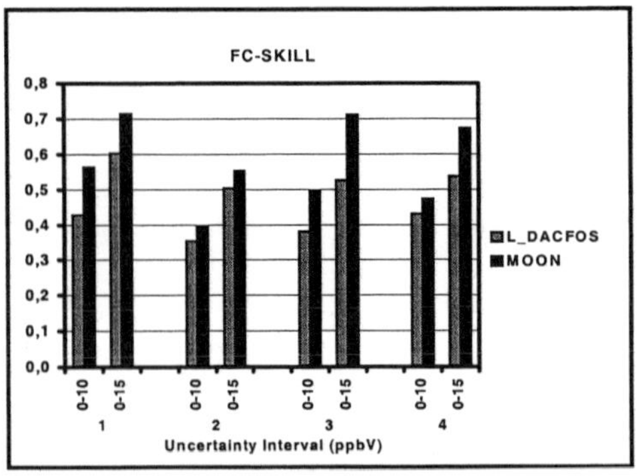

Figure 4: FC-skill for the 19 locations grouped according to the four classes for the period August 11 to August 24, 1995.

[0; 15] ppbV. We find that the FC-skills obtained by the MOON model are between 4% and 19% higher than those obtained by the DACFOS model.

CONCLUSION

The results from the two models show that the MOON model is a numerically more robust model than the L_DACFOS model, and that the use of the Gear algorithm and modeled photolysis compared to fast solvers such as the QSSA and parameterized photolysis is highly preferable. The general trends from all the simulations are that the MOON model gives higher concentration levels of surface ozone than the L_DACFOS model. But the results also show that both model have a too low diurnal cycle compared with the measurement data. Thus, they give at present too small AOT40 and AOT60 values.

In conclusion, the results presented in this paper show that the MOON model outperforms the L_DACFOS model in modeling surface ozone. For a more comprehensive validation and verification of the MOON model the reader is referred to (Gross, 2000) and (Gross et al., 2000a and b).

REFERENCES

Builtjes, P.J.H., 1998, Model validation of AOT40. *Proceeding from 2nd GLOREAM Workshop*, Madrid, Spain.

Gross, A., 1999, Surface Ozone and Tropospheric chemistry with applications to regional air quality modeling. (PhD thesis). *DMI Scientific Report 00-03*, Danish Meteorological Institute, Copenhagen, Denmark.

Gross, A, and Stockwell, W.R., 2000, Comparison of the EMEP, RADM2 and RACM mechanisms. *Atmospheric Environment*. Submitted.

Gross, A., Sørensen, J.H., and Stockwell, W.R., 2000a, A multi-trajectory vectorized Gear model: model formulation and process comparison. *Atmospheric Environment.* To be submitted.

Gross, A., Sørensen, J.H., and Stockwell, W.R., 2000b, A multi-trajectory vectorized Gear model: 3-D simulations and model validation. *Atmospheric Environment.* To be submitted.

Hesstvedt, E., Hov, Ø., and Isaksen, I.S.A., 1978, Quasi-steady-state approximations in air pollution modeling: comparison of two numerical schemes for oxidant prediction. *International Journal of Chemical Kinetics.* 10: 971.

Jacobson, M.Z., 1995, Computation of global photochemistry with SMVGEAR II. *Atmospheric Environment.* 29: 2541.

Jacobson, M.Z., 1998, Improvement of SMVGEAR II on vector and scalar machines through absolute error tolerance control. *Atmospheric Environment*, 32: 791.

Jacobson, M.Z., and Turco, R.P., 1994, SMVGEAR: a SMVGEAR-matrix, vectorized Gear code for atmospheric models. *Atmospheric Environment*, 28: 273.

Jensen, M.H., Rasmussen, A.., Svensmark, H., and Sørensen, J.H., 1996, Danish Atmospheric Chemistry FOrecasting System (DACFOS). *DMI Technical Report 96-03*, Danish Meteorological Institute, Copenhagen, Denmark.

Kiilsholm, S., Rasmussen, A., and Sørensen, J.H., 2000, Validation of DACFOS surface ozone forcasts 1996-98 - description ot the new verification system and model improvement. *DMI Technical Report 00-05*, Danish Meteorological Institute, Copenhagen, Denmark.

Kuhn, M., Builtjes, P.J.H., Poppe, D., Simpson, D., Stockwell, W.R., Andersson-Skøld, Y., Baart, A., Das, M., Fiedler, F., Hov, Ø., Kirchner, F., Makar, P.A., Milford, J.B., Roemer, M.G.M., Ruhnke, R., Strand, A., Vogel, B., and Vogel, H., 1998, Intercomparison of the gas-phase chemistry in several chemistry and transport models. *Atmospheric Environment.* 32: 693.

Müller, J.F., and Brasseur, G., 1995, IMAGES: a three-dimensional chemical transport model of the global troposphere, *Journal of Geophysical Research.* 100: 16445.

Simpson, D., Andersson-Sköld, Y., and Jenkin, M.E., 1993, Updating the chemical scheme for the EMEP MSC-W oxidant model: current status. *EMEP MSC-W Note 2/93*, Norwegian Meteorological Institute, Oslo, Norway.

Stockwell, W.R., 1986, A homogeneous gas phase mechanism for use in a regional acid deposition model, *Atmospheric Environment.* 20: 1625.

Stockwell, W.R., Kirchner, F., Kuhn, M. and Seefeld, S., 1997, A new mechanism for regional atmospheric chemistry modeling. *Journal of Geophysical Research.* 102: 25847.

DISCUSSION

R. YAMARTINO: Could you please comment on the time step you used in the QSSA solver, as its performance is very sensitive to its time step relative to the stiffness?

A. GROSS: The initial time step used in the QSSA solver is 900 seconds. An iteration loop for each time step is included in the numerical integration in order to increase the accuracy of the method. This iteration loop follows recommendations by Hesstvedt, Hov and Isaksen (1978. *International Journal of Chemical Kinetics*. X: 971).

A number of tests of the initial time step used in the QSSA solver have been made. These tests show that a decrease of the initial time step does not change the output results for ozone significantly. Similar results are obtained by Hertel, Berkowitz, Christensen and Hov (1993. *Atmospheric Environment*. 27A: 2591).

OBSERVATIONS AND MODEL STUDIES OF SOME RADIATIVE EFFECTS OF MINERAL DUST

Frédéric Pradelle,[1] Guy Cautenet,[1] Olivier Chomette,[2]
Michel Legrand,[2] Gilles Bergametti [3] and Béatrice Marticorena [3]

[1]LaMP - Aubière - France
[2]LOA - Lille - France
[3]LISA - Paris - France

INTRODUCTION

Mineral dust uptake and long-range transport have been monitored since many years and for long periods. Remote sensing over oceans (namely with satellite imagery) is now widely used, and allows for instance the characterization (Husar et al., 1997) or the climatology (Moulin et al., 1997) of tropospheric aerosols. Over land, observations are based upon IR (Legrand et al., 1994) or UV (Herman et al., 1997). The direct radiative effect of mineral aerosol is relatively weak when compared to the anthropic aerosol, reported to the average masses in the atmosphere: the former contributes for less than 20% to the total aerosol optical depth, whereas the anthropogenic sulfates contribute for 20 %, for a total mass ten times lower (Andreae, 1994). If we focus now on indirect effects, observations of desert dust plumes outbreaking off continental shores (Tropical Eastern Atlantic, Mediterranean, Indian Ocean, Pacific, ...) show that mixing of mineral particles with liquid phase or atmospheric chemicals compounds is likely to occur, and the particles are not removed by this process because clouds often evaporate before precipitation, which releases the dust within the atmosphere. Even though mineral dust is not hygroscopic at its native state, its interaction with clouds (wet capture) or other chemicals could modify this property: observations (Zhang et al., 1994; Levin et al., 1996) report strong chemical alterations ("aging" process) at the particle surface (formation of sulfates, for instance). The radiative parameters (diffusion and absorption of SW flux), but also the physico-chemical properties (such as CCN capability) of desert dust may be modified to a large extent during its wet stage, so that the global aerosol-cloud-earth surface forcing could be modified in the solar band. It follows also that a good survey of the direct and indirect effects of mineral dust requires, among others, an accurate modeling of all the aspects of its atmospheric cycle: uptake, transport, spectrum evolution, dry removal, and realistic schemes as regards radiation and scavenging (captures between droplets and particles) as well.

We first present an analysis of 6 years of daily VIS and IR Meteosat data over the Eastern Tropical Atlantic. We propose an analysis to interpret the apparent modifications in cloud albedo due to mineral dust presence, either inside or above water clouds. This study is based also on numerical modeling in microphysics and radiation, which are briefly described.

In a second part, we present a mesoscale study of the dust cycle which helps to support and interpret some aspects of the above mentioned observations over the Eastern Tropical Atlantic.

SATELLITE OBSERVATION OF DESERT DUST RADIATIVE EFFECTS OVER THE SEA

This section deals with the remote sensing of the radiative interaction between the maritime stratiform cloud cover and the desert aerosol. This study is based upon a 6-year (1991-1996) of ISCCP Meteosat B2 imagery series at 1130-1200 UTC over Eastern Tropical Atlantic. The visible satellite signal is expressed as an equivalent

Air Pollution Modeling and Its Application XIV, Edited by
Gryning and Schiermeier, Kluwer Academic/Plenum Publishers, New York, 2001

reflectivity according to Jankowiak and Tanré (1992), corrected of the water vapor absorption in the atmosphere (Tanré et al., 1990). Discrimination between a cloud-filled (or partially filled) pixel and a pixel with heavy dust content, which may have the same reflectivity in the visible spectral band, is performed by a method developed by Sèze and Desbois (1987) using both VIS and IR data. A more complete description of our method may be found in Pradelle et al. (2000).

The monthly averages of reflectivity for each pixel are presented on figure 1 (a to f) for three different months: January , April and July respectively. The upper figures show the mean number of dusty days, expressed as a percentage. We note a significant number of dusty days in any season, with a maximum in April: a large dark area (> 50%) is visible between 6°N and 15°. We note also a Northward shift of the occurrence areas from winter (between 5°N and 10°N in January) to summer (a large band between 15°N and 30°N in July). Minimum values of dusty days numbers are found at the ITCZ location, because most of the time the pixels are cloudy in this area, so it is difficult to see dust particles on satellite imagery, and also because dust is likely to be removed by wet scavenging. All these results (location, maximum values, shift and extent of the maximum area of dusty days number) are consistent with those found by Jankowiak and Tanré (1992), and by Guelle (1998) from AVHRR data.

The lower figures (d, e, f) in figure 1 display the average cloud visible apparent reflectivity of the surface-atmosphere (with cloud and dust layer) system for the same months, for the cloudy pixels. We remark a minimum in reflectivity (dark) over the Eastern Atlantic, near the NW African coast, from 5°N to 30°N. The lowest reflectivity values are found in April (a decrease of 20% as compared to the reflectivity values observed near 30°N), that is to say when outbreaks of dust particles over the Atlantic are the most frequent. This decrease in albedo reaches about 10% in January and 15% in July. Moreover the seasonal trend of these minimum areas is similar to the trend experienced by the maximum dust occurrence areas mentioned above.

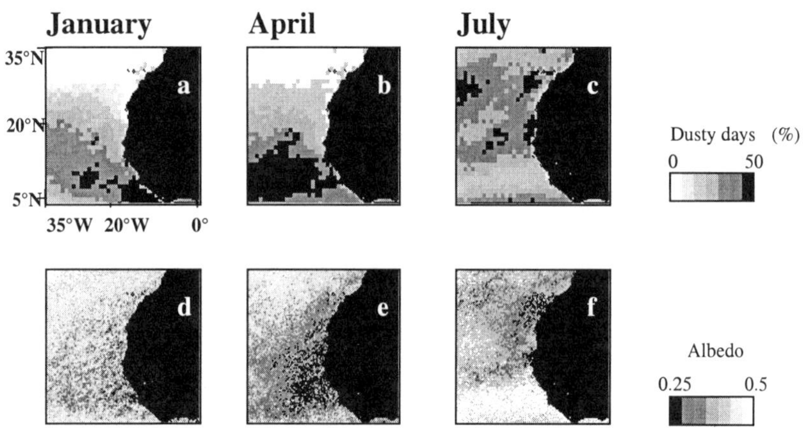

Figure 1. Upper row: % of dusty days for three months over the 6-year period.
Lower row: cloud albedo during the same period.

ANALYSIS OF THE POSSIBLE REASONS OF THE ALBEDO DECREASE

A minimum value in albedo of the system surface-water cloud is observed when dust plumes are present, which may be important on the regional radiative budget. Two hypotheses are invoked now: a direct effect, i.e. the dust plume above the water cloud enhances the SW absorption; an indirect effect, i.e. the dust particles interact with cloud droplets. Both effects are examined below.

Radiation and microphysics tests: the models.

The so-called "EC3" version of the ECMWF radiation model (Fouquart and Bonnel, 1980; Morcrette and Fouquart, 1986; Morcrette, 1989) is used here. This code is derived from high resolution models developed at the LOA via realistic simplifying assumptions. It is weakly CPU consuming and quite accurate, according to IRCCM tests (Inter-comparison of Radiation Codes for Climate Models; Fouquart et al., 1991). We have modified it so as to take the particular radiative properties of the dust particles into account (Chomette, 1999). We therefore can test our hypotheses: (i) what is the SW reflectance effect of a dust plume suspended above a marine stratocumulus? (ii) what is the SW reflectance effect of an interaction between mineral dust and cloud droplets?

The second question need a complementary scheme, i.e. a microphysics model able to describe the effect of an additional amount of partly soluble matter on a droplet cloud distribution. This model is the ExMix code, developed by Wobrock (1988) in order to study the growth of externally mixed condensation nuclei in clouds. This air-parcel model allows at all times to follow the individual activated aerosol particles through the droplet spectrum, as it considers only one number density for all aerosol particles and droplets together. In our version, five types of condensation nuclei can be distinguished, differing in solubility fraction, density, molecular weight and number of free ions. Therefore, the advantage of ExMix is that we can attribute to each type of aerosol particles a different chemistry, and so study the effect of each type of particles on the droplet distribution of the cloud. Comparison of ExMix simulations with results of field experiments can be found in Schell et al (1997) and Laj et al (1997).

Seasonal features of dust transport.

In view of subsequent discussions, we must specify the general scheme of mineral dust transport over NE tropical Atlantic (Chiapello et al., 1995): dust travels mainly under 1 km in winter (in the trade winds layer), and at 3-5 km in summer (in the Saharan Air Layer or SAL, above the trade winds inversion , which lies about 1.5 km). It follows that in summer, cloudy stratiform layers are topped by dust plumes, whereas the contrary occurs in winter, or at least dust and cloud may exist in the same layers. In that second case, microphysical interactions may occur between cloud and dust. These features are confirmed by mesoscale modeling (see below).

Model results: no microphysics interaction.

In this part, we study the direct radiative effect of a dust layer on a sea surface-cloud water system using the EC3 radiation model. We assume a typical marine stratocumulus cloud layer of 200 m thick, with an effective radius r_e of 8 μm and a liquid water content LWC=0.4 gkg^{-1}. If no dust is present, and for a normal zenith angle, we obtain an albedo α_0 of 0.42, with a surface albedo (ocean) of 0.05. Figure 2 represents the albedo α of the system (normalized by the reference value α_0) against the SW (550 nm) optical thickness τ of the dust layer. Two configurations have been studied: a summer case where the dust layer is above the cloud field, at the SAL level, and a winter case where the dust plume lies in the trade wind layer, i.e from sea surface to the cloud top.

The direct radiative effects of a dust plume when a stratiform cloud layer is present may be described as follows:

- if dust is above the cloud (summer case), albedo of the surface-cloud dust system decreases relatively strongly with the dust optical thickness. This decrease reaches about 17 % for τ = 0.5, the mean value observed in this region. It can exceed 30% during strong outbreaks corresponding to an optical depth greater than the unity.
- in winter case, where dust is found inside and beneath the cloud, the decrease is very slight: about 1% for τ = 0.5, 2.5% for τ = 1 (a slight increase may even be found if dust is quite under the cloud not displayed here).

It follows that during the summer season, the observed albedo minimum over the Eastern Atlantic Ocean may be explained by the presence of dust above the cloud deck. However, in winter, the direct radiative effect caused by the dust particles is weak and cannot alone explain the decrease of apparent albedo also observed during this season. The next subsection shows that another explanation may be invoked.

Model results: microphysics interaction.

First we assume a distribution composed of the aerosol particles types (except dust) observed over the Eastern Atlantic Ocean, mainly small organic particles, non sea salt sulfate and sea salts. From a typical dynamical field encountered in the tropical cloud layer, we simulate the growth of these particles using ExMix. After 1800 s, they form a stratocumulus cloud of 200m thick with a liquid water content of 0.4 gkg^{-1} and an effective radius of 8 μm at the top of the cloud. As previously, it corresponds to an apparent albedo α_0 equal to 0.42. In a second time, we add to the previous particles distribution a coarse mode of dust particles. After the same time of simulation we note the cloud parameters value and calculate the associate albedo α with the radiative model described above. For three soluble fractions (10, 20 and 30%) of this fourth aerosol type, we plot on the figure 3 the albedo α normalized by the reference value α_0 against the dust concentration.

All the curves are showing a decrease of the cloud albedo. This decrease can be explained by the fact that a modification of the aerosol particles distribution doesn't change significantly the liquid water content value. Therefore, an addition of large particles (median radius of 0.6 μm against about 0.1 μm for the sulfate particles) leads to form a cloud with less but larger droplets, i.e droplets with a larger effective radius and so a weaker cloud albedo. The observed reflectivity decrease depends strongly on the dust mass load and on her soluble fraction. During strong outbreaks, when the dust particles concentration can reach 600 μgm^{-3}, the albedo decreases from 3% for a soluble fraction equal to 5%, to 11% for a soluble fraction equal to 30%. Large uncertainties remain as regards to the solubility of the dust particles. Initially assumed as no soluble, dust particle should acquire a significant soluble fraction (and so act as CCN) by dry coagulation with other aerosol particles, by chemical processes on his surface, and especially by in-cloud processes. According to Levin (1996), this soluble fraction could take values ranging from 5 to 15%. However, in average, dust concentration takes weaker values, namely between 100 and 200 μgm^{-3}. It leads to an albedo decrease ranging from 0.5 to 5%, depending on the solubility. Added to the direct effect, the studied indirect effect couldn't totally explain the minimum of reflectivity observed in winter over the Eastern Atlantic Ocean. Transfer of a part of the particulate sulfate from the accumulation mode to the coarse mode (on the dust particles surface by chemical and microphysical processes mentioned above) could also act on the microphysics of the cloud, decreasing his reflectivity. Finally, the presence of black carbon particles over the cloud deck between 5 and 15°N in latitude could also change the reflectivity of the system. These particles, emitted by the biomass burning in Africa during this season, should absorb a part of the SW radiation and so decrease the albedo like the dust particles in summer case.

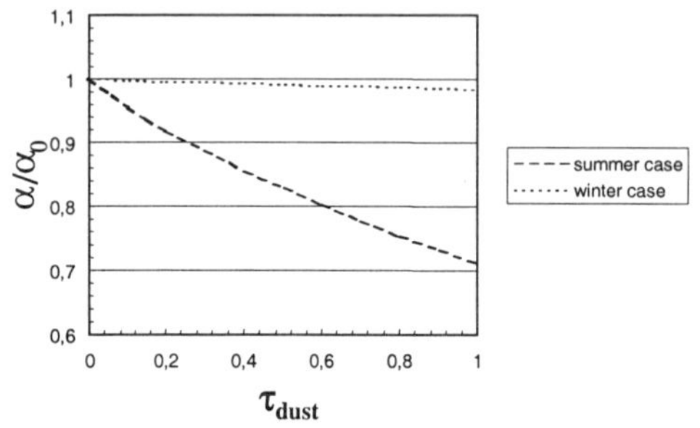

Figure 2: cloud albedo decrease due to dust (no microphysics)

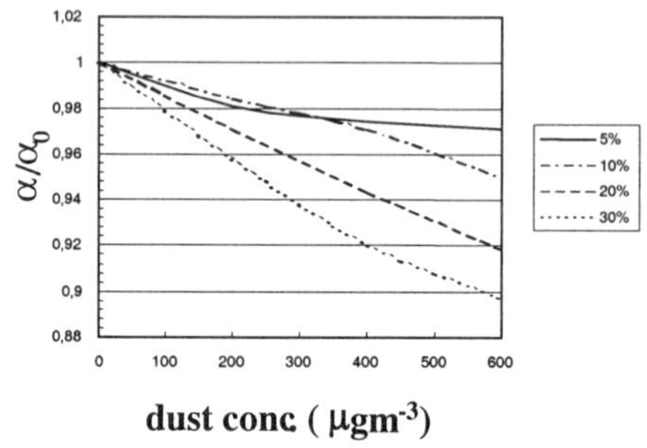

Figure 3: cloud albedo decrease due to microphysical effects

MESOSCALE MODELING OF THE DESERT PARTICLES ATMOSPHERIC CYCLE

We have simulated two typical situations (winter and summer) in view to control the levels dust may be encountered according to the season.

The mesoscale model

We use the RAMS model of the Colorado State University (Regional Atmospheric Modeling System, Pielke et al., 1992), in its non-hydrostatic "3b" version: it is a primitive equation, prognostic mesoscale model. Our model domain broadly ranges from 40°W to 15°E, and from 5°N to 40°N. We use a single grid with the horizontal grid step at 100 km, in agreement with the dust source mapping (see below) which has a space step of 1 squared degree. The vertical is divided into 30 steps, increasing with height, with 10 levels in the1500 m above surface. The ECMWF analysis database is used to initialize the model. The native radiation scheme has been replaced by the "EC3" mentioned above. Finally, we have added a complete scheme describing the atmospheric cycle of mineral dust.

The dust scheme

We have coupled to RAMS the dust source scheme developed by Marticorena and Bergametti (1995), where the aeolian dust flux is governed by the size distribution of the soil surface erodible particles, with a variable saltation threshold, and by the distribution of the surface roughness which controls the availability of the wind energy to be transferred to the soil particles. These authors have mapped a great part of the desert areas, including Sahara and Sahel. This model has been qualified with various experimental data, including Meteosat IR information (Marticorena et al., 1997). This emission algorithm is coupled on-line with the wind field from RAMS (Cautenet et al., 2000-a). More details may be found in the referenced literature. Dust advection and diffusion are controlled by a RAMS built-in scheme, whereas we added scavenging schemes (dry and wet removal). We use a spectral representation, because the dust removal processes (dry or wet) need a spectral approach, and the model incorporates the dust effect on radiative transfer, which requires a spectral framework also. The size distribution is a 3-lognormal spectrum derived from Alfaro et al. (1998), with mean radii ranging from 0.3 to 13 μm.

Modeled winter and summer dust transports

We have modeled two typical dust outbreaks (winter and summer). The first dust event (1995) took place between January, 9 and January, 14. The second (1996), between August, 16 and August, 21. We can partly control the quality of the simulations using satellite VIS pictures, or surface measurements for winter event. At Sal Island (Cape Verde, 23°W, 16°N), modeled and measured (B. Chatenet, personal communication) surface mass concentrations are in rather good agreement (figure 4a), and so are optical depths (figure 4b) as well (available from PHOTON/AERONET website http://aeronet.gsfc.nasa.gov:8080 , by courtesy of D. Tanré, PI). Surface data are relevant for winter events (due to the dust travel in the lower levels). In summer, we better present satellite pictures (figure 5): the modeled summer dust outbreak agrees with observation (recall that over land, VIS data are not used, because of the poor contrast). We see that dust concentrations profiles (figure 6) agree with the climatology data, i.e. dust travels in the lower levels in winter, and mainly at upper levels in summer.

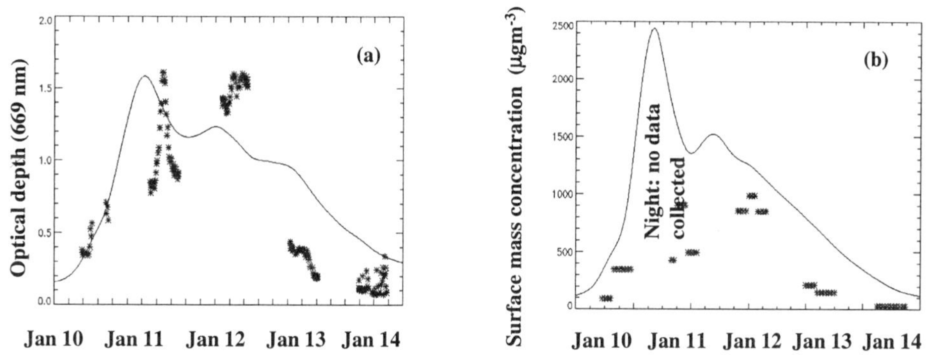

Figure 4 : Comparison of model data (solid lines) versus observations (stars) at Sal Island near surface during the winter dust event (January 1995) - (a): optical depth; (b): mass concentration.

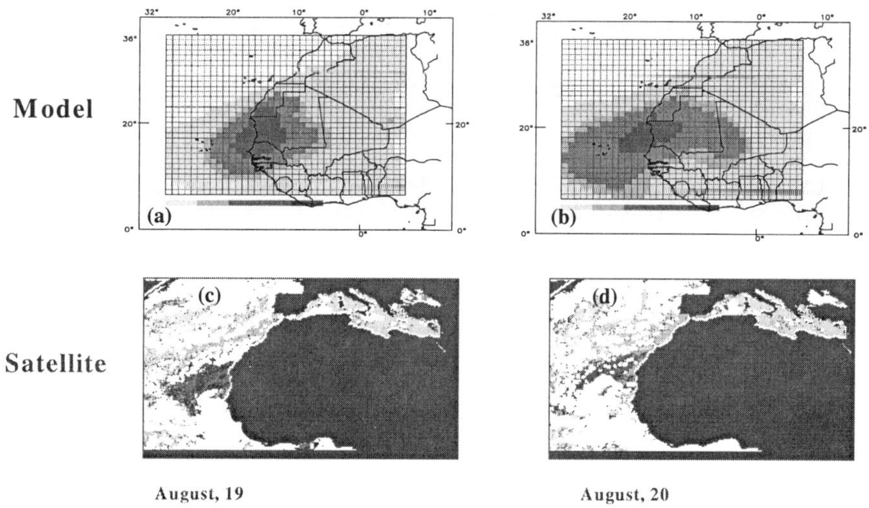

Figure 5: Comparison of model data (vertically integrated dust load, arbitrary units, a and b) versus Meteosat VIS imagery (c and d; water clouds in white; continental mask in black; heavy dust in dark) off the West African coast during the summer dust event (August 1996).

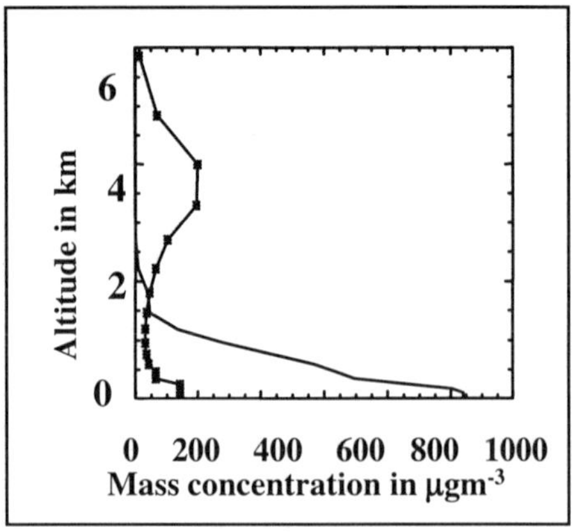

Figure 6: Modeled mass concentration profiles at Sal Island during the winter event (solid lines) and the summer event (solid line plus stars).

Modeled capture of dust particles by stratiform (non-precipitating) clouds.

We have developed an in-cloud spectral capture scheme (Cautenet et al., 2000-b). Here we assume that aerosol particles have not CCN properties, so that we do not consider any nucleation process: only a mechanical capture phenomenon of particles by droplets. The dust surface alteration may be triggered by such a process, and may lead to enhanced CCN properties for dust. All the main processes are considered: brownian processes, thermophoresis, diffusiophoresis, turbulence and gravity impaction (Pruppacher and Klett, 1997). We have calculated a general "capture kernel" K which depends on parameters (humidity, turbulence dissipation rate, temperature, concentrations, ...) currently handled by the RAMS model. An automatic look-up procedure is possible, which is quite fast. In figure 7, we show an example of the expected modification of a vertical aerosol profile (expressed here in number concentration: particles per cubic meter) after a 10-h residence within a typical stratocumulus field. About 50% of the mineral matter is captured by the liquid water phase. As a large part of the cloud water will evaporate, the mineral particles will surely undergo surface alterations before being released, so that a true chemical processing may have taken place.

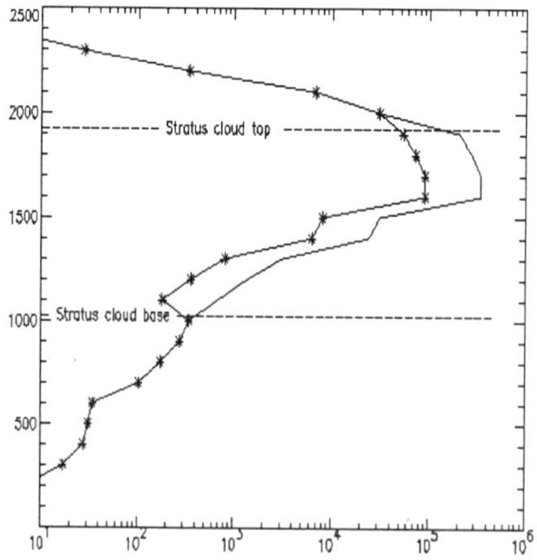

Figure 7: decrease of dust concentration when a stratocumulus is present (particles capture by droplets). Solid line: initial profile; solid line + stars: after 10 h.

CONCLUSION AND PERSPECTIVES

Our results merely illustrate some aspects of the complexity of the radiative role of aerosols, in particular mineral dust. Although this is generally considered as not hygroscopic, this may be no longer true after some processing, e.g. after a wet stage in a non-precipitating stratocumulus. Our observations over Eastern Tropical Atlantic suggest that when dust plumes and stratiform clouds are present in the same region, the net effect seems to be a slight but significant albedo decrease (on a statistical basis), of the order of some %. Two complementary explanations are proposed: an enhancement of SW absorption when dust is above cloud (no microphysics), or a microphysical interaction when dust and cloud are at the same level which leads to a growth in droplet size with the same final effect (albedo decrease). This latter case requires that some soluble material be present on the dust particle, which corresponds to observations of "aged" dust. Radiative and microphysics numerical tests reinforce these explanations, whereas mesoscale tests corroborate the dust outbreak climatology off West African coast, in particular the relative position of dust plume and stratiform clouds. However, field campaigns are quite necessary in view to disentangle the various mechanisms giving mineral aerosol its complex climatic role.

Acknowledgements

The authors wish to thank I. Jankowiak (LOA, Lille, France) and W. Wobrock (LaMP, Clermont-Ferrand, France) for their assistance. This work is supported by the Programme National de Chimie Atmosphérique (CNRS). It makes use of the RAMS model, which was developed under the support of the National Science Foundation (NSF) and the Army Research Office (ARO).

REFERENCES

Andreae, M. O., Climate effects of changing atmospheric aerosol levels, *World Survey of Climatology*, **XX**, Future Climate of the World, Henderson-Sellers Ed., 1994.

Alfaro, S., A. Gaudichet, L. Gomes and M. Maillé, Mineral aerosol production by wind erosion: aerosol particle sizes and binding energies, *Geophys. Res. Lett.*, **25**, 991-994, 1998.

Cautenet, G., F. Guillard, G. Bergametti, F. Dulac, B. Marticorena and J. Edy. Modeling of a Saharan dust event, *Contributions to Atmospheric Physics* (in press), 2000-a.

Cautenet,G. , F. Pradelle , M. Legrand, O. Chomette, G. Bergametti and B. Marticorena Mesoscale modeling of the desert aerosol transport in the Saharo-Sahelian and eastern Atlantic zones. Part I: methodology and preliminary tests, in rev., *J. Geophys. Res.*, 2000-b

Chiapello, I., G. Bergametti, L. Gomes, B. Chatenet, F. Dulac, J. Pimenta and E. Santos Soares, An additional low layer transport of Sahelian and Saharan dust over the North-Eastern Tropical Atlantic, *Geophys. Res. Lett.*, **22**, 3191-3194, 1995.

Chomette, O., Modélisation et analyse méso-échelle du cycle de l'aérosol désertique. Aspects radiatifs et dynamiques,. *Thèse de doctorat*, Université de Lille, France, 1999.

Fouquart, Y. and B. Bonnel, Computation of solar heating of the earth's atmosphere: a new parameterization, *Beitr. Phys. Atmosph.*, **53**, 35-52, 1980.

Fouquart, Y., B. Bonnel and V. Ramaswany, Intercomparing shortwave radiation codes for climate studies, *J. Geophys. Res.*, **96**, 8955-8968, 1991.

Guelle, W., Comparaison des simulations globales de ^{210}Pb et d'aérosols sahariens aux observations de surface et satellitales. *Thèse de doctorat,* Université de Paris 7, Paris, France, 1998.

Herman, J.R., P. K. Bhartia, O. Torres, C. Hsu, C. Seftor and E. Celarier, Global distribution of UV-absorbing aerosols from Nimbus 7/TOMS data, *J. Geophys. Res.*, **102**, 16911-16922, 1997.

Husar, R.B., J.M. Prospero and L.L. Stowe, Characterization of tropospheric aerosols over the oceans with the NOAA advanced very high resolution radiometer optical thickness operational product, *J. Geophys. Res.*, **102**,16889-16909, 1997.

Jankowiak, I. and D. Tanré, Satellite climatology of Saharan dust outbreaks. *J. Climate,* **5**, 646-656, 1992.

Laj, P., S. Fuzzi, M.C. Facchini, G. Orsi, A. Berner, C. Kruisz, W. Wobrock, A. Hallberg, K.N. Bower, M.W. Gallagher, K.M. Beswick, R.N. Colvile, T.W. Choularton, P. Nason, and B. Jones, Experimental evidence for in-cloud production of aerosol sulphate. *Atmos. Environm.*, **31**, 2503-2514.

Legrand, M., C. N'Doumé and I. Jankowiak, Satellite-derived climatology of the Saharan aerosol, *Passive Infrared remote sensing of clouds and the atmosphere II: proc. SPIE*, **2309**, 127-135, D.K. Lynch ed., 1994.

Marticorena, B. and G. Bergametti, Modeling the atmospheric dust cycle: 1- Design of a soil-derived dust emission scheme. *J. Geophys. Res.,* **100**, 16415-16430, 1995.

Marticorena, B., G. Bergametti, B. Aumont, Y. Callot, C. N'Doumé and M. Legrand, Modeling the atmospheric dust cycle: 2- Simulation of Saharan dust sources. *J. Geophys. Res.,* **102**, 4387-4404, 1997.

Morcrette, J. J., Description of the radiation scheme in the ECMWF model, *Tech. Memo. 165*, 26 pp, Res. Dep., European Center for Medium Range Weather Forecasts, Reading, England, 1989.

Moulin, C., C E. Lambert, F. Dulac and U. Dayan, Control of atmospheric export of dust from North Africa by the North Atlantic Oscillation, *Nature*, **387**,691-694.1997.

Levin, Z., E. Ganor and V. Gladstein, The effects of desert particles coated with sulfate on rain formation in the eastern Mediterranean, *J. Appl. Meteor.*, 35, 1511-1523, 1996.

Pielke, R. A., W. R. Cotton, R.L. Walko, C. J. Tremback, W. A. Lyons, L. D. Grasso, M. E. Nicholls, M. D. Moran, D. A. Wesley, T. J. Lee and J. H. Copeland, A comprehensive meteorological modeling system - RAMS. *Meteorol. Atmos. Phys.*, **49**, 69-91, 1992.

Pradelle, F., G. Cautenet and I. Jankowiak, Remote sensing of some radiative or microphysical interactions between marine stratocumulus clouds and Saharan dust: observations and climatology, submitted to *J. Geophys. Res.*, 2000.

Pruppacher, H.R and J. D. Klett, Microphysics of clouds and precipitation, Kluwer Academic Publishers, 1997.

Schell, D., W. Wobrock, R. Maser, M. Preiss, W. Jaeschke, H-W Georgii, M.W. Gallagher, K.N. Bower, K.M Beswick, S. Pahl, M.C Facchini, S. Fuzzi, A. Wiedensohler, h-C. Hansson and M. Wendisch, The size-dependant chemical composition of cloud droplets. *Atmos. Environm.*, **31**, 2561-2576.

Sèze, G., and M. Desbois, Cloud cover analysis from satellite imagery using spatial and temporal characteristics of the data, *J. Climate Appl. Meteor.*, **26**, 287-303, 1987.

Tanré, D., C. Deroo, P. Duhaut, M. Herman, J. J. Morcrette, J. Perbos and P. Y. Deschamps, Description of a computer code to simulate the satellite signal in the solar spectrum: The 5S code, *Int. J. Remote Sens.*, **11**, 659-668, 1990.

Wobrock, W., Numerische Modellsimulationen vonStrahlungsnebelepisoden unter Berücksichtigung spektraler Wolkenmikrophysk, *PhD Thesis*, available from Johannes-Gutenberg Universität, Frankfurt, Germany, 1988.

Zhang, Y., Y. Sunwoo, G. R. Carmichael and V. Kotamarthi , Photochemical processes in the presence of dust: an evaluation of the effect of dust on particulate nitrate and ozone formation, *J. Appl. Meteor.*, 33, 813-824, 1994.

CLIMATE EFFECTS OF SULPHATE AND BLACK CARBON ESTIMATED IN A GLOBAL CLIMATE MODEL

Trond Iversen, Alf Kirkevåg, Jón Egill Kristjánsson, and Øyvind Seland

Department of Geophysics, University of Oslo
P. O. Box 1022 Blindern, 0315 Oslo, Norway

INTRODUCTION

Aerosols influence processes in the climate system in a variety of ways. Particulate matter interacts with electromagnetic radiation by scattering and absorption, thus exerting *direct* radiative forcing. By extraction of moisture from ambient air or by their similarity to ice-crystals, particles affect the microphysics of clouds in ways that determine the efficiency of precipitation processes and the general cloudiness. Since clouds interact efficiently with solar and terrestrial radiation, particles thus exert *indirect* radiative forcing (Twomey, 1977; Albrecht, 1989).

Natural aerosol particles are ubiquitous in the atmosphere both primary as particles of marine (sea-salt) and continental (soil, clay) origins, and secondary as the result of gas-to-particle processes in the air. Important are also emissions of carbonaceous particles from forest fires. Anthropogenic aerosols contribute locally and regionally to adverse health effects and acidification. In order to influence the climate system, anthropogenic particles need to stay airborne for several days, and to disturb the natural interactions between particles and climate. Increasing evidence indicates that anthropogenic particulate matter may influence the climate system efficiently, but to what extent is not determined. Indirect effects of anthropogenic aerosols are particularly uncertain (IPCC, 1996).

This paper gives estimates of radiative forcing by sulphate and black carbon (BC). A scheme for production, dispersion and deposition is implemented in a version of the NCAR Community Climate Model 3 (CCM3) (Kiehl *et al*, 1998) with the cloud formulation of Rasch and Kristjánsson (1998). The model-clouds interact with radiation and precipitation, and the scheme is well suited for estimating indirect effects. The sulphate- and BC-scheme was tested in a hemispheric scale dispersion model by Seland and Iversen (1999), as was the scheme for size-distribution and optical parameters (Boucher *et al.*, 1998; Kirkevåg *et al.*, 1999; Kirkevåg and Iversen, 1999). Preliminary results are found in Iversen *et al.* (1998). Our work should not be confused with the schemes developed at NCAR (Barth *et al.*, 2000; Rasch *et al.*, 2000; Kiehl *et al.*, 2000).

GLOBAL DISTRIBUTION OF SULPHATE AND BC

The life-cycle scheme for the airborne concentration and deposition of the aerosol species is an extension of that presented in Seland and Iversen (1999). A mechanistic approach tags the

Air Pollution Modeling and Its Application XIV, Edited by
Gryning and Schiermeier, Kluwer Academic/Plenum Publishers, New York, 2001

particulate matter according to chemical and physical production mechanisms. This enables off-line reconstruction of the size-distributions.

Explicitly calculated sulphate precursors are emissions of DMS and SO_2 but a minor fraction (~2%) of the sulphur emissions are sulphate. One third of the DMS is assumed to yield MSA as an end product and two third SO_2 by oxidation with OH or NO_3 (Yin *et al.*, 1990). SO_2 is oxidised by OH in gas phase to sulphuric acid of which f=10% is assumed to produce new nucleation/aitken mode particles of pure sulphuric acid and water together with the directly emitted sulphate. Then 90% is assumed to condensate on pre-existing accumulation mode particles and produce internal mixtures. The sulphate budgets are negligibly influenced by variations in f (Seland and Iversen, 1999). In the lower troposphere sulphate is often efficiently produced in aqueous phase by oxidation with H_2O_2 and O_3. Oxidation with O_2 catalysed by Mn and Fe can be important at low oxidant levels. When H_2O_2 levels are high, the effective oxidation rate is determined by availability of clouds and the replenishment-rate of SO_2- and H_2O_2-rich air in clouds. Much sulphate produced in clouds is scavenged directly. However, detrainment and evaporation also leads to sulphate that is internally mixed with particulate matter already present in the droplets. The oxidation rate is considerably reduced in clouds dominated by ice. Seland and Iversen (1999) showed that the sulphur budget is moderately sensitive to the existence of non-precipitating stratocumulus clouds, and considerably sensitive to how in-cloud oxidation rates are reduced as clouds become glaciated. Externally mixed nucleation-mode sulphate is converted to internally mixed accumulation mode particles by coagulation in clear and cloudy air. Sulphate is tagged by four process-specific components, three of which are internally mixed with other particulate matter.

BC is emitted as primary particles by incomplete combustion. Fossil fuel emissions are assumed emitted as 90% nucleation-mode particles and 10% accumulation-mode conglomerates. This assumption affects considerably the rate at which BC turns from hydrophobic to hydrophilic, which is determined by coagulation in clear and cloudy air. Half of the BC emitted from biomass burning is assumed to be hydrophilic accumulation-mode particles due to coagulation with primary organic carbon (OC) in the fire plumes. The atmospheric residence-time for biomass burning BC is thus considerably shorter than for fossil fuel BC given that other conditions are similar. There are three production-tagged BC-components: nucleation- and accumulation-mode externally mixed BC, and accumulation-mode internally mixed BC.

The different components are dry and wet scavenged in accordance with particle size and degree of hygroscopicity. Dry deposition uses the resistance analogy, and is the same as in the sulphur-scheme developed at NCAR (Rasch *et al.* 2000). Wet deposition is calculated in full interaction with the cloud and precipitation scheme in the model. Seland and Iversen (1999) found that sulphate and BC budgets are highly sensitive to the formulation of below-cloud scavenging.

For oxidation the model uses prescribed fields of OH, H_2O_2 and O_3 from Berntsen and Isaksen (1997). Oxidation rates are temperature and pressure dependant (e.g. Seinfeld and Pandis, 1998). Coagulation rates are estimated by assuming standard size-distributions. Emission data are taken from Spiro *et al.* (1992), Barret and Berge (1996), Saeger *et al.* (1989) and Tarrasón *et al.* (1995) for sulphur, and from Cooke and Wilson (1996) for BC. S from natural sources are DMS (12Tg(S)/a) and SOx (6.5Tg(S)/a). BC from biomass burning are 8.1 Tg(C)/a. S from anthropogenic sources are SOx (103 Tg(S)/a). BC from fossil fuel combustion are 6.0 Tg(C)/a.

Figure 1 and Table 1 summarize results for the global distribution sulphur and BC. A full comparison with measurements is not ready. Table 1 compares global budgets with those of validated models. Our turnover time for SO_2 is in the mainstream whilst that for sulphate is slightly shorter than others. The effective oxidation and the fraction oxidised in clouds rate are smaller than others. The turnover time for BC is closer to that of Liousse *et al*(1996) than to Cooke and Wilson's (1996) and in agreement with Cooke *et al.* (1999), and we have a swift transfer from hydrophobic to hydrophilic particles. The quality of the column burdens is difficult to assess since measurements in the free troposphere are sparse. Comparing with the quite comprehensive validation made by Barth *et al.*(2000) and Rasch *et al.*(2000) for their sulphur scheme in CCM3, we have a smaller relative global burden of sulphate due to more efficient deposition (NB: our emissions are 50% larger). Whether this reflects a reduced overestimation in the free troposphere in our model remains to be seen. The underestimations of ground surface concentrations in the Arctic are pronounced both for sulphate and BC.

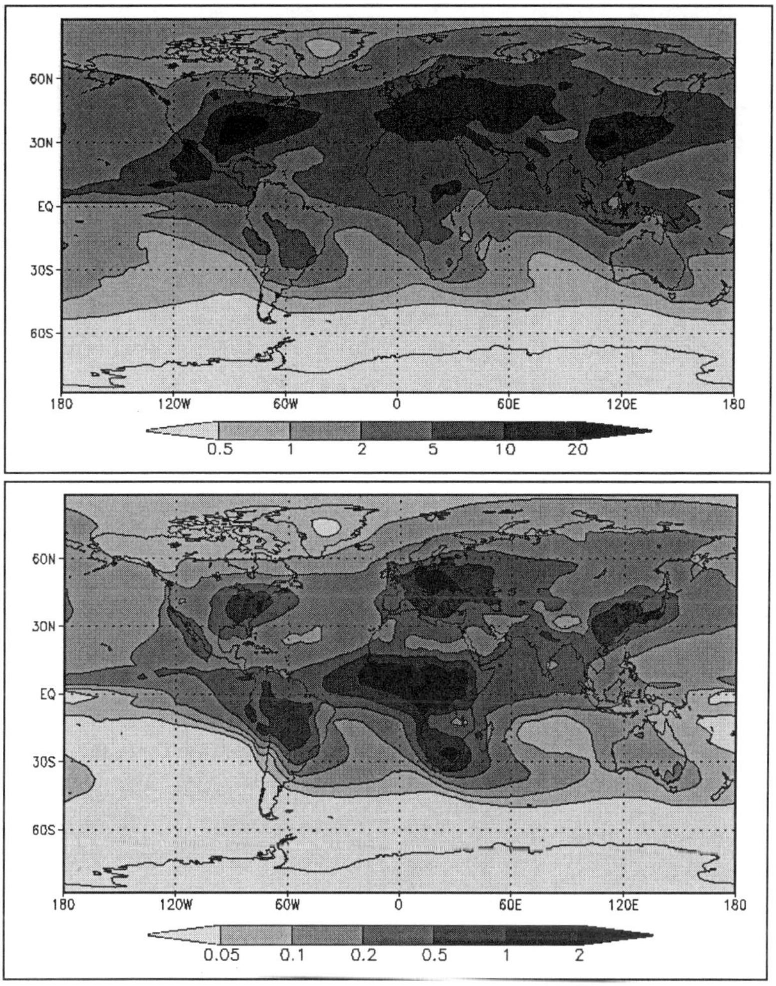

Figure 1. Model calculated vertical column burden of sulphate (upper panel, global mean: 4.3) and black carbon (lower panel, global mean 0.38). Units: mg m^{-2} as sulphate and carbon, respectively.

DIRECT RADIATIVE FORCING ESTIMATE

In order to calculate the radiative forcing by sulphate and BC as calculated above, bulk aerosol optical parameters (specific extinction coefficient, single scattering albedo, and asymmetry factor) due to particles in each grid volume at any time must be estimated. Since the particle size is close to the wavelength of visible light for which the solar radiance varies considerably with wavelength, these parameters should ideally be given for a wide range of wavelengths and particle diameters. Given that the particle composition may vary in time and space, the optical parameters must be calculated for every model time-step. Such calculations are computer demanding, even when assuming spherical and homogeneous particles so that Mie-theory can be applied. We have therefore developed tables with entries being background aerosol type, process-tagged concentrations of sulphate and BC, and relative humidity. Background aerosols are primary particles of natural origins, and we separate between continental, marine and polar aerosols w.r.t. size-distribution and composition. Based on d'Almeida *et al.* (1991), Covert *et al.*(1996) and Maenhout *et al.*(1996), the background aerosol may consist of 10 different modes. In developing the tables a wide selection of tagged sulphate and BC concentrations is assumed, and a size-

Table 1. Global turnover times (T) for total DMS, SO_2, sulphate and BC; and gross oxidation rate (K) for SO_2 to sulphate, and the fraction (w) of the chemical production taking place in aqueous phase.

	T_{DMS}(days)	T_{SO2}(days)	T_{SO4}(days)	T_{BC}(days)	$K(10^{-6}s^{-1})$	w
Other work:						
LR.91	3	1.2-1.8	4.6-5.6	-	5.3-2.3	0.84
P.95	0.9	0.6	4.7	-	9.8	0.90
F.96	2.1	1.6	4.4	-	3.7	0.67
C.96	1.0	1.3	3.9	-	4.6	0.85
Ch.97	-	1.3	4.1	-	4.3	0.87
Le.97	0.5	2.3	5.3	-	3.6	0.77
Re.98	5.0	2.0	4.5	-	3.8	0.78
Ra.00	1.4	1.9	4.0	-	4.3	0.83
CW.96-i	-	-	-	9.8	-	-
CW.96-ii	-	-	-	7.9	-	-
CW.96-iii	-	-	-	5.6	-	-
Li.96	-	-	-	3.9	-	-
Co.99	-	-	-	5.3	-	-
This work						
year "1989"	1.6	1.6	3.6	5.1	3.6	0.76
seasons: DJF	1.8	1.9	2.6	4.6	3.0	0.87
MAM	1.9	1.3	3.1	4.7	4.5	0.80
JJA	1.3	1.6	4.9	6.4	3.7	0.62
SON	1.4	1.6	3.7	4.7	3.5	0.76

LR.91=Langner and Rodhe(1991); P.96=Pham *et al.*(1995); F.96=Feichter *et al.*(1996); C.96=Chin *et al.*(1996); Ch.97=Chuang *et al.*(1997); Le.97=Lelieveld *et al.*(1997); Re.98=Restad *et al.*(1998), Ra.00=Rasch *et al.*(2000); CW.96=Cooke and Wilson (1996); Li.96=Liousse *et al.*(1996);Co.99=Cooke *et al.*(1999).

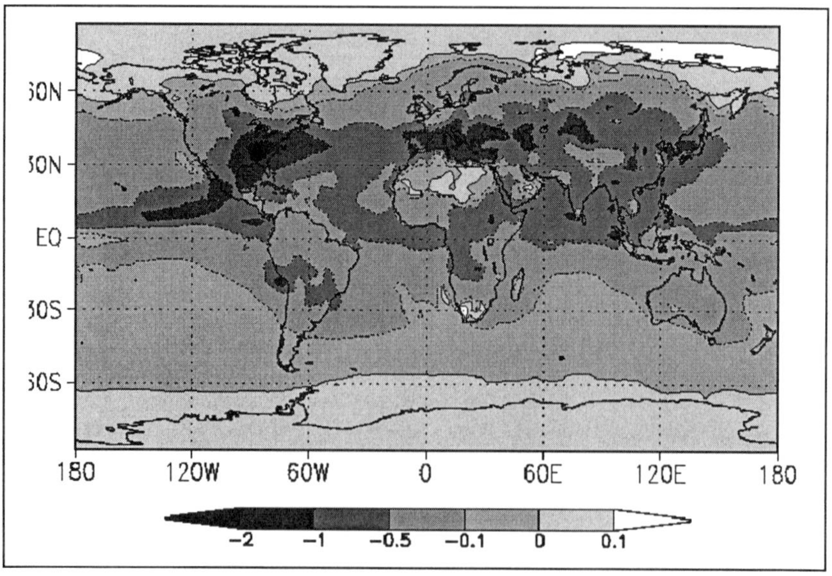

Figure 2. Model estimated top-of-the-atmosphere radiative forcing of anthropogenic sulphate and BC form fossil fuel combustion using NCAR CCM3 off-line with the dynamics. Global average is - 0.31. Unit: W m^{-2} .

resolved internally mixed aerosol is calculated for each case by solving a continuity-equation in the particle diameter space. Particle swelling due to hygroscopicity is calculated using data for water activity and solving the Köhler-equation. Finally, externally mixed particles are added. Details about this procedure can be found in Kirkevåg *et al.* (1999) and Kirkevåg and Iversen (1999).

The optical parameters are estimated from the tables for wavelengths in the radiative transfer scheme of NCAR CCM3. Without significant loss of accuracy for tropospheric aerosols, we have reduced the number of spectral bands for the aerosol optical parameters to 7, ranging from 3 in the ultraviolet (0.2-0.35:m), 1 in the visible (0.35-0.7:m), and 3 in the near infrared (0.7-4:m). In order to only estimate the top-of-the-atmosphere radiative forcing, we have run the standard CCM3-dynamics with monthly-averaged aerosol-fields. These forcings are so far not entered into the thermodynamic energy equation. The use of monthly averages lead to some regional errors, but the global forcing is not significantly influenced. Figure 2 shows the estimated direct radiative forcing due to anthropogenic sulphate and fossil fuel BC. The global average is -0.31W m^{-2}. The picture is heterogeneous with considerable negative forcing over polluted regions at mid-latitudes of the northern hemisphere. Positive forcing due to the absorptive properties of BC is estimated over high-albedo surfaces in the Arctic and in deserts. The latter is even more pronounced when BC from biomass burning is taken into account due to contributions from man-made forest fires in the tropics. The net effect of biomass burning BC and natural sulphate is slightly positive (0.03 W m^{-2}).

INDIRECT RADIATIVE FORCING ESTIMATE

We have estimated the two indirect effects of the model-calculated sulphate and BC on the radiative balance by changing cloud properties. As for the direct effect, only the radiative forcing at the top of the atmosphere is estimated, and not its response onto the dynamics of the atmosphere. For the first indirect effect caused by changing the radiative properties of cloud droplets, this is clearly possible to do. However, for the second indirect effect enforced by an increased amount of clouds, this is not as straightforward since it by nature takes more than a time-step to develop cloud-fields. Thus two cloud-fields are developed in parallel, one for the standard CCM3 control run and one controlled by the parameterised aerosol properties. Also here we have used monthly averaged concentration fields.

In order to include the indirect aerosol effects, the model's cloud microphysics scheme must be linked to some aerosol property as input (e.g. CCN), and to precipitation efficiency and radiation as output. The scheme of Rasch and Kristjansson (1998) has just these properties. Similar to other studies (Lohmann *et al.*, 1999; Kiehl *et al.*, 2000), indirect effects are imposed for warm clouds only, even though there are reasons to believe that the second indirect effect could be efficient also for mixed-phase clouds. The cloud albedo is linked to the cloud droplet size through an effective radius (r_e) directly proportional to the mean physical droplet radius (r). The proportionality coefficient varies from cloud type to cloud type, but is actually quite close to 1. The droplet radius is linked to the cloud liquid water content (water mass per air volume) q_{lwc}, which is a prognostic variable in the mode. The relation is $r = [3q_{lwc}/ 4B\kappa_w N]^{1/3}$ where N is the droplet number concentrations and κ_w is the water density. It can be shown that an increase in cloud albedo is largest for clouds with albedo around 0.5 and is proportional with the *relative* increase in droplet concentrations. Also the auto-conversion of cloud water to precipitation in warm clouds is proportional to the relative increase in cloud droplet number concentrations, but with a negative coefficient = -P/3, where P is the auto-conversion precipitation rate. For newly formed clouds, the number concentration equals the activated cloud condensation nuclei (CCN), and as a first approximation this number is used for N. CCN concentrations are tabulated for fixed super-saturations in similar ways as the optical parameters above, using the Köhler equation.

Figure 3 show the estimated total indirect effect due to the calculated sulphate and BC. The global average is – 1.5 W m^{-2}. Sensitivity tests indicate that the first indirect effect (due to increases cloud albedo) is about a factor of 4 larger than the second indirect effect. However, the two effects are not purely additive. Also the indirect effects appear to be strongly sensitive to some uncertain assumption in the parameterisation, such as the realised super-saturation. The super-saturation should ideally be estimated in the model and not be prescribed. More investigation is needed before more certain estimates can be given.

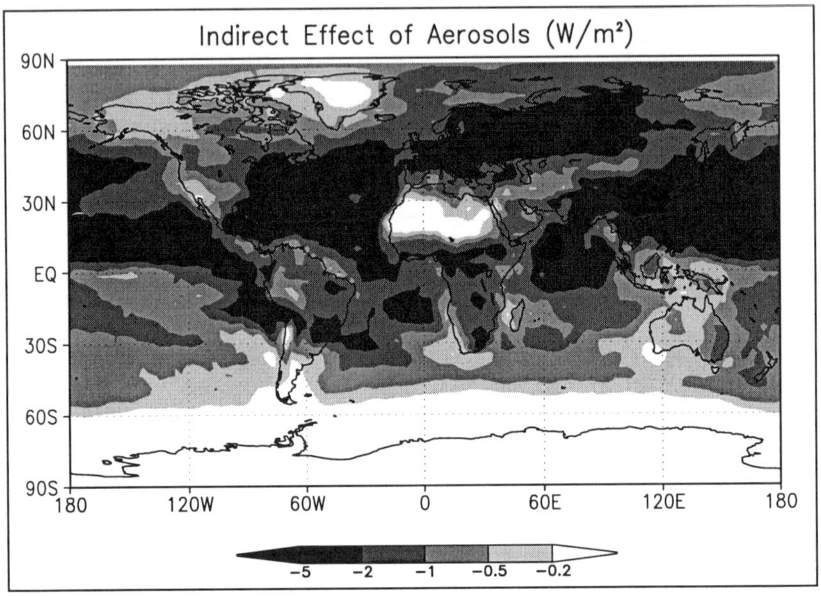

Figure 3. Estimated top-of-the-atmosphere radiative forcing due to the indirect effect caused by anthropogenic sulphate and BC from fossil fuel combustion. Global mean is –1.5. Unit. W m^{-2} .

SUMMARY

The work on climatic effects of aerosols has not come far enough to draw any firm conclusions, mainly due to uncertainties w.r.t. indirect effects. The concentration calculations as well as the forcing estimates need validation. Also organic carbon (OC), a large group of primary and secondary organic particles, should be included. Penner *et al.*(1996) estimated much larger indirect effects of OC to than the of sulphate, because OC-particles are not produced in clouds as opposed to sulphate. On the other hand, the indirect effects of sulphate estimated by Lohmann and Feichter (1997) is probably too high, since they did not take into account that much sulphate is produced in pre-existing droplets. The anthropogenic OC-burden in the troposphere is poorly known. Table 2 summarises our forcing estimates along with results by others. Despite the uncertainties it appears quite certain that the aerosols' impacts on radiative forcing is comparable to greenhouse gases. The pronounced regional patterns may affect the climate system more efficiently than indicated by the net global forcing. Climate scenario simulations are needed to reveal this.

Table 2. Top of the atmosphere global radiative forcing due to aerosol components estimated in this work as well as in several others. Unit: W m^{-2}. Also the burdens of sulphate, BC or OC are given when applied (unit mg m^{-2}.

	SO_4	BC	OC	Direct	Indirect
Other work:					
Jones and Slingo (1996)	1.8	0	0	not calc.	-1.5 - -0.5
Haywood *et al.* (1997) (BC/SO_4=7.5%)	1.8	0.13	0	-0.18 - -0.02	not calc.
Feichter *et al.*(1997)	2.23	0	0	-0.35	not calc.
Chuang *et al.* (1997)	3.23	0	0	-0.43	-1.24 - -0.62
Lohmann and Feichter (1997)	4.1	0	0	not calc.	- 4.3 - - 1.4
Haywood & Ramaswamy(1998) (external)	1.76	0.25	0	-0.42	not calc.
Myhre *et al.*(1998) (BC/SO_4=7.5%)	1.90	0.14	0	-0.16 - +0.10	not calc.
Penner *et al.*(1998) (external SO_4+BC+OC)----		not given	----	-0.81 - -0.53	not calc.
Cooke *et al.* (1999) (external BC+OC)	0	0.143	0.341	0.149	not calc.
Kiehl *et al.* (2000) (method II)	2.23	0	0	- 0.56	- 0.4
This work:	3.49	0.161	0	- 0.31	- 1.5

Acknowledgements

This work is a part of the nationally co-ordinated project "Regional Climate Development under Global Warming", financed by the Research Council of Norway under the programme on changes in climate and the ozone layer. The project receives support from the programme for supercomputing through a grant of computing time. Emission data for BC were provided by Dr. W. Cooke, and oxidant concentrations by Dr. T. Berntsen. The authors are grateful for advice by Dr. P. Rasch, and the opportunity to use parts of NCAR's developed model code for sulphur.

REFERENCES

Albrecht, B.A. (1089) Aerosols, cloud microphysics and fractional cloudiness. *Science,* **245**, 1227-1230.

D'Almeida, G.A., Koepke,P. and Shettle,E.P. (1991) *Atmospheric aerosols: global climatology and radiative characteristics.* Deepak publishing, Hampton, Virginia, USA.

Barret, K. and Berge,E, (1996). EMEP/MSC-W Report No.1/96. The Norwegian Meteorological Institute, Oslo, Norway.

Barth, M.C., Rasch, P.J., Kiehl, J.T., Benkowitz, C.M. and Schwartz, S.E.(2000) Sulfur chemistry in the National Center for Atmospheric Research Community Climate Model: Description, evaluation, features, and sensitivity to aqueous chemistry. *J. Geoph. Res.*, **105**, 1387-1415.

Berntsen, T.K. and Isaksen, I.S.A. (1997) A global threedimensional chemical transport model for the troposphere. 1, Model description and CO and ozone results. *J. Geoph. Res.*, **102**, 21,239-21,281.

Boucher, O., Schwartz, S.E. and 28 other authors (1998) Intercomparison of models representing shortwave radiative forcing by sulfate aerosols. *J. Geoph. Res.*, **103**, 16,979-16,998.

Chin, M, Jacob, D.J., Gardner, G.M: Foreman-Fowler, M.S. and Spiro, P.A. (1996) A global three-dimensional model of tropospheric sulphate. *J. Geoph. Res.*, **101,** 18,667-18,690.

Chuang, C.C, Penner, J.E., Taylor, K.E., Grossman, A.S. and Walton, J.J. (1997) An assessment of the radiative effects of anthropogenic sulphate. *J. Geoph. Res.*,**102,** 3761-3778.

Cooke, W.F. and Wilson, J.J.N. (1996) A global black carbon aerosol model. *J.Geophys.Res.,* **101**, 19395-19409.

Cooke, W.F., Liousse, C., Cachier, H. and Heichter, J. (1999) Construction of a 1x1 fossil fuel emission data set for carbonaceous serosol and implementation and radiative impact in the ECHAM4 model. *J. Geoph. Res.* **104,** 22,137-22,162.

Covert, D.S., Wiedensohler,A., Aalto, P., Heintzenberg, J., McMurry, P.H. and Leck, C. (1996) Aerosol number size distributions from 3 to 500 nm diameter in the Arctoc marine boundary layer during summer and autumn. *Tellus,* **48**, 197-212.

Feichter, J., Kjellström, E., Rodhe, H., Dentener, F., Lelieveld, J., and Foelofs, G.-J. (1996) Simulation of the tropospheric sulfur cycle in a global climate model. *Atmos.Env.*, **10**, 1693-1707.

Feichter. J, Lohmann, U. and Schult, I. (1997) The atmospheric sulphur cycle in ECHAM-4 and its impact on the shortwave radiation. *Clim. Dyn.,* **13,** 235-246.

Haywood, J.M., Roberts, D.L., Slingo, A., Edwards, J.M. and Shine, K.P. (1997) General circulation model calculations of the direct radiative forcing by anthropogenic sulphate and fossil fuel soot aerosol. *J. Clim.*, **10**, 1562-1577

Haywood, J.M. and Ramaswamy, V. (1998) Global sensitivity studies of the direct radiative forcing due to anthropogenic sulphate and black carbon aerosols. *J. Geoph. Res.*, **103**, 6043-6058.

IPCC (1996) *Climate Change 1995*. Ed. by J.T.Houghton et al., Intergovernmental Panel on Climate Change, J.T. Houghton et al , Cambridge University Press.

Iversen, T., Kirkevåg, A. and Seland, Ø. (1998) Hemispheris-scale modelling of sulphate and black carbon and thei direct radiative effects. *Air Pollution Modeling and its Application, XII.* Eds: S-E. Gryning and N. Chaumerliac, Plenum Press, New York, 477-487.

Jones, A., and Slingo, A. (1996) Predicting cloud-droplet effective radius and indirect sulphate aerosol forcing using a general circulation model. *Q. J. Roy. Met. Soc.*, **122**, 1573-1595.

Kiehl, J. T., J. J. Hack, G. B. Bonan, B. A. Boville, D. L. Williamson, and P. J. Rasch (1998) The National Center for Atmospheric Research Community Climate Model: CCM3, *J. Climate*, **11**, 1131--1149.

Kiehl, J.T., Schneider, T.L., Rasch, P.J., and Barth, M.C. (2000) Radiative forcing due to sulfate aerosols from simulations with the National Center for Atmospheric Research Community Climate Model, Version 3. *J. Geoph. Res.*, **105**, 1441-1457.

Kirkevåg, A, Iversen, T. and Dahlback, A. (1999) On radiative effects of black carbon and sulphate aerosols. *Atmos. Environ.*, **33**, 2621-2635.

Kirkevåg, A. and Iversen, T. (1999) Parameterization of Optical Properties of composite tropospheric aerosols in GCMs. Institute Report Series, No. 106. Dep. of Geophysics, Univ. of Oslo, Norway.

Langner, J. and Rodhe, H. (1991) A global three-dimensional model of the global sulfur cycle. *J..Atmos.Chem*, **13**, 255-263.

Lelieveld, J., Roelofs, G.-J., Ganzeveld, L., Feichter, J. and Rodhe, H. (1997) Terrestrial sources and distribution of atmospheric sulphur. *Phil. Trans. R. Soc. Lond.*, **352**, 149-158.

Liousse, C., Penner, J.E., Chuang, C., Walton, Eddleman, H. and Cachier, H. (1996) A global three-dimensional study of carbonaceous aerosols. *J. Geoph. Res.*, **101**, 19.411-19.432.

Lohmann, U., Feichter, J., Chuang, C. C., and Penner, J. E., (1999): Prediction of the number of cloud-droplets in the ECHAM GCM. *J. Geophys. Res.*, **104**, 9169-9198.

Lohmann, U. and Feichter, J. (1997) Impact of sulfate aerosols on albedo and lifetime of clouds: A sensitivity study with the ECHAM4 GCM. *J. Geoph. Res.*,**102,** 13,685-13,700.

Maenhout, W., Cornille, P., Pacyna, J.M., and Vitols, V. (1989) Trace element composition and origin of the atmospheric aerosol in the Norwegian Arctic. *Atmos. Environ.*, **23**, 2551-2570.

Myhre, G., Stordal, F., Restad, K. and Isaksen, I.S.A. (1998) Estimates of the direct radiative forcing due to sulphate and soot aerosols. *Tellus*,. **50B**, 463-477.

Penner, J.E., Chuang, C.C. and Liousse, C. (1996) The contribution of carbonaceous aerosols to climate change. In: *Nucleation and atmospheric aerosols, 1996*. M. Kulmala and P.E. Wagner (eds.). 759-769

Penner, J.E., Chuang, C.C. and Grant, K. (1998) Climate forcing by carbonaceous and sulfate aerosols. *Clim. Dyn.*, **14,** 839-851.

Pham, M., Müller, J.F., Brasseur, G.P., Granier, C., and Megie, G. (1995) A three dimensional study of the tropospheric sulfur cycle. *J. Geoph. Res.*, **100**, 26,061-26,092.

Rasch, P.J. and Kristjánsson, J.E. (1998) A comparison of the CCM3 model climate using diagnostic and predicted condensate parameterizations. *J. Climate.*, **11**, 1587-1614.

Rasch, P. J., Barth, M. C., Kiehl, J. T., Schwartz, S. E., and Benkovitz, C. M., (2000): A description of the global sulfur cycle and its controlling processes in the National Center for Atmospheric Research Community Climate Model, Version 3. *J. Geoph. Res.*, **105**, 1367-1385.

Restad, K., Isaksen, I.S.A. and Berntsen, T. (1998) Global distribution of sukphate in the troposphere. A three-dimensional model study. *Atmos. Environ.*, **32,**3 593-3609.

Saeger, M, and 12 co-authors (1989) The 1985 NAPAP emissions inventory (version 2): development of the annual data and modeler's tape. Rep. No. EPA-600/7-80-012a, US EPA, USA.

Seinfeld, J.H. and Pandis, S.N. (1998) *Atmospheric Chemistry and Physics. From air pollution to climate change.* John Wiley and Sons, Inc., USA. 1326 pp.

Seland Ø. and Iversen T. (1999) A cheme for black carbon and sulphate aerosols tested in a hemispheric scale, Eulerian disperion model. *Atmos. Environ.* **33**, 2853-2879.

Spiro. P.A., Jacob, D.J., Logan, J.A. (1992) Global inventory of sulfur emissions with 1x1 resolution. *J. Geoph. Res.*, **97,** 6023-6036.

Tarrasón, L., Turner, S. M. and Fløisand, I. (1995) An estimation of seasonal DMS fluxes over the North Atlantic Ocean. and their contribution to European pollution levels. *J. Geoph. Res.*, **100,** 11,623-11,639.

Twomey, S. (1977) The influence of pollution on the shortwave albedo of clouds. *J.Atm.Sci.*, **34**, 1149-1152.

Yin, F, Grosjean, D. and Seinfeld, J.H. (1990) Photooxidation of dimethyl sulphide and dimethyl disulphide. (II) Mechanism evaluation. *J. Atmos. Chem.*, **11**, 365-399.

NEW DEVELOPMENTS

chairpersons:	E. Batchvarova
	G. Carmichael
	S. Hanna
	S. LeDuc
	J. Weil
rapporteurs:	P. de Haan
	P.-E. Johansson
	C. Mensink

What did we learn from the ETEX Experiment?

W. Klug

Mittermayerweg 21, D-64289 Darmstadt, Germany.

1.0 Introduction

Mathematical models of atmospheric transport and chemistry generally improve our understanding of the atmosphere and are important tools in predicting developments in atmospheric composition over a wide range of time scales. In particular for accidental atmospheric releases the development of predictive and diagnostic capabilities is essential in order to be able to take appropriate measures and to issue timely warnings. Unfortunate examples of such accidents in the recent past are the spills of highly toxic chemicals in Seveso (Italy), Bhopal (India) and, in the case of radioactivity, Three Miles Island (USA). On a larger scale we recall also the release of radioactive material at Windscale (UK) and Chernobyl (USSR). The burning oilfields in Kuwait during the Iraq conflict can be considered also to be a major release into the atmosphere of chemical substances with potentially detrimental effects both in the immediate vicinity of the sources and on a regional or even global scale.

Having these facts in mind and following the Chernobyl accident in April 1986 an Atmospheric Transport Model Evaluation Study (ATMES) was initiated and jointly sponsored by the European Commission (EC), the World Meteorological Organisation (WMO) and the International Atomic Energy Agency (IAEA) (Klug et al., 1992). The goal of the ATMES-study was to compare simulated concentrations and depositions with measured values. The resonance to this study and especially the unexpected large number of participants showed the need for such exercises among the scientific community. However, there were several drawbacks connected with ATMES:

1. The release characteristics of the radioactive material at Chernobyl remained - even after considerable effort to reconstruct them post factum - uncertain and created an additional uncertainty when comparing model results with observational data;
2. The evaluation of the model results took place several years after the event, leaving ample time to adjust models according to measured data;

3. The measurement data base was very inhomogeneous with regard to sampling methods and intervals;
4. The model results were obtained with analysed (*not real time forecasted!*) meteorological data.

In view of the accidents mentioned before there is nowadays - in the frame of emergency response - a need to test the ability of meteorological services to forecast in real time the meteorological and concentration fields over distances of thousands of kilometres and up to several days. No verification of this ability has been shown so far and no experimental data were available to check the performance of the long range transport and diffusion models in connection with meteorological forecast models.
In the later stages of ATMES, a member of the Steering Committee, Dr. L. Machta, NOAA, suggested after the successful completion of the ANATEX (Across North America Tracer Experiment) and CAPTEX (Cross - Appalachian Tracer Experiment) to conduct a similar experiment in Europe with the aim to simulate an emergency situation of release of harmful material into the atmosphere avoiding the drawbacks mentioned before. In this case the release of the tracer material is announced only shortly before the event, the release characteristics (source strength and height, source location etc) are given to the modellers at this moment. The modellers task is then to forecast the evolution of the tracer cloud in real time and send in their forecasted concentration and deposition fields at prescribed time intervals to a central institution.
This idea was taken up by the Steering Committee and the new experiment ETEX (European Tracer Experiment) was carefully planned and executed. After detailed discussions the objectives of ETEX were defined and can be summarised as follows:

- to conduct a long-range atmospheric tracer experiment involving controlled releases under well-defined conditions together with co-ordinated environmental sampling of surface concentrations at distances approaching 2000 km.
- to provide notification of the releases to institutes responsible for producing rapid forecasts of atmospheric dispersion, these institutes then being required to produce such forecasts as quickly as possible.
- to evaluate the validity of these forecasts in the light of the subsequent environmental measurements.
- to assemble a data base of surface concentrations and meteorological data which will allow the evaluation of other present or future models of long-range atmospheric dispersion.

2.0 The Experiments

ETEX consisted of three major activities:

1. the tracer experiments as such, preceded by preparatory activities, such as 'dry runs' (for testing communications and forecasting procedures) and background measurements;
2. long range dispersion model calculations in real-time;
3. model results' evaluation.

ETEX was sponsored by the European Commission, the World Meteorological Organisation and the International Atomic Energy Commission. 17 countries were involved in the tracer experiment as such with 168 surface sampling stations distributed over their territories. 24 institutes from 20 countries participated in the real-time modelling activities with 28 models. It goes without saying that this five-year project is the result of the work and dedication of many individuals and institutions. Their activities are gratefully acknowledged. A detailed overview on the experiment is given in: "The European Tracer Experiment", 1998, EUR 18143 EN, Joint Research Centre, European Commission. A short summary of the work follows here.

2.1 The Releases

ETEX consisted of two tracer (perfluorocarbon) releases into the atmosphere sampled for three days after the begin of the emission using a sampling network spread over most of Europe. When the emission started about 30 modelling groups all over the world were informed about the details of the release. The modellers subsequently predicted in real-time surface concentrations over a period of 60 hours and sent their results to the reference centre at the Joint Research Centre of the EC at Ispra, Italy.

2.2 The Sampling

The sampling network consisted of 168 surface sampling stations in Western and Eastern Europe hosted by the synoptic stations of the National Weather Services. Three samplers were located on oil or gas platforms in the North Sea. The average spacing between the sampling stations was about 80 km. Only active samplers were used in ETEX. These samplers suck a known air volume through an absorber that quantitatively removes all PFC material from the air. The surface tracer concentration were averaged values over 3 hours.

The samples were shipped to JRC Ispra and chemically analysed. The release rate and the total mass released were chosen so that the expected concentrations even at large distances from the source would be well above background. - Three aircraft were employed to obtain the vertical structure of the cloud. They sampled the air on adsorption tubes and bags for subsequent laboratory analysis.

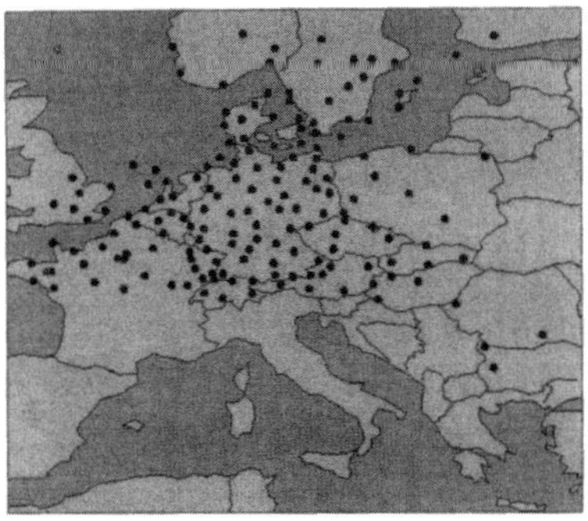

Fig. 1: Location of release site (North-west France) and sampling stations

2.3 Meteorological Measurements

In addition to the routine meteorological observations as obtained by the synoptic network of the Weather Services ground level and upper-air measurements at the release site were performed. Constant level balloons were launched to follow air trajectories departing from the release site during the 12 hours of release.

2.4 Meteorological Studies

A preliminary study before the experiment revealed that from a statistical point of view the most suitable weather situations with regard to a successful experiment would be those with westerly flow over Central Europe. This immediately led to the conclusion to have the release site in Western France and the release time either in fall or spring.

2.5 Time and Location of the Releases

Two tracer experiments were done: the first on the 23. October 1992, starting at 16:00 UTC and the second on the 14. November starting at 15:00 UTC, both lasting 12 hours. The tracer was Perfluoromethylcyclohexane (PMCH) which is inert and non-depositing. The release location was Monterfil, Brittany, France.

2.6 The Alert Procedure

The alert plan is triggered 48 hours before the beginning of the release after the weather situation in the light of the 48 to 96 hours numerical forecasts of the weather services were promising. If the situation was favourable a preliminary message was sent out to the co-ordinators of the release, the sampling and air craft crews.
The following days the development of the synoptic situation was checked and the co-ordinators and others were constantly informed. 24 hours before a possible release a pre-alert message was sent out containing only a warning. On the release day morning the alert message was sent out and in the afternoon the GO signal was transferred to the co-ordinators and at the same time the modellers were notified about the details of the release.
This procedure was tested in three so-called dry runs and the performance increased from one run to the next.

3.0 The Data Evaluation

The model results were submitted in real-time and afterwards by Fax, FTP or disks to the Evaluation Team at Ispra, Italy. The modellers sent by Fax concentration contours at 24, 48 and 60 hours immediately after release. Theses contours were plotted on a map of Europe. Nearly all modellers sent these results within the first 6 hours after release. This information allowed the Evaluation Team to ascertain shortly after the start which models gave qualitatively satisfying results.

3.1 Statistical evaluation

The large number of participating models made it impossible to evaluate each model on its own merits. With the limited resources available only statistical and graphical evaluation of the data were possible. This was done several months after the experiment
The statistical analysis was split into time, space and overall (global) analysis. The statistical parameters evaluated are the same as defined in Klug et al., 1992, p. 18-28. Here it suffices to name these quantities:

- Mean values, standard deviation, bias, normalised mean square error,
- Pearson's correlation coefficient,
- Fraction of calculated values which are within a factor of 2 or 5 of the observed ones,
- Figure of merit in time and in space,
- Box plots.

Experience showed that if one of the values above showed a good model performance the others did too, although each of them looked at another aspect of the data. Time of arrival and cloud duration of the travelling cloud as well as the peak concentration were computed and compared with the measured data.

4.0 Experiment 1

4.1 Details of the Release

The release was performed at Monterfil, Brittany, France on the 23/24 October 1994 from 1600 UTC to 0350 UTC. 340 kg of Perfluoromethylcyclohexane (PMCH) were released which is equivalent to a release rate of 7.98 g/sec. The source was at the surface.

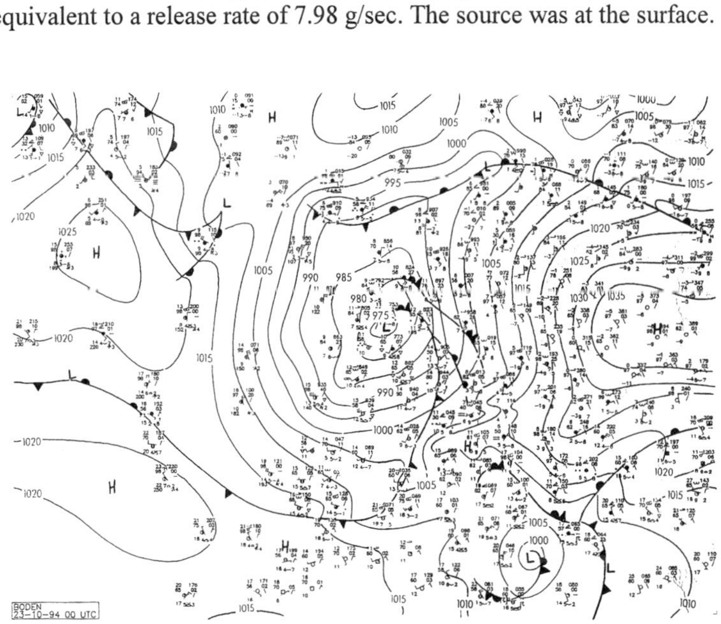

Fig. 2: Synoptic situation on 23 October 1994, 00 UTC

4.1.1 Synoptic Situation

The synoptic situation was dominated by a moderately strong low pressure area of 975 hPa over the British Isles which led to a westerly air flow over North-western France and Central Europe. This situation continued over the next 60 hours with a slight shift in the wind direction to Southwest.

4.1.2 Upper-Air Measurements

A SODAR was operating at the release site during the first experiment. Profiles of horizontal windspeed and direction were measured continuously up to 600 m above ground. Five pairs of constant level balloons were released during the experiment.

4.2 The Evolution of the Cloud during Experiment 1

The reconstruction of cloud position and concentrations at the surface showed that even a relatively simple meteorological situation produced unexpected cloud behaviour. The cloud moved during the first hours after the release in north-easterly direction. After 12 hours, the cloud could be detected not only in France, but also at a few places in Germany. The cloud crossed Germany and after about 60 hours travel time the eastward movement was stopped and the cloud was split into two parts, one moving into north-north-westerly, the other into south-south-easterly direction. This development was predicted by most of the meteorological models.

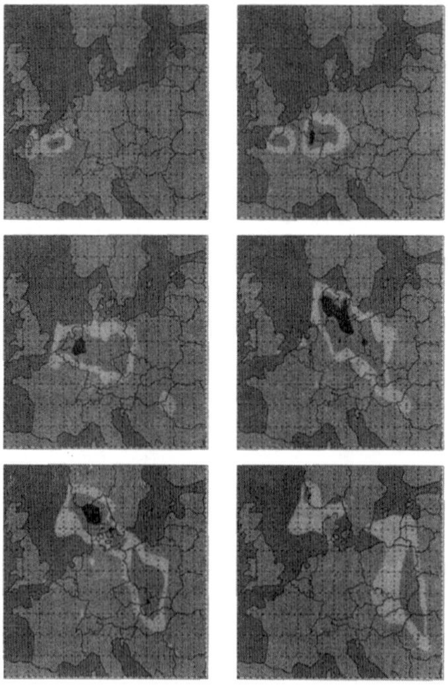

Fig. 3: Position of the measured tracer cloud during experiment #1 after 12, 24, 36, 48, 60 and 72 hours

4.3 Statistical Results

In order to facilitate the understanding of the time analysis 11 locations were selected forming two arcs on the centre cloud trajectory, the first one placed at a distance of about 600 km, the second approximately 1200 - 1400 km from the source. The first arc extended from Holland to Western Germany, consisting of 4 stations, the second from Northern Denmark over Germany, Czech Republic to Hungary, consisting of 7 stations.
The box plots show the frequency distributions of the predicted values (from 28 models) of time of arrival of the cloud, duration of the cloud and peak concentration. The station where concentrations were best forecasted is the Dutch station NL1. The station where most of the models failed was the Hungarian one.
The statistical analysis revealed that a group of models (6-8) were able to forecast **in real time** the cloud position and its horizontal extent. This is especially true for the first 48 hours. When the concentration values at the various locations are considered even the best models were not capable of always describing the concentrations in a satisfying way.
In order to get a summary of the models' behaviour it was decided to define indices of performance in terms of NMSE, bias, PCC, FMT and cumulated concentrations. Without discussing their values in detail it turned out that there were 3 groups of models:

- A first group of six models which showed excellent performance with respect to almost all of the above mentioned criteria;
- A second group of eight which had a few excellent and some average performances;
- Finally, a group of 4 with intermediate results.

Details of the statistical results for this experiment are given in a report by Graziani et al, 1998.

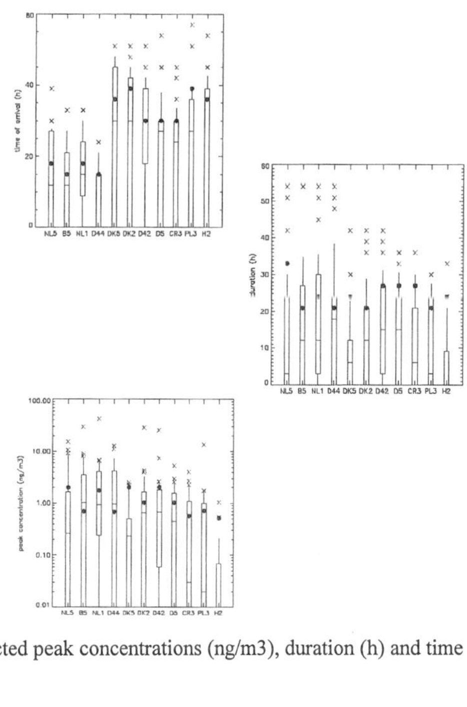

Fig. 4: Box plots of predicted peak concentrations (ng/m3), duration (h) and time of arrival (h), experim. #1

5.0 Experiment 2

5.1 Details of the Release

The release was performed at Monterfil, Brittany, France on the 14/15 November 1994 from 1500 UTC to 0245 UTC. 490 kg of Perfluoromethylcyclopentane (PMCP) were released which is equivalent to a release rate of 11.58 g/sec. The source was at the surface.

5.1.1 Synoptic Situation

A weak frontal system moved from the Atlantic to the Continent on the 13. November. On the 14. November 0000 UTC the low with its centre near Iceland was nearly stationary but had developed a wide warm sector extending from the Azores to the middle of Europe. The coldfront connected to this warm sector was still on the Atlantic but moved in rapidly and crossed Brittany while the experiment was on-going. This led to strong winds and showers during the passage of the front. After the passage the wind decreased considerably.

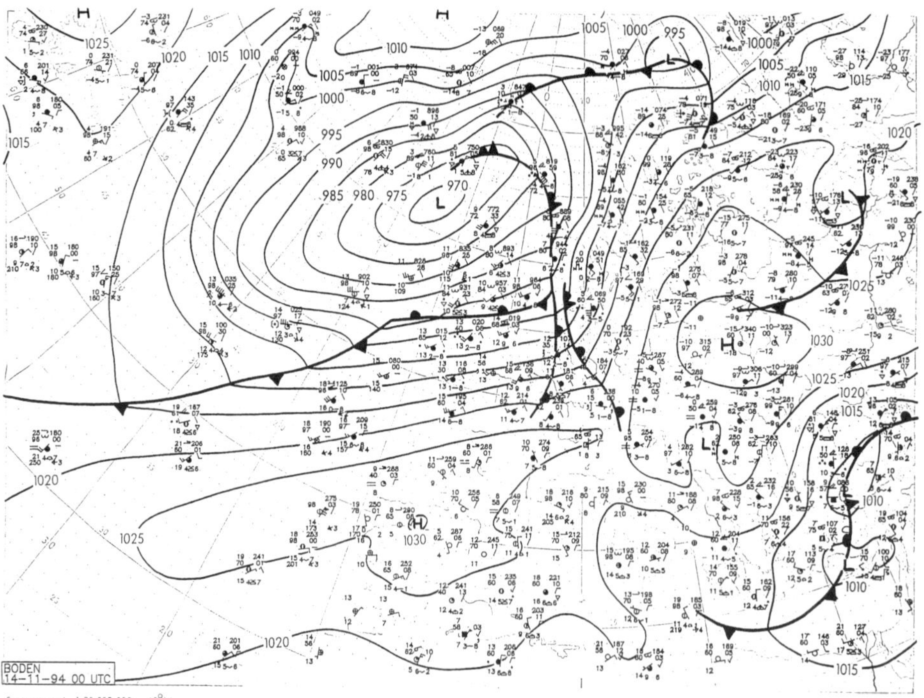

Fig. 5: Synoptic situation on 14 November 1994, 00 UTC

5.1.2 Upper-Air Measurements

A SODAR was operating at the release site also during the second experiment. However, showers and rain affected these measurements at times. A Sonic anemometer measured micrometeorological characteristics in the Prandtl-layer: friction velocity, surface heat flux and turbulence. 4 constant level balloons were successfully launched during the experiment.

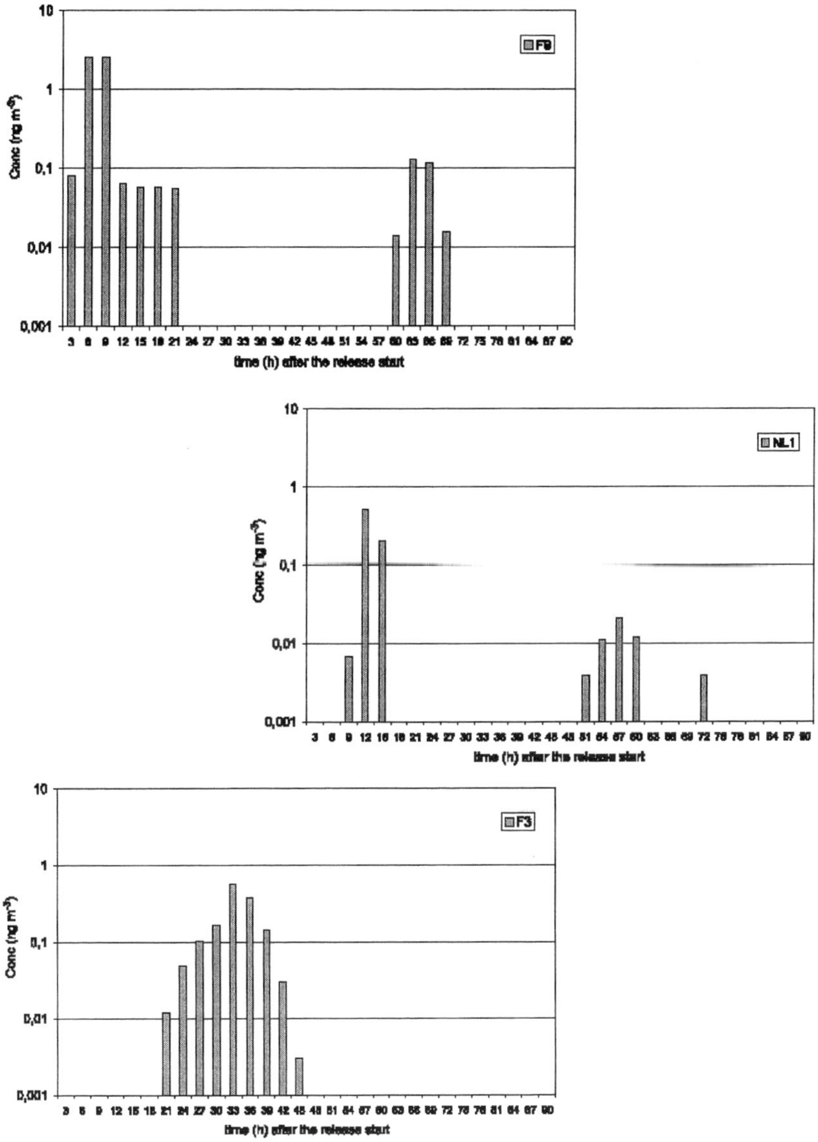

Fig. 6: Time series of measured concentrations at 3 stations, two in France, one in Holland, experiment # 2

5.2 The Evolution of the Cloud during Experiment 2

Due to the high wind speeds the tracer cloud crossed Europe very rapidly in a East-north-east direction. The observations revealed a split of the cloud. This might be real, but it might also be an artefact by the relatively large distances between sampling stations.
24 hours after the begin of the release the cloud was still split into two parts, one part in France, the other extended towards western Poland. The western part was displaced more to the Southwest, following the shift in wind direction towards the end of the release. The tracer cloud did not have a regular shape like in the first experiment.
At many sites the tracer was detected at intervals rather than continuously. At a few stations there was a double peak in the concentrations.
The concentration analysis of the second experiment left many questions open. First of all, it was not possible to arrive anywhere near a tracer mass balance. Only 75 % of the released tracer could be accounted for. Moreover, the cloud was split up into various fractions. Even more surprising was that a second peak in the measured concentrations showed up at some stations not far from the source after approximately 60 hours.
With these peculiarities in the measured concentration field it was not unexpected that the diffusion models had difficulties to reproduce the measured concentrations.

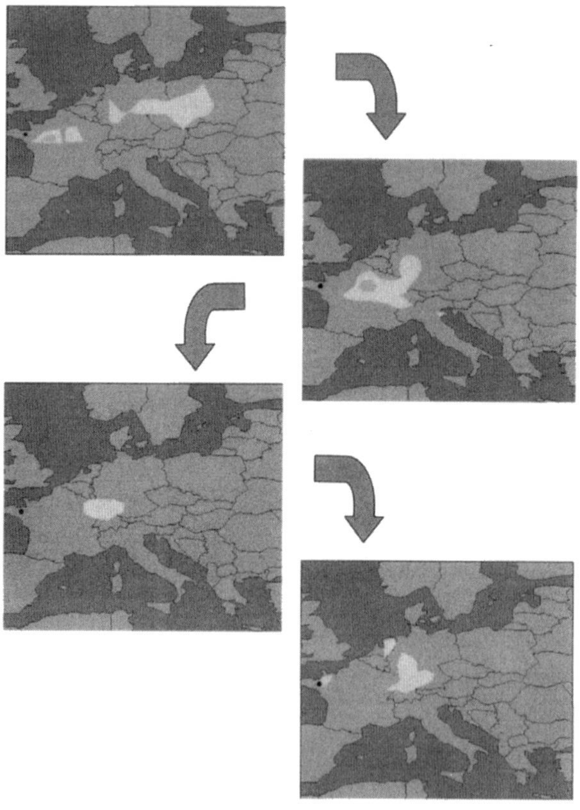

Fig.7: Position of measured tracer cloud during experiment # 2 after 24, 36, 48 and 60 hours

5.3 Statistical Results

Due to the limited number of non-zero measurements , the statistical comparison was limited to a few parameters. The presence of the ground level tracer cloud over the sampled area was limited to the first 24 hours after the beginning of the release. Even for that period none of the participating models was able to simulate correctly the ground level concentrations. The general trend was that the models greatly over-predicted the measured concentrations. After 24 hours the simulation of all models deteriorated.

6.0 The ATMES-II Exercise

The ETEX modelling was performed simultaneously with the experiment. The capability of providing these predictions in real time was considered to be an essential factor, as well as the performance of the model itself. It is obvious that only these those modellers could participate who had access to the results of a meteorological prediction model.

Almost 2 years after the experiment the ATMES-II modelling exercise - similar to the first ATMES project (see above) - was launched. Participants were asked to simulate the concentration fields of the **first** ETEX experiment using ECMWF **analysed** meteorological data as input to their own dispersion model. Here the differences between measured and predicted concentration fields could be more directed to the dispersion calculations since the same meteorological input was used. However, the different methods of using the meteorological information in the model might create differences.

A general, substantial improvement in the models' performance was observed for the common statistical parameters for those models which took part in the real-time exercise and the ATMES-II evaluation. This can be explained by the better resolution of the meteorological models, the availability of the tracer data that allowed participants to "tune" their models in light of the available data and also that in the two years between the real-time experiment and the ATMES-II project improvements in the modelling capacities took place.

The details of the ATMES-II results can be found in Mosca et al., 1998.

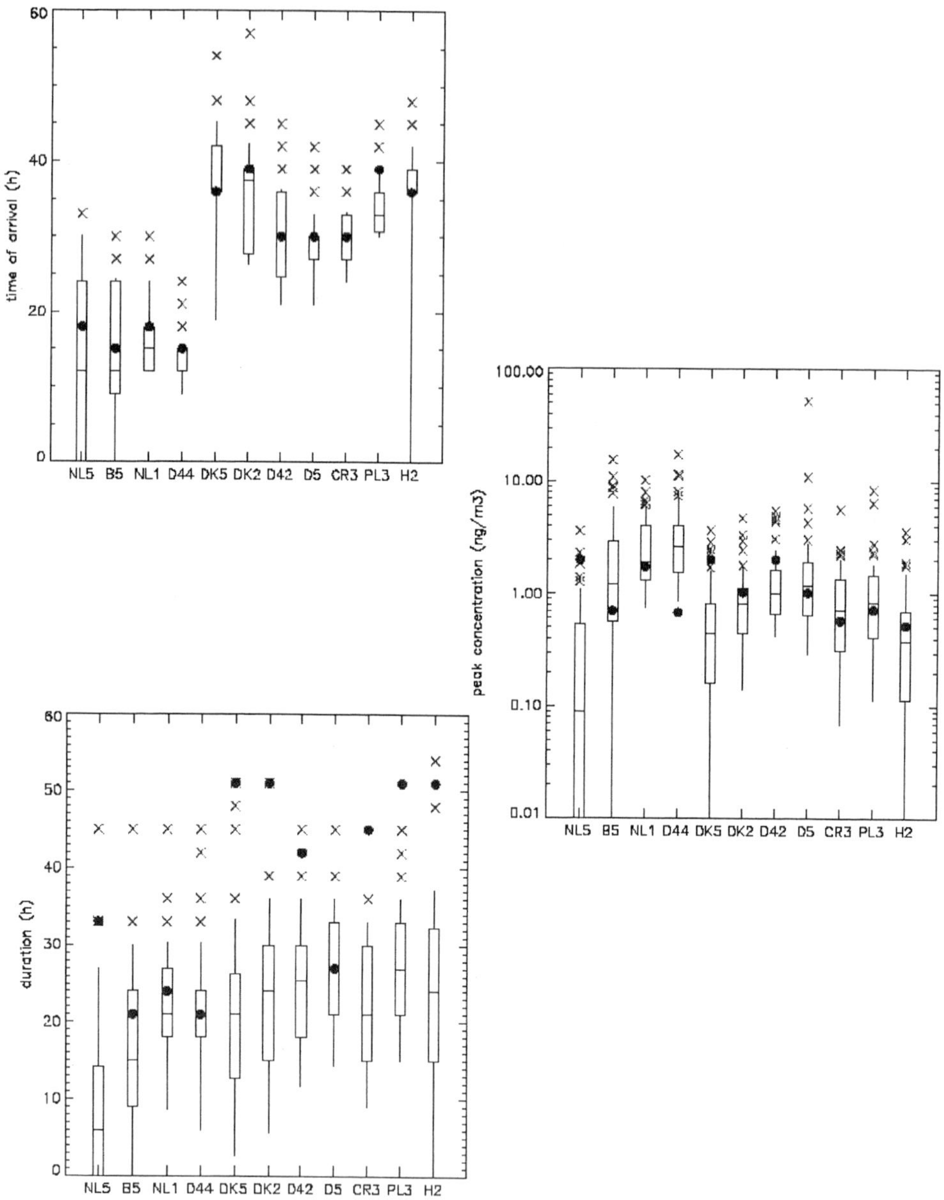

Fig. 8: Box plots of peak concentrations (ng/m3), duration (h) and time of arrival (h) with ATMES-II

7.0 What did we learn?

Before I come to this decisive chapter I want to define what is meant by "we": it means first of all Giovanni Graziani (JRC-Ispra), Stefano Galmarini (JRC-Ispra), Sonia Mosca (Enviroware) and myself. We assume that most of the modellers and the Steering Committee of ETEX will share our views, but there might also be differences.

- We learnt and showed that it is possible - after careful planning - to conduct a continent scale tracer experiment across Europe using the PFC tracer technique.
- We learnt that a large number of scientists and technicians from many countries across Europe can be brought together for one large experiment.
- We learnt that there is a large - more than we expected - number of institutes (than we expected) which can predict the dispersion of material ejected into the atmosphere in real-time and in a short time span.
- We learnt that the quality of the predictions can be described by differences between simulated and measured values of 3 to 6 hours in arrival time of the cloud at a certain location and by a factor of 3 in the maximum surface concentrations. These values can achieved by the "best" models under favourable conditions. It is not quite clear under which circumstances these models supply values which are less satisfying (see Release 2).
- Furthermore, we learnt that micro- and/or mesometeorological parameters in the surroundings of the source might influence the behaviour of the cloud even at long distances.
- We learnt that the model results differ considerably as can be seen by the box-plots graphs. This shows that not all modellers have the same prediction capability as others. This can easily be improved by systematic investigations on **why** some models behave "better" than others.
- We learnt that the way in which the meteorological information is used for dispersion purposes is still a topic of investigation for the future.
- We learnt that the prediction of transport and diffusion of atmospheric admixtures is still in its initial phase and should be improved by intensive studies in that area.

My hope is that the next generation of modellers will have interest and means enough to work in this fascinating field of meteorology and will improve our understanding of the atmosphere.

References

Graziani G., Klug, W., Mosca, S.,1998. *Real Time Long-Rang dispersion Model Evaluation of the ETEX First release*. EUR 17754, Luxembourg: Office for official publications of the European Communities.

Klug W., Graziani, G., Grippa, G., Pierce, D. and Tassone, C., 1992. *Evaluation of Long Range Atmospheric Models using Environmental Radioactivity Data from the Chernobyl Accident.* Elsevier Science Publishers, Barking, England.

Mosca S., R. Bianconi, R. Bellasio, G. Graziani, W. Klug, 1998. *ATMES-II -Evaluation of Long-range Dispersion Models using Data of the 1st ETEX Release.* EUR 17756 EN, Joint Research Centre , European Commission.

DISCUSSION

E. GENIKHOVICH: Do you not think that the problems with performance of the models in ETEX are particularly due to the fact that they had to predict an individual event rather than statistically stable characteristics? Did you consider the use of ensemble forecasts?

W. KLUG: Yes, the performance of the models would have been different if we attempted to make ensemble forecasts. For this reason the Ispra Research Center is planning a research project with the name "ENSEMBLE", which explains its purpose.

B. FISHER: Can you tell me the height of these ground-level releases and which local factors might influence the dispersion of a cloud released over a period of 12 hours?

W. KLUG: The release height was 4 m above ground. The gases were released from the top of the chimney at a temperature of 70-90 °C. The possible local factors which might influence the dispersion of the cloud released over a period of 12 hours are - apart from changing wind direction and speed with time - vertical wind shear and stability.

A. VENKATRAM:: Were there any models that were consistently "bad"? If so, what were the reasons for their poor performance?

W. KLUG: There were only 3 sets of data to evaluate : the two experiments plus the non-real-time model results of experiment # 1. Therefore, no general conclusion could be drawn. However, we looked into the question whether Lagrangian or Eulerian Models gave better results. No significant differences were found. It was left to the individual modeller to investigate the particular behaviour of their own model.

S. HANNA: If we were at a specific location in Europe and there was another accidental release of radioactive material similar to Chernobyl, what do the results of ETEX tell me about

1. the expected accuracy of the model predictions of cloud arrival time, dosages and deposition and
2. which of the 20+ models are likely to be the most reliable?

W. KLUG: Your question is difficult to answer: first of all the greatest uncertainty in case of an accident are the source characteristics. When we tried to simulate the Chernobyl cloud these data were very uncertain even after the event. For the first release most of the models showed differences of 3-6 hours in cloud arrival time and a factor of 3 in ground-level maximum concentrations. Deposition was not considered because the tracer was inert. The second experiment showed larger discrepancies between model results and observations. The reply to your second question is that from these two experiments under specific meteorological conditions a basis for statements on reliability is not given.

ADJOINT IMPLEMENTATION OF ROSENBROCK METHODS APPLIED TO VARIATIONAL DATA ASSIMILATION

Dacian Daescu,[1] Gregory R. Carmichael,[1] and Adrian Sandu[2]

[1]Center for Global and Regional Environmental Research
The University of Iowa
Iowa City, IA 52242
[2]Department of Computer Science
Michigan Technological University

INTRODUCTION

In the past decade the variational method has been successfully applied in data assimilation problems for atmospheric chemistry models [5, 6]. In 4D-var data assimilation a cost function is defined as the distance between model predictions and observations over the assimilation window. A minimization algorithm is then used to find the set of control variables which minimizes the cost function. Typical choices for the control variables are the boundary values, initial state of the model, emissions and deposition rates. Using the adjoint method, the gradient of the cost function can be computed fast, at the expense of a few additional function evaluations, making the optimization process very efficient.

We consider an atmospheric transport-chemistry model given by the system of differential equations:

$$\frac{\partial}{\partial t}c_i = -\nabla(\mathbf{u}c_i) + \mathrm{div}(K\nabla c_i) + f_i(t, x, c_1, \ldots, c_S) + E_i(t, x),\ i = 1 \ldots S. \tag{1.1}$$

on the bounded spatial domain $x \in \Omega \subset R^3$ and the time interval $t \in [t_0, T]$. The initial condition is $c(t_0, x) = c_0(x)$ and appropriate boundary values are prescribed. The solution $c(t, x) \in R^S$ of problem (1.1) represents the concentration vector of the chemical species in the model, where $\mathbf{u}(t, x)$ is the wind field, $K(t, x)$ is the diffusion tensor, and $E_i(t, x)$ represents the source and sink terms. The chemical reactions are modeled by the nonlinear stiff terms $f_i(t, x, c)$ which introduce coupling among species. Next we formulate the 4D-var data assimilation problem associated with (1.1) with the set of control variables given by the initial state of the model, c_0.
If space discretization is applied to problem (1.1) on a grid (N_x, N_y, N_z), the resulting ODE system of dimension $N = S \times N_x \times N_y \times N_z$ is:

Air Pollution Modeling and Its Application XIV, Edited by
Gryning and Schiermeier, Kluwer Academic/Plenum Publishers, New York, 2001

$$\begin{cases} \frac{dc}{dt} = F_A(c) + F_D(c) + F_R(c) \\ c(t_0) = c_0 \end{cases} \tag{1.2}$$

where F_A represents the advection and horizontal diffusion, F_D is the vertical diffusion, and the reaction and the source/sink terms are included in F_R. Under suitable assumptions, problem (1.1) has an unique solution and we can view this solution as a function of the initial conditions, $c = c(t, x, c_0)$. We assume that a previous analysis provides a "background estimate" c_b of c_0 with the error covariance matrix $\mathbf{B}$, and measurements $M_k, k = 1, m$ of the concentrations at moments t_k are scattered over the interval $[t_0, T]$. The errors in measurements and model representativeness are given by the covariance matrices $\mathbf{R}_k, k = 1, m$. In practice the matrices $\mathbf{B}$ and $\mathbf{R}_k$ are taken diagonal which corresponds to the assumption that background errors are uncorrelated and measurement errors are uncorrelated in space and time. This assumption is made in our next analysis. The 4D-var data assimilation finds an initial state c_0 that minimizes the distance between the model predictions and observations expressed by the cost function:

$$\mathcal{F}(c_0) = \frac{1}{2}(c_0 - c_b)^T \mathbf{B}^{-1}(c_0 - c_b) + \frac{1}{2}\sum_{k=1}^{m}(c_k - M_k)^T \mathbf{R}_k^{-1}(c_k - M_k) \tag{1.3}$$

The quality of the assimilation results is determined by various factors: availability and spatial-temporal distribution of measurements, accuracy of the background estimate, assimilation window, errors in measurements and model representativeness. Such an analysis is beyond the goal of this paper, and details can be found in [5, 11] and the references therein.

Most of the powerful optimization techniques require the evaluation of the gradient $\nabla_{c_0}\mathcal{F}$ of the cost function. In a comprehensive atmospheric chemistry model the dimension of the vector c_0 in (1.2) can easily be of order 10^6 which make the optimization a very expensive computational process.

In the variational approach one computes the gradient of the functional $\mathcal{F}$ by using the "adjoint method". The theory of adjoint equations for nonlinear problems is described in detail in [12] and the derivation of the adjoint model for the continuous and discrete cases are given in [6, 15].

Below we outline the basic ideas. The gradient of the cost function is:

$$\nabla_{c_0}\mathcal{F}(c_0) = \mathbf{B}^{-1}(c_0 - c_b) + \sum_{k=1}^{m}\left(\frac{\partial c_k}{\partial c_0}\right)^T \mathbf{R}_k^{-1}(c_k - M_k) \tag{1.4}$$

which can be computed with the following algorithm:

Step1. Initialize $gradient = 0$

Step2. $for\ k = m, 1, -1\ do$

$$gradient = \left(\frac{\partial c_k}{\partial c_{k-1}}\right)^T \left[\mathbf{R}_k^{-1}(c_k - M_k) + gradient\right]$$

Step3. $gradient = \mathbf{B}^{-1}(c_0 - c_b) + gradient$

The main advantage of the adjoint method is that explicit computation of the Jacobian matrices $\frac{\partial c_k}{\partial c_{k-1}}$ is avoided and the matrix-vector products can be computed directly at Step 2. For the theory and actual implementation of the adjoint computations the reader should consult [8] and the references therein. The algorithm described above requires the values of c_k in reverse order such that these values need to be stored from a previous run or recomputed. Moreover, in practice the measurements are usually sparse and the value of c_k is obtained from c_{k-1} with a sequence of steps $c_{k-1} \to c_k^1 \to \ldots \to c_k^s \to c_k$. The computational trade-off is then between allocating large amounts of memory to store the states of the system during the forward run, or frequent recomputations that increase the running time of the code. If an explicit numerical method is used to solve the stiff chemistry part of problem (1.2) then the "trajectory" from c_{k-1} to c_k may become very long, increasing the cost of the adjoint code. On another hand, if an implicit method is used then the adjoint computations may become complicated. Ideally one would like a method capable of taking large stepsize and an efficient adjoint implementation.

OPERATOR SPLITTING

A popular way to solve problem (1.2) is to use operator splitting, which has the advantage that different processes can be treated with different numerical methods. In a second order accurate Strang splitting approach with the time step $h = t_{n+1} - t_n$, the solution c_{n+1} is obtained from c_n as follows:

$$c(t_{n+1}) = \overline{F}_A\left(t_{n+\frac{1}{2}}, \frac{h}{2}\right)\overline{F}_D\left(t_{n+\frac{1}{2}}, \frac{h}{2}\right)\overline{F}_R(t_n, h)\overline{F}_D\left(t_n, \frac{h}{2}\right)\overline{F}_A\left(t_n, \frac{h}{2}\right)c(t_n) \quad (2.1)$$

where the operators $\overline{F}$ represent the numerical method used to solve each process. If $\overline{J}$ denotes the $N \times N$ Jacobian matrix associated to $\overline{F}$, the adjoint algorithm to compute the gradient (1.4) of the cost function requires products of the form $\overline{J}^T u$, with u an arbitrary seed vector. Since for large systems constructing the adjoint code by hand can be a frustrating process, automatic tools have been developed [8]. Automatic implementation allows also for flexibility, such that if the model is modified minimal user intervention is required.

Usually $\overline{F}_A$ is defined by an explicit method and may be nonlinear (if a flux-limiter is applied for positivity), $\overline{F}_D$ is linear, defined by a (semi-) implicit method. The products $\overline{J}_A^T u$ and $\overline{J}_D^T u$ can be then efficiently computed using an automatic adjoint compiler. The operator $\overline{F}_R$ is highly nonlinear, given by a stiff numerical method and the implementation of $\overline{J}_R^T u$ needs special consideration. Since the adjoint method requires several integrations of the direct model, the storage of (part of) the forward trajectory and the $(jacobian)^T \cdot vector$ products, the performance of the adjoint model is dominated by the implementation of the direct and adjoint method used in the chemistry integration which takes in practice as much as 90% of the CPU time. *Fisher and Lary*[6] show the adjoint computations for the adaptive-timestep *Bulirsch-Stoer* method, and *Elbern and Schmidt*[5] use the adjoint model for the QSSA scheme.

If the sparse structure of the models is carefully exploited Rosenbrock methods have been shown to be reliable chemistry solvers [13, 14] due to their outstanding stability properties and conservation of the linear invariants of the system. In the next sections we present the adjoint computations for a general 2-stage Rosenbrock method and an efficient implementation which is suitable for automatization. The L-stable method $ROS2$ we obtain as a particular case was applied for the chemistry integration in the forward 3D model LOTOS [2] for various types of operator splitting and as a W-method. Extension to a general s-stage method [9] is straightforward and the most popular methods used in practice are for $s \leq 4$.

i) Derivation of the Adjoint Formulas. We consider now the problem

$$\begin{cases} \frac{dc}{dt} = f(c) \\ c(t_0) = c_0 \end{cases} \tag{3.1}$$

with $c(t), c_0 \in R^n$ and $f : R^n \to R^n$, $f = (f_1, f_2, \dots f_n)^T$.
One step from t_0 to t_1 with $h = t_1 - t_0$ of a 2-stage Rosenbrock method as presented in [9] reads:

$$(\frac{1}{\gamma_{11}h}I - J_0)k_1 = f(c_0) \tag{3.2}$$

$$(\frac{1}{\gamma_{22}h}I - J_0)k_2 = f(c_0 + \alpha k_1) + \frac{\beta}{h}k_1 \tag{3.3}$$

$$c_1 = c_0 + m_1 k_1 + m_2 k_2 \tag{3.4}$$

where J_0 is the Jacobian matrix of f evaluated at c_0, $J_0 = (\frac{\partial f_i}{\partial c_j})_{ij}|_{c=c_0}$, I is the identity matrix, and the coefficients $\gamma_{11}, \gamma_{22}, \alpha, \beta, m_1, m_2$ are chosen to obtain a desired order of consistency and numerical stability. Since of special interest are the methods that require only one LU decomposition of $\frac{1}{\gamma_{ii}h}I - J_0$ per step, we consider the case when $\gamma_{11} = \gamma_{22} = \gamma$.
For the adjoint computations we have from (3.2), (3.3):

$$\left(\frac{\partial k_1}{\partial c_0}\right)^T = \left(J_0^T + (\frac{\partial J_0}{\partial c_0} \times k_1)^T\right)\left((\frac{1}{\gamma h}I - J_0)^T\right)^{-1} \tag{3.5}$$

$$\left(\frac{\partial k_2}{\partial c_0}\right)^T = \left((I + \alpha\frac{\partial k_1}{\partial c_0})^T J_1^T + \frac{\beta}{h}(\frac{\partial k_1}{\partial c_0})^T + \left(\frac{\partial J_0}{\partial c_0} \times k_2\right)^T\right)\left((\frac{1}{\gamma h}I - J_0)^T\right)^{-1}$$

where J_1 is the Jacobian evaluated at $c_0 + \alpha k_1$, and the terms $(\frac{\partial J_0}{\partial c_0} \times k_i)$, $i = 1, 2$ are $n \times n$ matrices whose j column is $(\frac{\partial J_0}{\partial c_0^j})k_i$, $i = 1, 2$. We want to stress here the fact that these matrices are not symmetric and we will return to the computation of these terms later. Using (3.4), for an arbitrary seed vector $u \in R^n$ we have:

$$\left(\frac{\partial c_1}{\partial c_0}\right)^T u = u + m_1\left(J_0^T + (\frac{\partial J_0}{\partial c_0} \times k_1)^T\right)\left((\frac{1}{\gamma h}I - J_0)^T\right)^{-1} u +$$
$$+ m_2\left(\left(I + \alpha(\frac{\partial k_1}{\partial c_0})^T\right)J_1^T + \frac{\beta}{h}(\frac{\partial k_1}{\partial c_0})^T + \left(\frac{\partial J_0}{\partial c_0} \times k_2\right)^T\right)\left((\frac{1}{\gamma h}I - J_0)^T\right)^{-1} u$$

Step 1. Solve for v the linear system $(\frac{1}{\gamma h}I - J_0)^T v = u$. Then,

$$\left(\frac{\partial c_1}{\partial c_0}\right)^T u = u + m_1\left(J_0^T + (\frac{\partial J_0}{\partial c_0} \times k_1)^T\right)v + m_2 J_1^T v + m_2(\frac{\partial k_1}{\partial c_0})^T(\alpha J_1^T + \frac{\beta}{h}I)v$$
$$+ m_2(\frac{\partial J_0}{\partial c_0} \times k_2)^T v \tag{3.6}$$

Step 2. Compute $\omega = J_1^T(m_2 v)$; $\omega_1 = \alpha\omega + \frac{m_2\beta}{h}v$.
Using (3.5) we get next:

Step 3. Solve for θ the linear system $(\frac{1}{\gamma h}I - J_0)^T\theta = \omega_1$.
After replacing in (3.6) and arranging the terms,
Step 4. Compute

$$\left(\frac{\partial c_1}{\partial c_0}\right)^T u = u + \omega + J_0^T(m_1 v + \theta) + (\frac{\partial J_0}{\partial c_0} \times k_1)^T(m_1 v + \theta) + m_2(\frac{\partial J_0}{\partial c_0} \times k_2)^T v \quad (3.7)$$

We now focus on the terms of the form $(\frac{\partial J_0}{\partial c_0} \times k)^T v$ whose evaluation dominate the computational cost of the algorithm given by *Step* 1-4. Here k , v $\in R^n$ are arbitrary constant vectors. For the i component we have:

$$\left((\frac{\partial J_0}{\partial c_0} \times k)^T v\right)_i = \left(\frac{\partial J_0}{\partial c_0^i} k\right)^T v = k^T \left(\frac{\partial (J_0^T v)}{\partial c_0^i}\right) = \left(\frac{\partial (J_0^T v)}{\partial c_0^i}\right)^T k \quad (3.8)$$

Consider now the function $H : R^n \to R^n$, $H(c_0) = J_0^T v$. Observe that the Jacobian matrix of H is symmetric, since for the (i,j) entry we have:

$$\frac{\partial H_i(c_0)}{\partial c_0^j} = \sum_{l=1}^{n} \left(\frac{\partial^2 f_l}{\partial c_0^i \partial c_0^j}\right) v_l = \sum_{l=1}^{n} H_{f_l}(i,j) v_l$$

where H_{f_l} is the Hessian matrix of the function $f_l : R^n \to R$.

Then $\frac{\partial H(c_0)}{\partial c_0} = \sum_{l=1}^{n} H_{f_l} v_l$, so $\frac{\partial H(c_0)}{\partial c_0}$ is symmetric. Using now (3.8) it results :

$$\left(\frac{\partial J_0}{\partial c_0} \times k\right)^T v = \left(\frac{\partial H(c_0)}{\partial c_0}\right) k \quad (3.9)$$

ii) Implementation of the Adjoint Code. The forward integration of problem (3.1) using implicit methods together with the performance analysis is given in [13, 14], proving that when the sparsity of the system is efficiently exploited Rosenbrock methods can be very efficient chemistry solvers. Implementation is done in the symbolic kinetic preprocessor KPP environment [4] which generates the sparse matrix factorization LU required in (3.2, 3.3) with a minimal fill-in [13] and the routine to backsolve the linear systems without indirect adressing. It is important to notice that the LU decomposition accounts for most of the CPU time of the code, and there is no need to repeat it during the backward adjoint integration.
One step of the adjoint code (from t_1 to t_0) for the chemistry integration requires a forward run from t_0 to t_1 given by the formulas (3.2-3.4) followed by the pure adjoint computations given by *Step*1-4. With the LU decomposition of $(\frac{1}{\gamma h}I - J_0)$ available from (3.2), *Step 1* reads : $U^T L^T v = u$. We extended the KPP abilities such that a loop free routine *tsolve* is generated for forward-backward solving this system in sparse format avoiding indirect addressing. The computational cost of *Step* 1,3 can be then compared with the corresponding part from (3.2) and (3.3) .
Step 2 requires evaluation of the product $J_1^T v$ which is automatically generated using sparse multiplications by a new routine *jactrvect.* This introduces some extra work (J_1 is evaluated at $c_0 + \alpha k_1$), but its cost is relatively cheap. The efficiency of the adjoint code is then dominated by the implementation of *Step* 4, given by the formula (3.7).
Using (3.9), we can rewrite (3.7) as:

$$\left(\frac{\partial c_1}{\partial c_0}\right)^T u = u + \omega + H_1(c_0) + \left(\frac{\partial H_1(c_0)}{\partial c_0}\right) k_1 + m_2 \left(\frac{\partial H_2(c_0)}{\partial c_0}\right) k_2 \quad (3.10)$$

with $H_1, H_2 : R^n \to R^n$, $H_1(c_0) = J_0^T(m_1 v + \theta)$, $H_2(c_0) = J_0^T v$.
In (3.10) we have then to compute the *jacobian·vector* products for the functions H_1, H_2 which can be done by *forward* automatic differentiation [1, 8] of the functions generated via routine *jactrvect*. The cost is then 2-3 times the cost of evaluating $H_1(c_0), H_2(c_0)$ and remains low due to the sparse structure of J_0. Automatic differentiation for H_1 provides also the value $H_1(c_0)$, so there is no need to compute it separately. In addition, these computations are independent allowing parallel implementation.

PERFORMANCE OF THE ADJOINT MODEL

The implementation presented in the previous section has the benefit that the adjoint part of the chemistry integration is generated completely automatically, taking full advantage of the sparsity of the system. This allows the user to easily move from one model to another and makes it very attractive compared with the hand written codes which construction for large models can be a difficult process. Moreover since symbolic computations are used, rounding errors are avoided and the accuracy of the results goes up to the machine precision. Implementation in the KPP context has also the advantage that avoids the introduction of auxiliary adjoint variables (routines *tsolve, jactrvect*) which has direct impact on the performance of the code both in terms of memory usage and CPU time. As a particular case we consider the 2nd order 2-stage Rosenbrock method $ROS2$ which is obtained from (3.2-3.4) by taking $\alpha = \frac{1}{\gamma}$, $\beta = -\frac{2}{\gamma}$, $m_1 = \frac{3}{2\gamma}$, $m_2 = \frac{1}{2\gamma}$. Choosing $\gamma = 1 \pm 1/\sqrt{2}$ the method is L-stable and the numerical experiments presented in this section were performed with $\gamma = 1 + 1/\sqrt{2}$.

In order to test the performance of the implementation we consider a 1-D horizontal model associated to (1.1). The chemistry part is based on the Carbon Bond Mechanism IV (CBM-IV, [7]) with 32 chemical species involved in 70 thermal and 11 photolytic reactions. The wind field and the diffusion coefficient are taken constant, $\mathbf{u} = 10$ km/hour (left-to-right), $K = 10^{-3}\text{Km}^2/\text{sec}$. Second order Strang splitting is applied

$$c(t_{n+1}) = \overline{F}_A\left(t_{n+\frac{1}{2}}, \frac{h}{2}\right)\overline{F}_R(t_n, h)\overline{F}_A\left(t_n, \frac{h}{2}\right)c(t_n)$$

with a splitting interval $t_{n+1} - t_n = 15$ min. The advection operator is discretized using a limited $k = 1/3$ upwind flux interpolation as presented in [10], and the diffusion operator using central differences formula. Together they define F_A. The numerical method defining $\overline{F}_A$ is the explicit trapezoidal rule. Dirichlet conditions are prescribed at the left boundary (x=0) and at the right boundary we consider $\frac{\partial c}{\partial n} = 0$. With the spatial domain [0,500] Km and a uniform grid $\Delta x = 5$ Km, the dimension of the corresponding (1.1) problem is 3200. A highly polluted region is considered between $100 - 200$ Km, with initial concentrations and emissions as for the urban scenario and for the rest of the domain rural concentrations and emissions are provided [14]. Interpolation is done between the center (150 Km) and the urban limits. In order to allow the system to equilibrate, box models (chemistry only) are integrated for one day over the whole grid. The results are the "true" initial conditions. "Measurements" are then generated every 15 min. by a 6 hours transport-chemistry run.

The data assimilation problem is set using the "twin experiments " method, with the background term dropped and the logarithmic form of (1.3). Taking the logarithm of the concentrations has the advantage that the positivity constraint is eliminated and scales the system. The minimization routine used is the Quasi-Newton limited memory L-BFGS algorithm [3], anticipating extension to large scale models.

For consistency with the implementation of the adjoint code for large-scale models where the storage of the entire forward trajectory is not a realistic option, we adopted a checkpointing strategy. First a full forward run is used to store the states of the model after each operator splitting interval. Second a forward run stores the states within a splitting interval (1+ all chemistry steps). This requires a full chemistry integration and a half transport integration (last stage is not needed). Third, a forward-backward integration is performed, where the forward part provides the intermediate states within a step. This requires simplified transport and chemistry computations. The adjoint code for the advection- difussion equations is automatically generated by the adjoint model compiler TAMC [8], while the adjoint code for the chemistry follows the algorithm given by *Step* 1-4, using (3.10).

For the assimilation problem, perturbations of the form $c_0^p(x) = c_0(x) * (1 + \alpha r(x))$ were introduced for 10 species in the model, where $r(x)$ are random numbers in the interval (0,1). The perturbed species and the corresponding value of α are presented in Table 1. Since observations can not be provided for all species, one of the goals of the data assimilation is to improve the analysis of unobserved species (like radicals) using mesurements for only few species. Assimilation starts at sunrise over a six hour interval (4:30-10:30 LT) and "mesurements" are provided *only* for ozone and NO2 every 15 min. Optimization stops when $\mathcal{F}_{opt}/\mathcal{F}_{init} < 10^{-4}$ or after 100 iterations. Figure 1 shows the configuration of the reference and perturbed initial and final concentrations for O3.

Two experiments are presented: in Run 1 measurements are at all grid points, in Run 2 measurements are 10 grid points apart (every 50 Km). The performance of the adjoint code and the optimization process is outlined in Table 1. It can be seen that the average ratio between the CPU time required to compute the gradient (and the cost function value) and the CPU time of a forward run is less than four, which makes our implementation very efficient. The results of the assimilation for O3 in Run 1 are shown in Figure 1 with solid dots. The quality of the assimilation is evaluated by the relative reduction in a monitor function $\mathcal{F}^i$ defined for each of the species in a similar way to the cost function, and which is updated every 15 min., *at all grid points.* The relative reduction $\mathcal{F}^i_{opt}/\mathcal{F}^i_{init}$ for 12 species is presented in Table 1. Success can be claimed for all presented species in both experiments, but it can be seen that sparsity of data can drastically affect the results of the analysis.

Table 1. Performance of the adjoint code and optimization process [1]

Run	No. iter.	~time/iter (sec)	$\sim$ t($\mathcal{F} + \nabla\mathcal{F}$)/ t($\mathcal{F}$)	$\mathcal{F}_{opt}/\mathcal{F}_{init}$
1.	43	7.5	3.8	9.9e-5
2.	23	7.9	3.8	9.8e-5

Species	α	$\mathcal{F}^i_{opt}/\mathcal{F}^i_{init}$ Run 1	$\mathcal{F}^i_{opt}/\mathcal{F}^i_{init}$ Run 2	Species	α	$\mathcal{F}^i_{opt}/\mathcal{F}^i_{init}$ Run 1	$\mathcal{F}^i_{opt}/\mathcal{F}^i_{init}$ Run 2
O	0.0	9.2e-5	6.4e-3	HO2	1.0	4.9e-1	6.3e-1
O3	0.3	1.4e-4	2.5e-3	N2O5	2.0	2.8e-3	5.0e-2
OH	2.0	9.9e-3	4.7e-2	C2O3	0.0	1.5e-2	1.6e-1
NO	2.0	1.8e-3	1.1e-1	HNO3	0.3	4.6e-1	5.0e-1
NO2	1.0	9.5e-5	7.8e-2	HONO	1.0	2.5e-1	2.9e-1
NO3	1.0	6.8e-3	3.3e-2	HCHO	1.0	6.6e-1	8.8e-1

[1] All the computations were done on a HP-UX B.10.20 A 9000/778 machine. The time to read-write data to files is not considered.

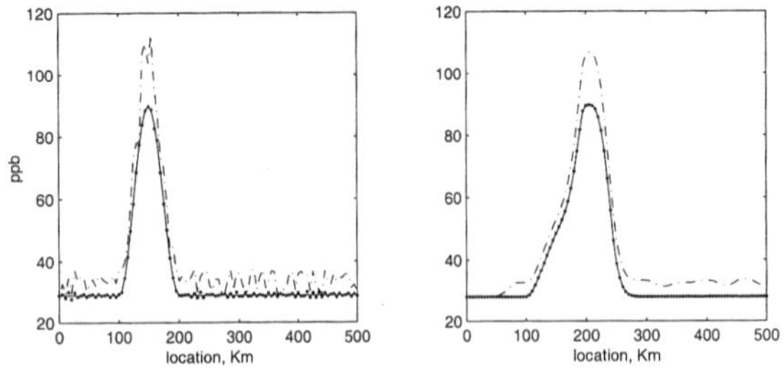

Figure 1. Distribution of O3 concentrations.Left- initial (LT= 4:30), right- final (LT= 10:30). Reference run with continuous line, perturbed run with dotted line, assimilation result for Run 1 with solid dots.

We introduced the adjoint computations and an efficient implementation of the 2-stage Rosenbrock methods which is suitable for automatization and parallel coding. The algorithm and the properties we described can easily be generalized to $s-$stage methods and it is of interest to analyse how this implementation can be extended to SDIRK and IRK methods [9]. Further work includes testing on comprehensive models, implementation in the context of W-transformation and different types of operator splitting as well as the possibility to use approximate gradients.

References

[1] Bischof, C., Carle, A., Khademi, P., Mauer, A.: The Adifor 2.0 system for the automatic differentiation of FORTRAN 77 programs. *Tech. Rep., Mathematics and Computer Science Division, Argonne National Laboratory, Argonne, Illinois*, 1994

[2] Blom, J.G., Verwer, J.G.: A comparison of integration methods for atmospheric transport-chemistry problems *CWI, Report MAS-R9910*, 1999

[3] Byrd, R., Liu, P., Nocedal, J., Zhu, C.: A limited memory algorithm for bound constrained optimization. *Tech. Rep. NAM-08, Northw. Univ., Evanston, Ill.*, 1994

[4] Damian-Iordache, V., Sandu, A.: KPP-A symbolic preprocessor for chemistry kinetics-User's guide. *Tech. Rep., Univ. of Iowa, Department of Mathematics*, 1995

[5] Elbern, H., Schmidt, H.: A four-dimensional variational chemistry data assimilation scheme for Eulerian chemistry transport modeling. *J. of Geophysical Research,* **104-D15**, 18583-18598, 1999

[6] Fisher, M., Lary, D.J.: Lagrangian four-dimensional variational data assimilation of chemical species. *Q.J.R. Meteorol. Soc.,* **121**, 1681-1704, 1995

[7] Gery, M.W., Whitten, G.Z., Killus, J.P., Dodge, M.C.: A photochemical kinetics mechanism for urban and regional scale computer modeling. *J. of Geophysical Research,* **94**, 12925-12956, 1989

[8] Giering, R.: Tangent linear and Adjoint Model Compiler, Users manual 1.2 . *"http : //puddle.mit.edu/ ~ ralf/tamc"*, 1997

[9] Hairer, E., Wanner, G.: Solving Ordinary Differential Equations II. Stiff and Differential-Algebraic Problems. *Springer-Verlag, Berlin*, 1991

[10] Hundsdorfer, D., Koren, B., Loon, M., Verwer, G.J.: A positive finite-difference advection scheme. *J. of Comput. Physics,* **117**, 35-46, 1995

[11] Khattatov, V.B. *et al.*: Assimilation of photochemically active species and a case analysis of UARS data. *J. of Geophysical Research,* **104-D15**, 18715-18737, 1999

[12] Marchuk, G.I., Agoshkov, I.V., Shutyaev, P.V.: Adjoint Equations and Perturbation Algorithms in Nonlinear Problems. *CRC Press*, 1996

[13] Sandu, A., Potra, F.A., Carmichael G.R., Damian, V.: Efficient implementation of fully implicit methods for atmospheric chemical kinetics. *J. of Comput. Physics,* **129**, 101-110, 1996

[14] Sandu, A., Verwer, J.G., Loon, M., Carmichael, G.R., Potra, A.F., Dabdub, D., Seinfeld, J.H.: Benchmarking stiff ODE solvers for atmospheric chemistry problems I: implicit vs explicit. *Atmos. Environ.,* **31**, 3151-3166, 1997

[15] Talagrand, O., Courtier, P.: Variational assimilation of meteorological observations with the adjoint of the vorticity equations. Part I. Theory. *Q.J.R. Meteorol. Soc.,* **113**, 1311-1328, 1987

DISCUSSION

B. E. A. FISHER: How will the adjoint approach, which has only been run on test cases, be adapted to real practical situations? This requires observations of a (large) number of primary and secondary chemical species. Have you considered how much observational data would be required for this?

D. DAESCU: Observational data is required not only in the adjoint approach, but in any type of data assimilation. In the past years the amount of data received from satellites and aircraft missions was increasing at a high rate. Nimbus 7 and UARS satellites provide a large set of data for chemical species such as O_3, NO_2, CH_4, H_2O, HNO_3, ... In comprehensive models the amount of data used may range from the MB to GB order. A detailed description may be found at the Data Assimilation Office, NASA Goddard Space Flight Center (DAO GSFC) web page: http://dao.gsfc.nasa.gov/

E. GENIKHOVICH: Because of nonlinearity of the governing equations, you should iterate your Jacobian. How many iterations did you do?

D. DAESCU: A close examination of the adjoint method shows that the construction of the adjoint code doesn't require an iterative process. An exact adjoint of the discrete forward code is constructed corresponding to the numerical method used in the forward integration.

THE DEVELOPMENT OF THE AUSTRALIAN AIR QUALITY FORECASTING SYSTEM: CURRENT STATUS

G. D. Hess[1], M. E. Cope[2,3], S. Lee[2], P. C. Manins[2], G. A. Mills[1], K. Puri[1], and K. Tory[1]

[1]Bureau of Meteorology Research Centre, Melbourne, Victoria, Australia
[2]CSIRO Atmospheric Research, Aspendale, Victoria, Australia
[3]CSIRO Energy Technology, North Ryde, NSW, Australia

INTRODUCTION

The Australian Air Quality Forecasting System (AAQFS) is being developed with funding from the Air Pollution in Major Cities Program (sponsored by Environment Australia). The project has the short-term objective of developing, validating and trialing an accurate, next-day (24-36 hour) numerical air quality forecasting system for a three-month demonstration period in Sydney, which includes the 2000 Olympics. Currently AAQFS forecasts are produced in both Melbourne and Sydney. After the Olympics, we hope to extend AAQFS to forecasting health- and visibility-related air quality metrics in the other major population centres of Australia. The principal partners of the project are the Bureau of Meteorology (BoM), CSIRO, Environment Protection Authority of Victoria (EPA-VIC) and the Environment Protection Authority of New South Wales (EPA-NSW).

The project has a number of specific goals: to provide the ability to generate 24-36 hour air quality forecasts twice per day (available 9 am and 3 pm); provide forecasts for a range of air pollutants including oxides of nitrogen (NO_x), ozone (O_3), sulfur dioxide (SO_2), benzene (C_6H_6), formaldehyde (CH_2O) and particulate matter (PM10 and PM2.5); provide forecasts at a resolution sufficient to consider suburban variations in air quality; and to provide the ability to generate simultaneous forecasts for a 'business-as-usual' emissions scenario and a 'green emissions' forecast. The latter scenario may correspond to minimal motor vehicle-usage and will be used to indicate the reduction in population exposure that could result from a concerted public response to a forecast of poor air quality for the next day.

In this paper, we provide a brief description of the AAQFS and present case studies of photochemical smog events in Melbourne and Sydney.

DESCRIPTION OF THE FORECAST SYSTEM

The AAQFS consists of five major components (Fig. 1): a numerical weather prediction (NWP) system, an emissions inventory module, a chemical transport module

(CTM) for air quality modelling, an evaluation module, and a data archiving and dissemination module (data package).

LAPS (Limited Area Prediction System), the NWP system, is a hydrostatic model with state-of-the-art numerics and physics packages, and has been used by BoM to generate operational meteorological forecasts since July 1996 (Puri et al. 1998). Meteorological forecasts will be provided at a horizontal resolution of 0.05° (LAPS05). Special attention will be paid to the resolution and treatment of surface processes in an effort to improve representation of local and meso-scale flows and boundary-layer growth. Accurate representation of these processes is crucial for realistic, high-resolution forecasting of air pollution dynamics.

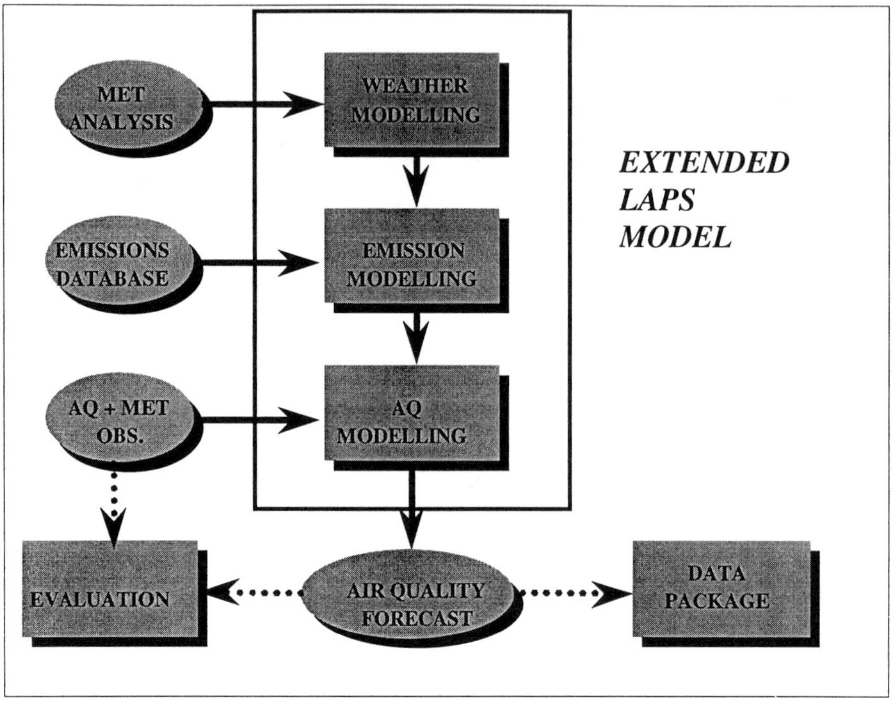

Figure 1. Flow diagram of the AAQFS. Note that modules located within the bounding-box labeled 'EXTENDED LAPS MODEL' correspond to components that will be run online during a meteorological forecast.

The aim is to extend LAPS to include the air quality modelling and the emissions modelling, in addition to the NWP component, by the time of the demonstration period. This will replace running the air quality model offline, and will give higher temporal and spatial resolution for the air quality transport and dispersion calculations. The online code is currently being implemented and tested.

EPA-VIC and CSIRO, with support from EPA-NSW, are undertaking emission inventory development. Presently emissions processing is undertaken offline with a resolution of 3-6 km with no allowance made for week/weekend or seasonal/local meteorological dependencies. The new system will use size-fractionated and speciated particle emissions, 0.01° gridded area sources over the densely populated regions and meteorologically dependant emissions that are generated online during LAPS operation.

A power-based vehicle emissions model, being developed at CSIRO, will be used to generate road-specific vehicle emission fluxes for the purpose of near-road impact modelling.

The Carnegie Mellon/California Institute of Technology (CIT) photochemical airshed model comprises the CTM (Harley et al. 1993; Cope et al. 1998, 1999). A notable modification is the implementation of the Generic Reaction Set (GRS) photochemical mechanism (Azzi et al. 1992), a highly condensed (7 species and 7 reactions) photochemical transformation mechanism featuring minimal computational overheads. The domain is divided into 10 non-uniform levels in the vertical (extending to 2000 m above ground level). The offline system has been used to generate 24-hour air quality forecasts (NO_x, O_3 and SO_2) using the 1100 UTC (2100 EST) LAPS05 forecasts. In the online system, the transport fields are updated at 5-10 minute intervals (60-minute intervals for the offline system). Note that the online CTM simulations will use a 0.05° outer grid, with nested 0.01° inner grids for major urban areas. Photochemical smog production will be simulated using an enhanced version of the GRS mechanism, which is under development, and particle transformation will be modelled using a modal-based particle scheme. A more comprehensive treatment of both processes will also be available in an offline version of the CTM.

Both the meteorological and air quality forecasts are the subject of on-going and case-specific validation. This is done through a comparison of LAPS meteorological fields with METAR/SYNOP (near-surface) and AMDAR (vertical profile) data and meteorological observations from the EPA monitoring networks. Air quality forecasts are compared to 1-hour EPA observations for NO_x (both as NO and NO_2) and O_3. This will be expanded by the time of the demonstration period to include SO_2, PM10, PM2.5, CO and, where available, non-methanic hydrocarbons. Critical to the validation process has been the availability of EPA data sets by the end of each forecast period, enabling the on-going validation to be substantially automated.

Data archiving is evolving from use of native system formats to unified NetCDF data packets (NetCDF is already used in the case of LAPS), which will be accessible via GUI-driven Q&A software. Sufficient information will be available in a data packet to enable the CTM to be run offline at a later time. The EPAs will have access to the daily forecasts via the AAQFS Web Site and will manage the dissemination of the forecast data.

PHOTOCHEMICAL SMOG EVENT IN SYDNEY

For Sydney where the coastline is essentially north-south (Fig. 2a), the primary meteorological task is to predict the onset and strength of the Tasman sea breeze whose penetration is often blocked or modified by a coastal range of mountains. In Fig. 3a we show a comparison of the forecast sea breeze with observations for a significant photochemical smog event in Sydney for the 1998/99 summer ozone season. The meteorological forecast was made using the latest version of LAPS with the Viterbo-Beljaars land-surface scheme and vertical diffusion, and direct calculation of the Monin-Obukhov surface fluxes (Viterbo and Beljaars 1995). In addition, a number of higher resolution modifications were made: topography was changed from 0.1° to 30-sec resolution; sea surface temperature from 1° (spatial resolution) and weekly to 0.25° and daily; soil texture and hydrological properties from constant values to 0.05° resolution; momentum roughness length and fractional vegetation coverage from 1° to 0.05° resolution. On the last day of this multi-day event the meteorological dynamics were complicated by the arrival of a wind surge up the coast, associated with a cold front known as a "Southerly Buster" (Fig. 3b). The arrival of these strong southerly synoptic-scale winds dominated over the mesocale sea-breeze circulation. The pollutants were concentrated at the head of the surge and transported to the northwest of Sydney.

Critical factors in the formation of the photochemical smog for the multi-day event were the presence of high ambient temperatures, low mixing depths and ventilation rates, and a low-level flow pattern, which sometimes resulted in the recirculation of the previous day's pollutants over Sydney.

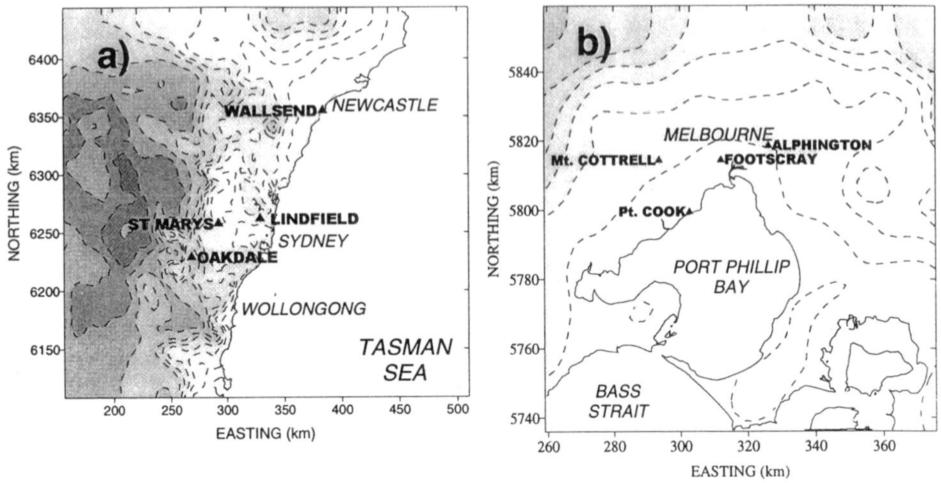

Figure 2. Map showing the topography and the location of the monitoring stations (in bold) sited in (a) the Sydney area and (b) the Melbourne area.

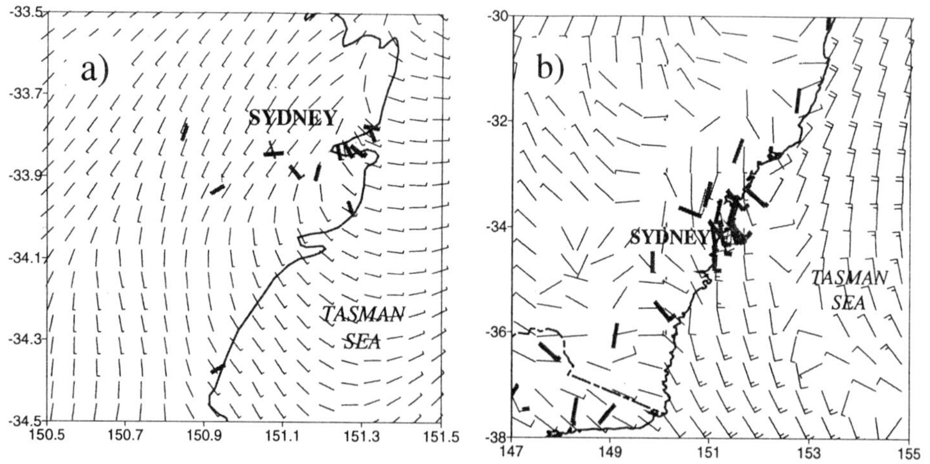

Figure 3. Comparison of the forecast 10-m wind field for Sydney with observations (in bold) for (a) 0100 UTC 13 December 1998; (b) Southerly Buster wind surge up the coast at 0500 UTC 13 December 1998 (note the difference in map scale; every 10th grid point shown.)

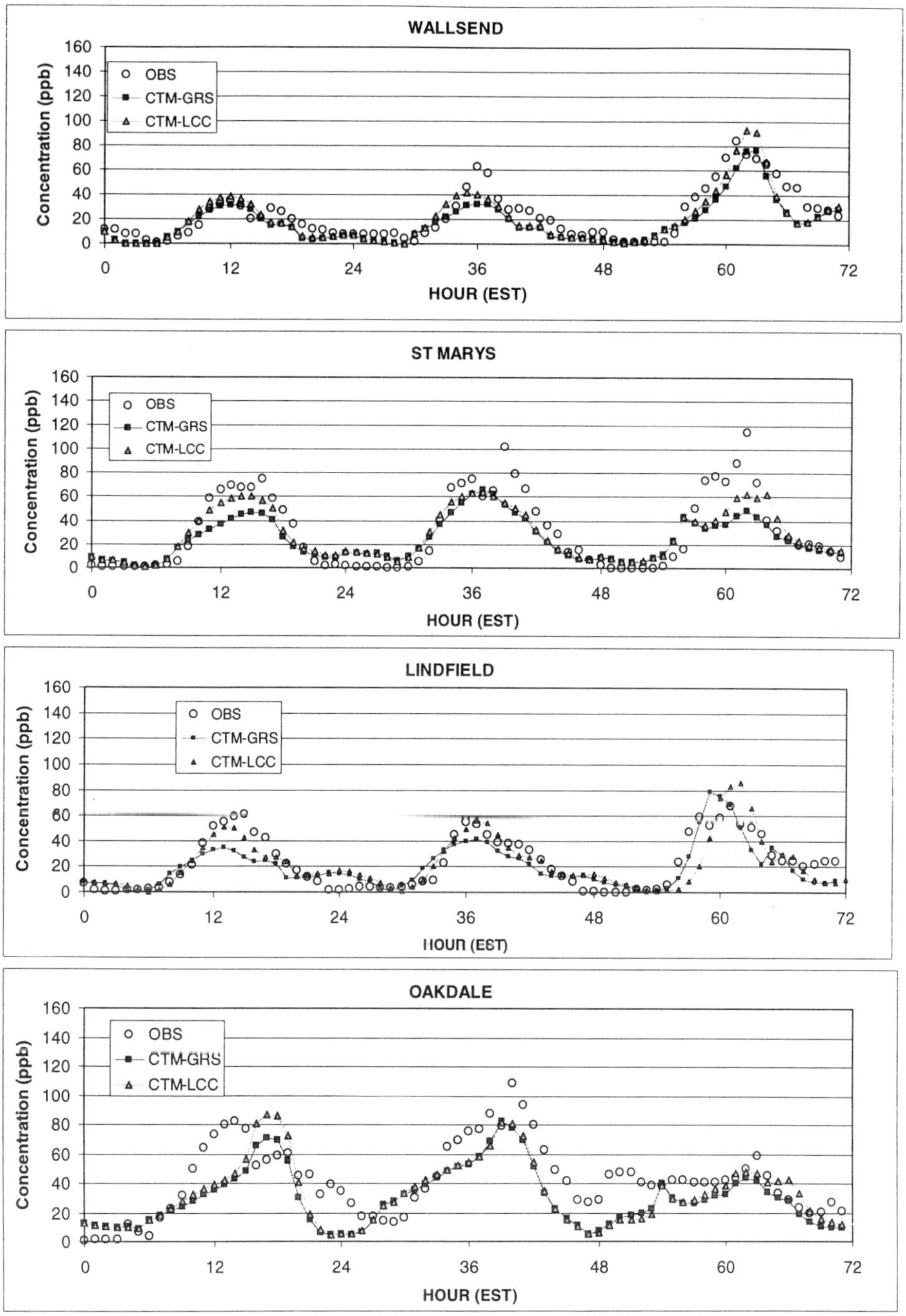

Figure 4. Comparison of predicted and observed 1-hour ozone concentration for multi-day photochemical smog event in Sydney, for the GRS and LCC (see text) photochemical mechanisms (hours 0-23 corresponds to 11 December 1998, hours 24-47 to 12 December 1998 and hours 48-71 to 13 December 1998).

The base-case ozone forecasts were generated by the offline forecasting system using the GRS photochemical mechanism (CTM-GRS). When the forecasts were initialised with horizontally homogenous pollutant fields, which are representative of continental clean-air conditions (a 'cold start'), the peak ozone concentrations for the period 9-14 December 1998 were markedly underpredicted. However, the forecasts for 11-13 December were considerably improved when the CTM was initialised using the forecast pollution field for the previous day (a 'warm start').

This is particularly apparent for 13 December 1998 where the forecast ozone concentration peak increased from 90 ppb to 124 ppb (compared to 127 ppb observed). Warm start forecasts were also generated by replacing the simple GRS mechanism with a comprehensive photochemical transformation mechanism (CTM-LCC; Lurmann, Carter, Coyner mechanism; Lurmann et al. 1987). The LCC mechanism (35 species, 106 reactions) is able to realistically simulate a wider range of conditions than GRS, however it requires considerably greater computational resources.

In Fig. 4 plots of observed and forecast (for both CTM-GRS and CTM-LCC) 1-hour ozone time series are shown for four of the EPA-NSW monitoring sites, representing different regions of the domain (see Fig. 2a). The agreement in reproducing the high peak in the Newcastle region (Wallsend) on 13 December is a very positive outcome because the peak appears to be caused by a combination of local photochemical production and transport of the Sydney plume within a meso-low. In the southwestern region (Oakdale) the system correctly predicts that the high peaks occur earlier (11-12 December) and are associated with the sea breeze. CTM-LCC, in general, performs better than CTM-GRS, which sometimes strongly underpredicts the ozone peak. The underprediction of both mechanisms on 13 December at St. Mary's is probably due to slight errors in determining the plume trajectory in the complex meteorological situation of that day and for flow over complex terrain.

PHOTOCHEMICAL SMOG EVENT IN MELBOURNE

The highest concentration of ozone in the Melbourne Airshed (see Fig. 2b) during the 1998/99 summer ozone season was observed on 10 December 1998. A maximum concentration of 122 ppb was observed at Mt Cottrell with 107 ppb observed at Point Cook (i.e. the maximum concentrations occurred on the western side of Port Phillip Bay). The meteorological situation this day was typical of many pollution events for Melbourne. There was an anticyclone located southeast of Melbourne in the Tasman Sea. During the 24-hour period the centre of the anticyclone drifted northeastward and then eastward. In the early morning, the near-surface winds over the Melbourne Airshed were light northerlies. However as the morning heating increased a Port Phillip Bay breeze developed, producing winds perpendicular to the bay shoreline and opposing the larger-scale northerly flow. By mid-afternoon the Bass Strait sea breeze with its southeasterly winds replaced the bay breeze. It is this delicate balance between opposing wind fields that must be accurately modelled; this includes accurately modelling the timing, direction and penetration of the bay breeze and the sea breeze.

In Fig. 5, we present wind field forecasts generated using the latest version of LAPS. A bay breeze develops in good agreement with the observations in direction and penetration. The timing of the onset (not shown) is also in good agreement. The point of maximum horizontal divergence is clearly visible. The sea breeze also shows good agreement with the observations.

The forecast air quality is illustrated in Fig. 6 where observed and forecast concentration time series of O_3 are shown for selected monitoring stations in the EPA-VIC network. The location of the monitoring stations is shown in Fig. 2b. A major O_3 peak occurs in mid-afternoon due to the photochemical transformation and recirculation of O_3 precursors in the bay breeze and this is seen in the model results. However, it is also clear that further improvements should be sought, given that peak ozone concentrations are still

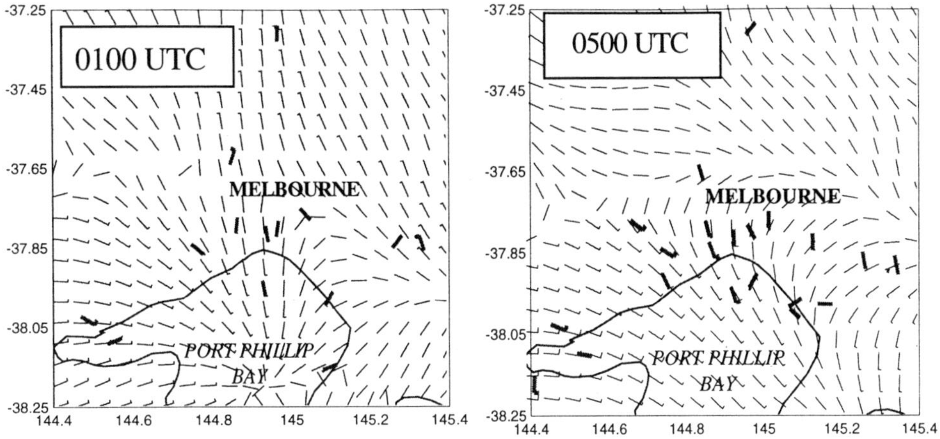

Figure 5. Comparison of 10-m forecast wind fields for Melbourne with observation (in bold). At 0100 UTC 10 December 1998 the flow pattern shows the bay breeze; at 0500 UTC the pattern shows the sea breeze.

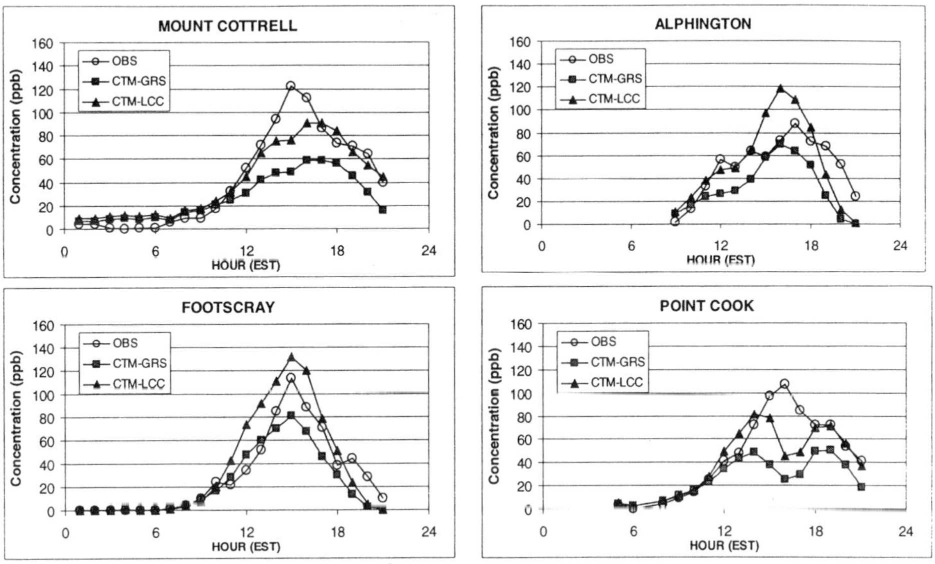

Figure 6. Comparison of predicted and observed 1-hour ozone concentration for the photochemical smog event in Melbourne, 10 December 1998 for GRS and LCC photochemical mechanisms.

underpredicted by the forecasting system at some stations on the western side of Port Phillip Bay and overpredicted at stations to the north and east sides of the Bay using the LCC mechanism. The predicted peak at Footscray (130 ppb) is close to the airshed maximum of 122 ppb found at Mt. Cottrell. The GRS mechanism systematically underpredicts the peak. A secondary peak is also observed in the early evening (and is shown in the model at Point Cook). This peak is associated with transport of pollution within the Bass Strait sea breeze.

There is also an observed NO_2 peak at hour 12 (not shown; see Hess et al. 1999), which is associated with bay-breeze activity and generally is well simulated. A second peak in NO_2 is observed near midnight local time. This peak is due to the recirculation of the aged air mass (with primary and secondary NO_2 concentrations), and possibly due to the titration of the O_3 plume by local emissions of NO. The model also forecasts a second peak, although the amplitude is too high and the phase is, in general, too early.

Given the demonstrated performance of the meteorological forecast, emphasis has now shifted to investigating the relative performance of the photochemical transformation mechanism, and the emissions inventory.

CONCLUSIONS

The AAQFS is routinely providing high spatial resolution air quality forecasts for guidance for the EPAs in Melbourne and Sydney. Case studies of photochemical smog events in Melbourne and Sydney have given encouraging agreement with observations. Meteorologically the two airsheds present different challenges: in Melbourne it is important to predict the onset and strength of the Port Phillip Bay breeze and the Bass Strait sea breeze; in Sydney is it important to predict the onset and strength of the Tasman sea breeze, the pollution plume trajectory for flow over complex terrain and the effects of local photochemical smog production and inter-regional transport. In the case studied for Sydney there was an additional complication of a synoptic-scale wind surge called the Southerly Buster. For both airsheds, the interaction between synoptic-scale forcing and mesoscale circulations can strongly influence the characteristics of an air pollution event and thus the meteorological model must be able to accurately simulate these interactions. In general, the LCC photochemical mechanism gave better predictions of the 1-hour ozone peak than the GRS mechanism. Improvements to the GRS mechanism and emissions inventory and online modelling of emissions and photochemistry are being developed and implemented. Work on the meteorological model to improve surface winds, soil moisture analysis and boundary-layer height also continues. We have yet to establish the limits of predictability of the system.

REFERENCES

Azzi, M., Johnson, G. J. and Cope, M., 1992, An introduction to the Generic Reaction Set photochemical smog mechanism, *Proc. 11th International Clean Air Conference*, Brisbane, 5-10 July 1992, 451-462.

Cope, M. E., Manins, P., Hess, D., Mills, G., Puri, K., Dewundege, P., Tilly, K. and Johnson, M., 1998, Development and application of a numerical air quality forecasting system, *Proc. 14th International Conf. Clear Air & Environment*, Melbourne, 18-22 October 1998, Clean Air Society of Australia and New Zealand, 353-358.

Cope, M., Hess, D., Lee, S., Azzi, M., Carras, J., Wong, N. and Young, M., 1999, Development of the Australian Air Quality Forecasting System: Current Status, *Proc. International Conf. Urban Climatology*, Sydney, 8-12 November 1999, in press.

Harley R.A., Russell A.G., McRae G.J., Cass G.R. & Seinfeld J.H., 1993, Photochemical modelling of the southern California air quality study, *Environ. Sci. Technol.*, **27**:378-388.

Hess, G. D., Cope, M. E., Lee, S. and Tory, K.,1999, LAPS and the Australian Air Quality Forecasting System, *Abstracts of Presentation at the Eleventh Annual BMRC Modelling Workshop*, 9-11 November 1999, BMRC Report No. 75, Bureau of Meteorology, Melbourne, 35-40.

Lurmann, F. W., Carter, W. P. and Coyner, L. A., 1987, A surrogate species chemical reaction mechanism for urban scale air quality simulation models, Final Report to US Environmental Protection Agency, Contract No. 68-02-4104.

Puri, K., Dietachmayer, G., Mills, G. A., Davidson, N. E., Bowen, R. A. and Logan, L. W., 1998, The new BMRC Limited Area Prediction System, LAPS, *Aust. Met. Mag.*, **47**: 203-233.

Viterbo, P. and Beljaars, A. C. M.,1995, An improved land surface parametrization scheme in the ECMWF model and its validation, *Tech. Report 75*, Research Department, ECMWF, Shinfield Park.

DISCUSSION

J. M. BALDASANO: Have you evaluated the influence of BVOC from the forests near to Sydney in the ozone production in the events considered.

M. COPE: We have run test case scenarios in which the biogenic emissions of isoprene have been set to zero. Peak ozone concentrations were reduced by 20-30%. Also, the spatial extent of the ozone footprint was reduced. It may be concluded that the modelled ozone formation is relatively sensitive to the emissions of BVOC.

J. M. BALDASANO: Have you also evaluated the influence of topography in the transport circulatory patterns, especially their relation with the sea breeze.

M. COPE: The Sydney airshed is bounded to the west by the Blue Mountains which peak at over 1000 m. According to the model, the mountains block the north-easterly sea breeze and cause the transported pollutants to remain in the airshed to be recirculated on the following day. This strongly enhances photochemical smog production on the second and third days of the event.

INVERSE MODELLING WITH A LAGRANGIAN PARTICLE DISPERSION MODEL: APPLICATION TO POINT RELEASES OVER LIMITED TIME INTERVALS

Petra Seibert

Institute of Meteorology und Physics
University of Agricultural Sciences Vienna
Türkenschanzstr. 18, A-1180 Wien, Austria
E-Mail: petra.seibert@boku.ac.at

INTRODUCTION

Inverse modelling in the context of air pollution modelling means to determine the sources of a substance emitted into the atmosphere from measurements of the ambient concentration (or deposition) of the substance, using a dispersion model which provides the relationship between the source and the ambient concentrations. There are many different methods which can be used, and they are briefly discussed in the next section. Here, a novel approach is presented. It uses an explicit source-receptor matrix for a linear relationship generated by a backward-running Lagrangian particle dispersion model (LPDM). It is shown how the elements of the source-receptor matrix can be derived from the gridded concentration fields which are the standard output of the LPDM. The method is then applied to the point release of the first European Tracer Experiment (ETEX) release. Earlier results have been presented by Seibert and Stohl (2000).

APPROACHES TO INVERSE AIR POLLUTION MODELLING

It is out of the scope of this paper to review all the work that has been done so far. However, the new method suggested shall be put in the context of previous approaches. A good overview of inverse modelling in the context of global biogeochemical cycles can be found in Kasibhatla et al. (1999). A longer discussion of previous work on inverse air pollution modelling is given in Seibert (1999).

Optimisation Methods

Inverse modelling is a kind of optimisation problem: the sources are adjusted so that the model output fits the observations best. As many practical problems are underdetermined and/or ill-conditioned, some a-priori knowledge is usually introduced. This can be through a formulation of the source reducing its dimension (e.g., prescribing a vertical profile, a horizontal distribution, or a temporal evolution), or by adding additional constraints, e.g. the minimisation of the deviation from a first guess. Adding such additional terms to the cost function is called a regularisation. Thus, in the general, the inverse modelling involves the minimisation of a cost function J which depends on a source vector $\boldsymbol{x}$ via a model $\mathcal{M}(\boldsymbol{x})$. [1] By calculating the derivatives $\partial J/\partial x_i$ and a suitable descend algorithm, the minimum of J and the associated solution $\boldsymbol{x}$ can be found iteratively. This method has been applied mainly for nonlinear problems, using an adjoint of the model (Elbern and Schmidt, 1999; Robertson and Langner, 1998).

The other approach, mainly used in linear problems, is the explicit calculation of the source-receptor relationships. In the linear case, each receptor value y_l (e.g., ambient concentration corresponding to a measurement) is a linear function of all source elements (m_{il} being the corresponding element of the source-receptor matrix):

$$y_l = \sum_i (m_{il} x_i)$$

If also the regularisation term is linear with respect to $\boldsymbol{x}$, the minimum of the cost function can be found by solving a linear system of equations. This is the approach adopted here.

Models: Eulerian or Lagrangian, Forward or Backward

In principle, all kinds of dispersion models can be used. Important distinctions are between Eulerian and Lagrangian models, and between forward (source-oriented) and backward (receptor-oriented) approaches. It has been recognised that generally for problems with a number of sources exceeding the number of receptors, a backward approach is computationally more efficient, and vice versa. The forward approach is conceptually simple: one runs the regular model with each source emitting one species, and records the concentrations at the receptors in order to obtain the source-receptor relationship. In the backward approach, the method of the adjoint is generally invoked, and a backward calculation of the adjoint model is made with the observed receptor values as input (each as a separate species, if the source-receptor matrix is to be calculated), and the sources as control variables, which means that the "adjoint tracer" (a quantity proportional to the observations) is evaluated at the source locations (Robertson and Langner, 1998). This approach has been applied to Eulerian models up to now. A rigourous construction of the adjoint of a LPDM appears difficult as in such models particles are usually emitted in source locations, a process that cannot be described by derivatives.

In the following section it is described how a source-receptor matrix can be derived running a LPDM backward without formally invoking the adjoint concept. The LPDM has two advantages compared to a Eulerian model in the context of inverse modelling. Firstly, it is considered to be more accurate as it does not suffer from errors associated with finite-difference advection schemes, e.g. numerical diffusion. Secondly, Eulerian models can represent well only sources that are at least of the dimension of a grid cell, and thus they have problems with point sources. In the backward mode, a point observation acts like a point source and will thus introduce additional errors in a backward Eulerian model. The LPDM does not have this kind of problem.

[1] The notation follows Kasibhatla et al., 1999.

DERIVATION OF THE SOURCE-RECEPTOR MATRIX WITH A BACKWARD-RUNNING LAGRANGIAN PARTICLE DISPERSION MODEL

General method

In the forward mode, particles are released from the source areas, and transported with the mean and turbulent velocity field. They carry a mass depending on source strength and particle release rate which can be altered by processes such as deposition or decay. To obtain concentrations, all the masses associated with the particles in a grid cell are summed up and then divided by the cell's volume.

In the backward mode, the same formalism and the same computer model are applied, but the particle trajectories are integrated backward in time, using a negative time step. However, there is a new interpretation: the values associated with the particles are now mixing ratios $\chi \equiv c/\rho$ (where c is the mass concentration and ρ the air density) in an infinitesimal volume of air. In our Lagrangian frame of reference, χ can change only by sinks such as deposition or decay and by uptake of emissions. Without sinks (in this paper, we are dealing with an invariant tracer), the individual rate of change of the mixing ratio is

$$\frac{d\chi}{dt} = S^{+}, \tag{1}$$

where S^{+} denotes the rate of change of χ due to sources. Thus, at any time t and location $\boldsymbol{r}_o$,

$$\chi(\boldsymbol{r}_o, t) = \int_0^t S^{+}[\boldsymbol{r}(t'), t']\, dt' + \chi(0), \tag{2}$$

where $\{\boldsymbol{r}(t'),\ t' \leq t\}$ is the back trajectory arriving in point $\boldsymbol{r}_o$ at time t, and $\chi(0)$ is the initial mixing ratio (background; already subtracted in the ETEX data). Eq. (2) is valid for an instantaneous mixing ratio, affected by turbulent fluctuations. The mean mixing ratio is obtained as an ensemble average, the normal LPDM procedure. In ETEX (and many other practical applications), measured concentrations represent a point in space and an average in time (ETEX: sampling interval 3 h). Thus, we need to average in time. If this averaging time exceeds substantially the time scale of turbulent mixing ratio fluctuations χ', then the temporal average and the temporal average of the ensemble mean are equal: $\overline{\chi'} = \overline{<\chi'>}$

We now introduce a discretisation in time (index j) with J back trajectories in equal intervals between t_1 and t_2, and index i for the temporal and spatial variation of S^{+} ("gridded in space and time"):

$$\overline{\chi} \equiv \frac{1}{t_2 - t_1} \int_{t_1}^{t_2} \chi(\boldsymbol{x}, t) dt \approx \sum_i \left[S_i^{+} \left(\frac{1}{J} \sum_j \Delta t'_{ij} \right) \right] \tag{3}$$

The term in parenthesis on the right-hand side is equal to the average residence time of all J trajectories in a spatio-temporal cell i contributing to the measurement under consideration, to be denoted by $\Delta\tau_i$.

Let us now consider that we have not just one measurement, but many (at different sites and times), denoted by l. Then we can write

$$\overline{\chi}_l = \sum_i (S_i^{+} \Delta\tau_{il}) \tag{4}$$

We see that the elements of the desired source-receptor matrix $\boldsymbol{M}$ are

$$M_{il} \equiv \frac{\partial \overline{\chi}_l}{\partial S_i^{+}} = \Delta\tau_{il} \tag{5}$$

If we rename the vectors of observations, $\overline{\chi}_l$, and emissions, S_i^+ as $\boldsymbol{y}$ and $\boldsymbol{x}$, respectively, to be consistent in notation with Kasibhatla et al. (1999) we arrive at the (yet unregularised) LSE to be solved for the inverse modelling:

$$\boldsymbol{y} = \boldsymbol{M}\boldsymbol{x}. \tag{6}$$

The key assumptions made to arrive at this result were that only volume sources are considered and no area, line or point sources, and that particles carry mixing ratios rather than masses. If the density variations between the receptors and the sources are small, χ and S^+ may be substituted by the mass concentration c and the emitted mass per volume and time $\dot{q}$. Otherwise, the mixing ratio values can easily be transformed to mass values by multiplying them with the air's density ρ_a at the receptor or source, respectively, whereby climatological values of ρ_a should be sufficient in most cases.

Implementation

The LPDM used here is the model FLEXPART (Stohl et al., 1998). It calculates concentration fields $\bar{c}_i$ similar to the method described above (one particle is counted for four horizontal grid cells using a uniform kernel of the size of grid cell):

$$\bar{c}_i = \frac{\mu_{tot}}{V_i\, NJ} \sum_{n=1}^{N} \sum_{j=1}^{J} f_{ijn}, \tag{7}$$

where V_i is volume of the grid cell, N the number of instants in time used to form a temporal average over an interval ΔT, μ_{tot} the total released mass, and f_{ijn} the fraction of the mass of a particle attributed to a specific grid cell, considering that all J particles have the same mass because there are no sinks.

If the density correction recently introduced in FLEXPART (Stohl and Thomson, 1999) is switched off, the required residence times $\Delta\tau_{il}$ can be obtained by simple scaling from the gridded output $\bar{c}_i$ of FLEXPART using a unit emission of a unique species for each measurement site and interval. This is because $\Delta\tau_i$ during ΔT can be calculated as

$$\Delta\tau_i = \frac{\Delta T}{NJ} \sum_{n=1}^{N} \sum_{j=1}^{J} f_{ijn}. \tag{8}$$

Combining the last two equations, we obtain the desired result

$$\Delta\tau_i = \bar{c}_i\, \frac{V_i\, \Delta T}{\mu_{tot}}. \tag{9}$$

A similar implementation should be possible also with other LPDMs. If the temporal behaviour of a point source with known location is sought, the parabolic kernel method implemented in FLEXPART could be used as well.

APPLICATION TO ETEX

The ETEX Experiment

During the first release of the European Tracer Experiment (ETEX, see the special issue of *Atmospheric Environment*[2] and ETEX, 1998) an inert, nondepositing gaseous tracer was released from a site in northwestern France during almost 12 hours from 23 Oct 16:00 UTC until 24 Oct 3:50 UTC, 1994. The release rate was constant with 7.98 gs^{-1}. The tracer was sampled during 90 hours by a network of 168 stations covering most of western and

[2] Volume 32, Issue no. 24, 1998

central Europe with a resolution of 3 hours. A forward simulation of this release with the FLEXPART model (Stohl et al., 1998) yielded a correlation coefficient between observations and simulated concentrations or 0.59 and a fractional bias of 0.01 and was thus among the best models tested on this experiment.

Model Set-Up

FLEXPART was used in the backward simulation with the same input as in the above-mentioned forward simulation, namely 3-hourly ECMWF fields with 1.0° horizontal resolution and 31 levels. For each of the 30 measurement intervals, a separate run was performed. In each run, at each of the 168 stations a unit emission of a stations-specific pseudo-species was released during the sampling interval (including those measurements where the observed tracer concentration was zero). The concentrations were sampled on a 41×21 grid with 1° resolution and 9 vertical cells during 35 time intervals of 3 h; later, another run was performed with hourly sampling during 105 intervals as it turned out that this improves the inversion considerably.

Inversion

The model set-up yields a source-receptor matrix with about $1.4 \cdot 10^9$ elements (3-hourly output) corresponding to 300 MByte of data even with a sparse matrix storage strategy. In order to reduce the dimension of the matrix to be inverted, only the lowest layer was considered. This appears to be justified because the pseudotracer concentrations arriving at the true source location had a rather uniform vertical distribution within the boundary layer. Furthermore, up to now it has also been assumed that either the source location is known (then the unknowns are the emissions during the 35, or, respectively, 105 temporal intervals), or the source time (then the unknowns are the release amounts in each of the 861 horizontal grid elements). In the first case, the source-receptor matrix for the release point is obtained by linear interpolation from the four surrounding cells.

A regularised inversion minimising the cost function

$$J = ||\boldsymbol{y} - \boldsymbol{M}\boldsymbol{x}||^2 + \theta_1^2||\boldsymbol{x}||^2 + \theta_2^2||\boldsymbol{D}\boldsymbol{x}|| \tag{10}$$

was then carried out, similar to Seibert (1997, 1999). The three contributions to the cost function are the misfit between observations $\boldsymbol{y}$ and modelled values, the variance of the solution and its "roughness", measured by a numerical implementation of the Laplacian, represented by the linear operator $\boldsymbol{D}$ (see Seibert, 1999). The θ values are the regularisation parameters, regulating the trade-off between the different conditions to be fulfilled. They were determined subjectively. The third term was not used ($\theta_2 = 0$) in some of the solutions shown below.

Results and Discussion

Fig. 1 shows the first result for the temporal evolution obtained with given source location, 3-hourly source resolution and $\theta_2 = 0$. The start of the release was reconstructed well, but its end is not well defined and tends to be too late. We can also see that a single outlier station (Mannheim) can have a visible negative impact. The underestimation of the total release amount of 35% may partly be attributed to the regularisation (Seibert and Stohl, 2000).

Fig. 2 shows the same but with 1-hourly source resolution. We see more clearly than in Fig. 1 that the inversion tends to create two separate emission peaks. The end of the release is captured somewhat better than with the 3-hourly resolution. If the minimisation of the Laplacian of the solution (smoothness) is added to the regularisation, a more realistic shape of the curve is obtained, though the principal problem (the fuzzy end of the release) remains. This is obviously caused by observed non-zero concentrations at a time when the modelled tracer cloud (in a forward run) had already passed the respective sites.

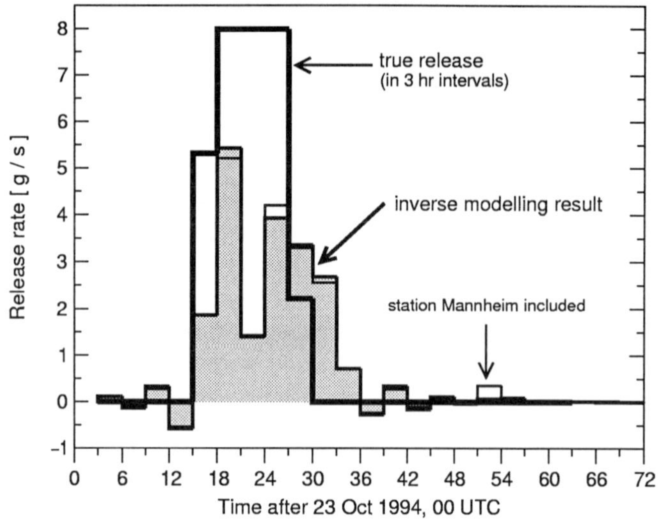

Figure 1. Temporal evolution of the true (heavy line) and the reconstructed emission (filled) based on 3-hourly source resolution. The thin line indicates the reconstructed emissions if one outlier station (Mannheim) is removed. Regularisation is based only on minimisation of variance. (Reprinted from Seibert and Stohl, 2000.)

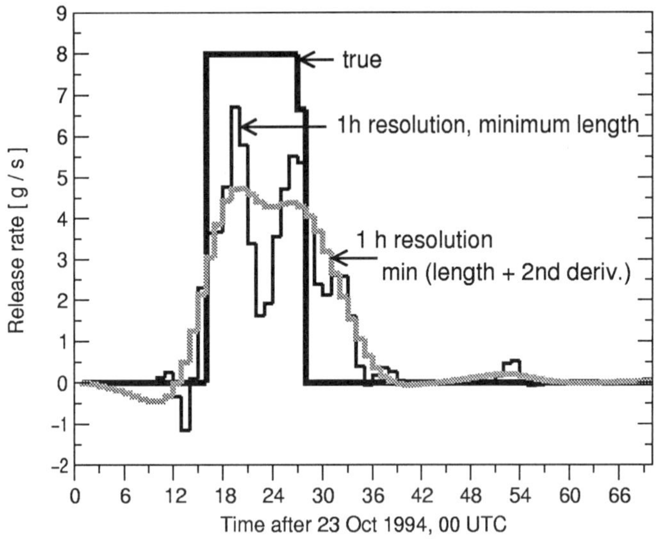

Figure 2. Temporal evolution of the true (heavy line) and the reconstructed emission based on 1-hourly source resolution. The two reconstructed emissions differ in the regularisation (see text).

An inversion with respect to the horizontal location of the source is shown in Fig. 3. As an effect of the regularisation requiring a smooth field, the source is smeared out spatially over several grid elements with the maximum about 2 grid elements ($\approx$ 150 km) west of the real source. Only relative values are given here because the interesting quantity, namely the total release, would be the integral over the whole three-dimensional domain. This is the first result of this kind presented for ETEX. Other regularisation techniques could be used to avoid the smearing of the peak, but these techniques would involve nonlinear constraints and

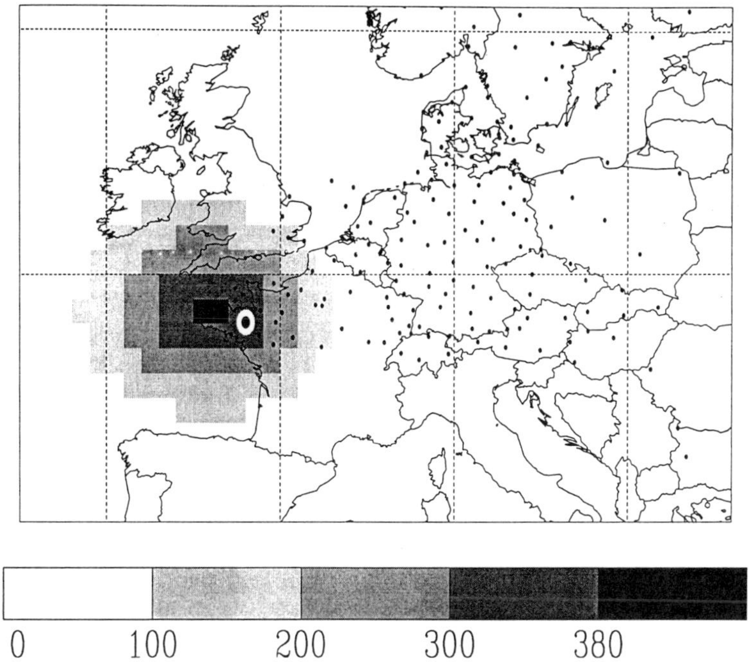

Figure 3. Horizontal source distribution (rel. units) for given temporal evolution (1 h resolution) and inversion with minimisation of both length and Laplacian (requiring a smooth field). The real source location is marked by a black dot in a white circle, and the measurement sites by small black dots.

thus require iterations. This does not mean, however, that they would be too costly in terms of computer time, because we only need to apply the source-receptor matrix to the source values of the previous iteration, compute the regularisation terms, and redo the inversion. It is not necessary to repeat the whole simulation as in the case of the adjoint modelling technique without explicit source-rector matrix. In such iterations, outliers among the measurements could be given a lower weight, so that a more accurate solution would be obtained (reduction of the horizontal location error).

Robertson and Langner (1998) have applied an adjoint Eulerian model to the same problem, however, without explicit calculation of source-receptor matrices. Their inversion was of a more limited scope, prescribing the horizontal source location and a narrow time window (not much larger than the real release time). They show also geographical distributions of the adjoint tracer cloud drifting over the release location but did not attempt to reconstruct a source location. Also Pudykiewicz (1998), in his inverse modelling exercise directed towards the detection of a possible nuclear explosion from a suggested radionuclide monitoring network for the control of the comprehensive nuclear weapons test ban, limited himself to a simulation of the adjoint tracer cloud with a Eulerian model, and simply fixed the source location at the point of the maximum tracer concentration for the source time estimated from nuclide ratios. The present paper goes beyond these approaches and has a good potential for mathematically well-based refinements.

CONCLUSIONS

A new method for inverse modelling of atmospheric trace species emissions, using a backward-running LPDM, has been introduced. It has been successfully applied to the first

ETEX tracer release, and could be used to reconstruct the timing or the spatial location of the release. However, regularisation methods and weighting of observations need to be further developed, including nonlinear regularisation terms, in order to improve the reconstruction of non-smooth shapes and to eliminate the detoriating influence of observations that are either wrong or cannot be explained well by the dispersion model. Introducing different weights for the observations is desirable also because of the correlation among the measurements. However, possibilites to estimate covariance matrices of both measurements and their errors are limited as compared to 4DVAR in numerical weather prediction, where a large climatological data set is available.

For point releases, other strategies for finding the minimum of the cost function (e.g., searching all possible grids as a source while using the inversion to determine the optimum temporal evolution for this grid) should be tried.

It is planned to extend the method to cases with deposition or decay. Indeed, the scaling presented here should be valid in this more general case, too, it appears only to be more difficult to provide the proof.

Potential applications include nuclear accidents and nuclear bomb testing (verification of the Comprehensive Test Ban Treaty[3]). Furthermore, it may be applied to any source determination problem (point or area source) where nonlinearities are not too important.

For updates of this work, please visit http://boku.ac.at/imp/envmet/invmod.html.

Acknowledgements. This work is a contribution to EUROTRAC-2 subprojects GLOREAM and GENEMIS-2, funded by FWF under P1295-GEO. Meteorological data from ECMWF have kindly been provided through ZAMG (Vienna). Special thanks go to Andreas Stohl who ran the FLEXPART model for the ETEX case and introduced some changes for better performance in this case with many species each emitted from exactly one source. Arnold Neumaier from the Mathematical Institute at the University of Vienna provided his advise concerning regularisation.

REFERENCES

Elbern, H. and Schmidt, H., 1999, A four-dimensional variational chemistry data assimilation scheme for Eulerian chemistry transport modeling, *J. Geophys. Res.* 104:18,583.

ETEX. The European Tracer Experiment, 1998, European Communities, Publ. No. EUR 18143EN, ISBN 92-828-5007-2, Luxembourg.

Kasibhatla, P., Heimann, M., Rayner, P., Mahowald, N., Prinn, R. G., and Hartley, D. E., eds., 1999, *Inverse Methods in Global Biogeochemical Cycles*, AGU Geophysical Monograph 114, ISBN 0-87590-097-6, Washington.

Pudykiewicz, J. A., 1998, Application of adjoint tracer transport equations for evaluating source parameters. *Atmos. Environ.* 32:3039.

Robertson, L., and Langner, J.,1998, Source function estimate by means of a variational data assimilation applied to the ETEX-I tracer experiment, *Atmos. Environ.* 32:4219.

Seibert, P., 1997, Inverse dispersion modelling based on trajectory-derived source-receptor relationships, in: Air Pollution Modeling and its Application XII, S.E. Gryning, N. Chaumerliac, eds., Plenum, New York.

Seibert, P., 1999, Inverse modelling of sulfur emissions in Europe based on trajectories, in: *Inverse Methods in Global Biogeochemical Cycles*, P. Kasibhatla et al., eds., AGU Geophysical Monograph Vol. 114, Washington.

Seibert, P., and Stohl, A., 2000, Inverse modelling of the ETEX-1 release with a Lagrangian particle model, in: *Proceedings of the 3rd GLOREAM workshop, Ischia, Italy*, University of Naples, in print. On-line at http://boku.ac.at/imp/envmet/glor3.html.

Stohl, A., Hittenberger, M., and Wotawa, G., 1998, Validation of the Lagrangian particle dispersion model FLEXPART against large scale tracer experiments, *Atmos. Environ.* 32:4245.

Stohl, A., Thomson, D.J., 1999, A density correction for Lagrangian particle dispersion models, *Bound.-Layer Meteor.* 90:155.

[3]for more information, see the web page of the CTBTO at http://www.ctbto.org/.

DISCUSSION

D. ANFOSSI: In performing your analysis, did you include the samplers which detected zero concentration?

P. SEIBERT: Yes, they contain important information.

D. SYRAKOV: As seen from W. Klug's presentation, a vertical spread of the source is quite possible. Have you made attempts to estimate this spread instead of the horizontal one?

P. SEIBERT: The pseudotracer released from the sampling locations was in general well mixed within the mixing layer when it arrived at the release site. This means that it doesn't contain information on the vertical distribution anymore. This relates to the sum of all contributions. Maybe if one would look at the nearby samplers and during night-time, one could derive more, but I have not done that.

H. van DOP: Did you apply the inverse analysis also to the second release of the ETEX experiment?

P. SEIBERT: No, I have not done that. All the models performed so badly on this release, including FLEXPART. In this case, inverse modelling would try to adjust the source so that it would compensate for the model error, so that no good results can be expected. But maybe one could derive where the mass has been "lost" from the simulated tracer cloud by a proper set-up of the inversion and thus help to understand the situation better.

AN ANALYTICAL AIR POLLUTION MODEL: EDDY DIFFUSIVITIES DEPENDING ON THE SOURCE DISTANCE

Gervásio[2] A. Degrazia, Davidson M. Moreira[1], Marco T. Vilhena[1] and Angela B. Moura[1]

[1]Universidade Federal do Rio Grande do Sul, Porto Alegre -Brasil
[2]Universidade Federal de Santa Maria , Santa Maria - Brasil

INTRODUCTION

The Eulerian dispersion model concept has been largely applied for estimating ground-level concentrations due to low and tall stack emission and it is usually suitable for regulatory use in air quality models. In principle, from advection-diffusion equation it is possible to obtain a practical model of dispersion from a continuous point source given appropriate boundary and initial conditions plus a knowledge of the time and space fields of U (mean wind vector) and K_α (eddy diffusivities; with $\alpha = x, y, z$)[1].

Much of the turbulent dispersion researches are related to the specification of concentration turbulent fluxes in order to allow the solution of the averaged advection-diffusion equation: this parameterization is called as the closure of the turbulent transport problem. The principal scheme for closing the equation is to relate turbulent concentration fluxes to the gradient of the mean concentration by K eddy diffusivities, which are properties of the turbulent flow but not of the fluid; i.e., first order closure. Even more important, the eddy diffusivities are not constant like the kinematic viscosity, but may vary in space. Hence the aforementioned scheme closes the advection-diffusion equation only to a certain degree: the eddy diffusivities still have to be determined[2].

The aim of this paper is to present an analytical solution for the advection-diffusion equation considering vertical eddy diffusivities varying with the distance from the source and related with the turbulence properties (inhomogeneous turbulence)[3]. The model performance is evaluated against ground-level concentrations using the convective runs of the Copenhagen and Prairie Grass dispersion experiments.

Air Pollution Modeling and Its Application XIV, Edited by
Gryning and Schiermeier, Kluwer Academic/Plenum Publishers, New York, 2001

MODEL DEVELOPMENT

Following Vilhena et al.[4,5], the concentration turbulent fluxes are often assumed to be proportional to the mean concentration gradient. This assumption, along with the equation of continuity, leads to the advection-diffusion equation. For a cartesian coordinate system in which the x direction coincides with that one of the average wind, the steady state advection-diffusion equation reads as[6]:

$$U\frac{\partial \bar{c}}{\partial x}=\frac{\partial}{\partial x}\left(K_x\frac{\partial \bar{c}}{\partial x}\right)+\frac{\partial}{\partial y}\left(K_y\frac{\partial \bar{c}}{\partial y}\right)+\frac{\partial}{\partial z}\left(K_z\frac{\partial \bar{c}}{\partial z}\right) \tag{1}$$

where $\bar{c}$ denote the average concentration, U the mean wind speed in x direction and K_x, K_y and K_z the eddy diffusivities. The cross-wind integration of the equation (1) (neglecting the longitudinal diffusion) leads to:

$$U\frac{\partial \overline{c^y}}{\partial x}=\frac{\partial}{\partial z}\left(K_z\frac{\partial \overline{c^y}}{\partial z}\right) \tag{2}$$

subject to the boundary conditions of zero flux at the ground and Convective Boundary layer (CBL) top, and a source with emission rate Q at height H_s:

$$K_z\frac{\partial \overline{c^y}}{\partial z}=0 \qquad \text{at } z=0,\, Z_i \tag{3}$$

$$U\overline{c^y}(0,z)=Q\delta(z-H_s) \qquad \text{at } x=0 \tag{4}$$

where $\overline{c^y}$ represents the average cross-wind integrated concentration and Z_i is the height of the CBL.

Bearing in mind the dependence of the K_z coefficient and wind speed profile U on variable z, the height Z_i of a CBL is discretized in N sub-intervals in such a manner that inside each interval $K_z(z)$ and $U(z)$ assume the average value:

$$K_n=\frac{1}{z_n-z_{n-1}}\int_{z_{n-1}}^{z_n}K_z(z)dz \tag{5}$$

$$U_n=\frac{1}{z_n-z_{n-1}}\int_{z_{n-1}}^{z_n}U_z(z)dz \tag{6}$$

Therefore the solution of problem (2) is reduced to the solution of N problems of the type:

$$U_n\frac{\partial}{\partial x}\overline{c_n^y}=K_z\frac{\partial^2}{\partial z^2}\overline{c_n^y} \quad \text{with } z_{n-1}\le z\le z_n \tag{7}$$

for $n = 1{:}\ N$, where $\overline{c_n^y}$ denotes the concentration at the nth sub-interval. To determine the *2N* integration constants the additional *(2N-2)* conditions namely continuity of concentration and flux at interface are considered:

$$\overline{c_n^y} = \overline{c_{n+1}^y} \quad n = 1,2,...(\text{N-1}) \tag{8}$$

$$K_n \frac{\partial \overline{c_n^y}}{\partial z} = K_{n+1} \frac{\partial \overline{c_{n+1}^y}}{\partial z} \quad n = 1,2,...(\text{N-1}) \tag{9}$$

Applying the Laplace transform in equation (7) results:

$$\frac{\partial^2}{\partial z^2} \overline{c_n^y}(s,z) - \frac{U_n S}{K_n} \overline{c_n^y}(s,z) = -\frac{U}{K_n} \overline{c_n^y}(0,z) \tag{10}$$

where $\overline{c_n^y}(s,z) = L\{c_n^y(x,z); x \rightarrow s$, which has the well-know solution:

$$\overline{c_n^y}(s,z) = A_n e^{-R_n z} + B_n e^{R_n z} + \frac{Q}{2R_a}\left(e^{-R_n(z-H_s)} - e^{R_n(z-H_s)}\right) \tag{11}$$

where

$$R_n = \pm\sqrt{\frac{U_n S}{K_n}} \quad \text{and} \quad R_a = \pm\sqrt{U_n S K_n}$$

Finally, applying the interface and boundary conditions we come out with a linear system for the integration constants. Henceforth the concentration is obtained inverting numerically the Laplace transformed concentration $\overline{c^y}$ by Gaussian quadrature scheme[7] yielding:

$$\overline{c_n^y}(x,z) = \sum_{j=1}^{M} A_j . \frac{P_j}{x}\left(A_n \exp\left(-\sqrt{\frac{P_j U_n}{xK_n}}.z\right) + B_n \exp\left(\sqrt{\frac{P_j U_n}{xK_n}}.z\right)\right) \tag{12}$$

$$\overline{c_n^y}(x,z') = \sum_{j=1}^{Ni} A_j . \frac{P_j}{x}\left[A_n e^{-\sqrt{\frac{P_j U_n}{xK_n}}.z} + B_n e^{\sqrt{\frac{P_j U_n}{xK_n}}.z} + \frac{1}{2}\frac{Q}{\sqrt{\frac{P_j K_n U_n}{x}}}\left(e^{-(z'-H_s)\sqrt{\frac{P_j U_n}{xK_n}}} - e^{(z'-H_s)\sqrt{\frac{P_j U_n}{xK_n}}}\right)\right] \tag{13}$$

The solution (12) is valid for layers that no contain the contaminant source. On the other hand, the solution (13) can be used to evaluate the concentration field in the layer that contains the continuous source. Here A_n and B_n are integrations constants which are determined applying the boundary and interface conditions; A_j and P_j are the weights and roots of the Gaussian quadrature scheme and are tabulated in the book by Stroud and Secrest[8].

FIELD EXPERIMENT AND TURBULENT PARAMETERIZATION

The model performance is tested by comparison with the cross-wind integrated concentration measured during Copenhagen and Prairie Grass field experiments. In the Copenhagen experiment, the tracer SF_6 was released without buoyancy from a tower 115 m high, and collected at the ground-level positions in up to three crosswind arcs of tracer sampling units. The sampling units were positioned 2-6 km from the point of release. The site was mainly residential with a roughness length of the 0.6 m. On the other hand, in the Prairie Grass experiment, the tracer SO_2 was released without buoyancy at a height of ~ 0.5 m, and collected at a height of 1.5 m at five downwind distances (50, 100, 200, 400 and 800 m). The Prairie Grass site was quite flat and much smooth with a roughness length of 0.6 cm.

To simulate the Copenhagen dataset (tall stack), the following vertical eddy diffusivity as function of downwind distance and of the turbulence characteristics is utilised in eq. (5) and (6):

$$\frac{K_z}{w_* Z_i} = \frac{0.21\psi^{1/3}(z/Z_i)^{4/3}}{q_w^{4/3}} \int_0^\infty \frac{sin\left(\dfrac{2.35\psi^{1/3}q_w^{2/3}n'X}{(z/Z_i)^{2/3}}\right)}{(1+n')^{5/3}} \frac{dn'}{n'} \tag{14}$$

while to simulate the Prairie Grass dataset (low stack) are used eddy diffusivities generated both by thermal and mechanical forcing mechanisms so that these two effects are modelled by the following vertical eddy diffusivities:

$$\frac{K_{zb}}{w_* Z_i} = \frac{0.24(z/Z_i)^{4/3}X^{1/2}}{q_w^{4/3}} \int_0^\infty \frac{sin\left(\dfrac{2.6q_w^{2/3}n'X^{3/2}}{(z/Z_i)^{2/3}}\right)}{(1+n')^{5/3}} \frac{dn'}{n'} \tag{15}$$

$$\frac{K_{zs}}{u_* z} = 0.26(1-z/Z_i)X^{1/2} \int_0^\infty \frac{sin\left(5.34(1-z/Z_i)n'X^{3/2}\right)}{\left(1+n'^{5/3}\right)} \frac{dn'}{n'} \tag{16}$$

where w_* and u_* are respectively, the convective and friction velocity scales, $X = w_* x/U$, $X' = u_* x/U$, q_w is related to the vertical profile of the more energetic convective eddies and z is the height above the surface. Subscripts b and s stand for buoyancy and shear forcings, respectively. The dissipation function ψ is given by[9]:

$$\psi^{1/3} = \left[\left(1-\frac{z}{Z_i}\right)^2\left(\frac{Z_i}{-L}\right)^{-2/3} + 0.75\right]^{1/2} \tag{17}$$

where L is the Monin- Obukhov length in the surface layer.

The vertical eddy diffusivity as given by eq. (14) is based on spectral properties and Taylor's statistical diffusion theory while the eqs. (15) and (16), valid to low stacks, take the variation of the vertical velocity standart deviation σ_w with z (inhomogeneous strong turbulence)[10] into account.

The wind speed profile is parameterized following the Similarity Theory of Monin-Obukhov and OML model[11] namely:

$$U = \frac{u_*}{k}\left[\ln(z/z_0 - \Psi_m(z/L) + \Psi_m(z_0/L)\right] \qquad \text{if } z \le z_b \tag{18}$$

$$U = U(z_b) \text{ if } z > z_b \tag{19}$$

where $z_b = \min\left[|L|, 0.1z_i\right.$, and Ψ_m is a stability function given by[12]:

$$\Psi_m = 2\ln\left[\frac{1+A}{2}\right] + \left[\frac{1+A^2}{2}\right] - 2\tan^{-1}A + \frac{\pi}{2} \tag{20}$$

$$A = (1-16z/L)^{1/4} \tag{21}$$

$k = 0.4$ is the Von Karman constant, u_* is the friction velocity and z_0 roughness length.

MODEL EVALUTION

The new parameterization for eddy diffusivities (eqs. 14, 15 and 16) is tested and compared with a parameterization derived from Taylor's statistical diffusion theory for large diffusion travel times[13]:

$$\frac{K_z}{w_* Z_i} = 0.22\left(\frac{z}{Z_i}\right)^{1/3}\left(1-\frac{z}{Z_i}\right)^{1/3}\left[1-\exp\left(-\frac{4z}{Z_i}\right) - 0.0003\exp\left(\frac{8z}{Z_i}\right)\right] \tag{22}$$

using the crosswind integrated concentration and meteorological data from Copenhagen and Prairie Grass field campaigns. Here, is utilised in eqs. (12) and (13), the news eddy diffusivities (eqs. 14, 15 and 16) and the old eddy diffusivity (eq. 22) as well as the wind field profile described in section 3.

Table 1. Statistical indices evaluating the model performances with different eddy diffusivities.

	Nmse	Fa2	Cor	Fb	Fs
Copenhagen (eq 22)	0.07	1.00	0.90	0.059	0.231
Copenhagen (eq. 14)	0.06	1.00	0.89	0.012	0.001
Prairie Grass (eq. 22)	0.09	0.61	0.98	-0.205	0.140
Prairie Grass (eqs. 15, 16)	0.08	0.86	0.97	0.061	0.152

Analysis based on statistical indices (Table 1), points out that the results obtained with the eddy diffusivities depending on the source distance are better than the ones reached with assymptotic eddy diffusivity valid only for the far field of a continuous point source.

CONCLUSIONS

This paper describes the development and testing of an analytical model that simulates the dispersion of contaminants into a convective boundary layer. The model is based on the advection-diffusion equation which is solved by the Laplace transform technique. This model assumes vertical eddy diffusivities varying with the distance from the source and related with the turbulence properties (inhomogeneous turbulence). The inclusion of eddy diffusivities as function of downwind distance has been often cited as something to be included in analytical dispersion models in order to take into consideration the different physics aspects of near-source and far-source dispersion. To investigate the turbulence memory effect for low and tall stacks a numerical comparison, using the well-known Copenhagen and Prairie Grass datasets, is also performed with results derived from a simulation using a vertical eddy diffusivity valid for large diffusion time and that is only dependent on the turbulence properties. It is also important to emphasize that the concentrations simulated by the eddy diffusivities depending on the source distance are better than the ones simulated by the asymptotic eddy diffusivity valid only for the far field of a continuos point source.

O presente trabalho foi realizado com o apoio do CNPq e CAPES, entidades do Governo Brasileiro voltadas ao desenvolvimento científico e tecnológico.

REFERENCES

1. S.R. Hanna, *Applications in Air Pollution Modelling, Atmospheric Turbulence and Air Pollution Modelling*, edited by F.T.M. Nieuwstadt and H. van Dop (Reidel, Boston), 1982.
2. A.A.M. Holtslag and C.H. Moeng , *J. Atmos. Sci.*, 48, pp. 1690, 1991.
3. G.A. Degrazia, U. Rizza, C. Mangia and T. Tirabassi. Validation of a new turbulent parameterization for dispersion models in a convective boundary layer, *Boundary Layer Meteorology*, 85, pp. 243-254, 1997.
4. M.T Vilhena, U. Rizza, G.A. Degrazia, C. Mangia, D.M. Moreira and T. Tirabassi, An analytical air pollution model: development and evalution, *Contribution to Atmospheric Physics*, pp. 315-320,1998.
5. D.M. Moreira, G.A. Degrazia and M.T. Vilhena, Dispersion from low sources in a convective boundary layer: an analytical model, *Il Nuovo Cimento*, pp. 685-691, 1999.
6. P.S. Arya, J. Appl. Met., 34, pp. 1112, 1995.
7. M. Heydarian and N. Mullineaux, *Appl. Math. Modelling*, 5, pp. 448, 1989
8. A.H. Stroud and D. Secrest, *Gaussian Quadrature Formulas* (prentice-Hall, Englewood Cliffs) 1966.
9. J. Hojstrup, Velocity spectra in the unstable boundary layer, *J. Atmos. Sci.*, 39, 2239-2248, 1982.

10. J.C. Weil, Dispersion in the convective boundary layer, *Lectures on air pollution modelling*, edited by A. Venkatram and J.C. Wyngaard, American meteorological Society, Boston, 1988
11. R.R. Berkowicz, H.R. Olesen and U. Torp, *The Danish Gaussian Air Pollution Model*, (OML): Description, Test and Sensivity Analysis in View of Regulatory Applications, Proceedings of the 15th International Techinical Meeting on Air Pollution Modelling and its Applications, April 15-19 (1985) St. Louis, USA (Plenum Press).
12. C.A.Paulsen, *J. Appl. Met.*, 9, pp. 857, 1975.
13. G.A. Degrazia, H.F. Campos Velho and J.C. Carvalho, Nonlocal exchange coefficients for the convective boundary layer derived from spectral properties, *Beitr. Phys. Atmos.*, pp. 57-64, 1997.

NUMERICAL TREATMENT OF AQUEOUS–PHASE CHEMISTRY IN ATMOSPHERIC CHEMISTRY–TRANSPORT MODELLING

Ralf Wolke, Oswald Knoth and Hartmut Herrmann

Institute for Tropospheric Research
Permoserstr. 15, D–04303 Leipzig, Germany,
e–mail: wolke@tropos.de

1. INTRODUCTION

Multiphase processes, such as gas scavenging by clouds, are of increasing importance for the comprehension of atmospheric processes. The gas scavenging by cloud drops leads to a transfer of chemical species between the gaseous and aqueous phases. This phase transfer and chemical reactions modify the concentrations of the species in both phases. The complexity of the processes involved has discouraged investigators from simultaneously treating all aspects of multiphase chemistry and microphysics with equal rigour. The description of cloud processing in most currently available box models and Eulerian grid models focuses either on detailed microphysics or complex multiphase chemistry. The chemical conversions in the liquid phase are described only in a few aggregated drop classes, or strongly simplified chemical mechanisms are used. Chemical conversions within cloud drops are essentially determined by the mass transfer between gaseous and liquid phase. It has been shown that these phase transitions must be described dynamically (Audiffren et al, 1998; Chaumerliaq et al., 2000). Furthermore, the gas uptake depends strongly on resolution of the drop spectrum (Wurzler, 1998).

In the paper we propose several numerical approaches for treating such processes in box and multidimensional chemistry–transport models. The droplets are subdivided into several classes with a mean droplet radius. This multifractional distribution and the transfer rates of liquid water between the different droplet classes are prescribed a priori by a microphysical cloud model. The phase transfer between the gaseous–phase and the aqueous–phase species in each class is described by the resistance model of Schwartz (1986). The very fast dissociations in the aqueous–phase chemistry are treated as forward and backward reactions. The pH value is not prescribed a priori. The H^+ concentration as part of the chemicl system is computed for each droplet class dynamically.

In contrast to the processes in the nature which perform in a coupled manner, these are decoupled in the numerical approach using the "operator splitting" scheme. The splitting error can be kept small only in the case of small time steps. The modelling of chemical conversions leads to stiff systems of ordinary differential equations. The equations describing liquid phase systems are in general much stiffer than those describing reactions systems in the gas phase. For stiff systems, explicit integration methods (QSSA, explicit Euler method) used in many models are only stable for very small step sizes (Sandu et al., 1997). Stiffness as well as operator splitting lead, using very small step sizes, to a high numerical expense or, if this expense is not invested, to instabilities or uncontrollable coupling errors in the time integration.

Therefore, we propose implicit–explicit schemes for the time integration of the resulting extremely stiff systems of ordinary differential equations (ODEs). The aqueous–phase and gas–phase chemistry, the mass transfer between the different droplet classes among themselves and with the gas phase as well as, in the multidimensional case, all vertical transport processes are integrated in an implicit and coupled manner by higher order BDF methods. For this part we apply a modification of the code LSODE (Hindmarsh, 1983) with an adapted step size control and a special linear system solver. This direct sparse solver exploits the special structure of the equations. In this approach we use an approximate matrix factorization which is related to operator splitting between chemistry and vertical transport at the linear algebra level (Verwer et al., 1998). The sparse Jacobians are generated explicitly (not by finite differences) and stored in a sparse form.

The efficiency and accuracy of our time–integration scheme is compared for the box case with the original LSODE Code of Hindmarsh (1983). For the tests in the box case, four different multiphase chemistry systems of different complexity are used. The numerical experiments indicate that the proposed integration scheme works stable and efficient. The results are discussed with respect to the use of this techniques for multidimensional chemistry–transport and parcel models with a complex cloud–chemistry.

2. MODEL DESCRIPTION

Mass Balance Equations in the 3D Case. The model equations base on mass balances described by systems of time–dependent, three–dimensional advection–diffusion–reaction equations

$$\begin{aligned}\frac{\partial \mu_{\mathbf{l},\mathbf{k}}}{\partial \mathbf{t}} &= -\frac{\partial}{\partial x}(u\mu_{l,k}) - \frac{\partial}{\partial y}(v\mu_{l,k}) - \frac{\partial}{\partial z}(w\mu_{l,k}) + \frac{\partial}{\partial z}(\rho K \frac{\partial \mu_{l,k}/\rho}{\partial z}) - \frac{\partial}{\partial z} w_s^k \mu_{l,k}, \\ &\quad + \sum_{\mathbf{i}\neq\mathbf{k}} \mathbf{T}_{\mathbf{k},\mathbf{i}}(\mathbf{t}, \mu_{\mathbf{l},\mathbf{k}}, \mu_{\mathbf{l},\mathbf{i}}) + \mathbf{R}_{\mathbf{l}}^{\mathbf{k}}(\mu_{\mathbf{1},\mathbf{k}}, \ldots, \mu_{\mathbf{N}_{\mathbf{Spc}},\mathbf{k}}, \mathbf{t}) + \mathbf{Q}_{\mathbf{l}}^{\mathbf{k}}, \qquad (1) \\ &\qquad l = 1, \ldots, N_{Spc}; k = 0, \ldots, N_{Class}\end{aligned}$$

where the following notation is used

$$\mu_{\mathbf{l},\mathbf{k}} = \mathbf{L}_{\mathbf{k}} \mathbf{c}_{\mathbf{l}}^{\mathbf{k}} \qquad (2)$$

and

$c_l^k = c_l^k(x,y,z,t)$	...	concentration of the l-th species in the k-th class,
$L_0 = 1$	...	gas phase (or mass fraction from "aerosols"),
$L_k = L_k(x,y,z,t)$	...	volume fraction $[V_k/V_{cell}]$ of the k-th droplet class,
(u,v,w),ρ,K	...	wind vector, density, vertical diffusion coefficient,
$Q_l^k = Q_l^k(x,y,z,t)$	...	natural and anthropogenic sources,
$R_l^k = R_l^k(\mu_{1,k},\ldots,\mu_{N,k})$	...	chemical reactions,
$T_{k,i}(t,\mu_{l,k},\mu_{l,i})$	...	mass transfer between the k-th and i-th class,
	...	(including phase interchange with the gas phase),
w_s^k	...	sedimentation velocity of the k-th droplet class .

In system (1) the horizontal diffusion is neglected. In addition, initial conditions as well as boundary conditions (including deposition at the earth's surface) have to be numerically treated.

Note that the number of species in the gas phase must not necessarily be the same as the number of aqueous species in all droplet classes. The decomposition of the droplet spectrum into classes is based on their droplet size and the amount of scavenged material inside the drops, respectively.

The species within one class are coupled only through the chemical reaction system. The mass transfer couples only the l-th species of different classes. The transport processes produce no coupling between different species c_l^k. For the following description of the mass transfer and the chemistry we consider a box model. The notation for gas and aqueous phase species is separated too.

Mass Transfer and Chemistry. The dynamics of the aqueous–phase and gas–phase chemistry within one grid cell ("box") can be described by a system of ODEs, see e. g. Lelieveld and Crutzen [4],

$$\frac{dc_l^G}{dt} = R_l^G(t,\mathbf{c^G}) - \kappa_l \sum_k L_k \cdot k_t^{kl} \cdot \left[c_l^G - \frac{c_l^k}{H_l}\right] \tag{3}$$

$$\frac{d\mu_{l,k}}{dt} = \frac{d(L_k c_l^k)}{dt} = L^k \cdot R_l^A(t,\mathbf{c^k}) + \kappa_l \cdot L_k \cdot k_t^{kl} \cdot \left[c_l^G - \frac{c_l^k}{H_l}\right] + \sum_i \left[T_{ik}c_l^i - T_{ki}c_l^k\right] \tag{4}$$

where c_l^G and c_l^k, $k = 1,\ldots,N_{Class}$, are concentrations of gas–phase and aqueous–phase species in the k-th droplet class. H_l are Henry's law constants for the l-th species. The mass transfer coefficient

$$k_t^{kl} := \left(\frac{r_k^2}{3D_g} + \frac{4r_k}{3\nu\alpha_l}\right)^{-1}$$

depends on the droplet size r_k, the gas diffusion coefficient D_g, the molecular speed ν and the mass accommodation coefficient α_l of the l-th species, Schwartz (1986). The indicator κ_l is equal to 1 if the species is soluble. In the other cases the second term will be dropped in both equations. The third term (4) describes the mass transfer between different droplet classes in linearized form. The mass transfer with the aerosol phase is not included in the model equations (4).

The reaction term R_l^A in the aqueous–phase contains fast dissociations. For instance

$$HNO_3(aq) \underset{k_-}{\overset{k_+}{\Longleftrightarrow}} H^+ + NO_3{}^-$$

gives the term

$$\frac{d\,HNO_3(aq)}{dt} = \ldots - k_+ \cdot HNO_3(aq) + k_- \cdot H^+ \cdot NO_3{}^- + \ldots = \ldots - k_- \cdot (\mathbf{k_e} \cdot HNO_3(aq) + H^+ \cdot NO_3{}^-) + ..$$

Such fast dissociations are considered as forward and backward reactions. The treatment as equilibria leads to large systems of differential algebraic equations (DAEs).

Table 1: Characteristics of tested reaction systems.

reaction mechanism	number of reactive species total	gas	aqua	number of reactions total	gas	Henry	disso	aqua
CHAUMERL	46	21	25	66	32	12	7	15
RADM2 / MM	116	62	54	270	158	20	17	75
RACM / CAPRAM 2.3	162	82	80	508	237	34	27	210
RACM / CAPRAM 2.4	220	83	137	654	237	34	55	328

Both approaches and different formulations of the DAEs are discussed in an earlier paper (Wolke and Knoth, 1996). The numerical costs for the implicit integration of the ODE and DAE system are comparable. Remark that in many cases only the equilibrium constant $\mathbf{k_e}$ is known. For such dissociations, k_- is set to a "large" prescribed value, 10^8 in our tests.

Chemical Mechanism. The modification or substitution of the chemical reaction system (gas and aqueous phase, phase transfer by Schwartz) can be done very easily, because the mechanism is read from a file. The syntax to describe the system is very easy and allows great flexibility. For the tests, four different multiphase chemistry systems of different complexity are used. Some characteristics of these mechanisms and corresponding references are given in Table 1.

Note that the photolytical reactions G1 in the mechanism CHAUMERL (Chaumerliac et al., 2000) is replaced by the reaction block given in the original system of Lelieveld and Crutzen (1991). Furthermore, all photolytical rates are time-dependent. The second mechanism RADM2–MM is more complex. Here the aqueous–phase mechanisms of Möller and Mauersberger (1995) is coupled with the RADM2 mechanism. The Chemical Aqueous Phase Radical Mechanism (CAPRAM) of Herrmann et al. (2000) contains a detailed treatment of oxidation of organic compounds with one and two carbon atoms, an explicit description of S(IV) oxidation and the reactions of OH, NO_3, Cl_2^-, Br_2^- and CO_3^-. The versions 2.3 and 2.4 are coupled with the gas phase mechanism RACM (Stockwell et al., 1997).

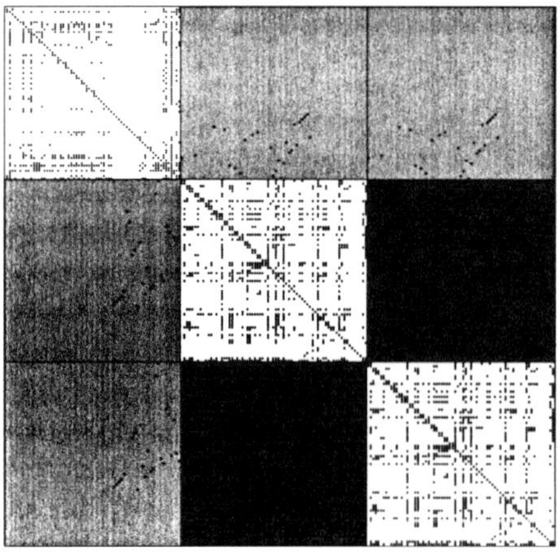

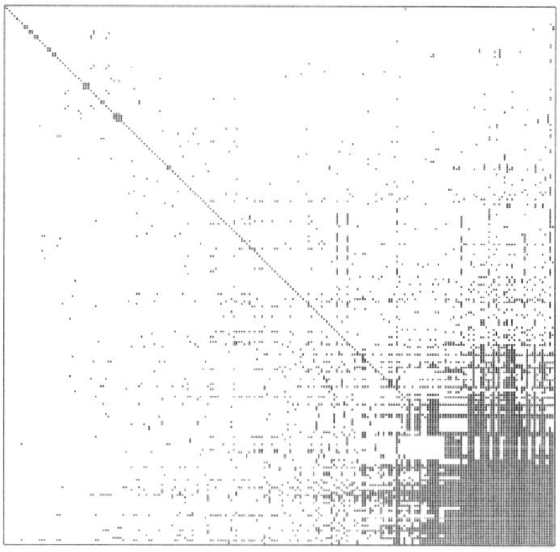

Figure 1: Sparse structure of Jacobian and LU decomposition

Table 2: Number of non-zeros in the Jacobian and the sparse LU decomposition for coupling and non-coupling between droplet classes.

mechanism	number of classes	number of variables	non-coupled Jacobian	non-coupled LU	non-coupled fill-in	coupled Jacobian	coupled LU	coupled fill-in
CHAUMERL	0	21	101	108	1.07			
	1	46	262	380	1.45			
	2	71	423	732	1.73	473	882	1.81
	3	96	584	1 005	1.72	734	1 615	2.28
	5	146	906	1 529	1.69	1 406	3 821	3.24
	10	271	1 711	2 894	1.69	3 961	13 299	4.75
RADM2 / MM	0	62	577	667	1.16			
	1	116	1 033	1 467	1.42			
	2	170	1 489	2 574	1.73	1 597	2 971	1.86
	3	224	1 945	3 587	1.84	2 269	5 485	2.42
	5	332	2 857	5 418	1.90	3 937	13 071	3.32
	10	602	5 137	9 274	1.81	9 997	47 470	4.75
RACM / CAPRAM 2.3	0	82	950	1 080	1.14			
	1	162	1 994	2 848	1.43			
	2	242	3 038	5 522	1.82	3 198	5 688	2.06
	3	322	4 082	8 564	2.10	4 562	12 199	2.67
	5	482	6 170	13 781	2.23	7 770	29 941	3.85
	10	882	11 385	26 967	2.37	18 585	110 837	5.96
RACM / CAPRAM 2.4	0	83	950	1 080	1.14			
	1	220	2 594	3 898	1.50			
	1	220	2 594	3 898	1.50			
	2	357	4 242	8 341	1.97	4 516	10 295	2.06
	3	494	5 890	13 102	2.22	6 712	20 246	2.67
	5	768	9 186	21 381	2.33	11 926	49 742	3.85
	10	1453	17 426	42 583	2.44	29 756	191 569	5.96

3. TIME INTEGRATION

Implicit-Explicit Methods. There is an ongoing effort to improve the numerics applied in chemistry-transport modelling. Special tuned implicit solvers for the time integration of the chemical kinetics system are used nowadays (e.g. Jacobson and Turco, 1994). They are superior in comparison to the classical QSSA schemes (Sandu et al., 1997). Such a solver with a step size control is included in the 3D chemistry-transport code MUSCAT (e.g. Wolke and Knoth, 1999), where the gas phase chemistry is integrated together with all vertical transport processes with the second order BDF method. In mesoscale applications large time steps can be obtained which are often equal to the horizontal advection time step. These step sizes can be achieved because horizontal advection and all other modelled processes are not coupled by the standard operator splitting method. They are treated by a new implicit-explicit integration scheme with full error control (Knoth and Wolke, 1998). Using these methods, several terms of the governing equations are treated explicitly, the remaining terms implicitly.

After the spatial discretization of transport terms, the mass balance equations (1) can be described as a large system of ODEs

$$c' = f_{Hor}(t,c) + f_{Vert}(t,c) + f_{TFrac}(t,c) + f_{Hen}(t,c) + f_{Chem}(t,c), \qquad (5)$$

where $f_{Hor}(t,c)$ represents the horizontal advection, $f_{Vert}(t,c)$ the vertical transport processes (including sedimentation), $f_{TFrac}(t,c)$ the mass transfer between different droplet classes, $f_{Hen}(t,c)$ the phase interchange with the gas phase and $f_{Chem}(t,c)$ includes the chemistry. It will be investigated for the aqueous phase: Whether and which processes can be treated explicitly and, furthermore, which loss of accuracy can be expected? For the gas phase, only $f_E := f_{Hor}$ is treated explicitly. The other terms

$$f_I := f_{Vert} + f_{TFrac} + f_{Hen} + f_{Chem} \tag{6}$$

in the right hand side of (5) are integrated implicitly.

BDF Scheme. In any implicit multi-step method the main computational task is the solution of a non–linear equation of the form

$$F(c) := c^{n+1} - C^n - \beta \Delta t_n f(t_{n+1}, c^{n+1}) = 0 \tag{7}$$

where $\beta > 0$ is a parameter of the integration method and C^n is a linear combination of previous values. If equation (7) is solved by a Newton–like method the new burden is to design a solution method for a linear equation system of the form

$$(I - \beta \Delta t J)\Delta c = b \tag{8}$$

where J is an approximation to $\partial f(t,c)/\partial c$. Normally the matrix J is held constant for several time steps and is recomputed either if the convergence is slow or if a fixed number of time steps has been reached. It is clear that the choice of a suitable method should be guided by the properties of the Jacobian.

An approximation of the Jacobian of the implicit part $f_I(t,c)$ is given as

$$J := J^{Vert} + J^{Box}$$

where the matrix

$$J^{Box} := J^{TFrac} + J^{Hen} + J^{Chem}$$

includes all chemical and mass transfer processes within one grid box, see Figure 1. The blocks in the diagonal are the Jacobians of the gas phase and aqueous phase reaction terms, respectively. The left and upper boundary blocks represent the phase interchange according to Schwartz. The "darker" diagonal matrices include the coupling terms resulting from the mass transfer between the droplet classes. The very sparse matrix J^{Box} as well as the band matrix J^{Vert} are generated explicitly and stored in a sparse form.

Sparse Solver for the Linear System (8). In MUSCAT the sparse linear system (8) is solved by Gauss–Seidel iterations. This technique has worked well for solving problems in atmospheric gas phase chemistry. In such applications three iterations are sufficient for the solution of the linear system (8). However, for aqueous chemistry mechanisms (e.g. WET in Sandu et al., 1997) the Gauss–Seidel iteration converges only slowly or even fails. In Wolke and Knoth (1999) an alternative approach is suggested, see also Verwer et al. (1998). In this scheme the splitting between chemistry and vertical diffusion is performed at the linear algebra level. The idea is to use an approximate factorization of the matrix $(I - \beta \Delta t J)$ into

$$(I - \beta \Delta t J) \approx (I - \beta \Delta t J^{Vert})(I - \beta \Delta t J^{Box}). \tag{9}$$

Then the linear system (8) can be solved by two sequential linear system solutions, one for the chemistry and one for the vertical transport:

$$(I - \beta \Delta t J^{Vert})b^* = b, \tag{10}$$

$$(I - \beta \Delta t J^{Box})\Delta c = b^*. \tag{11}$$

The linear system (10) is solved by a band solver without pivoting. The factorization of the transport Jacobian J^{Vert} is not stored. The factorization of the matrix and the

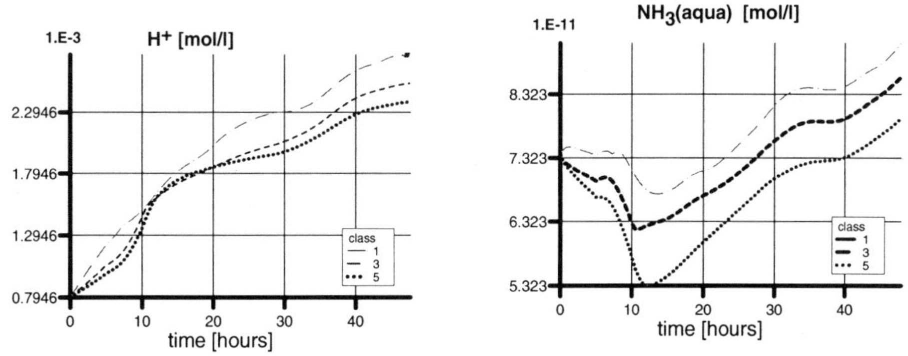

Figure 2: H^+ and $HN_3(aq)$ concentration in different droplet classes.

Table 3: The numerical efficiency for different prescribed relative tolerances *rtol* and maximal orders *maxord* of the BDF code.

chemical	BDF control		2 days simulation			30 s init.	
mechanism	maxord	rtol	steps	fcn	CPU [sec]	steps	fcn
CHAMERL	5	1.e-8	1475	1724	7.0	465	647
		1.e-5	904	975	6.6	133	202
		1.e-3	805	826	6.4	54	85
	2	1.e-3	651	658	6.2	51	79
RADM2 / MM	5	1.e-8	2522	3196	11.4	663	647
		1.e-5	1220	1490	8.5	294	417
		1.e-3	988	1224	8.2	183	274
	2	1.e-3	993	1219	8.1	187	262
RACM / CAPRAM 2.3	5	1.e-8	2198	2907	15.8	570	758
		1.e-5	1167	1451	11.2	251	337
		1.e-3	1013	1248	10.5	163	209
	2	1.e-3	942	1228	10.3	154	208
RACM / CAPRAM 2.4	5	1.e-8	5901	9427	46.4	2192	3795
		1.e-5	1326	1680	14.8	664	202
		1.e-4	1080	1381	13.1	436	723

solution of (10) is performed for each corrector iteration. For the direct solution of the sparse system (11) the diagonal Markowitz solver is implemented (e.g. Sandu et al., 1996). The sparse factorization is stored and only performed if the chemistry Jacobian J^{Box} has to be renewed. The fill-in by the sparse solver is investigated for the system with mass transfer between the droplet classes as well as for the system without this additional coupling, see Table 2. In the non-coupling case, the fill-in is moderate. If ten or more droplet classes are directly coupled then the fill-in is large. In this case an additional splitting of the Jacobian J^{Box} into J^{TFrac} and the matrix $J^{Hen} + J^{Chem}$ seems to be possible.

4. NUMERICAL TESTS

The described BDF solver is tested for the box case and different numbers of droplet classes. Four chemical mechanisms of different complexity are used, see Table 1. For the test scenario, time–constant microphysical values (liquid water content, droplet radius

Table 4: The numerical efficiency for a different number of droplet classes.

mechanism	linear solver	droplet classes	non-coupled steps	non-coupled fcn	non-coupled CPU [sec]	coupled steps	coupled fcn	coupled CPU [sec]
CHAMERL	sparse	1	759	778	6.3			
(rtol=10^{-3} maxord=2)		2	651	658	6.5	629	636	6.3
		3	706	723	6.7	706	727	6.7
		5	727	744	6.9	729	746	7.3
		10	734	746	7.7	732	742	9.6
		20	735	747	8.2	739	769	20.6
	full	2	697	703	6.9	697	709	6.9
		5	731	748	13.9	739	768	22.3
RADM2 / MM	sparse	1	933	1116	7.3			
(rtol=10^{-3} maxord=2)		2	993	1219	8.0	1020	1230	8.3
		3	1017	1246	8.6	957	1183	9.1
		5	967	1193	9.4	1013	1254	12.2
		10	955	1122	11.5	955	1105	25.2
		20	915	1029	14.7	929	1043	121.2
	full	2	1000	1190	10.7	989	1186	10.9
		5	1070	1340	24.9	1047	1286	36.0
RACM / CAPRAM 2.3	sparse	1	774	930	8.1			
(rtol=10^{-3} maxord=2)		2	960	1228	10.5	922	1167	10.9
		3	1090	1482	13.0	1036	1470	14.5
		5	1215	1788	19.6	1099	1574	25.3
		10	1375	2117	35.7	1352	2080	116.9
		15	1481	2365	43.0	1468	2120	219.2
	full	2	996	1264	18.3	939	1177	18.6
		5	1189	1687	78.7	1175	1656	121.3
RACM / CAPRAM 2.4	sparse	1	989	1128	9.7			
(rtol=10^{-4} maxord=5)		2	913	1030	11.3	917	1053	12.3
		3	954	1149	13.8	910	1046	15.6
		5	927	1077	17.1	929	1080	28.0
		10	867	989	26.1	910	1050	116.7
	full	2	939	1096	34.8	933	1057	33.7
		5	922	1056	187.5	922	1073	337.0

and liquid water fluxes between the droplet classes) are derived from cloud model data. The photolysis rates, the temperature and the humidity depend on the time of day. Time–constant emissions and dry deposition velocities are prescribed only for the gas phase species. At first an initialization run is performed over a time interval of 30 seconds. Then the two days simulation starts with the "balanced" final concentration of the initialization. In Table 3, the numerical costs are given for different prescribed relative tolerances $rtol$ and maximal BDF orders $maxord$. Here fcn denotes the number of function evaluations. The CPU time is measured on a RISC 6000 workstation with 512 MByte memory. Note that the benefit of higher order schemes becomes apparent especially for strong accuracy requirements.

The numerical efficiency for a varying number of droplet classes is compared in Table 4. For ten or more droplet classes, the coupling between the classes leads to large CPU times. This fact corresponds with the increase in the fill-in, see Table 2. The full matrix solver is not practicable for larger numbers of droplet classes. Figure 2 illustrates the differences in the aqueous phase chemistry for three droplet classes. The pH value for small droplets (class 1) is much higher than that of larger droplets (class 5). This fact results from the different mass transfer coefficients in the Schwartz approach.

5. CONCLUSIONS AND FURTHER INVESTIGATIONS

The BDF scheme with the sparse matrix solver is very efficient in the case of non–coupling between the droplet classes. In the coupling case an additional splitting in the approximation of the Jacobian seems to be a better approach. The choice of a suitable prescribed tolerance *rtol* is important for the efficiency of the scheme. But weak accuracy requirements can produce the breakdown of the integration scheme. The behaviour of this approach within multidimensional models will be investigated.

REFERENCES

N. Audiffren, M. Renard, E. Buisson and N. Chaumerliac (1998), Deviation from the Henry's law equilibrium during cloud events: A numerical approach of the mass transfer between phases and its specific numerical effects, *Atmospheric Research* **49**, 139–161.

N. Chaumerliac, M. Leriche and N. Audiffren (2000), Modeling of scavenging processes in clouds: Some remaining questions about the partitioning of gases among gas and liquid phases, *Atmospheric Research* **53**, 29–43.

H. Herrmann, B. Ervens, H.-W. Jacobi, R. Wolke, P. Nowacki and R. Zellner (2000), CAPRAM2.3: A Chemical Aqueous Phase Radical Mechanism for Tropospheric Chemistry, *J. Atmos. Chem.*, **36**, 231–284.

A.C. Hindmarsh (1983), ODEPACK, A systematized collection of ODE solvers, in: R.S. Steplman et al., Eds., *Scientific Computing*, (North–Holland, Amsterdam), 55-74.

M.Z. Jacobson and R.P. Turco (1994), SMVGEAR: A sparse-matrix, vectorized Gear code for atmospheric models, *Atmos. Environ.* **28**, 273–284.

O. Knoth and R. Wolke (1998), Implicit–explicit Runge–Kutta methods for computing atmospheric reactive flows, *Appl. Numer. Math.* **28**, 327–341.

J. Lelieveld and P.J. Crutzen (1991), The role of clouds in tropospheric chemistry, *J. Atmos. Chem.* **12**, 229-267.

D. Möller and G. Mauersberger (1995), Aqueous phase chemical reaction system used in cloud chemistry modelling, in: A. Flossmann, et al., Eds., *Clouds: Models and Mechanisms*, ISS, Garmisch–Partenkirchen, 77–93.

A. Sandu, J.G. Verwer, M. van Loan, G.R. Charmichael, F.A. Potra, D. Dabdub and J.H. Seinfeld (1997), Benchmarking stiff ODE solvers for atmospheric chemistry problems I. Implicit versus explicit, *Atmos. Environ.* **31**, 3151–3166.

A. Sandu, F.A. Potra, G.R. Charmichael and V. Damian (1996), Efficient implementation of fully implicit methods for atmospheric chemical kinetics, *J. Comput. Phys.* **129**, 101–110.

W.R. Stockwell, F. Kirchner, M. Kuhn and S. Seefeld (1997), A new mechanism for regional atmospheric chemistry modeling, *J. Geophys. Res. D22*, **102**, 25847–25879.

J.G. Verwer, W.H. Hundsdorfer and J.G. Blom (1998), Numerical time integration for air pollution models. *CWI Report MAS-R9825*, Centre for Mathematics and Computer Science, Amsterdam. To appear in Surveys for Mathematics in Industry.

R. Wolke and O. Knoth (1996), Numerical solution of air pollution models: Aqueous chemistry, *Zeitschrift für Angewandte Mathematik und Mechanik (ZAMM)*, Special Issue ICIAM95, **5**, 551-552.

R. Wolke and O. Knoth (1999), Implicit–explicit Runge–Kutta methods applied to atmospheric chemistry–transport modelling, *Environmental Modelling and Software*, in press.

S. Wurzler (1998), The scavenging of nitrogen compounds by clouds and precipitation: Part II. The effect of cloud microphysical parameterization on model predictions of nitric acid scavenging by clouds, *Atmospheric Research* **47–48**, 219–233.

A MODEL FOR TROPOSPHERIC MULTIPHASE CHEMISTRY: APPLICATION TO ONE CLOUDY EVENT DURING THE CIME EXPERIMENT

M. Leriche,[1] D. Voisin,[2] N. Chaumerliac,[1] and A. Monod[3]

[1]LaMP/OPGC, CNRS, Université Blaise Pascal
24 Avenue des Landais 63177 AUBIERE Cedex FRANCE
Tel. 4 73 407367, e-mail: leriche@opgc.univ-bpclermont.fr
[2]LGGE, CNRS
BP 96, 38402 Saint Martin d'Hères Cedex, FRANCE
[3]LCE, Université de Provence
3 place Victor Hugo, 13003 Marseille, FRANCE

INTRODUCTION

Tropospheric multiphase chemistry is still poorly understood, because the interactions between trace gases and the condensed phase are quite complex (Ravishankara, 1997). Clouds can interact with the tropospheric chemistry through different processes. They can remove gaseous species dissolved in the droplets by sedimentation or they can detrain them at various altitudes. Clouds can modify the actinic flux and the photochemistry is, then, more intense around the cloud top and less intense below (Madronich and Flocke, 1999). After their evaporation, clouds modify trace gas composition of the troposphere (Huret et al., 1994). Clouds can also indirectly perturb the gas chemistry through the uptake of soluble gases by cloud droplets and ice crystals, through aqueous chemical reactions (Lelieveld and Crutzen, 1991) and through microphysical redistribution (Grégoire et al., 1994).

There are two reasons for the lack of knowledge concerning such complex interactions. Measurements provide integrated results over several ten minutes especially in cloud water (see for example: Jaeschke et al., 1998) and do not provide detailed time resolved information about the scavenging process. Recent continuous measurements done with a CVI (Counterflow virtual Impactor) (Noone et al., 1988) introduce biases due to its principle of separating cloud phases. So, there are still a number of problems of chemical measurements in cloud phases (interstitial air, cloud water and particulate). The second problem is that multiphase model validation is still very limited due to the lack and the uncertainties in in-situ measurements. Sophisticated models often use theoretical scenarios rather than observations (Bott, 1999) or are based upon unrealistic microphysical conditions (Herrmann et al., 1999). Others are considering size-resolved cloud chemistry but most of the time limit their study to rather simple chemistry (Colvile et al., 1994; Wells et al., 1997).

Air Pollution Modeling and Its Application XIV, Edited by
Gryning and Schiermeier, Kluwer Academic/Plenum Publishers, New York, 2001

Nevertheless, those models need to be applied for studying processes and interactions and for supporting the interpretation of the multiphase measurements.

In this study, we present a box model for tropospheric multiphase chemistry that is based on an explicit chemical mechanism in both gas and aqueous phase. This model has been used for the modeling of a cloud event (Voisin et al., 2000) during the European CIME (Cloud Ice Mountain Experiment) experiment. Focusing on this particular event, a detailed analysis of the cloud chemical regime is presented with special emphasis on radical chemistry and on S(IV) to S(VI) conversion. The sensitivity of some volatile organic compounds oxidation pathways by radicals to the variable liquid water content is also examined.

DESCRIPTION OF THE MULTIPHASE BOX MODEL

The model is based on a chemical box model developed by Madronic and Calvert (1990) for the gas phase. The gaseous mechanism is explicit and has been adapted to a rural environment which includes 101 reactions and 48 gaseous species describing the chemistry of methane, sulfur, NOy and ammonia. The initial model has been extended to include the exchange of chemical species between gas and aqueous phase of the cloud (Audiffren et al., 1998), which is parameterized by the mass transfer kinetic formulation developed by Schwartz (1986). The chemical mechanism in aqueous phase is explicit with detailed chemistry of HOx, of chlorine, of carbonate, of N-species, of organic species with one carbon atom and of S-species. It includes 156 aqueous-phase reactions and 42 aqueous phase species. It has been developed based upon the work of Jacob (1986) updated with the latest kinetics and thermodynamic data available.

The mathematical formulation of the set of differential equation is written as:

$$\frac{dC_g}{dt} = P_g - D_g C_g + \frac{k_t C_w}{H^* RT} - L k_t C_g$$

$$\frac{dC_w}{dt} = P_w - D_w C_w + L k_t C_g - \frac{k_t C_w}{H^* RT}$$

where C_g and C_w are respectively the gaseous and aqueous concentrations in molec/cm^3 ; P_g and P_w, and D_g and D_w are respectively the gaseous and aqueous production (in molec/cm^{3}s) and destruction (in s^{-1}) terms ; k_t describes the mass transfer between gas and aqueous phases in s^{-1} (Schwartz, 1986) ; L is the liquid water content in vol/vol, H^* is the Henry's law effective constant in M/atm and R = 0.08206 atm/MK.

The set of differential equation is solved with the Gear's solver (Gear, 1971). The pH of the droplet is calculated at each time step by solving the ionic balance equation.

MODEL INITIALISATION

The multiphase box model is initialized with multiphase measurements made during the European CIME experiment by Voisin et al. (2000) during one cloud event. As a first step study, we used this experiment to evaluate the main reaction pathways in aqueous phase that we considered in our chemical model and to attempt some qualitative comparisons with real environmental data. In parallel, we use these real data to discuss possible discrepancies with more theoretical work as in Herrmann (1999).

The CIME experiment took place at the top of the Puy de Dôme (1465m), a station located in the center of France during winters 1997 and 1998. This station continuously measures temperature, intensity and horizontal wind direction, ozone concentration, SO_2 and

NO_x. The station is equipped with a wind tunnel that allows measurements of liquid water content, ice water content and effective radius. The multiphase measurements were performed by Voisin et al. (2000) during December 1997 in presence of mixed phase clouds and liquid clouds composed of supercooled droplets. They were measured in both gas and aqueous phases the concentrations of formic acid, nitric acid, nitrous acid, hydrochloric acid, sulfur dioxide and ammonia, and sulfate and pH value in aqueous phase. The minimum sampling time for aqueous phase was 15 minutes and was 30 minutes for the gas phase.

Table 1. Initial chemical conditions of the simulation.

Species	N_2	O_2	H_2O	O_3	NO	NO_2	CH_4	CO	HNO_3
Gas (ppbv)	$7,8.10^8$	$2,1.10^8$	$5,7.10^7$	37	0.9	6.2	1600	100	0.15
Aqueous (M)	0	0	cf. LWC	0	0	0	0	0	$5,45.10^{-4}$
Species	HNO_2	H_2O_2	HCHO	CH_3OH	SO_2	HCOOH	NH_3	HCl	S(VI)
Gas (ppbv)	0.054	0.07	6	2.25	0.45	0.28	0.13	0.07	0
Aqueous (M)	$2,59.10^{-6}$	0	0	0	$6,03.10^{-7}$	$4,89.10^{-6}$	$2,28.10^{-4}$	$1,40.10^{-4}$	$1,27.10^{-4}$

On December 13th, there was no ice and the meteorological situation was greatly stable with the presence of strato-cumulus made of supercooled droplets without rain. The origin of the air mass this day was north-north-east, a polluted origin. The simulation starts at 12.15 a.m. and ends at 3 p.m. local time. During this time, the actinic flux does not vary very much and the photolysis coefficients are taken constant and equal to the values at noon. The photolysis coefficients in aqueous phase are calculated from the data of Graedel and Weschler (1981) and Zellner et al. (1990). The temperature is taken constant equal to –3°C, the pressure is equal to 850hPa corresponding to the altitude of top of the Puy de Dôme. The droplet radius is 10 μm, typical value for a strato-cumulus cloud (Pruppacher and Klett, 1997). The liquid water content is taken variable following the measurements: its starting value is 0.26 $g.m^{-3}$, then, it decreases to 0.15$g.m^{-3}$ and increases to 0.22 $g.m^{-3}$ at 1.23 p.m. and finally decreases to 0.09 $g.m^{-3}$.

Table 1 presents the initial chemical conditions in both gas and aqueous phases. All values are taken from multiphase measurements performed by Voisin et al. (2000) or from continuous measurements at the Puy de Dôme station except for some gas concentrations. The values for CH_4 and CO are taken from Sillman (1990). For methanol and for H_2O_2, values are taken from measurements of Leibrock and Slemr (1997) and of Noone et al. (1991) respectively. Initial HCHO is taken from the CIME campaign (P. Laj, personal communication, 1999) on 15th of February 1997 in a polluted air mass with northern origin and in the presence of strato-cumulus without ice.

CLOUD CHEMSITRY DURING THE EVENT

The pH of the simulated cloud is very acidic between 3.3 and 2.7. In this section, model results will be analyzed and chemical regimes will be compared to the theoretical work of Herrmann et al. (1999), which includes more species (transition metals and C2 chemistry) but neglects some pathways such as $OHCH_2O_2$ reactivity. Also, in case of Herrmann's et al. (1999) the liquid water content is held constant throughout the simulation and the droplet radius is much smaller (1μm instead of 10μm in this work).

Radical chemistry

Radicals are very reactive species, responsible for the oxidizing capacity of the atmosphere. OH is the most important radical both in gas phase and aqueous phase. The sources of OH radical in aqueous phase is mainly transfer from the gas phase (60%), followed by the OH_{aq} production by ozone reaction with the superoxide ion (25%) and by nitrate ion photolysis (15%). Sinks for the OH_{aq} radical are due to the oxidation of the hydrated formaldehyde by OH (80%), leaving behind minor contributions of others reactions. In our scenario, a mean gas phase concentration of 2.10^6 molec.cm^{-3} and a mean aqueous phase concentration of about 2.10^{-14}M are found whereas in Hermann et al. (1999c) larger values are obtained in aqueous phase, of the order of 2.10^{-12} M. There are several reasons for these discrepancies. First, in this study, the net transfer rate of OH in aqueous phase is two orders of magnitude lower than the one in Herrmann et al. (1999) and can be explained by the different values of the droplet radius and of the liquid water content. As a result, the term: Lk_tC_g in Herrmann et al. (1999) is larger than in our case. Since transition metal chemistry is not considered in this work, a source of OH in aqueous phase is neglected from the reaction between hydrogen peroxide and Fe^{2+}/Cu^+.

Following Herrmann et al. (1999), for nitrate radical, the main source is also the transfer to aqueous phase and concentration levels are around 1.10^{-14}M.

For Cl_2^- radical, we obtain a comparable concentration in aqueous phase than Herrmann et al., around 2.10^{-13}M. Dichloride anion is formed from a fast equilibrium between hydrochloric ion and chlorine atom so that the sources for Cl_2^- correspond to those of the Cl atom. The Cl atom is formed for 85% from $ClOH^-$, formed by fast equilibrium between hydrochloric ion and hydroxyl radical, or for 15% from reaction of hydrochloric ion with NO_3. The dominant sink for this radical is due to reactions with the hydroperoxyl radical (90%) and it base, the superoxide ion (10%). Herrmann et al. (1999) do not find a significant sink from HO_2 radical because the presence of transition metals induces an important destruction of HO_2 radical.

Relative importance of radicals in the oxidation of volatile organic compounds

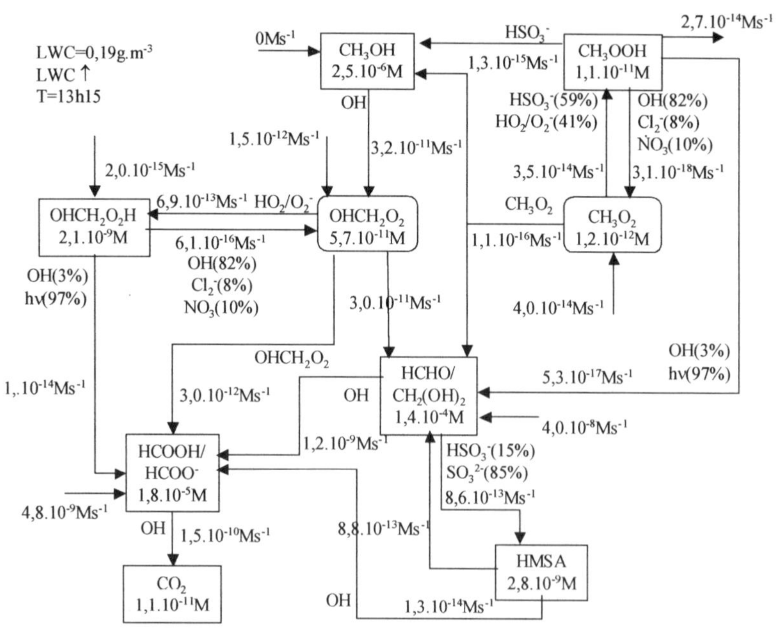

Figure 1. Concentration/flux diagram for oxidation pathways of COV at 1.15 p.m. and at LWC=0.19g.m^{-3}. Transfer fluxes of 0 M.s^{-1} (methanol) indicate Henry's law equilibrium.

Fig. 1 exhibits oxidation pathways of volatile organic compounds after one hour of simulation for a liquid water content of 0.19 gm^{-3}. At this time of the simulation, the cloud is in its growing stage. In the squares, concentrations of volatile organic compounds are given in M while in rounded boxes radical concentrations are provided. For each organic volatile compound, production and destruction fluxes are indicated with arrows, in order to give an estimate of their relative importance. For instance, methyl hydroperoxide CH_3OOH is destroyed by sulfite ion forming methanol and by various radicals forming CH_3O_2. This second destruction occurs with a relative contribution of reaction with OH dominating with 82% and with minor contributions of both Cl_2^- (8%) and NO_3 (10%). Same percentages are found in the oxidation of $OHCH_2O_2H$ since the same mechanism, similar to the hydrogen peroxide one is used. $OHCH_2O_2H$ is not considered in Herrmann et al. (1999), and neither is the intermediate radical $OHCH_2O_2$ in the oxidation of methanol. The lack of these two missing species can introduce an other source of formic acid as described in Monod and Carlier (1999). In another respect, our chemical model does not include volatile organic compounds with carbon atom greater than 1. This results in a missing source in aqueous phase from the oxidation of acetic acid of methyl peroxyl radical CH_3O_2. In Fig. 1, it is clearly seen that transfer from gas phase is the dominant source for volatile organic compounds, except for methyl hydroperoxide for which net aqueous phase production in the droplets is faster than its gaseous production. This transfer operates as converting gaseous alcohols into aqueous formic acid and aldehydes that further oxidize into formic acid.

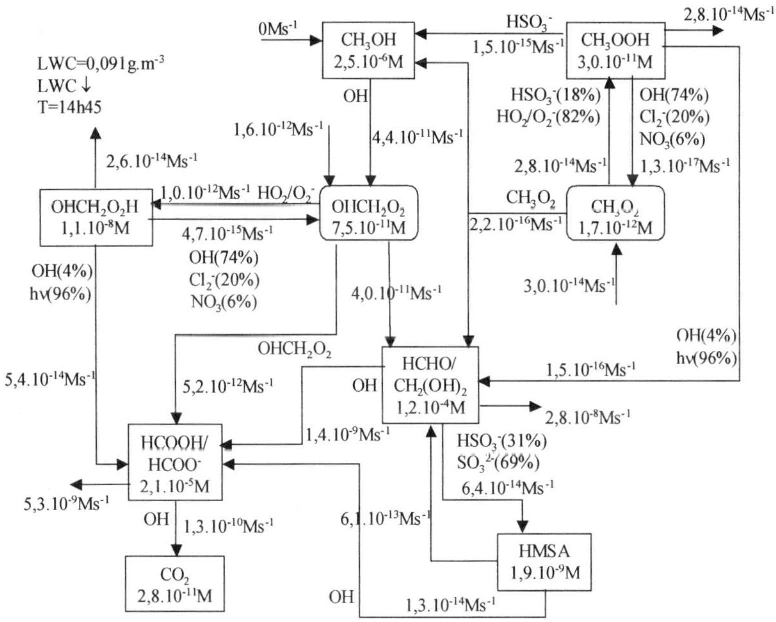

Figure 2. Same as Figure 1 but at 2.45 p.m. and at LWC=0.091$g.m^{-3}$.

Fig. 2 is similar to Fig. 1, except that concentration/flux diagram is presented later on in the simulation. At this time, the liquid water content is smaller and is decreasing and gaseous material is less available than at the start of the simulation since the gas phase is not replenished by any dynamical or microphysical processes. Due to this difference, we observe a general decrease in the percentages of oxidation pathways of organic volatile compounds by radicals. Compared to the case where the cloud water content was increasing, we notice that hydrated formaldehyde is not anymore in equilibrium with hydroxy methane sulfonate since there is less sulfite available to form HMSA. Also with decreasing liquid water content, all organic volatile compounds are degassing with negative transfer rates, contrasting with the situation previously observed in Fig. 1. When the cloud dissipates with a decrease of its liquid water content, droplets are more concentrated in aqueous phase.

Some oxidation pathways are clearly different in Fig. 1 versus Fig. 2. For instance, conversion from methyl peroxyl radical to methyl hydroperoxide is dominated either by reaction with HSO_3^- (Fig. 1) or by reaction with HO_2/O_2^- (Fig. 2). Finally, even if volatile organic compounds concentrations are in the same order of magnitude than in Herrmann et al. (1999), we observe slightly smaller values directly related to the fact that larger droplets are considered in our model, leading to a reduced total droplet surface associated with a reduced droplet number concentration.

Conversion of S(IV) to S(VI)

The conversion from S(IV) to sulfate is partly at the origin of the acidification of the cloud. The main path is through reaction with H_2O_2 at low pH and with O_3 at high pH (Graedel and Weschler, 1981). The importance of metal ions, as catalysers has also been stressed in the literature (Grgic et al., 1991). In Fig. 3, production pathways of sulfate are drawn and in our conditions peroxonitric acid is the main oxidizing agent of S(IV) into S(VI). The only reference mentioning the possible importance of peroxonitric acid is Amels et al. (1996). In order to explain such a phenomena, we has examined its gaseous budget. In clear air, this acid is in equilibrium with HO_2+NO_2. With the presence of clouds, this equilibrium is disrupted because of the high solubility of peroxonitric acid. Due to its net gaseous transfer destruction, the fraction that enters the aqueous phase is very significant to oxidize S(IV), especially in the present conditions of low H_2O_2 concentrations, increased by the fact that H_2O_2 is not refilled by any entrainment or other process that will renew its gas phase concentration. In order to support such an hypothesis, a sensitivity test has been performed with initializing the chemical model with 1ppb of H_2O_2. In this case, we retrieve more classical conclusions. H_2O_2 is the main oxidant, but is closely followed by the peroxonitric acid to a less extent. At the end of the simulation, other oxidation pathways appear with the reactions of Cl^- and H_2O_2 with the sulfate radical anion, due to the artifact of the box model that neglects cloud entrainment of fresh H_2O_2 oxidant.

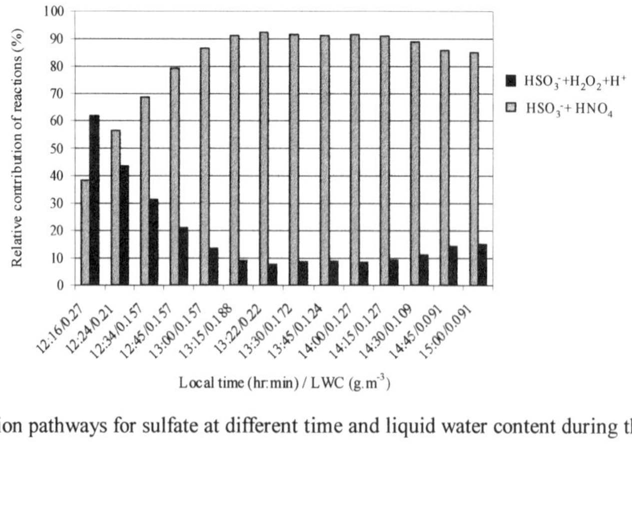

Figure 3. Production pathways for sulfate at different time and liquid water content during the simulation.

During the CIME experiment, the high NO_x regime and the low H_2O_2 concentration, combined with unrealistic dynamical conditions in the chemical box model lead to the possible contribution of very soluble peroxonitric acid in the conversion of S(IV) to S(VI).

Moreover, this result highlights the fact that peroxonitric acid can contribute significantly to S(IV) to S(VI) conversion and can compete with H_2O_2 oxidation of S(IV) in aqueous phase and could partly explain degassing of H_2O_2 as observed by Laj et al. (1997).

CONCLUSION

In this study, we have used a multiphase box model which takes into account an explicit chemistry mechanism for both gas and aqueous phase for a rural environment and the kinetic of mass transfer between phases (Schwartz, 1986). The model is, then, initialized with multiphase measurements performed by Voisin et al. (2000) during the CIME experiment. The 13th December 1997, the cloud chemistry is mainly governed by high NO_x and high formaldehyde levels and by an acidic pH in the droplets. A comparison of the model results is performed versus recent theoretical results from Herrmann et al. (1999), who proposed a slightly different chemical scheme, including C2 chemistry and transition metal chemistry but neglecting some reaction pathways, such as the one involving $OHCH_2O_2$ radical and using unrealistic microphysical cloud conditions (cloud duration, constant liquid water content, small droplet radius).

On the basis of this confrontation with theoretical results from Herrmann et al. (1999), a detailed analysis of radical chemistry and of the relative importance of those radicals in the oxidation of volatile organic compounds is performed. Depending on the stage of evolution of the cloud, either growing or dissipating, a different partitioning of organic volatile compounds is observed between the gas phase and the aqueous phase. These differences lead to different oxidation pathways that should be accounted for in simulating multiphase chemistry.

Then, evidence for the possible role of aqueous peroxonitric acid in the conversion of S(IV) to S(VI), already underlined by Amels et al. (1996) has been found in case of low hydrogen peroxide concentration levels associated to the high NO_x regime observed during the experiment.

These results show the capability of such a chemical box model, including exhaustive and explicit multiphase chemistry of highlighting wild variability in reaction and oxidation pathways, that cannot be directly put in evidence through actual measurements. They also indicate the need for lacking observations of gases such as formaldehyde that would greatly help this kind of approach.

ACKNOWLEDGMENTS

This work was supported by the "Programme National de Chimie Atmosphèrique" (PNCA) of the INSU (Institut des Sciences de l'Univers). Computer resources were provided by I.D.R.I.S (Institut du développement et des Ressources en Informatiques Scientifique), project n°000187.

REFERENCES

Amels, P., Elias, H., Götz, U., Steinges, U., Wannowius, K.J., 1996, Kinetic investigation of the stability of peroxonitric acid and of its reaction with sulfur(IV) in aqueous solution, In: *Heterogeneous and*

Liquid Phase Processes, Transport and Chemical Transformation of Pollutants in the Troposphere Vol. 2, P. Warneck, ed., Springer, Berlin.

Audiffren, N., Renard, M., Buisson, E., and Chaumerliac, N., 1998, Deviations from the Henry's law equilibrium during cloud events: a numerical approach of the mass transfer between phases and its specific numerical effects, *Atm. Res.* 49:139.

Bott, A., 1999, A numerical model of the cloud-topped planetary boundary-layer: chemistry in marine stratus and the effects on aerosol particles, *Atmos. Environ.* 33:1921.

Colvile, R.N., et al., 1994, Computer modelling of clouds at Kleiner Feldberg, *J. Atmos. Chem.* 19:189.

Gear, C.W., 1971. Numerical initial value problems in ordinary differential equation. pp. 158-166. Prentice-Hall, Englewod Cliffs, NJ.

Graedel, T.E., Weschler, C.J., 1981, Chemistry within aqueous atmospheric aerosols and raindrops, *Rev. Geophys. Space Phys.* 19:505.

Grégoire, P.J., Chaumerliac, N., and Nickerson, E.C., 1994, Impact of cloud dynamics on tropospheric chemistry: advances in modeling the interactions between microphysical and chemical processes, *J. Atmos. Chem.* 18:247.

Grgic, I., Hudnik, V., Bizjak, M., Levect, J., 1991, Aqueous S(IV) oxidation I. Catalytic effects of some metal ions, *Atmos. Environ.* 25A:1591.

Herrmann, H., Ervens, B., Jacobi, H.-W., Wolke, R., Nowacki, P., Zellner, R., 1999, CAPRAM2.3: A chemical aqueous phase radical mechanism for tropospheric chemistry, *J. Atmos. Chem.* in press.

Huret, N., Chaumerliac, N., Isaka, H., and Nickerson, E.C., 1994, Impact of different microphysical schemes on the dissolution of highly and less soluble non-reactive gases by cloud droplets and raindrops. *J. Appl. Meteor.* 33:1096.

Jacob, D.J., 1986, Chemistry of OH in remote clouds and its role in the production of formic acid and peroxymonosulfate, *J. Geophys. Res.* 91:9807.

Jaeschke, W., Dierssen, J.P., Günther, A., Schumann, M., 1998, Phase partitioning of ammonia and ammonium in a multiphase system studied using a new vertical wet denuder technique, *Atmos. Environ*. 32:365.

Laj P., et al., 1997, Cloud processing of soluble gases, *Atmos. Environ.* 31:2589.

Leibrock, E., Slemr, J., 1997, Method for measurements of volatile oxygenated hydrocarbons in ambient air, *Atmos. Environ.* 31:3329.

Lelieveld, J., and Crutzen, P.J., 1991, The role of clouds in tropospheric photochemistry, *J. Atmos. Chem.* 12:229.

Madronich, S., Calvert, J.G., 1990, The NCAR Master Mechanism of the gas phase chemistry, NCAR technical Note, TN-333+SRT, Boulder Colorado.

Madronich, S., Flocke, S., 1999, The role of solar radiation in atmospheric chemistry, In: *Environmental Photochemistry*, P. Boule, ed., Springer, Berlin.

Monod, A., Carlier, P., 1999, Impact of clouds on the tropospheric ozone budget: Direct effect of multiphase photochemistry of soluble organic compounds, *Atmos. Environ.* 33:443.

Noone K.J., Charlson, R.J., Covert, D.S., Ogren, J.A., Heintzenberg, J., 1988, Design and calibration of a counterflow virtual impactor for sampling of atmospheric fog and cloud droplets, *Aerosol Sci. Technol.* 8:235.

Noone K.J., Ogren, J.A., Birgitta Noone, K., Hallberg, A., Fuzzi, S., Lind, J.A., 1991, Measurements of the partitioning of hydrogen peroxide in a stratiform cloud, *Tellus* 43B:280.

Pruppacher, H.R., Klett, J.D., 1997, Microphysics of Clouds and Precipitation 2nd Ed., Reidel, Dordrecht.

Ravishankara, A.R., 1997, Heterogeneous and multiphase chemistry in the trposphere, *Science* 276:1058.

Schwartz, S., 1986, Mass-transport considerations pertinent to aqueous phase reactions of gases in liquid water clouds, in: *Chemistry of mulitphase systems*, W. Jaeschke, ed., Springer-Verlag, Berlin.

Sillman, S., Logan, J.A., Wofsy, S.C., 1990, The sensivity of ozone to nitrogen oxides and hydrocarbons in regional ozone episodes, *J. Geophys. Res.* 95:1837.

Voisin, D., Legrand, A., and N. Chaumerliac, 2000, Investigations of the scavenging of acidic gases and ammonia in mixed liquid solid water clouds at the Puy de Dôme mountain (France), *J. Geophys. Res.* 105:6817.

Wells, M., et al., 1997, The reduced nitrogen budget of an orographic cloud, *Atmos. Environ.* 31:2599.

Zellner, R., Exner, M., Herrmann, H., 1990, Absolute OH quantum yields in the laser photolysis of nitrate, nitrite and dissolved H_2O_2 at 308 and 351 nm in the temperature range 278-353K, *J. Atmos. Chem.* 10:411.

DISCUSSION

F. MÜLLER: How did you treat the photolysis rates inside the droplet?

M. LERICHE: The actinic flux inside the droplets is equal to the actinic flux in the interstitial air multiplied by a factor 1.7 (Madronich and Flocke, 1999). The photolysis rates in the droplets are then calculated with data from Graedel and Weschler (1981) and Zellner et al. (1990).

F. MÜLLER: For your simulations, you used a fixed radius of 10μm but you varied the liquid water content. Did you solve extra equations for the number density?

M. LERICHE: We did not solve any extra equation for the number density because the microphysical processes are not taken into account in the first approach presented in this paper. So, we have been considering here a monodisperse cloud with droplets of 10μm and with an evolving liquid water content, taken from measurements. During the cloud event, the effective droplet radius has a value around 11μm with small variations around +/- 2μm as measured by a Gerber instrument. So, we used a constant value for the droplet radius corresponding to the typical value for a stratocumulus cloud without rain, as a mean radius close to the effective radius measured. The coupling of the chemical box model with a module of quasi-spectral microphysics is underway and we still have to couple this new model with an air parcel model.

PROGRESS IN COUNTER-GRADIENT TRANSPORT THEORY

Han van Dop[1] and Gé Verver[2]

[1] Institute for Marine and Atmospheric Research
Utrecht University
p.o. box 80005
3508 TA Utrecht, The Netherlands
[2]Royal Netherlands Meteorological Institute
p.o. box 201
3730 AE De Bilt, The Netherlands

INTRODUCTION

Convective boundary layers are characterised by almost uniform, monotonous mean profiles of temperature etc. Deardorff (1972) was one of the first who noted that the upper parts of these profiles had small positive gradients and he proposed to modify the usually applied flux-gradient relationship in the K-theory approach according to

$$\overline{w\theta} = \phi^h = K\left(\frac{\partial\Theta}{\partial z} - \gamma\right), \tag{1}$$

where K is the eddy diffusion coefficient for heat and γ the countergradient part of the mean temperature profile. The observations suggested that the order of magnitude for γ was approximately $0.7Kkm^{-1}$(Deardorff (1966)). Based on the Reynolds averaged second moment equations, Deardorff (1966) gave theoretical support for this term.

More recently, in studies to improve vertical flux parameterizations in GCM's, and facilitated by much more evidence and knowledge of the CBL structure, new schemes and parameterizations have been developed (e.g., Wyngaard and Brost (1984); Fiedler and Moeng (1985); Holtslag and Moeng (1991); Holtslag and Boville (1993); Schumann (1993); Hamba (1993), Robson and Mayocchi (1994), Zilitinkevich et al. (1999), van Dop and Verver (2000)). The apparent asymmetry in the vertical turbulent transport of the CBL - upward transport takes place in somewhat smaller but highly turbulent thermals, while downward transport is dominant in more 'quiet' areas of a CBL - plays a major role in these studies. This asymmetry can be described by adopting distribution functions for the vertical velocity which may be gaussian but depending on height, or by adopting skewed (uniform) distribution functions (or both). In addition 'mass flux' approximations proved to be useful in CBL applications (Randall et al. (1992))

Based on the analysis in van Dop and Verver (2000), a simple generalisation of the various counter-gradient approaches (Deardorff (1972) (D), Holtslag and Moeng (1991) (HM) and Wyngaard and Weil (1991) (WW)), is proposed, leading to a modified diffusion equation, similar to one suggested by Wyngaard and Moeng (1990) for the transport of a passive scalar.

THEORY

It is easily demonstrated that (1) and similar expressions are a special case of:

$$(1 + s(z)\frac{\partial}{\partial z} + \tau(z)\frac{\partial}{\partial t})\phi = -K(z)\frac{\partial C}{\partial z}, \tag{2}$$

(van Dop and Verver (2000)), where ϕ denotes the scalar flux $\overline{wc}$. Here, $s(z)$ and $\tau(z)$ are height dependent length and time scales, which, in addition to $K(z)$, further characterize the diffusion. With the mean equation for the scalar C,

$$\frac{\partial C}{\partial t} + \frac{\partial \phi}{\partial z} = 0, \tag{3}$$

we may eliminate ϕ and obtain for constant τ, K and s a modified diffusion equation:

$$\tau\frac{\partial^2 C}{\partial t^2} + \frac{\partial C}{\partial t} + s\frac{\partial^2 C}{\partial z \partial t} = K\frac{\partial^2 C}{\partial z^2}, \tag{4}$$

which is consistent with 'Taylor diffusion' and can be transformed into the 'Telegraph equation' (see van Dop and Verver (2000)).

REYNOLDS EQUATIONS

The equations for the first three moments of a passive scalar in horizontally homogeneous conditions are

$$\frac{\partial C}{\partial t} = -\frac{\partial \phi}{\partial z} \tag{5a}$$

$$\frac{\partial \phi}{\partial t} = -\overline{w^2}\frac{\partial C}{\partial z} - \frac{\partial m}{\partial z} - \frac{\phi}{\tau} \tag{5b}$$

$$\frac{\partial m}{\partial t} = -\overline{w^3}\frac{\partial C}{\partial z} - 2\overline{w^2}\frac{\partial \phi}{\partial z} - \phi\frac{\partial \overline{w^2}}{\partial z} - \frac{m}{\tau}, \tag{5c}$$

where $\phi \equiv \overline{wc}$ and $m \equiv \overline{w^2 c}$. We have introduced the timescale τ by

$$\frac{\phi}{\tau} \approx \frac{g}{T}\overline{\theta c} - \overline{c\frac{\partial p}{\partial z}}$$

The fourth moment $\overline{w^3 c}$ was approximated by $3\,\overline{w^2}\phi$. The constant of proportionality follows from the assumed Gaussianity of the dynamics of the CBL. The mass flux approximation for the CBL also provides an estimate for this relationship:

$$\overline{w^3 c} = \frac{1 - 3\sigma + 3\sigma^2}{\sigma(1-\sigma}\overline{w^2}\phi, \tag{6}$$

where σ defines the updraft area fraction of the CBL. The latter varies between 0.35 and 0.5 (Petersen (1999)), yielding for the constant of proportionality between the moments

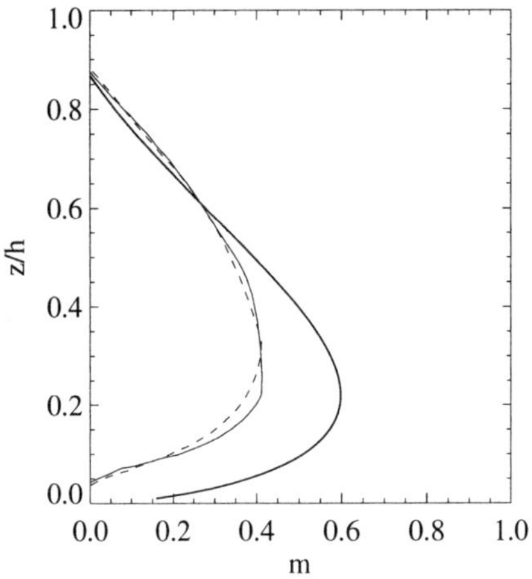

Figure 1: Vertical profiles of the third moments $m \equiv \overline{w^2c}$. The thick line depicts our result, the thin line is an LES result from Moeng and Wyngaard (1989) and the dashed line is the parametrisation of Zilitinkevich et al. (1999)

in (6) a considerably lower value between 1 and 1.5. In the rest of the paper, however, we shall adopt the value 3, since it agrees well with experimnetal CBL dispersion characteristics (de Baas et al. (1986)). Equations (5a,b,c) were also considered by Zilitinkevich et al. (1999) who proposed an advection/diffusion parameterisation for the third order correlation. In their parameterisation, however, two additional adjustable constants were introduced facilitating a better fit through the data.

In the stationary, zero skewness case (5c) simplifies to an algebraic equation in m. Upon substitution in (5b) and solving for the concentration gradient we arrive at an equation of the form (2) with for $K(z)$ and $s(z)$

$$K = \frac{\overline{w^2}\tau}{1 - \tau\frac{\partial}{\partial z}\left(\tau\frac{\partial \overline{w^2}}{\partial z}\right)} \tag{7}$$

$$s = -\frac{2\tau\frac{\partial \overline{w^2}\tau}{\partial z} + \tau^2\frac{\partial \overline{w^2}}{\partial z}}{1 - \tau\frac{\partial}{\partial z}\left(\tau\right)\frac{\partial \overline{w^2}}{\partial z}}, \tag{8}$$

respectively. This gives some theoretical support for the proposed equation (4). In the more general case, with skewness $S \neq 0$, a more complicated expression for the concentration profile follows.

RESULTS

We have examined the above formulations for the dispersion of a passive scalar, in a 'standard' CBL, using a heat entrainment ratio -0.25. This enables comparison with (LES) data.[1] We have used a large number of experimental data in order to determine

[1]Hereafter, all variables are expressed in dimensionless units through combinations with w_*, θ_*, c_*, and h.

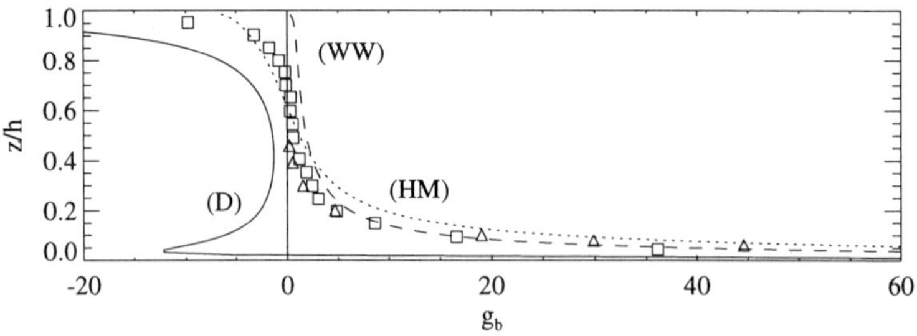

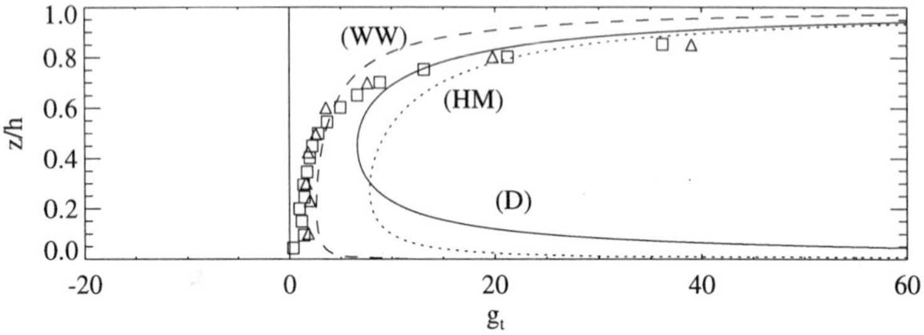

Figure 2: g_t and g_b functions for the D, HM and WW case. Full line (D), dotted (HM) and dashed (WW). The symbols refer to LES data from Moeng and Wyngaard (1989).

standard profiles for the timescale and second and third moments of the vertical velocity (summarized in Moeng and Wyngaard (1989) and Garratt (1992)).

First we have solve m from Eq.(5a,b,c) using as boundary condition $m(1) = 0$. The resulting CBL-profile (cf. figure 1) of m overestimates the LES simulations (the shape is similar but the maximum is 0.4 whereas here it is 0.6). This casts some doubts on the Gaussian approximation made in (5c). However, we have done the similar evaluation for the moisture fields from AMTEX and GATE and got in this case much better predictions for $\overline{w^2q}$.

Next we have compared various counter-gradient approaches (D, HM and WW) with LES data. K and s functions were derived and dimensionless profiles, g_t and g_b, evaluated. We have plotted g_t and g_b in Figure 2 together with the data from Moeng and Wyngaard (1989) for two resolutions of their LES model. The D-parametrization compares poorly with the large eddy results. Results for the HM and WW parametrizations are not very conclusive: in the lower part of the CBL, WW fits well to the LES data, both in the bottom-up and top-down case. In the upper part of the boundary layer LES data and HM agree fairly. The agreement in the top-down case is generally less.

Finally the role of the skewness was determined by comparing the full numerical solution of (5a,b,c) with the one for $S = 0$ (Figure 3). We observe that the latter gives a much better fit to the LES data, suggesting that the inclusion of the third moment improves the prediction.

CONCLUSION

We have reformulated the diffusion equation in a way that includes counter-gradient

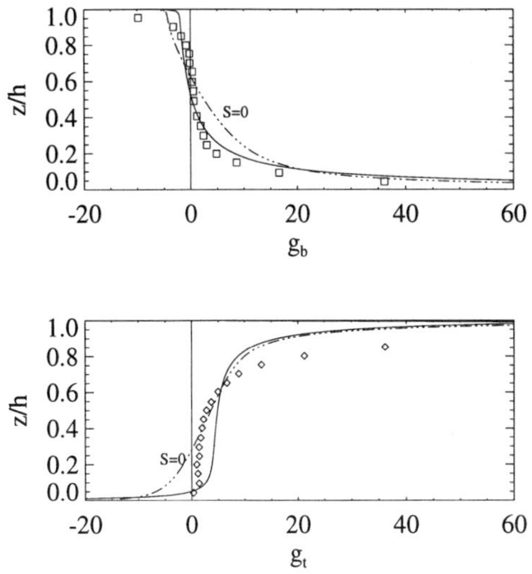

Figure 3: Dimensionless profiles g_t and g_b. solid line: full Reynolds averaged approximation (5a,b,c); dashed line: the same for $S = 0$. The symbols refer to LES data from Moeng and Wyngaard (1989).

effects. The Reynolds-averaged equations up to third order gives some support (in appropriate approximations) for this formulation.

The general shape of the CBL profiles is largely determined by the inhomogeneities in the turbulence profiles (variance, timescale etc.) but the extended diffusion equation (2) provides a small but essential counter-gradient contribution.

In a simulation using the third-order Reynolds-averaged equations, both bottom-up and in top-down diffusion are counter-gradient. In the top-down case the counter-gradient range is, however, limited to the lowest 5% of the CBL. The fit with the LES data is fair both for the top-down and bottom-up diffusion.

The impact of the skewness can be nicely demonstrated with the Reynolds averaged model (5a,b,c) by comparing runs with and without skewness. Though the profiles are qualitatively the same, the agreement with the data improves considerably when the skewness is taken into account (see figure 3). This conclusion is in line with the findings of Cuijpers and Holtslag (1998).

References

Cuijpers, J. W. M. and Holtslag, A. A. M. (1998). Impact of skewness and nonlocal effects on scalar and buoyancy fluxes in convective boundary layers. *J. Atmos. Sci.*, 55:151–162.

de Baas, A. M., van Dop, H., and Nieuwstadt, F. (1986). An application of the langevin equation for inhomogeneous conditions to dispersion in a convective boundary layer. *Quart.J.Roy.Meteorol.Soc.*, 112:165–180.

Deardorff, J. (1966). The counter-gradient heat flux in the lower atmosphere and in the laboratory. *J.Atmos.Sci.*, 23:503–506.

Deardorff, J. (1972). Theoretical expression for the countergradient vertical heat flux. *J.Geophys.Res.*, 77:5900–5904.

Fiedler, B. and Moeng, C.-H. (1985). A practical integral closure model for mean vertical transport of a scalar in a convective boundary layer. *J.Atmos.Sci.*, 42:359–363.

Garratt, J. R. (1992). *The Atmospheric Boundary Layer.* Cambridge University Press. 316 pp.

Hamba, F. (1993). A modified first-order model for scalar diffusion in the convective boundary layer. *J.Atmos.Sci.*, 50:2800–2810.

Holtslag, A. A. M. and Boville, B. (1993). Local versus nonlocal boundary-layer diffusion in a global climate model. *J.Clim.*, 99:99–99.

Holtslag, A. A. M. and Moeng, C.-H. (1991). Eddy diffusivity and countergradient transport in the convective atmospheric boundary layer. *J.Atmos.Sci.*, 48:1690–1698.

Moeng, C.-H. and Wyngaard, J. (1989). Evaluation of turbulent transport and dissipation closures in second-order modeling. *J.Atmos.Sci.*, 46:2311–2330.

Petersen, A. (1999). *Convection and chemistry in the atmospheric boundary layer.* PhD thesis, IMAU, Utrecht University.

Randall, D., Shao, Q., and Moeng, C.-H. (1992). A second-order bulk boundary-layer model. *J.Atmos.Sci.*, 49:1903–1923.

Robson, R. and Mayocchi, C. (1994). A simple model of countergradient flow. *Phys. Fluids*, 6:1952–1954.

Schumann, U. (1993). Transport asymmetry in skewed convective circulations. *J.Atmos.Sci.*, 50:116–119.

van Dop, H. and Verver, G. (2000). Countergradient transport revisited, submitted to the *J. Atmos. Sci.*

Wyngaard, J. and Brost, R. A. (1984). Top-down and bottom-up diffusion of a scalar in the convective boundary layer. *J.Atmos.Sci.*, 41:102–111.

Wyngaard, J. and Moeng, C.-H. (1990). Large-eddy simulation in geophysical turbulence parameterization. 99 999, workshop on Large-Eddy Simulation: where do we stand?, St. Petersburg, Florida.

Wyngaard, J. and Weil, J. (1991). Transport asymmetry in skewed turbulence. *Phys. Fluids*, A(3):155–162.

Zilitinkevich, S., Gryanik, V., Lykossov, V., and Mironov, D. (1999). Third-order transport and nonlocal turbulence closures for convective boundary layers. *J.Atmos.Sci.*, 56:3463–3477.

DISCUSSION

R. BORNSTEIN: How was the moisture treated in the formulation?

H. van DOP: All the turbulence data were based on observed characteristics of the (quasi-stationary) dry CBL, with an entrainment ratio of -0.25. Moisture was treated as a passive scalar diffusing in this CBL. However, here we have an extra degree of freedom since the upper boundary condition of the moisture flux can be chosen more or less arbitrarily, depending on the water vapour concentration difference between the top of the CBL and the overlying layer. We have done simulations with entrainment ratios for moisture of 0.5 and 1, respectively. Our simulation tends to underestimate the data, which show, however, an enormous scatter.

E. GENIKHOVICH: It is known that the telegraph equation cannot guarantee positiveness of its solution in the 3-D case. Have you checked this problem in your approach?

H. van DOP: No, I did not. Ianushets (Z. Angew. Math. Phys., **46**, 445-458, 1994) notes that the telegraph equation guarantees positive definite solutions only in the 1-D case. This is the case I have addressed here. It would indeed be interesting, also in view of practical applications, to investigate the conditions where also the 3-D telegraph equation yields positive solutions, given proper initial conditions.

D. W. BYUN: In your presentation you did not show how the variance of the vertical velocity component is parameterized. Where did you get $\overline{w^2}$?

H. van DOP: I used empirical profiles to characterize the CBL turbulence.:

$$\overline{w^2} = 1.3\ z^{2/3}\ (1 - 0.65z)^2$$

$$\overline{w^3} = S\,\sigma_w^3$$

(all parameters in dimensionless form, using w_* and h). S is the skewness parameter, which in this study was put to 1.

PARAMETERISATION OF FLUXES OVER A SPARSE BOREAL FOREST AT HIGH LATITUDE

Ekaterina Batchvarova[1,2], Sven-Erik Gryning[1], and H. A. R. de Bruin[3]

[1]Risø National Laboratory, Denmark
[2]Bulgarian Academy of Sciences, Bulgaria
[3]Wageningen University, the Netherlands

INTRODUCTION

Measurements carried out in Northern Finland on radiation and turbulent fluxes over a sparse, sub-arctic boreal forest with snow covered ground were analysed. The measurements represent winter conditions characterised by low sun angles. At low solar angles the forest shades most of the snow surface. Therefore an important part of the radiation never reaches the snow surface but is absorbed by the forest. The sensible heat flux above the forest was fairly large, reaching more than 100 Wm^{-2} during the afternoon. The measurements of sensible heat flux within and above the forest revealed that the heat flux from the snow surface is negligible and the heat flux above the forest stems from warming of the trees. A simple model for the surface energy balance of a sparse forest is presented. The model treats the diffuse and direct global radiation separately. It introduces a factor that accounts for the shading of the ground at low solar angles, and a parameter that deals with the transmittivity of the forest.

SITE AND INSTRUMENTATION

Meteorological measurements of energy fluxes were carried out in March 1997 at the Sodankylä Meteorological Observatory (67° 22' N, 26° 38' E) at Tähtelä in Finnish Lapland. The observatory is located in a sparse coniferous forest of typically 6-8 meter tall trees. The area is typical for the subarctic Northern Finland with coniferous forests and large open mires dominating the landscape. The area is rather flat both on small and large scales, with hills reaching 500 meters height within 20 km. During the experimental period the ground was covered with snow, the lakes were frozen, and the trees most of the time were without snow cover. Day and night were approximately equally long.

The instrumentation of the site during the experiment was rather comprehensive. Wind and temperature fluctuations were measured with a frequency of 10 Hz by the use of

Air Pollution Modeling and Its Application XIV, Edited by
Gryning and Schiermeier, Kluwer Academic/Plenum Publishers, New York, 2001

ultrasonic anemometers (Solent Research 3D sonic anemometers) in and above the sparse forest, mounted on a mast at heights of 2, 6, 12 and 18 meters. In addition measurements of humidity fluctuations were performed at 18 meters height by use of an OPHIR optical hygrometer. From these measurements half-hourly values of wind speed and direction, fluxes of momentum and sensible heat at the four levels and the latent heat at the upper level were determined. The measurement programme started on March 13 and ended on March 24, 1997. As part of the continuous observation programme at the Sodankylä Meteorological Observatory, hourly averaged measurements of global and diffuse (by use of a shading ring) short wave radiation were performed with pyranometers, mounted at 16 meters height well above the forest to obtain a free horizon. Hourly observations of cloud cover were performed as part of the standard synoptic observation programme.

ANALYSIS OF MEASUREMENTS

In order to illustrate the behaviour of the daily variation of the temperature and sensible heat flux in and above the forest, the diurnal variation of the temperature and sensible heat flux, averaged over the experimental period, was calculated. Figure 1 (upper panel) shows the daily variation of the averaged temperature for the period 13 to 23 March, based on half-hourly temperature measurements.

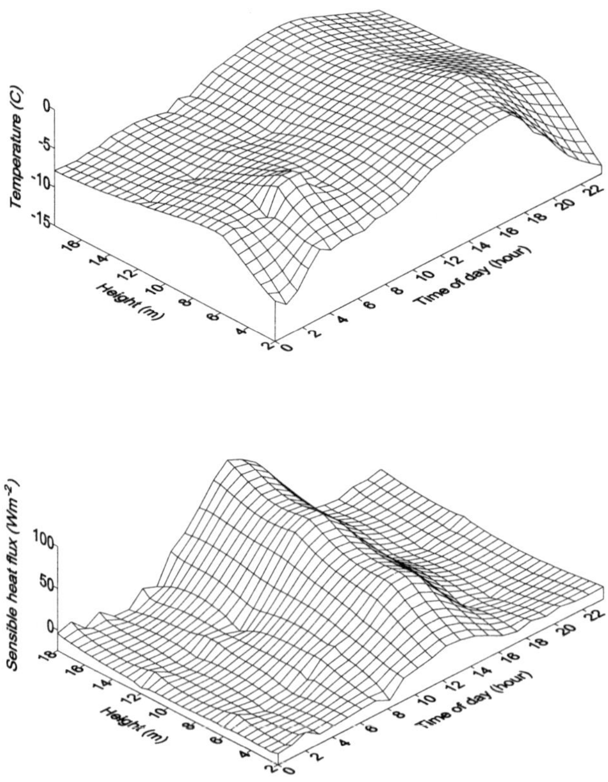

Figure 1. Average temperature (upper panel) and sensible heat flux (lower panel) shown as function of height and time of the day. The plots are based on half hourly measurements at 2, 6, 12 and 18 meters height for the period 13 to 24 March 1997.

Near the forest floor the temperature has its minimum approximately an hour past midnight, then it increases slowly reaching a maximum at 16 LST (local standard time), when the temperature starts to fall rapidly. Above the forest (18 meters) the minimum temperature during the night is larger and the maximum temperature during the day is smaller than inside the forest. Thus the temperature amplitude inside the forest is larger than above the forest. It is also characteristic that generally the temperature stratification inside the forest is stable and it is neutral or unstable above the forest. The sensible heat flux, plotted in the same way, is shown in Figure 1 (lower panel). It is seen that the heat flux is negative (downward directed) during the night. It is also characteristic that the heat flux during daytime increases between the forest floor and the top of the forest. It reaches a maximum value at noon of 100 Wm^{-2} above the forest as compared to only 10 Wm^{-2} near the forest floor. This shows that the upward directed heat flux above the forest originates mainly from warming of the trees.

SIMPLE MODEL OF THE FLUXES ABOVE THE FOREST

At low solar angles the canopy absorbs an important portion of the direct solar radiation. Since the canopy is sparse it is transparent for radiation. However, the ground is covered by snow, which has a very high albedo. So, most of the solar radiation reaching the snow will be reflected and the canopy will absorb a part of this reflected radiation also. That implies that for direct solar radiation the apparent vegetation cover of the forest is much greater than the actual one (defined as the portion of the ground covered by vegetation projected vertically). In most operational Soil-Vegetation-Atmosphere Transfer models used in numerical meteorological models, these features typical for a sparse sub-arctic are not incorporated (Melas *et al.,* 2000).

Here we present a simple parameterisation of the energy balance of this sparse sub-arctic coniferous forest. This parameterisation differs from existing approaches, primarily in the description of the short-wave radiation balance.

Simple expression for the shading factor

In order to describe the effect of low solar angles leading to an apparently high vegetation cover for direct radiation a *shading factor* f_{sh} is introduced. If the ground is shaded entirely then f_{sh} = 1, and when the shading is only directly under the trees corresponding to a sun at zenith $f_{sh} = \sigma_f$. For instance if f_{sh} = 0.8, 80% of the ground is shaded and 20% receives direct sunshine. The variation of the shading factor is restricted to the range $\sigma_f \leq f_{sh} \leq 1$. In addition, we introduce a *transmittancy factor* τ describing the fact that the forest canopy is not opaque for short-wave radiation. If $\tau = 1$ the canopy is entirely transparent, if $\tau = 0$ the canopy is opaque.

The shading factor, f_{sh}, represents the part of the ground that is covered by trees or lies in the shade of trees. Considering a tree with height h and typical crown diameter d, it covers an area of:

$$\frac{\pi}{4} d^2$$

and casts it shade over an area of:

$$\frac{hd}{tg(\alpha)},$$

where α is the elevation of the sun. The total area of ground per tree is:

$$\frac{\pi\, d^2}{4\,\sigma_f}.$$

At the critical sun elevation angle, α_c, the ground is entirely shaded by the trees:

$$\frac{\pi}{4}d^2 + \frac{hd}{tg(\alpha_c)} = \frac{\pi\, d^2}{4\,\sigma_f},$$

which can be solved to yield

$$tg(\alpha_c) = \frac{4h}{\pi d}\left(\frac{\sigma_f}{1-\sigma_f}\right).$$

These conditions correspond to $f_{sh} = 1$. When the elevation of the sun is higher than α_c, only a fraction f_{sh} of the ground is shaded:

$$\frac{\pi}{4}d^2 + \frac{hd}{tg(\alpha)} = \frac{\pi\, d^2}{4\,\sigma_f} f_{sh},$$

which can be solved

$$f_{sh} = \sigma_f\left(1 + \frac{4h}{\pi\, d\, tg(\alpha)}\right) \qquad \alpha \ge \alpha_c\,.$$

to yield the expression for f_{sh}. It can be seen that the variation of the shading factor falls within the range $\sigma_f \le f_{sh} \le 1$.

Short wave radiation

It is proposed here to distinguish between *direct* and *diffuse* incoming solar radiation

$$K^{\downarrow} = D + I\sin\alpha,$$

in which D is the diffuse component of the total global radiation $K^{\downarrow}$ and I the incoming solar radiation at the top of the vegetation across a plane perpendicular to the solar beam. Since the path through the atmosphere depends on the solar angle α, in general D and I are expected to be a function of the solar angle α.

The short-wave radiation received by the canopy consists of three terms. The first refers to the diffuse solar radiation. The diffuse radiation is treated as in Deardorff (1978), which means the canopy is assumed to be opaque for diffuse radiation and receives σ_f of the total incoming diffuse radiation. The second term refers to the direct radiation received by the canopy and the third term describes the reflected short-wave radiation received by the canopy from the snow cover. So the gain (superscript G) of short-wave radiation for the canopy is:

$$K_c^G = \sigma_f D + f_{sh}(1-\tau)I\sin\alpha + \sigma_f a_s (1-\tau) K_s^G .$$

In the last term K_s^G is the short-wave radiation received by the snow cover:

$$K_s^G = (1-\sigma_f)D + f_{sh}\tau I\sin\alpha + (1-f_{sh})I\sin\alpha .$$

Multiple reflected radiation (e.g. from snow to vegetation and next reflected back to snow) is ignored, because the albedo of the vegetation is rather low.

The loss (superscript L) terms for the canopy and the snow are given by:

$$K_c^L = a_c K_c^G$$

and

$$K_s^L = a_s K_s^G$$

in which a_c and a_s are the albedo for the canopy and the snow respectively. For the net short-wave radiation of the canopy and the snow we get:

$$K_c^n = K_c^G - K_c^L$$

and

$$K_s^n = K_s^G - K_s^L .$$

Long wave radiation

For the parameterisation of the long-wave radiation it is assumed that a flat horizontal plate can replace the canopy with emissivity similar to the canopy. The plate emits radiation towards the atmosphere and towards the snow surface. The one-sided loss of long-wave radiation of the canopy per unit (projected) area of the canopy, is in this way:

$$L_c^O = \varepsilon_c \sigma T_c^4 ,$$

in which ε_c is the emissivity of the canopy, σ the Stephan-Boltzmann constant and T_c the absolute canopy temperature. The gains of long-wave radiation for the canopy, L_c^G and the snow surface, L_s^G are:

$$L_c^G = \sigma_f \varepsilon_c L_{atm}^{\downarrow} + \sigma_f \varepsilon_c L_s^L$$

$$L_s^G = (1-\sigma_f)\varepsilon_s L_{atm}^{\downarrow} + \sigma_f \varepsilon_s L_c^O ,$$

where $L_{atm}^{\downarrow}$ is the incoming long-wave radiation from the atmosphere. The long wave radiation emitted from the canopy and snow surface, L_c^L and L_s^L are:

$$L_c^L = 2\sigma_f L_c^O$$

$$L_s^L = \varepsilon_s \sigma T_s^4,$$

where ε_s is the emissivity of the snow and T_s the absolute temperature of snow surface. Therefore, the long-wave radiation balances for the canopy, $L_c^n = L_c^G - L_c^L$, and snow surface, $L_s^n = L_s^G - L_s^L$ read:

$$L_c^n = \sigma_f \varepsilon_c L_{atm}^{\downarrow} + \sigma_f \varepsilon_c \varepsilon_s \sigma T_s^4 - 2\sigma_f \varepsilon_c \sigma T_c^4$$

$$L_s^n = (1-\sigma_f)\varepsilon_s L_{atm}^{\downarrow} + \sigma_f \varepsilon_c \varepsilon_s \sigma T_c^4 - \varepsilon_s \sigma T_s^4 .$$

Energy balance

The energy balance for the canopy is simplified. It is assumed that the trees are not able to transpire because the canopy temperature is well below zero. In addition, the biomass of the trees is small, so the heat storage in the vegetation is ignored. This means that the canopy the net radiation $Q_c^n = K_c^n + L_c^n$, equals the sensible heat flux:

$$Q_c^n = H_c .$$

The energy balance of the snow cover reads:

$$Q_s^n = H_s + \lambda E_s + G_s ,$$

in which $Q_s^n = K_s^n + L_s^n$ is the net radiation for the snow surface, H_s is the sensible heat flux for the snow surface, E_s the snow evaporation, λ the latent heat of sublimation and G_s the heat flux inside the snow.

Having determined the radiation balance of the forest canopy it is possible to estimate the temperatures T_s, T_c and the fluxes within and above the forest by use of a simple approach based on a network of resistance's suggested by Deardorff (1978) and Shuttleworth and Wallace (1985). Details are given in Gryning *et al.* (2000).

VALIDATION OF THE MODEL

The model was used to simulate the measurements from Sodankylä. The resistances suggested by Shuttleworth and Wallace (1985) were applied. As input were used the measurements of wind speed and temperature at 18 meters above the forest, the measured global and diffuse radiation above the forest, the humidity of the air at 1.5 meter height measured in a clearing of the forest and observations of the total cloud cover. Gryning *et al.* (2000) summarises the parameters that were used to describe the site. Figure 2 (upper panel) shows the modelled and measured sensible and latent heat fluxes for the entire measuring period. During daytime the agreement between measurements and simulations is generally good for both the sensible and the latent heat fluxes. During night-time good agreement between model predictions and measurements of the sensible heat flux are observed for a cloud-covered sky (15/16 March). At reported clear skies, however, the simulated sensible heat flux is twice lower than the measured (16/17 and 17/18 March) and near constant. When repeating the simulation assuming a cloud covered sky, the agreement between model and measurements for the sensible heat flux during the nights of 16/17 and 17/18 March was good. Thus the performance of the model during night-time depends on the accuracy of the cloud cover observations, which are known to be flimsy (Fig.2, lower panel). The latent heat flux during the night is predicted rather well.

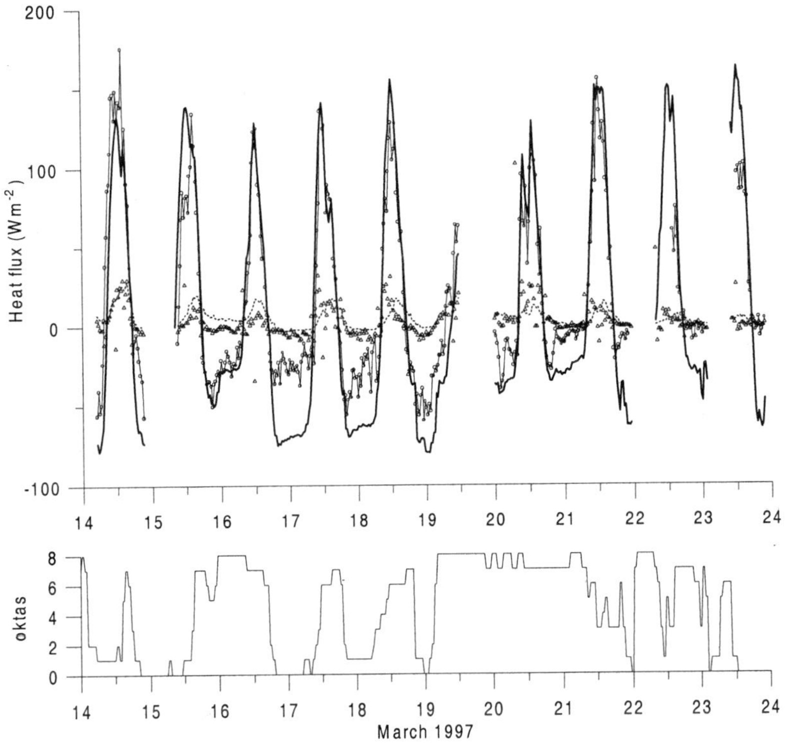

Figure 2. Simulated sensible (full line) and latent heat flux (dashed line) above the forest. Measurements of sensible and latent heat fluxes are shown as (o) and (Δ) - upper panel. Observed total cloud cover - lower panel.

DISCUSSION

The meteorological conditions with rather long days, a snow covered ground surface and bare pine or spruce trees are typical for the Northern boreal forest during the last part of the winter (March to May), but hardly characteristic for the whole winter period. The measurements show that the forest has a pronounced effect on the local meteorology. The forest induced sensible heat flux is comparable to those found in middle Europe, and is able to form a convective mixed-layer that can reach depths of typically 1000 meters in the late afternoon. The convection is caused by warming of the trees by solar radiation. The trees are bare (not snow covered) with low albedo and thus an efficient absorber of short-wave radiation. The ground is covered with snow that reflects short wave radiation. The upward directed heat flux above the forest originates mainly from the warming of the trees, while the heat flux from the forest floor remains small. While the sensible heat flux can reach more than 100 Wm^{-2} during the day, the latent heat flux, which originates mainly from sublimation of snow, is 10 times lower, typically 10-20 Wm^{-2}.

Under low solar angles the canopy is apparently fully covering the ground, implying that it absorbs an important part of the direct radiation, and only a minor part reaches the snow surface. Therefore methods based on the distribution and relative coverage of land surface types that do not consider the shading effect of the trees at low solar angles (i.e. present methods based on satellite measurements of surface temperature and classifications of land use) are likely to fail under cloudless conditions. In a similar way the albedo that can be derived from satellite and air plane measurements is not the albedo that controls the absorption of short wave radiation under cloud free conditions with low solar angles, because the effective albedo is not only a function of the surface coverage but also the solar angle. For cloudy conditions, the direct radiation will be small compared to diffuse radiation, the shading effect of the trees will be negligible and use of traditional methods based on land use classifications can be better justified.

The shading effect is not only confined to a sparse forest. Similar effects can be expected in an urban environment, where the sensible heat flux above the urban area will be controlled by the thermal characteristics of the urban canopy and to a lesser degree by the thermal characteristics of the surface.

Acknowledgements

We are grateful to Martti Heikinheimo (the Finnish Meteorological Institute) and Rigel Kivi (the Sodankylä Meteorological Observatory) for providing the radiation and synoptic data, and to Per-Erik Johansson (the Swedish National Defense Research Establishment) for flux data. The hospitality and help of the staff at the Sodankylä Meteorological Observatory and the assistance of John Hansen and Lars Christensen are acknowledged. The European Union supported the study, contract number ENV4-CT96-0324.

REFERENCES

Deardorff, J.W.,1978, Efficient prediction of ground surface temperature and moisture, with inclusion of a layer of vegetation. *J. Geo. Res.,* 83 C4, 1889-1903.

Gryning, S.E., Batchvarova, E. and de Bruin, H.A.R., 2000, Energy balance of sparse coniferous high-latiude forest under winter-conditions. Submitted to *Boundary-Layer Meteorology.*

Melas, D., Persson, T., de Bruin, H.A.R., Gryning, S.E. and Batchvarova, E., 2000, Numerical model simulations of boundary-layer dynamics during winter conditions'. Submitted to *Theoretical and Applied Climatology.*

Shuttleworth, J. and Wallace, J. S., 1985, Evaporation from sparse crops - an energy combination theory, *Quart. J. R. Meteorol. Soc.* 111, 839-855.

DISCUSSION

G. SCHAYES: When considering the diffuse radiation, did you separate between overcast sky and clear sky. They have different properties, overcast sky radiation is isotropic, but clear sky diffuse radiation is not, particularly when the sun is low.

E. BATCHVAROVA: The idea of separating direct and diffuse radiation is introduced in this study in order to model the details of shading. The canopy is considered opaque for the diffuse radiation. Based on a 3-month period (March-May 1997) of hourly observations of total and diffuse radiation at Sodankylä, we have parameterised the ratio between total and diffuse incoming radiation. This work is submitted for publication in Boundary-Layer Meteorology.

D. STEYN: It would seem that the fluxes should be related to the needle temperature. Were such temperatures measured?

E. BATCHVAROVA: No, during this pilot experimental campaign there were no measurements of canopy temperature. Definitely, such data are needed also to verify the model presented here. We model the canopy temperature based on the canopy resistances suggested by Shuttleworth and Wallace, 1985 (Quarterly J. R. Meteorol. Soc. 111, 839-855). The Sodankylä Meteorological Observatory became a centre for other EU projects on winter conditions and such measurements could be a part of future studies.

DEPENDENCE OF TURBULENT VELOCITY VARIANCES ON SCALE AND STABILITY

L. Mahrt,[1] Erin Moore,[1] Dean Vickers,[1] and N. O. Jensen[2]

[1]College of Oceanic and Atmospheric Sciences
Oregon State University
Corvallis, OR 97331 USA

[2]Meteorology and Wind Energy Dept.
Risø National Laboratory
DK 4000 Roskilde, Denmark

INTRODUCTION

Turbulent variances and kinetic energy have been successfully approximated by turbulence similarity theory, although differences exist between studies. Discrepancies are greatest for very stable conditions. Existing formulations poorly describe mesoscale variations. Smedman (1988) finds that mesoscale motions do not obey Monin-Obukhov similarity theory and their inadvertent inclusion in the turbulence variances degrades the performance of similarity theory.

Separation between turbulence and mesoscale variances is required to understand the behavior of turbulence in very stable conditions and formulate these variances. One application of formulated variances is the modeled relationship between dispersion and the standard deviation of the crosswind velocity (Kristensen et al., 1981; Hanna, 1983; Skupniewicz, 1987; Lyons and Scott, 1990), one of the goals of this study.

THE DATA

This study analyzes data from an old aspen forest in Central Canada (Blanken et al., 1997, 1998; Mahrt et al., 2000), a beech forest site as a part of Euroflux in Denmark (Pilegaard et al., 2000) and a second beech forest site in Denmark on Falster Island, a grassland in Central USA (Sun, 1999), a heather site in Denmark and a pine site in complex terrain in Western USA. The data are described in more detail in Mahrt et al., (2000).

Velocity fluctuations are computed as

$$\phi' = \phi - \overline{\phi} \qquad (1)$$

where the overbar represents averaging over a time period of τ. Modeling variances as a function of averaging time τ is a fundamental goal of this study. For application of turbulence similarity theory, τ must be chosen large enough to include most of the turbulence yet exclude significant mesoscale motion. Instantaneous variances are then averaged over a longer averaging scale to produce

$$[\phi'^2] \tag{2}$$

[~] represents the longer term average, chosen to be four hours in this study. The dependence of the variance on the averaging time τ can be examined in terms of any arbitrary set of averaging scales.

For each site, the variances are composited for different stability classes, ($z/L < -0.2$), (-0.2, -0.05), (-0.05, 0.04), (0.04, 0.1), (0.1, 0.5), (> 0.5). The choice of these classes are based on the frequency distribution of the observations.

CROSSWIND VARIANCE

This study emphasizes the behavior of the crosswind velocity variance, $[v'^2]$ (τ), which is critical input for dispersion models. The crosswind variance is more erratic and more difficult to model compared to the along-wind and vertical velocity variances, examined in Section 5. We initially focus on the Microfronts grassland site.

The velocity variances, composited for individual stability classes (Figure 1), increase rapidly with averaging time at small averaging times due to the inclusion of more turbulence with increased averaging time. Recall that the averaging time τ defines the fluctuations, while the variances of the fluctuations are then averaged over four-hour periods. The gap between turbulent and mesoscales is indicated by the intermediate scale regime where the variance is essentially constant or increases only slowly with increasing averaging time (Figure 1), corresponding to a minimum in the spectral density (not shown). The gap is best defined for stable conditions where the turbulence is confined to smaller scales. The very slow increase of the variance with increasing averaging time in the gap region implies that the variance is not sensitive to the choice of the averaging time τ, provided that τ is within the gap region. Evaluation of turbulence similarity theory much choose averaging times corresponding to the gap region. However, the time scales corresponding to the gap region decrease substantially with increasing stability, implying that the averaging time defining turbulent fluctuations should decrease substantially with increasing stability. The turbulence extends up to only about $\tau = 100$ s in very stable conditions.

CROSSWIND VARIANCE MODEL

Our approach here is to model the turbulence part of the horizontal variance and then model the residual from this model as mesoscale motion. With this approach, mesoscale and turbulence motions may overlap in scale, as opposed to assuming all motions smaller (greater) than a certain scale are turbulent (mesoscale). If successful, this approach would reduce the contamination of the turbulence variances due to inadvertent inclusion of mesoscale motion in the turbulent velocity variance with fixed averaging times (see Introduction). The separation of the mesoscale and turbulent contributions to the variance is motivated by the different roles of turbulent diffusion and mesoscale motions on transport.

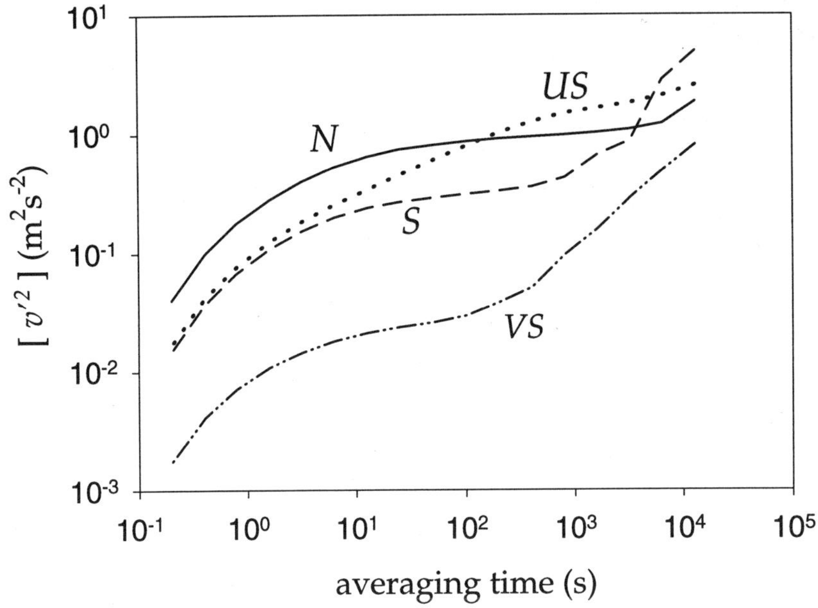

Figure 1. Composited crosswind variances for the grassland site for the stability classes: very stable (VS) $z/L > 0.5$, stable (S) $0.1 < z/L < 0.5$, near-neutral (N) $-0.05 < z/L < 0.04$ and unstable conditions $z/L < -0.2$.

The variance of the crosswind component is reasonably approximated by the function

$$[v'^2](\tau) = Cu_*^2\left(1 - \exp([\tau/\tau*]^n)\right) + [v'^2(\tau_r) - Cu_*^2](\tau/\tau_r)^p \tag{3}$$

where the first term on the right hand side models the turbulence contribution and the second term models the mesoscale contribution. The time scale $\tau*$ is the governing time scale for the turbulent flow. The coefficient C describes the amplitude of the turbulence variance relative to the surface friction velocity. For large averaging time with this model, the turbulence contribution to the variance asymptotes to Cu_*^2. For dispersion models, Cu_*^2 would be the input for the turbulent crosswind velocity variance. The reference time scale, τ_r, will be chosen as 4 hours in the analysis below. This defines the largest scale included in the mesoscale calculation. The variance $[v'^2(\tau_r) - Cu_*^2]$ along with the exponent p determines the magnitude of the mesoscale variance.

The choice of the values of the coefficients for the turbulence contribution to the variance were guided by systematically varying n, C and $\tau*$ and plotting the error in C - $\tau*$ space and in n - $\tau*$ space for fixed values of the third variable.

The coefficient C can be interpreted as a nondimensional turbulence velocity variance$[v'^2]/u_*^2$. The best fit value of C increases significantly with increasing instability

(Figure 2) and increases more slowly with increasing stability. A subjectively determined fit to all of the data sets (Figure 2, solid line) is expressed as

$$C = C_0\left(1 - a\frac{z}{L}\right)^q \tag{4}$$

where $C_o = 2.6$, $q = 2.0$ for unstable conditions and 1.0 for stable conditions and $a = 1.0$ for unstable conditions and 0.2 for stable conditions. Again, C is defined in terms of the estimated turbulent part of the velocity variance so that it is not exactly equal to $\lfloor v'^2 \rfloor / u_*^2$ for any given averaging time

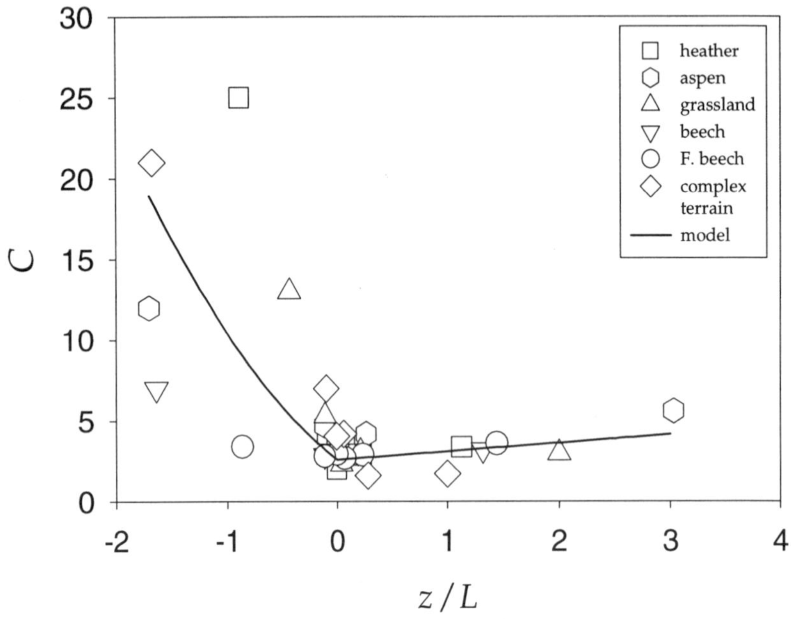

Figure 2. Dependence of the coefficient C (Eq. 3) on stability and the formulation for $C(z/L)$ (solid line; Eq. 4).

The time scale τ^* increases with increasing instability corresponding to large eddies for very unstable conditions (Figure 3), and decreases to very small values for strongly stable conditions corresponding to small eddies. Here, we model the variance in terms of averaging time τ, which can be more directly utilized in applications such as dispersion models. The dependence of $\tau*$ on z/L (Figure 3) is modeled as

$$\tau^* = \tau_0 \exp(-bz/L) \tag{5}$$

where $\tau_o = 20$ s and $b = 1.75$ for unstable conditions and 0.95 for stable conditions.

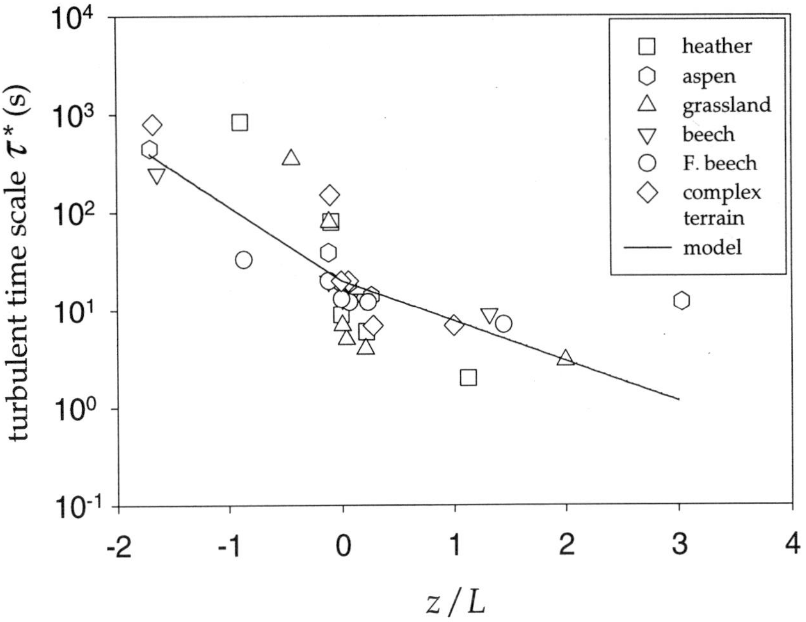

Figure 3. Dependence of the turbulence scaling time scale, τ^* (Eq. 3) and the modeled value (Eq. 5, solid line) on stability.

The observations suggest a simple formulation for the turbulent exponent, where n = 0.5 for unstable conditions and n = 0.7 for stable conditions. The smaller value of the exponent for unstable conditions corresponds to a more gradual approach of the turbulence variance to the gap regime.

For inclusion of height-dependence into Eq. 3, we recommend that the first term on the right hand side of this equation be generalized to

$$C(z/L)u_*^2\left(1-\exp\left([\tau/\tau^*]^n\right)\right)f(z/h) \tag{6}$$

where $f(z/h)$ is $(1 - z/h)^{3/2}$ for stable conditions and unity for unstable conditions. Our data analysis did not find a clear dependence of the mesoscale variance on height and the second term on the right hand side of Eq. 3 remains independent of height.

MODEL FOR THE THREE VELOCITY VARIANCES

The behavior of the along-wind and vertical velocity variance is more similar between sites compared to the crosswind variance. The model parameters for the along-wind and vertical velocity components are determined in the same way as that outlined for the crosswind variance in the previous sections. The total model is Eqs. 3, 4, and 5, with the coefficients listed in Tables 1-2 below.

TABLE 1: Values of the model coefficients for the three turbulence velocity variances. C_o is the scaled amplitude of the turbulence, a (unstable/stable) and q (unstable/stable) are coefficients for the stability dependence, τ_o is the characteristic time scale of the turbulence for neutral conditions, expressed in seconds, b (unstable/stable) is the coefficient for variation of the characteristic time scale with stability, n (unstable/stable) is the exponent coefficient for the turbulence.

variable	C_0	a	q	τ_0	b	n
$[v'^2]$	2.6	1.0/0.2	2.0/1.0	20	1.75/0.95	0.7/0.5
$[u'^2]$	4.7	1.0/0.2	1.0/1.0	35	1.75/0.95	0.7/0.6
$[w'^2]$	1.5	3.0/3.0	0.67/0.67	5	1.75/0.95	0.65/0.55

TABLE 2: Values of the model coefficients for the three velocity components for the mesoscale velocity variances where p is the mesoscale exponent and *meso-amp* is the mesoscale amplitude $([\phi'^2](4hours) - Cu_*^2)$, expressed in m^2s^{-2}.

Variable	p	meso-amp
$[v'^2]$	0.8	0.6
$[u'^2]$	0.7	0.6
$[w'^2]$	na	0.0

Notice that the amplitude of the variance and the time scale are both largest for the along-wind variance and both smallest for the vertical velocity component, consistent with traditional spectra (e.g., Kaimal and Finnigan (1994)). The model coefficients for the scaled vertical velocity variance is based on Panofsky et al. (1977) and worked well for the short vegetation but tended to over predict the scaled variance for the tall vegetation. The discontinuity of the exponents, n and the height-dependence z/h at neutral stability is undesirable. However, more complex relationships must await further collection of data.

As with the crosswind variance, the height-dependence for levels, which are not small compared to the boundary-layer depth, can be included by replacing the first term on the right hand side of Eq. 3 with Eq. 6 for the along-wind variance, except using the coefficients for the along-wind variance from Table 1. The same height-dependence is used for the vertical velocity variance for stable conditions. For the height-dependence for the vertical velocity component for unstable conditions, one can use (e.g., Garratt, 1992),

$$f(z/h) = (z/h)^{1/3}(1 - z/h)^{1/3} \tag{7}$$

The parameters for the mesoscale term in Table 2 are estimated in Mahrt et al. (2000). The mesoscale vertical velocity variance was small and subject to significant relative errors in all of the data sets and is set to zero in the model.

Acknowledgments

This material is based upon work supported by Grant DAAD19-9910249 from the Army Research Office and Grant 9807768-ATM from the Physical Meteorology Program of the National Science Foundation.

REFERENCES

Blanken, P. D., Black, T.A., Yang, P.C., Neumann, H.H., Nesic, Z., Staebler, R., den Hartog, G., Novak, M.D., and Lee, X., 1997, Energy balance and canopy conductance of a boreal aspen forest: Partitioning overstory and understory components. *J. Geophys. Res.*, 102:28,915-28,927.

Blanken, P. D., Black, T.A., Neumann, H.H., den Hartog, G., Yang, P.C., Nesic, Z., Staebler, R., Chen W., and Novak, M.D., 1998, Turbulent flux measurements above and below the overstory of a boreal aspen forest. *Bound.-Layer Meteor.*, 89:109-140.

Garratt, J. R., 1992, *The Atmospheric Boundary Layer*. Cambridge University Press.

Hanna, S. R., 1983, Lateral turbulence intensity and plume meandering during stable conditions. *J. Climate Appl. Meteor.*, 22:1424-1430.

Kaimal, J. C. and J. J. Finnigan, 1994, *Atmospheric Boundary Layer Flows: Their Structure and Measurement*. Oxford Press.

Kristensen, L., N. O. Jensen, and E. L. Peterson, 1981, Lateral Dispersion of Pollutants in a Very Stable Atmosphere; The Effect of the Meandering. *Atmos. Environ.*, 15:837-844.

Lyons, T. and B. Scott, 1990, *Principles of Air Pollution Meteorology*. CRC Press. Boca Raton.

Mahrt, L., X. Lee, A. Black, H. Neumann and R. M. Staebler, 2000, Vertical mixing in a partially open canopy. To appear *Ag and For. Meteorol.*

Mahrt, L., E. Moore, D. Vickers and N.O. Jensen, 2000, Dependence of turbulent velocity variances on scale and stability. Submitted to *J. Appl. Meteor.*

Panofsky, H. A., H. Tennekes, D. H. Lenschow, and J. C. Wyngaard, 1977, The characteristics of turbulent velocity components in the surface layer under convective conditions. *Bound.-Layer Meteor.*, 11:355-361.

Pilegaard, K., P. Hummelsjøj, N. O. Jensen and Z. Chen, 2000, Contrasting the results from the first two seasons of continuous CO_2 eddy-flux measurements over a Danish beech forest. Submitted to *Agric. For. Meteorol.*

Skupniewicz, C. E., 1987, Measurements of over water diffusion: the separation of relative diffusion and meander. *J. Climate Appl. Meteor.*, 26:949-958.

Smedman, A. S, 1988, Observations of a multi-level turbulence structure in a very stable atmospheric boundary layer. *Bound.-Layer Meteor.*, 44:231-253.

Sun, J., 1999, Diurnal variation of thermal roughness height over a grassland. *Bound.-Layer Meteor.*, 92:407-427.

DISCUSSION

A. VENKATRAM: Is the model for the variance in the "mesoscale motion" region purely empirical in that it is a functional fit to data? Is the time scale in the equation related to a length scale and a velocity scale of the mesoscale motions?

L. MAHRT: The fit is purely empirical. We were unable to identify universal length and velocity scales that could describe the wide variety of contributions to the mesoscale regime. Even in the better behaved turbulence region, we were unable to find velocity and length scales that were suitable for a range of stability conditions.

S. HANNA: There may be another way to approach the problem of estimation of the wind speed horizontal velocity variance (σ_u^2 or σ_v^2) for averaging times of several hours or more. Observations of σ_y (lateral dispersion) at travel times ranging from one hour to many days from many mesoscale and regional experiments consistently show σ_y approximately between 0.5 and 1.0 (ms^{-1}) times travel time (s). At these large averaging times, the clouds should be dispersed vertically through the boundary layer. Since σ_y grows like σ_v t where t is the travel time, then it results that σ_v is approximately 0.5 to 1 ms^{-1} averaged over the depth of the boundary layer. Gifford states in his Langevin model description, that this σ_v includes many contributions, such as terrain effects, mesoscale eddies, and so on. I was wondering whether your CASES and other data suggest a generalized σ_v of about 1 ms^{-1} over long averaging times and over the full vertical range of tower observations.

L. MAHRT: We always see σ_v increasing with averaging time up to the 4-hour periods that we analyzed. The values for 4-hours are typically in the range you suggested. However integrating spectra in the literature for the mesoscale-synoptic regime would produce an indefinite increase with increasing averaging time. As the averaging time increases, a larger range of motions are captured in the σ_v calculation. This suggests that the observed σ_v may not be appropriate for estimation of σ_y at large averaging times? However, this is outside my area of expertice and I need to talk to you about this more.

NEW DEVELOPMENTS IN DISPERSION EXPERIMENTS FOR THE CONVECTIVE BOUNDARY LAYER

J.C. Weil[1], W.H. Snyder[2], R.E. Lawson, Jr.[3], and M.S. Shipman[4]

[1] Cooperative Institute for Research in Environmental Sciences
University of Colorado
Boulder, CO 80309

[2] 7312 Grist Mill Rd.
Raleigh, NC 27615

[3] Atmospheric Sciences Modeling Division
Air Resources Laboratory
National Oceanic and Atmospheric Administration
Research Triangle Park, NC 27711

[4] Geophex, Ltd.
Raleigh, NC 27603

INTRODUCTION

Dispersion from elevated sources usually is most rapid in a convective boundary layer (CBL) with light winds due to the large turbulent eddies within the CBL. For buoyant releases, plume segments can be brought to the surface when a convective downdraft overcomes the plume rise. For highly buoyant plumes, the maximum ground-level concentrations (GLCs) often are found during lofting followed by downward mixing, i.e., when the plume rises to the CBL top, is trapped there by the elevated inversion capping the boundary layer, and then disperses downwards (Briggs, 1988; Weil, 1988).

Our understanding of buoyant plume dispersal in the CBL was advanced significantly by Willis and Deardorff's (1983, 1987) experiments in a laboratory convection tank. The laboratory plumes were visually similar to their full-scale counterparts. In addition, the experiments demonstrated the importance of the dimensionless buoyancy flux F_* in characterizing the relative effects of buoyancy and ambient convection far downwind. The F_* is given by

$$F_* = \frac{F_b}{U w_*^2 z_i} , \tag{1}$$

where F_b is the stack buoyancy flux, U is the mean wind speed in the CBL, w_* is the

Air Pollution Modeling and Its Application XIV, Edited by
Gryning and Schiermeier, Kluwer Academic/Plenum Publishers, New York, 2001

convective velocity scale, and z_i is the CBL depth. For $F_* = 0.03$, the plume behaved similarly to a nonbuoyant plume after some initial rise, but for $F_* = 0.11$, the plume rose to the CBL top, where it lofted and then gradually mixed downwards. Willis and Deardorff (1987) found a systematic variation in the mean surface concentrations with F_* and downwind distance.

These experiments were revolutionary but were limited by the experimental measurement techniques and the small sample sizes. For example, in the highly-buoyant plume studies, the limited repetitions resulted in uncertainty in the mean concentration values near the surface and underestimates of the lateral plume spread (Weil et al., 2000).

The above limitations were overcome in convection tank experiments conducted at the Environmental Protection Agency's Fluid Modeling Facility. The new apparatus consisted of three major components: 1) a convection tank patterned after that used by Willis and Deardorff, 2) a laser-induced fluorescence system for obtaining non-intrusive measurements of tracer concentrations, and 3) an automation system for controlling the experimental operation and obtaining a large number of repeated measurements; see Snyder et al. (2000) for a detailed description.

This paper reports on new measurements of dispersion in highly-buoyant plumes ($F_* \gtrsim 0.1$), which have presented a major problem to dispersion modeling (see Weil, 1988). We briefly present key aspects of the mean concentration field including the vertical and lateral spatial statistics, plume meander, etc. and some new data on concentration fluctuations; see Weil et al. (2000) for a more detailed summary. We believe that this data set offers one of the more complete descriptions to date on the dispersion characteristics of buoyant releases in the CBL.

EXPERIMENT DESCRIPTION

The experimental arrangement was similar to that of Willis and Deardorff (1987). The convection tank was about 124 cm on a side, was filled with water to a depth of 34 cm, and had an initial stratification aloft of 1°C/cm. The convection was driven by an electrically-heated bottom surface that produced a $z_i \simeq 20$ cm and $w_* = 0.74$ cm/s at the time of the measurements. A mean wind was simulated by towing a model stack just above the tank floor at a speed U of 2.07 cm/s. The stack emitted a water - ethanol mixture to simulate the buoyancy, and the mixture contained a small amount of Rhodamine dye, which fluoresced when excited by laser light.

In an approach different from that of Willis and Deardorff, a laser was mounted on a movable table alongside the tank and towed at the stack speed in order to illuminate a y - z (crosswind, vertical) plane at a fixed distance x downstream of the stack. Pictures of the fluorescent dye were taken from a video camera viewing the illuminated plane at a right angle; the light intensity was digitized, stored, and converted to concentration in $y \times z$ intervals of 1.8 mm × 2.3 mm. In each tow, 59 cross-sectional images were digitally recorded at 0.8 s intervals as the laser sheet traversed the tank. The tows were repeated 6 to 7 times for a total of 354 to 413 realizations, which is an unprecedented data volume. The averaging time corresponding to these repeats was 10.5 to 12.2 z_i/w_*, where z_i/w_* (= 27 s) is the eddy turnover time; for a typical field value of z_i/w_* ($\sim$ 500 s), the full-scale averaging time would be $\sim$ 1.5 hr.

Four experiments were performed, each with a different value of F_*, but the same effluent speed, U, and CBL variables; the $F_* = 0, 0.1, 0.2$ and 0.4, with $F_* = 0$

serving as a reference case. The stack conditions that were essentially fixed were the effluent speed (9.9 cm/s), stack radius (0.16 cm), and stack height ($z_s = 3$ cm). The dimensionless stack height and towing speed, $z_s/z_i = 0.15$ and $U/w_* = 2.79$, were close to the values used by Willis and Deardorff (1987). In each experiment, cross sections were sampled at each of 8 equally-spaced distances; see Weil et al. (2000) for further details.

RESULTS

In the following, we discuss profiles of the mean crosswind-integrated concentration (CWIC), dispersion parameters, and surface values of the CWIC and fluctuating concentrations; the CWIC C^y is defined by $C^y = \int_{-\infty}^{\infty} C(x, y, z)dy$, where C is the ensemble mean concentration. The plume variables are presented as a function of the dimensionless downstream distance

$$X = \frac{w_* x}{U z_i} , \qquad (2)$$

which is the ratio of the travel time (x/U) to the eddy turnover time.

Vertical Profiles of CWIC

Figure 1 shows vertical profiles of the dimensionless CWIC as a function of X for the moderately-buoyant plume ($F_* = 0.1$); the CWIC is nondimensionalized by the well-mixed value Q/Uz_i, where Q is the source strength. Note that the small enhanced C^y region near the surface ($z < 0.05z_i$) is an anomaly caused by refracted light near the tank floor. Due to this anomaly, we choose the "surface" C^y and C values as those at $z = 0.05z_i$.

The profiles in Fig. 1 show that the surface C^y values are low near the source and that an elevated C^y maximum is maintained above z_i. For $X \geq 2.5$, the mixed-layer profile is essentially uniform, but the C^y is lower than the well-mixed value (Q/Uz_i) due to the elevated C^y maximum. For $X > 4$, we expect that the elevated maximum would decrease and the mixed-layer value of C^y would increase due to entrainment of the plume aloft. For $F_* = 0.2$ and 0.4, the profiles (not shown) are qualitatively similar to those in Fig. 1, but the elevated maximum in C^y is larger and the mixed-layer and surface values are smaller than for $F_* = 0.1$.

A key difference between these results and those of Willis and Deardorff (1987) is the vertical C^y distribution within the mixed layer over the range $3 \leq X \leq 4$. As can be seen, our profiles are quite uniform, whereas those deduced from the CWIC contours of Willis and Deardorff (their Figs. 2 to 6) are not. Their profiles show large positive gradients ($\partial C^y/\partial z$) for $z/z_i < 0.8$, which suggest that source buoyancy affects the mixed-layer distribution over the range $3 \leq X \leq 4$. Moreover, Willis and Deardorff (1987) state that for $F_* \gtrsim 0.2$, the vertical distribution is not well mixed for X as large as 5. We believe that their findings are erroneous and result primarily from insufficient sampling.

Dispersion Parameters

The "total" dispersion parameters in the lateral and vertical directions—σ_y and σ_z—are presented together with the root-mean-square (rms) meander and relative dispersion. These are some of the first measurements of all components—meander, relative dispersion, and total dispersion—for dispersing plumes in a convection tank.

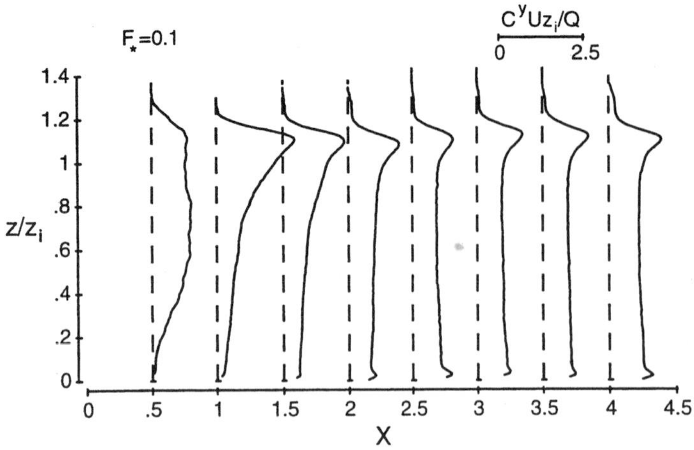

Figure 1. Vertical profiles of dimensionless crosswind-integrated concentration as a function of X for $F_* = 0.1$ and $z_s/z_i = 0.15$.

In homogeneous turbulence, σ_y is related to the meander σ_{ym} and relative dispersion σ_{yr} through $\sigma_y^2 = \sigma_{ym}^2 + \sigma_{yr}^2$, assuming that σ_{ym} and σ_{yr} are independent. For a passive plume, the meander should follow $\sigma_{ym} = \sigma_v t$ if $t \ll T_{Ly}$ and $d\sigma_{ym}^2/dt = 0$ if $t \gg T_{Ly}$, where t is time, σ_v is the rms lateral turbulence component, and T_{Ly} is the Lagrangian integral time-scale (Csanady, 1973). Thus, for large t, we expect σ_{ym} to tend to a constant.

Figure 2 gives an overview of the lateral dispersion components—σ_{ym}, σ_{yr}, and σ_y—and demonstrates that all generally increase in a systematic manner with X and F_*. The σ_{ym} follows the general behavior predicted by Csanady (1973) except that the data do not extend to a sufficiently small t $(= x/U)$ to verify the short-time limit. This limit is shown by the dashed line, corresponding to $\sigma_{ym}/z_i = 0.51X$, with the σ_v $(= 0.51w_*)$ found from analysis of the σ_y data for $F_* = 0$. Figure 2a suggests that σ_{ym} approaches a constant, long-time asymptote but the asymptotic value depends on F_* (see Weil et al., 2000 for a discussion).

By comparing Figs. 2a to 2c and noting the scale changes, one can see that σ_{yr} is the main contributor to the total dispersion. The meander is at most $\sim 0.5\sigma_{yr}$, which means that it only contributes about 12% to σ_y.

Figure 3 compares our σ_y/z_i data with the mean σ_y/z_i from field observations of buoyant plumes (Weil and Corio, 1985; where $F_* \leq 0.06$) and shows that the tank data for $F_* = 0$ to 0.2 are in reasonable agreement with these observations. Our measurements also are much more consistent with the field observations than are the earlier data of Deardorff and Willis (1984) (filled squares), which are only about 60% as large. We suspect that the earlier data are low because of an insufficient detection limit that prevented the resolution of low concentration tails.

Figure 3 shows that σ_y is slightly enhanced for $F_* = 0.2$, relative to the lower buoyancy cases, and clearly enhanced for $F_* = 0.4$ as also suggested by field observations (Briggs, 1985). Briggs likened the buoyant plume spread at an elevated inversion to the enhanced lateral spread of a buoyant line thermal in a water tank halted vertically by a horizontal plate. Based on analysis of field data, he proposed the following σ_y expression

$$\frac{\sigma_y}{z_i} = c_1 F_*^{1/3} X^{2/3} \tag{3}$$

with $c_1 = 1.6$. Our data show that the above functional dependence is followed for $F_* = 0.2$ and 0.4 (dashed lines) but the $c_1 \simeq 0.8$. The smaller c_1 in the laboratory may be partially explained by other effects (crosswind shear, mesoscale variability, etc.) that are present in the field but absent in the convection tank.

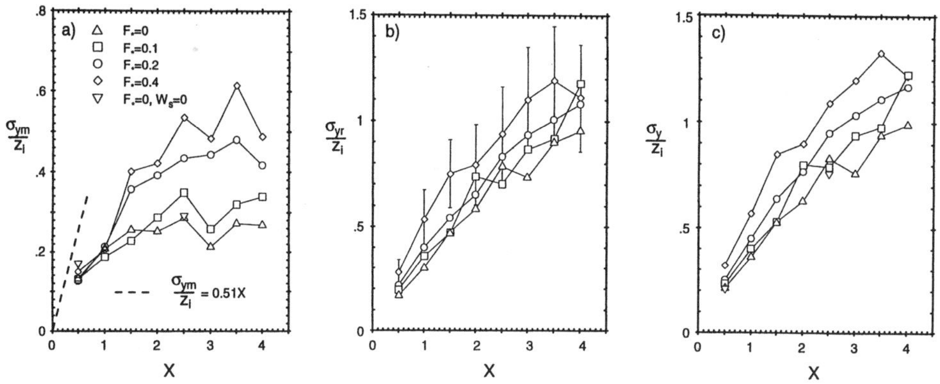

Figure 2. Dimensionless values of crosswind dispersion components as function of X and F_*. a) Root-mean-square (rms) meander, b) relative dispersion, and c) total dispersion. Bars in (b) show typical rms deviation of relative dispersion.

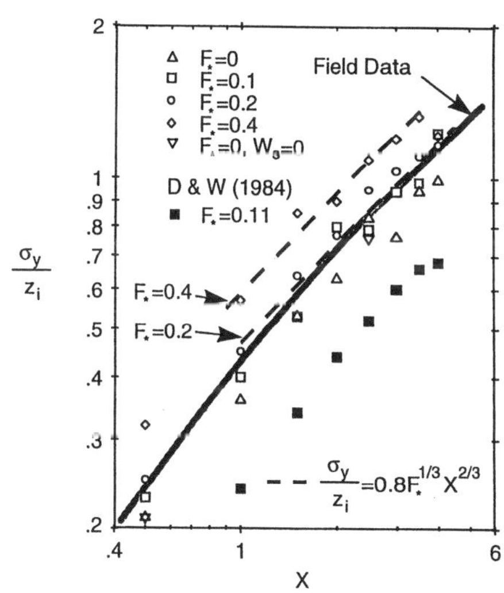

Figure 3. Dimensionless crosswind dispersion versus X showing: a) present laboratory data (open symbols), b) Deardorff and Willis (1984) data, c) field data (Weil and Corio, 1985), and d) Eq. (3) with $c_1 = 0.8$.

Figure 4 presents the vertical dispersion components as functions of X and F_*. It shows that the meander (σ_{zm}) peaks at $X = 0.5$ and generally decreases to small values for $X \geq 3$. In analogy with the σ_{ym}, we expect that a linear growth of σ_{zm}/z_i with X occurs for $X < 0.5$ as suggested by the dashed line (Fig. 4a), where σ_w is taken as $0.6w_*$; however, further data are necessary to confirm this. At $X = 0.5$, the decrease in the meander with F_* is attributed to: 1) the increased plume rise and greater resistance to downward dispersion with increasing buoyancy, and 2) the confinement imposed by the elevated inversion. In summary, we find that σ_{zm} differs significantly from Csanady's predictions.

The relative dispersion (σ_{zr}) arises from plume-generated turbulence and ambient turbulence in the inertial subrange. For $X \leq 2$ and $F_* \leq 0.2$, the σ_{zr} is approximately independent of buoyancy (Fig. 4b), but it is appreciably smaller for $F_* = 0.4$ over the same X range. In the latter case, the plume vertical dimension is "squashed" as the rapidly-rising plume is halted by the elevated inversion. As $X \to 4$, the σ_{zr} tends to an asymptotic value due to plume trapping and homogenization within the CBL.

Figure 4c shows that the absolute dispersion (σ_z) differs from σ_{zr} primarily for $X < 2$ where the meander is most significant; there is little difference between σ_z and σ_{zr} farther downstream.

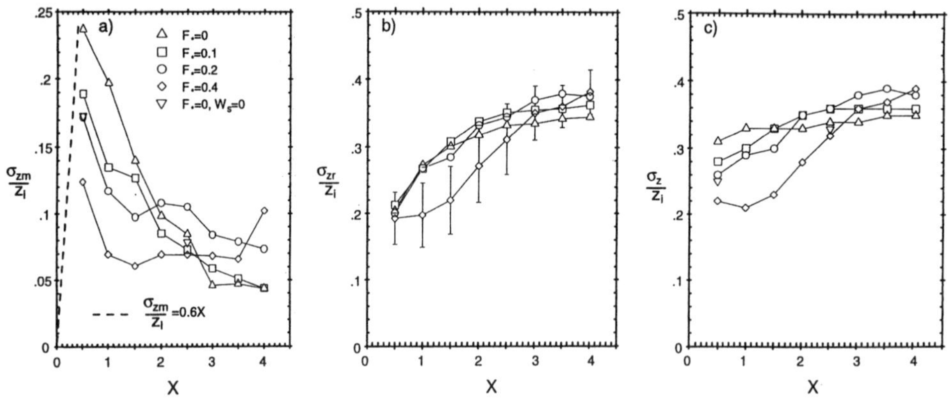

Figure 4. Dimensionless values of vertical dispersion components as function of X and F_*. a) Root-mean-square (rms) meander, b) relative dispersion, and c) total dispersion. Bars in (b) show typical rms deviation of relative dispersion.

Surface CWIC and RMS Concentration Fluctuation

The dimensionless CWIC near the surface ($z/z_i = 0.05$) is presented in Fig. 5a along with data from the Deardorff and Willis (1984) and Willis and Deardorff (1987) experiments. Our data shows that the addition of buoyancy significantly reduces the CWIC near the source ($X < 2$) by comparison with the CWIC for the nonbuoyant plume (open triangles). Further systematic reductions in the CWIC occur as F_* increases to 0.4 due to more of the plume becoming trapped in the inversion layer ($1 \leq z/z_i \leq 1.25$).

A similar trend of the CWIC with F_* was found by Willis and Deardorff (1987) and Deardorff and Willis (1984). As can be seen, their results are in general agreement with ours for $F_* = 0.1$ but are substantially lower than our data for $F_* \geq 0.2$. The latter finding is probably due to: 1) an inadequate detection limit in their experiments that precluded the measurement of low concentration tails, and 2) insufficient repetitions in their experiments as suggested in their paper (Willis and Deardorff, 1987).

A quantity of much interest in air quality is the near-surface concentration fluctuation intensity σ_c/C along the plume centerline. Figure 5b shows this as a function of X and F_*. At $X = 0.5$, the data clearly demonstrate a systematic increase in σ_c/C with F_*, a result which was not attained in the earlier Deardorff and Willis experiments for the broad F_* range shown here. The increase is due to the increase in the elevated plume centerlines for the more buoyant releases and to a more intermittent plume at the surface. Although there is a significant variation with F_* at short range, the σ_c/C exhibits a more gradual variation with X for $X > 1.5$. For $0 \leq F_* \leq 0.2$, the σ_c/C collapses to a nearly universal distribution which is attributed to the greater homogenization of the plume within the mixed layer as X increases.

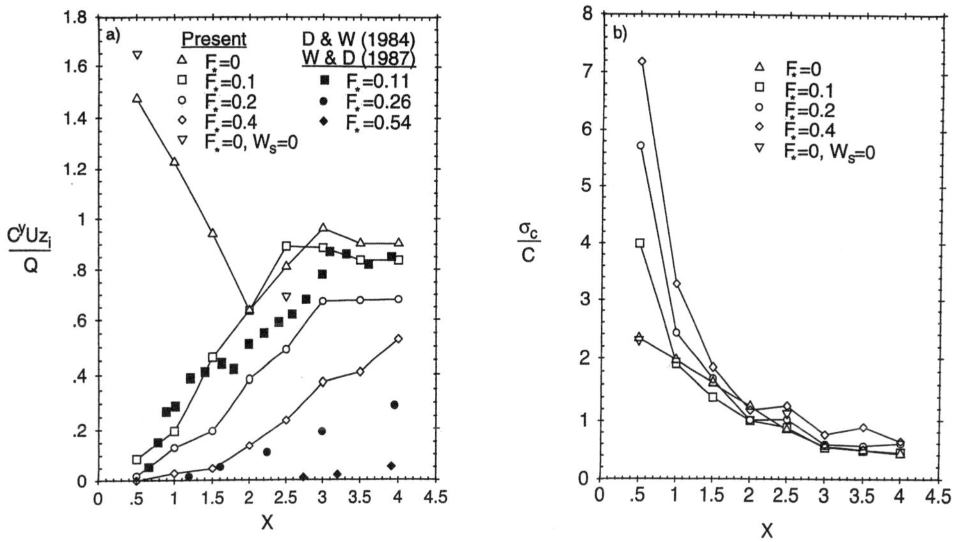

Figure 5. a) Dimensionless crosswind-integrated concentration at the surface ($z/z_i = 0.05$), and b) concentration fluctuation intensity along the plume centerline at the surface, both as a function of X and F_*.

SUMMARY AND CONCLUSIONS

Results from experiments on buoyant plume dispersion in a laboratory convection tank were presented with emphasis on highly-buoyant plumes ($F_* \gtrsim 0.1$), which have posed a major problem to dispersion modeling. For the mean concentration field, some results showed trends similar to those found by Willis and Deardorff (1987), but our results went beyond theirs.

Our data showed the following. 1) The CWIC fields including the surface values exhibited systematic dependencies on X and F_*. Far downstream ($X \geq 3$), the plumes were characterized by a vertically well-mixed distribution below z_i with a C^y maximum aloft, in contrast to the nonuniform profiles obtained earlier by Willis and Deardorff over the same X range. 2) The lateral and vertical dispersion parameters including the rms meander, relative dispersion, and total dispersion followed systematic trends with X and F_*. These data were the first laboratory measurements to show a buoyancy-enhanced σ_y, with $\sigma_y/z_i \propto F_*^{1/3}X^{2/3}$ for the two most buoyant cases ($F_* = 0.2, 0.4$), and a qualitative agreement of σ_y with field observations. 3) Measurements of the concentration fluctuation intensity σ_c/C along the plume centerline showed that close to the source the near-surface σ_c/C increased systematically with buoyancy due to the greater concentration intermittency. For $F_* \leq 0.2$, the surface σ_c/C decreased steadily with distance and attained a nearly universal distribution with X for $X > 1.5$; this was attributed to the strong vertical mixing in the mixed layer.

ACKNOWLEDGMENTS

This work was supported by the Strategic Environmental Research and Development Program of the U.S. Department of Defense, Department of Energy, and Environmental Protection Agency (EPA) through a cooperative agreement between EPA and The Pennsylvania State University with J.C. Wyngaard as the project director. J.C. Weil was supported under a subcontract to the University of Colorado.

DISCLAIMER

This paper has been reviewed in accordance with the U.S. Environmental Protection Agency's peer and administrative review policies and approved for presentation and publication. Mention of trade names or commercial products does not constitute endorsement or recommendation for use.

REFERENCES

Briggs, G.A., 1985, Analytical parameterizations of diffusion: the convective boundary layer, *J. Climate Appl. Meteor.*, 24:1167.

Briggs, G.A., 1988, Analysis of diffusion field experiments, in: *Lectures on Air Pollution Modeling*, A. Venkatram and J.C. Wyngaard, Eds., Amer. Meteor. Soc., Boston.

Csanady, G.T., 1973, *Turbulent Diffusion in the Environment*, Reidel, Dordrecht.

Deardorff, J.W., and G.E. Willis, 1984, Ground-level concentration fluctuations from a buoyant and a non-buoyant source within a laboratory convectively mixed layer, *Atmos. Environ.*, 18:1297.

Snyder, W.H., R.E. Lawson, Jr., M.S. Shipman, and J. Lu, 2000, Fluid modeling of atmospheric dispersion in the convective boundary layer, *Bound.-Layer Meteor.* (submitted).

Weil, J.C., 1988, Dispersion in the convective boundary layer, in: *Lectures on Air Pollution Modeling*, A. Venkatram and J.C. Wyngaard, Eds., Amer. Meteor. Soc., Boston.

Weil, J.C., and L.A. Corio, 1985, Dispersion formulations based on convective scaling, Maryland Power Plant Siting Program, Maryland Dept. of Natural Resources, Annapolis, MD, Rept. No. PPSP-MP-60.

Weil, J.C., W.H. Snyder, R.E. Lawson, Jr., and M.S. Shipman, 2000, Experiments on buoyant plume dispersion in a laboratory convection tank, *Bound.-Layer Meteor.* (submitted).

Willis, G.E., and J.W. Deardorff, 1983, On plume rise within the convective boundary layer, *Atmos. Environ.*, 17:2435.

Willis, G.E., and J.W. Deardorff, 1987, Buoyant plume dispersion and inversion entrapment in and above a laboratory mixed layer, *Atmos. Environ.*, 21:1725.

DISCUSSION

R. BORNSTEIN: Comment. During a summer field study in Sacramento, CA, USA, we were unable to get an instrumented balloon to penetrate into a strong elevated subsidence inversion. It just flattened out at the inversion base in a pattern just like your gravity current/dispersion figure.

A. VENKATRAM: The horizontal spread caused by self-induced (buoyancy-induced) turbulence is similar in form to that caused by the "gravity current". How do you distinguish between the two?

J. WEIL: The gravity current model pertains to the lateral spread of a highly-buoyant plume that lofts at the top of the CBL. The model is based on a constant vertical thickness h of the current or plume with $h \sim \Delta z_i$, $(=0.25 z_i)$ where Δz_i is the thickness of the inversion layer - the upper part of the entrainment zone (see Weil et al., 2000). The gravity current model leads to an expression of the form $\sigma_y = c_1 F_b^{1/3} x^{2/3} / U$, where Briggs (1985) finds $c_1 = 1.6$ and our laboratory data show $c_1 = 0.8$. Your point is that this σ_y expression is of the same form as that for a round, freely-rising plume within the CBL, i.e., before the plume "feels" or reaches the inversion. The standard deviation or spread σ_r of the round plume is related to the plume radius r as: $\sigma_r = r/\sqrt{2} \approx 0.45 F_b^{1/3} x^{2/3} / U$. The difference between the two expressions is that the round plume applies to two-dimensional dispersion or spreading about the plume axis, whereas the gravity-current model applies only to one-dimensional spread, i.e., in the lateral direction; the vertical spread is assumed to be constant. Thus, the plume cross sectional area would increase in proportion to $x^{4/3}$ for the round plume but only as $x^{2/3}$ for the "lofting" plume.

B. FISHER: Have you looked at situations when the plume has sufficient buoyancy to penetrate the inversion? It would be interesting to investigate how the plume fractionates or partitions.

J. WEIL: We have considered but not investigated the higher buoyancy case where the plume can penetrate the inversion top. I agree that this would be an interesting case to explore, but it must be done with care. There are two ways to achieve a greater penetration: 1) reduce the density gradient in the undisturbed stable layer above the entrainment zone, and/or 2) increase the source buoyancy. As a first step, we chose a situation in which the CBL depth increased very little (~ 3%) over the course of the experiment. This was the case with the experiments reported here and was done to simplify the data interpretation, i.e., we assumed a constant z_i. With a sufficiently weak density gradient aloft to achieve a greater plume penetration, the unsteadiness in z_i would have to be taken into account. We will consider weaker density gradients for future experiments.

ANALYSIS AND POLLUTION IMPLICATIONS OF THE $E - \epsilon$ TURBULENCE MODEL PREDICTIONS OF THE NEUTRAL ABL

Frank R. Freedman

Environmental Fluid Mechanics Laboratory
Department of Civil and Environmental Engineering
Stanford University
Stanford, CA 94305-4020

INTRODUCTION

A common turbulence parameterization used in Reynolds-averaged flow solvers is the $E - \epsilon$ model. In this model, turbulence fluxes are represented by traditional gradient-diffusion relationships with turbulent diffusivities diagnosed from prognostically solved values of turbulent kinetic energy (E) and its dissipation rate (ϵ). The primary advantages of this model over traditionally used mixing-length approaches are the maintenance of some of the non-equilibrium effects on turbulence through the inclusion of unsteady and transport terms in the E and ϵ equations, and the removal of the need to specify mixing length l, since this can be modeled as $l \propto E^{3/2}/\epsilon$. This latter advantage makes the $E - \epsilon$ model attractive, for example, for computation of complex flow, in which the manner to prescribe l is unclear.

In spite of its advantages, use of the $E - \epsilon$ model in linked meteorological flow and air-pollution dispersion computations is not widespread. This seems in part due to apparent errors in the model's predictions of prototypical steady-state, one-dimensional atmospheric boundary layers (ABLs), for which accurate model predictions are routinely assured prior to final application (e.g., Andrén, 1990; Hurley, 1997). Here, this issue is revisited for the horizontally homogeneous, barotropic, neutrally stratified ABL (NABL) in light of mathematical analysis of the $E - \epsilon$ model solution near the NABL top (edge). Results of NABL computations using the $E - \epsilon$ model are then interpreted based on this analysis. From this, it is seen that the oft-reported failure of the $E - \epsilon$ model in application to the NABL can be traced to the inability of the model's analytic edge solution to collapse to that consistent with a steady-state diffusion-dissipation TKE balance when standard values of closure constants are maintained. A constraint on the ratio of closure constants is proposed based on the analysis to attain such consistency. Computations with the constants chosen to satisfy this constraint are shown to rectify this predictive failure of the $E - \epsilon$ model.

Air Pollution Modeling and Its Application XIV, Edited by
Gryning and Schiermeier, Kluwer Academic/Plenum Publishers, New York, 2001

EQUATIONS AND ANALYSIS

For the NABL, the mean momentum equations take the form

$$\frac{\partial \overline{u}}{\partial t} = f(\overline{v} - v_g) - \frac{\partial \overline{u'w'}}{\partial z} \tag{1}$$

$$\frac{\partial \overline{v}}{\partial t} = -f(\overline{u} - u_g) - \frac{\partial \overline{v'w'}}{\partial z}, \tag{2}$$

where all symbols have standard definitions. The $E - \epsilon$ model for turbulence fluxes in (1) and (2) is given by

$$\overline{u'w'} = -K_m \frac{\partial \overline{u}}{\partial z} \tag{3}$$

$$\overline{v'w'} = -K_m \frac{\partial \overline{v}}{\partial z}, \tag{4}$$

with turbulent momentum diffusivity (K_m) given by

$$K_m = c_\mu E^2/\epsilon. \tag{5}$$

The following prognostic equations for E and ϵ are then included to close the system

$$\frac{\partial E}{\partial t} - \frac{\partial}{\partial z}\left(\sigma_e K_m \frac{\partial E}{\partial z}\right) = -\overline{u'w'}\frac{\partial \overline{u}}{\partial z} - \overline{v'w'}\frac{\partial \overline{v}}{\partial z} - \epsilon \tag{6}$$

$$\frac{\partial \epsilon}{\partial t} - \frac{\partial}{\partial z}\left(\sigma_\epsilon K_e \frac{\partial \epsilon}{\partial z}\right) = -\frac{c_{\epsilon 1}}{\tau}\left(\overline{u'w'}\frac{\partial \overline{u}}{\partial z} + \overline{v'w'}\frac{\partial \overline{v}}{\partial z}\right) - \frac{c_{\epsilon 2}}{\tau}\epsilon, \tag{7}$$

where $\tau \equiv E/\epsilon$ is a turbulence time scale. The standard values for closure constants $[c_\mu, c_{\epsilon 1}, c_{\epsilon 2}, \sigma_\epsilon, \sigma_e = 0.09, 1.44, 1.92, 0.77, 1.00]$ are set to give reasonably accurate predictions of several engineering benchmark flows in comparison to laboratory data. A constraint invoked in the determination of these constants is that (6) and (7) must yield the proper value of the Von Karman constant (k) through the relationship

$$k^2 = \frac{c_\mu^{1/2}(c_{\epsilon 2} - c_{\epsilon 1})}{\sigma_\epsilon}, \tag{8}$$

derived by inserting classic surface layer similarity expressions into the steady form of (7). Plugging the standard constants into (8) gives $k \approx 0.43$. The more commonly accepted value $k = 0.40$ can be achieved by resetting $\sigma_\epsilon = 0.90$.

In an analysis of mean and turbulence equations near the NABL top (edge), Deaves (1981) found that the only primary turbulent kinetic energy (TKE) balance in this region consistent with the search for a steady-state solution and satisfactory of a series of reasonable upper boundary conditions was between diffusion and dissipation. Assuming such a balance, "local" (at the edge) power-law solutions of the form $E \sim \eta^p$ and $\epsilon \sim \eta^q$ are proposed subject to the upper boundary conditions $E = 0$ and $\epsilon = 0$ at $\eta = 0$. Here, $\eta \equiv 1 - z/h$ is a normalized height variable in terms of height h, at which the flow laminarizes with respect to TKE. Insertion of these relationships into (6) and (7) assuming a steady-state diffusion-dissipation balance can then be shown to yield

$$(6-3\kappa)p^2 - 7p + 2 = 0 \tag{9}$$

$$q = 3p/2 - 1, \tag{10}$$

where

$$\kappa = \frac{c_{\epsilon 2}\sigma_e}{\sigma_\epsilon}. \tag{11}$$

The details of the derivation leading to (9) and (10) are given in Freedman and Jacobson (2000). Valid solutions for p and q are those satisfying the upper boundary conditions with finite derivatives at $\eta = 0$. It can be shown that such solutions result from the positive solution branch of (9) so long as $0.625 < \kappa < 2$. A narrower range

$$1 \leq \kappa \leq 1.45, \tag{12}$$

however, is proposed so that $\epsilon \to 0$ as fast or faster than $E \to 0$ as $\eta \to 0$ (the lower bound), and so that the decrease of E as $\eta \to 0$ is not excessively rapid (the upper bound). The former is justified from classical arguments of TKE decay, whereby small dissipation bearing eddies are extinguished before large energy containing eddies. The latter is based on qualitative observation of the Coleman (1999) DNS data of the NABL, discussed further in the next section. From (9) and (10), it is seen that the bounds of (12) correspond to the approximate ranges $2 < p < 4$ and $2 < q \leq 5$.

Insertion of the standard constants (with σ_ϵ equal to either 0.77 or 0.90) into (11) yields $\kappa > 2$, outside the range of consistency with the NABL edge solution sought above. In another analysis of the $E - \epsilon$ model solution at the edge of turbulent shear layers, Cazalbou et al. (1994) found solutions in the form of diffusive fronts based in essence on a TKE balance between material derivative and diffusion terms. Computations performed for a series of streamwise-varying shear flows showed edge behavior consistent with this proposed solution. Although the Cazalbou et al. (1994) edge solution is consistent with the search for a steady-state solution in streamwise-varying flows, where material derivative terms are manifested as horizontal mean advection, it is not consistent with such a search for flows without streamwise variation such as the NABL, since in this case the material derivative is manifested as unsteadiness. The existence of a steady-state NABL, however, gains wide support from theory (Blackadar and Tennekes, 1968), field observations (Lettau, 1962; Grant, 1986), and LES and DNS computations (Mason and Thomson, 1987, Andrén and Moeng 1993; Coleman, 1999). Any tendency for the $E - \epsilon$ model to collapse to an unsteady edge solution in application to the NABL should therefore be viewed as erroneous, and as a possible source of inaccuracy. To achieve the hypothesized-correct NABL edge solution sought above, it is proposed necessary to alter the values of the constants in some way so that (12) is satisfied. Because of complications involved with also satisfying (8), the simplest way to do this is to alter the value of σ_e. In the next section, $E - \epsilon$ model computations of the NABL employing the standard value $\sigma_e = 1$ and the value $\sigma_e = 0.6$, corresponding to $\kappa = 1.28$ and thereby satisfying (12), are compared in light of their different analytical edge solutions.

COMPUTATIONAL RESULTS

We now proceed to describe a series of numerical computations of the NABL using the $E-\epsilon$ model. These computations were performed by numerical integration of (1)-(7) to steady-state over four inertial periods, where the inertial period $\tau_f \equiv 2\pi/|f|$. A grid of 150 levels is employed up to a model top of $\approx$ 20 km, chosen to be far removed from the disturbed region of the flow. Roughly 100 of these levels reside below 5 km. Surface layer similarity relationships are employed for lower boundary conditions. Runs were made with different imposed values of surface Rossby number $\mathrm{Ro} = G/|f|z_0$ (Table 1), where G is the geostrophic wind speed, $f = 2\Omega \sin \phi$ the Coriolis parameter and z_0 the roughness length (entering the computations through the lower boundary conditions), so that consistency with the results of Rossby number similarity (Blackadar and Tennekes, 1968) can be checked. Results are further compared with the Coleman (1999, hereafter C99) DNS data of the NABL. These are chosen to ensure comparison with data reflective of purely neutral conditions, probably never completely present in field data, and because of its high spatial resolution throughout the NABL. As discussed in C99, the Reynolds number of the DNS is sufficiently high to achieve near Reynolds (or equivalently Rossby) number invariance in the outer region of the NABL. This invariance assures that comparison of outer layer profiles and bulk quantities predicted by the $E-\epsilon$ model over a rough surface with those computed by the DNS over a smooth surface is valid. Also, since the effects of the horizontal component of rotation on turbulence are not accounted for in the $E-\epsilon$ model, there are no interpretative complications with respect to this physics between the $E-\epsilon$ computations (with different imposed values of latitude ϕ - see Table 1) and the DNS (which assumed $\phi = 90°$). Further details on the design and results of these computations can be found in Freedman and Jacobson (2000).

Computations of non-dimensional boundary layer height $\tilde{h} \equiv h_\tau |f|/u_\star$ for different values of constants σ_e and σ_ϵ are shown in Fig. 1. Here, h_τ is defined as the height at which the magnitude of the momentum flux is 5% of the surface value $u_\star^2$. Rossby number similarity predicts that $\tilde{h}$ is a unique constant independent of Ro, an invariance reproduced by the $E-\epsilon$ computations at steady-state. The C99 DNS data, as well as the LES results of Andrén and Moeng (1993), report $\tilde{h} \approx 0.60$. Overpredictions with respect to this value result from use of the standard constants ($\tilde{h} \approx 0.85$); a moderate improvement ($\tilde{h} \approx 0.75$) is found when the value $\sigma_\epsilon = 0.90$ is employed to be consistent with $k = 0.40$. Similar overpredictions of NABL depth are reported in all other known published applications of the $E-\epsilon$ model to NABL (see Xu and Taylor, 1997 and references therein). Accurate collapse to $\tilde{h} \approx 0.60$, however, results when the value $\sigma_e = 0.6$ is employed. These results provide evidence supporting the hypothesis that proper prediction of the NABL by the $E-\epsilon$ model is dependent on the collapse of its analytic edge solution to that consistent with a steady-state, diffusion-dissipation TKE balance. When the standard value $\sigma_e = 1$ is maintained, thereby collapsing this solution to the erroneous (for the NABL) unsteady one of Cazalbou et al. (1994), correspondingly erroneous maintenance of TKE aloft (as will be shown below) and overpredicted $\tilde{h}$ result. This is consistent with the lack of dissipation aloft implied by this edge solution. Contrarily, when $\sigma_e = 0.6$ is chosen, collapsing the analytic edge solution to the proper steady-state one, such TKE maintenance aloft does not occur and $\tilde{h}$ is accurately predicted. This is consistent with the presence of dissipation aloft implied by this edge solution.

Table 1: Cases for $E - \epsilon$ NABL computations.

Case	G (ms^{-1})	ϕ (deg)	z_0 (m)	Ro
RO5	5	60.0	0.100	3.96×10^5
RO6	10	43.3	0.100	1.00×10^6
RO7	5	30.0	0.005	1.37×10^7
RO8	30	70.0	0.002	1.09×10^8

Evidence for proper representation of NABL edge behavior is further seen from computed profiles of non-dimensional TKE ($\tilde{E}$) and length scale ($\tilde{l}$) shown in Fig. 2. For computations using $\sigma_e = 1$, $\tilde{E}$ (Fig. 2a) is overpredicted in the upper region. As discussed above, this is an expected consequence of the unsteady analytic edge solution imposed from such constant selection. The lack of ϵ near the top of the NABL for this case is seen to have dramatic effects on computed length scales (Fig. 2b). Here, a large valued maximum in $\tilde{l}$ is achieved well above $\tilde{z} \approx 1$, the approximate point at which TKE laminarizes. The existence of such a maximum is intuitively unappealing, and is at great odds with LES deductions of Andrén and Moeng (1993) in which l approached zero near the NABL top. The dimensional value of the maximum for this case is $l \approx 500$ m, much larger than asymptotically approached values $l_\infty \approx 50 - 100$ m typically prescribed in Blackadar-type mixing length representations. In contrast, when the value of $\sigma_e = 0.6$ is set, excellent agreement with the DNS is found in $\tilde{E}$ aloft, with $\tilde{l}$ a maximum *within* the NABL approaching $\tilde{l} \approx 0$ at the TKE laminarization point $\tilde{z} \approx 1$. These length scale predictions seem physically more realistic, and qualitatively agree with Andrén and Moeng's (1993) LES deductions. The approach of l to zero at the TKE laminarization point h ($\eta = 0$) is furthermore consistent with (10), which implies $l \propto E^{3/2}/\epsilon \sim \eta$. This provides additional evidence for consistency of the edge solution computed by the $E - \epsilon$ model with $\sigma_e = 0.6$ with the correct one based on steady diffusion-dissipation TKE balance. Such consistency is not found when the standard constants are maintained, where l continues to grow well beyond h. [1]

To examine the air-pollution implications of the above findings, a plume of passive tracer with mean concentration $\bar{c}$ was introduced in the region $1 \leq \tilde{z} \leq 1.25$ at $\tilde{t} \equiv t/\tau_f = 3$. This region was chosen to be above h, where little tracer diffusion should occur. The turbulent flux of tracer was modeled in a manner analogous to that for momentum; i.e. $\overline{w'c'} = -K_c \partial\bar{c}/\partial z$ with $K_c = K_m$. Tracer concentrations at $\tilde{t} = 4$ (Fig. 3a) show that the plume remains essentially intact when the value of $\sigma_e = 0.6$ is set; only minor and qualitatively realistic erosion of the lower portion of the plume is computed. On the other hand, large unrealistic tracer diffusion is computed when $\sigma_e = 1$ is maintained, with the initially concentrated profile of $\bar{c}$ centered at $\tilde{z} = 1.125$ smeared-out nearly uniformly over a region ≈ 8 km deep. This is attributed to the large computed values of l and K_c (Fig. 3b) above h for this case, viewed as a consequence of the inability of the $E - \epsilon$ model with the standard constants to collapse to the proper NABL edge solution. When the value $\sigma_e = 0.6$ is set to obey (12), these quantities properly collapse to zero near h, consistent with the steady edge solution. This results in realistic tracer diffusion for the initial conditions specified in this case.

[1] The underprediction of TKE near the surface in both cases is a consequence of an imposed value $c_\mu = 0.09$ higher than that inferred from atmospheric surface layer measurements, which instead suggest values $c_\mu \approx 0.05$. Computations with this change in c_μ (not shown), although improving TKE predictions in the lower NABL, yielded no significant change to overall predictions elsewhere in the flow.

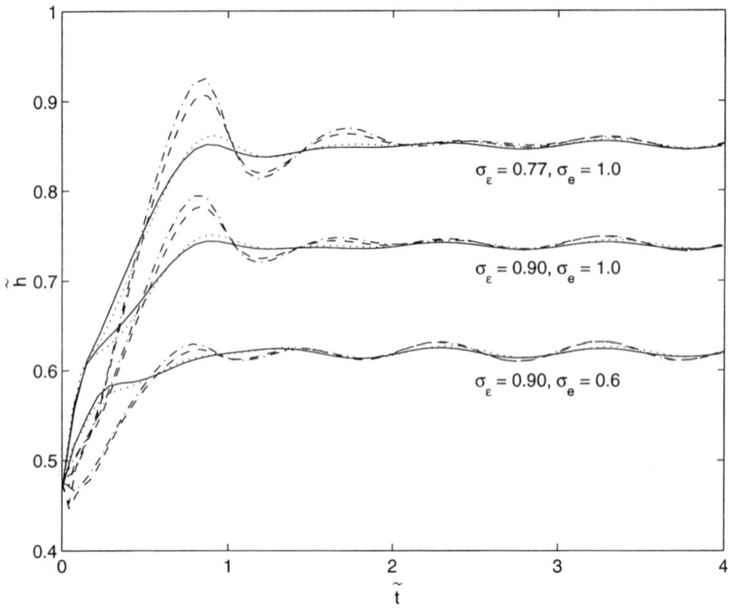

Figure 1: Plots of numerically computed non-dimensional boundary layer height $\tilde{h} \equiv h_T|f|/u_\star$ vs. non-dimensional time $\tilde{t} = t|f|/2\pi$ for different choices of closure constants and Ro (Table 1); solid line: RO5 simulation, dotted line: RO6 simulation, dashed line: RO7 simulation, dashed-dotted line: RO8 simulation. A value $\tilde{h} \approx 0.60$ is found from the C99 DNS.

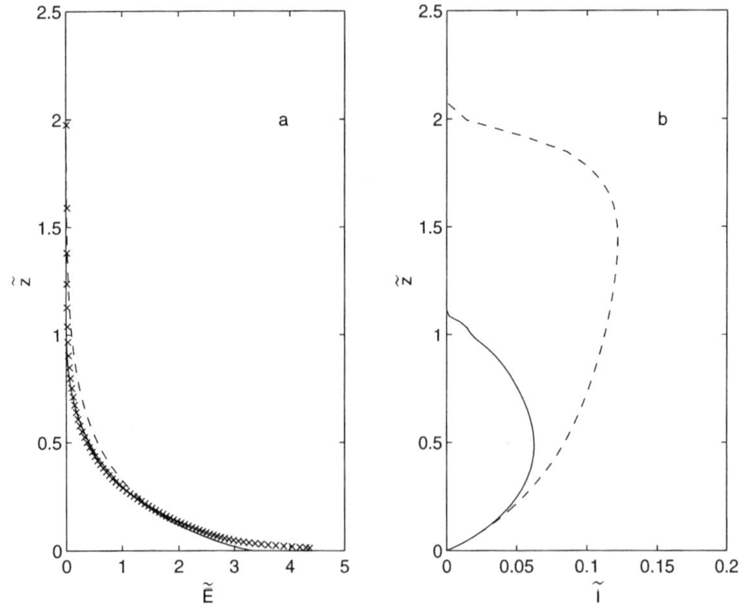

Figure 2: Plots of numerically computed a) non-dimensional turbulent kinetic energy $\tilde{E} \equiv E/u_\star^2$ and b) non-dimensional length scale $\tilde{l} \equiv (c_\mu^{3/4}\tilde{E}^{3/2})/\tilde{\epsilon}$ for case RO6 (Table 1), where non-dimensional dissipation $\tilde{\epsilon} \equiv \epsilon/u_\star^2|f|$. Solid line: $\sigma_\epsilon = 0.90$, $\sigma_e = 0.60$; dashed line: $\sigma_\epsilon = 0.90$, $\sigma_e = 1.00$. The C99 DNS $\tilde{E}$ data is plotted as symbol 'x', and has been truncated to only show data for $\tilde{z} > 0.01$.

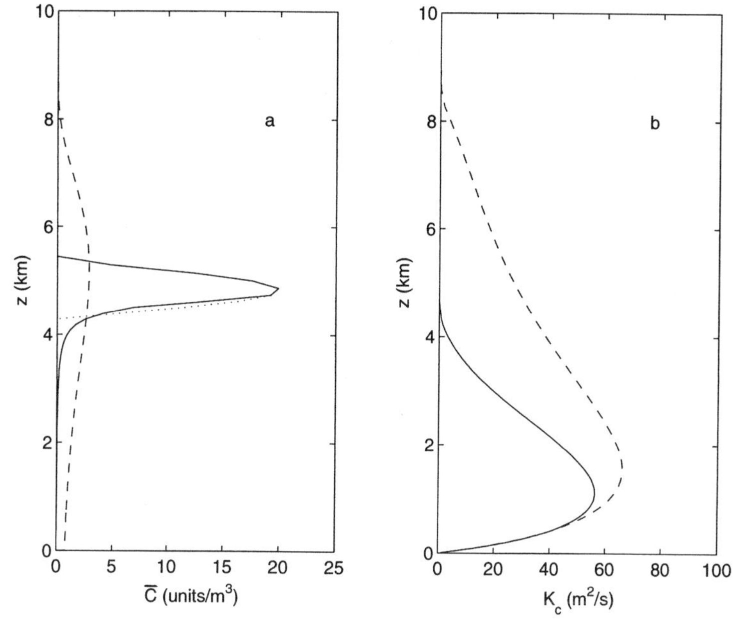

Figure 3: Plots of numerically computed profiles of a) mean tracer concentration $\bar{c}$ and b) mean tracer diffusivity K_c for case RO6 (Table 1). Solid line: $\sigma_\epsilon = 0.90$, $\sigma_e = 0.60$; dashed line: $\sigma_\epsilon = 0.90$, $\sigma_e = 1.00$. The dotted line in a) is the initial profile at $\tilde{t} = 3$.

CONCLUSION

The central result discussed in this paper was that condition (12) on closure constants should be satisfied in order for the $E - \epsilon$ turbulence model to properly represent the solution for turbulence quantities near the NABL top (edge). This condition was derived from mathematical analysis of the $E - \epsilon$ model equations near the NABL top assuming a steady-state balance between diffusion and dissipation terms in the E and ϵ equations. The accuracy of numerical predictions of the NABL by the $E - \epsilon$ model when the value of σ_e was changed from 1 to 0.6 to satisfy (12) supports the validity of this constraint. This stated, it should be emphasized that this change in the value of σ_e is *not* in and of itself suggested as a "fix" to the $E - \epsilon$ model for NABL prediction; the change was only made to satisfy (12) in attempts to demonstrate the predictive effects of the model's analytic collapse to the proper NABL edge solution. It remains for future work to examine the physical implications of (12), and as a result to hopefully arrive at a means of satisfying this constraint consistent with the physics of the flow. In the meantime, the alteration $\sigma_e = 0.6$ from the standard value $\sigma_e = 1$ is seen as a practical solution to the problem of satisfying (12), and therefore of rectifying the oft-reported failure of the standard $E - \epsilon$ model for NABL prediction.

Future research will be directed towards performing analyses and computations with the $E - \epsilon$ model for prototypical stratified ABLs similar to those described in this paper for the NABL. The outcome of such efforts should aid considerably in understanding and improving the model so that its use in linked meteorological flow and air-pollution dispersion computations can be more frequently and confidently carried out.

ACKNOWLEDGEMENTS

The author would like to express thanks to Dr. G. N. Coleman for supplying the DNS data used for comparison in this study.

REFERENCES

Andrén, A., 1990, Evaluation of a turbulence closure scheme suitable for air pollution applications, *J. Appl. Meteor.* 29:224.

Andrén, A., and C. H. Moeng, 1993, Single-point closures in a neutrally stratified boundary layer, *J. Atmos. Sci.* 50:3366.

Blackadar, A. K., and H. Tennekes, 1968, Asymptotic similarity in neutral barotropic planetary boundary layers, *J. Atmos. Sci.* 25:1015.

Cazalbou, J. B., P. R. Spalart, and P. Bradshaw, 1994, On the behavior of two-equation models at the edge of a turbulent region, *Phys. Fluids* 6:1798.

Coleman, G. M., 1999, Similarity statistics from a direct numerical simulation of the neutrally stratified planetary boundary layer, *J. Atmos. Sci.* 56:891.

Deaves, D. M., 1981, A note on the upper boundary conditions for turbulence models in the neutral atmosphere, *Bound. Layer Meteor.* 21:489.

Freedman, F. R., and M. Z. Jacobson, 2000, Reanalysis of the $E - \epsilon$ turbulence model predictions of the neutral ABL, to be submitted to *Bound. Layer Meteor.*

Grant, A. L. M., 1986: Observations of boundary layer structure made during the KONTUR experiment, *Quart. J. Roy. Meteor. Soc.* 112:825.

Hurley, P. J., 1997, An evaluation of several turbulence schemes for the prediction of mean and turbulent fields in complex terrain, *Bound. Layer Meteor.* 83:43.

Lettau, H. H., 1962: Theoretical wind spirals in the boundary layer of a barotropic atmosphere, *Beitr. Phys. Atmos.* 35:195.

Mason, P. J., and D. J. Thomson, 1987, Large-eddy simulations of the neutral static-stability planetary boundary layer. *Quart. J. Roy. Meteor. Soc.* 113:413.

Xu, D., and P. A. Taylor, 1997, An $E - \epsilon - l$ turbulence closure scheme for planetary boundary-layer models: the neutrally stratified case, *Bound. Layer Meteor.* 84:247.

DISCUSSION

A. VENKATRAM: Can you explain how a relatively small variation in the constants in the E-ε equation can make such a big difference in the length scale profiles? I suppose you can choose the constants so the ε goes to zero faster than E does.

F. FREEDMAN: Although, the modifications $\sigma_e = 0.6$ and $\sigma_\varepsilon = 0.9$ from their standard values $\sigma_e = 1$ and $\sigma_\varepsilon = 0.77$ seem minor, the *ratio* $\sigma_e / \sigma_\varepsilon$, which appears in κ, decreases by roughly a half. This is a more significant change, which strongly increases ε-transport relative to E-transport near the NABL top. As a result, the solution at the NABL edge shifts from one in which ε approaches zero faster than the $E^{3/2}$, to one in which ε approaches zero slower than $E^{3/2}$. Hence the profile for length scale (proportional to $E^{3/2}/\varepsilon$) decreases with height near the NABL top after the constant modification is made. This is an expected result based on the power-law analytic solution at the edge. In both the standard and modified cases, ε approaches zero faster than E (but slower than $E^{3/2}$ only in the modified case).

R. BORNSTEIN: What would need to be done in order to extend this work to diabatic ABLs?

F. FREEDMAN: From the literature, it appears that turbulence models using the standard ε-equation give poor predictions of diabatic as well as neutral ABLs. The work described here shows how to solve the predictive problem for the neutral case, providing the foundation on which to extend to thermally-stratified (diabatic) cases. To extend to these cases, one idea is to analyze the model equations for consistency with M-O similarity theory, that provides a firm basis to extend to diabatic from neutral ABLs. If inconsistent, the model can perhaps be altered to enforce consistency with this theory.

D. W. BYUN: Commenting on the extension of the algorithm, you could consider a boundary layer matching approach to link surface-layer similarity with the modification of the PBL wind with additional power-law description.

F. FREEDMAN: Yes, this should be helpful, since it is a way to extend the analyses of the model equations beyond the surface layer.

D. STEYN: "Data" is the plural of "Datum", and means known, fixed or reference points or values. You have used the term "DNS data". Could you explain how you elevate mere (DNS) model output to the status of data?

F. FREEDMAN: I agree that viewing computer-generated output from a DNS as "data" (with the definitional implications you noted) warrants explanation. The elevation of this output to data is based on the fact that a DNS, in spirit, is an exact solution to the full Navier-Stokes equations, and it is the turbulence statistics from this solution that we aim to replicate with turbulence models. In practice, the only inherent error present in DNS output are those associated with numerical discretization (as opposed to LES, there is no subgrid scale model in DNS). If the DNS is done well, these errors are both quantifiable and tolerable. Such is judged to be the case with the Coleman (1999) DNS used here. The other consideration needed to deem the DNS applicable to the atmosphere is that the Reynolds number of the simulation be sufficiently high so that the results are invariant to viscosity. Again, this is shown to an acceptable degree to be the case with the Coleman DNS. Hence, the Coleman DNS of the NABL is viewed to be NABL "data", and has the advantages of being purely neutral (as opposed to field measurements) and of very high vertical resolution.

PLPM: A NEW PHOTOCHEMICAL LAGRANGIAN PARTICLE MODEL. BASIC IDEAS AND PRELIMINARY RESULTS.

Gabriele Zanini[1], Rita Lorenzini[1], Luca Delle Monache[1], Sonia Mosca[2], Roberto Bellasio[2], Roberto Bianconi[2] and Sara Peverieri[1]

[1]ENEA/AMB/CAT/INAT – via Martiri di Monte Sole, 4 – 40129 Bologna, Italy
[2]Enviroware, Centro Direzionale Colleoni - Andromeda 1, 20041 Agrate Brianza, Italy

INTRODUCTION

Lagrangian particle models (LPMs) are the most advanced tools to simulate atmospheric dispersion of inert pollutants and they are more and more extensively used. Eulerian models (EMs) show limitations that LPMs can overcome, as discussed hereafter.

• In mesoscale applications EMs integrate equations over a numerical grid having cells as large as some kilometres, for efficiency. This causes the resolution of meteorological fields and orography to be limited. Also, emissions are uniformly distributed within the cell containing the source, which prevents detailing the concentration field close to strong sources as industrial plant chimneys. This problem is faced either reducing the grid size, or using the Lagrangian plume in grid technique in the initial phase, until the plume size can be compared to the grid extension.

Transport and diffusion in LPMs are instead independent from any grid and all the available meteorological information can be used. Also, nestings of higher resolution in regions with complex atmospheric circulation are possible.

• Diffusion coefficients are proportional to the concentration gradient (K-theory) in some EMs. Under convective meteorological conditions this is questionable since the null vertical concentration gradient produces a null turbulent vertical flux of mass. In reality, the average vertical well-mixing is due to a counterbalance between strong local upwards or downwards motions. The use of a higher-order closure scheme solves this problem but adds additional partial differential equations to an already complex non-linear system.

On the other hand, LPMs can describe convective local updrafts and downdrafts appropriately (e.g. Luhar and Britter, 1989) and thus correctly reproduce high stack emission descent and ground-level emission rise in the initial diffusion phase.

• Comparing mean concentration values predicted by EMs in a cell with values obtained by sampling just some liters of air is often problematic, especially where strong concentration gradients exist (for instance close to sources). Instead, a continuous concentration field can be computed in LPMs, using a density estimator or smoother.

• The space discretization in EMs introduces numerical diffusion that can be reduced but not totally eliminated. Numerical diffusion may even dominate over the physical one (Nguyen et al., 1997).

Apart from these advantages of LPMs vs EMs, until few years ago it was thought that LPMs were unsuitable for treating photochemical pollution. Chock and Winkler (1994) introduced a first photochemical LPM using a simplified chemical mechanism with 10 reactions for 10 species. Chock and Winkler computation of concentration values is made using the box-counting method, consisting in summing the particles' masses in the cells of a grid, and dividing by the cells' volumes. Results depend on the volume of the boxes and are discontinuous, thus making the model grid-dependent.

The model presented in this paper, named PLPM (Photochemical Lagrangian Particle Model) is a Lagrangian particle model that uses the complex chemical mechanism SAPRC90 (Carter, 1990).

The following sections present the basic ideas underlying the modelling approach and preliminary evaluation tests. It is thought to be worth to bring the results obtained so far to the attention of the air quality modelling community, together with the still-open questions that would benefit of the efforts of other researchers.

BASIC IDEAS OF PLPM

PLPM is a three-dimensional complex-terrain and multi-source photochemical LPM. Several source types (point, area, volume and linear) can be located anywhere within the simulation domain. Area and volume sources can have arbitrary extensions with various shapes. Initial conditions are treated as instantaneous volumetric sources activated at the beginning of the simulation. Boundary conditions are described through a number of area sources overlapping domain boundaries and can have any temporal variation. Emissions are preprocessed for lumping according to the condensed chemical mechanism that is selected.

Each released particle has a mass composed by different substances. The number of substances in a particle and its mass vary with time since chemical reactions result in the production or loss of species' mass.

PLPM directly reads the output file of CALMET (Scire et al., 1999) to load the two- and three-dimensional meteorological fields needed to move the particles.

Particles move independently in each direction following the flow. Their velocity is split into a mean component responsible for the transport and a fluctuating component representing the diffusion part. The fluctuating particle velocity follows the stochastic differential equation by Thomson (1987), integrated with variable time step depending on local turbulence.

THE KERNEL ESTIMATOR

The most critical aspect is the transformation of species within each particle. PLPM implements two ways for computing concentration values, the standard box counting and the kernel estimator.

The box counting is simple but, as already stated, sensitive to the grid size and to the number of particles used.

The kernel is grid-independent and gives local concentration values, however the re-assignment of masses to the particles is quite complex and still under study.

Up to now both approaches have been implemented and considered. Also, a mix of the two could be envisaged. Consequently, one line of development has been the identification and implementation of a kernel estimator that gives reliable concentration values with a limited number of particles. Several kernels have been considered (Silvermann, 1986; de Haan, 1999) and specific algorithms for the calculation of the bandwidth have been developed to deal with particles of different mass. While for equal masses the computation of the bandwidths is based on geometrical considerations, in case of variable masses the bandwidth in one direction is found as proportional to the maximum of the distances in that direction of the closest particles that bring a given fraction of the total mass of the species into the domain.

Two kernels have been selected:

$$\hat{f}(\vec{x}) = \frac{1}{n\, h_1 \ldots h_d} \sum_{i=1}^{n} \left\{ \prod_{j=1}^{d} K\left(\frac{x_j - X_{ij}}{h_j^x} \right) \right\}$$

$$\hat{f}(\vec{x}) = \frac{1}{n} \sum_{i=1}^{n} \left\{ \prod_{j=1}^{d} \frac{1}{h_{ij}} K\left(\frac{x_j - X_{ij}}{h_{ij}} \right) \right\}$$

where $\vec{x}$ represents the receptor coordinates and X_{ij} is the coordinate of the i-th particle along the j-th direction.

The first one uses bandwidths (h^x_j) calculated as the distance of the particles from the point where the concentration is computed. The bandwidths of the second kernel (h_{ij}) are based on the interparticle distances.

Upon sensitivity and optimization, some theoretical distributions have been reproduced satisfactorily in both cases (Peverieri, 2000).

INERT APPLICATION

In order to test the kernel ability to predict concentrations in a real situation a comparison has been made between PLPM and CALPUFF (Scire et al., 1999) through the simulation of an inert emission from a tall stack in the center of a domain with elevations ranging from 150 to more than 3000 m in less than 50 km. Satisfactory results for PLPM are obtained generating a small number of particles (360 particles per hour). Both models use the same meteorological input from CALMET. The qualitative patterns at different times are similar (Figure 1). The irregularity of the contours of PLPM not necessarily represents an inaccurate reproduction of the concentration field. In fact, due to the complexity of topography and atmospheric circulation channeling of pollutants into valleys takes place. The capability of the kernel to compute local concentrations focuses the distribution of the pollutant.

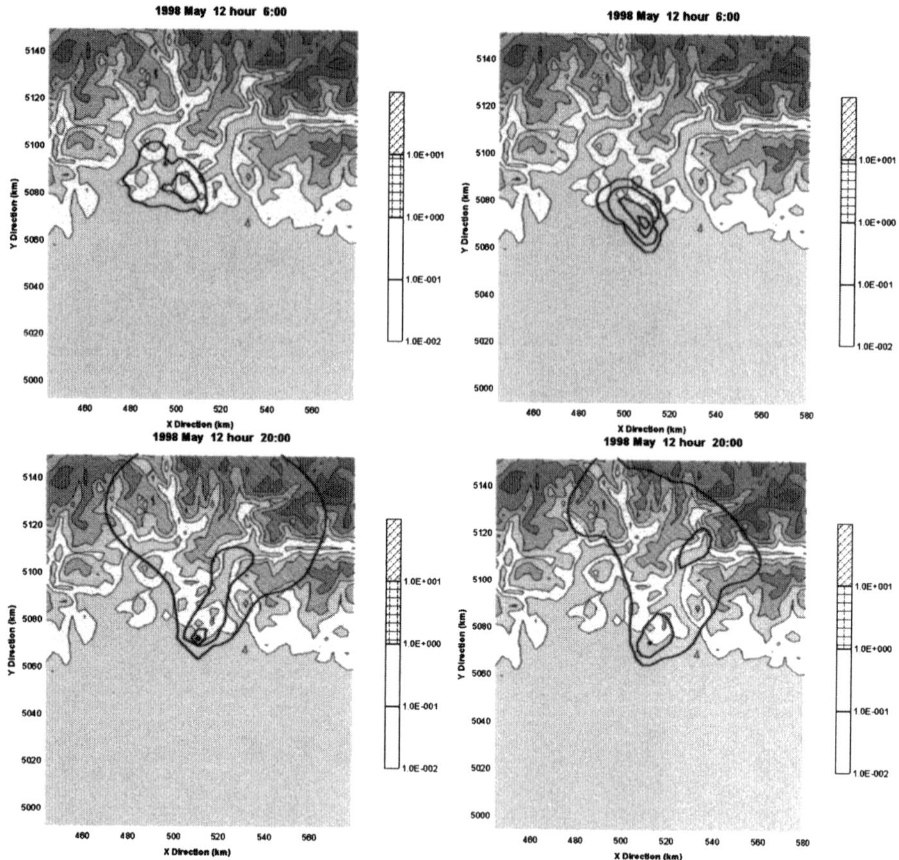

Figure 1. Results of CALPUFF (left) and PLPM (right) for an inert simulation.

The concentration trends at selected receptors are compared for the two models. A satisfactory agreement is observed (Figure 2), especially the first day of the run.

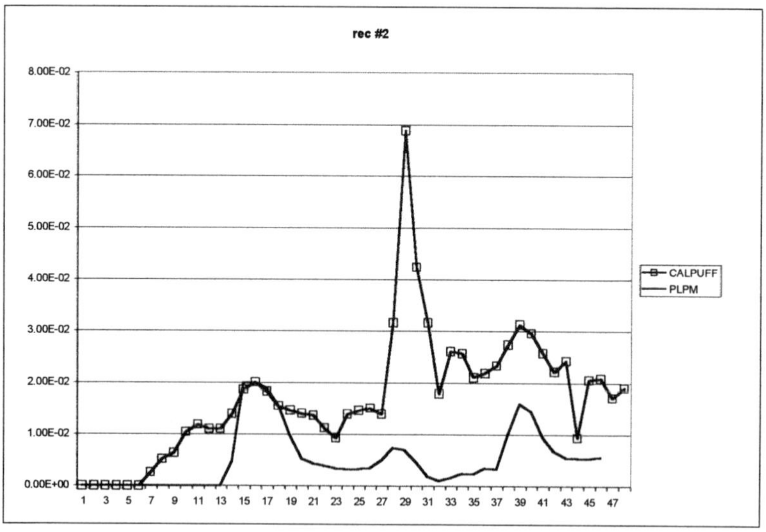

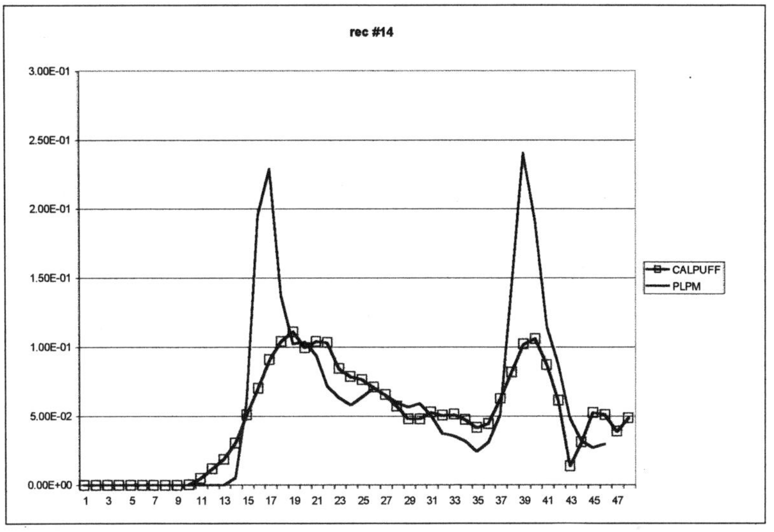

Figure 2. Concentration trends predicted by CALPUFF and PLPM at two selected receptors.

The magnitude of the concentrations predicted by the two models is the same at different receptor locations. The times at which maximum concentrations are reached are also the same or slightly shifted as, for example, in receptor 14.

PHOTOCHEMISTRY

In parallel with the kernel development, efforts have been put in the implementation of the SAPRC90 chemical mechanism into PLPM. One of the main difficulties for chemistry in LPMs, as mentioned above, is the particle mass reassignment after chemical transformations.

Photochemical transformation of masses within a particle is activated at discrete time steps. Let's consider a given particle, hereafter named receptor. Concentration of all species due to all other particles (donors) at receptor location is computed using the density estimator. These values are input to the chemical mechanism and new concentration values are received in output. Starting from these new concentration values, the mass of the substances within the receptor are redistributed with an inverse kernel procedure.

If m_j and C_j are mass and concentration of the j-th species in a receptor before the chemistry step and m_j' and C_j' are the values after chemistry, following Chock and Winkler (1994) it is imposed that $m_j' = m_j \; (c_j' / c_j)$. This equation allows masses of each species to be modified according to the variation of their concentration.

Two options are currently under evaluation: the pure box counting and the use of the kernel. Both schemes hold in the first tests: the gas chamber and the dispersion over complex terrain of an uniformly speciated mass released from a volume source in the same domain of the inert simulation introduced above.

The first set of tests consisted in running PLPM with initial conditions as discussed in Seinfeld (1986). Particles are generated uniformly in the domain and do not move: only concentration calculation, chemistry and re-assignment are activated, and the system is let to evolve for 10 hours, at a constant solar radiation. No active and boundary sources were activated and only four species were evaluated: nitrogen oxide (NO), nitrogen dioxide (NO_2), ozone (O_3) and formaldehyde (HCHO). Figure 3 gives the time evolution of these species, starting from initial concentration in ppm of 0.01, 0.1, 0.1 and 0, respectively.

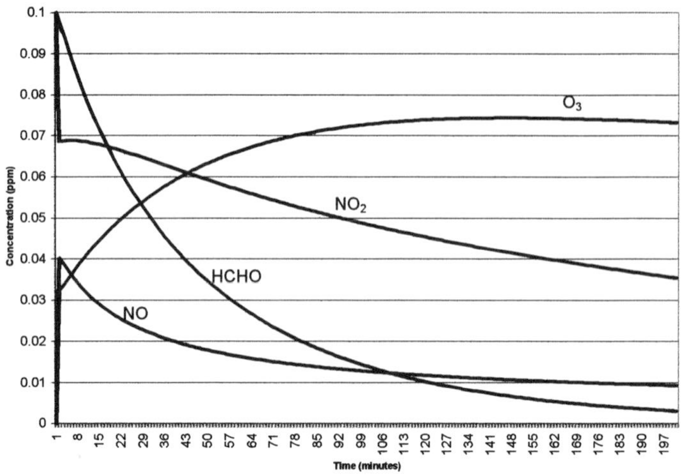

Figure 3. Concentration trends of NO, NO_2, O_3 and HCHO inside the "gas chamber" as predicted by PLPM.

Both with the box counting and the kernel PLPM correctly describes the evolution of the species, confirming the correct implementation of the SAPRC90 mechanism. Thanks to the modularity of the code, other mechanisms could be implemented as well.

FIGURE 4 shows the evolution of a multispecies mass over the domain of the inert test. A fictitious volume source emitting a mixture of pollutants has been placed approximately at the centre of the domain and was kept active for five hours: from 1 to 5 of 12 May 1998. Ozone concentration remains null until 6 a.m., while NO and NO2 concentrations increase due to the continuous emissions in these first hours. During the morning ozone levels increase and a maximum is observed over the lake of Como and toward the Valtellina valley. Ozone production is limited because the emissions last for only five hours during the night and then NO_2, formed by the oxidation of NO, is rapidly consumed. During the night of 13 May NO and O_3 are consumed to form NO_2. It can be seen (Figure 4) that at 4 a.m. there is only a small spot of NO at the North of the domain. On the contrary NO concentrations increase during the following morning due to the photo dissociation of NO_2.

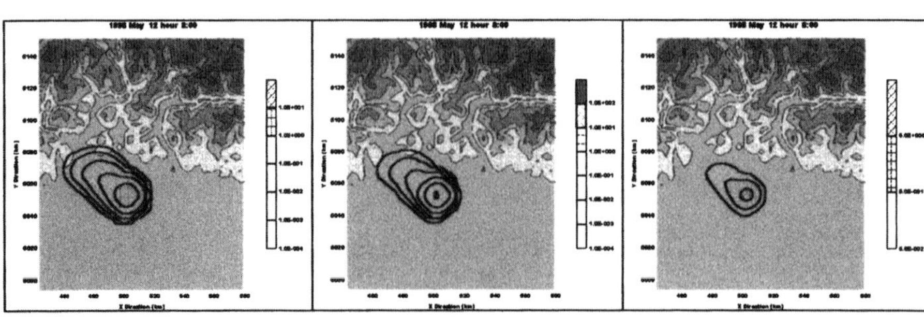

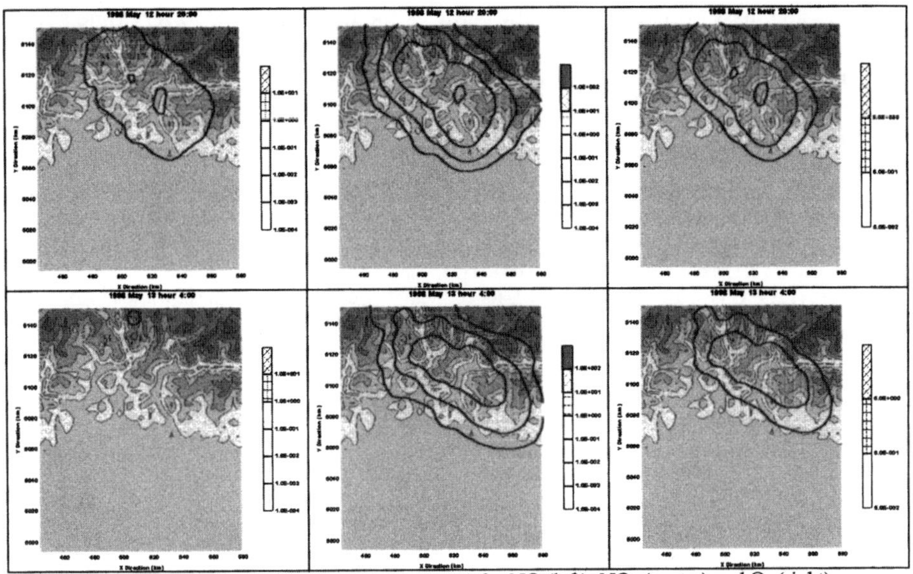

Figure 4. Selected maps of concentration (ppm) for NO (left), NO_2 (centre) and O_3 (right).

DISCUSSION

The test showed that LPMs are suitable to simulate photochemical transformations in a grid-free system using a complex chemical mechanism. However, many questions are still open.

It is here recognized that these performed so far are simple and ideal cases, where the composition of the particles is the same. The simulation of emissions with different species profiles raises the problems with the individuation of the correct and most consistent approach to redistribute masses among particles, or recompute fractions of species at single particles to produce a global behaviour that is coherent with the chemical transformation that has taken place.

In order to identify sources responsible for high concentration levels and acting for selective reductions, approaches of concentration calculation and redistribution of mass to particles maintaining particles' identity were also considered. In case of photochemical reactions, however, the status of a particle as substances mass distribution does not depend only on the source of origin, but also strongly on the chemical reactions undergone. Therefore, the dependence on the source of origin is lower than in case of primary pollution. It is thus necessary to investigate if it is better to privilege approaches conserving the identity of particles or not.

Due to the chemical transformation and the redistribution of mass some particles could lose mass becoming quite small or gain mass and become too large. Both these two opposite situations can be problematic since particles become no representative or too much representative of a volume of space. In order to avoid these problems some rules can be introduced to lump particles when they are too small and to split them when they are too big.

The choice that many pollutants compose each particle may be problematic to deal with substances that have a vertical stratification. Two possible solutions are a vertical redistribution of mass or a distinction between two types of particles: those representing light substances and those representing substances subject to vertical stratification.

Generating particles each representing only one substance would directly take into account a vertical stratification of heavy substances, but would involve many more particles and mass exchanges, and new particles should be generated every time a new substance is created.

Future work will address the above mentioned issues.

REFERENCES

Carter W. P. L. (1990) A detailed mechanism for the gas-phase atmospheric reactions of organic compounds Atmospheric Environment, 24A, No. 3, 481-518

Chock D. P. and Winkler S. L. (1994) A particle grid air quality modelling approach, 2. Coupling with chemistry, J. Geophys. Res. 99, D1, 1033-1041.

de Haan P. (1999) On the use of density kernels for concentration estimations within particle and puff dispersion models. Atmospheric Environment, 33, No. 13, 2007-2021

Gingold R. A. and Monaghan J. J. (1982) *Kernel estimates as a basis for general particle methods in Hydrodynamics.* Journal of Computational Physics, 46, 429-453.

Luhar A. K. and R. E. Britter (1989) A skewed homogeneous Larangian particle model for convective conditions, *Atmospheric Environment*, **27A**, 619-624.

Nguyen K. C., Noonan J. A., Galbally I. E. and Physick W. L. (1997) Predictions of plume dispersion in complex terrain: Eulerian versus Lagrangian models. Atmospheric Environment, 31, 7, 947-958.

Peverieri S. (2000) Kernel density estimators for the Photochemical Lagrangian Particle Model. Graduation thesis. University of Bologna, Italy.

Scire J.S., D.G. Strimaitis and R.J.Yamartino (1999) A User's Guide for the CALPUFF dispersion Model (Version 5.0), Earth Tech, Inc

Seinfeld J. H. (1986) Atmospheric chemistry and physics of air pollution. pp. 738. John Wiley and Sons, Inc., New York.

Silvermann B. W. (1986). Density Estimation for Statistics and Data Analysis. pp. 175. Chapman and Hall Ltd., London.

Thomson D. J. (1987) Criteria for the selection of stochastic models of particle trajectories in turbulent flows. J. Fluid Mech., 180, 529-556.

DISCUSSION

P. SEIBERT: What is the bandwidth of the kernel?

L. DELLE MONACHE: With the kernel approach, the concentration of one specie at one point (receptor), is computed taking in account the contribution of all particles (donator), spatially distributed around the receptor point, that carry mass of the specie under consideration. The bandwidth is the most important parameter of the kernel technique. It tell us how much is spread (large bandwidth) or peaked (small bandwidth) the contribution (that can be seen as a "bump" over the donator point) of each donator point. While for equal masses the computation of the bandwidth is based on geometrical considerations, in case of variable masses the bandwidth is found as proportional to the maximum of the distances in that direction of the closest particles that bring a given fraction of the total mass of the species into the domain.

H. van DOP: Do you realize that you make an unknown error by dispersing two (or more) chemical reactants independently from each other? In reality there will be some coherence in the motion of two reacting particles, certainly when they are closed together. See: (reference: Crone, van Dop, Dinar in Atmospheric Environment 1998/99).

L. DELLE MONACHE: Actually the particles doesn't react between each other. In our approach the particles can be seen as a representative sample of the entire population. With a statistical meaning in mind, it doesn't really matter how much two particles are closed together. More they are closed, and more, they will contribute (by kernel density estimation), to the amount of concentration located at the position of each particle. Then, after the concentrations are computed, the chemical reactions take place, independently on each particle location.

P. BUILTJES: Can you explain in which way the particles carry the chemistry. Does a particle contain all the chemistry in total, or carries a particle only NO, or NO_2 etc. and do they have to meet before the reaction can take place?

L. DELLE MONACHE: Each particle carry all the chemistry, that means that each particle can bring a fraction of mass of each specie. By the kernel estimation, we are able to compute, on a fixed point (usually a point where a particle is located) the concentrations of different species. Then, with those values, in that point, the reactions take place. After that, on the specific spatial position there will be new values of concentrations. From those values, with an inverse process respect to the kernel one, the masses of the substances within this point are redistributed.

ADAPTIVE DISPERSION MODELLING AND ITS APPLICATIONS TO INTEGRATED ASSESSMENT AND HYBRID MONITORING OF AIR POLLUTION

Eugene L. Genikhovich, Alexander D. Ziv, Elena N. Filatova
Air Pollution Modelling and Forecasting Laboratory
Main Geophysical Observatory
St. Petersburg 194021, Russia

INTRODUCTION

A significant progress in development of sophisticated dispersion models during several last decades was not accompanied by similar progress in technologies of application of these models. Nowadays as well as forty years ago dispersion models are mainly used independently from results of measurements after completing the process of their evaluation. The problem of data assimilation which is of major importance in, for example, the weather forecast has not been considered as an important one in the field of dispersion modelling. It was due partially to the fact that the existing air pollution monitoring networks are too sparse to provide reasonably detailed fields which could be used for initialisation of the dispersion models and partially to the fact that the existing dispersion models just have no "levers" which could be "manipulated" depending on the measured concentrations of air pollutants; in other words, they are not aimed for adaptive dispersion modelling.

The Kalman filter (Kalman, 1960) is, probably, the main tool widely used for adaptive modelling in engineering applications (e.g., Leondes, 1976). There were several attempts made to apply this technique to diffusion problems in the urban (Runca *et al.*, 1979), regional (Bultijes *et al.*, 2000) and global (Haas-Laursen *et al.*, 1996) scales as well as to use it in meteorological applications (Houtekamer, Mitchell, 1998). In the majority of theoretical and practical works on the meteorological weather forecast, however, the data assimilation is based on the theory of objective analysis developed by Gandin (1963) (see also Sasaki, 1970). The goal of the present publication is to discuss possible reasons for limited applicability of the Kalman filtering in dispersion modelling and certain approaches to adaptive dispersion modelling based on the methodology of the objective analysis of meteorological fields.

Adaptive dispersion modelling is considered here in connection with the problems of the integrated assessment and hybrid monitoring of air pollution in cities, *i.e.*, combined use of measured and calculated concentrations for assessing and monitoring the air pollution in cities. Fields of concentration of air pollutants in cities have very complex and irregular structure, which, in particular, reflects a complex and irregular pattern of the source distribution and possible singularities of concentrations in the vicinity of the sources. Because of economical and technical limitations, monitoring networks in cities are not dense enough to resolve such a fine structure. Therefore, the monitoring data can characterise mainly the air pollution at the monitoring sites and usually cannot be used to draw the map of the air pollution in the whole city. Standard techniques like, for example, weighted interpolation do not use the information about the sources of the pollution and, thus, cannot perform well in mapping the air pollution. On the contrary, computed concentration fields are usually based on the use of detailed emission data and have no formal limitations on their resolution. Formally, dispersion models can generate the values of concentration at the arbitrary receptor points. These values, however, include some inherited errors, in particular, because of possible uncertainties in the emission and meteorological data. It is here where enter the adaptive dispersion models.

THE KALMAN FILTER

Let us assume that the air pollution in the city is described by the n-dimensional vector $\underline{c}(t)$ of concentrations corresponding to the discrete set of the grid points and discrete time t. Its evolution is governed by the equation

$$\underline{c}(t+1)=F_t\underline{c}(t)+\underline{w}_t, \tag{1}$$

where F_t is the n x n transition matrix (evolution operator) and $\underline{w}_t$ is the model error (noise). The concentrations are monitored at k receptor point; so the results of the measurements can be described as follows:

$$\underline{y}(t)=H_t\underline{c}(t)+\underline{v}_t, \tag{2}$$

where $\underline{y}$ is the k- dimensional vector of measurements, H_t is the k x n measurement matrix (in the simplest case its elements are either 0 or 1), and $\underline{v}_t$ is the measurement errors. It is usually assumed that mean values of the noises $\underline{w}_t$ and $\underline{v}_t$ are equal to zero, their covariance matrixes are, respectively, Q and R, and that there is no cross-correlation between $\underline{w}_t$ and $\underline{v}_t$.

As soon as the initial state $\underline{c}(t-1)$ is known and the noise in Eq. (1) is omitted, this equation can be used to predict the "initial value" $c_{t|t-1}$ concentrations at the moment t:

$$\underline{c}_{t|t-1}=F_{t-1}\underline{c}_{t-1}\,. \tag{3}$$

Having substituted it into the right-hand side of Eq. (2), one can define the renewal process $\underline{\nu}_t$ as

$$\underline{\nu}_t=\underline{y}(t)-H_t\underline{c}_{t|t-1}\,. \tag{4}$$

The corrected value of concentration $c_{t|t}$ at the moment t is presented now as follows:

$$\underline{c}_{t|t}=\underline{c}_{t|t-1}+K_t\nu_t, \tag{5}$$

where K_t is the amplification matrix. Its optimal (corresponding to minimum of the residual dispersion of the estimates) value is determined from the following recurrent equations:

$$K_t = P_{t|t-1} H_t^T [H_t P_{t|t-1} H_t^T + R]^{-1}; \qquad (6)$$

$$P_{t|t} = P_{t|t-1} - K_t H_t P_{t|t-1}, \qquad (7)$$

$$P_{t+1|t} = F_t P_{t|t} F_t^T + Q, \qquad (8)$$

where H_t^T is the transposed matrix H_t, $P_{t|t-1}$ and $P_{t|t}$ are covariance matrixes of the residuals of estimates of concentrations with use of $\underline{c}_{t|t-1}$ and $\underline{c}_{t|t}$, respectively.

OPTIMAL INTERPOLATION

Let us consider a receptor point No. 0 We are looking for representation of the deviation of concentration $\underline{c}'_0$ from its mean value at this receptor points as a linear combination of the concentrations at the monitoring stations (not coinciding with the point "0"):

$$\underline{c}'_0 = \sum_{i=1}^{k} p_{0i} \underline{c}'_i, \qquad (9)$$

with certain coefficients p_{0i}. It is known (see, for example, Gandin, 1963) that the minimum of the residual dispersion corresponds to the coefficients determined form the following set of the linear equations:

$$\sum_{n=1}^{k} m_{in} p_{0n} = m_{0i}, \qquad (10)$$

where m_{ik} are second central moments of the concentrations. This kind of interpolation is used nowadays in graphical packages and widely known as kriging.

DETERMINATION OF THE SECOND MOMENTS

It is obvious that the optimal interpolation (in the form presented here) explicitly uses an information about "connectedness" of the concentrations only in space and the Kalman filter accounts for both, space and time correlations. From this point of view, the Kalman filtering can be considered as more efficient tool. It should be noted, however, that Eq. (6) and (7) were derived using ensemble averaging which cannot be used for the real atmosphere. In practical applications, therefore, it should be replaced with averaging over either time or space assuming that the process under investigation is either stationary or spatially homogeneous one. Unfortunately, neither of these assumptions is valid for the concentration fields in the city which reflect the spatial structure and temporal variations of the sources of pollutants as well as governing meteorological parameters.

The optimal interpolation works only with spatial covariances, and it greatly simplifies its practical applications. Still, methods of determination of the covariance matrixes in use should be specified here too. In this connection, the following representation of the short-term averaged concentration field $\underline{c}(t)$ in the city is assumed in the present paper (see also Genikhovich et al., 2000):

$$\underline{c}(t) = \underline{C}(t) + \underline{c}_b + \underline{\varepsilon}(t), \qquad (11)$$

where $\underline{C}(t)$ is the concentration field calculated with a dispersion model, $\underline{c}_b(t)$ is the background concentration field from the sources located outside the city. The noise $\underline{\varepsilon}(t)$ comprises all kind of errors including the errors of measurements, input data for the model (both, emissions and meteorology) and the model itself; its average value is assumed to be equal to zero. Eq. (11) yields

$$<\underline{c}>=<\underline{C}>+<\underline{c}_b>; \tag{12}$$

$$m_{ij}=M_{ij}+\sigma_b^2+\sigma_\varepsilon^2\delta_{ij}, \tag{13}$$

where $<\underline{c}>$, $<\underline{C}>$ and $<\underline{c}_b>$ are mean values of the actual, calculated with the model and background concentrations, m_{ij} and M_{ij} are covariances of actual and calculated concentrations, correspondingly, σ_b^2 and σ_ε^2 are dispersions attributed to the background concentrations and noise, and δ_{ij} is the Kronecker delta (= 1 if i=j and =0 otherwise). When deriving Eq. (13), it was assumed that three terms in the right-hand side of Eq. (11) correspond to different length scales and, therefore, their cross-correlations are equal to zero.

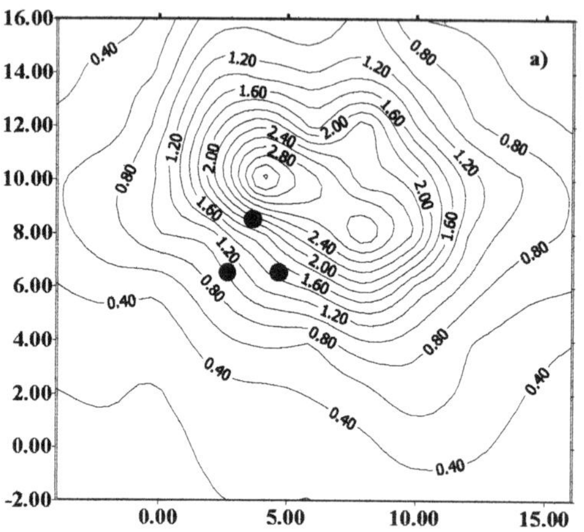

Fig 1. Mean annual calculated concentrations of SO_2.

The evolution operator F_t is explicitly included in the Kalman filter and can be defined in several different ways. Accordingly to Runca et *al.* (1979), for example, F_t is determined as a result of discretization of the advection-diffusion equation. Eq. (13) indicates that F_t can be used for the optimal interpolation too. However, we do not limit ourselves with differential operators or their matrix approximations. Any multiple-source dispersion model intended for applications in the urban scale can be used to generate the moments M_{ij}. Model test runs presented in this paper were carried out with the hybrid dispersion model introduced by Genikhovich *et al.* (2000), but they are used only for illustrations.

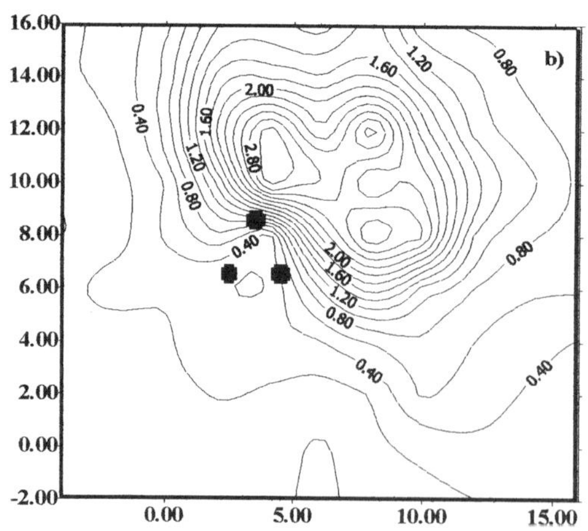

Fig 2. Mean annual corrected concentrations of SO_2.

The averaging procedure which is used with the optimal interpolation should be flexible depending on the relationship between the averaging time

T_a and characteristic time T_c of the variability of the concentration field, which in its turn depends on the variability of the governing meteorological parameters and emissions. If $T_a >> T_f$, as for example in the case of the annual averaging, the brackets in Eq. (11) mean the temporal averaging over the interval T_a. However, if these two time scales are either comparable or $T_a < T_f$, the brackets should indicate the conditional averaging corresponding to the "fixed" values of the governing parameters like the wind speed and wind direction, time of the day and so on (in practical applications these parameters should belong to certain intervals). The results of application of the optimal interpolation technique to mapping the mean annual concentrations of SO_2 (in 10 $\mu g/m^3$) in the city of Pskov, Russia, are shown on Fig. 1 and 2; they correspond to the fields calculated and corrected with Eq. (9) –(10), respectively. The information which has been used for the model calculations consists of emission data (154 point sources in the city) and measurement data from three monitoring stations (their locations are shown on Fig. 1 and 2 by dots). Measured concentrations correspond to the 20 min averaging time and were monitored three times a day in 1993.

DISCUSSION AND CONCLUSION

The aim of the presented paper is to outline certain approaches for constructing the adaptive dispersion models and for implementing the existing ones in adaptive regimes. The later option seems to be more effective because it could be applied to the majority of existing dispersion models including regulatory ones. As an example, one of the problems of the integrated air quality assessment is considered which is dealing with mapping of the air pollution in the city. It is obvious that, usually having a very limited number of monitoring stations in the city, one cannot use only these data to draw, for example, the map of concentrations of NOx in the city. Formally, modeling results can be used to generate as much detailed field of concentrations as one wish and, therefore, to draw the map, but it is very hard to say how correct such a map will be. The most effective way seems to be in the joint use of measured and calculated concentrations because they both carry certain information about the structure of the concentration field. The results of the corresponding mapping are presented in this paper. A new hybrid dispersion model developed at MGO was used in this calculations, but the general approach can be the same, for example, with a "standard" Gaussian-type dispersion model.

Adaptive dispersion models could become also a tool for hybrid monitoring of the air pollution. Until now, all precautions were taken to assure that monitoring networks are built on the "uniform" measuring equipment which provide compatible results of measurements. It has become obvious, however, that, due to economic and other limitations, the number of the monitoring stations in the cities is always insufficient and many harmful pollutants are not monitored at all. The solution of this problem could be in enhancing the existing networks with the combined use of comparatively simple instruments (for example, passive samplers) and results of "computational monitoring" based on the dispersion modeling. Such kind of the hybrid monitoring of the air pollution will not destroy the existing system but will require, however, the active use of the adaptive modeling.

REFERENCES

Bultijes P., van Loon, M., Segers, A., 2000, Data assimilation: a tool to shape the future, in: *Symposium*

2000 Eurotrac-2 Abstracts, GSF, Munich

Gandin, L.S., 1963, *Objective Analysis of Meteorological Fields*, Hydrometeoizdat, Leningrad (in Russian)

Genikhovich, E., Filatova, E., Ziv, A., 2000, A method for mapping the air pollution in cities with combine use of measured and calculated concentrations. *Int. Journ. Envir. Pollut.* (in press)

Genikhovich, E.L., Graheva, I.G., Groisman, P.Ya., Khurshudyan, L.G., 1999, A new Russian regulatory dispersion model MEAN for calculation of mean annual concentrations and its meteorological preprocessor. *Int. Journ. Envir. Pollut.* (in press)

Haas-Laursen, D.E., Harley, D.E., Prinn, R.G., 1996, Optimizing an inverse method to deduce time-varying emissions of trace gases, *Journ. Geophys. Res,* D17:22823

Houtekamer, P.L., Mitchell, H.L. , 1998, Data assimilation using an ensemble Kalman filter technique. *Mon. Wea. Rev.*, 126: 796.

Kalman, R.E., 1960, A new approach to linear filtering and prediction problems, *Trans. ASME, Journ. Basic Eng.* 83:146

Leondes, C.T. ,ed., 1976, *Control and Dynamic Systems*, Acad. Press, NY

Runca, E., Melli, P., Spirito, A., 1979, *Real Time Forecast of Sulphur Dioxide Concentrations in the Venetian Lagoon Region. P. 1. Advection diffusion model*, IIASA, Laxenburg

Sasaki Y. 1970. Some basic formalisms in numerical variational analysis. *Mon. Wea. Rev.* 98: 875

DISCUSSION

D. BYUN: To apply the optimal interpolation method, don't you need to obtain gradient (sensitivity) of the output to the input from the model?

E. GENIKHOVICH: Actually we do it when we estimate the covariance matrix using the model.

F. MÜLLER: You are not afraid to bias the model predictions over the model domain by using highly, locally influenced measurements at the surface?

E. GENIKHOVICH: The field of concentration in the city is very irregular and the length scale of its correlations is usually rather small. It means that the local effects can influence the model results only in the vicinity of the monitoring station.

D. SYRAKOV: In what time scale do you use the proposed technique, case-by-case or for the long-term averaged fields?

E. GENIKHOVICH: Numerical results presented here correspond to the annual averaging, but this technique can be applied also on the case-by-case basis.

J. BARTNICKI When you apply the data assimilation procedure to model results, the mass balance of pollutants is destroyed. How do you deal with this problem in practical applications?

E. GENIKHOVICH: We have generalized the technique of the optimal interpolation to take into account additional restrictions like the mass balance. The results obtained will be published separately.

ARTIFICIAL NEURAL NETWORK-BASED ENVIRONMENTAL MODELS

Marija Zlata Božnar and Primož Mlakar

Jožef Stefan Institute
Jamova 39
SI-1000 Ljubjana
Slovenia

ABSTRACT

Artificial neural network-based air pollution prediction models have become very popular. The paper describes feature determination and pattern selection strategies that help to improve significantly the performance of neural-network based models.

The most important methods of feature determination are preprocessing, heuristic determination, feature extraction and feature selection. The most important methods for pattern selection are meteorological knowledge-based cluster determination and Kohonen neural network-based cluster determination.

To summarize the explanation of the proposed methods, their influence on model behaviour is discussed.

Keywords: Artificial neural networks, air pollution, prediction model, feature determination, pattern selection, cluster.

INTRODUCTION

Recently artificial neural network-based air pollution prediction models became very popular. In the last ten years we devoted much effort to development of a neural network-based SO_2 air pollution prediction model (Božnar, 1993). SO_2 and O_3 pollution prediction are the most popular topics, but not the only problems where neural networks are being used in the environmental field (Božnar, 1995). Although the number of applications increased significantly in the last few years, most authors start with the basic usage of neural networks. This is reflected in the efficiency of the models and in the "black box" treatment of them.

To overcome these problems, we developed methods that help to improve significantly the results of neural network-based air pollution prediction models (Božnar, 1998, 1997[1], Mlakar, 1997[1]). It is also important that these methods help the modeller to understand the resulting behaviour of the model obtained.

STUDIED PROBLEM: SO_2 AIR POLLUTION PREDICTION

The feature determination and pattern selection methods were developed for the case of an SO_2 air pollution prediction model. The model that we developed (Mlakar, 1997[2]) is based on a Perceptron neural network and predicts SO_2 concentrations for half an hour (one measuring interval) in advance for a chosen air pollution measuring station. We studied the case of SO_2 pollution around Slovenia's largest thermal power plant at Šoštanj (ŠTPP). ŠTPP is located in the complex orography of the Velenje basin which is characterized by very low winds and winter thermal inversions that emphasize the pollution of the TPP surroundings. The ŠTPP is the major source of SO_2 pollution in the area. An efficient air pollution prediction model can help to "run" the power plant in such a way to reduce or prevent at least the most harmful pollution episodes. This is possible because of a local community law that forces the TPP to reduce the emission of SO_2 in case of high pollution episodes.

ŠTPP has a modern environmental information system that measures meteorological and air pollution data at six stations in the surroundings influenced by the TPP and an emission station that measures the emission of pollutants from the three TPP stacks and some operational parameters (Božnar, 1994).

The huge data base from all these measurements from 1990 to 1994 (half hour average values) serves as a basis for developing the air pollution prediction model.

NEURAL NETWORK

As a basis for our air pollution prediction model we chose a multilayer Perceptron neural network with two hidden layers and a sigmoid transfer function in the hidden layer and output neurons (Rumelhart, 1986). For training a back propagation agorithm was used. This type of neural network configuration is highly non-linear and proven to be universal approximator (Cybenko, 1989), and on that account it is among the most suitable types of neural networks for numerical prediction of multivariable time series.

For input features we used meteorological measurements (ground level wind speed and direction, temperature, relative humidity) and ambient SO_2 concentrations at some of the measuring station for the current half hour interval and for the previous intervals. There was only one output feature – the ambient SO_2 concentration (real value) at a selected measuring station for the following half hour interval (Mlakar, 1997[2]).

A pattern is defined as a vector, compounded from values of the input features and the value of the output feature. For each half hour interval within the four year data base available for research purposes (over 70,000 half hour intervals with measured values), one pattern can be constructed for the model examined.

There are several good and user friendly commercially available packages that implement the above mentioned neural networks.

PROBLEMS THAT SHOULD BE SOLVED IN ORDER TO CONSTRUCT AN EFFICIENT PREDICTION MODEL

When developing neural network-based prediction models, modellers usually take all the available data base and then concentrate on adjusting the learning algorithm parameters (such as learning rate and momentum in the backpropagation algorithm) in order to achieve best model performance. This adjustment certainly influences the model's performance, but it is not an important problem. Some guidelines to solve this problem can be found in our papers: Mlakar, 1997[2], 1994[1,2].

The most important problem in the model construction is selection of the proper information (relevant data) with which to train the neural network. The modeller should find relevant measurements that give essential information for concentration prediction. If the neural network is trained with low-information data, the prediction results will also be poor.

The data used for neural network training (learning) should be free of patterns and features with low or no information relevant to pollution prediction.

If we define the problem of pollution prediction at a chosen station as a highly nonlinear function in n-dimensional space, than the modeller should assure sampling the defined range of the studied function (its indepentent variables) homogenously and in sufficient density. Only if this rule is satisfied, will the model predict the output values of the unknown (during the learning process) patterns well enough.

In order to solve the model construction problems, we developed feature determination and pattern selection strategies that provide the guidelines for finding the most relevant information in the data base available for model construction.

Determination of model generalizing capabilities

To measure the improvement of the model due to feature determination and pattern selection strategies, a suitable form of measure (measurement criterion) should be defined. In the domain of air pollution forecasting forms of measure such us the correlation coefficient, normalized mean square error or fractional bias are used. But none of these functions directly reflects the most important capability of an air pollution prediction model – the prediction of high concentrations. In our case measured high concentrations (output feature) usually represent only about 5% of all concentrations. Concentrations usually rise and fall very quickly and an episode of high concentration lasts only one or two half hour intervals, only rarely longer (Božnar, 1996). From the user's point of view, it is important to predict high concentrations properly and not to create false alarms by predicting high concentrations when the actual ones turn out to be low.

To fulfill these requirements for our modelling puposes we defined a new form of measure–p^6, the probability of successful prediction of high concentrations (Mlakar, 1997[3]). It is defined as the number of intervals with successfully predicted high concentrations divided by the sum of the number of intervals with false alarms or actual high concentrations.

FEATURE DETERMINATION

In the process of feature determination the patterns are transformed from high dimensional "measurement space" into "feature space" of much lower dimension (Mlakar 1997[3]). Modern environmental measuring stations measure several parameters, not all of them relevant for a particular pollution forecasting problem.

The main reasons for the transformation of patterns into much lower dimensional space are:

- Extraction of the most relevant information from patterns in the measuring space – the features should represent only the most significant aspect of the input data.
- Reduction of the number of features simplifies the model and enables more efficient and quicker training.
- Reduction of the number of learning patterns: more features require more patterns.

Preprocessing and heuristic determination of features

Preprocessing and heuristic determination of features highly depend on the specific application. In our problem knowledge about air pollutant dispersion is relevant. The expert determines the features that are the most probable predictors of the observed phenomenon. Air temperatures measured at several stations are not directly correlated with SO_2 pollution. But vertical temperature difference between two properly selected stations at different elevations is a good indicator of atmospheric stability and that is one of the basic indicators of diffusion mechanisms.

Feature extraction

Feature extraction is usually a highly non-linear transformation of the subset of measuring space into one feature.

As an example of such a transformation we developed unsupervised classification schemes for ground level wind fields classification. If the measurements of ground level wind speed and direction at the six stations around the ŠTPP are used, the neural network-based model does not perform well. But it can be assumed that the wind fields above the basin can be classified into a finite number of clusters. Each cluster contains very similar wind fields that cause similar pollution patterns.

A very efficient tool for cluster determination is the Kohonen neural network (Kohonen, 1995). This is a special neural network (totally different from the Perceptron neural network) that takes the patterns (in our case vectors of all wind measurements) and classifies them into a finite number of clusters (the number is selected by the modeller in advance). When classification is done, the variability of each cluster should be determined with a suitable measure of distance. In our case the weighted mean standard deviation of cluster wind speed and weighted mean standard deviation of cluster wind direction was calculated. The process should be repeated for several numbers of clusters. The natural, meaningful number of clusters (that suits the problem best) is the smallest one that does not significantly increase the overall measure of distance (to determine it a XY plot of distance measure versus number of clusters for several cases is used).

As an input feature for the prediction model, the cluster number (from classification into the natural number of clusters) is used. The results prove that wind field cluster number is a very important feature for SO_2 forecasting in the Šoštanj case (Mlakar, 1996).

Feature selection

Feature selection is a procedure in feature space (with the previously described methods already reducing the measuring space) which searches for the best subset of n features. In the best subset the features from the original space that are redundant or do not contain enough relevant information are eliminated.

It is apparent that all possible subsets cannot be tested. Some optimal or suboptimal search algorithms can be used (an example is the Branch and bound algorithm, Narendra, 1977).

But the Perceptron neural network itself gives a very good but nonoptimal indication of the relevancy of each feature.

When the Perceptron neural network is trained, the sum of the network connection absolute weights leading from a particular input feature to the hidden layer neurons is a direct indicator of the observed feature's importance. The modeller should calculate this value for all the features. The desired model performance should be balanced with the number of features taken into account. Features of high importance should be included in the final subset of features and features of no importance should be excluded. But it is a subjective decision which or how many features of low importance can be included.

There are some modifications and enhancements of the above procedure given by Ruck, 1990 and Belue, 1995, and modified by Mlakar, 1997[3]. Ruck defined the saliency metric as the average derivative of the output with respect to a specific input. Ideally, each input should be independently sampled over its expected range of values. Because the derivative evaluations increase tremendously with the number of features and input points, Ruck suggested using transformed training vectors. For each one, each input is sampled over its range and the other input values are determined by the training vector currently being evaluated. To determine salient features a white noise variable as a known irrelevant feature should be added to the input vector. In the case of the Šoštanj air pollution data base only the actual patterns (training set) were used as starting points. For a three layer Perceptron neural network with a sigmoid transfer function a simplified version of the saliency metric Λ_j for an input feature j is defined as:

$$\Lambda_j = N^{-1} \sum_n \sum_i \left| \frac{\partial y_i}{\partial x_j}(x,w) \right|$$

where

$$\frac{\partial y_i}{\partial x_j}(x,w) = y_i(1-y_i) \sum_m w^2_{mi} x^1_m (1-x^1_m) w^1_{jm}$$

is the partial derivative of the ith output with respect to the jth input. N indicates the number of training patterns, n is the index of the pattern, y_i are the outputs, w^1_{jm} represents the weighted connection of the jth input and the mth hidden neuron, w^2_{mi} represents the weighted connection of the mth hidden neuron and the ith output, and x^1_m is the output of the mth hidden neuron.

To determine significant features (Mlakar, 1997[3]), firstly the network should be trained with training patterns. Then the saliency metrics should be computed for all features. The procedure should be repeated with slightly different learning parameters to obtain several results. The mean value and 97.5% confidence interval (a Student T distribution is assumed) for each feature's saliency metrics should be computed. The relevant features are the ones whose average saliency metrics are above the confidence interval for the mean value of the saliency for the noise feature.

PATTERN SELECTION STRATEGIES

A certain number of features of the input vector require some minimal number of training patterns which is not easy to determine. The problems described in the literature usually do not have enough training patterns, so the problem of their selection is not so obvious. But environmental measuring systems usually provide an extensive data base.

If a huge data base of possible patterns is available (over 70,000 in the case of Šoštanj case), patterns should be selected in order to achieve:

- improved generalizing capabilities of the model and hence improved prediction quality;
- lower computational cost during the learning process.

The whole process of building a neural network-based prediction model is based on the assumption that meteorological and TPP operating situations that cause particular pollution patterns are repeated if the overall situation is observed through a long enough historical data base. Each pollution pattern has a similar antecedent in the observed history. If the neural network is trained with the proper patterns from history, it will generalize well on new patterns.

The process of selecting the proper learning patterns receives almost no attention in the literature. In our opinion, this problem is as relevant as feature selection and therefore we developed several strategies (Božnar, 1997[2], 1997[1]). The two most successful ones are described in this paper.

Subset determination

The pattern selection strategies should give the guidelines for the division of the universal set of patterns (i.e. all the patterns possible in the data base available according to the selected features) into several independent subsets (sets without common elements). A TRAINING set is used for adjustment of the neural network's connection weights during the process of training using the backpropagation algorithm. A TESTING set is used for periodical testing of the network's prediction capabilities during training to determine the optimal network (not overlearned). A PRODUCTION set is the one used for testing the final model (optimal network) in order to determine its performance. If a possibility exists to test the model in real time, then the patterns constructed in this way form the ON-LINE set. Patterns from the universal set of data which do not belong to any of the above described sets form the REMAINING set. A common name for the union of the training and testing set is a LEARNING set.

The learning set patterns are responsible for the performance of the model. The production set is responsible for proper assessment of model performance.

Reference models

A model trained with unselected learning patterns is the first reference model. This is the model built with the largest possible learning set of patterns. The patterns are not selected. All the patterns from a sufficiently long time interval are taken.

The second reference model is a naive predictor. This is a "model" that predicts an unchanged concentration for the following half hour interval. Models should perform better than this one.

Pollution situation versus clear air patterns ratio

A detailed examination (Božnar, 1997[2]) shows that it is very important how the output feature (predicted concentration) is distributed in the learning set of patterns. If a model is built with a learning set containing too high percentage of patterns with high output concentration it tends to overpredict the actual values. If a model is built with a learning set containing too low percentage of patterns with high output concentration it tends to underpredict the actual values and loses its prediction capabilities and starts to behave just like a naïve predictor. The correct ratio should be determined experimentally for the particular case and should be kept in all further model development.

Meteorological knowledge – based pattern selection strategy

The patterns selected for use in model training should represent meteorological and pollution situations typical of the observed measuring station. It must be ensured that the learning set of patterns contains essential information needed for recognizing unknown pollution situations which appear during the use of the model on new data.

The main rules can be summarized as follows:

- First the typical pollution mechanisms for the observed measuring station should be determined according to knowledge of air pollution dispersion and the study of the polluted area.
- An equal number of patterns with high output concentration should be selected for each typical mechanism. The number of each mechanism's patterns is constrained by the

number of patterns of the mechanism having the smallest number of representative patterns.

- The patterns selected representing situations with no pollution should also be equally distributed between the main typical meteorological situations.
- The proper overall ratio between patterns with high output concentrations and patterns with low output concentrations should be taken into account.

These rules should be used both for training and testing sets.

The model should be trained with selected training and testing sets and its final performance should be evaluated with an appropriate form of measure (p^6).

Kohonen neural network – based pattern selection strategy

The idea of this strategy is to make a similar selection as in meteorological knowledge – based strategy, but to avoid the need for a detailed knowledge of the model domain. Instead of a detailed expert study a Kohonen neural network (Kohonen, 1995), capable of unsupervised learning, is used to determine the clusters of patterns. Each cluster represents meteorological and pollution situations that have a certain degree of similarity. The idea of using a Kohonen neural network is described in the Feature extraction paragraph.

- For the clustering process we used the whole learning set as used for training the reference model with unselected patterns. It is extremely important that for the Kohonen neural network input features all the Perceptron neural network input features plus its output feature are used.
- First a meaningful number of clusters should be found. Clustering should be performed for different numbers of clusters. For each division the quality of clustering should be calculated using a cost function that measures the dispersion of patterns within each cluster. The value of the cost function against the number of clusters should be plotted. The natural, meaningful number of clusters occurs at the point where the line's slope changes. For further evaluation this number of clusters should be used.
- When the clusters are obtained the procedure is then the same as with the meteorological knowledge-based pattern selection strategy.

TESTING PATTERN SELECTION STRATEGIES USING THE BEST POSSIBLE FEATURES

The final tests were made for all models with the best possible features.

Both pattern selection strategies were tested with several models constructed according to their rules. Both strategies show about a 20% relative increase of average p^6 value on the production set of about 1000 patterns from the most interesting time interval with many pollution situations. When tested with almost all the patterns left for a possible production set (over 7,000 patterns), the relative increase was about 8%.

Details of the tests (figures and other numbers) can be found in Božnar, 1997[2], Mlakar, 1997[2], 1997[3].

CONCLUSIONS

The problems of feature determination and pattern selection are the essential ones that should be solved in order to obtain a good neural network-based air pollution prediction model. In this paper the strategies suggested give general guidelines for solving this problem. These guidelines help modellers to understand the behaviour of neural networks.

REFERENCES

Beleue, L. M., Bauer, K. W. Jr., 1995, Determing input features for multilayer perceptrons, *Neurocomputing*, vol. 7, 2, 111-121

Božnar, M., Lesjak, M., Mlakar, P., 1993, A neural network-based method for short-term predictions of ambient SO2 concentrations in highly polluted industrial areas of complex terrain. *Atmos. environ., B Urban atmos.*, vol. 27, 221-230.

Božnar, M., Brusasca, G., Cavicchioli, C., Faggian, P., Finardi, S., Mlakar, P., Morselli, M. G., Sozzi, R., Tinarelli, G., 1994, Application of advanced and traditional diffusion models to en experimental campaign in complex terrain. V: Baldasano, J. M. , ed., *Air Pollution II. Volume 1, Computer simulation*. Southampton; Boston: Computational Mechanics Publications, 159-166.

Božnar, M., Mlakar, P., 1995, Neural networks - a new mathematical tool for air pollution modelling. V: Power, H. ed., Moussiopoulos, N. ed., Brebbia, C. A. , ed., *Air pollution III. Volume 1, Air pollution theory and simulation*. Southampton; Boston: Computational Publications, 259-266.

Božnar, M., Mlakar, P., 1996, Analysis of ambient SO_2 concentrations in the surroundings of the Šoštanj thermal power plant. V: Caussade, B. ed., Power, H. ed., Brebbia, C. A. , ed., *Air pollution IV : monitoring, simulation and control*. Southampton; Boston: Computational Mechanics Publications, 727-736.

Božnar, M., Mlakar, P., 1997[1], Pattern selection strategies for SO_2 forecasting models. V: Power, H. (Ur.), Tirabassi, T. ed., Brebbia, C. A. , Ed., *Air pollution V : Šmodelling, monitoring and management].* Southampton; Boston: Computational Mechanics Publications, 547-556.

Božnar, M., 1997[2], Pattern selection strategies for a neural network - based short term air pollution prediction model. V: ADELI, H. , ed., Intelligent Information Systems IIS'97, Grand Bahama Island, Bahamas, December 8-10, 1997. *Proceedings*. Los Alamitos, California: IEEE Computer Society, 340-344

Božnar, M., Mlakar, P., 1998, Improvement of air pollution forecasting models using feature determination and pattern selection strategies. V: Gryning, Sven-Erik ed., Chaumerliac, Nadine. *Air pollution modeling and its application XII*, (NATO challenges of modern society, vol. 22). New York; London: Plenum Press, 725-726.

Cybenko, G., 1989, Approximation by superposition of a sigmoidal function, *Mathematics of control, signals and systems,* 2, 303-314

Kohonen, T., 1995, Self-organising maps, Springer - Verlag, Berlin

Mlakar, P., Božnar, M., 1994[1], Short-term air pollution prediction on the basis of artificial neural networks. V: Baldasano, J. M. , ed., *Air Pollution II. Volume 1, Computer simulation*. Southampton; Boston: Computational Mechanics Publications, 545-552.

Mlakar, P., Božnar, M., Lesjak, M., 1994[2], Neural networks predict pollution. V: GRYNING, Swen-Erik ed., MILLÁN, Millán M. , ed., *Air pollution modeling and its applications* (NATO challenges of modern society, 18). New York; London: Plenum Press, 531-532

Mlakar, P., Božnar, M., 1996, Analysis of winds and SO_2 concentrations in complex terrain. V: Caussade, B. ed., Power, H. ed., Brebbia, C. A. , ed., *Air pollution IV : monitoring, simulation and control*. Southampton; Boston: Computational Mechanics Publications, 455-464.

Mlakar, P., Božnar, M., 1997[1], Feature determination for air pollution forecasting models. V: POWER, H. ed., Tirabassi, T. (Ur.), Brebbia, C. A. , ed., *Air pollution V : Šmodelling, monitoring and management]*. Southampton; Boston: Computational Mechanics Publications, 577-586.

Mlakar, P., Božnar, M., 1997[2], Perceptron neural network - based model predicts air pollution. V: ADELI, H. , ed., Intelligent Information Systems IIS'97, Grand Bahama Island, Bahamas, December 8-10, *Proceedings*. Los Alamitos, California: IEEE Computer Society, 345-349.

Mlakar, P., 1997[3], Determination of features for air pollution forecasting models, V: ADELI, H. , ed., Intelligent Information Systems IIS'97, Grand Bahama Island, Bahamas, December 8-10, *Proceedings*. Los Alamitos, California: IEEE Computer Society, 350-354.

Narendra,P. M., Fukunaga, K., 1977, A branch and boudn algorithm for Feature subset selection, *IEE trans. Comput.*, vol. 26, 917-922

Ruck, D. W., Rogers, S. K., Kabrisky, M., 1990, Features selection using multilayer perceptron, *Neural network comput.* 20, 40-48

Rumelhart, D. E., McClelland, J. L., 1986, Parallel distributed processing 1,2, MIT Press, Cambridge, MA

DISCUSSION

P. SEIBERT: Could you explain what the numbers (0.35, 0.4) in the results mean exactly?

M. BOZNAR: The p^6 measure is defined as probability of successful prediction of high concentrations. First we define a cost function J^6. J^6 equals 1 (correct prediction) if the measured concentration is high (more than 0.15 mg/m^3) and if at the same time the absolute error of prediction is less than 0.1 mg/m^3 or if the relative error is less than 20%. Otherwise J^6 equals 0. It equals 0 also if there is false alarm: predicted concentration over 0.25 mg/m^3 and no measured high concentration. It is important that the correct classifications and classification errors are not counted through the whole production set. Correct classifications and classification errors are counted if there is high SO_2 concentration or if there is a false alarm. The number of this situations (both) is N. If there is low measured concentration and low predicted concentration, the pattern is not counted. The probability of correct classification p^6 is defined as:

$$p^6 = \mathrm{sum}(J^6_i) / N$$

where i is from 1 to N.

P. SEIBERT: Is it right that the model makes a prediction for one averaging interval?

M. BOZNAR: Yes, the model, that we report the results in this paper, predicts pollution for one averaging interval (30 minutes). But it is of course possible to built a model that would predict pollution for more averaging intervals in advance. The results would probably be worse.

P. SEIBERT: Can you compare the performance of the neural-network based model to that of a physically based (dispersion) model?

M. BOZNAR: A direct comparison can not be made for the Šoštanj case, because the neural network calculates concentrations for next half hour interval, the dispersion model would need a forecast of meteorological parameters for that. We do not have such a short term precise forecast. But of course we compared the results to actual measurements (taken automatically on the measuring station). The results are shown (see the detailed explanation of p^6 measure).

D. SYRAKOV: As your model predicts for 0.5 hour, what will happen if the TPP change the coal or the regime of work?

M. BOZNAR: If only the percent of sulfur in the coal is changed, this would cause different SO_2 concentrations in the TPP stacks. There the concentrations are measured on-line and the measurements are input features to the model. This way the model would »notice« the change and adjust the predictions. The problem occurs, if the emissions are so constant (this can happen in reality) that the model is not sensitive to this input feature any more. This can happen if there is no patterns with varying load of the TPP block (only full load or no load). In this case or in the case of previously unknown regime of work, the model should be trained again with new patterns (describing the new regime) included in the learning set.

UNSTEADY BEHAVIORS OF GAS-AEROSOL INTERACTIONS CAUSED BY INTRA-PHASE MASS TRANSFER LIMITATIONS – UNSTEADY GAS AEROSOL MODEL (UGAM)

S.Y. Cho[1] and G.R. Carmichael[2]

[1]Department of Environmental Engineering
Inha University
Incheon, Korea
[2]Department of Chemical and Biochemical Engineering
University of Iowa
Iowa City, IA 52242

INTRODUCTION

Phase equilibrium calculations have important applications in many areas of science and engineering. General purpose computer software such as MTDATA and Therm Calc are commercially available and a vast amount of thermodynamic data containing several thousand organic and inorganic species have been compiled (Hack, 1996). In the atmospheric aerosol area, the importance of equilibrium relations between gas and aerosol phases was first reported by Stelson et al.(1979). Since then, significant efforts have been made to develop comprehensive equilibrium models that calculate the partitioning of acidic and basic chemical species into gas, aqueous and solid phases (Basett and Seinfeld, 1983; Pilinis and Seinfeld, 1987, Wexler and Seinfeld, 1991; Kim et al., 1993; Meng et al. 1995; Jacobson, 1999).

There are two general kinds of numerical methods used in solving the gas-aerosol equilibrium relations. One method is to minimize Gibb's free energy (Wexler and Seinfeld, 1991). Although this method is numerically robust and has been successfully used in many applications, it requires large computation times to be used in multi-dimensional air quality models (Kim and Seinfeld, 1993). The second method is designed to enhance numerical efficiency by solving aerosol equilibrium relations directly (Kim and Seinfeld, 1993; Jacobson et al., 1996). However, the algebraic equations arising from the gas-aerosol equilibrium relations are numerically ill-conditioned due to inherent discontinuity in the solution domain. In addition to these numerical efficiency issues, equilibrium may not be achieved between gas and aqueous aerosol due to inter-phase mass transfer limitations (Wexler and Seinfeld, 1990). Furthermore, the equations derived from gas-aerosol equilibrium relations may yield multiple solutions if multiple aerosol size bins are considered (Wexler and Seinfeld, 1990).

All of these problems stem from assuming equilibrium between the gas and aerosol phases. One way to overcome these problems is to take a kinetic approach. In this paper, we treat all the gas-aerosol interactions including solid forming processes kinetically. Under this approach, a set of first order differential equations can be derived and incorporated into differential equations describing gas and aqueous phase chemical kinetics. Therefore, there is no need to solve non-linear algebraic equations describing aerosol equilibrium relations separately. More importantly, this approach fully resolves the unsteady behavior of gas-aerosol interactions caused by inter-phase mass transfer limitations.

SOLUTION PROCEDURES FOR EQUILIBRIUM RELATIONS

In the atmosphere, acids react with bases to generate salts, which constitute aerosol. If the relative humidity is lower than the deliquescence points of all the salts present in the aerosol, the aerosol may stay in the solid state. Otherwise, water vapor condenses onto the aerosol surface to make aqueous aerosol. Then, gaseous acids and bases dissolve into the water film surrounding the aerosol and are subsequently ionized. In addition, acids and bases may be produced by aqueous reactions inside the aerosol. The ionized acids and bases may react to generate solid salts again.

In this paper, a kinetic approach is employed to overcome numerical problems associated with the equilibrium approach. One advantage of a kinetic approach is that it can be easily implemented using existing gas phase chemistry solvers. In order to implement the kinetic approach, the following differential equations describe the reactions between acids and bases:

$$\frac{d[A_i]}{dt} = \sum_{j=1}^{j=n}\left(-a_{i,j}k_f[A_i]^{a_{i,j}}[B_j]^{b_{i,j}} + a_{i,j}k_b[A_{i,a_{i,j}}B_{j,b_{i,j}}]\right). \quad \text{(1-a)}$$

$$\frac{d[B_j]}{dt} = \sum_{i=1}^{i=m}\left(-b_{i,j}k_f[A_i]^{a_{i,j}}[B_j]^{b_{i,j}} + b_{i,j}k_b[A_{i,a_{i,j}}B_{j,b_{i,j}}]\right). \quad \text{(1-b)}$$

$$\frac{d[A_{i,a_{i,j}}B_{j,b_{i,j}}]}{dt} = k_f[A_i]^{a_{i,j}}[B_j]^{b_{i,j}} - k_b[A_{i,a_{i,j}}B_{j,b_{i,j}}]. \quad \text{(2)}$$

where $[A_{i,a_{i,j}}B_{j,b_{i,j}}]$ is the fictitious molal concentration of the solid phase chemical species defined as the number of moles per unit volume of solvent. In addition, k_f and k_b are fictitious forward and backward reaction rate constants with the following properties:

$$k_f = 0 \text{ and } k_b > 0 \quad \text{if } [A_i]^{a_{i,j}}[B_j]^{b_{i,j}} < K_{eq,i,j} \quad \text{(3-a)}$$

$$k_f > 0 \text{ and } k_b = 0 \quad \text{if } [A_i]^{a_{i,j}}[B_j]^{b_{i,j}} > K_{eq,i,j} \quad \text{(3-b)}$$

$$k_f = 0 \text{ and } k_b = 0 \quad \text{if } [A_i]^{a_{i,j}}[B_j]^{b_{i,j}} = K_{eq,i,j} \quad \text{(3-c)}$$

The reaction rates k_f and k_b should be a function of chemical species concentrations and should be continuous for well-behaved numerical solutions. Furthermore, the first order derivatives of k_f and k_b should also be continuous if numerical methods such as Newton's method are used.

In the present study, k_f and k_b are devised as follows to satisfy these criteria:

$$k_f = \frac{k_c}{K_{eq,i,j}} \times fs(\frac{[A_i]^{a_{i,j}}[B_j]^{b_{i,j}}}{K_{eq,i,j}} - 1) \tag{4-a}$$

$$k_b = k_c \times fs(1 - \frac{[A_i]^{a_{i,j}}[B_j]^{b_{i,j}}}{K_{eq,i,j}}) \tag{4-b}$$

where k_c is a characteristic reaction rate constant. In equation 4-a, k_c is divided by $K_{eq,i,j}$ to normalize the forward reaction rate as well as to make the units consistent. The exact value of k_c is not important as long as the acid-base equilibrium is achieved. In the present study, k_c is set equal to 10 sec^{-1}. Then, the half-lives of acids, bases and resulting salts become less than 0.1 sec assuming pseudo first order reactions for the forward and backward reactions. Numerical experiments have found that k_c larger than 1 sec^{-1} ensures equilibrium for all the cases tested in section 4.

The function fs in equation 4 is defined as:

$$fs(x) = 0 \quad \text{if} \quad x \le 0 \tag{5-a}$$

$$fs(x) = x^2 \quad \text{if} \quad 0 < x \le 2 \tag{5-b}$$

$$fs(x) = 2x \quad \text{if} \quad 2 < x \tag{5-c}$$

Equation 5 c keeps k_f and k_b from becoming too large even when $[A_i]^{a_{i,j}}[B_j]^{b_{i,j}}$ greatly exceeds $K_{eq,i,j}$. This kinetic treatment of equilibrium applies only to solid forming equilibrium relations. Ionic dissociations of acids and bases are calculated from explicit algebraic equations with an electron neutrality constraint. The electron neutrality constraint is given in a form of the algebraic equation.

$$\sum_{all\ anions} v_i\{C_i\} - \sum_{all\ cations} v_i\{C_i\} = 0 \tag{6}$$

where the parenthesis {} denotes a concentration and v_i is the electronic charge. The above equation can be easily solved for the hydrogen ion concentration by a simple direct substitution or Newton iteration method. However, it can be costly because thermodynamic variables should be re-evaluated at each iteration step. Because the present kinetic approach intends to eliminate all the computationally costly iterations, the following differential equation is introduced as an electronic neutrality constraint:

$$\frac{d\{H^+\}}{dt} = k_i\{H^+\}(\sum_{all\ anions} v_i\{C_i\} - \sum_{all\ cations} v_i\{C_i\}) \tag{7}$$

The above equation decreases the H^+ concentration if cation concentrations are larger than anion concentrations and increases the H^+ concentration otherwise until the cation concentrations equal the anion concentrations. Here k_i is a fictitious rate constant that satisfies the electronic neutrality condition satisfied. k_i should be sufficiently large to ensure satisfying the electron neutrality condition, but an excessively large number may introduce numerical stiffness. In the present study, k_i is chosen as 1,000 based on numerical experiments.

The present kinetic approach eliminates discontinuous non-linear algebraic equations and accelerates the convergence. Any number of equilibrium relations can be accommodated without modifying the solution procedure. Furthermore a variety of user interface software is available to interpret chemical mechanisms and to calculate Jacobian matrices (Kumar et al., 1995; Sandu et al. 1996). Therefore, adding or deleting chemical species and equilibrium relations can be done without modification of the computer code. In multi-dimensional atmospheric models, the differential terms derived by the present kinetic approach can be simply added to the gas phase and aerosol phase chemical species equations. Therefore, it can be easily implemented in the existing atmospheric models without modifying the solution procedure.

MATHEMATICAL MODEL FORMULATIONS FOR AEROSOL CHEMISTRY AND THERMODYNAMIC RELATIONS

Both chemical reactions and inter-phase mass transfer should be considered in addition to equilibrium relations to fully describe aerosol interactions with gaseous chemical species in the atmosphere. For simplicity all the mathematical formulations presented here assume a single size bin but the analysis can be easily extended to multi-size bin as noted in section 5.

Assuming the gas phase is well mixed, the gas phase and solid aerosol phase equations for solid aerosol are as follows.

gas phase:

$$\frac{dC_{g,i}}{dt} = R_{g,i} + E_{g,i} \tag{8}$$

solid aerosol phase:

$$\frac{dC_{s,i}}{dt} = E_{s,i} \tag{9}$$

where subscript g and s denote a gas and solid aerosol phase, respectively, and subscript i indicates the i-th chemical species. R is the chemical reaction rate and E represents an equilibrium related flux defined in equations 1, 2 and 3. A produced solid aerosol species may diffuse onto existing aerosol, but this process is not included here for simplicity.

For aqueous aerosol, gaseous chemical species dissolve into the aerosol surface. The dissolved chemical species may react with each other to form solid salts. Assuming the gas and aerosol phases are well mixed, the chemical species equations of each phase are:

gas phase:

$$\frac{dC_{g,i}}{dt} = R_{g,i} - h_i (C_{g,i} - H_i C_{\ell,i}) + E_{g,i} \tag{10}$$

aqueous aerosol phase:

$$\frac{dC_{\ell,i}}{dt} = R_{\ell,i} + \frac{h_i}{w_\ell}(C_{g,i} - H_i C_{\ell,i}) + E_{\ell,i} \tag{11}$$

solid aerosol phase:

$$\frac{dC_{s,i}}{dt} = E_{s,i} \tag{12}$$

where subscript ℓ denotes an aqueous aerosol phase, w_ℓ is the water content, and H is the Henry's law constant. C represents a chemical species concentration in the gas and solid phases and an activity in the aqueous phase, and h is the mass transfer rate constant defined as follows:

$$h_i = \frac{4\pi r D_i}{1 + f(Kn,\alpha)Kn} n(r) \tag{13}$$

$$f(Kn,\alpha) = \frac{1.333 + 0.71Kn^{-1}}{1 + Kn^{-1}} + \frac{4(1-\alpha)}{3\alpha} \tag{14}$$

where D is the molecular diffusion coefficient, Kn is the Knudsen number, α is the accommodation coefficient, and $n(r)$ is the number of aerosols with the radius of r. In addition to the above differential equations, the ion dissociation equations with electron neutrality constraints and the water content equations are solved along with estimations of activity coefficients. A detail discussion of these equations is found in Kim et al. (1993). Furthermore, a general aerosol dynamic equation may also be coupled to complete the aerosol equations but only the growth equation is considered here for simplicity.

The above set of first order differential equations can be readily solved by various numerical methods. One common method is to split the chemical reaction terms and inter-phase mass transfer terms to decouple the gas and aqueous phase chemical species equations (Carmichael et al., 1991; Jacobson, 1997). Numerical accuracy and efficiency of the time splitting method depend on discrepancy of characteristic time scales of the splitted equations (Kim and Cho, 1998). Because the characteristic time scales of chemical reaction rates and inter-phase mass transfer rates do not differ significantly, a small time splitting interval ranging 5–15 seconds is typically used, and this requires frequent restarts of the ODE solvers (Carmichael et al., 1991; Jacobson, 1997). More importantly, the time splitting method may introduce non-negligible numerical errors.

In this paper, these equations are solved as a coupled system without employing any decoupling methods in order to obtain a highly accurate solution. Of course, decoupling may improve the computational efficiency greatly but may degrade the computational accuracy. The present work focuses on developing a numerically robust and accurate gas-aerosol model which a more numerically efficient model.

EVALUATION OF UNSTEADY GAS-AEROSOL MODEL PERFORMANCE

Description of the Unsteady Gas-Aerosol Model (UGAM)

An unsteady gas-aerosol model (hereafter denoted as UGAM) was developed here based on equations 8-12. Binary and multi-component activity coefficients were calculated by the Kusik and Messner's method (Kusik and Messner, 1978). Additional thermodynamic data such as water activities, Henry's constants, and equilibrium constants, deliquescence points were adopted from Kim et al. (1995) and Jacobson (1999), where the procedures for calculating thermodynamic properties of aqueous aerosol are presented in detail.

Because a solid salt does not exist at a relative humidity above its deliquescence point, UGAM solves only the salts identified to exist under a given relativity humidity. In the present work, UGAM includes 6 cations (H^+, Ca^{++}, Mg^{++}, K^+, Na^+, NH_4^+) and 6 anions ($C\ell^-$, $CO_3^=$, HCO_3^-, OH^-, HSO_4^-, NO_3^-, $SO_4^=$) and all the corresponding salts. In addition, the Lurman Atkinson Lloyd Winge's mechanism (Lurman et al., 19860) was employed to describe the gas phase chemistry.

UGAM was also applied to the more general case specified in Table 1. The initial gaseous chemical species concentrations represent a moderately polluted area and the aerosol compositions include elements not only from sea salt aerosol but also from mineral aerosol. These gaseous chemical species concentrations and aerosol compositions were chosen to form solids inside aqueous aerosol to evaluate the model's ability of simulating solid forming processes. Furthermore, the given initial gaseous concentrations were designed to produce *in-situ* a non-negligible amount of acids to test the model's ability of handling temporally varying gas and aerosol compositions. Solar radiation fluxes were calculated every 15 minutes to update photo-dissociation rates and therefore LSODES was reinitialized every 15 minutes.

Table 1. The gas phase concentrations and aerosol compositions used for model simulations.

initial gas phase concentrations(ppb)		initial aerosol composition($\mu g/m^3$)	miscellaneous conditions
NO2=20	NO=20	$CaCO_3$=20	starting date = 9 A.M. July 1(GMT)
SO2=20	O3=30	$NaC\ell$=10	longitude = 80 E, latitude = 40 N
alkane=20	alkene=20	K_2CO_3=10	accommodation coefficient = 0.1
HNO_3=5	NH_3=10	$MgC\ell_2$=5	initial aerosol radius = 0.5 μm
sulfate=5			initial aerosol mass = 100 $\mu g/m^3$
			relative humidity = 65 %

All the gas phase concentrations and aerosol compositions other than specified above were set equal to zero.

Figure 1 shows temporal variations of the major chemical species under the conditions given in Table 1. Gaseous HNO_3 drops sharply for the first thirty minutes or so as it dissolves into the aqueous aerosol and then steadily increases as the photochemical reactions proceed. Aqueous NO_3^- and sulfate also steadily increases as the initially present or photo-chemically produced gaseous HNO_3 and sulfuric acid are incorporated into the aqueous aerosol. The water content also increases for the first thirty minutes as the water absorbing acids/bases are incorporated into the aerosol.

A minor decrease in the water content is calculated between 11 and 12 hours because the formation of solid $CaSO_4$, $NaNO_3$, and KNO_3 removes the corresponding ions from the aqueous phase. However, as the photo-chemically produced HNO_3 and sulfate are further added into the aerosol, the water content increases again to dissolve the solid $NaNO_3$ and KNO_3 as shown at 11 hour. Solid $CaSO_4$ is finally ionized to increase the water content at hour 14.

CONCLUDING REMARKS

A kinetic approach was taken to develop a new gas-aerosol interaction model named UGAM. UGAM calculates aerosol compositions under temporally varying atmospheric conditions accurately and efficiently. The simulation results in this study illustrate that aerosol compositions may vary significantly with time due to inter-phase mass transfer limitations as well as gas phase unsteadiness over a long period of time.

The kinetic approach proposed in this work successfully incorporates complex aerosol related processes into the chemical kinetic equations. Therefore, aerosol related processes can be easily implemented into any existing multi-dimensional Eulerian chemistry/transport/deposition models. Although an implicit ODE solver is used in UGAM to obtain a highly accurate solution, more approximate methods such as semi-implicit method, which is widely employed in multi-dimensional Eulerian models, may be used. Furthermore, time splitting may be applied to decouple the gas and aqueous phase equations.

All the analyses in this work are limited to aerosol with a single size bin. However, the proposed kinetic approach can be easily extended to aerosol with multi-size bins. The numerical stiffness may not increase significantly by adding more size bins but the number of differential equations increases accordingly. A special structure of the Jacobian matrix associated with multi-size bins may be exploited to achieve a high numerical efficiency. More study is required on this issue.

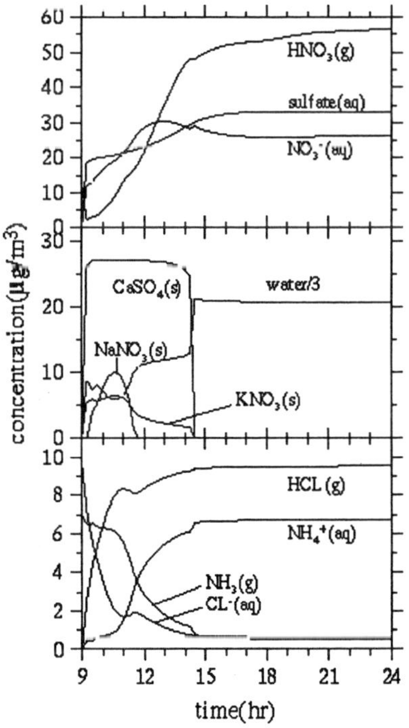

Figure 1. Calculated temporal variations of major chemical species with the conditions given in Table 1.

Acknowledgements – This work was supported in part by grants from the DOE-ACP program and NASA.

REFERENCES

Bassett, M.E. and Seinfeld, J.H., 1983, Atmospheric equilibrium model of sulfate and nitrate aerosol, *Atmos. Environ.*, 17, 2237-2252.

Bischof. C., Carle, A., Corliss, G., Griewank A., and Hovland, P., 1992, ADIFOR - Generating derivative codes from FORTRAN programs, *Scientific Programming*, 1, 1-29.

Brown, P.N., Byrne, G.D., and Hindmarsh, A.C., 1989, VODE: A variable coefficient ODE solver, *SIAM J. Sci. Stat/Comput.*, 10, 1038-1051.

Carmichael, G.R., Peter, L.K., and Saylor, R.D., 1991, The STEM II regional scale acid deposition and photochemical model, *Atmos. Environ.*, 25A, 2077-2092.

Hack, K., 1996, *The SGTE/Casebook - Thermodynamics at Work*, Material Modeling Series, The Institute of Materials, London.

Hindmarsh, A.C., 1983, *Scientific Computing*, Edited by R. Stepleman, IMACS, North Holland, Amsterdam.

Jacobson, M.K., Tabazadeh, A. and Turco, R.P., 1996, Simulating equilibrium within aerosols and non-equilibrium between gases and aerosols, *J. of Geophys. Research,* 101, 9079-9091.

Jacobson, M.K., 1997, Development and application of a new air pollution modeling system-II: Aerosol module structure and design, *Atmos. Environ.*, 31, 131-144.

Jacobson, M.K., 1999, Studying the effects of calcium and magnesium on size-distributed nitrate and ammonium with EQUISOLV-II, *Atmos. Environ.*, 33, 3635-3649.

Kim. J.H. and Cho, S.Y., 1997, Computational accuracy and efficiency of the time splitting method in solving atmospheric transport/chemistry equations, *Atmos. Environ.*, 31, 2215-2224.

Kim, Y.P., Seinfeld, J.H., and Saxena, P., 1993, Atmospheric gas-aerosol equilibrium I. Thermodynamic model, *Aerosol Sci. Tech.*, 19, 157-181.

Kim, Y.P. and Seinfeld, J.H., 1995, Atmospheric gas-aerosol equilibrium III: Thermodynamics of crustal elements Ca^{2+}, K^{+}, and Mg^{+2}, *Aerosol Sci. Technol.*, 22, 93-110.

Kumar, N., Lurmann, F.W., Carter, W.P.L., 1995, *Development of the Flexible Chemical Mechanism Version of the Urban Airshed Model,* Final report prepared for the California Air Resources Board. Sacramento, CA by Sonoma Technology Inc., Santa Rosa, CA. STI-94470-1508-FR, August.

Kusik, C.L., and Meissner, H.P., 1978, Electrolyte activity coefficients in inorganic processing. *AIChE Symp. Ser.*, 173, 14-20.

Meng Z., Seinfeld, J.H., and Saxena, P., 1995, Gas/aerosol distribution of formic and acetic acids, *Aerosol Sci. Tech.*, 23, 561-578.

Nenes, A., Pandis, S.N., and Pilinis, C., 1999, Continued development and testing of a new thermodynamic aerosol module for urban and regional air quality models, *Atmos. Environ.*, 33, 1553-1560.

Pilinis, C. and Seinfeld, J.H., 1987, Continued development of a general equilibrium model for inorganic multi-component atmospheric aerosols, *Atmos. Environ.*, 21, 369-371.

Sandu, A., Potra, F.A., Carmichael, G.R. and Damian, V., 1996, Efficient implementation of fully implicit methods for atmospheric chemical kinetics, *J. of Comput. Phys*, 129, 101-110.

Song, C.H., 1999, Tropospheric aerosols in east Asia: A model study of the evolution processes of dust and sea salt particles during long range transport, PhD thesis, Univ. of Iowa, Iowa City, Iowa.

Stelson, A.W., Friedlander, S.K. and Seinfeld, J.H., 1979, A note on the equilibrium relationship between ammonia and nitric acid and particulate ammonium nitrate, *Atmos. Environ.*, 13, 369-371.

Wexler, A.S. and Seinfeld, J.H., 1990, The distribution of ammonium salts among a size and composition dispersed aerosol, *Atmos. Environ.*, 24A, 1231-1246.

Wexler, A.S. and Seinfeld, J.H. 1991, Second generation inorganic aerosol model, *Atmos. Environ.*, 25A, 2731-2748.

Zang, Y., Seigneur, C., Seinfeld, J.H., Jacobson, M.Z., Clegg, S.L., Binkowski, F., 1999, A comparative review of inorganic aerosol thermodynamic equilibrium modules: similarities, differences, and their likely causes, *Atmos. Environ.*, in press.

DISCUSSION

E. GENIKHOVICH: Referring to the UGAM formulation: if you take the logarithm from both sides of your equations you will have a system of linear equations and inequalities, which can be directly solved using linear programming. Why not use that approach?

G. R. CARMICHAEL: That is an interesting idea and we will evaluate it. Our initial efforts were directed to casting the problem as an ODE system (eliminating the costly algebraic components) so that we could make efficient use of our new implicit ODE solvers.

A. VENKATRAM: It appears that at the end of every model step you still need the equilibrium solution. This means that your unsteady solution method is simply a numerical solution for the ionic reactions in the aerosol?

G. R. CARMICHAEL: No. We do not need to explicitly calculate the equilibrium concentration (at each step or at anytime in the calculation). Rather the way the system is cast, the production/destruction terms are formulated (along with electro-neutrality) so that the equations look exactly like the gas phase chemistry equations and thus we never have to explicitly compute the equilibrium concentrations. Of course as time progresses or for fast equilibrium steps, the solution will converge to the equilibrium concentrations.

SPATIAL-TEMPORAL VARIABILITY OF AEROSOL OPTICAL PROPERTIES SIMULATED IN CAM/GCMIII

J. P. Huang[1], S. L. Gong[1], L. A. Barrie[2] , J.-P. Blanchet[3]

[1]Meteorology Service of Canada, Environment Canada
4905 Dufferin Street, Downsview, Ontario M3H 5T4 Canada
[2]Pacific Northwest National Laboratory (PNNL)
902 Battelle Boulevard, P.O. Box 999
Richland, WA 99362, USA
[3]Earth Sciences Department, UQAM, P.O. Box 8888,
Station "Centre Ville", 201 President Kennedy Ave.
Montreal, Que., H3C 3P8 Canada

INTRODUCTION

Radiative forcing by aerosol is recognized as an important contributor to climate change [IPCC 1995]. Aerosol influence radiative forcing directly by scattering/absorption of solar radiation, and indirectly by altering cloud droplet size distribution and concentration. Therefore, aerosols exert a cooling influence on climate. This cooling influence is thought to be offsetting a substantial fraction of anthropogenic greenhouse warming, but its magnitude is highly uncertain. This uncertainties are thought to represent the greatest contribution to uncertainty in climate forcing over the industrial period [IPCC, 1995; Schwartz and Andreae, 1996]. Estimates of global or hemispheric average radiative forcing by anthropogenic sulfate aerosol are uncertain by somewhat more than a factor of 2 [Nemesure et al., 1995]. A major consequence of this uncertainty is that it limits the ability to empirically detect the climate change due to increased greenhouse gases or to quantitatively infer the sensitivity of climate to greenhouse gas forcing. It also precludes evaluation of performance of global climate model simulations of climate change over the industrial period. It is therefore necessary to resolve this largest scientific uncertainty in climate forcing.

The principal reason of this uncertainties are due to the lack of information about the aerosol spatial and temporal distribution as a function of particle size and composition [IPCC, 1995]. Additionally aerosol microphysical properties are not a universal constant, but depend on sources and composition and evolve as a consequence of chemical and physical processes occurring in the atmosphere. The mass loading, composition, and the microphysical properties of aerosols such as number concentration and size distribution directly affect their direct and indirect radiative forcing of climate. Kiehl and Briegleb [1993] showed that the geographical pattern of climate forcing changes dramatically when anthropogenic sulfate aerosols are included, while Taylor and Penner [1994] showed that both the pattern of climate response and the globally averaged climate sensitivity are highly dependent upon the geographical pattern of forcing. Though not explicitly stated, these conclusions apply to the temporal pattern of climate forcing as well, which is markedly variable for aerosols on diurnal, synoptic, and seasonal scales [Boucher 1995]. In the case of aerosols stemming from anthropogenic sulfur emissions, geographical and temporal variation are certainly caused by variations in local mass concentration [Charlson et al., 1991; Kiehl and Briegleb, 1993], but could also arise from variation in the optical properties of sulfate aerosols. The optical properties depend fundamentally on aerosol size and chemical composition. The size dependent aerosol optical properties are, in turn, a result of synoptic scale meteorology and air mass sources [Quinn et al. 1996].

Reducing the uncertainty in aerosol radiative forcing will require a major effort both in characterizing the present distribution and properties of aerosols and in developing understanding required to represent the processes controlling loading and properties of tropospheric aerosols in numerical models. In this study, the third-generation Canadian Global Climate Model (GCM III) coupled with CAM (Canadian Aerosol Module) was used to simulate the global distribution of size-segregated sea-salt and sulfate. From the simulated aerosol distributions we estimate optical depth and focus on their global and vertical distribution. The purpose of this paper is to assess of the sensitivity of optical properties to aerosol size distribution and composition. It is hoped that the assessment so obtained can lead to a better understanding of the uncertainty in climate forcing.

MODEL AND OPTICAL DEPTH COMPUTATION

Using the sectional approach, the Canadian Aerosol Module (CAM) was developed to be used in atmospheric models (climate or real-wind driven) for investigating the role of aerosols in the atmosphere [Gong et al., 1997; 2000]. It includes major aerosol processes in the atmosphere: production, growth, coagulation, nucleation, condensation, dry deposition, below-cloud scavenging, activation, an explicit micro-physical cloud module to treat the aerosol-cloud interactions and chemical transformation of sulfur species in clear air and in clouds. Emissions to the atmosphere of anthropogenic and natural aerosols are simulated for sea-salt and sulfate. In order to simulating the size distribution, the size-segregated multi-component aerosol algorithm has been introduced into CAM. The global sea-salt results were obtained using Canadian GCMIII at a horizontal resolution of 96×48 Gaussian grid (~3.75 °), vertical resolution of 32 levels (from surface to 1 hPa pressure level) and 20 minute integration time step. The size spectrum of sea-salt aerosols (dry) from 0.005 μm to 20.48 μm in radius was divided into 12 size bins. The model ran for four winter months (Jan. to April) and the averaged results for the four months were presented in this paper.

These optical properties can be calculated from standard Mie theory once the index of refraction (which depends on the chemical composition) and size distribution of the aerosol have been determined. In this study, the model directly calculates the size distribution that are used to estimate optical depth. The single aerosol optical depth $\tau_i(\lambda)$ is calculated from the specific extinction coefficient of the chosen aerosol type i (i=sea-salt; sulfate):

$$\tau_i(\lambda) = \int_{z_1}^{z_2} \beta_{e,i}(\lambda, z)dz \qquad \textbf{(1)}$$

Here, $\beta_{e,i}(\lambda)$ is the single aerosol specific extinction for a given wavelength λ with considering the relative humidity (RH) (Blanchet and List, 1983). Increased RH results in large particles, which scatter more radiation.

RESULTS

The global distribution of four month averaged optical depth for sea-salt and sulfate are shown in Fig. 1(a) and Fig. 1(b), respectively. It shows the highest optical depth of sea-salt values between 40°S-60°S latitude in all Southern Hemisphere Ocean and in the tropical Pacific and Atlantic Ocean. The steepest horizontal gradients of sea-salt optical depth are found in middle latitude due to the influence of the land-sea distribution. To compare with sea-salt, the optical depth of sulfate (Fig. 1(b)) are much lower in the ocean regions and the highest optical depth are found in the East Asia, West Europe and East U.S.. Figure 2 show the relative contribution of sea-salt and sulfate to the total optical depth as estimated from CAM/GCMIII, expressed as percentage of total two aerosol extinction optical depth. The sea-salt dominates the aerosol extinction over the ocean. It contribute more than 75% over Southern Pacific due to the absence of other aerosol species. In tropical Pacific and Atlantic region, sea-salt also contribute more than 60%. These results are greater than the modeling results by Tegen et al. (1995) and similar to the observation results by Quinn et al. (1996). Sulfate dominate the extinction over source region. The vertical distribution of extinction coefficients for different locations are shown in Fig. 3. These profiles are representative of typical locations in global. As indicated in the figure, the vertical profile can be divided into three distinct types. For high latitude station, Alert and Heimey, the vertical profile is dominated by a maximum near 650 hPa. While for the middle station, Oahu, it shows two peak. One is around 970 hPa and other is 250 hPa. However, for tropical and Southern Hemisphere, the maximum of sea-salt optical depth is located in surface and decreases with height rapidly.

CONCLUSIONS AND DISCUSSIONS

The results show significant spatial-temporal variability of aerosol optical properties. The dominant spatial pattern associated with changes in emissions sources. Any representation of global aerosol by a single aerosol type or by several types are fixed in space and time would be insufficient for studies of aerosol effects on climate change. While many studies on global aerosol effects presently concentrate only on sulfate aerosol, the contribution of sea-salt aerosol to the aerosol optical depth is important for studies climate change. The vertical distribution also shows strong spatial variability and it is crucial also for estimating the climate effect.

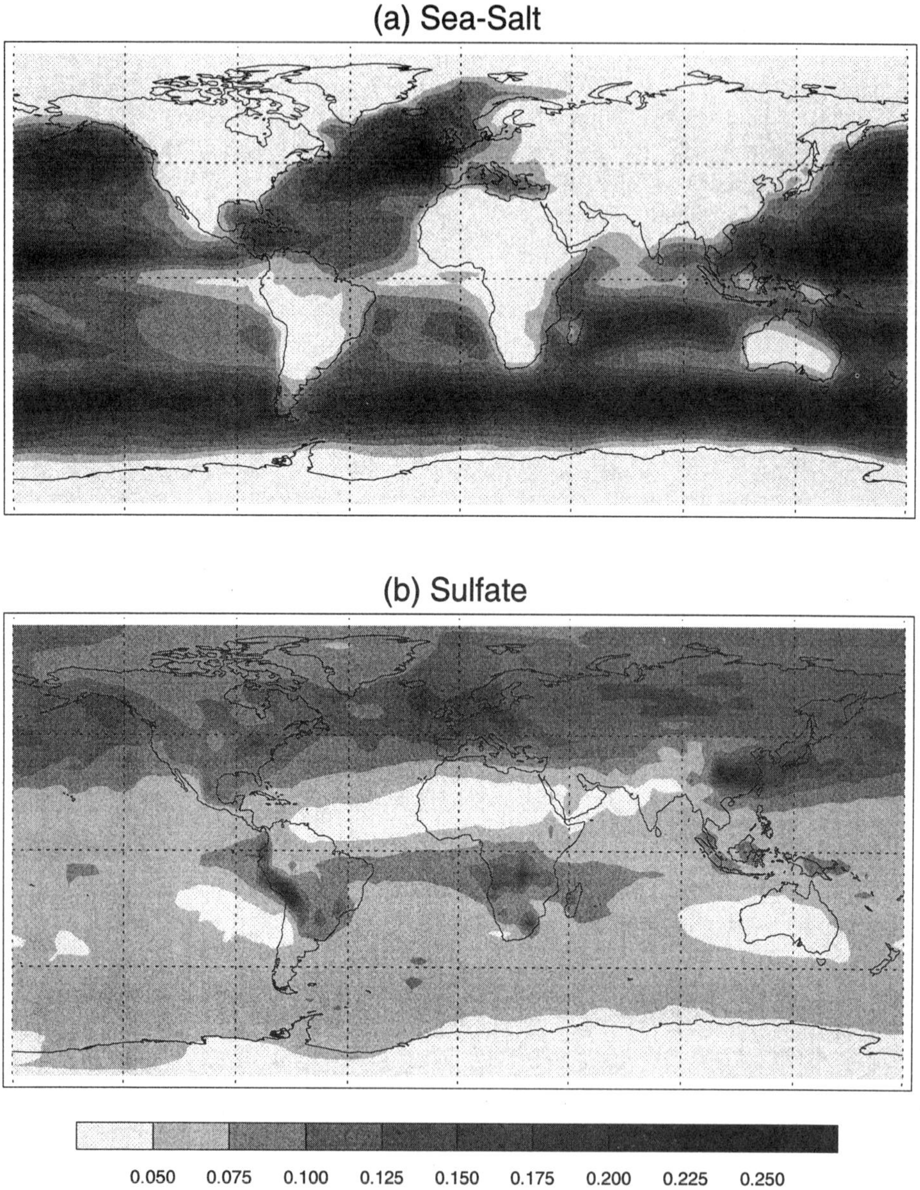

Fig.1 Global distribution of optical depth for (a) Seasalt; and (b)Sulfate

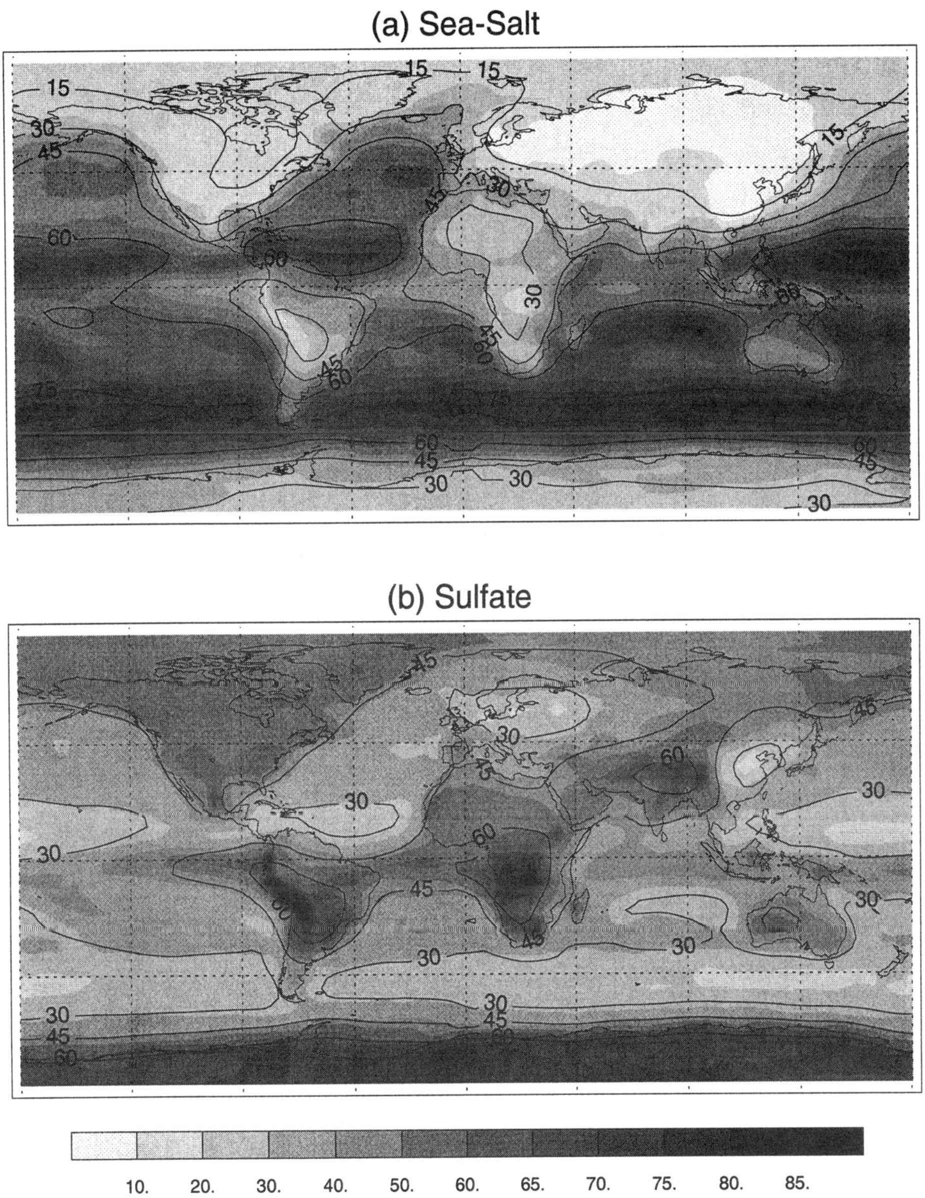

Fig.2 Relative contribution (x100%) for (a) Seasalt; and (b)Sulfate

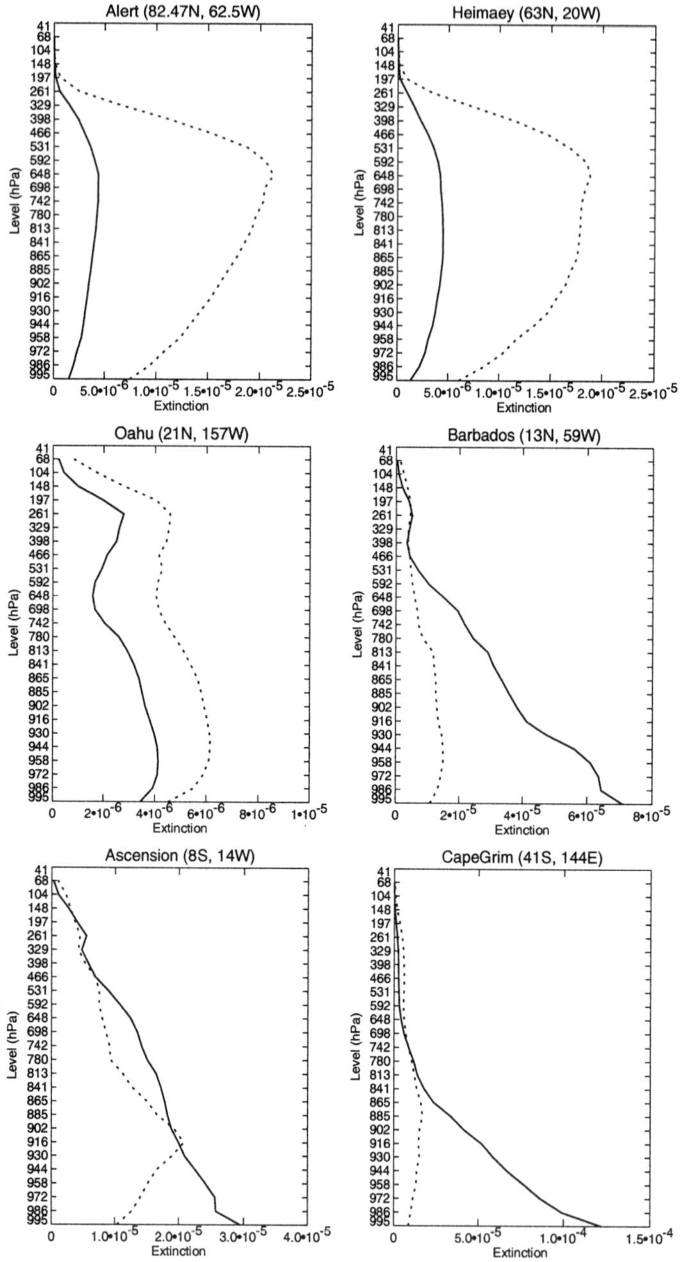

Fig.3 Vertical distribution of extinction coefficient. Solid line is for sea-salt and dash line for sulfate

REFERENCES

Blanchet, J.-P. and R. List, Estimation of optical properties of Arctic Haze using a numerical model, **21**, *Atmosphere-Ocean,* 444-465, 1983

Boucher, O. and T. L. Anderson, General circulation model assessment of the sensitivity of direct climate forcing by anthropogenic sulfate aerosols to aerosol size and chemistry. *J. Geophys. Res*, **100**, 26117-26134, 1995.

Charlson, R. J., J. Langner, H. Rodhe, C. B. Levoy, and S. G. Warren, Perturbation of the northern hemisphere radiative balance by backscattering from anthropogenic sulfate aerosols. *Tellus*, **43Ab**, 152-163, 1991.

Gong, S.L., L.A. Barrie, and J.-P. Blanchet, Modeling sea-salt aerosols in the atmosphere. Part 1: Model development, *J. Geophys. Res.*, **102 (D3)**, 3805-3818, 1997.

Gong, S.L., L.A. Barrie, J.-P. Blanchet, K. von Salzen , U. Lohmann, G. Lesins, D.Lavoué, J. Jiang, H. Lin, E. Girard, R. Leaitch , H. Leighton, P. Chylek and L. Spacek, CAM:Treatment of the size segregated atmospheric aerosols for climate and air quality models 1. Module Development, Submitted to *J. Geophys. Res*.

IPCC, Climate Change 1995, The Science of Climate Change, Contribution of Working Group I to the Second Assessment Report of the Intergovernmental Panel on Climate Change, J. T. Houghton, L. G. Meira Filho, B. A. Callander, N. Harris, A. Kattenberg, and K. Maskell (Eds.), Cambridge University Press, UK, 1996.

Kiehl, J. T., and B. P. Briegleb, The relative roles of sulfate aerosol and greenhouse gases in climate forcing. *Science*, **260**, 311-314, 1993.

Nemesure, S., R. Wagener and S. E. Schwartz, Direct shortwave forcing of climate by the anthropogenic sulfate aerosol: Sensitivity to particle size, composition, and relative humidity. *J. Geophys. Res.*, **100**, 26105-26116, 1995.

Quinn, P. K., V. N. Kapustin, T. S. Bates and D. S. Covert, Chemical and optical properties of marine boundary layer aerosol particles of the mid-Pacific in relation to sources and meteorological transport. *J. Geophys. Res.*, **101**, 6931-6951, 1996.

Schwartz, S. E., and M. O. Andreae, Uncertainty in climate change caused by anthropogenic aerosols, *Science*, **272,** 1121-1122, 1996.

Taylor, K. E. and J. E. Penner, Response of the climate system to atmospheric aerosols and greenhouse gases. *Nature,* **369,** 734-737, 1994.

Tegen, I., P. Hollrig, C. Chin, I. Fung, J. Daniel, and J. Penner, Contribution of different aerosol species to the global aerosol extinction optical thickness: estimates from model results, *J. Geophys. Res.*, **102 (D20)**, 23895-23915, 1997.

DISCUSSION

J. A. van JAARSVELD: You showed size-distributions for the different species. I suppose that these are the initial distributions. Where do you get them, from observations? If so, then I must say that measured distributions are in fact 'aged' distributions, while you probably need 'fresh' distributions. Can you comment?

S. GONG: First of all, the size-distributions for sea-salt and sulphate aerosols I presented were from our model simulations not from observations. Since the distribution was expressed as the mass concentration, it is hard to distinguish between 'fresh' and 'aged' aerosols from these figures. A better conclusion would be drawn from the number size distribution for 'fresh' or 'aged' aerosols.

INTERCOMPARISON OF PHOTOCHEMICAL MECHANISMS USING RESPONSE SURFACES AND PROCESS ANALYSIS

Gail S. Tonnesen[1] and Deborah Luecken[2]

[1]University of California
Center for Environmental Research and Technology
Riverside, CA 92521

[2]U.S. Environmental Protection Agency
Research Triangle Park, NC 27711

INTRODUCTION.

Air Quality Models (AQMs) are used to simulate the emissions and transport of trace gases into urban and regional atmospheres and the subsequent photochemical reactions that produce secondary pollutants such as ozone (O_3) and fine particulates (PM2.5). The results of these simulations are used by policy makers to decide whether to emphasize reductions in emissions of volatile organic compounds (VOC) or nitrogen oxides (NO_x) for attaining ambient air quality standards for O_3 and PM2.5. The choice of control strategy also has implications for certain toxic compounds such as formaldehyde for which the major source in urban atmospheres is secondary photochemical production. The economic implications of air quality policy-making are significant; it is estimated that the annual cost of compliance with regulations for managing urban and regional O_3 in the U.S. exceeds US$ 1 billion (USEPA, 1997).

Because of the large economic and social costs of decisions affecting O_3 and other secondary pollutants, we wish to avoid potential errors by using AQMs to provide accurate predictions of the type and the amount of emissions control necessary to meet mandated air quality goals. There are large uncertainties, however, in the inputs to AQMs and corresponding uncertainties in the usefulness of model simulations. Furthermore, there is a large variety of both Lagrangian and Eulerian AQMs that are used for developing control strategies for O_3 and PM. Given the large uncertainties in the models' input variables, it is possible that different AQMs may indicate contradictory control strategies. An important step in the evaluation and validation of AQMs is the comparison of the performance and predictions of the diverse AQMs currently in use.

In the U.S., regulatory agencies require the use of Eulerian models for developing air quality attainment strategies because these model are presumed to have more complete representation of turbulent dispersion and transport processes and, consequently, greater physical realism than do Lagrangian models. A number of difficulties arise, however, in performing intercomparisons of Eulerian models due to the complex interactions and feedback effects among the transport and chemical processes. When AQMs are run in their "native mode" (i.e., in which each model implements its unique emissions, meteorology, chemical mechanism and solver algorithms) it can be difficult to determine which of the models' component processes control the differences in model predictions. This occurs because differences in emissions, meteorology and chemistry can each have similar effects on predicted O_3 concentrations (e.g., Dennis et al., 1999; Jang et al., 1995). As an alternative, models can, in principle, be operated using "harmonized inputs" (i.e., each model is operated with identical emissions, meteorology, chemistry, etc). In practice, harmonizing the models input is not feasible (Hass et al., 1997), and furthermore, it would not provide a true comparison of the models as they are actually used in specific applications. Hass et al. (1997) recommend an approach of native mode intercomparisons of the models combined with intercomparisons of the models sub-modules using harmonized inputs. This paper describes a new approach for performing comprehensive intercomparisons of one important sub-module: the photochemical mechanisms which control the production of secondary pollutants.

METHODS

There are two major difficulties in performing mechanism intercomparisons: First, the results of the comparison can vary significantly depending on both the absolute levels of VOC and NO_x emissions and on the ratio of VOC/NO_x. Mechanism comparisons are typically performed for a few different sets of VOC and NO_x emissions, for example, Kuhn et al. (1998) in the most comprehensive mechanism intercomparison to date used five sets of VOC-NO_x combinations, only two of which corresponded to urban conditions. Second, model predicted concentrations of O_3 and other species are non-linear functions of the model's independent variables, and furthermore, some secondary products can be produced and removed by multiple reaction pathways and other sink or source processes. As a result, model predictions of species concentration can be poor indicators of important differences in the chemical processes that control formation and removal of important trace species.

To address these problems, we have modified a research version of a widely used Lagrangian model, the Ozone Isopleth Plotting Package (OZIPR) (Gery and Crouse, 1991) to facilitate the creation of time-dependent, isopleth diagrams, or response surfaces, for each of the trace species concentrations for a wide range of VOC and NO_x emissions. In addition, we have written algorithms that use the integrated rates of chemical reactions to calculate and output important diagnostic terms that describe the chemical budgets of radicals, total odd nitrogen (NO_y) and odd oxygen (O_x). We have implemented these algorithms for several widely used photochemical mechanism, and we describe below the results of a comparison of the SAPRC99 (Carter, 1990) and the CB4-99 (Adelman, 1999). Other recent mechanisms are being evaluated and results are reported by Tonnesen and Luecken (2000).

The simulation conditions used here are derived from an Atlanta, GA scenario developed by the U.S. EPA (Baugues, 1991). We note, however, that the response surface is generated for a large range of VOC and NO_x emissions scenarios, and so this analysis is not intended to characterize actual conditions in Atlanta. Instead, this analysis is intended to be illustrative of a new modeling tool for performing process-based evaluations and intercomparisons of photochemical mechanisms. The scenario conditions were "harmonized" for the two mechanisms by using a identical photorates for the inorganic species and for formaldehyde (HCHO). Because different procedures are used for partitioning emitted and ambient VOC into each mechanisms unique "model" VOC species, it is also necessary to develop VOC splits from an identical set of emitted/ambient VOC. Here we used VOC splits created for the ambient data from a canister study of 39 cities in the U.S. (Jeffries and Sexton, 1994).

The modified OZIPR code creates three new output files which contain the time-varying response surfaces for: concentration for each model species; integrated rates of each chemical reaction; and key chemical processes describing the budgets of HO_x, NO_y and O_3. The new output files are written in a machine independent, binary data format that can be analyzed using PAVE, a public domain scientific visualization package designed for performing model evaluations (Thorpe et al., 1996). The OZIPR model is designed to facilitate model simulations for a wide range of VOC and NO_x emissions levels so that the "response surface" of peak O_3 concentration and other parameters can be viewed as a function of the precursor emissions levels. The OZIPR model automates the process of creating response surface plots by allowing the user to specify the maximum range of the initial VOC and NO_x concentrations, and the model will perform a matrix of 11 by 11 simulations in which initial concentrations and emissions are scaled proportionally from zero to 100% of the specified maximum VOC and NO_x range. Typically the results of these simulations are displayed as contour surfaces showing the dependent variables as a function of VOC emissions on the x-axis and NO_x emissions on the y-axis.

RESULTS

First we compare the mechanisms in terms of their predicted peak O_3 concentrations, sensitivity of O_3 to precursor emissions, and the relative effectiveness of VOC and NO_x control strategies. Figure 1 (top-left) shows the O_3 response surface (also known as the O_3 isopleth diagram) for the SAPRC99 mechanism, where the contour lines show the peak O_3 concentration predicted for the 11 by 11 matrix of 12 hour trajectory model simulations, and Figure 1 (top-right) shows the percent difference in the peak O_3 concentration for the CB4.99 mechanism, where for this and subsequent plots the difference is calculated as the percent change in CB4.99 relative to the SAPRC99 values.

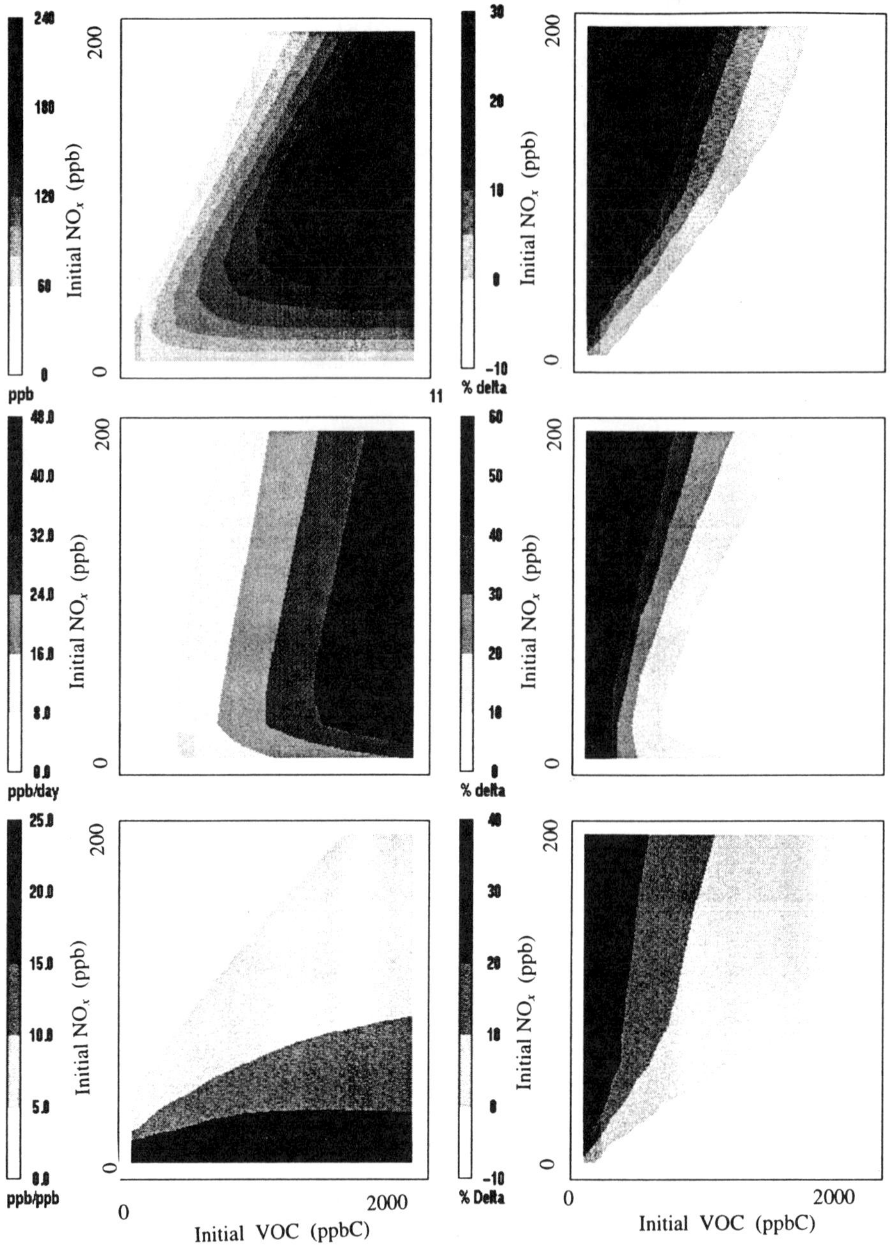

Figure 1. Left-hand side shows response surfaces for SAPRC99 and right-hand side shows % change in response surface for CB4.99 calculated as 100(CB4.99/SAPRC99–1): (top) peak O_3 concentration; (middle) cumulative radical initiation of HO_2 and RO_2; and (bottom) O_x production efficiency per NO_x. VOC emissions increase along the x-axis and NOx emissions increase along the y-axis

The two mechanism give very similar predictions for O_3 at conditions near the ridgeline of maximum O_3, but there are three notable differences in the predicted O_3 concentrations for the two mechanisms:

1) CB4.99 predicts higher O_3 levels for radical-limited conditions (*i.e.*, high NO_x and low VOC) above the ridgeline;
2) SAPRC99 predicts higher O_3 levels for NO_x-limited conditions (*i.e.*, high VOC and low NO_x) below the ridgeline;
3) There is considerable temporal variability between the two mechanisms where production of O_3 is much more rapid in SAPRC99 than in CB4 from 8 to 10 AM.

Figure 2 compares the sensitivity of peak O_3 to changes in emissions of VOC (dO_3/dE_{VOC}) and emissions of NO_x (dO_3/dE_{NOx}). The CB4.99 has greater dO_3/dE_{VOC} than does SAPRC99 for low VOC/NO_x ratios and slightly lower dO_3/dE_{VOC} for high VOC/NO_x ratios. The comparison is more complex for dO_3/dE_{NOx}; SAPRC99 has greater dO_3/dE_{NOx} than does SAPRC99 both at low and very high VOC/NO_x ratios, while CB4.99 has greater dO_3/dE_{NOx} at VOC/NO_x ratios from 5 to 20 ppbC/ppb. It is difficult to interpret the overall effect of these differences in O_3 sensitivity, and a more direct comparison of the mechanisms can be performed using the Empirical Kinetic Modeling Approach (EKMA) (EPA 1983) to estimate the relative effectiveness of VOC versus NO_x reductions strategies for the two mechanisms. The EKMA method was used here to evaluate the necessary reduction in VOC and NO_x for attaining the current U.S. national air quality one-hour standard of 120 ppb for O_3.

Table 1 shows the results of the EKMA analysis. The CB4.99 required higher levels of VOC reductions and lower levels of NO_x reductions than did SAPRC99. Thus, the SAPRC99 would tend to make VOC reduction strategies appear relatively more effective, while the CB4.99 would make NO_x reduction strategies appear relatively more effective.

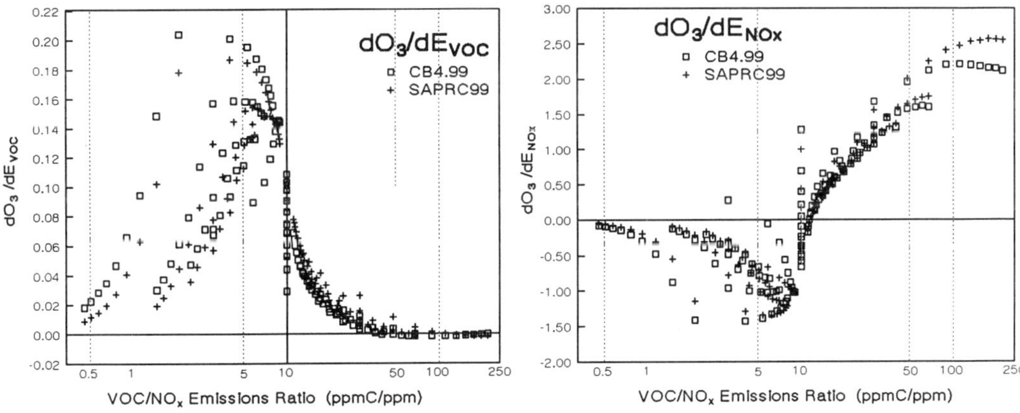

Figure 2. O_3 sensitivity to changes in precursor emissions of VOC (left) and NO_x (right), calculated as the centered difference of ±2% changes in emissions for each of the simulation used to form the O_3 response surface in Figure 1.

Table 1. Comparison of simulated VOC and NO_x emissions reductions needed to attain the U.S. air quality standard for O_3 of 120 ppb one-hour-average. Predicted controls are calculated for the SAPRC99 and CB4.99 photochemical mechanisms.

VOC/NO_x Ratio (ppmC/ppm)	% VOC Control		% NO_x Control	
	SAPRC99	CB4.99	SAPRC99	CB4.99
10.0	30	36	72	66
12.5	42	49	65	61
20.0	59	67	57	53
30.0	70	77	53	49

A mechanistic explanation for the differences in the control strategy effectiveness between the mechanisms can be obtained by analyzing the mechanism performance in terms of the budgets of radicals and NO_y. For radical-limited conditions (i.e., low VOC/NO_x ratios above the O_3 ridgeline), O_3 production is limited by the production of new radicals. Conversely, for NO_x limited conditions (i.e., high VOC/NO_x ratios below the O_3 ridgeline), production of O_3 is primarily limited by the O_3 production efficiency per NO_x.

To analyze the differences in radical initiation, it is necessary to describe the temporal differences in the rate of O_x and O_3 production. The SAPRC99 was much "hotter" in the first few hours of the simulation; O_x was produced at a maximum rate of 80 ppb in the first hour for SAPRC99 compared to only 27 ppb for CB4.99. Given that the scenario conditions are radical-limited in the early morning and that the initial O_3 concentration in this system was nearly zero, differences in the early morning reactivity must be due to differences either in the initiation of new HO_2 and RO_2 radicals from organic chemistry or in the propagation efficiency of radicals. Both effects contribute here, as described next.

First, SAPRC99 had a higher rate of OH attack on aromatics (i.e., an increase in propagation reactions). For the simulation with maximum reactivity in the first hour, OH attacked aromatics in SAPRC99 at a rate of 6.6 ppb/hr in the first hour compared to 2.6 ppb/hr in CB4.99. This occurred both because the initial aromatic concentrations were higher in SAPRC99 (59 ppb in SAPRC99 compared to 55 ppb in CB4.99), and because the aromatic OH rate constants for aromatic reactions were about 10% higher in SAPRC99 than in CB4.99.

Secondly, SAPRC99 has a much richer description of RO_2 radicals than does CB4.99. In SAPRC99 the radical species that contributed most significantly in the early morning were methyl peroxy, acetyl peroxy and RO_2R radicals, with the most significant initiation reactions being OLE_2+O_3 and photolysis of MGLY, BACL, DCB_2, and DCB_3 (where these intermediates were primarily produced by OH attack on aromatics). The CB4.99 uses an operator species (XO_2) and prompt formation of HO_2 to represent implicitly RO_2 formation. This implicit treatment of RO_2 complicates the comparison of radical chemistry between the mechanisms. Nonetheless comparisons of total peroxy radical initiation (HO_2+RO_2) shows that, in the most reactive simulation, SAPRC produced 7.3 ppb new peroxy radicals in the first hour compared to only 3.0 ppb in CB4.99. The primary cause of this difference was that nearly 100% of the SAPRC99 aromatic reaction products (BACL, DCB_2 and DCB_3) photolyzed to produce two new radicals, while much lower fractions of the CB4.99 aromatic reaction intermediates produced new radicals. For example in CB4.99, only 30% of aromatic reaction product TO_2 reacted with NO to produce OPEN (which subsequently decomposes to produce two radicals) while 70% of TO_2 decomposed in a simple propagation reaction producing no new radicals.

In summary, SAPRC99 was much more reactive than CB4.99 in the early morning because intermediate products of the OH attack on aromatics had considerably higher yields of new peroxy radicals and because SAPRC99 had higher radical propagation efficiencies (due to the higher rate of OH+VOC reactions) in the early morning. This created positive feedback effects in which more rapid production of HCHO and O_3 led to greater morning radical initiation in SAPRC99.

While the analysis of the temporal dynamics provides insights into the controlling mechanistic processes of the two mechanisms, the resulting differences in the simulated peak O_3 concentrations can be understood more readily in terms of the cumulative HO_x initiation and O_x per NO_x production efficiency where these terms are integrated over the full simulations time. Figure 1 (middle-left) shows for SAPRC99 the cumulative initiation of new peroxy radicals (HO_2+RO_2), where RO_2 represents all organic peroxy radicals. Figure 1 (middle-right) shows the percent change in HO_2+RO_2 initiation for CB4.99 relative to SAPRC99. The CB4.99 had greater radical initiation for the radical limited regime which contributed to the greater O_3 production in CB4.99 above the O_3 ridgeline (Figure 1 top-right). The higher total rate of radical initiation in CB4.99 resulted in part from a larger fraction of HCHO photolyzing to produce new HO_2 radicals, and in part because the slower photochemistry in CB4.99 resulted in higher production rates of carbonyls later in the midday period when photolysis rates were maximized.

Figure 1 (bottom-left) shows for SAPRC99 the average O_x production ($P(O_x)$) efficiency per NO_x converted to inert forms of odd nitrogen ($P(NO_z)$). Figure 1 (bottom-right) shows the percent change in $P(O_x)/P(NO_z)$ for CB4.99. Comparison of Figure 1 (top-right) and Figure 1 (bottom-right) shows that the lower O_3 concentrations in CB4.99 for NO_x limited conditions can be explained in terms of the 10% lower $P(O_x)/P(NO_z)$ values in CB4.99 at the lowest NO_x levels. Above the ridgeline in the radical limited regime, differences in $P(O_x)/P(NO_z)$ did not affect O_3 production because there was excess NO_x for radical limited conditions.

Finally, the differences in the control strategy effectiveness as shown in Table 1 for the two mechanisms can be understood in terms of the HO_x and NO_y budgets: VOC controls were relatively less effective in CB4.99 because it produced more radicals for the radical limited conditions, while NO_x controls were less effective in SAPRC99 due to its higher values of $P(O_x)/P(NO_z)$.

CONCLUSIONS

As recommended by Kuhn et al. (1998) a comprehensive mechanism evaluation or intercomparison requires that the mechanisms be evaluated for a large number of VOC and NO_x scenarios, and, ideally, for the full response surface. Key difficulties in performing such an evaluation include: the large number of VOC/NO_x scenarios to be performed; the large number of species concentrations and reaction rates to be analyzed; and the temporal variability in the evolution of species concentrations. We have demonstrated here modifications to the widely used OZIPR model that facilitate mechanism evaluations. New model outputs are created allowing the PAVE visualization package to be used to perform sophisticated, time-dependent analyses for multiple data sets. In addition, outputs that describes the budgets of HO_x, NO_y and O_x allows for convenient evaluation of the important component chemical processes. Our example analysis here shows that there are important differences in the CB4.99 and the SAPRC99 mechanisms, particularly in the O_x production efficiency per NO_x for NO_x-limited conditions and in the HO_x budget for radical-limited conditions. It is not possible to conclude which mechanism is more correct, but the model tool demonstrated here will allow mechanism developers to more readily assess the effects of future revisions to these photochemical mechanisms, and to determine which chemical process require better elucidation in future chamber and kinetics experiments.

Disclaimer: This paper has been reviewed in accordance with the U.S. Environmental Protection Agency's peer review policies and approved for presentation and publication. Mention of trade names or commercial products does not constitute endorsement or recommendation for use.

REFERNCES

Adelman, Z.E., 1999, A reevaluation of the Carbon Bond-IV photochemical mechanism, Masters Thesis, University of North Carolina, Chapel Hill, NC. 27499.

Baugues, K., 1990, Personal communication. U.S. Environmental Protection Agency, Research Triangle Park, North Carolina.

Carter, W. P. L., 1999, Documentation of the SAPRC-99 Chemical Mechanism for VOC Reactivity Assessment, Draft report to the California Air Resources Board, Contracts 92-329 and 95-308, 9/13/99.

Dennis, R.L., Arnold, J.R., Tonnesen, G.S., Yonghong Li, 1999, A new response surface approach for interpreting Eulerian air quality model sensitivities, *Computer Physics Communications*, 117: 99-112.

Gery M. and Crouse, R., 1990, *User's Guide for Executing OZIPR*, Order No. 9D2196NASA, U.S. Environmental Protection Agency, Research Triangle Park, NC, 27711.

Jang J.C.C., Jeffries H.E. and S. Tonnesen. 1995. Sensitivity of ozone to model grid resolution -- II: Detailed process analysis for ozone chemistry. *Atmos. Environ.*, 29:3101-3114.

Jeffries, H. E. and Sexton, K. G., 1994, The relative ozone forming potential of methanol fueled vehicles and gasoline fueled vehicles in outdoor smog chambers. Final Report, Coordinating Research Council, Atlanta, GA, Project ME-1, 1994.

Kuhn, et al., 1998, Intercomparison of the gas-phase chemistry in several chemistry and transport models, *Atmos. Environ.*, 32:693-709.

Hass, H., Builtjes, P.J.H., Simpson, D., Stern, R., 1997, Comparison of model results obtained with several European regional air quality models, *Atmos. Environ.*, 31:3259-3279.

Thorpe, S., J. et. al., 1996, The Package for Analysis and Visualization of Environmental Data. Proc., Computing in Environmental Resource Management, 2-4 December, Research Triangle Park, NC, Air & Waste Management Assoc., 241-249.

Tonnesen and Luecken, 2000, A comprehensive intercomparison of three new gas-phase photochemical mechanisms, manuscript to be submitted to *Atmos. Environ.*

U.S. EPA, 1983, *Proceedings of the Empirical Kinetic Modeling Approach (EKMA) Validation Workshop.* EPA Report No. EPA-600/9-83-0141, Office of Research and Development, Research Triangle Park, NC, 27711.

U.S. EPA, 1997, Regulatory Impact Analyses for the Particulate Matter and Ozone National Ambient Air Quality Standards and Proposed Regional Haze Rule. USEPA OAQPS, Research Triangle Park, NC.

DISCUSSION

P. BUILTJES: Your results show a clear influence whether you use a box model or a run with entrainment. Do you agree that an intercomparsion of chemical mechanisms in a 3-D Eulerian grid model would be better, and that these results would show smaller differences?

G. TONNESEN: The model simulations did show greater differences in peak ozone concentration among the various photochemical mechanisms depending on whether the scenarios were constant volume or included mixing and entrainment. The pattern of differences was similar in simulations with constant volume and with entrainment, but the magnitude of the differences were smaller with entrainment. For example, in comparing the RACM mechanism to the RADM2 mechanism, the RACM peak O_3 ranged from 27% lower to 14% greater than for RADM2 depending on the VOC to NO_x ratio. For constant volume simulations, the range in peak O_3 was 33% lower to 40% greater in RACM compared to RADM2.

This difference is certainly due in part to the effect of entrainment of O_3. The scenarios with entrainment included a constant regional background O_3 concentration of 40 ppb, and the height of the mixed layer increased from 250 m to 2100 m, so entrained O_3 contributed significantly to the peak O_3 levels in this scenario. It is also probable that mixing and entrainment affected the range of values of the VOC/NO_x ratio perhaps reducing extreme values.

In comparisons with an Eulerian model, background O_3 may vary somewhat due to differences in chemical production among the mechanisms. Hence, it is possible that results in an Eulerian model would be intermediate between the constant volume and entrainment scenarios examined here.

It will be necessary to perform intercomparisons in Eulerian models to determine how the response may change. The value of performing comparison in the one-dimensional model is that it is possible to systematically examine the system response as a function of VOC and NO_x emissions magnitudes and ratios, making it possible to anticipate expected differences for various photochemical regimes in the Eulerian model.

AN EVALUATION OF TWO ADVANCED TURBULENCE MODELS FOR SIMULATING THE FLOW AND DISPERSION AROUND BUILDINGS

S.T. Chan and D.E. Stevens

Lawrence Livermore National Laboratory
Livermore, CA 94551

INTRODUCTION

Numerical modeling of airflow and pollutant dispersion around buildings is a challenging task due to the geometrical variations of buildings and the extremely complex flow created by such surface-mounted obstacles. The airflow around buildings inevitably involves impingement and separation regions, building wakes with multiple vortices, and jetting effects in street canyons. The interference from adjacent buildings further complicates the flow and dispersion patterns. Thus accurate simulations of building-scale transport phenomena requires not only appropriate physics submodels but also significant computing resources. We have developed an efficient, high resolution CFD model for simulating chemical and biological releases around buildings. Our primary goal is to support incident response and preparedness in emergency response planning and vulnerability analysis.

Our model, FEM3MP, solves the three-dimensional, time-dependent Navier-Stokes equations. The numerical algorithm uses an innovative finite element approach to accurately represent complex building shapes and a fully implicit projection method for efficient time integration (Gresho and Chan, 1998). For turbulence parameterization, we have implemented, in addition to a k-ε turbulence submodel, a nonlinear eddy viscosity (NEV) and a large eddy simulation (LES) turbulence submodels. Other model physics, including UV radiation decay, aerosols, surface energy budget, and tree canopy effects, have also been implemented. Our model has been developed to run on both the serial and massively parallel computer platforms.

Model validation is a fundamental component of establishing the credibility of this model for use in the assessment of flow and dispersion of hazardous agents around buildings. We have performed model validations using, among others, the tow-tank experimental data obtained from flow and dispersion past a cubical building (Zhang et al., 1996) and also similar data around a 2-D array of buildings in a wind tunnel (Brown et al., 2000).

In this paper, we will describe briefly the salient features of our model, present and discuss results from a model-data comparison study. Due to space limitations, only results for the cubical building, with a focus on the performance of the NEV and LES turbulence submodels, are presented in this paper. Additional results will be presented at the conference.

NUMERICAL MODEL

Governing Equations

The equations modeled in this validation study are a subset of the generalized anelastic equations available to FEM3MP. These equations, written in Cartesian tensor form, are:

$$\frac{\partial}{\partial t}u_i + u_j\frac{\partial u_i}{\partial x_j} = -\frac{\partial p}{\partial x_i} + \nu\frac{\partial^2 u_i}{\partial x_j \partial x_j} + \frac{\partial}{\partial x_j}(-\overline{u_i' u_j'}) \tag{1}$$

$$\frac{\partial}{\partial x_j}u_j = 0 \ , \tag{2}$$

$$\frac{\partial c}{\partial t} + u_j\frac{\partial c}{\partial x_j} = \nu\frac{\partial^2 c}{\partial x_j \partial x_j} + \frac{\partial}{\partial x_j}(-\overline{u_j' c'}) \tag{3}$$

In the above equations, u_i is the i-th component of the mean velocity, c is the mean tracer concentration, p is the dynamic pressure, ν is the kinematic viscosity, $\overline{u_i' u_j'}$ are turbulent stresses, and $\overline{u_j' c'}$ is the mass flux of the tracer. The RANS (Reynolds-Averaged Navier-Stokes) and LES approaches are used to parameterize the turbulent stresses and mass fluxes.

The above set of governing equations are firstly discretized in space by the Galerkin finite element method, with piecewise constant representation for pressure and trilinear approximations for other field variables, to obtain a couple system of nonlinear first-order ordinary differential equations. In order to solve cost-effectively large three-dimensional problems, we employ the fully implicit projection method developed by Gresho and Chan (1998). With the method, the coupled system of equations are segregated and solved sequentially for each of the field variables via efficient iterative solvers.

RANS Turbulence Submodel

In our RANS approach, we used a nonlinear eddy viscosity (NEV) turbulence submodel developed by Suga (1995). This 3-equation turbulence submodel has many desirable properties, including anisotropy, a cubic constitutive law, and no need for wall functions. Since the model has been derived from the Reynolds stress closure models, it retains many of their attributes—but at a significantly reduced computational cost.

The crux of the submodel involves the following three fairly complex, coupled equations, with details available in Suga (1995), Gresho and Chan (1998):

$$\frac{\partial k}{\partial t} + u_j\frac{\partial k}{\partial x_j} = \frac{\partial}{\partial x_i}\left[\left(\nu\delta_{ij} + 0.22 f_g \overline{u_i' u_j'}\frac{k}{\tilde{\varepsilon}}\right)\frac{\partial k}{\partial x_j}\right] + P_K - \varepsilon, \tag{4}$$

$$\frac{\partial \tilde{\varepsilon}}{\partial t} + u_j\frac{\partial \tilde{\varepsilon}}{\partial x_j} = \frac{\partial}{\partial x_i}\left[\left(\nu\delta_{ij} + 0.18 f_g \overline{u_i' u_j'}\frac{k}{\tilde{\varepsilon}}\right)\frac{\partial \tilde{\varepsilon}}{\partial x_j}\right] + \left(1.2 P_K - C_\varepsilon\tilde{\varepsilon}\right)\frac{\tilde{\varepsilon}}{k} + P_\varepsilon, \tag{5}$$

$$\frac{\partial A_2}{\partial t} + u_j\frac{\partial A_2}{\partial x_j} = \frac{\partial}{\partial x_i}\left[\left(\nu\delta_{ij} + 0.22 f_g \overline{u_i' u_j'}\frac{k}{\tilde{\varepsilon}}\right)\frac{\partial A_2}{\partial x_j}\right] + \frac{a_{ij}}{k}\left[P_{ij} - \varepsilon_{ij} + \phi_{ij}\right] - \frac{2A_2}{k}P_K. \tag{6}$$

In the above equations, k is the turbulent kinetic energy (TKE), $\tilde{\varepsilon}$ is the isotropic dissipation rate, and A_2 is the second invariant of the dimensionless anisotropic Reynolds stress tensor.

The Reynolds stresses are defined by a cubic constitutive relationship

$$\begin{aligned}\overline{u_i' u_j'} &= \frac{2}{3}k\delta_{ij} - \nu_T S_{ij} + \frac{\nu_T k}{\tilde{\varepsilon}}\{c_1(S_{ik}S_{kj} - \frac{1}{3}S_{k\ell}S_{k\ell}\delta_{ij}) \\ &+ c_2\left(\Omega_{ik}S_{kj} + \Omega_{jk}S_{ki}\right) + c_3(\Omega_{ik}\Omega_{jk} - \frac{1}{3}\Omega_{\ell k}\Omega_{\ell k}\delta_{ij}) \\ &+ \frac{k}{\tilde{\varepsilon}}\left[c_4\left(S_{ki}\Omega_{\ell j} + S_{kj}\Omega_{\ell i}\right)S_{k\ell} + c_5 S_{ij}S_{k\ell}S_{k\ell}\right] + c_6 S_{ij}\Omega_{k\ell}\Omega_{k\ell}\} \ ,\end{aligned} \tag{7}$$

in which S_{ij} is the mean strain rate tensor and Ω_{ij} is the mean rotation tensor. The isotropic turbulent eddy viscosity is defined as

$$\nu_T = c_\mu f_\mu k^2/\bar{\varepsilon}, \tag{8}$$

where c_μ is a turbulent viscosity parameter and f_μ is a wall damping function, both being functions of the turbulence variables k, $\bar{\varepsilon}$, and A_2.

LES Subgrid-Scale Modeling

The role of the turbulence submodel in an LES simulation is to parameterize the effect of motions that cannot be resolved by the computational mesh. The one used here was developed by Smagorinsky (1963), with the subgrid TKE, e, and the eddy viscosity given by,

$$e^{1/2} = \frac{c_s^2}{c_m} l\left|S_{ij}\right|, \qquad K_m = c_s^2\ l^2\left|S_{ij}\right|. \tag{9}$$

Here c_s and c_m are empirical constants with values of 0.1 and 0.3 in this study. The subgrid scale fluxes are related to the gradients of the mean variables as,

$$\overline{u_i' u_j'} = -K_m S_{ij}, \qquad \overline{u_j' c'} = -K_m \frac{\partial c}{\partial x_j}, \tag{10}$$

Although simpler conceptually, the properties of the flow are more difficult to diagnose for an LES than a RANS simulation. For example, TKE is computed explicitly in a RANS model. For an LES calculation, one must first generate a mean velocity field and the deviations from this field, square and average again to generate the resolved TKE. Averaging is also needed for the sub-grid TKE, since the sum of these two quantities has to be used for comparison with RANS and the experimental results.

One problem associated with LES turbulence models is their inability to model near wall behavior. For this study, we used free-slip on the top and bottom boundary to maximize turbulent dispersion. Also, c_s = 0.1, instead of 0.23 (a value typically used in atmospheric LES), was used to enhance crosswind dispersion.

MODEL-DATA COMPARISON

This paper presents a validation study of flow and dispersion around a cubical building, using both the RANS and LES approaches. There have been many laboratory experiments and numerical studies performed on this flow. The fidelity of our model with respect to the flow field is judged by comparison with the laboratory measurements of Martinuzzi and Tropea (1993) and the LES results of Shah (1998). For dispersion assessment, we use the results from a study by Zhang et al. (1996) for a continuous, ground-level tracer release behind the cube.

The computational domain used in our numerical simulations is 10H by 7H by 2H, H being the cube height, in the longitudinal, lateral, and vertical directions, respectively. The origin is situated at the center of the cube, with the inflow at x = -3.5H.

Since the RANS assumes a steady state solution and the geometry is symmetric, only one-half of the domain was simulated. This enables the RANS to use a small mesh of 96x37x30 nodal points. No slip boundary condition was used along the channel walls and on surfaces of the cube. The mean velocity of a fully-developed channel flow of average velocity of 0.6 m/s was specified on the inlet plane. The entire simulation took about 25 hours on 2-3 processors of a shared-memory multi-processor DEC ALPHA to integrate 1200 time steps.

The LES simulation used a computational mesh of 129x65x65 nodal points and integrated for 1920 time steps. This generated a time series of data for 57.6 nondimensional time units, t* = U_o t/H. A uniform inflow velocity U_o= 0.6 m/s was used to match the inflow characteristics of Zhang et al. (1996). The LES calculation requires roughly an order of magnitude more effort than the RANS simulation. However, due to the use of parallel computing using 64 IBM processors, this simulation took only 18 hours to complete.

The LES simulation was integrated for 40 non-dimensional time units before averaging starts. At this point, the mean concentration in the domain has stabilized and the flow has settled down in alternately shedding pairs of vortices behind the cube. Our simulation generated a pair of vortices every 7.5 non-dimensional time units, which compares reasonably well with an estimated value of 8.5 by Shah's LES model.

In Fig. 1, time-averaged streamlines are compared. The main features of all flow fields are similar. They include the separation zones in front of the cube, on the roof and the two sides (not shown), a primary recirculation zone in the wake, and a pair of counter-rotating vortices on the horizontal plane (not shown). The predicted reattachment length is 1.85 by the RANS model and is 1.55 by the LES model. These values agree well with the value of 1.68 measured by Martinuzzi and Tropea (1993), and with the value of 1.64 predicted by Shah (1998). The present model results compare much better than the value of 2.85 predicted by the simpler k-ε turbulence model (Chan and Lee, 1999) and the values between 2.68 and 3.4 predicted by variants of two-layer k-ε models (Lakehal and Rodi, 1997).

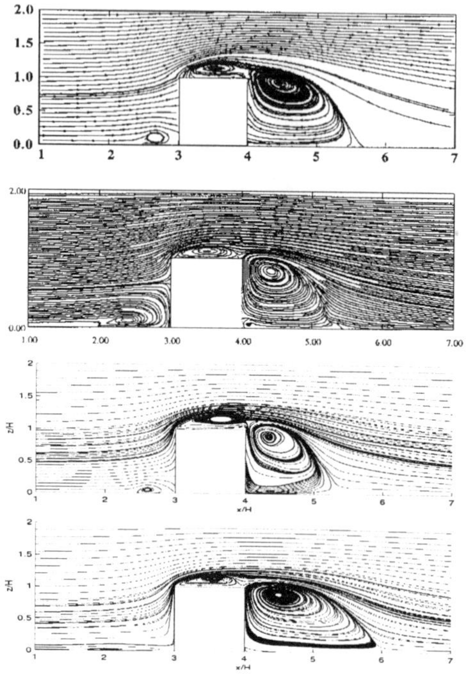

Figure 1: Comparison of time-averaged streamlines along the symmetry plane for the experimental results of martinuzzi and Tropea (top), Shah (top middle), FEM3MP LES (bottom middle), and RANS (bottom).

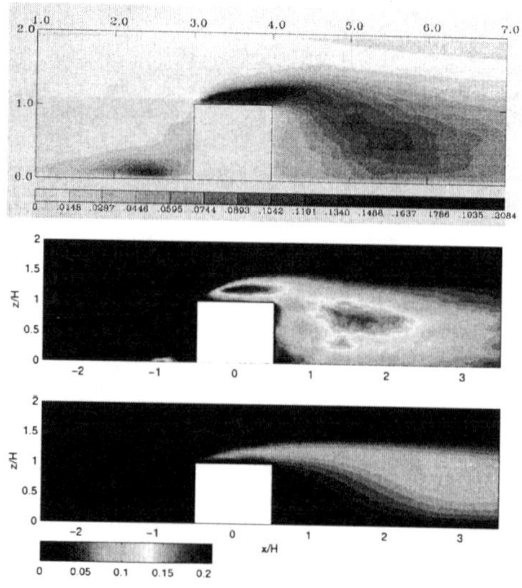

Figure 2: Comparison of normalized TKE along the symmetry plane for Shah's LES (top), FEM3MP LES (middle), and RANS (bottom).

In Fig. 2, the predicted TKE fields on the vertical plane of symmetry are compared with that obtained by Shah (1998), who used an LES model, a finer grid of 192x64x96, and a fully developed channel flow as inlet boundary condition. The TKE fields from both LES models tend to be larger and extend closer to the roof than that from the RANS. Our LES tends to predict higher TKE behind the cube, while Shah's has a bigger region of large TKE values in front of the cube. Besides the difference in inlet velocity profiles, Shah used no slip boundary conditions, while we used free-slip boundary conditions.

In Fig. 3, we compare the predicted turbulent kinetic energy profiles at selected locations with the data of Martinuzzi and Tropea (1993). These profiles indicate that both turbulence submodels are able to predict the bulk flow features observed in the experiment. Overall, the LES predicts the TKE better than the RANS. The small discrepancies near the top and bottom of the domain from the LES model are generated by the free-slip boundary conditions. Both models predict peak TKE values at approximately the correct locations.

The results of Zhang et al. (1996) are used herein to assess the accuracy of concentration patterns predicted by our model. These results were created with a tow-tank in which a cube of 0.1m high mounted to a plate was moved at 0.6 m/s through a water tank of 18m long and 1m x 1m in cross section. The dispersion was generated by a ground-level, continuous source placed at 0.25H behind the cube. The experiment showed that the tracer is entrained into the recirculating eddy behind the cube and dispersed downstream. The plume quickly becomes much wider than the cube width within a few block heights downstream.

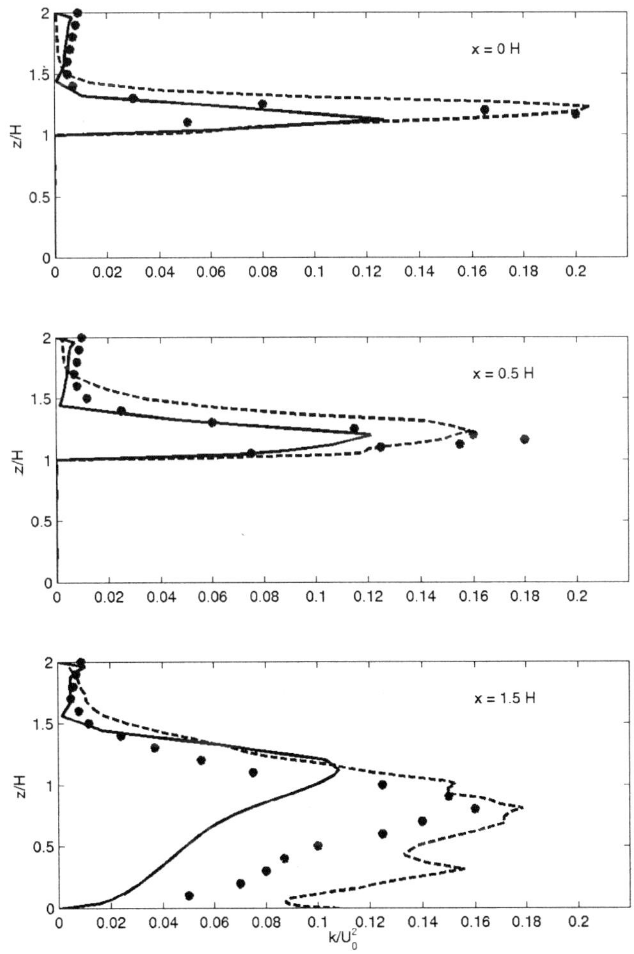

Figure 3: Comparison of normalized TKE for experiment (dots), LES (dashed line), and RANS (solid line) at three vertical profiles along the symmetry plane.

In Figs. 4 and 5, concentration patterns on the floor and the plane of symmetry are displayed. The contours are normalized concentration $\chi = CU_o^2H/Q$, where C is concentration and Q is the source flux. An interesting feature of these plots is that the instantaneous plume of LES looks very similar to the RANS plume in terms of width and height, except for a small horizontal translation downstream. The higher concentration contours are very similar among the three panels, indicating that both the RANS and LES simulations are performing well for the bulk of the dispersion. However, the RANS simulation is missing the wavy structure of the dispersion patterns shown in Fig. 4 and the eddy mixing at the top edge of the plume shown in Fig. 5. In addition, the LES plume has a wider spread in mean concentration, which was also observed in the experiment as a result of vortex shedding generated by the cube.

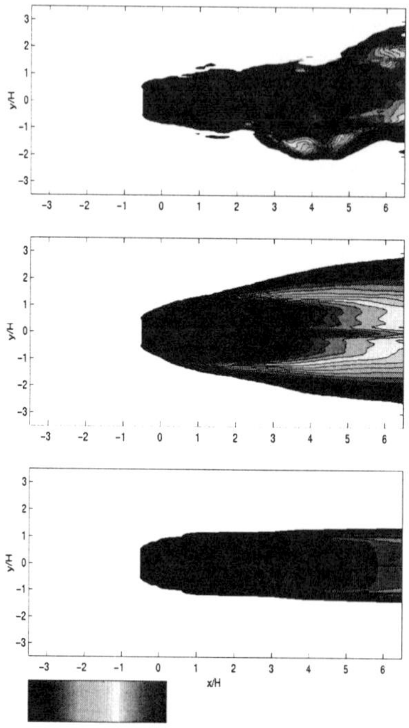

Figure 4: Tracer concentrations along the floor plane showing the instantaneous pattern form the LES (top), the mean from the LES simulation (middle), and from the RANS simulation (bottom).

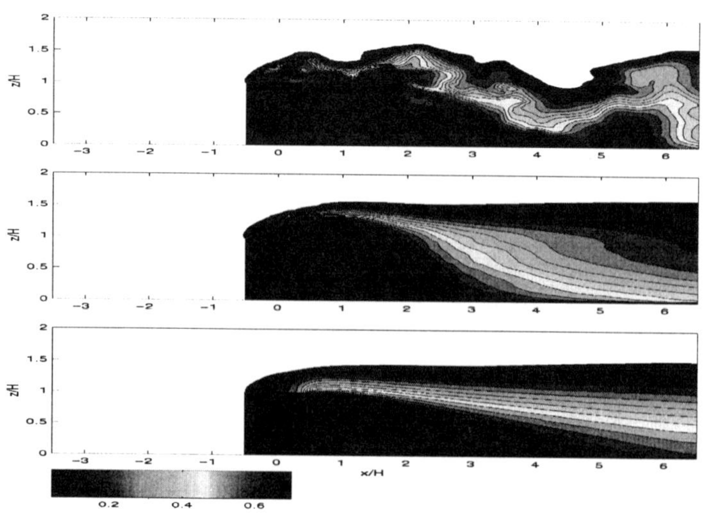

Figure 5: Tracer concentrations along the symmetry plane showing the instantaneous pattern from the LES (top), the mean from the LES simulation (middle), and from the RANS simulation (bottom).

In Fig. 6, we compare the predicted concentration profiles with measurements (shaded area) along three selected lines. The top plot shows the LES capturing the horizontal dispersion at the outflow as observed in the experiment. The middle plot shows the LES and RANS agreeing with experiments along the floor of the symmetry plane. Both of these plots support our observation that the RANS plume is similar to a realization of the LES plume. The bottom plot shows the vertical concentration profile centered in the middle of the outflow plane. This plot further supports this idea, since both RANS and mean values of LES have similar vertical extent as they agree towards the top of the profile. However, the lack of vortex shedding in the RANS simulation has resulted in higher concentrations near the bottom of the profile.

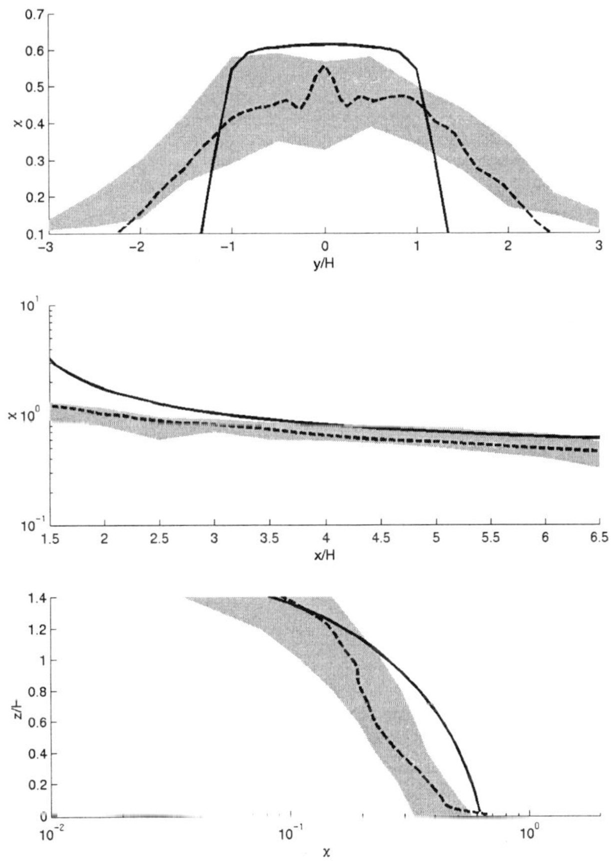

Figure 6: Comparison of the mean concentrations along three lines: the horizontal line at outflow (top), the horizontal line at the bottom of the symmetry plane (middle), and the vertical line at outflow (bottom). LES (dashed line), RANS (solid line).

SUMMARY

We have presented a validation for our model through simulating the flow and dispersion around a cubical building. It was found that the LES model yields more accurate results but with roughly an order of magnitude increase in computational cost. Several conclusions can be drawn from this study:

1. Instantaneous snapshots of concentration from the LES model have similar vertical depth and width as the RANS. This indicates RANS is producing a reasonable plume, but without the spatial fluctuation associated with vortex shedding in the experimentally observed wake.
2. The RANS and LES yield similar mean velocity fields near the building. If mean velocity fields are sufficient and decoupled from the advected species, the RANS solution is the more economical method.

3. The LES can capture turbulent variations that the RANS does not model. For situations in which the source lasts sufficiently long, LES simulates more accurately the width of the dispersion than RANS does. This is useful for situations where accurate species concentrations in both space and time are important.

ACKNOWLEDGMENTS

This work was performed under the auspices of the U.S. Department of Energy by University of California Lawrence Livermore National Laboratory under contract number W-7405-ENG-48.

REFERENCES

Brown, M.J., R.E. Lawson, D.S. DeCrox, and R.L. Lee, 2000, Mean flow and turbulence measurements around a 2-D array of buildings in a wind tunnel, 11th J. AMS/AWMA Conf. on the Appl. of Air Poll. Met., Long Beach, CA., 35-40.

Chan, S.T.and R.L. Lee, 1999, A model for simulating airflow and pollutant dispersion around buildings: *Air Pollution VII*, WIT Press, 39-48.

Gresho, P.M. and S.T. Chan, 1998, Projection 2 goes turbulent - and fully implicit, *Int. J. of Comp. Fluid Dyn.* 9:249-272.

Lakehal D. and W. Rodi, 1997, Calculation of the flow past a surface-mounted cube with two-layer turbulence models, *J. Wind Eng. Ind. Aerodyn.* 67 & 68:65-78.

Martinuzzi, R. and C. Tropea, 1993, The flow around surface-mounted, prismatic obstacles placed in a fully developed channel flow, *J Fluids Eng.* 115:85-92.

Shah, K.B., 1998, Large eddy simulations of flow past a cubic obstacle. Ph.D. thesis, Stanford University.

Smagorinsky, J., 1963, General circulation experiments with the primitive equations: 1. The basic experiment, *Mon. Wea. Rev.* 121:1847-1859.

Suga, K., 1995, Development and application of a nonlinear eddy viscosity model sensitized to stress and strain invariants. Ph.D. thesis, UMIST, Manchester, UK.

Zhang, Y.S., S. Arya, and W. Snyder, 1996, A comparison of numerical and physical modeling of stable atmospheric flow and dispersion around a cubical building, *Atm. Env.* 30:1327-1345.

DISCUSSION

R. BORNSTEIN: What was the stability of the RANS simulations? Why does the plume behind the cube seem somewhat narrow?

D. STEVENS: All of the calculations used neutral stability. The narrow spread of the plume is due to the fact that the RANS does not capture vortex shedding. The LES by capturing this vortex shedding had a much wider plume. One cannot fix this problem in the RANS by merely increasing the horizontal viscosity while keeping the Schmidt number constant. An increase in horizontal eddy viscosity causes the velocity field to be more homogenous and hence less able to cause horizontal dispersion via advection.

C. MENSINK: Is there a possibility to solve the RANS equation in such a way that one would obtain or allow a transient solution?

D. STEVENS: The RANS equations are derived by a Reynold's Average over dynamic timescales. One could alter this timescale to achieve much lower eddy viscosities. However, this would effectively create an LES turbulence model as Taylor's hypothesis links spatial and temporal scales. Lowering the timescale would result in a lengthscale that is resolvable on the mesh and would then be related to the mesh spacing.

A. BAKLANOV: Does the RANS model correctly simulate the eddy viscosity over the shear layer of the ABL?

D. STEVENS: This RANS model was developed for wind engineering scale flows and hence was tuned to the scales of interest to Mechanical Engineers. This is appropriate for simulating wind effects around individual building complexes. For a larger scale flow, it will probably need some of the modifications that were discussed at this conference.

A. BAKLANOV: Is the computing cost of the nonlinear three-equation RANS model much higher in comparison with the standard K-ε model?

D. STEVENS: This three equation model does not need any wall functions through the use of a third equation involving an invariant of the Reynolds' Stress tensor. This equation automatically senses the presence of a boundary and automatically and consistently varies K-ε in the presence of a wall. This is something that is very difficult to do with standard wall functions. Unfortunately, due to high gradients near the ground and building surfaces, there are very sharp gradients in this quantity. This lowers the convergence rate of this equation (implicitly integrated) and slows the computation down greatly. The standard K-ε takes around 1/10 the time of the LES. The three-equation model takes roughly the same amount of time.

EMISSIONS INVENTORY ESTIMATION IMPROVEMENTS USING A FOUR-DIMENSIONAL DATA ASSIMILATION METHOD FOR PHOTOCHEMICAL AIR QUALITY MODELING

Alberto Mendoza-Dominguez and Armistead G. Russell

School of Civil and Environmental Engineering
Georgia Institute of Technology
Atlanta, GA 30332

INTRODUCTION

Four-dimensional data assimilation (FDDA) is a method to directly, or indirectly, incorporate temporally and spatially resolved observational data in a deterministic model by modifying the model inputs or parameters. This has not yet been developed for three-dimensional air quality models (AQMs), in part due to the lack of an appropriate approach, though it has had wide use in the meteorological modeling community. Original FDDA applications in meteorological models were oriented to the global scale, but the necessity for its use on mesoscale applications was soon appreciated (Stauffer and Seaman, 1990). Meteorological models play an important role in air quality modeling since the reliability of the meteorological fields provided by those models can affect substantially the performance of the AQMs. Urban and regional (multiscale) FDDA approaches have been produced to help the meteorological models perform better (Seaman, 1992; Stauffer and Seaman, 1994). Several applications have been documented on the use of FDDA in air quality studies (e.g. Seaman *et al.*, 1995; Lyons *et al.*, 1995; Mueller *et al.*, 1996). However, in all of these studies FDDA has been applied as part of the meteorological modeling portion, not to the air quality model.

Outputs from meteorological applications, when fully processed, are used either in Eulerian AQMs-as the studies mentioned above-or Lagrangian random-particle statistical models (e.g. Yamada *et al*, 1989) to investigate the transport and fate of pollutants. The meteorological inputs for AQMs derived using FDDA-driven meteorological models are usually relatively reliable. This is typically not the case with other AQM inputs, especially emissions inventories (National Research Council, 1991). Here an FDDA approach for AQMs has been used to suggest improvements in the emissions inventory for a particular application to the Atlanta, Georgia nonattainment area. The approach used here differs from typical Newtonian Relaxation (or nudging) methods (Seaman, 1992) implemented in meteorological models in that the solution obtained from nudging is not mathematically optimized, while the emission corrections derived from the AQM-FDDA approach taken here are optimal in a least square sense. The theory behind the AQM-FDDA approach as applied in this case is that emissions are the main reason for the discrepancy between model-derived concentrations and observations and, assuming that the AQM (and other inputs and parameters) have significant lower uncertainty, minimizing this discrepancy by adjusting the emissions inventory will result in better estimates of real emission strengths, composition and spatial and temporal patterns.

METHODS

The AQM-FDDA method consists of an integrated modeling approach that combines a three-dimensional photochemical AQM, direct sensitivity analysis, and a receptor model. The objective of the modeling effort is to derive emission correction estimates of major source categories based on ambient concentration of primary and secondary species. The discrepancy between modeled and observed concentrations and the sensitivity of the modeled concentrations to the sources of interest drive the calculation of the FDDA module. In essence, the AQM provides the concentration fields while the direct sensitivity analysis module provides sensitivity fields. These two then are used to estimate emission scaling factors of a base case emissions inventory using a receptor modeling approach, in this case, ridge regression (Draper and Van Nostrand, 1979). The receptor model in the FDDA approach solves the following algebraic linear system:

$$\mathbf{f} = \left(\mathbf{G}^{T}\mathbf{W}_{c}\mathbf{G} + \mathbf{W}_{f}\right)^{-1}\mathbf{G}^{T}\mathbf{W}_{c}\mathbf{d} \tag{1}$$

In equation 1, **f** represents a vector of estimated scaling factors (one scaling factor for each source involved in the assimilation process), each component of vector **d** is the difference in the concentration measured and modeled of ozone and its precursors, **G** is a matrix of weighted sensitivity coefficients (which indicate how the modeled concentrations vary as emissions are decreased or increased from a given base value). Each observation is weighted by means of matrix $\mathbf{W}_c$ in order to account for estimated uncertainties (and other properties) of the measurements. Finally, the matrix $\mathbf{W}_f$ provides further information of the *a priori* characteristics of the emission strengths. This method is discussed in greater detail elsewhere (Mendoza-Dominguez and Russell, 2000).

Emission corrections were calculated for the emissions inventory created for the Atlanta, Georgia historic ozone episode of August 9-10, 1992. Formation and transport of photochemical pollutants was followed using the CIT Airshed Model (Russell *et al.*, 1988; Harley *et al.*, 1993) extended with DDM-3D (direct decoupled method for three-dimensional models) to calculate sensitivity coefficients (Yang *et al.*, 1997). The CIT and DDM-3D provide information to the **d** vector and **G** matrix, respectively. Ridge regression calculations (including the estimation of the $\mathbf{W}_f$ matrix) are based on the work of Hoerl and Kennard (1976) and Aldrin (1997). Details in the setup of the modeling domain and description of the inputs to the AQM can be found elsewhere (Chang *et al.*, 1997; Mendoza-Dominguez, 2000)

The AQM-FDDA procedure makes no assumption on the number of sources that can be used. This is a function of computational resources available and the extent of the observations database. Here two assimilation simulations were conducted, both focused on estimating corrections to the emissions inventory for the second day of the episode. The first one consisted in computing emission corrections factors for domain-wide emissions of mobile, point and stationary NO_x (NO_x = NO + NO_2), and of mobile, stationary and biogenic volatile organic compounds (VOCs). In the second simulation, emissions of individual VOCs for which observations were available were treated separately (e.g.

isoprene was an observed species that is treated individually in the emissions inventory). In this case, the assimilation handled 24 sources: NO_x from anthropogenic surface sources (mobile and stationary combined), NO_x from point sources, 20 individual VOC species, other non-observed anthropogenic VOCs (as a lumped class) and other non-observed biogenic VOCs.

As emissions are corrected by the FDDA module the sensitivity coefficients change due to variations in the relative amounts of VOCs and NO_x present in different parts of the modeling domain. Thus, the emission corrections from an AQM-FDDA run serve as the input to subsequent model runs, and iterations are stopped until convergence in the estimated scaling factors is obtained. This process yields new strengths for all sources involved. Of particular interest are the corrections applied to VOC emissions and the possible changes in ambient VOC reactivity generated by the change in the emissions inventory suggested by the FDDA method. This was investigated computing absolute and normalized peak and spatial exposure reactivities (Bergin *et al.*, 1995), and comparing them to the Maximum Incremental Reactivity (MIR) scale (Carter, 1994).

RESULTS AND DISCUSSION

Table 1 presents the scaling factors computed for domain-wide emissions of NO_x and VOCs. The FDDA results imply that NO_x emissions from stationary sources need to be decreased by almost half their base case value. However, the uncertainty in this calculation is high as the $\pm$ 1σ value of 0.49 suggests. Correction estimates for the other two NO_x sources appear to be more reliable, and in both cases emissions tend to be increased. These results illustrate that the emission changes suggested by the FDDA approach can have limitations. In the particular case of NO_x sources, observations of nitrogen-containing species are limited in the modeling domain: only two urban sites measured NO_2 and peroxyacetyl nitrate (PAN), and three rural sites measured NO_y (reactive odd nitrogen). Even though corrections for NO_x emissions are influenced by all 27 observed species present in the observations database (including O_3, NO_x, NO_y, NO, NO_2, PAN, total organic carbon, and 20 individual VOC species), the sensitivity of nitrogen-containing species to NO_x emissions are the strongest. Thus, in this case a poorly spatially resolved observations field is influencing the FDDA results. The impact is more severe for stationary sources in comparison to mobile sources because they are more evenly distributed around the domain, while mobile source are concentrated in the central part of the domain where observations of nitrogen-containing species are more abundant. VOC scaling factors for anthropogenic sources appear to be more consistent with each other. The increase by a factor of two for the anthropogenic VOC emissions is in agreement with other studies (e.g. Harley *et al.*, 1993) where VOC emissions were suspect of being underestimated. Biogenic VOCs were practically unchanged.

Table 1. Domain-wide base case emission scaling factors ($\pm$ 1σ).

Emissions Category	Scaling factor
Anthropogenic NO_x area source	0.53 ± 0.49
Anthropogenic NO_x mobile source	1.45 ± 0.14
Anthropogenic VOC area source	2.02 ± 0.03
Anthropogenic VOC mobile source	2.36 ± 0.04
Biogenic VOC emissions	1.13 ± 0.06
Anthropogenic NO_x point source	1.61 ± 0.08

Further detail in the evaluation of the VOC base case emissions inventory was accomplished calculating emission corrections for individual VOC species. This tests possible biases in the speciation of VOCs. Figure 1 presents the estimated scaling factors for all anthropogenic (mobile + stationary) VOC emissions. Eighteen specific VOC species are present, plus an additional VOC(A) class which represents all other VOCs that were not present in the observations database (e.g. alcohols, ketones, other olefins and parafins, etc.). Fourteen sources were increased from their base value, and overall the anthropogenic VOC mass emitted increased by a factor of about 1.75. Biogenic VOC emissions were separated into three categories: isoprene, α-(and β-) pinene and VOC(B) (other biogenic VOC emissions). Isoprene and the pinenes were scaled upwards less than 5%, while the VOC(B) class was scaled by 40%. Overall, biogenic VOCs mass emitted increased by 10%. Area source (mobile + stationary) anthropogenic NO_x emissions for this run remained at their base case level (scaling factor of 0.98 ± 0.04, while point source NO_x emission correction was in accordance to the results presented in Table 1 (scaled upwards by 54% ± 12%).

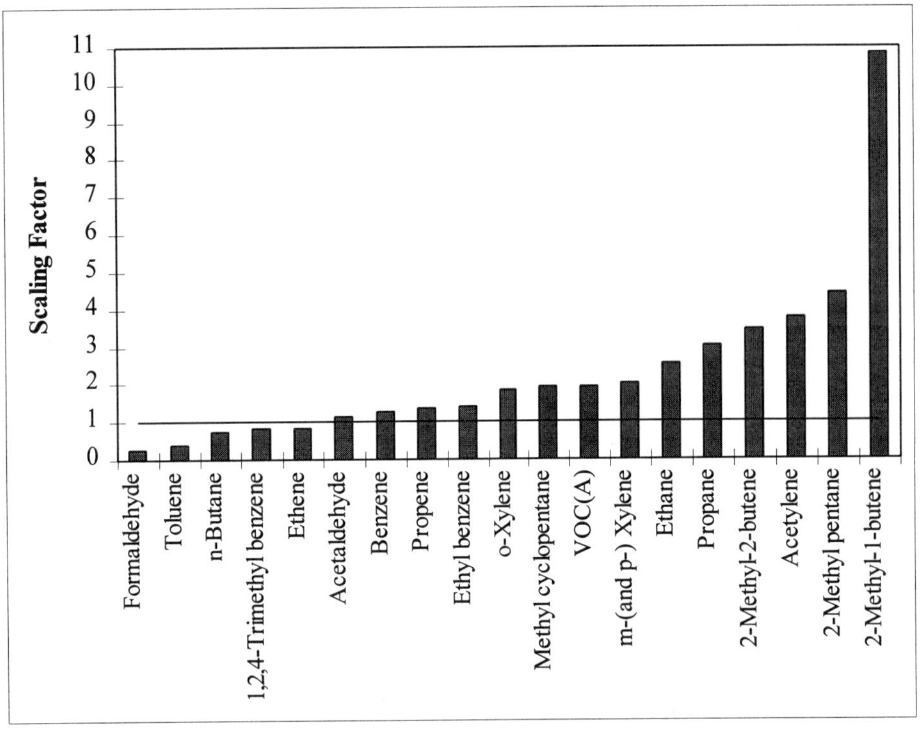

Figure 1. Emission scaling factors for specific anthropogenic VOC species with respect to the base case emissions inventory.

Ozone performance for the second day of simulation for the base case and using the corrections for the assimilation simulation that derived the results from Figure 1 is presented in Table 2. Overall, domain-wide ozone statistics changed little from one run to the other, while peak ozone performance improved using the FDDA-derived emissions. Even though the ozone performance was acceptable using the base case inventory, ozone fields were computed using an emissions inventory that was likely too low in anthropogenic VOC emissions, and possibly low n point source NO_x. While the ozone field had little variation when altering emissions, sensitivities do change and this has direct implications on control strategies that might be suggested from the AQM simulation. For example, Table 3 lists ozone sensitivities before and after an assimilation run. FDDA emission corrections are those presented in Table 1. Sensitivities are provided for the time and location of the observed ozone peak and as a daily spatial exposure. Spatial exposure sensitivity for source *j* was computed as the sum of the sensitivities of source *j* over all grid cells and over all hours (for the second day of simulation) where simulated ozone concentration was greater or equal to 40 ppb. These results indicate, for example, that a reduction of mobile VOC emissions by one metric ton will have less impact in the domain-wide ozone production than what is expected from the base case scenario if the FDDA results are used.

Table 2. Ozone model performance statistics for the second day of simulation.

Statistics	Base case	FDDA
Normalized bias, %	10.6	12.5
Mean bias, ppb	5.4	6.9
Normalized error, %	20.1	21.3
Normalized Mean Square Error (NMSE)	0.05	0.06
Root Mean Square Error (RMSE), ppb	17.4	18.0
RMSEs (systematic RMSE), ppb	10.8	11.1
RMSEu (unsystematic RMSE), ppb	13.6	14.2
Daily Index of Agreement	0.95	0.94
Peak estimation accuracy (%)		
Paired in space and time	-18.5	-14.9
Paired in space	-18.2	-12.1
Paired in time	-9.2	-2.6
Unpaired	-4.3	2.9

Table 3. Peak and exposure ozone sensitivity comparison between the AQM run using the base case emissions inventory and the FDDA-corrected inventory.

	Peak ozone[1]		Spatial exposure by percent[2]		Spatial exposure by mass[3]	
Source[4]	Base case	FDDA	Base case	FDDA	Base case	FDDA
NO_x(A)	5.9×10^{-2}	4.7×10^{-2}	0.577	0.440	0.596	0.845
NO_x(M)	4.5×10^{-2}	1.1×10^{-1}	0.910	1.227	0.516	0.480
VOC(A)	4.0×10^{-2}	5.1×10^{-2}	0.170	0.246	0.085	0.061
VOC(M)	3.9×10^{-2}	6.1×10^{-2}	0.175	0.305	0.098	0.073
VOC(B)	1.5×10^{-1}	1.2×10^{-1}	1.553	1.523	0.088	0.077
NO_x(P)	4.5×10^{-4}	5.3×10^{-4}	0.064	0.112	0.016	0.018

[1] Sensitivity at the time and location of the observed ozone peak (ppb/% increase in emissions).
[2] Spatial exposure sensitivity by one percent increment in emissions (ppm/% increase in emissions-day).
[3] Spatial exposure sensitivity by unit mass increment in emissions (ppm/metric ton-day).
[4] A: stationary; M: mobile; B: biogenic; P: point.

The results in Table 3 indicate that the absolute ozone-forming potential (reactivity) of the VOC emissions before and after the application of the AQM-FDDA approach varies.

Now, it is of interest if the relative reactivity also changes or remains approximately the same. This again has implications on control strategy design (Russell *et al.*, 1995). This point was investigated comparing the normalized reactivities (NRs) of the emissions of specific VOC species given by the base case inventory and the scaled emissions provided by the AQM-FDDA run that led to the results presented in Figure 1. The computed NRs are illustrated in Figure 2, and are presented in a similar way as the sensitivities of Table 3. Spatial threshold exposure NR of species *j* refers to the ratio of the spatial exposure sensitivity of source *j* and the spatial exposure sensitivity of a composite VOC mixture (Bergin *et al.*, 1995). For this calculation, the spatial exposure sensitivities were converted to units of ppm of O_3 per ppmC of species *j* and the threshold was set to 90 ppb of O_3. (NRs for a threshold of 40 ppb of O_3 were also computed but no major difference was observed in the resulting values and trends). The normalizing composite was taken as the average composition of the mobile VOC emissions. Peak normalizing reactivity was simply computed as the ratio between the ozone sensitivity at the time and location of the simulated ozone peak (in a O_3 ppm/ppbC basis) and the normalizing composite. NRs based on the MIR scale of Carter (1994) are also included in Figure 3 in order to compare the AQM-FDDA derived values with box model results.

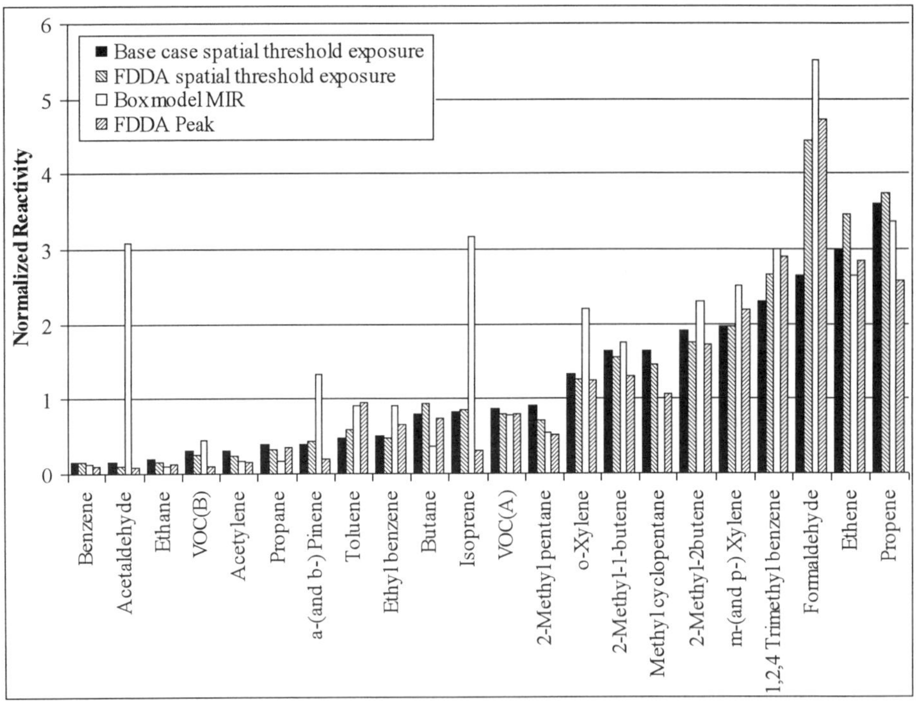

Figure 2. Comparison of normalized reactivities obtained from AQM simulations using the base case and FDDA-derived emissions inventory, as well as MIRs computed from Carter (1994).

Figure 2 indicates that the relative reactivity scale (from spatial exposure sensitivities) computed from the base case emissions inventory is very similar to the one calculated from the AQM-FDDA results. The only major differences are in the NRs for formaldehyde. However, previous analyses have indicated discrepancies with the formaldehyde predictions of AQM simulations in the Atlanta domain (Sillman et al., 1995) that might be clouding the NR derived for the base case inventory. When comparing the spatial exposure NRs to the MIR-NRs, some differences are evident. In general, species that tend to form PAN and PAN-like compounds (e.g. acetaldehyde, isoprene, ethylbenzene, etc.) occupy a lower position in the AQM relative reactivity scale than the one predicted by the MIR scale. This is due to the chemical conditions in the atmosphere of the Atlanta area that tend to conduct to higher levels of organo-nitrates. It is in some instances also customary to calculate peak NR and compare these to the exposure ones. Peak NRs from the FDDA application are also presented in Figure 2, and in general compare well to the exposure NRs.

CONCLUSIONS

Emissions inventories used in typical three-dimensional air quality model applications have significant uncertainties, and in many cases suffer major deficiencies in correctly approximating real source strengths and compositions. A four-dimensional data assimilation method for air quality models was used to suggest emission correction factors to the base case inventory used in modeling Atlanta's August 9-10, 1992 ozone episode. Total VOC mass emissions from anthropogenic sources in the base case inventory are suspect of being underestimated by about a factor of two, while area source (stationary and mobile source combined) anthropogenic NO_x emissions appear to be well estimated in the base case inventory. Point source NO_x was estimated to be underpredicted by about 50 to 60%. Finally, biogenic VOC emission corrections were less that 10%. Limitations in the procedure due to a poorly resolved observations field were evident while trying to compute correction factors for NO_x from stationary and mobile sources (separately). Corrections for mobile-source NO_x were more certain than for area-source due to the concentration of nitrogen-containing species measurements around the city's downtown area, precisely where the mobile sources are more abundant. Further analysis of the VOC emissions indicates that the speciation profiles used to partition the total anthropogenic VOC mass need to be revised. The AQM-FDDA approach provides not only information of the strength, uncertainty and bias of the base case emissions inventory, but also provides revised information for source contribution analyses and control strategies design. Here it was demonstrated that the ozone response to a similar mass reduction of emissions is different if the base case or corrected inventory is used. However, on a relative scale, VOCs reactivities are very similar no matter what inventory is used.

ACKNOWLEDGMENTS

This work was supported by the National Science Foundation (contract no. BES-9613729) and Georgia Power. A. Mendoza-Dominguez acknowledges the *Consejo Nacional de Ciencia y Tecnologia* (Mexico) for partial financial support during his research stay at Georgia Tech. The authors thank M.E. Chang for providing the UAM-IV input files used in this work.

REFERENCES

Aldrin, M. (1997) Length modified ridge regression, *Computational Statistics & Data Analysis*. 28:377.

Bergin, M.S., Russell, A.G., Milford, J.B. (1995) Quantification of individual VOC reactivity using a chemically detailed, three-dimensional photochemical model, *Environ. Sci. Technol*. 29:3029.

Carter, W.P.L. (1994) Development of ozone reactivity scales for volatile organic compounds, *J. Air & Waste Manage. Assoc*. 44:881.

Chang, M.E., Hartley, D.E., Cardelino, C., Haas-Laursen, D., Chang, W.-L. (1997) On using inverse methods for resolving emissions with large spatial inhomogeneities, *J. Geophys. Res.* 102(D13), 16023-16036.

Draper, N.R., Van Nostrand, R.C. (1979) Ridge regression and James-Stein estimation: Review and comments, *Technometrics*. 21:451.

Harley, R.A., Russell, A.G., McRae, G.J., Cass, G.R., Seinfeld, J.H. (1993) Photochemical modeling of the Southern California Air Quality Study, *Environ. Sci. Technol.* 27:378.

Hoerl, A.E., Kennard, R.W. (1976) Ridge regression: Iterative estimation of the biasing parameter, *Commun. Statis.-Theor. Meth.* A5(1):77.

Lyons, W.A., Tremback, C.J., Pielke, R.A. (1995) Applications of the Regional Atmospheric Modeling System (RAMS) to provide input to photochemical grid models for the Lake Michigan Ozone Study (LMOS), *J. Appl. Meteor.* 34:1762.

Mendoza-Dominguez, A., Russell, A.G. (2000) An iterative Four Dimensional Data Assimilation method for photochemical air quality modeling based on inverse modeling and direct sensitivity analysis, *Environ. Sci. Technol.* Submitted.

Mueller, S.F., Song, A., Norris, W.B., Gupta, S., McNider, R. (1996) Modeling pollutant transport during high-ozone episodes in the Southern Appalachian Mountains, *J. Appl. Meteor.* 35:2105.

National Research Council (1991) *Rethinking the ozone problem in urban and regional air pollution*, National Academy Press, Washington D.C.

Russell, A.G., McCue, K.F., Cass, G.R. (1988) Mathematical modeling of the formation of nitrogen-containing air pollutants. 1. Evaluation of an Eulerian Photochemical Model. *Environ. Sci. Technol.* 22:263.

Russell, A., Milford, J., Bergin, M.S., McBride, S., McNair, L., Yang, Y., Stockwell, W.R., Croes, B. (1995) Urban ozone control and atmospheric reactivity of organic gases, *Science*. 269:491.

Sillman, S., Al-Wali, K.I., Marsik, F.J., Nowacki, P., Samson, P., Rodgers, M.O., Garland, L.J., Martinez, J.E., Stoneking, C., Imhoff, R., Lee, J.H., Newman, L., Weinstein-Lloyd, J., Aneja, V.P. (1995) Photochemistry of ozone formation in Atlanta, GA-Models and measurements, *Atmospheric Environment*. 29:3055.

Stauffer, D.R., Seaman, N.L. (1990) Use of four-dimensional data assimilation in a limited-area mesoscale model, *Mon. Wea. Rev.* 118:1250.

Seaman, N.L. (1992) Meteorological modeling applied to regional air-quality studies using four-dimensional data assimilation, in: *Environmental Modelling*, P. Melli and P. Zannetti, eds., Computational Mechanics Publications, Southampton.

Stauffer, D.R., Seaman, N.L. (1994) Multiscale four-dimensional data assimilation, *J. Appl. Meteor.* 33:416.

Seaman, N.L., Stauffer, D.R., Lario-Gibbs, A.M. (1995) A multiscale four-dimensional data assimilation system applied in the San Joaquin Valley during SARMAP. Part I: Modeling design and basic performance characteristics, *J. Appl. Meteor.* 34:1739.

Yamada, T., Kao, C.-Y. J., Bunker, S. (1989) Airflow and air quality simulations over the western mountainous region with a four-dimensional data assimilation technique, *Atmospheric Environment.* 23:539.

Yang, Y.-J., Wilkinson, J.G., Russell, A.G. (1997) Fast, direct sensitivity analysis of multidimensional photochemical models, *Environ. Sci. Technol.* 31:2859.

DISCUSSION

R. BORNSTEIN: What are the implications, as the method actually produced increased biases?

A. MENDOZA-DOMINGUEZ: The air quality model run that used the adjusted emissions gave a larger bias for ozone model performance compared to the bias obtained using the base case emissions inventory. This implies that adjusting emissions is not enough to bring correspondence between measurements and predictions. Here we have assumed that emissions are the sole source of this discrepancy, however we also acknowledge that other inputs and model parameters introduce errors in the calculation of simulated concentrations. Taking these other processes (e.g. meteorology, chemistry, etc.) in account is expected to give a better agreement between observations and predictions.

M. BOZNAR: Have you done any verification of the dispersion model before you have used it in your procedure?

A. MENDOZA-DOMINGUEZ: The CIT model has undergone extensive verification and diagnostic evaluation. The model uses state-of-the-science descriptions for the chemical mechanism, transport schemes, dry deposition and numerical routines. The model has been applied to several urban (e.g. Los Angeles, Atlanta, Mexico City, Athens) and regional (e.g. Northeast United States, Southeast United States, Swiss Plateau, Mexico-U.S. border area) domains with very different characteristics. This has provided the opportunity to use a variety of databases to evaluate the model.

R. SAN JOSE: Do you have any reasons why biogenic emissions were not so sensitive to your technique? Also, would it be possible to improve somehow the quality of biogenic emissions by using your technique?

A. MENDOZA-DOMINGUEZ: The reason why biogenic emissions were not adjusted as much is that the model used to obtain those emissions (U.S. EPA's BEIS2) has been extensively used and tested against data collected in the Atlanta area. A high degree of confidence existed, before we applied our FDDA method, that biogenic emissions were well characterized in this area. Our results proved valid that expectation.

During the presentation I did not addressed the level of resolution that one could obtain for adjustments in biogenic emissions; the paper discusses this further. In essence, one can obtain a scaling factor for the entire biogenic VOC emissions, or obtain separate scaling factors for different species (e.g., isoprene, α-(and β-) pinene, and other biogenic VOCs). One could also address the issue of adjusting for different subdomains. This is an issue that we are currently investigating. However, one has to acknowledge that greater refinement in the adjustments implies greater amount of observational data needed that in some cases might not be available.

G. CARMICHAEL: The modifications of the biogenic hydrocarbon emissions were small. This is somewhat surprising to me and raises the question of how robust these results are- i.e., will the modified emissions vary greatly from episode to episode?

A. MENDOZA-DOMINGUEZ: Again, the biogenic emissions inventory, at least for the Atlanta area, has been subject of extensive research. The biogenic emissions model used has been tested against actual measurements in Atlanta, making the biogenic inventory a highly reliable one for this area. So, it is really not that surprising that the biogenic hydrocarbon emissions were not changed that much. In fact, the expectation was that they should not be adjusted a lot. However, we agree that for other areas it might be the case that the biogenic inventory could be very uncertain, but we are confident that our technique is robust enough to identify possible flaws in the emission estimates in those areas (as done here for the anthropogenic emissions).

ADAPTIVE GRIDS IN AIR POLLUTION MODELING: TOWARDS AN OPERATIONAL MODEL

M. Talat Odman[1], Maudood N. Khan[1], and D. Scott McRae[2]

[1] School of Civil and Environmental Engineering
Georgia Institute of Technology
Atlanta, GA 30332-0512, U.S.A.
[2] Department of Mechanical and Aerospace Engineering
North Carolina State University
Raleigh, NC 27695-7910, U.S.A.

INTRODUCTION

The uncertainty in the predictions of air quality models (AQMs) is attributed to various sources. Inadequate resolution of the numerical grid can be an important source of this uncertainty. A large grid size could lead to significant errors in the concentrations of pollutant species, especially those that are formed via non-linear chemical reactions. To address this issue, nested and multi-scale modeling techniques have been developed (Odman and Russell, 1991; Odman et al., 1997). These techniques use finer grids in areas that are presumed to be of interest (e.g., cities) and coarser grids elsewhere (e.g., rural locations). Limitations include loss in accuracy due to grid interface problems and inability to adjust to dynamic changes in resolution requirements. Adaptive grids are not subject to such limitations and do not require a priori knowledge of where to place finer resolution. Through a process called grid clustering, they allocate finer resolution over areas of interest, automatically. Adaptive grid AQMs can capture the details of atmospheric dynamics and chemistry more efficiently than their fixed grid counterparts. The vicinities of emission sources are of particular interest with their large pollutant concentration gradients and rapid chemical kinetics. Dynamic adaptation to such regions assures optimal use of computational resources at all times during the simulation.

In this paper, we report on the status of the development of an operational, regional scale, adaptive-grid AQM. After a brief description of the adaptive grid algorithm, we discuss its performance in simplified test cases that emphasize the treatment of urban and point source plumes. We compare both its accuracy and efficiency to classical fixed grid models. Finally, we discuss the development of more efficient algorithms for the processing of emissions and meteorological data.

MODEL FORMULATION AND ADAPTIVE GRID METHODOLOGY

In this section, the critical parts of the adaptive grid AQM formulation will be highlighted. The adaptive grid methodology is based on the Dynamic Solution Adaptive Grid Algorithm (DSAGA) of Benson and McRae (1991), which was developed for aerospace applications. DSAGA employs a constant number of grid nodes that, in two-dimensional space, partition a rectangular domain into N by M quadrilateral grid cells. The nodes move throughout the simulation but the grid structure remains the same. In other words, each interior node is still connected to the same four nodes (and boundary nodes to the same two or three nodes), but the length of the links and the area of the grid cells change. The movement of the nodes is controlled by a weight function whose value is proportional to the error in the solution. The nodes are clustered around regions where the weight function bears large values, thereby increasing the resolution where the error is large. Since the number of nodes is fixed, refinement of grid scales in regions of high interest is accompanied by coarsening in other regions where the weight function has smaller values. This yields a continuous multiscale grid where the scales change gradually. Unlike nested grids, there are no grid interfaces, which may introduce numerous difficulties due to the discontinuity of grid scales. The availability of computational resources determines the number of grid nodes that can be afforded in any AQM. By clustering grid nodes automatically in regions of interest, DSAGA makes optimal use of computational resources throughout the simulation.

The time integration of the atmospheric diffusion equation on a moving grid is performed as a two-step operation. First, the equations are solved on a fixed grid. In the second step, the grid nodes are moved and the solution is redistributed to the new locations of the nodes.

Fixed Grid Solution of Atmospheric Diffusion Equations

This step is very similar to the solution of the equations in fixed grid AQMs (e.g., Odman and Ingram, 1996; Byun et al., 1999). The equations are written in generalized coordinates $x^i, i = 1,2,3$, as:

$$\frac{\partial(Jc_n)}{\partial t} + \frac{\partial\left(Jv^i c_n\right)}{\partial x^i} + \frac{\partial}{\partial x^i}\left(JK^{ii}\frac{\partial c_n}{\partial x^i}\right) = JR_n + JS_n \tag{1}$$

where c_n is the concentration of species *n*, v^i are the contravariant velocities or components of the wind vector in each x^i direction, K^{ii} is the contravariant turbulent diffusivity tensor assumed to be diagonal, R_n and S_n are the chemical reaction and emission source terms for species *n*, and *J* is the Jacobian determinant of a series of coordinate transformations. First, there are transformations due to the use of a conformal map projection in the horizontal and a terrain-following σ coordinate in the vertical. Then, there is one additional transformation in the horizontal plane:

$$\xi = \xi(x, y); \eta = \eta(x, y) \tag{2}$$

This transforms the adaptive grid that is non-uniform in the (x, y) space of map coordinates to a uniform grid in the (ξ, η) space. At this point, all the solution algorithms of existing AQMs that assume uniform grid spacing become usable to solve the transformed equations:

$$\frac{\partial(Jc_n)}{\partial t}+\frac{\partial(Jv^{\xi}c_n)}{\partial \xi}+\frac{\partial(Jv^{\eta}c_n)}{\partial \eta}+\frac{\partial(Jv^{\sigma}c_n)}{\partial \sigma}+\frac{\partial}{\partial \xi}\left(JK^{\xi\xi}\frac{\partial c_n}{\partial \xi}\right)$$
$$+\frac{\partial}{\partial \eta}\left(JK^{\eta\eta}\frac{\partial c_n}{\partial \eta}\right)+\frac{\partial}{\partial \sigma}\left(JK^{\sigma\sigma}\frac{\partial c_n}{\partial \sigma}\right)=JR_n+JS_n \quad (3)$$

where

$$J=\frac{1}{m^2}\frac{\partial z}{\partial \sigma}\left(\frac{\partial x}{\partial \xi}\frac{\partial y}{\partial \eta}-\frac{\partial y}{\partial \xi}\frac{\partial x}{\partial \eta}\right) \quad (4)$$

and m is the map scale factor. Due to space limitation, the expressions for the contravariant velocities and the diffusivity tensor are omitted here.

Note that we have restricted the movement of the grid nodes to the horizontal plane. Adaptation in the vertical direction is also possible but significantly more complicated. Assuming there is already sufficient vertical resolution in current AQMs, which use 15 to 30 unequally spaced vertical layers, we did not attempt to use adaptations in the vertical. One limitation is that the horizontal grid would be the same in each vertical layer. Conceptually, there may be a need for clustering nodes in different regions across vertical layers. This issue is left for future research.

Grid Movement and Solution Redistribution

After the solution is obtained over the fixed grid, the grid nodes are moved using a weight function along with a center-of-mass repositioning scheme. The weight function $w(\xi,\eta,t)$ must be such that its value is large in regions where grid nodes need to be clustered in order to achieve a more accurate solution. To assure an accurate solution in the next time step, the resulting grid must be free of highly sheared or skewed cells, and there must be a smooth transition from small to large cells with no voids in regions where the solution is relatively uniform. Laflin and McRae (1996) developed a weight function that satisfies these requirements and is very easy to compute. We use a similar weight function defined as:

$$w=A^{1+e}\sum_n \alpha_n \nabla^2 c_n + w_{\min} \quad (5)$$

where ∇^2 is a discrete approximation to the Laplacian, which represents the error between the computed value of c_n and the value obtained from the interpolation of c_n values in the neighboring cells. A relatively small value of $\nabla^2 c_n$ for any cell indicates that the grid can support relatively accurate interpolations of the solution in the neighborhood of that cell. The parameter e controls weighting with respect to the cell area A (a positive value gives more weight to larger cells) and promotes the generation of an orthogonal (not skewed) and smooth grid. A minimum weight, $w_{\min}$, inhibits evacuation of grid nodes from regions of uniform concentration. Note that w is a linear combination of the errors in the concentrations of various chemical species. The mechanisms used in AQMs to represent the chemistry of the atmosphere usually have a large number of species. Each species may have very different resolution requirements. This is, in part, due to the non-homogeneous distribution of emissions and disparate residence times in the atmosphere. Determining α_n, or how each species would contribute to the weight function, such that ozone concentrations can be estimated most accurately is one of the objectives of this research. Initially, we will give the largest weight to NO_X species. Recall that the horizontal grid will be the same for all vertical layers. We choose to use the surface layer concentrations in the weight function.

Repositioning of the grid nodes is accomplished by a *center-of-mass* scheme. In this scheme, the weight function is viewed as a mass distribution. Consider a new cell definition whose vertices are the original grid cell centers and centroid is the grid node itself. The new position of the grid node, $\vec{P}^{new}$, is defined as the center-of-mass of this newly defined cell:

$$\vec{P}^{\,new} = \frac{\sum_{i=1}^{4} w_i \, \vec{P}_i}{\sum_{i=1}^{4} w_i} \tag{6}$$

where $\vec{P}_i$ are the position vectors of the original cell centers and w_i are the values of the weight function at those locations.

After repositioning of grid nodes, the solution must be redistributed to the new grid through interpolation. For high-order accurate and non-oscillatory interpolation, we use the advection-interpolation equivalence (Alapaty et al., 1998) and the monotonic advection scheme of Collela and Woodward (1984) known as the *Piecewise Parabolic Method* (PPM). This latter method is also used for the solution of the advection terms in Equation (3).

Finally, a check is made for grid node movement. If the maximum movement of all grid-nodes is below a preset tolerance, the grid is considered to have resolved the solution sufficiently. Otherwise, the weight function is recomputed and the grid adaptation and solution redistribution procedure is reiterated. Upon meeting the movement tolerance, Equation (3) can be integrated to advance the solution through the next time step. However, note that repositioning the nodes leads to a non-uniform grid in the (ξ,η) space. Reestablishing a uniform grid requires computation of a new transformation. In other words, the metric terms, i.e., the Jacobian and the contravariant terms, must be recalculated. In addition, since the locations of the grid nodes in the physical domain have changed, meteorological variables and emissions inputs must be re-gridded. Another objective of this research is to develop efficient algorithms to reduce the overhead involved in all these operations.

PRELIMINARY APPLICATIONS

The adaptive grid algorithm was applied to problems with increasing complexity and relevance to air quality modeling. First, it was applied to pure advection tests (Srivastava et al., 2000). In a rotating cone test where the PPM was used for advection, the adaptive grid solution was more accurate than the fixed uniform grid with the same number of grid nodes. The error in maintaining the peak was only 13% compared to 39% with the fixed grid: an accuracy that could only be achieved by using 22 times more grid nodes with a fixed uniform grid. In a second test, four cones were rotated to measure the performance of the algorithm in following multiple features. Once again, the peak accuracy was better than the fixed uniform grid alternative. The peak error increased to 23% since the grid nodes, whose number was the same as in the previous test, were being clustered around four cones instead of one. Srivastava et al. (2000) conducted a third test with concentric conical puffs of NO_X and VOCs reacting in a rotational wind field. The parameters of this problem are such that, after a certain time, ozone levels drop below the background near the base of the conical puffs but they peak near the vertex of the cones. This feature was resolved by the adaptive grid solution while it was completely missed by the uniform fixed grid solution. To resolve this feature with the fixed grid would require approximately 22 times more grid nodes. The overhead involved in adaptive grid computations is much smaller than the time spent in computing the chemical kinetics for these additional nodes, overall, the adaptive grid algorithm is about 10 times more efficient than the uniform grid alternative for the same level of accuracy.

Srivastava (1998) reported two more tests: one simulating a power plant plume and another with multiple sources. A two dimensional plume with a VOC/NO_x emission ratio of 14% was advected with uniform winds and diffused over a background with a VOC/NO_x ratio of 35. Other parameters were chosen to make the dispersion as realistic as possible. After about 12 hours of simulation, the composition of the plume was analyzed taking cross sections at various downwind distances. At 10 km downwind, the adaptive grid solution showed a NO_x rich, but ozone deficient core. This feature was completely missing in the uniform grid solution, which artificially diffused the NO_x and displayed highest ozone levels at the core of the plume. The adaptive grid, on the other hand, had ozone bulges developing near the plume edges. At a downwind distance of 30 km, these bulges continued to grow as NO_x diffused slowly from the core to the edges (at a rate more in line with physical diffusion) and radicals were entrained into the plume. This plume structure started disappearing after about 80 km. At a downwind distance of 135 km, the plume was fully *matured* with an ozone peak at the center. The peak ozone concentration was larger than one predicted by the fixed uniform grid. A similar evolution of the plume was observed in the fixed uniform grid solution when the number of grid nodes was increased by a factor of 9. However, this solution was about 5 times more expensive than the adaptive grid solution.

In the multiple source test, a city is placed at the center of a domain with diagonal winds. The city consists of an inner circular source representing urban conditions and an outer circle representing suburban conditions. The VOC/NO_x emission ratios are 5 and 9, respectively. There are two identical point sources with a VOC/NO_x emission ratio of 14%:

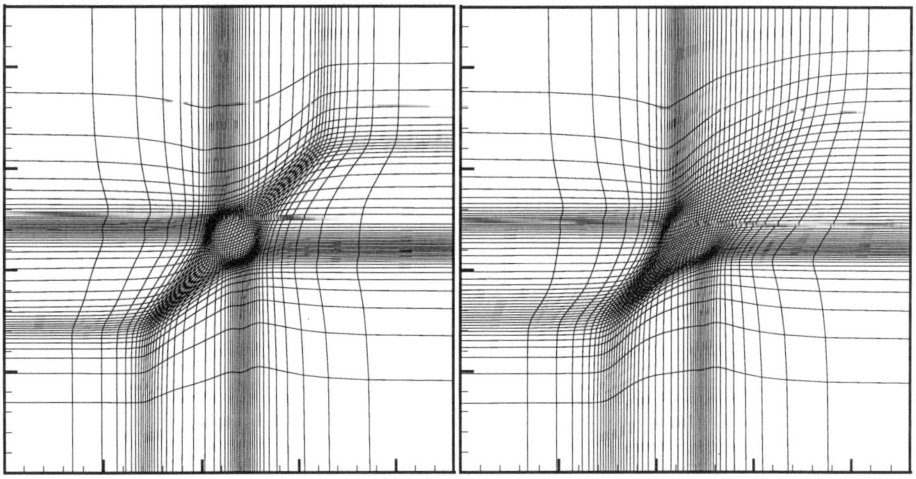

Figure 1. Adaptive grid for the multiple source test: initial (left) and after 12 hours of simulation (right). Reproduced from Srivastava, 1998.

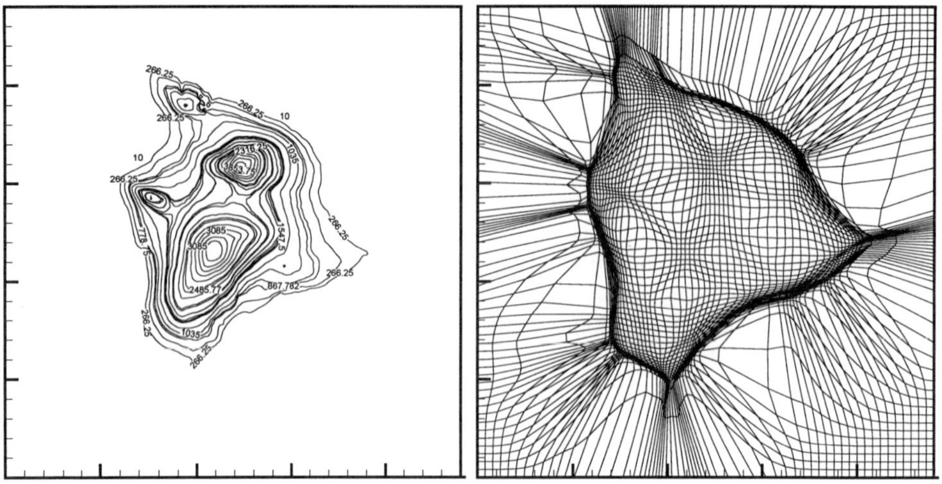

Figure 2. Surface elevation for the island of Hawaii (left) and the grid adapted to the terrain features (right).

one is located upwind and the other downwind from the city. The background has a VOC/NO_x ratio of 10. Figure 1 shows the adaptive grid at the beginning of the simulation and 12 hours later. The locations of the sources are clearly visible in the initial grid. Note that the grid nodes are clustered around the point sources and at the transitions from the background to the outer suburban circle and from the suburban to the urban circles. Once the simulation started, the nodes continued to cluster around the upwind source but moved away from the downwind source towards the upwind end of the city. This behavior was mostly due to a more uniform ozone field downwind of the city.

Finally, in order to evaluate the performance of the adaptive grid algorithm in resolving complex features, we conducted a test where we used the elevation data for the island of Hawaii. Figure 2 shows the terrain features of this geographic region at 1-km resolution. This elevation data was mapped on an initially uniform, 8-km resolution grid. Then, this grid was adapted to capture the details of the terrain. As shown in Figure 2 there is good agreement between the regions where the grid nodes are clustered and the prominent terrain features. This result suggests that the adaptive grid methodology described here can also capture the complexity of pollutant concentration fields in the atmosphere.

AIR QUALITY MODEL DEVELOPMENT

Among the most important components of the adaptive grid AQM are the input data processors, especially those for emissions and meteorological data. In fixed grid AQMs, the input data are gridded once, before the simulation. In adaptive grid AQMs on the other hand, the grid nodes move during the simulation. Since this movement is governed by the solution, there is no a priori knowledge of where the nodes would be located at any given time. Therefore, gridding of data must be performed during the simulation, and repeated after each adaptation step. Data processing must be efficient not to slow the computations.

Processing of Emissions

Emission processing makes extensive use of geographic information systems (GISs). The gridding of emission sources is known as the *intersection* problem in GIS. Most intersection algorithms are developed for general applications and they are not efficient enough to be used in the adaptive grid AQM. We have developed new algorithms for the specific problem at hand. An important element of these algorithms is that they take advantage of the topology and structure of the grid. Recall that although the locations of the grid nodes change after each adaption, the grid structure is maintained. In other words, a grid cell will have the same neighboring cells before and after adaptation. This property yields significant savings during the search and intersection operations.

Algorithms were developed for the processing of point, mobile, and area source emissions. A point source may not lie in the same grid cell before and after adaptation. Thus, the grid cell containing each point source must be re-identified. However, since the source would be located in close proximity to the grid cell where it was located before, the search is started from that cell and continues with its neighbors and the neighbors' neighbors and so on. The local nature of this search provides significant savings over more general, global searches.

The problem is more complex for mobile and area sources where the fractional contribution of sources in each grid cell needs to be computed. Mobile sources are identified as lines with emissions specified in mass per unit length. A line source can lie wholly or partially within a grid cell. Intersecting the line with the grid cell identifies the appropriate case. Once the length of the line that falls within the grid cell is calculated, it can be multiplied with the per-unit-length emissions to yield the contribution of this line source to the cell. Summing the contributions from all line sources that intersect with the grid cell yields the total emissions for the cell. The area source emissions are typically reported over a fixed uniform grid. Each cell on this emission grid represents a fixed emission flux from the geographic area that it covers. Consider an adapting grid cell G and a uniform emissions grid with cells E_i. The intersection of G with each E_i results in polygons S_i. The total mass of emissions in the cell G can be calculated by summing the products of the area of S_i by the emission rate of cell E_i.

Processing of Meteorology

The ideal solution is to have a meteorological model (MM) that runs in parallel with the AQM and operates on the same grid. The weight function used to move the grid nodes can be made a function of pollutant concentrations as well as some essential meteorological parameters such that the solution of both models are accurately resolved. After each adaptation step, the MM would calculate dynamically consistent inputs (i.e., wind components, temperatures, densities, mixing rations, etc.) at the grid node or cell center locations as needed by the AQM. This solution is left for future research. Currently, we are using inputs from a very high resolution fixed grid MM and interpolating to the adaptive grid. The techniques used for interpolation are similar to the intersection algorithms developed for emissions, with some differences. For mass conservation, the vertical wind components are readjusted as described in Odman and Russell (2000). There are also some additional requirements such as momentum and energy conservation.

CONCLUSIONS

We developed an adaptive grid algorithm for use in AQMs. It employs a structured grid where the nodes move throughout the simulation. The movement is controlled by a weight function whose value depends on a linear combination of the errors in various pollutant concentrations. The algorithm generates a continuous, multiscale grid where the scales change gradually, and makes optimal use of computational resources at all times. So far, we evaluated the algorithm in idealized model problems involving dispersion of power plant plumes and chemical reactions. The results are much more accurate than those achieved by uniform fixed grid models that use the same computational resources. The adaptation criterion we used so far gives equal weight to the errors in the concentrations of different pollutant species. We are developing criteria that would be more sensitive to reaction pathways that are more important in the formation of specific secondary pollutants, such as ozone.

The incorporation of the algorithm into an operational air pollution model is under way. The movement of the grid nodes necessitates reprocessing of meteorological and emissions inputs after each adaptation step. While the ideal solution would be to have a meteorological model running in parallel, over the same grid, since there is no such model at present, we are interpolating data from a nested grid model onto the adaptive grid. We developed efficient intersection algorithms to assure proper allocation of point and line source emissions into each grid cell every time the nodes move.

Acknowledgements

This research is supported by the U.S. Environmental Protection Agency under Grant Agreement No. R 827028-01-0 and, in part, by MCNC under Contract No. D96-7050-000. The authors wish to tank Dr. Ravi Srivastava of U.S. Environmental Protection Agency for permitting the use of material from his Ph.D. dissertation.

References

Alapaty, K., Mathur, R., and Odman, T., 1998, Intercomparison of spatial interpolation schemes for use in nested grid models, Mon. Wea. Rev. 126;243.

Benson, R.A, and McRae, D.S., 1991, A Solution adaptive mesh algorithm for dynamic/static refinement of two and three dimensional grids. In: Proceedings of the Third International Conference on Numerical Grid Generation in Computational Field Simulations, Barcelona, Spain, 185.

Byun, D.W., Young, J., and Odman, M.T., 1999, Governing equations and computational structure of the community multiscale air quality (CMAQ) chemical transport model. In: Science Algorithms of the EPA Models-3 Community Multiscale Air Quality (CMAQ) Modeling System, D.W Byun and J.K.S. Ching, (eds.), EPA/600/R-99/030, U.S. EPA, Washington, Chapter 6.

Collela, P., and Woodward, P.R., 1984, The piecewise parabolic method (PPM) for gas-dynamical simulations, J. Comput. Phys. 54; 174.

Laflin, K.R., and McRae, D.S., 1996, Three-dimensional viscous flow computations using near-optimal grid redistribution. In: Proceedings of First AFOSR Conference on Dynamic Motion CFD, Rutgers University, New Jersey.

Odman, M.T., and Russell, A.G., 1991, A multiscale finite element pollutant transport scheme for urban and regional modeling, Atmospheric Environment 25A; 2385.

Odman, M.T, Ingram, C., 1996, Multiscale Air Quality Simulation Platform (MAQSIP): Source Code Documentation and Validation, MCNC Technical Report ENV-96TR002, Research Triangle Park, North Carolina.

Odman, M.T., Mathur, R., Alapaty, K., Srivastava, R.K., McRae, D.S., and Yamartino, R.J., 1997, Nested and adaptive grids for multiscale air quality modeling. In: Next Generation Environmental Models and Computational Methods, G. Delic and M.F. Wheeler (eds.), SIAM, Philadelphia, 59-68.

Odman, M.T, and Russell, A.G., 2000, Mass conservative coupling of non-hydrostatic meteorological models with air quality models. In: Air Pollution Modeling and its Application XIII, S.-E. Gryning and E. Batchvarova (eds.), Kluwer Academic/Plenum Publishers, New York, 651-660.

Srivastava, R.K, 1998, An Adaptive Grid Algorithm for Air Quality Modeling, Ph.D. Dissertation, North Carolina State University, Raleigh, North Carolina.

Srivastava, R.K., McRae, D.S., and Odman M.T., 2000, An adaptive grid algorithm for air quality modeling, J. Comput. Phys. (revision in press)

DISCUSSION

R. BORNSTEIN: 1) Why not have the weights proportional to gradients instead of curvature, e.g., the very large number of grids placed on your Hawaiian coast?

2) What are the advantages of this method over "moveable fine mesh grids"?

T. ODMAN: 1) What makes a pollutant or terrain field complex is not the gradient (i.e., slope) but the change in gradient (i.e., curvature). A field with large but constant gradient can be represented with relatively fewer points than one with large curvature. For example, two points can capture a constant slope, no matter how large. However, more than two points would be necessary to capture any changes in this slope. The adaptive grid algorithm allocates the grid nodes such that the features of the field, whether terrain or pollutant, can be best captured. Therefore, it is better to use curvature as the clustering criteria. The grid-repositioning algorithm is an iterative process and continues until the node movement tolerance is met. In the example with the terrain field of the island of Hawaii, we used an absolute tolerance, which may be too strict leading to more clustering than desired near the coastline. We are planning to use a relative tolerance (i.e., we will stop the iterations when any node movement is less than a fraction of the minimum distance to neighboring nodes). This should enhance the quality of the grid by allocating fewer grid cells to places like the coastline where the curvature is large over a very narrow band.

2) The main advantage of the method over "moveable fine mesh grids" is its simplicity. The grid for this method is structured, i.e., all internal nodes are connected to four neighboring nodes. This results in a very simple data structure in the form of a matrix. The data structures for the moveable mesh grids are considerably more complex. Further, the grid for this method can be mapped onto a uniform grid by using the coordinate transformations described above. At that point, all the solution algorithms (e.g., advection) that exist for the uniform fixed grid AQMs become useable. Finally, this method changes the grid size gradually while the moveable fine mesh methods have an abrupt change in scales and a grid interface that may lead to numerical difficulties.

EFFECTS OF URBAN AND INDUSTRIAL ROUGHNESS OBSTACLES ON MAXIMUM POLLUTANT CONCENTRATIONS

Steven R. Hanna[1] and Rex E. Britter[2]

1. CSI MS 5C3, George Mason Univ., Fairfax, VA 22030-4444, USA
2. Dept. of Eng., Univ. of Cambridge, Cambridge, CB2 1PZ, UK

INTRODUCTION

The objective of this paper is to describe how structures or obstacles (such as buildings and storage tanks) in an urban or industrial site affect transport and dispersion and show how these effects can be parameterized in consequence models. The general point is made that the concentration downstream from a source is primarily influenced by wind speed (advection) and turbulence (diffusion). The urban or industrial structures affect both of these – above the obstacles, within the obstacles, and downstream of the obstacles. In most cases the effects generally enhance the dilution substantially.

The effects of surface features on transport and dispersion can be assessed in two ways – first, by attempting to resolve the flow around the individual surface obstacle, and second, by parameterizing the combined effect of groups of surface obstacles (Robins and Macdonald 1999, Hanna and Britter 2000). The first approach is appropriate for pollutant clouds released near buildings or other obstacles where there is interest in assessing the effects in the immediate neighborhood of the obstacle and the plume dimensions are smaller than the obstacle height. The second approach (the focus of the current paper) is appropriate for clouds that have dispersed to a size greater than the roughness obstacles.

The effects of parameterizations of surface features in dispersion models can be understood through analysis of a simple transport and dispersion model. For a continuous non-buoyant plume with emission rate, Q, released at the ground, the ground-level concentration, C, on the plume centerline predicted by the Gaussian plume formula is:

$$C/Q = (\pi u \sigma_y \sigma_z)^{-1}, \qquad (1)$$

where u is the wind speed and σ_y and σ_z are the lateral and vertical plume spreads. Next the simple assumption can be made that, for nearly-neutral conditions close to the source, both σ_y and σ_z are proportional to u^*t or u^*x/u, where t is travel time from source to distance x (Hanna et al., 1996). The proportionality factors are about 2 and 1.3 for σ_y and σ_z, respectively. The log wind profile law, applicable for neutral stability, is assumed:

Air Pollution Modeling and Its Application XIV, Edited by
Gryning and Schiermeier, Kluwer Academic/Plenum Publishers, New York, 2001

$$u = (u^*/0.4) \ln((z - d)/z_o) \quad (2)$$

Consequently, equation (1) simplifies to:

$$C/Q = (\ln((z\text{-}d)/z_o)) / (1.04\pi u^* x^2) \quad (3)$$

where d is the displacement height and z_o is the surface roughness length. This solution is valid for situations where the plume centroid height h is greater than the obstacle height H_r. For other conditions the same, as the surface roughness length, z_o, is increased in this equation, the numerator will decrease and the denominator will increase, because u* will be larger for a larger z_o. Therefore both terms act to decrease the normalized maximum ground-level concentration, C/Q, as surface roughness increases.

Equation (1) also reveals how it is possible, under certain conditions, for concentrations to not decrease as roughness obstacle heights increase in situations when the plume height is less than the obstacle height. Because the wind speed is located in the denominator of the equation, if the wind speed is markedly decreased down within the obstacles, it may sometimes dominate over the increase in turbulent velocities and cause the predicted C/Q to not decrease. This situation is most likely to happen for dense obstacle packing.

SURVEY OF RELEVANT EXPERIMENTS

There have been several relevant field and wind tunnel experiments concerning the effects of surface roughness obstacles on dispersion, as summarized below. The experiments are divided into two groups depending on whether the plume centroid height, h, is larger than or less than the height of the roughness obstacles, H_r.

The first case of interest concerns plumes with centroid heights larger than H_r, such that the bulk of the plume extends into the surface boundary layer. For these cases, the wind flow and the dispersion processes are explained by standard surface boundary layer theories. There is no need to resolve the flow around individual roughness obstacles and z_o and d are used to characterize the effects of the roughness obstacles.

Full-scale urban tracer experiments, such as the St. Louis tracer data (McElroy and Pooler, 1968), show that, for a given stability class, the dispersion coefficients for urban conditions are consistently larger than the dispersion coefficients for rural conditions. On a smaller scale, but still in the field, the Kit Fox field experiments involved ground-level releases of CO_2 gas within a large array of roughness obstacles (Hanna and Chang, 2000). It was intended that the field experiment be about a 1/10 scale representation of the roughness typical at an oil refinery. Near-ground concentrations were observed on monitoring arcs at downwind distances of 25, 50, 100, and 225 m. The vertical extent of the plume, h, was always observed to be about equal to or larger than H_r. Releases took place over three different surface roughness types, with z_o = 0.0002 m, 0.02 m, and 0.2 m. The maximum normalized concentrations, C/Q, during each of the three roughness types occurred during similar meteorological conditions – light-to-moderate wind stable conditions. A very clear trend was seen in the observed C/Q values, which were observed to decrease by about a factor of four as roughness increased from 0.0002 m to 0.02 m, and decrease by another factor of four as roughness further increased from 0.02 m to 0.2 m. Therefore there was a factor of 15 to 20 decrease in C/Q as roughness increased from 0.0002 to 0.2 m.

There have been some laboratory studies of flow and dispersion over arrays of roughness elements where the plume size exceeds H_r and therefore the elements can be parameterized through z_o. For example, Roberts et al. (1994) presented results that confirm that the concentration would decrease as the roughness increases, for all other conditions the same. Britter (1991) analyzed other laboratory data and concluded that it was appropriate to parameterize the effects of an industrial area on dispersion by means of a surface roughness length, z_o.

The second case of interest concerns plumes with heights less than H_r. Clearly the increased turbulence within the obstacle array will cause the dispersion coefficients, σ_y and σ_z, to increase, thus tending to decrease the maximum concentration, C/Q. It is expected that, for loosely-packed obstacles, where there is less of a decrease in u, the normalized concentration, C/Q, will decrease as the roughness element height increases. But for some configurations of densely-packed obstacles, where there is more of a decrease in u, the normalized concentration, C/Q, may not decrease and may possibly increase under certain scenarios.

There have been some small-scale field experiments with plumes whose centroid height is less than H_r. For example, Macdonald et al. (1997) used movable cubes in the field with dimensions of about 1.1 m to set up a variety of obstacle arrays, with different spacings and groupings in a flat grassy field. A point source plume was released at a height of ½ H_r and at a location about two H_r upwind of the array. Concentrations were observed at various points within the arrays. For comparison purposes, an identical experiment was carried out with no obstacles in place. The results showed little effect of the obstacles on the maximum ground-level concentration at a given downwind position within the obstacle arrays. For increased obstacle density, the plume dispersion increased, thus tending to decrease the concentration, but at the same time, the wind speed decreased, tending to increase the concentration. The two effects roughly cancelled out for this specific scenario involving a plume release 2 H_r upwind of the obstacle array.

Davidson et al. (1995) carried out a similar small-scale field experiment, but with 2 m cube-shaped obstacles, and used a similar source position (about two H_r upwind of the array). They reached similar conclusions – that the maximum concentration at a given downwind position was little affected by changes in the obstacle arrays.

There have been several wind tunnel experiments concerning plume releases within obstacle arrays. For example, Hall et al. (1996) and Davidson et al. (1996) carried out wind tunnel experiments concerning dispersion in regularly spaced obstacle arrays, nearly exactly matching the geometry of the small scale field experiments by these same groups that were discussed above (Macdonald et al., 1997, and Davidson et al., 1995, respectively). Not coincidentally, similar conclusions were reached – the presence of the obstacle arrays has little effect on the maximum ground-level concentration at a given downwind distance for plumes released two H_r upwind of the array at an elevation of ½ H_r.

The conclusions described above for the sets of field and laboratory experiments by Macdonald and Davidson and their colleagues have been influenced by the fact that the source was located upwind of the obstacles. Consequently the plume had already dispersed substantially before it encountered the front of the obstacle array. These experiments are quite different from the Kit Fox field experiments, where the source was at ground level in the middle of the obstacle array. The Kit Fox results have been confirmed by the Macdonald et al. (2000) wind tunnel experiments, which showed a decrease of concentration as roughness increased. Thus, when comparing the results of the various experiments, it is important to compare data from similar types of source scenarios.

Furthermore, all of the small-scale field studies and laboratory studies have made use of obstacles with uniform shapes and sizes. Consequently, these results should be extrapolated with caution to real urban and industrial areas, which consist of a mixture of obstacle shapes and sizes.

ESTIMATING BOUNDARY LAYER PARAMETERS

Earlier, H_r was defined as the average height of the roughness obstacles. Additional important scaling parameters are the ratio of obstacle plan area to ground area, λ_p, and the ratio of obstacle frontal area (as seen by the wind) to the ground area, λ_f (Grimmond and Oke, 1999). If we accept an uncertainty of a factor of two say in z_o then the recent review by Grimmond and Oke (1999) suggests that for λ_p between 0.2 and 0.6 or λ_f between 0.1 and 0.4, typical of urban and industrial areas, the rules of thumb: $z_o/H_r = 0.1$ and $d/H_r = 0.5$, could be assumed. Since the average building/obstacle height, H_r, is 10 m, it follows that $z_o = 0.1*10$ m = 1 m and d = 0.5*10 m = 5 m. The data in the references demonstrate that these simple constants are likely to be within a factor of two or three of the local "observed" values at individual field sites.

For the general situation where the roughness obstacles could have any types of spacing, it may be appropriate to account for the non-linear dependency of z_o and d on λ. For artificial wind tunnel and field experiments with uniform-sized obstacles, the ratio z_o/H_r is linear with λ at small values, reaches a plateau with $z_o/H_r = 0.15$, then decreases as λ approaches 1.0 (i.e., close packing). But in a real urban area or industrial plant, the ratio z_o/H_r does not decrease to zero as λ approaches 1.0, since real obstacles (e.g., buildings and tanks) are not cubes with uniform shapes, but have various individual heights and shapes and therefore have a relatively large effective roughness even when packed closely. Therefore a value of $z_o/H_r = 0.15$ is suggested at large values of λ. The ratio d/H_r increases monotonically from 0.0 to 1.0 as λ increases from 0.0 to 1.0. The following formulas provide reasonable fits to the data over the entire range of roughness obstacle spacing:

$$z_o/H_r = \lambda_f \qquad \text{for } \lambda_f < 0.15 \qquad (4a)$$
$$z_o/H_r = 0.15 \qquad \text{for } \lambda_f > 0.15 \qquad (4b)$$

$$d/H_r = 3\lambda_f \qquad \text{for } \lambda_f < 0.05 \qquad (5a)$$
$$d/H_r = 0.15 + 5.5(\lambda_f - 0.05) \qquad \text{for } 0.05 < \lambda_f < 0.15 \qquad (5b)$$
$$d/H_r = 0.7 + 0.35(\lambda_f - 0.15) \qquad \text{for } 0.15 < \lambda_f \qquad (5c)$$

Given the estimates of z_o/H_r and d/H_r from the above formulas, the friction velocity, u*, can be estimated from equation (2), given a wind speed observation or estimate, u, at some height, z, which is above H_r. Bentham and Britter (2000) showed that the average wind speed, U_c, below the obstacle heights is given by:

$$U_c/u^* = (z_o/2H_r)^{-1/2} \qquad z < H_r \qquad (6)$$

The log-linear wind profile is assumed to apply down to a height, z_{int}, such that $u(z_{int}) = U_c$. For obstacles with large spacings (i.e, λ approaching zero), z_{int} can become much less than H_r and iteration is required to estimate U_c.

The turbulent velocities are assumed to be given by the following relations at all heights, above and below H_r: $\sigma_u/u^* = 2.4$, $\sigma_v/u^* = 1.8$, and $\sigma_w/u^* = 1.3$. However, current research is likely to improve estimates of turbulent velocities below H_r.

It must be noted that the determination of the "roughness element height" or the "average building height", H_r, will produce the greatest uncertainty in application. The standard deviation of individual H_r among a grouping of real obstacles is found to equal to about the same order as the mean H_r (Grimmond and Oke, 1999).

VARIATION OF MAXIMUM CONCENTRATIONS WITH ROUGHNESS

Clouds with $h \gg H_R$: As shown above, as the surface roughness increases, the friction velocity u^* along with the components of the turbulence intensity increase. The increased turbulence intensity leads to increased turbulent dispersion and hence, as suggested by equation (3), a tendency towards reduced concentrations from a given source release, for all conditions the same (i.e., constant net radiation and constant free-stream wind speed). This conclusion is valid at heights within the atmospheric boundary layer such that the boundary layer has adjusted to the combined effects of individual roughness obstacles. It may be inferred that this conclusion will be true as long as the cloud depth is (much) greater than the obstacle depth, H_r.

Britter et. al. (2000) reviewed several analyses for vertical dispersion and also combined the results from three parallel studies in different wind tunnels and a reanalysis of the Prairie Grass data to find that the ground level concentration C from a continuous line source of strength q was given by

$$d(q/C)/dx = (0.65 \pm 0.05)\, u^* \tag{7}$$

showing a direct dependence upon u^* (and by implication z_o).

For negatively buoyant releases, Briggs et. al. (2000) studied three parallel experiments in different wind tunnels and found (somewhat surprisingly) that all the effects of surface roughness on vertical dispersion were through the variation of u^* with z_o.

The lateral dispersion, σ_y, for neutrally-buoyant releases will depend directly on the lateral turbulence intensity, which also relies, in part, on u^*. However the lateral dispersion is also dependent upon variations in wind direction. Additionally the size, geometry and spacing of the roughness elements will likely introduce a lateral spatial scale into the turbulence that can also affect the lateral dispersion and not be characterized within u^* or z_o. This latter effect will be less important when the lateral plume dimension is large compared with the lateral spatial scale of the turbulence.

This reduction in concentration with an increase in surface roughness length can be clearly seen in the Kit Fox experiments as described in Section 3.

Thus it appears through analysis and these experiments that increased surface roughness has produced an increased plume dilution. And that this increased dilution is arising due to the increase in u^*, and related z_o. However, the frontal area ratio λ_f was not very large in these experiments.

The surface roughness length, z_o, increases with greater obstacle density until the obstacle interference limits z_o. At even larger obstacle densities, z_o can decrease, particularly in laboratory studies using obstacles of constant height. It would seem reasonable to speculate that any constancy or reduction in z_o and u^* due to greatly increased λ_f would also be reflected in the change in maximum concentration.

Thus for clouds much deeper than the obstacle heights, H_r, there can be a wide range of obstacle densities for which the maximum cloud concentrations will not be sensitive to the obstacle density. But, as implied earlier, a further influence on lateral dispersion will be the setting of a spatial turbulence scale by the size, geometry and specific spatial arrangement of the obstacles. This might be interpreted as a "flow-channeling" (Roberts et al., 1994). Flow channeling however may be more an artefact of idealized experiments than would be expected in a real urban or industrial scenario.

Clouds with h < H_r: Accidental releases leading to pollutant clouds may be initially located within the array of roughness obstacles and must grow through the array. If the array is not too densely-packed, a dispersion model based on an advection-diffusion equation approach is plausible. For very dense arrays such an approach is less reasonable; an advection velocity may not be an appropriate description and diffusion may be from within the array to above the array, followed by advection by flow above the array and diffusion back down into the array.

In the former case, the solution should be of much the same form as for the case when h >> H_r. with an advection velocity, and cloud dimensions, and if necessary longitudinal dispersion. Predictions of maximum concentrations therefore rely on specifying the advection velocity and the growth of the lateral and vertical (and longitudinal) dimensions.

Bentham and Britter (2000) developed a simple correlation (equation (7)) for the advection velocity within the array: $U_c/u^* = (z_o/2H_r)^{-1/2}$, where u* is determined from z_o and from the flow above the roughness array.

Experiments on the growth of the cloud dimensions within the array are few. The Kit Fox field experiments clearly demonstrate the increase in vertical dimension growth rate caused by the obstacles in the array. The cloud (plume and puff) depth in the array had a growth rate larger than 0.16, this with $\lambda_f = 0.12$. This growth rate is considerably larger than the estimates of the growth rate of the clouds once they are larger than the depth of the array. In a laboratory experiment using two-dimensional bluff obstacles, Melia (1991) observed a linear growth rate of the centroid height, h, such that dh/dx = 0.16 for λ_f = 0.037 and 0.056. It is not yet clear how the growth rate of the vertical plume dimension increases with obstacle characteristics and spacing. However, it is likely that the growth rate is constant and larger than the growth rate for the cloud when h >> H_r. This may be a useful operational observation. Further, and particularly for λ_f not large, the growth rate will depend upon the turbulence levels within the obstacle array and this may be estimated semi-analytically.

The growth of the lateral plume dimension, σ_y, within the array is influenced by the turbulence levels generated within the array, the spatial scale of the turbulence within the array, and a "topological diffusion" related to the physical presence of the obstacles. Several experiments show that σ_y is less strongly dependent upon the obstacle density than might be expected. Macdonald et al. (2000) present data for dispersion within and above cube-shaped obstacles. Close to the dispersion source, where the cloud is principally within the array, σ_y is much the same for $0.05<\lambda_p<0.91$ with all these results being significantly larger than for $\lambda_p = 0.0$. In addition, the growth of σ_y with x is more parabolic than linear and this suggests that the diffusion might be characterized by a turbulent diffusivity based on a constant (with λ_p) turbulence intensity and a spatial scale determined by the obstacle dimension.

As stated earlier, the maximum concentration depends upon how the advection velocity and σ_y and σ_z vary with the obstacles. The ERP study indicated that these varied in such a way as to substantially decrease the concentration with $\lambda_f = 0.12$; by a factor of about 4 when comparing the ERP and the URA results at the first measurement arc.

The Macdonald et al. (2000) results, close to the source where the cloud is principally within the array, produce a reduction in concentration of a factor of around three to four for all $\lambda_f = \lambda_p$ between 0.05 and 0.40 over that for no obstacles. Larger $\lambda_f = \lambda_p$ reverses the trend with $\lambda_f = \lambda_p = 0.0$ and 0.69, 0.91 being similar.

Thus it is clear that over a wide range of λ_f or λ_p that the obstacles act to reduce the maximum concentrations; the analytical support for this is being developed.

CONCLUSIONS

The maximum ground level pollutant concentration in an urban or industrial area can be approximated given knowledge of the wind speed, u, at some height, z, above the surface roughness obstacle height, H_r, and a measure of the obstacle density parameter, λ. During nearly-neutral stabilities, these parameters allow the surface roughness length, z_o, the displacement length, d, the friction velocity, u*, and the entire wind profile above and below H_r to be prescribed. The turbulent dispersion coefficients, σ_y and σ_z, can then also be estimated from u*.

For plumes released at ground-level and with depths greater than H_r, the maximum ground-level concentration is always decreased as z_o increases. Typically the concentration decreases by about a factor of two or three for each order of magnitude increase in z_o.

For plumes released at ground-level and with depths less than H_r, and for small-to-moderate obstacle densities λ (between zero and about 0.6), the concentration behavior is similar to that for plumes with depths greater than H_r. That is, maximum ground-level concentrations decrease as z_o increases. For larger obstacle densities (λ slightly less than 1.0), the wind speed below H_r may decrease sufficiently that, even with enhanced diffusion, the maximum concentrations may not decrease as z_o increases.

The conclusions of many of the previous small-scale field and laboratory studies of this effect have been influenced by the fact that nearly all obstacle arrays that were constructed for the experiments consist of uniform shapes and sizes such as rectangular billboards or cubes. As these arrays with constant height become tightly packed (i.e., as λ approaches 1.0), their z_o becomes small and u* also decreases. However, for real urban and industrial areas, as λ approaches 1.0, there remains much variability in the obstacle heights, with a standard deviation in H_r approximately equal to the mean value of H_r. Consequently, z_o, u*, and turbulent dispersion are relatively large in real environments even when the obstacles have small spacing.

ACKNOWLEDGEMENTS

This research was supported by grants from the American Institute of Chemical Engineers and from the U.S. Defense Threat Reduction Agency.

REFERENCES

T. Bentham and R.E. Britter, Spatially-averaged flow velocity within large groups of obstacles. Submitted to *Atmos. Env.* (2000).

G.A. Briggs, R.E. Britter, S.R. Hanna, J.A. Havens, A.G. Robins and W.H. Snyder, Dense gas vertical diffusion over rough surfaces: results of wind tunnel studies. Submitted to *Atmos. Environ.* (2000).

R.E. Britter, The Development of Simple Models for Dense Gas Dispersion over Obstacles or Sloping Terrain. Task B: Releases within a Complex Array of Obstacles. Rep. FM31/90 prepared by CERC for British Gas PLC Midlands Research Station (1991)

R.E. Britter, G.A. Briggs, S.R. Hanna and A.G. Robins, The passive limit of short-range vertical dispersion of dense gases in a turbulent boundary layer, Submitted to *Atmos. Env.* (2000).

R.E. Britter and J.M. McQuaid, *Workbook on the Dispersion of Dense Gases*. HSE Report No. 17/1988, Health and Safety Executive, Sheffield, UK, 158 pages (1988).

M.J. Davidson, K.R. Mylne, C.D. Jones, J.C. Phillips, R.J. Perkins, J.C.H. Fung, and J.C.R. Hunt, Plume dispersion through large groups of obstacles – a field investigation, *Atmos. Environ.* 29: 3245-3256 (1995).

M.J. Davidson, W.H. Snyder, R.E. Lawson and J.C.R. Hunt, Plume dispersion from point sources upwind of groups of obstacles – wind tunnel simulations, *Atmos. Environ.* 30: 3715-3725 (1996).

C.S.B. Grimmond and T.R. Oke, Aerodynamic properties of urban areas derived from analysis of surface form, *J. Appl. Meteorol.* 38: 1261-1292 (1999).

D.J. Hall, R. Macdonald, S. Walker and A.M. Spanton, Measurements of dispersion within simulated urban arrays – a small scale wind tunnel study, BRE Client Report CR 178/96, Building Research Establishment, Garston, Watford (1996).

S.R. Hanna and R.E. Britter, The Effects of Roughness Obstacles on Flow and Dispersion at Industrial Facilities, AIChE, 3 Park Ave., New York, NY 10016 (2000).

S.R. Hanna and J.C. Chang, Kit Fox dense gas dispersion field experiment and HEGADAS model testing, Submitted to *Atmos. Env.* (2000).

S.R. Hanna, P.J. Drivas, and J.C. Chang, *Guidelines for Use of Vapor_Cloud Dispersion Models*, AIChE/CCPS, 345 East 47th St., New York, NY 10017, 285 pages (1996).

R.W. Macdonald, R.F. Griffiths and S.C. Cheah, Field experiments of dispersion through regular arrays of cubic structures, *Atmos. Environ.* 31: 783-795 (1997).

R.W. Macdonald, D.J. Hall and R.F. Griffiths, Scale model study of building effects on dispersion in the urban canopy at intermediate source distances, To appear in *J. Environ. Eng*. (2000).

J.L. McElroy and F. Pooler, St. Louis Dispersion Study Volume II – Analysis, US Dept. HEW, Arlington, VA (1968).

J. Melia, The dispersion of dense fluids through arrays of obstacles, Ph.D. Thesis, University of Cambridge (1991).

P.T. Roberts, R.E. Fryer-Taylor and D.J. Hall, Wind-tunnel studies of roughness effects on gas dispersion, *Atmos. Environ.* 28: 1861-1870 (1994).

A. Robins and R. Macdonald, A Review of Flow and Dispersion in the Vicinity of Groups of Buildings, Report No. ME-FD/99.93 (Draft) University of Surrey (1999).

DISCUSSION

E. GENIKHOVICH: Did you have concentrations measured below the displacement height? What is the ratio of concentrations above and below this height?

S. HANNA: During the Kit Fox field study, concentrations were measured at three levels below the displacement height and two levels above. These observations were made at a downwind distance of about ten obstacle heights, after the ground-level-released plume had passed by about three rows of obstacles. It should be noted that the displacement height was found to equal about 0.7 times the obstacle height. The concentrations below and just above the displacement height were nearly constant, since the tracer cloud appeared to be mixed uniformly in the vertical behind the obstacles at that downwind distance.

MODEL ASSESSMENT AND VERIFICATION

chairpersons:	C. Borrego D. Steyn I. Uno
rapporteurs:	M. Boznar P. de Haan W. B. Petersen

EVALUATION OF THE CONCENTRATION FLUCTUATION PREDICTIVE POWER OF THE KINEMATIC SIMULATION PARTICLE MODEL

R. J. Yamartino[1], D. G. Strimaitis[2] and A. Graff[3]

[1] 191 East Grand Avenue, Old Orchard Beach ME 04064 USA
[2]Atmospheric Sciences Group, Earth Tech, Concord MA 01742 USA
[3] Umweltbundesamt, Bismarckplatz 1, D-14193 Berlin, Germany

INTRODUCTION

The Kinematic-Simulation-Particle (KSP) model is a new type of atmospheric dispersion model having the intrinsic capacity to predict second-by-second concentration patterns and fluctuations that can be critical to the assessment of odor and hazardous chemical exposure problems.

KSP uses the mean flow fields and a realistic spectrum of turbulent eddies to transport smaller eddies and tracer particles. Just as in nature, the instantaneous distribution of these tracer particles yields concentration field 'snapshots' that may be analyzed to yield critical statistics (e.g., peak values, durations above thresholds, higher moments), as well as the more usual longer time averages and ensemble means. KSP represents a 'parallel universe' model of the atmosphere and can be used to investigate phenomena such as relative- versus absolute-diffusion as well as more applied problems. In addition, the user-friendly KSP model (Strimaitis et al., 1995) provides the capability to assess the consequences of air pollution emissions over a broad range of: spatial and temporal scales; terrain and land use environments; and non-steady meteorological land release conditions.

The KSP (Version 1.9) system has recently undergone an extensive evaluation process funded by the German EPA (the Umweltbundesamt), to determine biases and quantify model uncertainties using the Prairie Grass, Lillestroem, Copenhagen, and Kincaid field program data as well as wind tunnel results obtained at the University of Hamburg wind tunnel. These analyses are fully documented in Yamartino and Strimaitis (1998) and Schatzmann et al., (1998). The present paper reviews the model's basic formulation and some recent enhancements, and summarizes the key results of comparisons with data obtained from wind tunnel studies of mean concentrations and concentration fluctuations.

1 Email correspondence: rjy@maine.rr.com

2. MODEL FORMULATION AND FEATURES

The KSP model's sub-grid-scale (SGS) flow field generator relies on explicit yet evolving, periodic, and divergence-free functions to quantify the larger scale eddies responsible for pollutant dispersion. Beginning with the approach of Fung et al. (1992) for isotropic turbulence, the KS formalism was modified and extended to the more general atmosphere containing height dependent variations in wind velocity (i.e., shear) and turbulence plus turbulence anisotropies. This was accomplished by first transforming Fung's 3-d isotropic eddies into component 2-d eddies lying in the x-z, y-z, and x-y planes, respectively. The amplitudes of each of these components could then be independently varied to achieve the desired vertical and horizontal plume dispersion rates. The amplitudes of the x-z and y-z components were set equal and constrained so that the quadratic sum of amplitudes equaled the desired vertical velocity variance. Any excess of the horizontal velocity variance over the vertical velocity variance was then channeled into the x-y velocity meander eddy.

The original KSP model used five eddy size components, with wavelengths ranging from 12.5m to 3200m in wavelength multiples of four and with time scales in seconds equal to the wavelengths in meters, whereas an experimental, six-eddy version of the model considers the expanded wavelength range of 1m to 3125m in wavelength multiple steps of five) and with time scales in seconds equal to twice the wavelengths in meters. The smaller eddies of this latter version were thought necessary to model near-surface releases and short source-receptor travel times, but this need has not yet been convincingly demonstrated. In both versions, the relative admixture of the various wavelength components is determined from empirical power spectra profiles, nS(n,z), for the stable and convective boundary layers as published by Moraes (1988) and Hojstrup (1982), respectively. These spectral intensities are then combined with our characterization of the KS system's response to various input wavelengths and time scales to yield amplitudes for each component. Though only five or six discrete wavelengths are input, the fact that the amplitudes and phases of these various Fourier components of turbulence evolve according to Langevin equations causes the final spectrum of velocity fluctuations to reproduce the target spectra quite well and have other desirable features. For example, the traditional, independent-particle Langevin equation approach generally involves use of parameterized integral time-scale profiles, $T_L(z)$, which only implicitly recognize the fact that longer wavelength eddies have longer time scales than the shorter components. In the KS approach, each spectral wavelength component has a single, fixed time scale T_L; however, as the mix of spectral components varies with height, one obtains an effective $T_L(z)$ profile. Both versions of the model also incorporate a purely random, white noise velocity of 0.03 m/s to account for unresolved, small eddies

The various eddies are transported by the mean wind, as well as by all eddies larger than themselves, and combine to yield realistic particle dispersion. Under convective conditions, the larger two eddies become convective towers with a more intense, inner updraft zone surrounded by slower downdraft air. Horizontal plume meander is further enhanced via a 10km wavelength eddy having a Lagrangian time scale of 1200s

Several recent operational enhancements to KSP facilitate its application to episodic single-source releases, including tracer data sets. These enhancements include:

i) an option to select R,θ coordinates (i.e., in contrast to x-y grids) for the receptor array definition;

ii) an option to compute arc-integrated concentrations with an accuracy much greater than Δθ spacing of the receptor resolution and determine the arc-maximum concentration to the Δθ arc resolution; and

iii) an option to emulate point receptors through use of a triweight sampling kernel (de Haan, 1998) and integral averages of these kernels. Earlier model versions required definition of receptor boxes of significant volume to enable the counting of particles. It is worth noting that the 'sigma' x, y and z needed for the kernel estimation are computed internally by KSP and are equivalent to 'relative diffusion' estimates.

KSP also allows for the custom treatment of supplemental meteorological observations taken during tracer field programs and has the ability to automatically simulate multiple realizations of each scenario or trial in order to obtain an ensemble of predictions.

3. COMPARISON OF KSP WITH WIND TUNNEL RESULTS

KSP is generally applied to tracer experiments using the 'user defined' options for winds and turbulence. These routines basically invoke the micrometeorological profiles of the LASAT model (Janicke, 1995) to determine appropriate profile shapes for the given Monin-Obukhov length, L, and mixing height H, but then use profile rescaling to ensure that the winds and/or turbulence match specific measured values. Thus, the 'user defined' routines enable KSP to run in a mode that combines the strengths of the experimental database with existing knowledge and profiles of the PBL.

Several modifications to KSP were needed to create flow and turbulence fields appropriate to the University of Hamburg wind tunnel, scaled up by 1000:1. These included:

- choice of the micrometeorological parameter values: z_0=0.86m, u_*=0.3 m/s, H=603 m, and L=9999 m, to achieve an optimal fit to the measured wind tunnel profiles of u, σ_v and σ_w;
- replacement of the spectral shape algorithm with the Teunissen (1980) spectral function:

$$nS(n)/u_*^2 = A\, f / (\, 1 + B\, f^{5/3}\,)$$

  where n is frequency and f is the non-dimensional frequency, $f=nz/U=z/\lambda$, λ is wavelength, and B is a 'constant' that was found to display some z dependence;
- use of five KS wavelengths of 1m, 5m, 25m, 125m, and 625m to emulate the Teunissen vertical spectra;and,
- reduction of the 'meander' eddy wavelength to 2.5 km to yield lateral spectra that account for the sidewalls of the 2.5m wide wind tunnel.

Obtaining appropriate spectral parameters for KSP was more difficult as the position of the spectral peak f_m is found to vary with height in the tunnel, whereas analysis of the Teunissen spectral function indicates that $f_m = (1.5/B)^{0.6}$ is independent of z. Various height dependent B(z) were tested, but none was as satisfactory as using fixed values of B=23.33 for the z=60 mm release and B=8.44 at z=100 mm release. One notes that the latter value of B=8.44 corresponds to the neutral limit of the Moraes spectrum and yields a peak at f_m=0.355, whereas the lower z value of B=23.33 yields a reduced f_m=0.192. These f_m values correspond reasonably with the observed spectral peaks in the wind tunnel's vertical turbulence spectra and suggest that the lower portion of the wind tunnel is somewhat 'richer' in longer wave eddies than is encountered in a neutral atmosphere.

Mean Concentrations

An array of receptors in the wind tunnel enabled measurement of crosswind integrated and arc-maximum concentrations at 33 downwind locations ranging from 0.3m to 4.0m downwind and corresponding to full scale downwind distances of 300m to 4km. Tracer releases at 60mm and 100mm (i.e., 60m and 100m full scale) were conducted. The wind-tunnel-modified version of KSP was run for both the 60m and 100m releases at a

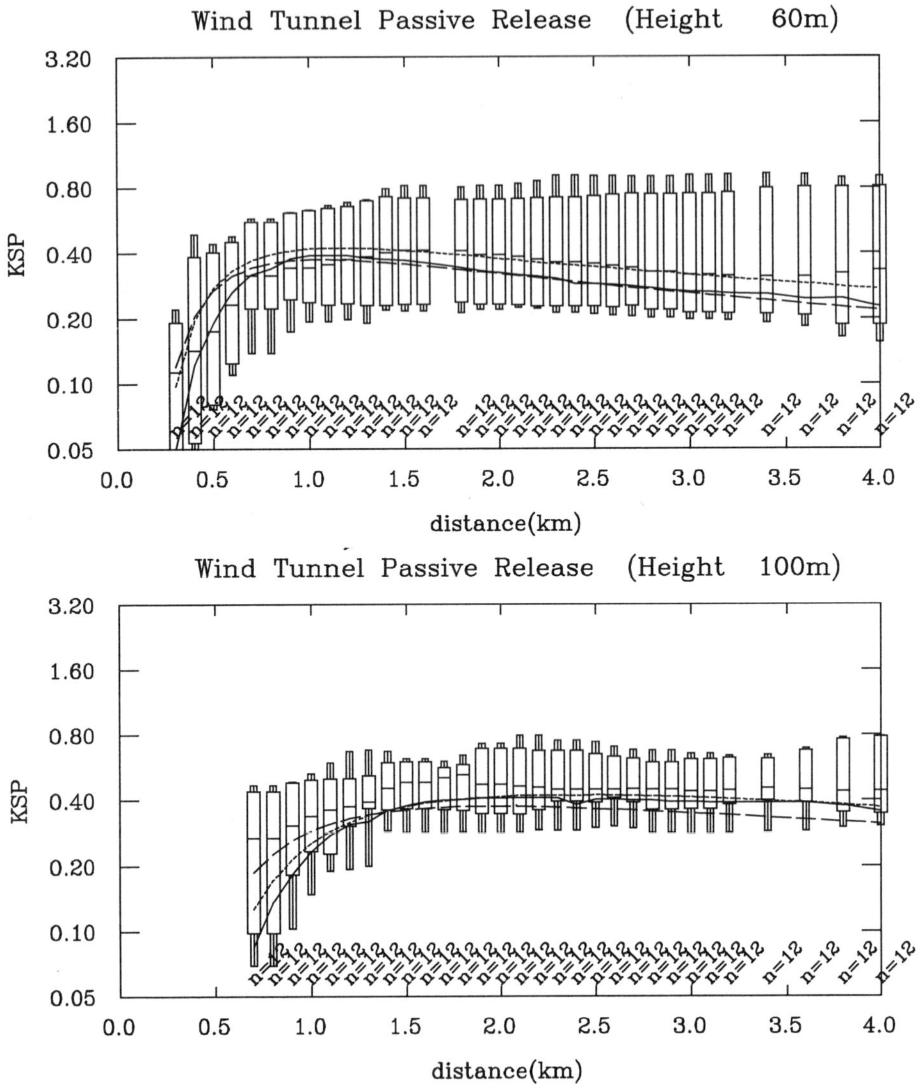

Figure 1. Non-dimensional crosswind-integrated concentrations from KSP(1.9) using 5 of 6 wavelength components and custom wind tunnel modifications to the turbulence and spectral profiles. Whiskers indicate the 10th, 25th, 50th, 75th and 90th percentiles of 12 replicate, 30-minute average runs. Accompanying curves are measurements from the UOH wind tunnel at 1000:1 scale (solid line) and model results from LASAT (dotted line) and Berljand (dashed line).

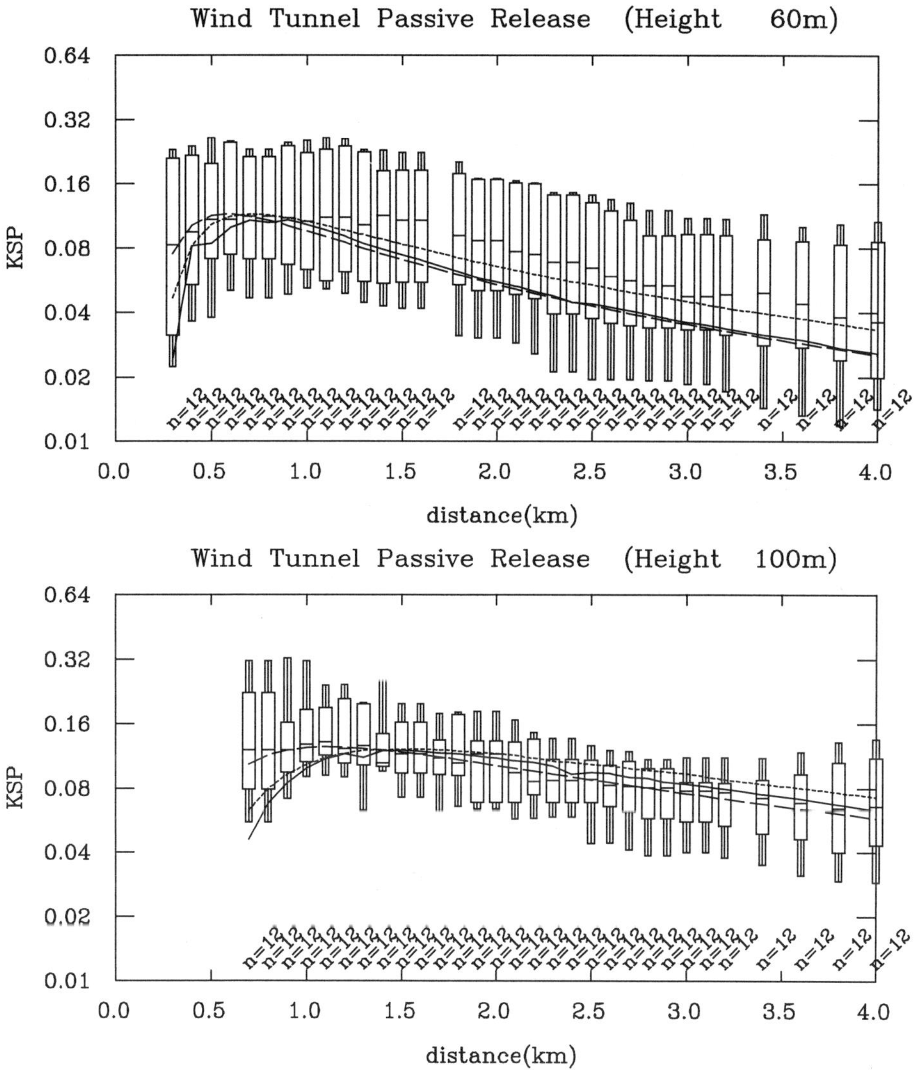

Figure 2. Non-dimensional peak concentrations from KSP(1.9) using 5 of 6 wavelength components and custom wind tunnel modifications to the turbulence and spectral profiles. Whiskers indicate the 10th, 25th, 50th, 75th and 90th percentiles of 12 replicate, 30-minute average runs. Accompanying curves are measurements from the UOH wind tunnel at 1000:1 scale (solid line) and model results from LASAT(dotted line) and Berljand(dashed line).

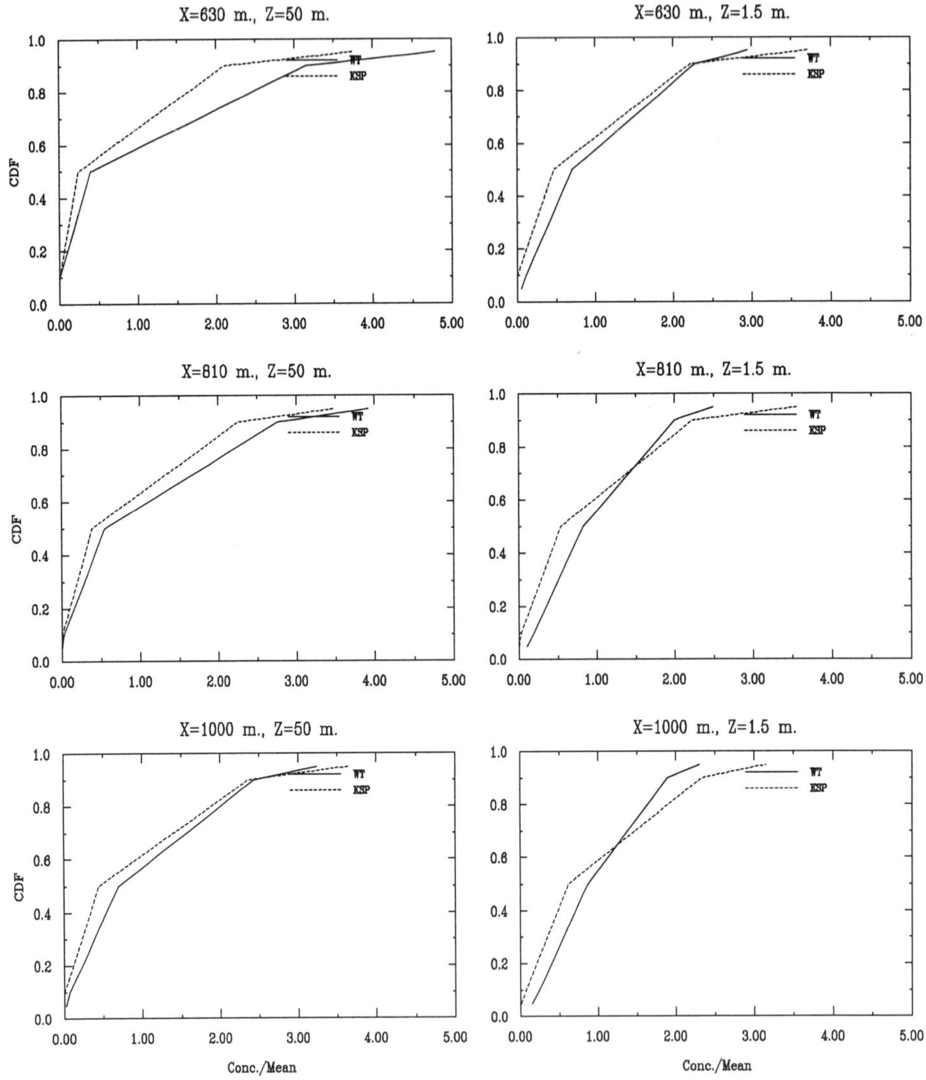

Figure 3. Cumulative distribution functions (CDF) of KSP(1.9) 1-second concentrations normalized by the 6-hr average concentration (dotted line) compared with the CDF of normalized concentrations from the wind tunnel (solid line).

time step of $\Delta t = 2s$ for six consecutive half-hour periods, with the first period discarded to avoid model spin-up effects. Resulting non-dimensional, CWICs (i.e., C^yUH_q/Q) and peak concentrations (i.e.,$C^yUH_q^2/Q$) are presented in Figures 1 and 2, respectively, along with corresponding wind tunnel data (solid line), LASAT results (dotted line) and Berljand model (1982) (dashed line) results. The KSP results are in good agreement with observations and the results of simpler models, but display a large amount of scatter. This is due to the fact that a half-hour is much too short an averaging time period to obtain results comparable to the ensemble-average means of the models or the long averaging period of the wind tunnel; however, 30 minutes was chosen as this is the averaging time basis of German regulations (TA Luft).

Concentration Time Series

High time resolution (i.e., 500 Hz sampling) concentration time series were obtained with the wind tunnel operating at a scale of 300:1. This scale shift necessitated making minor changes to the wind tunnel specific version of KSP. The principal changes were use of a smoother $z_0 = 0.254$m and a smaller $u_* = 0.2688$ m/s to match the wind of 3.6 m/s at the 50m release height, and a shift in the value of the spectral coefficient to B=70, corresponding to a w-spectrum peak at $f_m = 0.1$. The turbulence routines were also adjusted to yield the observed σ_w and σ_v, with reasonable turbulence profiles being again achieved through choice of the mixing height and Monin-Obukhov length as H=603 m and L=9999 m.. A KSP run at $\Delta t = 1s$ for 6 hours led to a sample of 21,600 concentrations at each receptor point. Though the scaled wind tunnel sampling rate of 300/500Hz=0.6s suggests that KSP should have been marched at a smaller time step, sub-second time stepping is not currently possible. Further, one notes that while the experimental sampling rate was 500Hz, spectral plots of the concentration data did not suggest significant power beyond 60Hz, thus, suggesting a minimum meaningful KSP time step of 300/60Hz=5s. Hence, KSP results were first processed through an 11-point Gaussian filter with an effective smoothing time constant of 6 seconds (i.e., approximately 300/60Hz). This smoothing had little impact on most percentile level concentrations, though it did reduce maximum values by up to a factor-of-two. Though this KSP data sample is less than one-tenth the wind tunnel sample of 300,000 points (i.e., 500Hz times the 600s sample duration), it is interesting to compare the cumulative frequency distributions (CFDs) of KSP predictions versus wind tunnel observations. These distributions, normalized by their respective run-length-average value, are shown in Figure 3 for three downwind distances and two sampling heights (i.e., the source height of 50m and a near ground level height of 1.5m). While these CFDs appear qualitatively similar, the changes in CFD slope with downwind distance and receptor height suggest that additional analyses with alternative spectral shapes be tried.

4. CONCLUSIONS

This evaluation has shown that the current version of KSP:

- produces a range of crosswind integrated and arc-maximum concentrations that encompass neutrally-stratified, elevated release wind tunnel measurements and the predictions of simple plume and Lagrangian particle models;
- predicts a variability in concentration estimates on the order of a factor-of-two to four for exposure durations of 30 minutes. That is, σ_c/C is of order unity and indicates that the natural variability of 30-minute average concentration values will always limit the level of agreement

one is able to achieve using ensemble average models; and

• generates cumulative frequency distributions (CFDs) of concentrations that have the same qualitative behavior as those measured in the wind tunnel; however, these modeled CFD shapes are very sensitive to the assumed turbulence power spectra and its profile, nS(n,z), and assumed time scales.

ACKNOWLEDGEMENT

This work was supported through a contract to the Meteorological Institute of the University of Hamburg by the German Federal Environmental Agency (Umweltbundesamt) as part of the Environmental Research Plan of the Federal Ministry of Environment, Nature Conservation, and Reactor Safety.

REFERENCES

Berljand, M.E., 1982: Moderne Probleme der atmosphärischen Diffusion und der Verschmutzung der Atmosphäre, Akademie-Verlag, Berlin.

Fung, J.C.H., J.C.R. Hunt, N.A. Malik and R.J. Perkins, 1992: Kinematic simulation of homogeneous turbulent flows generated by unsteady random Fourier modes. *J. Fluid Mech.*, **236**, 281-318.

Hojstrup, J., 1982: Velocity spectra in the unstable planetary boundary layer. *J. Atmos. Sci.*, **39**, 2239-2248.

Janicke, L., 1995: Ausbreitungsmodell LASAT: Referenzbuch zu Version 2.51, Primelweg 8, D-88662, Überlingen, FRG. (In German)

Moraes, O.L.L., 1988: The velocity spectra in the stable atmospheric boundary layer. *Boundary Layer Meteorol.*, **43**, 223-230.

Schatzmann, M., L. Janicke, W. Klug, S. Rafailidis, D. Strimaitis, R. Yamartino, 1998: Validierung des "KSP Model" für Anwendungen im Vollzug des BImSchG. UBA Final report 98-295-43-354.

Strimaitis, D., R. Yamartino, E. Insley, and J. Scire, 1995: A User's Guide for the Kinematic Simulation Particle Model. EARTH TECH Document 1274-2 prepared for the Free University of Berlin and the Umweltbundesamt, Berlin, FRG.

Teunissen, H.W., 1980: “Structure of Mean Winds and Turbulence in the Planetary Boundary Layer over Rural Terrain,” *Bound-Layer Meteorol* **19**, 187-221.

Yamartino, R., and D. Strimaitis, 1998. Refinement and Evaluation of the Kinematic Simulation Particle Model. Appendix A to UBA Final report 98-295-43-354.

DISCUSSION

P. BUILTJES: Can you comment on the relation between the time-scales in the wind-tunnel and your 30-minute averaged calculations?

R. J. YAMARTINO: The wind tunnel concentration measurements were averaged over 10 minutes. However, given the 1000:1 scaling, this corresponds to a very long, full-scale, time averaging period of 167 hours. In any case, the wind tunnel averaging period was long enough that the resulting concentrations were very stable, and thus can reasonably be considered as ensemble average values. The KSP model was run at full-scale for thirteen (13) half-hour periods. The first half-hour of concentration data was discarded, as it includes start-up effects, and the subsequent twelve, half-hour average values were used to construct the data whiskers shown in Figures 1 and 2. Unlike the single, ensemble-average concentrations provided by the other models or the wind tunnel, the significant spread of KSP values results from the fact that 30 minutes is too short an averaging time to yield an ensemble average concentration, and such variability in measured concentrations is to be expected. Nevertheless, this 30-minute averaging duration is important to consider, as it is the short-term averaging period considered in the German air quality regulations.

DEVELOPMENT OF A NEW OPERATIONAL AIR POLLUTION FORECAST SYSTEM ON REGIONAL AND URBAN SCALE

Jørgen Brandt, Jesper H. Christensen, Lise M. Frohn,
Gary Geernaert, and Ruwim Berkowicz

National Environmental Research Institute,
Department of Atmospheric Environment,
Frederiksborgvej 399, P.O. Box 358,
DK-4000 Roskilde, Denmark.

INTRODUCTION

Episodes with high levels of harmful air pollution concentrations, as e.g. nitrogen dioxide or ozone, can have damaging effects on human health and plants. In order to minimize the effects of these episodes it is useful to be able to predict them. An operational air pollution modelling system is an important tool that can be used for air pollution forecasting, monitoring and scenarios. Where critical limit values are exceeded, air pollution forecasts can be used to inform and warn the public. An operational system can be integrated into national urban and rural monitoring networks. Furthermore, it can be used to study the effects of emission reduction scenarios, where emissions from e.g. a part of the traffic is reduced in the model. Such scenarios can be used as guidelines for the authorities for decision making concerning restrictions on e.g. traffic, during episodes of high air pollution levels.

A new operational air pollution forecast system, the DMU-ATMI THOR air pollution forecast system, has been developed at the National Environmental Research Institute, Department of Atmospheric Environment, ATMI, Denmark. The integrated system consists of a numerical weather forecast model and a series of air pollution models, covering a wide range of scales (from European scale to street scale in cities) and applications. The system is designed to automatically produce 3 days air pollution forecasts of some of the most important air pollution species on a continuous basis. The system is able to resolve details of individually street canyons. The various models, the coupling/integration and the configuration of the models, and the real time performance on fast workstations with parallel architecture will be described. Examples of comparisons of model results with measurement are presented.

A schematic diagram of the different modules and the data flow chart of the DMU-ATMI THOR air pollution forecast system is shown in Fig. 1. Presently, the model system consists of a coupling of four models. A numerical weather forecast model, ETA, is applied

(Nickovic et al., 1998). This model is initialized with data from a global circulation model run at the National Centers for Environmental Prediction, NCEP, USA.

The weather forecast is subsequently used as input to a long-range transport air pollution model, the Danish Eulerian Operational Model, DEOM, producing air pollution forecasts on European scale. DEOM is an operational version of the Danish Eulerian Model, DEM (Zlatev et al., 1992), which has been run and validated by comparison to EMEP measurements over a period of 10 years. Meteorological data from ETA and air pollution concentrations from DEOM are used as input to the Background Urban Model, BUM (Berkowicz, 1999b) calculating the urban background pollution based on emission inventories with e.g. a 2 km x 2 km grid resolution covering the central city of Copenhagen. The output from BUM is subsequently used as input to the Operational Street Pollution Model, OSPM, (Berkowicz, 1999a) producing the air pollution concentrations at street level, at both sides of the street in cities. The weather forecasts, the European air pollution forecasts, the urban background forecasts and the street canyon forecasts are currently run operationally, four times every day, initiated with data at 00 UTC, 06 UTC, 12 UTC and 18 UTC. The system is fully automated – meaning that the entire procedure of receiving the data, running the models, producing the visualizations and sending the specified result to the end users is controlled by cron-jobs and Unix scripts. The whole system and the operational procedure have been run, tested and validated since August 1998.

The system produces huge amounts of output data in every run. These data are impossible to comprehend without fast and advanced visualization and animation techniques. Different visualization techniques have been developed and implemented in the system. For every forecast that is produced between two to four times a day, nearly 1000 visualizations and animations are produced and systemized to be presented using an Internet browser. In this way, it is easy to follow the temporal and spatial development of the air pollution levels and to discover whether critical air pollution limit values are to be exceeded. Visualizations of the daily maximum air pollution concentrations over an area provide an effective tool for monitoring warning or alert situations. In the case of a predicted exceedance of the critical limit values for e.g. ozone, the population can be informed and eventually alerted. A demonstration of the entire system including visualizations, animations and time series of the weather and air pollution concentrations is available at the web page: http://www.dmu.dk/AtmosphericEnvironment/thor.

THE DANISH EULERIAN OPERATIONAL MODEL, DEOM

The Danish Eulerian Operational Model, DEOM, calculates the regional background pollution levels on European scale. The operational version of the model calculates transport, dispersion, deposition and chemistry of 35 species. Three vertical layers are presently used in order to reduce computing time. The three layers are defined as the mixing layer (below the mixing height), the old mixed layer (between the present mixing height and the advected mixing height from the previous day) and the reservoir top layer located between the advected mixing height and the free troposphere. Experiments have been carried out with a fourth layer, the emission layer (close to the ground). However no major performance improvements were achieved. The emission data used in DEOM are a combination of data provided by the European Monitoring and Evaluation Programme (EMEP) (with a spatial resolution of 50 km x 50 km) and the Danish emission inventories, which have a resolution of down to 1 km x 1 km. The data include emissions of NO_x, SO_2, NH_3 and anthropogenic VOC emissions. Biogenic VOC emissions are calculated from land use data. The model is based on a system of partial differential equations (PDE's):

$$
\begin{aligned}
\frac{\partial c_s}{\partial t} = & -u\frac{\partial c_s}{\partial x} - v\frac{\partial c_s}{\partial y} \\
& + K_x\frac{\partial^2 c_s}{\partial x^2} + K_y\frac{\partial^2 c_s}{\partial y^2} \\
& - (\kappa_{1s} + \kappa_{2s})\, c_s \\
& + E_s \\
& + Q_s(c_1, c_2, \ldots, c_q) \\
& + V(c_s)\,, \quad s = 1, 2, \ldots, q
\end{aligned}
$$

where q is the number of the chemical species. The horizontal advection is represented by the first two terms on the right-hand-side of the equation. The third and the fourth terms describe the horizontal dispersion. The dry and wet deposition terms, $-(\kappa_{1s} + \kappa_{2s})c_s$, are linear, while the chemistry term, $Q_s(c_1, c_2, \ldots, c_q)$, introduces non-linearity. The concentrations are denoted by c_s, u, v are wind velocities, K_x, K_y are dispersion coefficients, assumed constant 10000 $m^2\ s^{-1}$, E_s describe the emission sources, κ_{1s} and κ_{2s} are dry and wet deposition coefficients, respectively, and the chemical reactions are denoted by $Q_s(c_1, c_2, \ldots, c_q)$. Different vertical exchange procedures, $V(c_s)$, which are functions depending on the number of layers have been developed, based on ideas in Pankrath (1988).

It is difficult to treat the system of equations directly. Therefore, the model is split into four sub-models. A simple splitting procedure, based on ideas from e.g. McRae et al. (1984), can be defined. The four sub-models include (1) horizontal transport, (2) horizontal dispersion (3) dry and wet deposition, and (4) emissions and chemistry. An Accurate Space Derivative algorithm is used to handle the transport terms in the first sub-model. Time integration for the advection term is performed with a predictor-corrector scheme with several correctors. For the horizontal dispersion, truncated Fourier series approximate the concentrations. It is then possible to find an analytical solution for each Fourier coefficient in the Fourier space. The deposition terms are solved directly. The chemical scheme used in the model is the CBM-IV scheme (Zlatev et al., 1992).

THE BACKGROUND URBAN MODEL, BUM

The BUM model is suitable for calculations of urban background concentrations when the dominating source is the road traffic. For this source the emissions take place at ground level, and a good approximation is to treat the emissions as area sources, but with an initial vertical dispersion determined by the height of the buildings. The applied emission data have to be provided on a grid with the same resolution as used in the model. The concentrations calculated by the model include NO, NO_2, NO_x and O_3. Contributions from the individual area sources, subdivided into a grid with a resolution of 2 km x 2 km, are integrated along the wind direction path assuming linear dispersion with the distance to the receptor point. Horizontal dispersion is accounted for by averaging the calculated concentrations over a certain, wind speed dependent, wind direction sector, centred on the average wind direction. Formation of the nitrogen dioxide due to oxidation of NO by ozone is calculated using a simple chemical model based on an assumption of photochemical equilibrium on the time scale of the pollution transport across the city area. This time scale governs the rate of entrainment of fresh rural ozone. The model is described in more detail in Berkowicz (1999b).

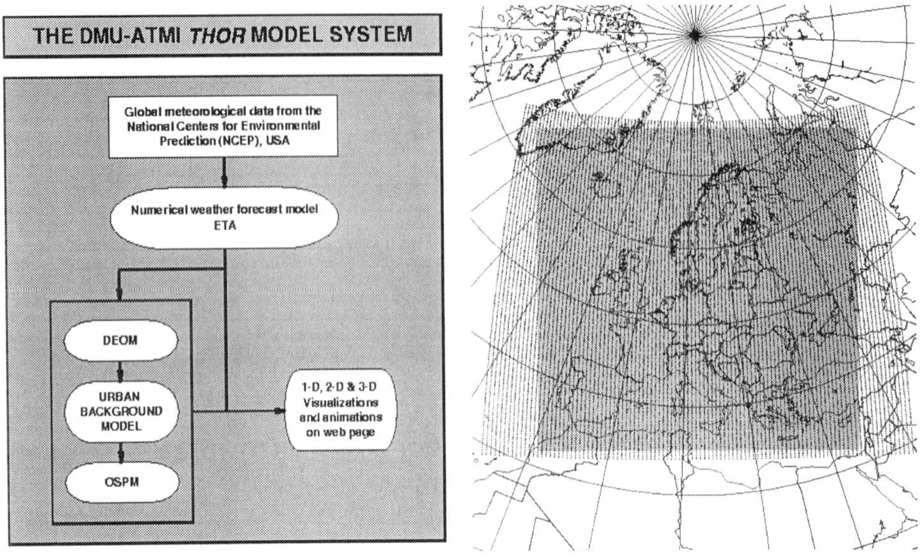

Figure 1. Left figure: a schematic diagram of the main modules and the data flow in the DMU-ATMI THOR air pollution forecast system. Right figure: the model domains of the ETA model (dots) and the Danish Eulerian Operation Model (shaded square). The center of the domains is at NERI, Roskilde, Denmark.

THE OPERATIONAL STREET POLLUTION MODEL, OSPM

The OSPM is a parameterized semi-empirical model making use of a priori assumptions about the flow and dispersion conditions in a street canyon. In the model, concentrations of exhaust gases are calculated using a combination of a plume model for the direct contribution and a box model for the recirculating part of the pollutants in the street. Parameterizations of flow and dispersion conditions in street canyons are deduced from extensive analysis of experimental data and model tests. Results from these tests have been used to improve the model performance, especially with regard to different street configurations and a variety of meteorological conditions. The model calculates air concentrations of NO, NO_2, NO_x, O_3, CO and benzene in the street canyon at both sides of the street. For a more detailed description of the model, see Berkowicz (1999a).

MODEL CONFIGURATIONS AND PERFORMANCE

When operational air pollution forecasts are produced, it is important that the different models and the model system are optimized on fast computers and that the system is fully automated, in order to ensure available prognoses early in the morning. The THOR system has been optimized to run operationally on two powerful workstations, both SGI Origin 200 with 4 processors. One machine is dedicated to run the models and the other to produce visualizations and animations.

The production time for one total forecast depends, of course, on the configuration of the models with respect to the number of grid points and the grid resolution. The ETA model is run on a staggered Arakawa E grid using a latitude/longitude system with shifted pole. The horizontal grid resolution is 0.25^o x 0.25^o corresponding to approximately 39 km x 39 km at 60° north. The number of horizontal grid points is 104 x 175 and the number of vertical layers is 32. The Danish Eulerian Operational Model is applied on a polar stereographic projection using an Arakawa A grid. The model is run with a horizontal grid resolution of 50 km x 50 km grid resolution at 60° north and a grid size of 96 x 96 horizontal grid points. Experiments have also been made with a grid resolution of 25 km and a grid size of 192 x 192 grid points. The domains of the two models are displayed in Fig. 1.

The ETA and DEOM models have been parallellized and are running with a speed-up of approximately 3.5 when 4 processors are used and with a real performance of approximately 350 MFLOPS corresponding to 20% of the peak performance. The visualizations and animations are calculated on another similar workstation. By the time the weather forecast is finished, the air pollution forecast will be running simultaneously with the production of the visualizations of the weather forecasts on the two different computers. The time spent on the different operational tasks is shown in Table 1. The total time required for calculating a 3 days forecast is around 3 hours, including the data downloaded from the internet to initialize the system (which takes around 30 min) and the production of the visualizations.

Table 1. Real time performance of the model system for a total forecast on SGI Origin 200 (225 MHz) with 4 processors.

Task	Approximate time
Data transfer of global meteorological data from NCEP	~ 30 min
ETA model	130 min (~ 40 km resolution, 32 layers)
Danish Eulerian Operational Model	9 min (~ 50 km resolution, 3 layers)
BUM and OSPM	< 1 sec
Production of visualization and animations	~ 30 min
Total time for a single forecast	~ 3 hours

EXAMPLES OF MODEL RESULTS AND COMPARISONS TO MEASUREMENTS

The model system is continuously being validated against measurements from the Danish urban monitoring network. For a description of the measurements and techniques, see Kemp and Palmgren (1999). Some examples given as scatter plots are shown in Fig. 2. The figure shows comparisons of measurements and the daily mean values of the 12 hours forecasted NO_x modelled concentrations (meaning that two forecasts are included in the daily mean) for the period August 1998 to September 1999.

Statistics for the plots have been calculated and given in the figures. These include the number of data samples, *N*, mean values, standard deviations, correlation coefficients with test for significance, figure of merit, *FM*, the bias with the 95% confidence interval, the fractional bias, *FB*, the fractional standard deviation, *FSD* and the normalized mean square error, *NMSE* with 95% confidence interval. These statistical parameters give a good indication of the system performance concerning both mean values and data variability. Detailed explanations for all these statistical parameters can be found in Brandt (1998) and Spiegel (1992).

The upper left figure includes comparison of measurements with model results from the long-range transport/chemistry model, DEOM, for a rural site named Lille Valby, located 30 km west of the city of Copenhagen, Denmark. As seen in the figure, calculated mean values are very close to the measured value, and with a highly significant correlation coefficient. The upper right figure shows similar results for a different site – an urban background station located at the top of a building in the center of Copenhagen (HCØ Institute, University of Copenhagen). The figure shows that the DEOM model results are not representative for the urban background pollution levels. Comparison to measurements shows a large bias and smaller correlation. However, when using the Background Urban

Model, BUM, which takes into account emissions from the city with higher resolution (2 km x 2 km), a much smaller bias and higher correlation are observed (lower left figure). The lower right figure shows comparison of measurements and OSPM model results at street level, at the eastern side of the street of Jagtvej, Copenhagen. The street of Jagtvej is located near to the urban background site (HCØ). Although, much higher mean concentrations is seen at street level compared to the rural background and the urban background, a small bias and a higher correlation coefficient of 0.83 are seen.

CONCLUSIONS AND FUTURE WORK

An air pollution forecast system has been developed. At present, the system includes weather forecasts and air pollution forecasts for the whole of Europe, urban air pollution forecasts for the urban background of Copenhagen, Denmark, and street canyon air pollution forecasts for an individual street in Copenhagen. The present system has been optimized and fully automated on local fast workstations. The system is presently being validated against measurements. Some comparisons of measurements and model results from the three models DEOM, BUM and OSPM have been shown. These results indicate good agreement with measurements for all the models – especially for the street level forecasts. The results also emphasize the importance of coupling models operating on different scales into a system.

The single models and the whole system are in the process of validation against operational and historical measurements. At present, validation of both the individual models and the whole model system is carried out. Measurements from the Danish urban and rural monitoring network are used. The 3-dimensional version of DEOM is presently under development. Experiments are made for different grid-resolutions and different number of vertical layers. Further species will be included in the model system – especially particles are important.

Two additional models will furthermore be implemented in the THOR system. These are 1) the Danish Eulerian Hemispheric Model, DEHM, (Christensen, 1995; Christensen 1997) which is used to describe the transport of SO_2, SO_4, Lead and Mercury on hemispheric scale and 2) the Danish Rimpuff and Eulerian Accidental release Model, DREAM, (Brandt 1998; Brandt et al., 1998) which is used to describe the transport, dispersion, deposition and radioactive decay from a point source, as e.g. the Chernobyl accident.

An important field for further development of the system is data assimilation of real time air pollution measurements. At present, data from the Danish monitoring stations are used for real time validation of the system. These measurements will be assimilated into the model for a better estimation of the initial fields in DEOM.

A new comprehensive high-resolution model, REGINA (REGIonal high resolutioN Air pollution model), is under development for the regional air pollution forecast. The model will be a variable scale model with nesting capabilities, using high-resolution meteorological and land use input data and able to resolve order of a few kilometers variabilities in heterogeneous terrain. The model includes 56 chemical species (only 35 species is presently used in DEOM), and include a full 3-D description of the transport, with an optional number of vertical layers.

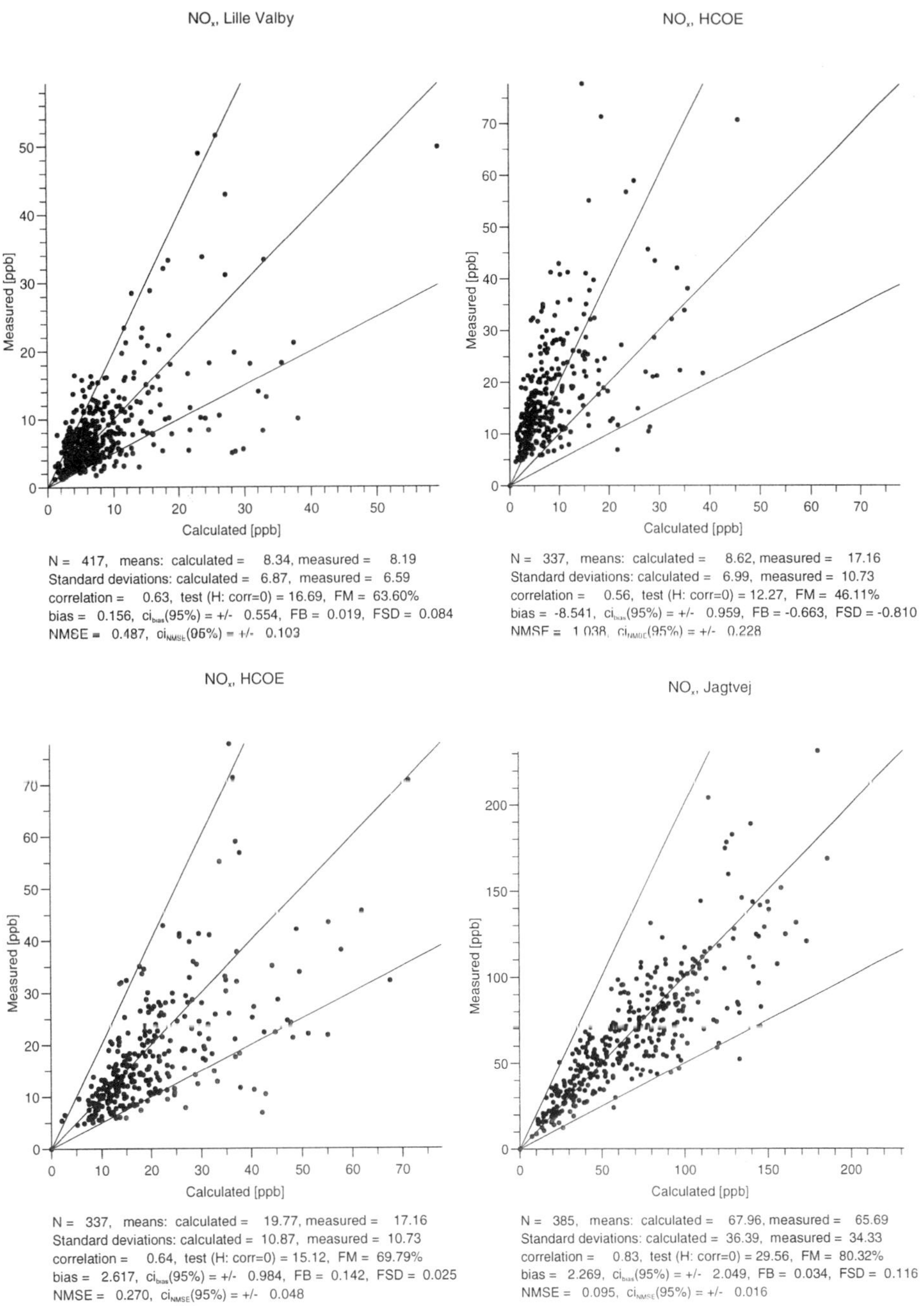

Figure 2. Comparisons of measurements and the daily mean values of the 12 hours forecasted NO_x modelled concentrations for the period August 1998 to September 1999. Upper left: DEOM results for Lille Valby – a rural site 30 km west of Copenhagen. Upper right: DEOM results for HCØ Institute (roof level) which is urban background over Copenhagen. Lower left: BUM results for HCØ Institute (roof level) which is urban background over Copenhagen. Lower right: OSPM results at street level for Jagtvej, Copenhagen, Denmark.

ACKNOWLEDGEMENTS

Dr. Kåre Kemp, National Environmental Research Institute, Denmark has kindly provided air pollution measurement data. Global meteorological data have been kindly provided by The National Centers for Environmental Prediction, NCEP, USA. Dr. Ole Hertel, National Environmental Research Institute, is acknowledged for useful comments to the manuscript.

REFERENCES

Berkowicz, R., 1999a. OSPM - A parameterized street pollution model, 2nd International Conference - Urban Air Quality, Measurement, Modelling & Management, 3-5 March, Madrid, pp. 8.

Berkowicz, R., 1999b. A simple model for urban background pollution, 2nd International Conference - Urban Air Quality, Measurement, Modelling & Management, 3-5 March, Madrid, pp. 8.

Brandt, J., 1998. Modelling Transport, dispersion and deposition of passive tracers from accidental releases, PhD Thesis. National Environmental Research Institute, Department of Atmospheric Environment, Frederiksborgvej 399, P.O. Box 358, DK-4000 Roskilde, Denmark, pp. 307.

Brandt, J., Bastrup-Birk, A., Christensen, J. H., Mikkelsen, T., Thykier-Nielsen, S., and Zlatev, Z., 1998. Testing the importance of accurate meteorological input fields and parameterizations in atmospheric transport modelling using DREAM - validation against ETEX-1, Atmospheric Environment, Vol. 32, No. 24, pp. 4167-4186.

Christensen, J. H., 1995. Transport of Air Pollution in the Troposphere to the Arctic, PhD Thesis. National Environmental Research Institute, Department of Atmospheric Environment, Frederiksborgvej 399, P.O. Box 358, DK-4000 Roskilde, Denmark, pp. 377.

Christensen, J. H., 1997. The Danish Eulerian hemispheric model - a three-dimensional air pollution model used for the arctic, Atmospheric Environment, Vol. 31, No. 24, pp. 4169-4191.

Kemp, K. and F. Palmgren, 1999. The Danish Air Quality Monitoring Programme. Annual Report for 1998. NERI Technical Report No. 296. Ministry of Environment and Energy, National Environmental Research Institute, November 1999, pp. 66.

McRae, G. J., Goodin, W. R., and Seinfeld, J. H., 1984. Numerical solution of the atmospheric diffusion equations for chemically reacting flows, Journal of Computational Physics, Vol. 45, pp. 1-42.

Nickovic, S., Mihailovic, D., Rajkovic, B., and Papadopoulos, A., 1998. The Weather Forecasting System SKIRON, Vol. II, Description of the model, Athens June 1998, pp. 228.

Pankrath, J., Stern, R., Builtjes P., 1988. Application of long-range transport models in the framework of control strategies: Example of photochemical air pollution. Environmental Meteorology. Proceedings of an International Symposium, Würzburg, 1987. K. Grefen, J. Löbel (Eds.), Kluwer Academic Publishers.

Spiegel, M. R., 1992. Theory and Problems of Statistics, 2. ed. Schaum's Outline Series. McGraw-Hill, INC. 6th print, pp. 504.

Zlatev, Z., Christensen, J., and Hov, Ø, 1992. An Eulerian air pollution model for Europe with nonlinear chemistry, Journal of Atmospheric Chemistry, Vol. 15, pp. 1-37.

DISCUSSION

R. SAN JOSE: We are running a similar operational air quality system in Madrid under a new project APNEE. My question is what are the reactions of the people when they visit the web site?

G. GEERNAERT: For scientific researchers, the responses to the THOR forecasts have been quite positive both in Denmark and abroad. The THOR system, however, has not been readily available to the public very long, and there have been no significant air pollution episodes since the THOR became operational with public access. We have therefore had very little feedback from the public so far. It is likely that after an air pollution episode is encountered in Denmark (normally a few times each summer), we could expect feedback from people accessing the web site.

EVALUATION OF THE CHEMISTRY–TRANSPORT MODEL MECTM USING TRACT–MEASUREMENTS - EFFECT OF DIFFERENT SOLVERS FOR THE CHEMICAL MECHANISM

Frank Müller[1], K. Heinke Schlünzen, Michael Schatzmann

Meteorological Institute, University Hamburg, Bundesstr. 55, Germany

INTRODUCTION

Recently, meso–scale air quality modeling systems gain more and more importance due to their application in regulatory and political decision processes. They are used e.g. for the prediction of the exceedance of trace gas threshold values, for the distribution of accidental releases, or for the calculation of trans–boundary transports of pollutants. The knowledge of the quality of the model predictions for a certain purpose, i.e. the knowledge of the variability of the model output with respect to internal and external model parameter variations, is essential for a sound judgment and application of the model results.

An essential part of an air quality modeling system is the numerical solver for the chemical reaction mechanism used for the determination of the trace gas concentrations. Two of widely applied solvers in air quality modeling are the QSSA–solver (Hesstvedt et al., 1978) and the HYBRID–solver (Young and Borris, 1977) which will be further tested in this study. Comparisons of these numerical solvers in box model simulations, which were frequently performed in the past (Kuhn et al., 1997, Sandu et al., 1998), have shown that the QSSA–solver is less mass conserving than the HYBRID–solver. This is especially true for fast reacting species (e.g. radicals) and for nitrogen dioxides (NO_x). Under moderate to polluted conditions the QSSA–solver tends to higher ozone (O_3) forecasts compared to the HYBRID–solver. For NO_x limited conditions the application of QSSA leads to an under–prediction of O_3. Projecting these results to 3D–air quality models one could expect an improvement of the NO_x forecast in the case of the HYBRID–solver. Consequently, this would improve ozone predictions, i.e. reduced ozone minima during the early morning hours (with relatively high NO values accumulated in the nocturnal boundary layer) and higher O_3 maxima develop during the early afternoon hours within the well mixed boundary layer.

Since the model result depend on both the processes regarded in the model (model equations, parmeterizations, etc.) and the model errors (discretization errors, operator splitting errors, parame–terization errors, etc.) the model output uncertainty is a result of the non–linear superposition of the uncertainties of the modeled processes. Additionally, the uncertainty of necessary input data used for the simulation (emissions, initial and boundary values, landuse, reaction rate, etc.) also contributes to the overall model output uncertainty. Therefore, the aim of this study is to investigate the impact of the numerical solution technique for the chemical reaction mechanism on the overall model performance taking into account all other 'model effects'. The significance of the change of the solver is checked against measurements.

---

1 Author to whom correspondence should be addressed. Present affiliation: Max–Planck–Institute for Meteorology, Bundesstr. 55, D–20146 Hamburg, Germany

Air Pollution Modeling and Its Application XIV, Edited by
Gryning and Schiermeier, Kluwer Academic/Plenum Publishers, New York, 2001

THE MODEL SYSTEM METRAS/MECTM

The applied non–hydrostatic meteorological model is METRAS which is described in detail in Schlünzen et al. (1996) and in Lüpkes and Schlünzen (1996). It is used to calculate all necessary meteorological fields with a time resolution of 15 minutes for MECTM. The chemistry–transport model MECTM is off–line coupled to METRAS. Simulation domain and resolution of both models are identical. The horizontal resolution was set to $5x5km^2$, while the vertical resolution changes from 20m at the ground to about 1000m at the top of the model domain which was chosen to be 11km. 15 vertical levels are allocated below 1000m above ground. For the discretization an Arakawa–C–grid with terrain following coordinates was applied. The size of the model domain was $300x300km^2$.

The chemical reaction mechanism employed in MECTM is a modified version of RADM2 (Stockwell, 1990). Beside O^{1d} and O^{3p}, which are treated in steady–state, all species are considered as prognostic. The numerical solver for the chemistry simultaneously solves chemical reactions, vertical turbulent diffusion, dry deposition, and emissions. Hence, only advective transport and ''extended chemistry' are treated in de–coupled mode. The transport time step, which is equal to the integration interval of the chemistry, is calculated based on the CFL for advection only. For the QSSA–solver the chemical time step was fixed to 1s. Since the integration intervals are always longer than 50s the number of integration cycles per integration interval is sufficiently large to ensure correct integration behavior of QSSA. The build–in time step control of the HYBRID–solver enables variable time steps which are arbitrarily bounded at a minimum time step of $\Delta t_{min}=10^{-5}s$ and at a maximum time step of $\Delta t_{max}=5min$. The METRAS/MECTM was one–way meteorologically as well as chemically nested into the EURAD modeling system MM5/CTM2 (Grell et al., 1994; Hass, 1991)

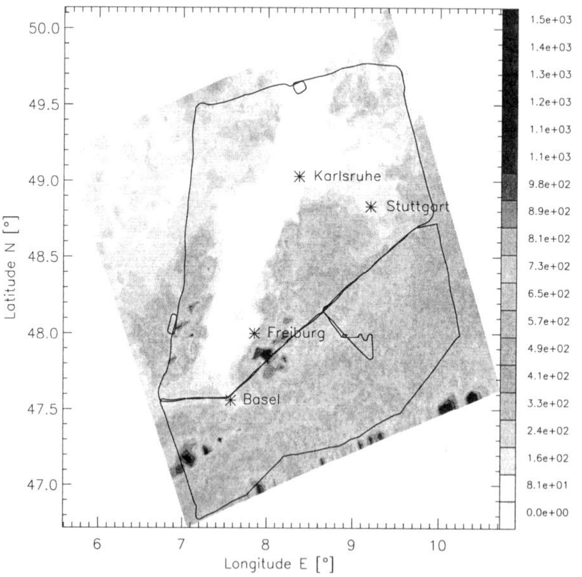

Figure 1: Orography of the model domain and flight paths of the aircraft measurements. Upper part of the flight path: North box, lower part: South box. Bar code: Orography height [m].

THE TRACT–EVALUATION CASE

For the evaluation of the model system the TRACT–intensive measurement campaign was chosen. It took place in the Upper Rhine Valley mainly in Germany in September 1992. A detailed description of the measurement campaign was given by Zimmerman (1995). For this study September, 16 was chosen where quality insured aircraft measurements of ozone, NO_x and some meteorological parameters are available from three different aircrafts. Since two of the flights had identical flight pattern the two complementary flights were used for the evaluation. Those flights took place between about 1pm and 5pm MEST on September, 16. This time period was also chosen as evaluation period. Figure 1 shows the modeling domain with topography and the corresponding flight pattern for the so–called North and South box, respectively. The meteorological situation was characterized by the passage of a week frontal system from North to South. Especially the northern part of the model domain was affected by this convergence zone by changing the wind direction and pollutant levels.

Generally, for a proper evaluation using measured data some requirements should be fulfilled: The accuracy of the used measurements has to be known. A significant number of measurements must be available which are, in addition, representative for the area of interest. Finally, the measured data should reflect as much as possible the overall situation, they should not predominantly represent local influences which are not resolved by the model. Since we use aircraft measurements, which were taken within the whole boundary layer and partly above, they do not reflect local effects which are predominantly related to the surface. With about 3500 measurements per flight also the amount of data seems to be sufficient for an evaluation. The quality of the aircraft measurements was reported by Kanter et al. (1996). The criteria for the calculation of the hit rates to compare the dependency of the model performance on the numerical solver were derived based on the reported measurement uncertainties. We used for the evaluation very high criteria: 10% of the median for the measured ozone concentration (5.1ppb), and 50% of the median for NO_x (1.6ppb).

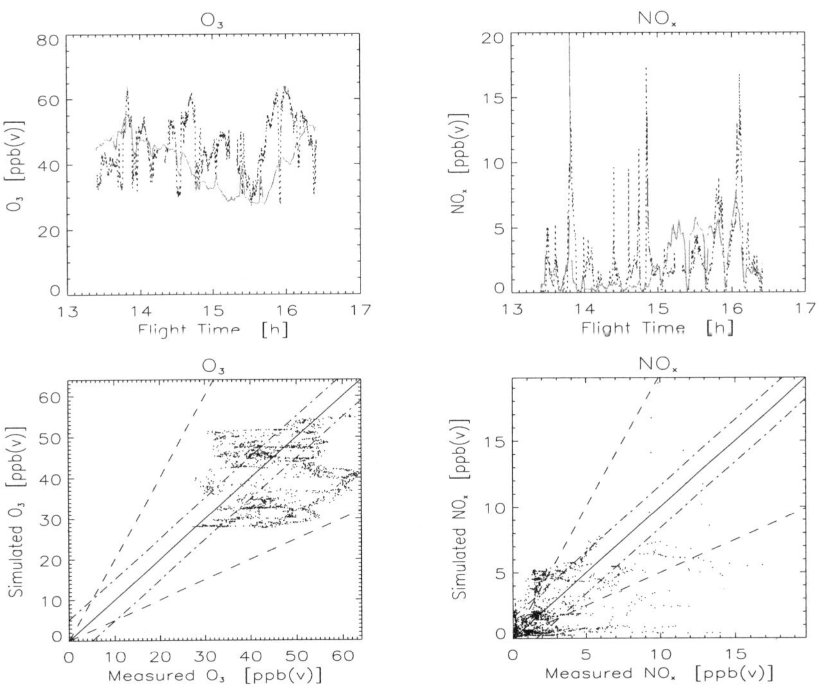

Figure 2: Comparison between aircraft measurements and model results for ozone and NO_x in the North box. Top panel: Time series (solid: model, dotted: measurement), lower panel: corresponding scatter plot. Dashed line: Factor 2–line, dotted dashed: Evaluation criteria.

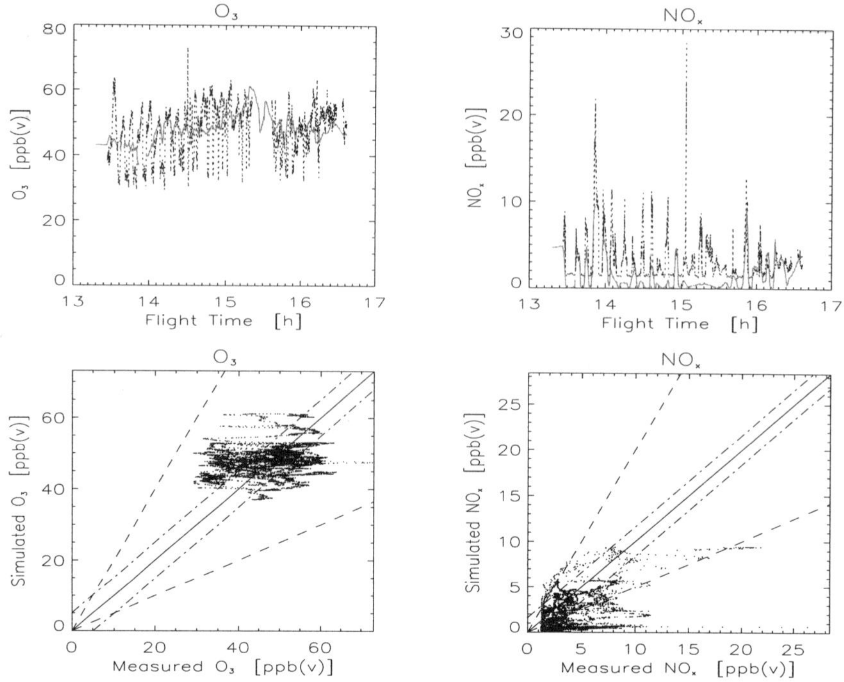

Figure 3: Same as Figure 2, but for the South box

RESULTS

For the comparison of the model results with the aircraft measurements the corresponding model output was interpolated to the flight path within the model domain. Figure 2 and 3 show the time series of measured and simulated O3 and NO_x concentrations (base case) together with the corresponding scatter plots for the North and South box, respectively. In the base case simulations the QSSA–solver was employed. It is evident from the figures that the measurements for both ozone and NO_x show more variability than the model results. This can be explained by the fact that the aircraft measurements can distinguish single plumes of elevated point sources whereas the model results represent only the volume averaged signal of all point sources located in one grid cell and their corresponding volume averaged ozone concentration. While the O_3 concentration level is reasonably well predicted in the South box, the ozone is increasingly underestimated in the North box beginning at about 2.30pm. This time the convergence zone is penetrating into the northern model domain. The passage of the convergence was predicted with a delay of about 2 hours by the larger scale model and, hence, also by the meteorological driver model METRAS. The effect of the delayed frontal passage can also be recognized for the NO_x concentration, especially for the time between 2.30pm and about 3.30pm. In the South box the predicted NO_x is in general much lower than the corresponding measurements, though the peak in the NOx concentrations are well predicted in time and, therefore, in space. The impact of the convergence line has not reached the area of the South box. Nevertheless, the NO_x measurements show a cut–off at about 2ppb(v) which contributes considerably to the bias in the comparison. This cut–off can be attributed to the detection limit of the measurement device.

The substitution of QSSA by the HYBRID–solver results in slightly higher NO_x and lower ozone concentrations as expected from the box model simulations. This behavior is documented in Figure 4 in which modeled NO_x and O_3 concentrations are compared for the used solvers for all model grid points within the planetary boundary layer. The correlation coefficients show similar values (0.9, Figure 4) for the selected species. This indicates that the simulated chemistry is not different using different solvers. However, the spread in the data, especially for NO_x, is caused by the different quality of the solver prediction predominantly close to the surface emissions.

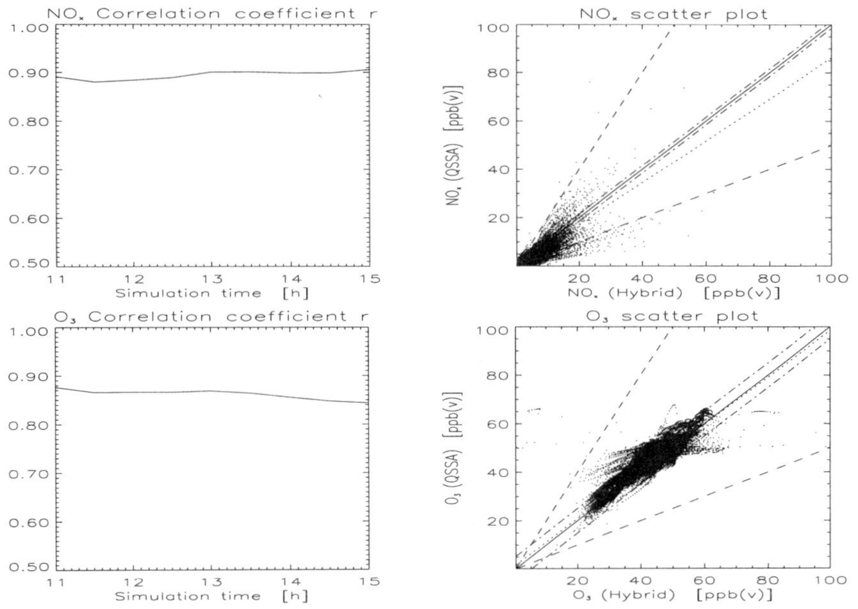

Figure 4: Sensitivity of the modeled NOx (top) and O3 (bottom) concentration to the applied solver for the chemical reaction mechanism. Data points are shown along both flight paths.

Statistical insight into the details of the differences between the measured and modeled concentrations is provided by Figure 5 for NO_x, and by Figure 6 for O_3. Shown are the cumulative frequencies and the probability densities for the concentration differences using the two solvers based on the aircraft measurements. The shaded areas depict the range of allowed uncertainties of model results based on the observation uncertainties. Values within this areas count as 'right' predicted for the calculation of the hit rate.

Figure 5 shows that the HYBRID–solver tends to decrease the marked under–prediction of NO_x and slightly increases the not very much pronounced over–prediction. The shift of the probability density function towards the center of the shaded area can also be seen for the corresponding cumulative frequency curves. The non–symmetric shape of the probability function for NO_x with higher weights for under–predicted values is an indication for the problem of the treatment of elevated point sources in the model as well as for uncertainties related to the transport. In MECTM elevated point emissions act as volume sources. They are not treated as separated plumes in the sense of Lagrangian modeling which could account for the observed spatial and temporal variability of pollutants in the boundary layer. Both over–predicted and under–predicted concentrations are the result of partly wrong transported plumes, which are resolved by the model, due to deficiencies in the wind prediction. Those plumes are generated by single big point sources or by a large number of smaller elevated point sources in one grid cell.

Figure 6 shows the statistical results for the comparison of the modeled and measured ozone concentrations. In this case the cumulative frequency curves do not show any improvement of the ozone prediction using the HYBRID solver instead of QSSA whereas the probability to predict correct O_3 values slightly increases for the whole model domain. The probability of under–prediction and over–prediction is not changed by the substitution of the numerical solver. This is likewise a strong indication for other reasons controlling the discrepancies between model results and aircraft observations like mode boundary values (Lenz et al., 2000).

Results of the comparison of aircraft measurements with smulated ozone and NO_x concentrations using the two different solvers are shown in Figure 7. The used data set is slightly decreased since only aircraft measurements within the planetary boundary layer are taken into account. Additionally, the flight paths were not distinguished. Therefore, these results are representative for the whole model domain. It follows from Figure 7 that the NO_x prediction could be slightly improved using the HYBRID solver.

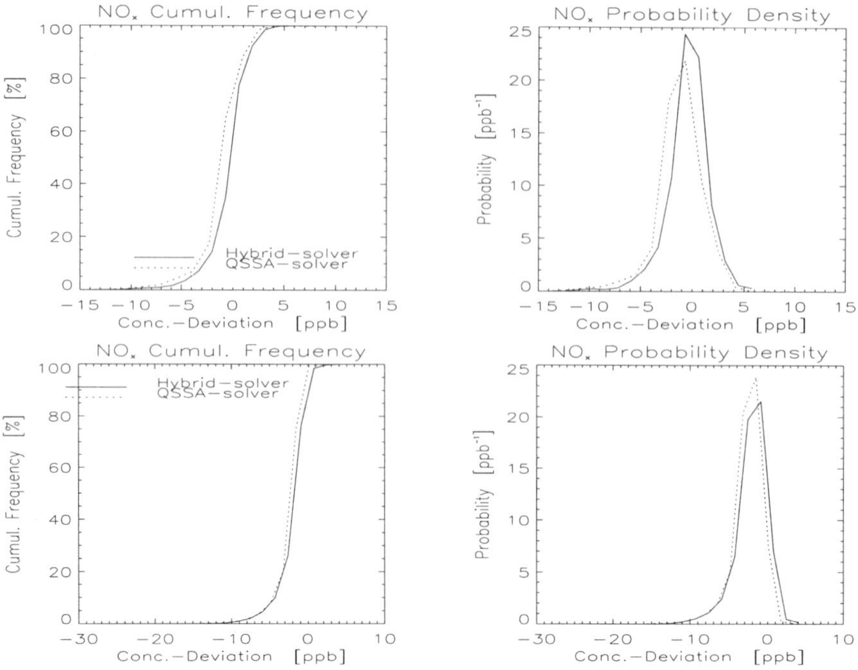

Figure 5: Cumulative frequency and probability density of the NO_x concentration differences between measurements and simulation results using two different solver. Top: North box, bottom: South box

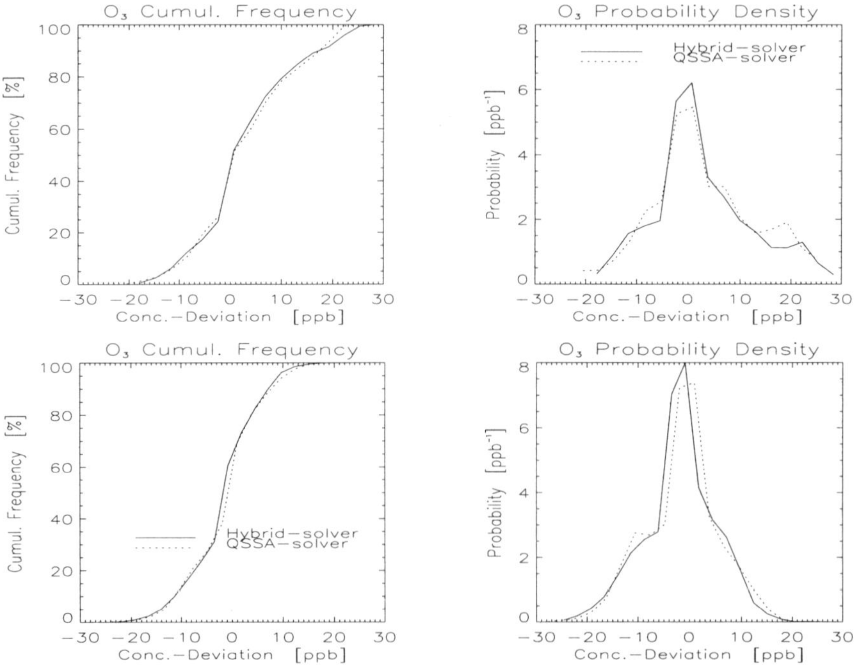

Figure 6: Same as Figure 5, but for the ozone concentration.

Median, 95%–percentile, and correlation coefficient increase. In the case of ozone (not shown here) the correlation coefficient slightly decreases. No impact on the median is seen, which coincides with the results of Figure 6. The 95% percentile is slightly increased in conformance with the increase of the O_3 probability density.

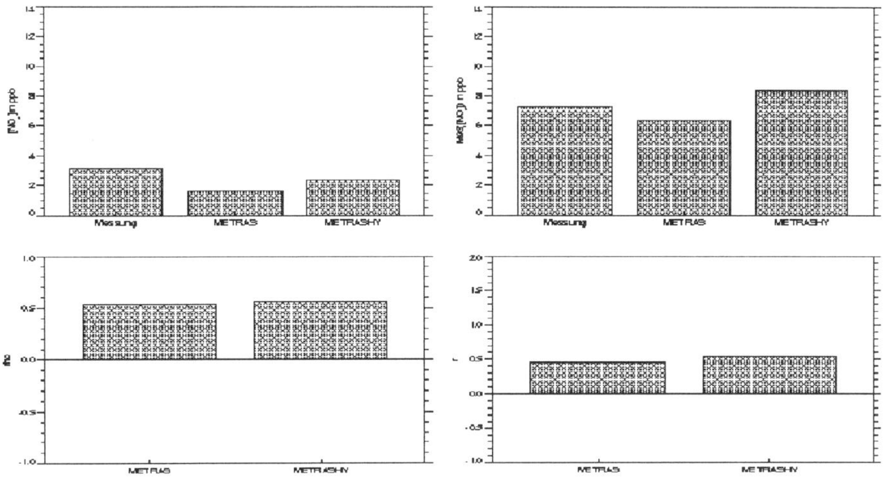

Figure 7: Statistical comparison of the NOx and O3 concentration applying the QSSA solver (indicated by METRAS) and the HYBRID–solver (indicated by METRASHY). Shown is the median (top right), 95% percentile (top left), correlation rho (bottom left), and regression r (bottom right).

CONCLUSION

This study deals with the determination of the uncertainty of the predicted concentration of selected key trace gases due to the applied numerical solver for the chemical mechanism in the lower troposphere. It could be seen that the numerical solution technique for the chemistry has an effect on the overall model result. The HYBRID–solver improves the simulated NOx concentration values while O3 concentrations are only slightly affected. The discrepancies between the model results and the corresponding measurements for the target quantities NO_x and O_3 could, however, only partly explained by the change of the numerical solver.

Since the chosen evaluation episode was transport dominated the investigated effect on the change of the pollutant concentrations is small compared to transport related changes. Therefore, the observed differences between measurements and model results can be explained to a large extend by inaccuracies of the lateral boundary values and of the predicted meteorological parameters such as wind direction and wind speed (Lenz et al., 2000) as well as by the treatment of elevated point emissions. The significance of the shown results are mainly reached by the employment of quality insured high resolution aircraft measurements.

ACKNOWLEDGMENT

As part of the Tropospheric Research Program (07TFS10/LT1–C.5) this study was funded by the German Ministry of Education and Science (BMBF). The authors gratefully acknowledge the support by both the EURAD group from the University Cologne and the Emission group at the IER Stuttgart, who have provided nesting and emission data, respectively, for the model simulations.

REFERENCES

Grell, G.A., Dudhia, J., Stauffer, D.R., 1994, A Description of the fifth generation Penn State/NCAR mesoscale model (MM5), National Center for Atmospheric Research, Boulder, Colorado, NCAR Technical Note

Hass, H., 1991, Description of the EURAD Chemistry–Transport–Model Version 2 (CTM2), Mitteilungen aus dem Institut für Geophysik und Meteorologie der Universität zu Köln, Heft 83

Hesstvedt, E., Hov, Ö., Isaksen, I.S.A., 1978, Quasi–steady–state approximations in air pollution modeling: Comparison of two numerical schemes for oxidant prediction, Int. J. Chem. Kinet., 10:971

Kanter, H.–J., Slemr, F., Mohnen, V., 1996, Airborne chemical and meteorological measurements made during the 1992 TRACT Experiment: Quality control and Assessment, J. Air and Waste Management Association, 46:710

Kuhn, M., Builtjes, P.J.H., Poppe, D., Simpson, D., Stockwell, W.R., Andersson–Skold, Y., Baart, A., Das, M., Fiedler, F., Hov, O., Kirchner, F., Makar, P.A., Milford, J.B., Roemer, M.G. M., Ruhnke, R., Strand, A., Vogel, B., Vogel, H., 1998, Intercomparison of the gas–phase chemistry in several chemistry and transport models, Atmos. Environ., 32:693

Lenz, C.–J., Müller, F., Schlünzen, K.H., 2000, The sensitivity of mesoscale chemisry transport model results to boundary values, Environmental Maintaining and Assessment, in press

Lüpkes, Ch., Schlünzen, K.H., 1996, Modelling the arctic convective boundary with different turbulence parameterizations, Boundary Layer Meteorol., 79:107

Sandu, A., Verwer, J.G., van Loon, M., Carmichael, G.R., Potra, F. A., Dabdub, D., Seinfeld, J. H., 1997, Benchmarking stiff ODE solvers for atmospheric chemistry problems. Part I: Implicit vs explicit, Atmos. Environ., 31:3151

Schlünzen, K.H., Bigalke, K., Lenz, C.–J., Lüpkes, Ch., Niemeier, U., von Salzen, K., 1996, Concept and realization of the mesoscale transport and fluid model METRAS, METRAS–Technical Report 5, Meteorological Institute, University Hamburg

Stockwell, W.R., Middleton, P., Chang, J.S., 1990, The second generation regional acid deposition model chemical mechanismfor regional air quality modeling, J. Geophys. Res. 95:16343

Young, T.R., Borris, J.P., 1977, A numerical technique for solving stiff ODEs associated with the chemical kinetcs of reactive flow problems, J. Phys. Chem., 81:2424

Zimmermann, H., 1995, Field phase report of the TRACT Field Measurement Campaign, EUROTRAC, Scientific Secretariat, Garmisch–Partenkirchen, Germany

INFLUENCE OF TURBULENCE PARAMETERIZATION ON THE MIXING LAYER HEIGHT PREDICTION WITH A MESOSCALE MODEL

Olaf Hellmuth and Eberhard Renner

Institut für Troposphärenforschung
Permoserstr. 15
D–04303 Leipzig, Germany

INTRODUCTION

The mixing layer height (MLH) is a critical parameter in determining air pollution concentrations. For example, studies performed by Rao et al. (1994) indicate that predicted ozone concentrations are extremely sensitive to errors and uncertainties in the MLH. Berman et al. (1997) show that the variability in predicted ozone concentrations due to uncertainties in the specification of the MLH evolution is comparable or greater than that due to different chemical mechanisms. Here, we present results from a case study dealing with the influence of turbulence parameterizations on the MLH prediction in the convective boundary layer (CBL) using a mesoscale model.

MODEL DESCRIPTION

We use the nonhydrostatic, three-dimensional **ME**soscale **TRA**nsport and Stream model METRAS (Schlünzen (1988), Schlünzen (1990), Schlünzen et al. (1996)). It is based on the primitive equations which ensure the conservation of mass, momentum and energy. The model equations are solved in a terrain-following coordinate system. METRAS is indended to apply to maximum areas of about 500 × 500 km^2 with horizontal grid increments between some meters and 5 km. Basic model assumptions are anelastic approximation and constant Coriolis parameter.

TURBULENCE PARAMETERIZATION

In the model, the parameterization of vertical fluxes of heat, moisture and momentum in the Prandtl layer is based on the Monin-Obukhov similarity theory, using the similarity functions of Dyer (1974). In the Ekman layer, different flux parameterizations are applied.

Profile closure after Dunst (1982) In this closure, fluxes for momentum, heat and humidity are proportional to the vertical gradients of the transported mean quantities, where the eddy diffusivity of momentum is a function of heigth, local gradient Richard-

Air Pollution Modeling and Its Application XIV, Edited by
Gryning and Schiermeier, Kluwer Academic/Plenum Publishers, New York, 2001

son number and friction velocity. Schlünzen (1990) modified this scheme to ensure the consistency of the eddy diffusivity at the top of the Prandtl layer from Prandtl layer similarity theory and from Ekman layer formulation as well as the continuity of its derivates with respect to stability. The eddy diffusivity for heat is determined from that for momentum taking into account atmospheric stability (Schlünzen et al. (1996), Lüpkes and Schlünzen (1996)).

Mixing length approach Here, fluxes are parameterized according to the downgradient approach. The eddy diffusivity for momentum is related to the mixing length for neutral stratification after Blackadar (1962) and the vertical gradient of the horizontal wind vector. Atmospheric stability is considered by the profile functions after Dyer (1974). A modified Herbert and Kramm (1985) scheme after Lüpkes and Schlünzen (1996) is used to ensure flux continuity at the top of the Prandtl layer. The eddy diffusivity for heat is derived from that for momentum considering atmospheric stability as a function of local gradient Richardson number (Schlünzen et al. (1996), Lüpkes and Schlünzen (1996)).

Countergradient scheme after Lüpkes and Schlünzen (1996) The CBL turbulence parameterization scheme proposed by Lüpkes and Schlünzen (1996) is similar to that after Troen and Mahrt (1986) and Holtslag and Moeng (1991). It considers both down- and countergradient transport terms. The eddy diffusivity for heat is parameterized using a modified empirical profile function after Zilitinkevich, which takes into account mechanical and buoyancy driven turbulence, and the countergradient flux term is that after Holtslag and Moeng (1991). Matching of Prandtl layer and CBL parameterization schemes ensures heat flux continuity at the top of the Prandtl layer. The eddy diffusivity for momentum is derived from that for heat using the Prandtl number. Nonlocal momentum fluxes are neglected. For stable stratification, the mixing length approach is taken.

TKE closure This scheme is based on the Prandtl-Kolmogorov relation for the eddy diffusivity (Kolmogorov (1942), Prandtl (1945)), after which the eddy diffusivity for momentum is in inverse proportion to the peak wavenumber of the energy density spectrum of wind velocity, and directly proportional to the square of the turbulent kinetic energy (TKE). The inverse of the peak wavenumber of the energy density spectrum can be interpreted as the mixing length, which is parameterized in terms of the mixing length for neutral stratification after Blackadar (1962) and the similarity function for momentum after Dyer (1974). For the eddy diffusivity for heat, the similarity functions for heat are used, accordingly. The TKE is determined from a forecast equation, which takes into account advective transport, shear production or loss, turbulent transport, buoyant production or consumption, humidity covariance, and loss of TKE due to dissipation. The pressure redistribution term is neglected, and the dissipation rate is determined diagnostically (for details see Schmittner and Lüpkes (1996)).

Modified countergradient scheme Following Holtslag and Moeng (1991) and Lüpkes and Schlünzen (1996), the turbulent heat flux in the CBL is splitted into a down- and a countergradient term:

$$\overline{w'\theta'} = -K_h \left(\frac{\partial \bar{\theta}}{\partial z} - \gamma_h \right) \quad , \quad \gamma_h = b \frac{w_*^2 \theta_f}{\overline{w'^2} z_i} \quad . \tag{1}$$

Here, $K_h = \tau \overline{w'^2}/2$ is the eddy diffusivity for heat, τ the return-to-isotropy time scale, $\overline{w'^2}$ the variance of the vertical velocity, γ_h the countergradient term reflecting nonlocal transports induced by large eddies, z_i is the mixing layer height, θ_f the convective temperature scale, w_* the convective velocity scale, and b the turbulence transport parameter, which is empirically given as $b = 2$. In the Prandtl layer, the heat flux is determined from the Monin-Obukhov theory. The continuity of the heat flux at the top of the Prandtl layer is ensured by matching the CBL and Prandtl layer parameterization schemes (Lüpkes and Schlünzen (1996)) leading to

$$K_h = \frac{\kappa u_* z_p}{\Phi_h|_{z_p} - \frac{\kappa z_p}{\theta_*}\gamma_h|_{z_p}} \left(\frac{\overline{w'^2}}{\overline{w'^2}|_{z_p}} \right) \quad , \tag{2}$$

where κ denotes the von Kármán constant, u_* the friction velocity, z_p the top of the Prandtl layer, Φ_h the similarity function for heat, and θ_* the kinematic temperature scale. For the variance of the vertical velocity we use a synthetic profile function, considering mechanical turbulence after Tassone et al. (1994) and buoyancy driven turbulence after Sorbjan (1989):

$$\overline{w'^2} = \left(\overline{w'^2}\right)_{mech} + \left(\overline{w'^2}\right)_{conv} \tag{3}$$

$$\left(\overline{w'^2}\right)_{mech} = u_*^2 \left(1.7 - \frac{z}{z_i}\right) \tag{4}$$

$$\left(\overline{w'^2}\right)_{conv} = w_*^2 \left[\left(\frac{z}{z_i}\right)^{2/3} \left(1 - \frac{z}{z_i}\right)^{2/3} + 0.5R^{2/3} \left(\frac{z}{z_i}\right)^{2/3} \left(1 - \frac{z}{z_i} + D\right)^{2/3} \right] \quad . \tag{5}$$

The entrainment heat flux ratio R is parameterized after Flamant et al. (1997) and the normalized entrainment layer thickness D after Physick (1997):

$$R = \frac{\left(\overline{w'\theta'_v}\right)_i}{\left(\overline{w'\theta'_v}\right)_0} \approx \frac{\kappa L}{z_i} - 0.2 \quad , \quad D = \frac{\Delta z_i}{z_i} = 0.9 \left(\frac{w_e}{w_*}\right)^{\frac{1}{3}} \tag{6}$$

with L being the Monin-Obukhov length and w_e the entrainment velocity. The eddy diffusivity for momentum is derived from that for heat using the Prandtl number (Lüpkes and Schlünzen (1996)). Nonlocal momentum fluxes are neglected. For stable stratification, the mixing length approach is taken.

APPLIED MIXING LAYER HEIGHT CRITERIONS

Heat flux minimum criterion: For convective conditions, z_i is determined diagnostically as the level, where the heat flux attains a minimum. In the case of indifferent and stable stratification, z_i is the first level above the height, where the vertical gradient of potential temperature becomes positive (Schlünzen et al. (1996), Lüpkes and Schlünzen (1996)).

Bulk Richardson number criterion: For stable, indifferent and convective conditions, z_i is determined as the level, where the bulk Richardson number exceeds a critical threshold of 0.25 (Holtslag et al. (1995)).

Modified air parcel method: In the CBL, the MLH is parameterized using a modified parcel method (MP), which is based on an idea proposed by Nelson et al. (1989). The MLH is considered to be the average $\bar{z}_{i,MP}$ (gridscale value) of an ensemble of heights z_i (subgridscale values), up to which individual buoyant air parcels with different sizes and Prandtl layer vertical velocities can rise. For each air parcel, the level of vanishing buoyancy z_i is determined by integrating balance equations for momentum, temperature and humidity. The frequency distribution of z_i depends on the occurence frequency of air parcel size and air parcel vertical velocity in the Prandtl layer, which are derived from empirical data. The model to compute $\bar{z}_{i,MP}$ is a simplified version of the approach described in Hellmuth and Renner (1999). For application in the mesoscale model, a number of CBL radiosounding profiles from the TRACT evaluation case study (see next chapter) has been used to parameterize $\bar{z}_{i,MP}$ in terms of the level z_i^{DPI} at which the dry adiabatic profile of potential temperature intersects the environmental temperature (**D**ry **P**arcel **I**ntersection method after Holzworth (1964)):

$$\bar{z}_{i,MP} = z_i^{DPI}(0.72 + 5.9 \cdot 10^{-5} z_i^{DPI}) \tag{7}$$

For stable and indifferent conditions, the bulk Richardson number criterion after Holtslag et al. (1995) is used.

EVALUATION RESULTS FROM A CASE STUDY

Model results are compared with aircraft measurements and radiosounding data from the TRACT episode 16.09.1992 (**TR**ansport of **A**ir Pollutants over **C**omplex **T**errain, 1992)(Binder (1997)), performed over complex terrain in the southwestern part of Germany, France and Swizerland. The size of the model domain is 300 × 300 km^2, topographic heights are in the range 100 m – 1400 m. The northern part of the model domain was influenced by a weak cold front with cloud development. Winds are from NW with $u < 5$ m/s in the planetary boundary layer and $u > 10$ m/s above. The grid resolution is $\Delta x = \Delta y = 5$ km, and $\Delta z = 20 - 230$ m. To force the model, we use forecast data from the model system EURAD-MM5 (Ebel et al. (1997)) with a mesh size of 25 km. The model is integrated over 24 h starting at 15.09.92, 1800 Local Time. MLH model results are compared to that from daytime radiosounding data and afternoon Do228 aircraft measurements (flight D-CALM and D-CFFU). Measured data are projected onto forecast data according to the pointer method (Schaller and Wenzel (1999)). The MLH from aircrafts measurements is determined as the mean value of lower and upper boundary of the temperature inversion. In the case of radiosounding data, the MLH determination depends on the applied MLH criterion in the model. Using the heat flux minimum criterion, the MLH is determined from the DPI method. Otherwise the same criterion is used as in the model. A "MLH hit rate" is determined considering forecast MLH to be a hit when its deviation from the observation is lower or equal 30 % (Schaller and Wenzel (1999)).
From Table 1 follows, that the correlation coefficients between modelled and observed mixing layer heights and the hit rates in the North-box (D-CALM) are generally higher than in the South-box (D-CFFU). Compared to the northern part which is largely covered by the wide Rhine valley and low mountain range, the southern part has a more complex orography with the foothills of the Alps and high mountain range reaching larger relief amplitudes. The synoptical analysis exhibits ascending air masses with formation of convective clouds and higher MLH's in the northern and northeastern part and descending air masses in the southern part of the model domain (Binder (1997)).

The comparison of different parameterization schemes and MLH criterions shows, that the prognostic TKE closure does not result in a generally higher MLH correlation coefficient and hit rate than the diagnostic schemes. The application of the MLH criterion (7) tends to improve the MLH hit rate at least with respect to the aircraft measurements. Note that the radiosounding data considered here, are taken from morning and afternoon soundings in the North- and the South-box. Therefore, verification results based on radiosounding data reflect the overall applicability of the investigated turbulence parameterizations. Figure 1 shows some selected scatter plots of modelled and observed MLH's for the mixing length approach after Herbert and Kramm (1985), the countergradient scheme after Lüpkes and Schlünzen (1996) and the TKE closure after Schmittner and Lüpkes (1996), each with the bulk Richardson number criterion and Equation (7). For all investigated parameterization schemes, the largest, unbiased scatter occure for the heat flux minimum criterion (not presented here), while the bulk Richardson number criterion tends to overestimate the MLH. Slightly better results are obtained with Equation (7). As already noticed, the TKE closure does not result in a better agreement between modelled and observed MLH's than the diagnostic schemes. The corresponding MLH forecast fields presented in Figure 2 show similar pattern for all parameterization schemes with higher MLH's using the bulk Richardson criterion instead of Equation (7). The simulations confirm the enlargement of the MLH in the northern part owing to convective activity associated with the weak cold front. The very low values in the southeastern part results from stable stratification over Lake Constance during the day. In Figure 3, results with the modified countergradient scheme are depicted, which show a rather good agreement between modelled and observed values. Correlation coefficients and hit rates are $r = 0.84$ and $T = 77$ % for D-CALM aircraft data, $r = 0.26$ and $T = 44$ % for D-CFFU aircraft data, and $r = 0.62$ and $T = 50$ % for the radiosounding data.

SUMMARY AND CONCLUSIONS

In the case study presented here, a maximal MLH hit rate of 88 % in the CBL over the northern part of the model domain is achieved. However, despite of its tolerant definition, the maximal MLH hit rate over the mountainous terrain in the South does not exceed 39 %. Possible explanations for the low hit rate may be shortcomings of the applied parameterizations schemes over complex, inhomogeneous terrain, grid resolution and model forcing. As the prognostic TKE closure does not provide basically better results than the countergradient schemes, we conclude that the characteristic velocity scale is properly considered in the countergradient schemes. Therefore, the low hit rate may be also adressed to the parameterization of the integral length scale. This is suggested by sensitivity studies with the TKE closure, showing a strong influence of the integral turbulence length scale on the MLH prediction. Nevertheless, the verification results presented here are similar to findings from Sørensen and Rasmussen (1997), who exemplary quote a correlation coefficient of $r \approx 0.4$ for their 24-h MLH forecast using the DMI-HIRLAM NWP model (Danish Meteorological Institute, High Resolution Limited Area Model) verified against ETEX radiosounding data (European Tracer Experiment 1994). Our findings are also in the range of evaluation results from Schaller and Wenzel (1999) for five regional atmospheric chemistry transport models showing correlation coefficients between 0.4 and 0.6 and overall hit rates between 49 and 69 % for a 24-h MLH forecast period of TRACT, 16.09.92.

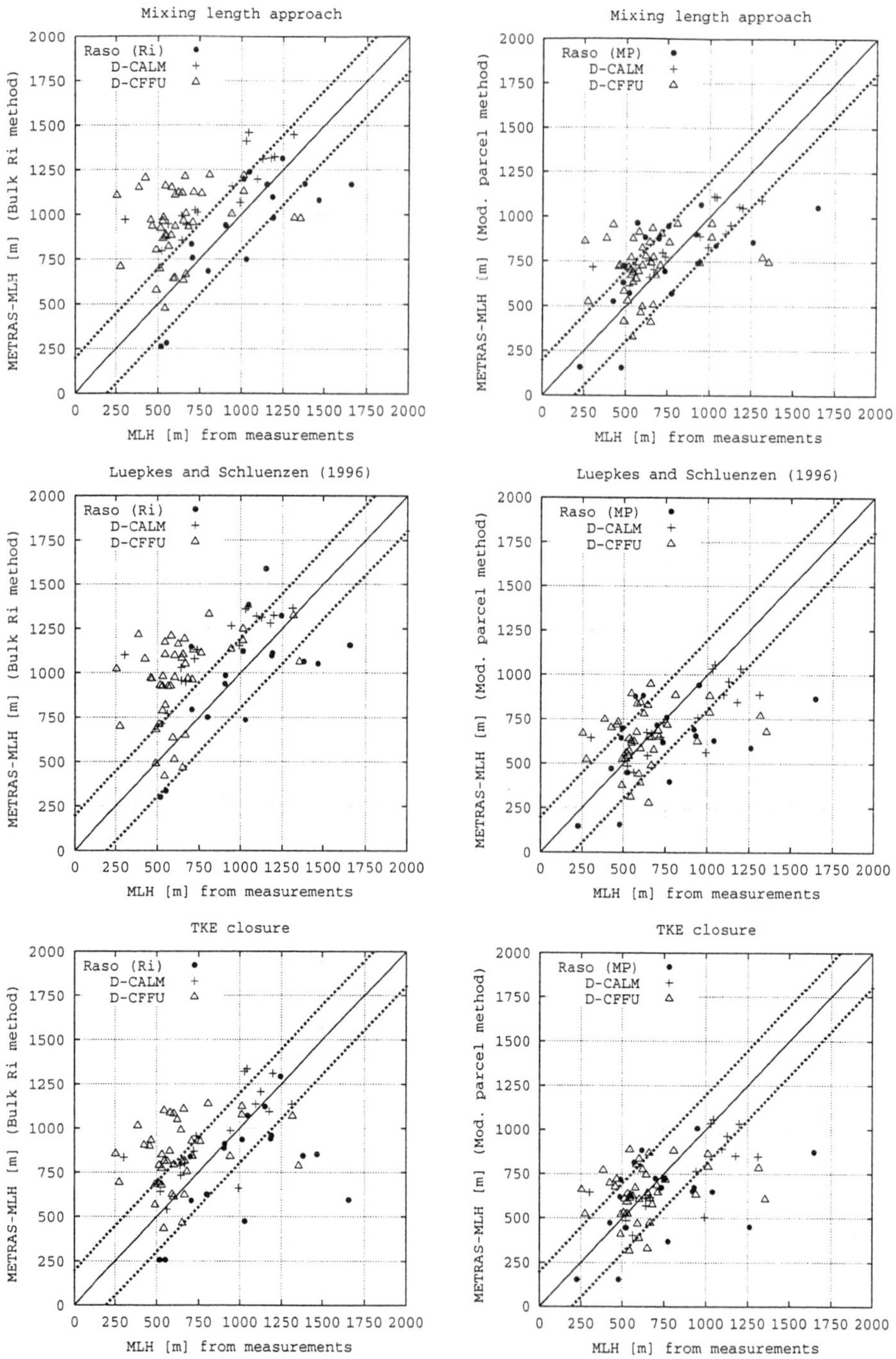

Figure 1. TRACT,16.09.92: Comparison of modelled and observed mixing layer heights. (Left: Bulk Richardson number criterion. Right: Modified parcel method.)

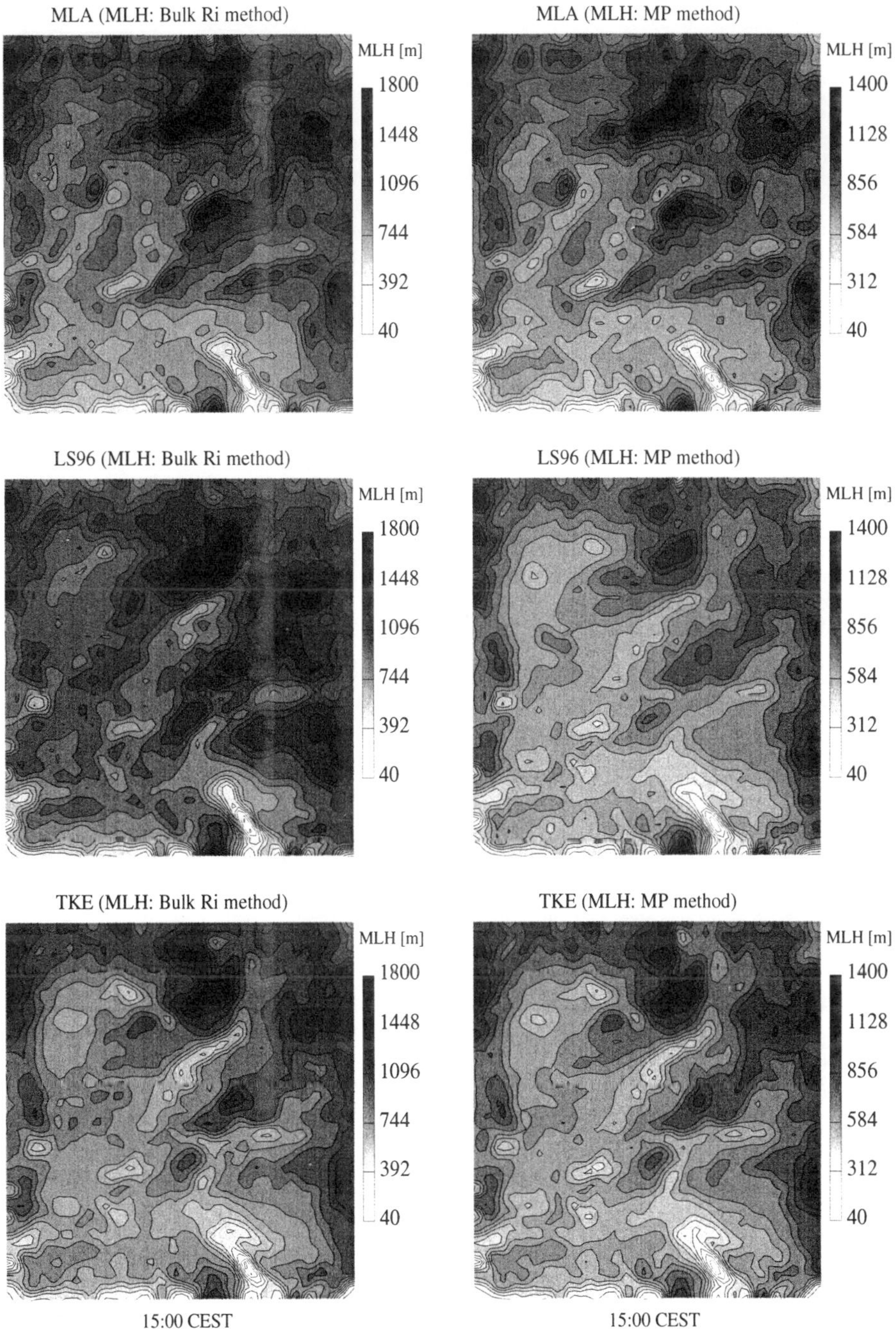

Figure 2. TRACT,16.09.92,15 CEST: Examples of MLH forecast. (Left: Bulk Richardson number criterion. Right: Modified parcel method. Top to bottom as in Figure 1.)

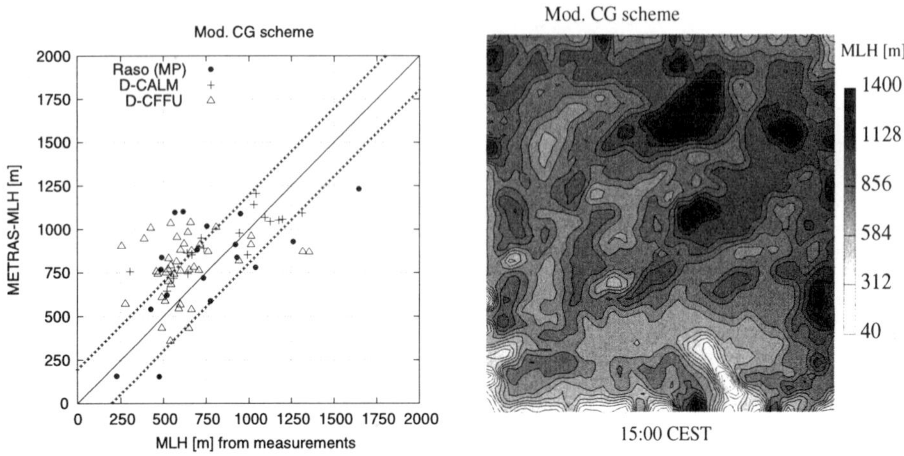

Figure 3. TRACT,16.09.92: Modified countergradient scheme with MLH according to Equation (7). (Left: Comparison of modelled and observed mixing layer heights. Right: Afternoon MLH forecast.)

Table 1. TRACT, 16.09.92: Correlation coefficient (**r**) and hit rate (**T**) of z_i (D-CALM: North-box, time 13.12-16.47 CEST; D-CFFU: South-box, time 13.26-16.38 CEST)

Turbulence scheme	**Data**	**MLH criterion**					
		$\left(\overline{w'\theta'}\right)_{z_i} = \min$		$(Ri_b)_{z_i} \geq 0.25$		$\bar{z}_{i,MPM}$	
		r	**T** [%]	**r**	**T** [%]	**r**	**T** [%]
Profile closure after Dunst (1982)	D-CALM	0.57	53	0.85	29	0.83	88
	D-CFFU	0.34	34	0.30	20	0.29	51
	Radiosonde	0.63	44	0.77	61	0.67	56
Mixing length scheme after Herbert and Kramm (1985)	D-CALM	0.67	35	0.87	35	0.87	88
	D-CFFU	0.30	37	0.23	22	0.23	56
	Radiosonde	0.74	44	0.79	56	0.66	56
CG scheme after Lüpkes and Schlünzen (1996)	D-CALM	0.70	71	0.81	35	0.76	71
	D-CFFU	0.08	46	0.39	24	0.27	58
	Radiosonde	0.64	56	0.61	44	0.48	44
TKE closure after Schmittner and Lüpkes (1996)	D-CALM	0.41	47	0.75	71	0.74	71
	D-CFFU	0.33	51	0.29	44	0.25	58
	Radiosonde	0.53	22	0.53	50	0.44	44

References

Berman, S. et al., 1997. Uncertainties in estimating the mixing depth - comparing three mixing-depth models with wind profiler measurements. *Atmos. Environ.*, 31 (18):3023–3039.

Binder, H.-J., 1997. *Tageszeitliche und räumliche Entwicklung der konvektiven Grenzschicht über stark gegliedertem Gelände.* PhD thesis, Universität (TH) Karlsruhe, Fakultät für Physik. 309 S.

Blackadar, A. K., 1962. The vertical distribution of wind and turbulent exchange in a neutral atmosphere. *J. Geophys. Res.*, 67.

Dunst, M., 1982. On the vertical structure of the eddy diffusion coefficient in the pbl. *Atmos. Environ.*, 16:2071–2074.

Dyer, A. J., 1974. A review of flux-profile relationship. *Boundary-Layer Meteorology*, 7:362–372.

Ebel, A. et al., 1997. Air Pollution Studies with the EURAD Model System (3): EURAD - European Air Pollution Dispersion Model System. Mitteilungen, Institut für Geophysik und Meteorologie der Universität zu Köln.

Flamant, C. et al., 1997. Parameterization of the depth of the entrainment zone above the daytime mixed layer. *Boundary-Layer Meteorology*, 83:247–284.

Hellmuth, O. and E. Renner, 1999. Diagnostic determination of mixing layer height and entrainment layer thickness in the convective boundary layer using a spectral entraining jet model. Contr. Atmos. Phys. Accepted.

Herbert, F. and G. Kramm, 1985. Trockene Deposition reaktionsträger Substanzen, beschrieben mit einem diagnostischen Modell der bodennahen Luftschicht. In Becker, K. H. and J. Löbel, editors, *Atmosphärische Spurenstoffe und ihr physikalisches Verhalten.*, page 264. Springer Verlag.

Holtslag, A. A. M. and C.-H. Moeng, 1991. Eddy diffusivity and countergradient transport in the convective atmospheric boundary layer. *J. Atmos. Sci.*, 48:1690–1698.

Holtslag, A. A. M., E. van Meijgaard and W. C. D. Rooy, 1995. A comparison of boundary layer diffusion schemes in unstable conditions over land. *Boundary-Layer Meteorology*, 76:69–95.

Holzworth, G. C., 1964. Estimates of mean maximum mixing depths in the contiguous United States. *Mon. Wea. Rev.*, 92:235–242.

Kolmogorov, A. M., 1942. Equations of turbulent motion of an incompressible fluid. *Izv. Akad. Nauk SSSR, Seria Fizicheska Vi.*, 1-2:56–58.

Lüpkes, C. and K. H. Schlünzen, 1996. Modelling the arctic convective boundary-layer with different turbulence parameterizations. *Boundary-Layer Meteorology*, 79: 107–130.

Nelson, E., R. Stull and E. Eloranta, 1989. A prognostic relationship for entrainment zone thickness. *J. Appl. Met.*, 28:885–902.

Physick, W. L., 1997. Recent developments in closure and boundary conditions for Lagrangian stochastic dispersion models. In Gryning, S.-E. and N. Chaumerliac, editors, *22nd NATO/CCMS International Technical Meeting on Air Pollution Modelling and its Application*, pages 214–227. Imprimerie des UFR Sciences, Clermont-Ferrand, France, Plenum Press.

Prandtl, L., 1945. Über ein neues Formelsystem für die ausgebildete Turbulenz. *Nachr. Akad. Wiss. Göttingen, Math.-Phys. Klasse.* S. 6.

Rao, S. T. et al., 1994. Sensitivity of the urban airshed model to mixing height profile. In *Proceedings of the 8th AMS/AWMA Joint Conference on the Applications of Air Pollution Meteorology, Nashville, TN.*

Schaller, E. and A. Wenzel, 1999. Evaluierung regionaler atmosphärischer Chemie-Transport-Modelle, Fall 1: TRACT, 16.9.1992, 13-17 Uhr MESZ. Technical report, Lehrstuhl für Umweltmeteorologie, Fakultät 4 - Umweltwissenschaften und Verfahrenstechnik, Brandenburgische Technische Universität Cottbus.

Schlünzen, K. H., 1988. Das mesoskalige Transport- und Strömungsmodell 'METRAS'– Grundlagen, Validierung, Anwendung. *Hamburger Geophysikalische Einzelschriften*, A88:139 pp.

Schlünzen, K. H., 1990. Numerical studies on the inland penetration of sea breeze fronts at a coastline with tidally flooded mudflats. *Contr. Atmos. Phys.*, 63:243–256.

Schlünzen, K. M. et al., 1996. Concept and realization of the mesoscale transport and fluid model METRAS. Technical report, Meteorologisches Institut der Universität Hamburg.

Schmittner, A. and C. Lüpkes, 1996. Parametrisierung subskaliger Flüsse in einem mesoskaligen Modell der arktischen atmosphärischen Grenzschicht. Master's thesis, Universität Bremen.

Sorbjan, Z., 1989. *Structure of the Atmospheric Boundary Layer.* Prentice Hall, New Jersey.

Sørensen, J. H. and A. Rasmussen, 1997. Mixing height derived from the DMI-HIRLAM NWP model, and used for ETEX dispersion modelling. In Gryning, S.-E., F. Beyrich and E. Batchvarova, editors, *The Determination of the Mixing Height - Current Progress and Problems*, EURASAP Workshop Proceedings, 1-3 October 1997, pages 41–44. Risø National Laboratory Roskilde, Denmark.

Tassone, C., S.-E. Gryning and M. W. Rotach, 1994. A random walk model for atmospheric dispersion in the daytime boundary layer. In Gryning, S.-E. and M. M. Millan, editors, *Air Pollution Modelling and its Application X*, pages 204–250. Plenum, New York and London.

Troen, I. and L. Mahrt, 1986. A simple model of the atmospheric boundary layer; sensitivity to surface evaporation. *Boundary-Layer Meteorology*, 37:129–148.

DISCUSSION

P. SEIBERT: I agree very much with your last conclusion, that turbulence measurements in the free atmosphere would be useful. I wonder, however, why you based your evaluation of the turbulence scheme only on mixing heights. They cannot be directly measured, the so-called measured MH must have been derived from the measurements with one of the methods - involving uncertainty. As the measurements give the temperature and humidity profiles throughout the whole BL and the entrainment layer, why don't you use the whole profiles for the comparison?

O. HELLMUTH: The results that had been presented here are obtained within the framework of Principal Subject (PS) 1 of the Tropospheric Research Program of the Federal Ministry for Education, Science, Research and Technology. The aim of PS 1 was to establish and improve the diagnostic and forecasting capabilities of atmospheric chemistry models and to test them in practice. In order to determine the accuracy and the limits of the applicability of present community chemistry transport models for the regional ozone forecast, a comprehensive model evaluation via simulation of test-cases, of which field data are available, has been performed. Because of its relevance for air pollution problems the mixing layer height has been selected as a meteorological key parameter in addition to chemical parameters. For the time being only aircraft measurements from two "sawtooth" flight pattern of the TRACT experiment, 16.09.92, were available. From these flight patterns the mixing layer heights are derived for the model verification. Due to the lack of corresponding ground-based measurements, several uncertainties result in determining the MLH. For example, missing heat flux and air parcel temperatures in the surface layer do not allow the application of the bulk Richardson number criterion from the aircraft measurements. To reduce the uncertainties mentioned above, available radiosounding data from the daytime boundary layer have been included later on. Despite of the remaining problems of "measuring the MLH", in the present approach all models and turbulence parameterizations have been verified against the same "defective MLH measurements". Doing this we believe that a predication of the relative capability of the models and parameteriation schemes is warrantable and a relative weighting of parameterization goodness using a "MLH hit rate" as a reference value should be reasonable. However, we agree that a validation of the MLH prediction in terms of absolute units is of limited value due to the difficulties in determining the MLH from experimental data. Going a step further, a more detailed analysis of the TRACT data using vertical profiles of the relative humidity, the potential temperature and horizontal wind velocity at preselected grid points has been performed in the frame of a diploma thesis by Eger (2000: Simulation der Mischungsschichthöhe über komplexem Terrain in Abhängigkeit von der Turbulenzparametrisierung. Master's theis, Universität Leipzig, Fakultät für Physik und Geowissenschaften). As the results do not show a uniform picture of the behavior of the different turbulence parameterizations we refer the reader for further details to Eger (2000), which can be provided on demand. Our main motivation to perform a comparison of modelled and "measured" MLH was to make our results comparable to findings from other model groups within the project framework and from the literature. Our present activities are directed on the evaluation of the new operative mesoscale model (LM - Lokal Modell) of the German Weather Service using top quality measurements from aircraft, radiosounding ascents, free balloon flights, windprofiler, Sodar/Rass and Lidar from the BERLIOZ ozone campaign from summer 1998.

H. van DOP: In the countergradient approach the temperature profile tends to be more stable than in the standard approach. As a consequence the temperature jump over the inversion, Δ, tends to be smaller. This leads to higher entrainment velocity estimates using: $w_e = (\overline{w\theta_i'})/\Delta$ although the inversion layer height z_i may therefore be larger than observed, in agreement with your observations.

O. HELLMUTH: Splitting the turbulent heat flux in the CBL into a down- and a countergradient part as described in the corresponding ITM paper with the countergradient term being always > 0 we find, that the "countergradient corrected" gradient of the potential temperature is lower compared with the standard gradient of the potential temperature. This means that the countergradient approach tends to destabilize the CBL (see Fig. 1 for illustration). Due to the reduction of the temperature gradient in the core of the CBL and in its upper part the point of intersection of the environmental profile of the potential temperature and the dry adiabatic profile is lifted which leads to an enhancement of the MLH. This argument is in accordance with our observations and corresponds to Van Dop's arguments.

Temperature profile in the CBL in the downgradient and countergradient approach

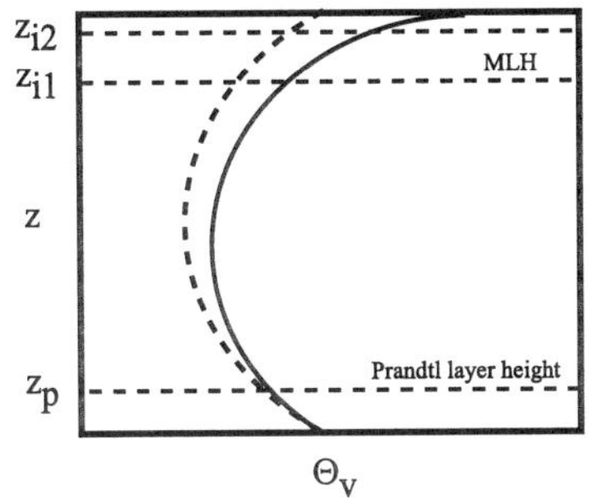

—— standard temperature profile (downgradient approach) (z_{i1})

- - - - "countergradient corrected" temperature profile (z_{i2})

Figure 1 Differences between the vertical profiles of the potential temperature in the standard approach and in the countergradient flux representation.

H. van DOP: Also the Holzworth method tends to overestimate z_i, since a parcel is moving vertically adiabatically. In reality the parcel will cool a bit due to mixing, leading to lower estimates of z_i, see figure below.

Influence of lateral mixing on MLH determination

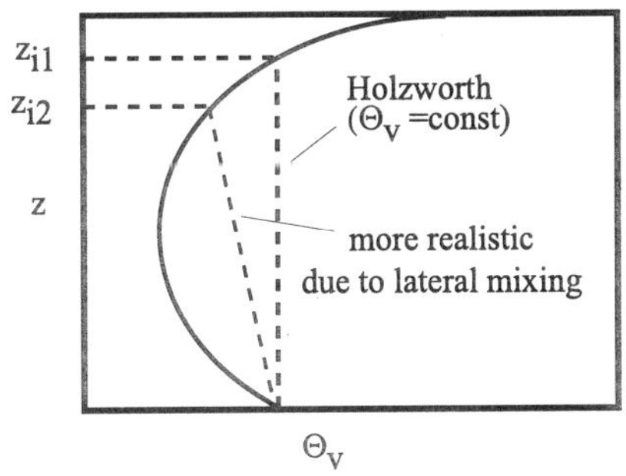

—— environmental temperature profile
- - - - individual air parcel temperature profile

O. HELLMUTH: To overcome the overestimation of the MLH according to the Holzworth method, a modified air parcel method has been implemented in our mesoscale model which does exactly consider the decreasing effect of lateral mixing on the air parcel buoyancy with the consequences mentioned by Van Dop. In this approach the MLH is considered to be the average (gridscale value) of an ensemble of heights (subgridscale values), up to which individual buoyant air parcels with different sizes and Prandtl layer vertical velocities can rise. For each air parcel, the level of vanishing buoyancy is determined by integrating balance equations for momentum, temperature and humidity using an entraining jet model which considers the lateral mixing of heat, humidity and momentum from the environment into the parcels. The frequency distribution of the height of vanishing buoyancy depends on the occurrence frequency of air parcel size and air parcel vertical velocity in the Prandtl layer, which are derived from empirical data. From the TRACT data we found, that new method provides MLHs which are nearly 30 percent lower compared with that from the Holzworth method. In the present case we performed simulations with both the countergradient flux representation leading to an enhancement of the MLH as well as the modified air parcel approach resulting in a substantial reduction of the MLH. As a net effect an overestimation of the MLH is observed. From this result we conclude, that the countergradient heat flux is overestimated in our approach.

EVALUATION AND FURTHER APPLICATION OF THE MICRO-SCALE PHOTOCHEMICAL MODEL MICRO-CALGRID

R. Stern[1] and R. J. Yamartino[2]

[1]Free University Berlin, Institute of Meteorology, Carl-Heinrich-Becker-Weg 6-10, D-12165 Berlin, Germany
[2]191 East Grand Avenue, Old Orchard Beach, ME 04064, USA

INTRODUCTION

This paper describes the first evaluation of the micro-scale photochemical model, MICRO-CALGRID, MCG, using full-scale, urban field programs conducted within street canyons in Berlin and Milan.

MCG is both a 2-d and 3-d photochemical model for high-resolution studies of complex urban environments, including street canyons and urban quarters, that:

- includes two modern photochemical reaction schemes, the SAPRC-93 and the latest CBM-IV chemical mechanism, and a simple NO-NO_2-O_3-chemistry scheme that reflects the short traveling times of pollutants within modelling domains extending over only one or just a few canyons;
- treats all individual buildings as obstacles with impenetrable surfaces;
- accepts the 3-dimensional, micro-scale, atmospheric flow and turbulence fields generated by the Navier Stokes flow model MISKAM;
- adds in the turbulence effects of multi-lane, volume and velocity dependent vehicle traffic;
- links to the emissions model MOBILEV to provide all traffic related emissions;
- considers all linear removal mechanisms and non-linear chemical transformations; and
- links to the urban scale photochemical grid model CALGRID, allowing integrated assessments of the urban- and micro-scales.

A more detailed model description can be found elsewhere (Stern and Yamartino, 1998). The development of the MCG micro-scale modelling system was initiated by the Federal Republic of Germany's (FRG) Environmental Protection Agency (Umweltbundesamt) as part of a programme designed to fulfil the requirements on air quality assessment defined in the new EU air quality framework directive 96/62/EU and its daughters directives. This paper gives an overview about an evaluation study carried out within the framework of the European Community's AUTO OIL 2 Programme, AOP2, (Skouloudis et al., 1999). The detailed results can be found in Stern (1999).

THE BERLIN-SCHILDHORNSTRASSE APPLICATION

Street geometry

The Schildhornstraße is a major inner city four-lane road in the Berlin district of Steglitz. It runs from north-west to south-east and serves as an access route to the city motorway. The Schildhornstrasse street canyon is 26 m wide and the height of the buildings on the left (north) side is 19 m, and 22 m on the right (south) side. A permanently operating measuring container (station 117) is located at the kerb on the south side and a measuring bus (station 088) is located in the parking lane on the other (north) side of the street. The latter station was only operating during a measurement programme lasting from Dec 4, 1994 through March, 10, 1995.

Meteorological data

For the AOP2-modelling period, Feb 21 (Tuesday)– Feb 25 (Saturday), 1995, meteorological data were taken from a meteorological measuring station located on a roof top in Berlin-Schoeneberg, approx. 3 km north-east of the Schildhornstrasse. Figure 1 shows wind speed and wind direction during the modelling period. During the first four days of the modelling period, the flow is dominated by wind direction between west and south. In the evening of Feb 24, the wind direction turns to north. On Feb 25, it turns toward west and south to easterly directions. Wind speed varies most of the time between 2 and 6 m/s. Very low wind speeds are measured in the night from Feb 24 to Feb 25.

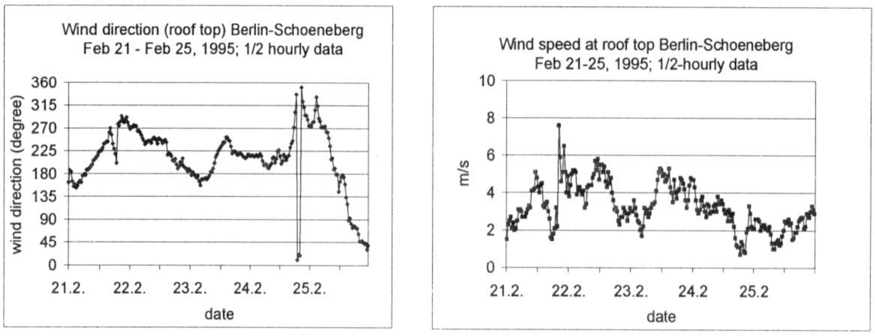

Figure 1 Wind direction and wind speed in the modelling period Feb 21 – Feb 25, 1995

Air quality data

The following pollutants were measured in the Schildhornstraße: CO, NO, NO_2 and TSP half-hourly at station 088 and station 117; SO_2 half-hourly at station 117; benzene as daily average at station 088. The measurements represent total concentrations, i.e. the pollution caused by the traffic in the Schildhornstrasse plus the background pollution caused by all other sources.

Model region

The modelling region is set up in such a way that the Schildhornstrasse runs parallel to the x-axis. The chosen grid size in x-direction along the Schildhornstraße is 12 m. The grid-size in y-direction is 2.9 m (width of a lane). Thus, the cross-section in the Schildhornstrasse street canyon is resolved by 9 grid cells. Overall, the horizontal extent of the modelling domain is approximately 1000 m in x-direction and 300 m in y-direction. The meteorological modelling region has been enlarged to make sure that the inflow conditions for the air quality modelling region are not disturbed by boundary effects. The atmosphere is resolved in the vertical up to 100 m, with grid sizes of 3 m in the lowest 24 m and increasing grid sizes above this height.

Modelling of the emissions

Hourly emissions for the modelling period were calculated with the emissions model MOBILEV (Fige, 1997). This model requires lane-specific information about the average daily traffic (ADT) and the percentage of heavy-duty vehicles as primary inputs. Additional supplemental input data include: a street characterization, a slope of the street, percentage of weekday, Saturday and Sunday traffic cycles, speed of the vehicles, temperature data, stop-and-go times, cold-start percentages and fleet compositions. Most of these data were available from the extensive traffic counting from Dec 1994 until March 1995; otherwise, default values were applied. Emissions for all 11 streets within the modelling domain were calculated with MOBILEV as hourly values for the species NOx, CO, benzene, SO_2 and TSP. The ADT for the Schildhornstraße was about 45.000 vehicles per day with approximately 6% heavy trucks.

Modelling of the flow conditions

The flow and turbulence conditions were calculated with MISKAM (Eichhorn, 1996). Major input data for MISKAM are the building geometry, and wind direction and wind speed upwind of the building geometry. The meteorological data were derived from the hourly measurements in Berlin-Schoeneberg (see Fig. 1). The results of the MISKAM model runs are stationary, three-dimensional wind- and turbulence fields for each simulation hour. These fields are input to MCG.

Determination of background concentrations

MCG needs hourly resolved background concentrations that describe the situation outside the modelling region. In the Schildhornstrasse application, these input data were derived from measurements. Urban background pollution data were available from several stations. Because of the rather small differences between the observations at these stations, the average of the concentrations was used as background for MCG.

MCG base run results

To initialize MCG, a pre-episode period from Feb 20 was run and the model was applied for the whole period without any further re-initialization. The CO, NO_2 and TSP time series shown in Figure 2 represent the concentrations calculated for the grid cells where the air is drawn into the monitoring devices of stations 117 and 088. The overall behaviour of the observations is modelled quite well at both stations. On the first three days of the episode, wind directions were in the south to west sectors (Fig. 1), so that the roof top inflow directions were effectively perpendicular to the orientation of the Schildhornstrasse. Hence, in this part of the episode, station 117 was always down-wind of the sources at the canyon ground and observed and calculated concentrations at station 117 were much higher than those at station 088. The morning and afternoon emission peaks are reflected in the observations and the calculations. In the evening of Feb 24, the wind direction turned towards north (Fig. 1), the direction of the vortex in the street canyon reversed and station 088 became the down-wind station for a few hours. In the evening of Feb 24, the wind speeds were very low and a rather high peaks were observed and calculated at station 088. Later on, the wind turned over west and south towards easterly directions. Therefore, on Feb 25, there was no clear down-wind or up-wind station. In general, the low observed concentrations at station 088 are overestimated. This might be due to the combined effects of the numerical, physical and vehicle induced diffusion that seems to lead to an overestimation of mass-transport to the upwind station. Figure 2 also shows the background concentrations used in the MCG run. It is obvious that most of the observed and calculated CO concentrations at the down-wind station are due to the emissions inside the canyon. For both, measured and calculated NO_2 concentrations, the differences between down-wind and up-wind concentrations are much smaller than for the primary pollutant

CO. This is partly due to the chemical formation of NO_2 and partly due to the much higher fractional role of the background in the canyon concentrations. As for CO, the observed low NO_2 peak concentrations are slightly overestimated at the upwind-station 088. The peak overestimation is more pronounced for NO_2 .This might be an indication that the estimated NO_2 background could be a little bit too high. Besides the few underestimated isolated peak TSP observations the agreement of measured and observed TSP concentration levels is also quite good.

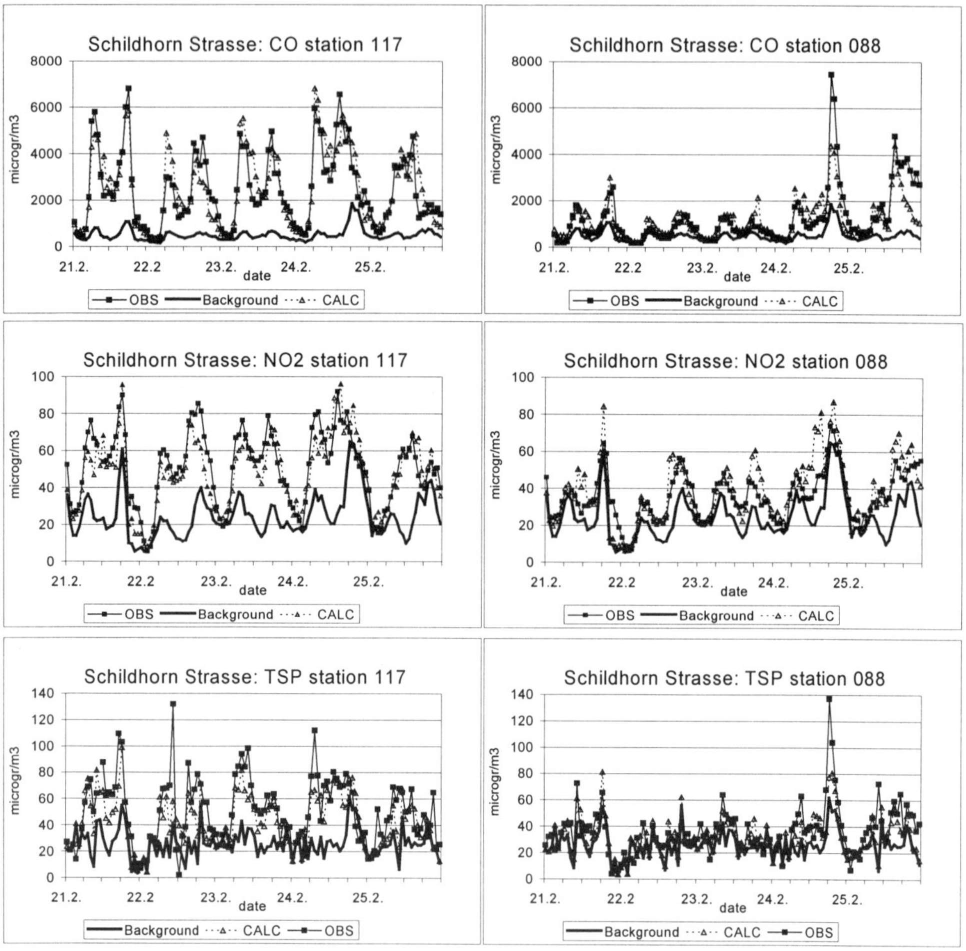

Figure 2 Time series of observed and calculated CO, NO_2 and TSP concentrations. The thick line indicates the observed background concentrations used in the calculations. Schildhornstrasse, Berlin, Feb 21-25 1995

The episode mean modelled concentrations lie in the following ranges of the observed episode mean concentrations: CO: –7% (station 117) to -3% (station 088), NO: -2% to +6% ;NO_2: -6% to +7% ;NO_x: 1% to +4% ; TSP:-15% to –10%; SO_2 -11% (station 117 only); benzene +10% (station 088 only). For all pollutants the correlation for station 117 is better than that for station 088 (Table 1). Station 117 is the down-wind station in most hours of the simulated period. Hence, the temporal concentration patterns reflect directly the temporal emission patterns and the high correlation is a hint that the modelled emission cycle agrees quite well with reality. Station 088 is much more influenced by the background values than station 117. Direct impact by the emissions of the Schildhornstrasse occurs only for a few hours. The very high observed concentrations at those hours are underestimated by the model whereas the low observed peak concentrations on the other days are overestimated (see Fig. 2). This provides an additional reason for the worse correlation. The best correlation is calculated for SO_2 at station 117. This has its basis in the low SO_2-emissions inside the canyon, thus the larger part of the concentrations is due to the background concentrations that are transported into the canyon from aloft. TSP shows the poorest performance for most the statistical measures. This might be due to the rather uncertain emission factors (both diesel exhaust and tire and break wear) and in the behaviour of the observed time series that exhibit large concentration amplitudes within a few hours (see Fig. 2).

Calculated daily mean benzene concentrations at the upwind station 088 are within approximately 20% of the measured values at four of the five simulated days (Fig. 3). On day 3, Feb 23, the observed daily mean benzene concentration is overestimated by 45%. The large relative deviations of the calculated daily benzene concentrations from the observations might be due to the rather uncertain benzene background concentrations, which were available as daily mean only and which contribute a considerable part to the concentrations calculated at the upwind station 088 inside the canyon. In this context it has also to be considered that the accuracy of the benzene measurements is less than that of the other pollutants and also that the emission factors for benzene are rather uncertain.

Measure	CO		NO		NO2		NOX (as NO2)		TSP		SO2
	117	088	117	088	117	088	117	088	117	088	117
O_{MEAN}	2461	1240	135.8	58.5	50.8	36.0	259.1	125.7	49.8	35.9	35.0
C_{MEAN}	2303	1224	144.5	57.8	48.1	38.6	269.7	127.3	42.6	32.5	31.1
$C_{BACKMEAN}$	507		8.2		24.9		37.5		26.1		20.1
σ_O	1599	1207	95.0	72.0	19.9	13.2	164.0	120.8	25.0	18.3	16.9
σ_C	1573	905	112.1	62.0	19.2	17.5	188.3	110.5	19.5	14.7	16.1
PCR	.866	.779	.888	.739	.879	.849	.892	.757	.745	.701	.94

Table 1: Statistical measures for the station 117 and station 088 in the Schildhornstrasse, Berlin. Episode mean values (O observed, C calculated) and standard deviations σ are in $\mu gr/m^3$. PCR Pearson correlation coefficient r. No lower or upper cutoff value. Modelling period Feb 21 –Feb 25, 1995, 120 hours.

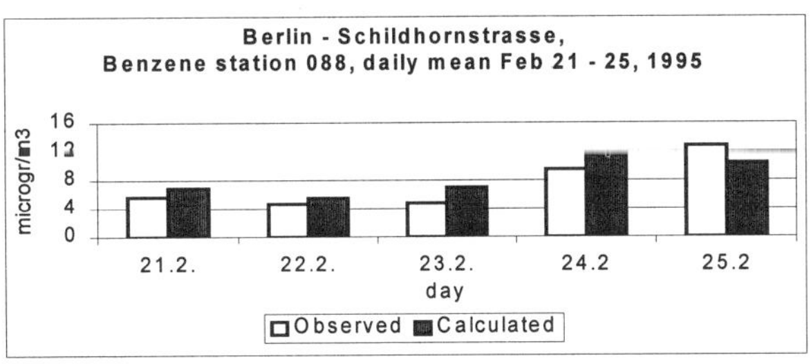

Figure 3 Calculated and observed daily mean benzene concentrations. Schildhornstrasse, station 088.

THE MILAN-VIALE MURILLO APPLICATION

Street Geometry

The Viale Murillo is part of the outer Milan ring road and runs from north to south. It is a four lane road with a wide strip between the north-, respectively south-bound lanes that serves as an parking area. In Viale Murillo, a measurement programme was carried out in

November and December 1997. The period from Nov 14 (Friday) to Nov 17 (Monday), 1997, was used for the MCG model evaluation. In the vicinity of the measuring cross section the Viale Murillo is 38,6 m wide and the height of the buildings on the left (west) side is 24 m, and 12 m on the right (east) side. The pavements are approximately 3 m wide. Each lane is 4 m wide. The parking strip in the middle of the street is 16,4 m wide. Measurements have been conducted at the edges of the pavements.

Meteorological data

Wind speed and wind direction were recorded at roof top of a building close to Viale Murillo. These hourly data (Fig. 4) were used as input for the flow model MISKAM. Wind speeds are generally low during the selected period. Only at Sunday, Nov 16, the measured wind speeds are higher than 2.5 m/s. Wind directions are quite variable; however during the first three days of the modelling period the flow is dominated by wind direction between west and north. In the evening of Nov 16, the wind direction turns toward south to easterly directions. On Nov 17, it turns from north toward west and south to north-easterly directions.

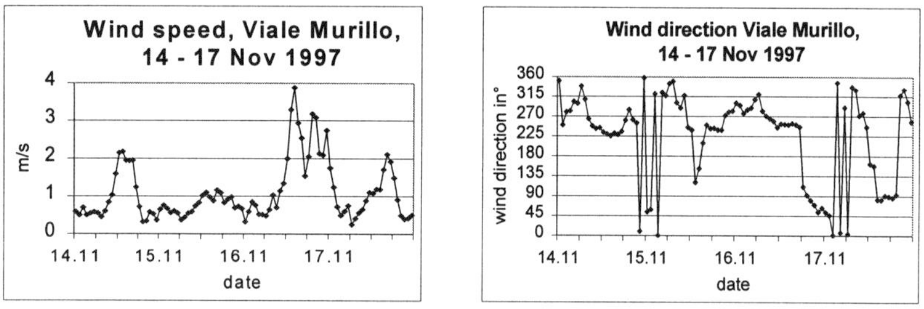

Figure 4: Roof top wind direction and wind speed in the modelling period , Milan, Viale Murillo

Air quality data

Only CO was measured at the kerbs of Viale Murillo and at a background station at the top of a building nearby. The CO data recorded at roof top level were used as background concentrations. The CO measurements at the receptors A (east side of Viale Murillo at 2.8 m height) and receptor D (west side of Viale Murillo at 2.8 m height) were used for model evaluation.

Traffic data

Detailed hourly traffic counts were available for each lane of Viale Murillo. The quite peculiar traffic pattern for the selected period is shown in Figure 5. Lane 1 on the west side of Viale Murillo (driving direction south) exhibits the lowest amount of traffic with peak values in the night of the workdays and on Sunday (November 16). The adjacent lane 2 is the most busy lane of Viale Murillo. In the workday peak traffic hours the counts gave more than 1200 cars/h. The two lanes in northerly direction (lane 3 and lane 4) have approximately equal amount of traffic with peak values of about 800 cars/h. In the night from Saturday to Sunday there is a secondary traffic peak on each lane. For the lanes in northern direction this peak is as high as the traffic peaks during the workdays. Averaged over all lanes and counting times, the percentage of light duty vehicles is 6.4%, that of heavy duty vehicles and busses is 4.6% and that of motorcycles is 3%. Hourly emissions for the modelling period have been calculated with the emissions model MOBILEV based on these measured traffic data.

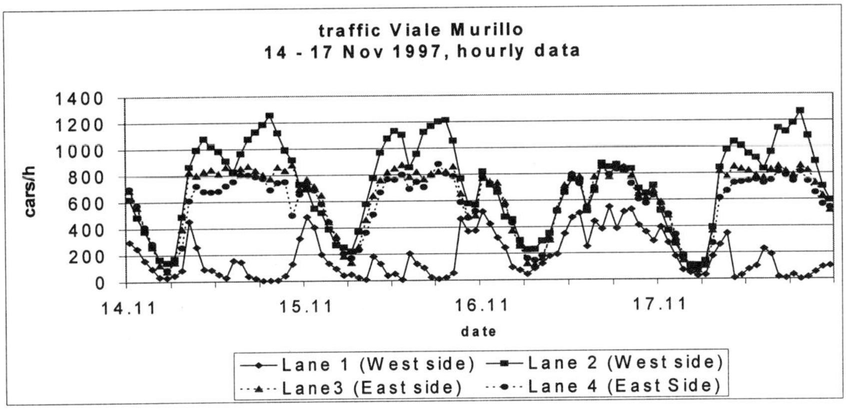

Figure 5: Number of cars/h for the four lanes of Viale Murillo. Modelling period from November 14 to November 17, 1997.

Model region

The selected modelling region has a horizontal extent of approximately 800 m in x-direction and 300 m in the y-direction with Viale Murillo running parallel to the x-axis. The modelling region includes Viale Murillo, the nearby Piazzale Zavattari and the neighbouring streets. The chosen grid size in x-direction along the Viale Murillo is 10 m. The gridsize in y-direction is 4 m for the area of the lanes and the strip in the middle of the street, 3 m for the area of the pavements and again 4 m outside of Viale Murillo. Thus, the cross-section in the Viale Murillo is resolved by 10 grid cells. As in the Schildhornstrasse application, the meteorological modelling region has been enlarged. The atmosphere is resolved in the vertical up to 100 m, with grid sizes of 2 m in the lowest 20 m and increasing grid sizes above this height.

MCG base run results

The last hour of a pre-episode run for Nov 13 was used to start the base run, and the model was applied for the whole period Nov 14 -17 without any further re-initialization. The CO time series shown in Figure 6 represent the concentrations calculated for the grid cells where the air is drawn into the monitoring devices of the receptors A and D at the kerbs of Viale Murillo. The overall behaviour of the observations is modelled quite well. Most of the observed peaks can also be found in the calculated time series. The highest CO concentrations as well as the lowest CO concentration occur about at the same time in the measured as in the calculated time series. In agreement with the observations the calculations give always lower concentrations for receptor D at the west side of Viale Murillo. CO concentration levels are quite high with a pronounced peak in the evening of the first day of the modelling period. The Milan background contribution to the concentrations inside the canyon is large and considerably higher than the respective Berlin background CO concentrations in the Schildhornstrasse. Compared to the Schildhornstrasse application, measured and calculated concentrations are much higher in Viale Murillo, although the 24 hour averaged weekday traffic amount is similiar. This can be explained by the higher background contribution, the much lower roof top wind speeds, and the specific traffic pattern of Viale Murillo. Calculated weekday morning emissions peak values for Viale Murillo are not very different from those of the Schildhornstrasse, but the unequal traffic distribution between the four lanes causes lane specific traffic disturbances, so that MOBILEV calculates higher emissions for Viale Murillo than for Schildhornstrasse, particularly for workday afternoons.

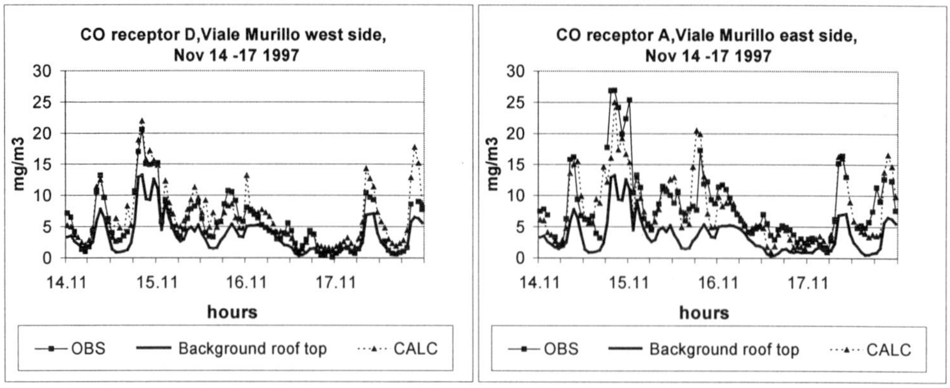

Figure 6: Observed and calculated CO concentrations, Nov 14 – Nov 17, 1997. Viale Murillo, Milan.

Measure	CO Receptor A	CO Receptor D
O_{MEAN}	8.59	5.57
C_{MEAN}	7.84	6.54
$C_{BACKMEAN}$	3.61	3.61
σ_O	5.73	4.26
σ_C	5.25	4.49
PCR	0.829	0.927

Table 2 Statistical CO measures for the receptors A and D in the Viale Murillo, Milan. Episode mean values (O observed, C calculated) and standard deviations σ are in mg/m^3. No lower or upper cutoff value. Modelling period Nov 14 –Nov 17, 1997, 96 hours.

CONCLUSIONS

The MICRO-CALGRID photochemical micro-scale model was applied to two street canyons for two winter episodes. The overall behaviour of all measured pollutants (CO, NO, NO_2, benzene, SO_2 and TSP in the Berlin canyon, CO in the Milan canyon) could be reproduced well in both canyons. Future work will focus on the application to summer time cases and longer time periods.

REFERENCES

Eichhorn, J., 1996. Validation of a microscale pollution dispersal model. Air Pollution Modeling and ist Application XI. Plenum Press

Fige, 1997. Mobilev – Documentation and user's guide (in german). F+R-Project 105 06 044. Umweltbundesamt, Berlin

Skouloudis, A., Suppan, P., Walker, B., 1999. The Autooil II Programme: Methodology and status of air-quality modelling. International Conference Air quality in Europe: Challenges for the 2000s. Venice, 19th-21st May 1999

Stern, R., 1999. Application of the MICRO-CALGRID photochemical model for high-resolution studies in urban street canyons of Berlin and Milan. IVU-Umwelt GmbH, Final Report under the EU contract No. 14342-1998-10-F1ED ISP DE commissioned by the Joint Research Centre Ispra.

Stern, R., and R. Yamartino, 1998. Development and Initial Application of the MICRO-CALGRID Photo-chemical Model for High-Resolution Studies of Urban Environments. Proceedings of the 23rd NATO/CCMS ITM on Air Polluting Modeling and its Application, Varna, Bulgaria, Sept. 28 - Oct. 2.

CHALLENGES OF AIR POLLUTION MODELING AND ITS APPLICATION IN THE NEXT MILLENNIUM

Akula Venkatram

College of Engineering
University of California
Riverside, CA 92521-0425

INTRODUCTION

In the opening talk of this conference, Peter Bultjes summarized the state-of-the-art in air pollution modeling. In this paper, I will discuss some of the challenges that face us, and make suggestions on how we might be able to address them in the future. A reasonable way of guessing at what the future holds is to extrapolate "data points" from the past. So, I will briefly examine the history of air pollution modeling to discern the trends that might determine the future.

As is normally done, air pollution models will be classified according to the scales of application. Short-range models apply to space scales of tens of kilometers, while urban and long-range transport models apply to larger scales. This classification is useful because, typically, models at these two scales are based on different mathematical frameworks. Short range models generally use a Lagrangian framework to focus on a single plume from a source. Long-range models use a Eulerian framework to account for interaction between plumes from multiple sources. While this distinction might be useful at this time, it may well disappear in the future for reasons I will discuss in the paper.

SHORT RANGE DISPERSION MODELS

The first short-range model was formulated by Pasquill (1961) about 40 years ago. This model was based on Gaussian concentration distributions in the vertical and horizontal, and was specifically formulated for surface releases. What made the model useful was that plume spreads, the scales of the concentration distributions, could be estimated from routinely available meteorological measurements such a wind speed and cloud cover. A set of meteorological measurements was mapped to a stability class, which in turn was related to the plume spreads, sigma "y" and sigma "z". This set of semi-empirical relationships was based on well-designed field experiments, the most notable of which was the Prairie Grass experiment conducted in Nebraska (Barad, 1958). This Pasquill scheme for calculating plume spread is still used in regulatory dispersion models such as the Industrial Source Complex (ISC) model.

The next improvement in short range models came with advances in micrometeorology during the 1970s (Haugen, 1970). The new understanding of the atmospheric boundary layer allowed better estimation of turbulence properties, which in turn were used to estimate plume spread using a variety of approaches. Because transport of heat and momentum was well characterized in the surface boundary layer, it was possible to estimate dispersion of surface releases using methods that essentially assumed that the eddy diffusivity of mass was the same as that of heat (Gryning et al., 1983; Horst, 1979; Venkatram, 1982).

The next set of advances came when Deardorff (1972) generated turbulent velocity fields in the CBL using large eddy simulation. A little later, Deardorff and Willis (1975) initiated a series of innovative water tank experiments that provided valuable insight into dispersion in the convective boundary layer (CBL). Lamb (1978, 1981) used the velocity fields from Deardorff's simulations to simulate particle motion in the convective boundary layer. These three developments laid the foundation for our current understanding of turbulence and dispersion in the convective boundary layer.

Over the past ten years, theoretical work by Sawford (1985), Thomson (1987) and others has led to dispersion models that rely on simulating particle motion in turbulent flows (For example, Luhar and Britter, 1989). The rapid increase in computational resources over this period has increased the popularity of the method, in which a continuous source is simulated by releasing a large number of particles into the flow. The concentrations are estimated by sampling these particles at receptors. The major advantage of these particle-tracing methods is that, in principle, they can be applied to model dispersion under complex flow conditions with non-uniform turbulent statistics. I suspect that, in the future, this computationally intensive method will become popular in the future. However, the particle model, as it is currently formulated, cannot handle non-linear chemistry between species from interacting plumes. Thus, its use is likely to be confined to species that are inert or undergo first order chemistry. The next section discusses models that use the Eulerian framework to provide a more comprehensive description of air quality.

URBAN AND LONG-RANGE TRANSPORT MODELS

Most of the currently available oxidant and acid deposition models belong to this class, which are referred to as comprehensive models. They have all evolved from the Urban Airshed Model (UAM, Reynolds et al, 1973), which first incorporated the governing processes in an Eulerian framework: the governing mass conservation equations are solved numerically using a fixed grid system. Since then, a large number of models, applicable to both urban and regional scales, have been developed. The increase in computing power has allowed the inclusion of large number of processes. The most recently developed models (CIT, Meng et al., 1997 for example) incorporate detailed gas phase and aqueous phase chemistry, aerosol physics and chemistry, and cloud and fog process.

Russell and Dennis (2000) provide a wealth of information on the performance of currently available comprehensive models. They make several observations that are relevant to the future of comprehensive modeling:

1. Increases in computing power have led to inclusion of more detailed treatments of processes.
2. The increased complexity (improvements if you want) of these models has not been accompanied by improved performance in describing ozone observations.
3. The comprehensive models perform poorly in estimating the concentrations of precursors, such as NO_x, of secondary pollutants such as ozone and aerosols.
4. The responses of secondary pollutant concentrations to changes in precursor emissions are sensitive to parameterizations of mixing as well as model inputs. In fact, for the same emission change in NO_x, the change in ozone concentrations at a receptor can be negative or positive depending on the specification of the mixed layer height (Li et al., 1998).

Russell and Dennis (2000) provide several possible reasons for the lack of discernable improvement in model performance in response to inclusion of more processes in the models. One reason they have not considered is related to the treatment of mixing in these comprehensive models. Current models have relatively coarse horizontal resolution of the order of kilometers, which has the effect of mixing pollutants and thus leading to chemical reactions that do not really occur. To see this, consider two pollutant sources separated by 1000 m. If the wind speed is 5 m/s and the horizontal turbulent velocity is 1 m/s, pollutants from these sources are not likely to mix for distances of at least 5000 m; on the other hand, the model falsely mixes these pollutants instantaneously over the grid square. There is little point in modeling the details of the chemistry of species that do not mix in the first instance. It is important to remember that mixing has to occur at molecular scales before chemistry can occur.

Enhanced horizontal diffusion associated with coarse grid resolution can have other effects such as converting a reaction that is oxidant limited to one that is not. Consider SO_2 oxidation limited by hydrogen peroxide. The amount of SO_2 converted to sulfate is limited by the concentration of hydrogen peroxide. Thus, close to the source, where the concentration of SO_2 is high relative to hydrogen peroxide, this conversion is slow. On the other hand, in a numerical model, the SO_2 is mixed instantaneously to falsely low concentrations giving rise to enhanced oxidation rates.

Several studies appear to indicate that more realistic treatment of large point sources (plume in grid models; Seigneur et al., 1983) does not make much difference to ozone estimates. This might be true only because these point sources might be making small contributions to overall ozone concentrations in the first place. Also, the plume in grid treatment could be wrong. The only way of finding out whether the proper treatment of horizontal mixing is important is to resolve the entire precursor emission inventory to the small meter scales at which sources really occur. Accurate treatment of horizontal mixing might resolve the problems with predictions of ozone precursors. At this point, it is difficult to place stock in ozone estimates from comprehensive models, even if they appear to be accurate at times, when the predicted precursor concentrations are consistently biased.

I believe that future progress in comprehensive modeling will have to pay more attention to mixing processes, which, after all, have to precede chemistry, however complex and impressive the chemistry might be. In principle, increases in computational resources and modeling will allow us to overcome the mixing problem, and models will approach reality. But is this a scenario that we should look forward to? If models do indeed become alternate realities, we have to deal with a system that is as messy as reality. It appears to me that the biggest challenge in the future is related to evaluating these models with observations, and interpreting model results. I discuss this next.

MODEL EVALUATION AND INTERPRETATION OF RESULTS

As models become more complex, unavoidable errors in models inputs and process parameterizations will make it more difficult to compare model estimates to corresponding observations. In response to a need for better model evaluation techniques, several new tools have been developed over the past few years. Process analysis (Jeffries and Tonnesen, 1994) attempts to trace the effects of individual processes on the concentrations, and thereby gauge the relative importance of errors in model inputs. Sillman (1995) has proposed the use of certain "indicator" species, such as NO_z and hydrogen peroxide to establish correspondence between the evolution of chemistry in the model and the real system. These are promising approaches that attempt to evaluate models without direct comparison between model estimates and observations. Irwin (1999) and others are developing parallel methods for short-range models.

The rich repertoire of responses from a comprehensive model also makes it difficult to draw broad generalizations about the system being modeled. This was brought home to me recently when I saw a NOVA TV program on climate. One of the scenes shows Tom Wigley of NCAR and his colleagues sitting in front of a large screen displaying beautiful pictures of model predicted temperature patterns over the globe. What made it interesting is that these scientists were wearing 3-D glasses to interpret these pictures (they were in pairs)! I am sure that air quality models will provide similar entertainment. But, it is not clear that all this high tech, even if it is a lot of fun, will lead to broad, simple generalizations about the system. While some might argue that such generalizations are not possible and not even necessary, I believe that they are at the heart of what understanding is. I will try to make my case in the following paragraphs.

One way of understanding some of the major responses of a complex model is through a semi-empirical model. I think of a semi-empirical model as a model of a system that is simpler than the system being modeled. Thus, the semi-empirical model is not a statistical fit to the data. It is a true model in that it is a description of a system using the framework of the relevant conservation equations; it is the system that is simplified rather than the governing equations. The degree of simplification depends on the model, but the adequacy of the model is ultimately determined by its ability to describe and explain chosen features of the real system. Comparison with reality will entail determining some of the parameters of the model. The model is empirical because most of its credibility derives from comparison of its responses with either those of the real system or those of the comprehensive model. The semi-empirical model is thus a parable that focuses attention on chosen aspects of the real system. Let me illustrate these ideas through some examples.

Dispersion from Surface and Elevated Releases

Consider dispersion from a surface release, which has been studied by several authors using a variety of methods. In a recent study, Shuming Du and I (Du and Venkatram, 1998) used a Lagrangian stochastic model to simulate the motion of particles released in the surface boundary layer. This approach is based on the best available understanding of dispersion. It is computationally intensive and is capable of producing "realistic" pictures of time-varying plumes. These are the hallmarks of current approaches to air pollution modeling.

Our simulations produced a mass of data, which was essentially meaningless until we interpreted it using a semi-empirical model derived earlier (Venkatram, 1992). We found that the cross-wind integrated concentration from the surface releases under unstable conditions could be described by

$$\overline{C}_*^y = \frac{1}{x_*\left(1+\alpha x_*^2\right)^{1/2}}, \tag{1}$$

where α is an empirical constant, and

$$\overline{C}_*^y = \frac{Q}{u_*|L|};\ x_* = \frac{x}{|L|}. \tag{2}$$

The semi-empirical model provides valuable generalizations of the near and far source behavior of the cross-wind integrated concentration:

$$\begin{aligned} \overline{C}_*^y &\sim x_*^{-1};\ small\ x_* \\ \overline{C}_*^y &\sim x_*^{-3/2};\ large\ x_* \end{aligned} \tag{3}$$

The "bouncing ball" model, proposed by Weil and Furth (1981) for dispersion in the convective boundary layer provides another example of semi-empirical modeling. By assuming that particles remember their velocities for an indefinite time after release from an elevated source, we can derive an extremely simple expression for the cross-wind integrated concentration as a function of height. The expression for the ground-level concentration is particularly simple (Venkatram, 1983):

$$\overline{C}^y = \frac{2Q}{x}\left[P(w_d|h_s) + P(w_i|h_s)\right]$$
$$where \qquad (4)$$
$$w_d = -uh_s / x$$
$$w_i = u(2 - h_s / z_i)/x$$

In equation (4), $P(w/h_S)$ is the probability density function of the vertical turbulent velocities at release height, and the rest of the symbols have their customary meaning. The performance of the model in explaining results from a particle model (Lamb, 1978) is shown in Figure 1. While model estimates are not precise reproductions of the results from the numerical model (Lamb, 1978), it is clear that the probability density function of turbulent velocities is crucial to explaining the ground-level concentration.

Emboldened by the performance of the model, I used it later to estimate the error associated with using a Gaussian concentration distribution in the vertical (Venkatram, 1993).

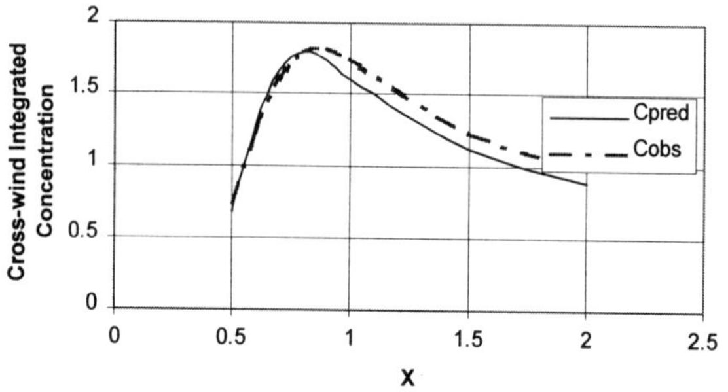

Figure 1: Comparison of ground-level concentrations from simple model and numerical model (Lamb, 1978), referred to as observed. Release height is 0.49 of mixed layer height.

Consider dispersion in complex terrain, which is governed by several complicated processes. Under unstable conditions, the plume is depressed towards the surface of the obstacle as it goes over it. The implied compression of the streamlines is associated with a speed-up of the flow and an amplification of vertical turbulence. Under stable conditions, part of the flow flowing towards an obstacle tends to remain horizontal, while the other part climbs over the hill. Experiments show (Snyder et al., 1983) that this tendency for the flow to remain horizontal can be described using the concept of the dividing streamline height, denoted by H_c. Below this height, the fluid does not have enough kinetic energy to surmount the top of the hill; a plume embedded in the flow below H_c either impacts on the hill or goes around it. On the other hand, the flow and hence the plume above H_c can climb over the hill. These and other effects are accounted for in the Complex Terrain Dispersion Model (CTDMPLUS, Perry, 1992) that attempts to provide accurate concentration estimates for plumes dispersing in complex terrain.

AERMOD (Cimorelli et al., 1996) is a new regulatory model designed to replace ISC. In developing the AERMOD, we incorporated a semi-empirical model (Venkatram et al. 1998) that mimics the major features of CTDM. The model is essentially an interpolation of knowledge of flow and dispersion in complex terrain in two asymptotic states. Under very stable conditions, the flow and hence the plume embedded in it, tends to remain horizontal when it encounters an obstacle. Under unstable conditions, the plume is more likely to climb over the obstacle. Thus, the very stable and very unstable conditions represent the two asymptotic states. Under stable or unstable conditions, the real flow is likely to lie between these two states.

AERMOD assumes that the concentration at a receptor, located at a position (x,y,z), is a weighted combination of two concentration estimates: one assumes that the plume is horizontal, and the other assumes that the plume climbs over the hill. The concentration associated with the horizontal plume dominates during stable conditions, while that caused by the terrain-following plume is more important during unstable conditions. These assumptions allow us to write the concentration, $C(x,y,z)$, as

$$C(x,y,z) = fC_f(x,y,z) + (1-f)C_f(x,y,z_e). \tag{5}$$

The first term on the left-hand side of Equation (5) represents the contribution of the horizontal plume, while the second term is the contribution of the terrain-following plume. The concentration, $C_f(x,y,z)$, is that associated with a plume that is unaffected by the terrain; the plume axis remains horizontal. In the first term, $C_f(x,y,z)$ is evaluated at the receptor height, z, to simulate a horizontal plume. In the second term, the concentration is evaluated at an effective height, z_e, which will be discussed later.

The formulation of the weighting factor, f, uses the observation that the flow below the critical dividing streamline height, H_c, tends to remain horizontal as it goes around the terrain obstacle (Snyder et al., 1983). This suggests the following formulation for f:

$$f = f(\phi) \tag{6}$$

where

$$\phi = \frac{\int_0^{H_c} C_f(x,y,z)\mathrm{d}z}{\int_0^{\infty} C_f(x,y,z)\mathrm{d}z}, \tag{7}$$

represents the fraction of the plume mass (assuming that the plume is horizontal) below the critical dividing streamline height. This fraction goes to zero under unstable conditions because H_c is zero. The weight, f, is a function of ϕ, is chosen to be

$$f = \tfrac{1}{2}(1+\phi) \tag{8}$$

$$z_e = (z - z_h), \tag{9}$$

where z_h represents the height of the terrain at (x,y), so that $(z - z_h)$ represents the height of the receptor above local terrain.

When ϕ goes to unity, the entire plume lies below H_c, and f goes to unity. Under these conditions, the hill concentrations are entirely determined by the horizontal plume. When ϕ goes to zero under unstable conditions, f becomes $\frac{1}{2}$. This means, that under unstable conditions, the concentration at an elevated receptor is the average of the contributions from the horizontal plume and the terrain-following plume.

Figure 2 shows the performance of the model relative to CTDM is explaining ground-level concentrations in a complex terrain site at Tracy, Nevada. We see that the simple semi-empirical model captures the essential features of the statistics of concentrations measured in complex terrain. The model has been tested at several other complex terrain sites, and we find that it performs at least as well as CTDM in the limited task describing concentration statistics.

Tracy SF6 1-Hr Q-Q Plot (Conc.)

Figure 2: Comparison of concentrations estimated by CTDMPLUS and AERMOD for the Tracy Complex Terrain site. Model estimates and observations are ranked before comparison.

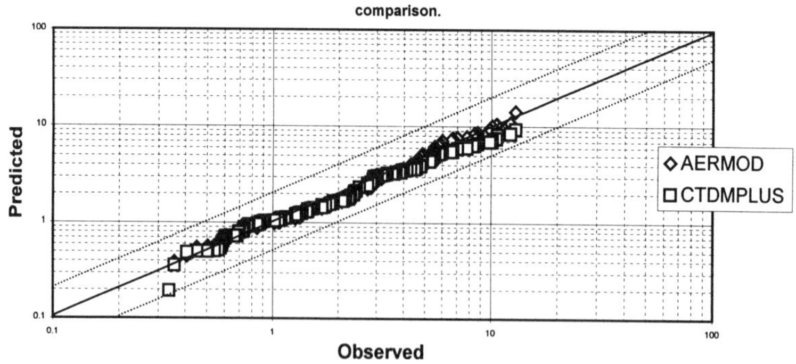

Semi-empirical models have an extensive history of application to long-range transport and deposition of sulfur (Fisher, 1978; Eliassen and Saltbones, 1975). The simplest possible model for long-range transport of sulfur considers two species: primary sulfur, S(IV) and the oxidized state, S(VI). Then, if we represent by the flux of primary pollutant by p and that of secondary pollutant by s, the evolution of the fluxes as a function of travel time, t, from the source can be written as

$$\begin{aligned}\frac{dp}{dt} &= -kp - \lambda p \\ \frac{ds}{dt} &= kp - \bar{\lambda} s\end{aligned} \tag{10}$$

where k is the primary to secondary pollutant conversion rate, and the λs represent removal rates by both dry and wet deposition. Equation (1) can be readily integrated to yield

$$\begin{aligned}&p = p_o e^{-(k+\lambda)t} \\ &and \\ &s = \frac{kp_o}{\left(k+\lambda-\bar{\lambda}\right)}\left(e^{-\bar{\lambda}t} - e^{-(k+\lambda)t}\right)\end{aligned} \tag{11}$$

where p_o is the initial flux of primary pollutant. The long-term concentration at a distance r can be estimated by assuming that plume is spread over an angle of 2π. Then, the concentration corresponding to the fluxes in Equations (3) are simply obtained by dividing the fluxes by $2\pi ruz_i$.

It turns out that a slightly modified form of Equation (3) can provide estimates of annually averaged sulfur concentrations in rain that compare well with both observations as well results from comprehensive models (Venkatram et al., 1990).

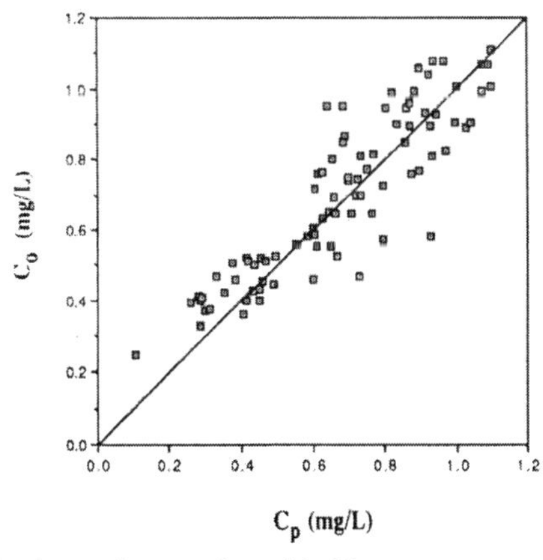

Figure 3: Comparison of model estimates from semi-empirical long-range transport model with observations of sulfur in rain averaged over 1982-1985. Data correspond to stations in the northeastern United States.

This semi-empirical model does not pretend to explain the details of the phenomena that govern sulfur deposition. However, it does show that long-term deposition of sulfur (and concentrations in rain) is constrained by simple mass balances. This has justified the use of the model in the analysis of alternative SO_2 emission reduction strategies (EPRI, 1991). In principle, the uncertainty in such applications can be examined by varying the parameters over ranges suggested by the comprehensive model.

Because the model has an analytical solution, it can provide useful insight into the behavior of the sulfur system. For example, it is easy to show that for travel times small compared to the scavenging time scale, the sulfate concentration is essentially independent of distance from the source, and is given by the simple expression:

$$C_{so_4} = \frac{kQ}{2\pi z_i u^2} \tag{12}$$

where Q is the SO_2 emission rate. This concentration happens to be the maximum value of the sulfate concentration.

A Semi-Empirical Model For Ozone

A semi-empirical model for ozone can be constructed (Simplified Ozone Modeling System or SOMS; Venkatram et al., 1994, 1998) by using the simple idea that ozone at a receptor can be represented as a function of essentially three variables: concentrations of NO_x and VOC, and the time (the "age") over which these chemical species are exposed to sunlight to produce ozone. This idea can be converted into a computational scheme by a providing a method to compute these three variables. SOMS assumes that the precursor concentrations are simply those in the absence of chemistry, and the age of each species is the concentration weighted travel time from the source to the receptor of concern.

This concept is best illustrated by considering the simplest possible case in which a source emits both VOC and NO_X. If the wind speed is assumed to be u, the age of the molecules at a distance x from the source is simply x/u. This simple idea can be generalized to more complex wind flows by formulating a conservation equation for species age. This conservation equation, derived elsewhere (Venkatram et al., 1998) is

$$\frac{\partial \phi}{\partial t} + u_i \frac{\partial \phi}{\partial x_i} = \frac{1}{\rho_a} \frac{\partial}{\partial x_i} \left(\rho_a K^i \frac{\partial \phi}{\partial x_i} \right) + q(\vec{x},t) \tag{13}$$

where

$$\phi = Aq \tag{14}$$

and A is the age and C is the corresponding concentration. u_i is the mean wind, K^i is the eddy diffusivity, ρ_a is the air density, and q is the mixing ratio given by

$$q = \frac{C}{\rho_a}. \tag{15}$$

Figure 4 shows the comparison of a comprehensive model that incorporates the age concept with the same model in which chemistry and transport are coupled in the usual manner.

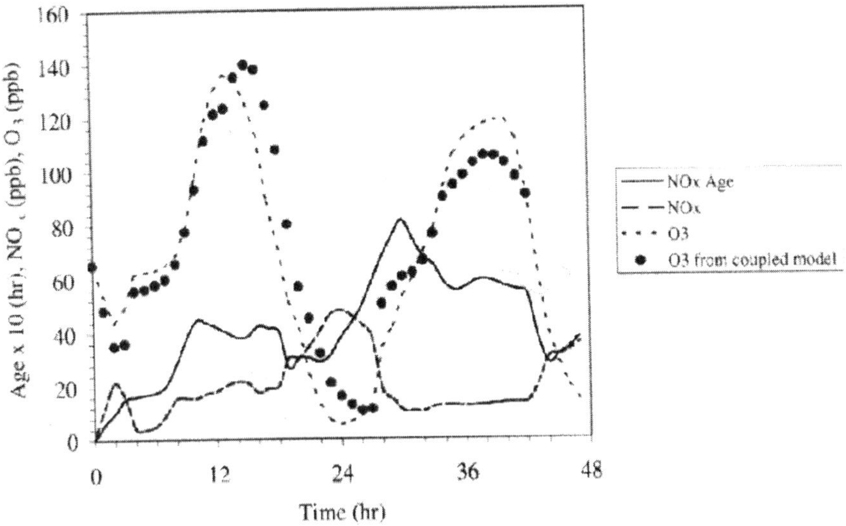

Figure 4: Comparison of ozone estimates from coupled and uncoupled (using the age concept) comprehensive models for a site in Wisconsin during the period June 26-27, 1991. Dotted line corresponds to uncoupled model.

The age concept is useful because it can be used to examine the impact of a chosen source of VOC or NO_x on ozone at a receptor. This is because the composite age, A, for the sum of mixing ratios, $\sum q_i$, resulting from several sources, E_i, is given by

$$A = \frac{\sum A_i q_i}{\sum q_i}, \tag{16}$$

where A_i corresponds to a particular mixing ratio distribution q_i. Because we can associate a q_i for every source in a modeled region, the age field can be viewed as a matrix, A_{ij}, that quantifies the contribution of a source at "*i*" to the composite age at receptor "j". This matrix can be used to compute the composite for any required distribution of sources. The linearity of the concentration and age equations also means that effect of any source can be estimated without recalculating the concentrations and ages associated with other sources.

A simplified version of SOMS has been used to estimate three-year statistics of daily maximum ozone concentrations at an urban site in Baltimore, Maryland (Vukovich et al.,

2000). The comparison with observations indicates that the SOMS concept is capable of describing the statistics of daily maximum ozone concentrations.

CONCLUSIONS

It is clear that the future will see more realistic models as improvements in computer resources are used to incorporate more details of governing processes into comprehensive models. As models become more comprehensive, it will become more difficult for individual scientists or even small organizations to participate in research related to modeling. I agree with Dennis and Russell (2000) that this will lead to the development of a few "community" models that can be accessed by interested scientists through the Internet. We will also see the integration of air quality models into larger frameworks that will not only include processes in all environmental media but economic models to understand the economic consequences of emission changes. Such integrated models are already being used in climate studies (Nordhaus, 1994).

I strongly believe that the development of comprehensive models will lead to better understanding of the processes that govern air quality. However, it should be accompanied by the development of simple semi-empirical models that allow us to gauge the relative importance of these processes. I have been careful about defining a semi-empirical model; it is a model of a simpler system, but it is a model, nonetheless, in the sense that it is based on mathematical equations that are consistent with the conservation laws.

In this paper, I make the case for using simple semi-empirical models to interpret both observations and results from comprehensive models. Semi-empirical models are also ideally suited for inclusion into Integrated Assessment Models. For example, the semi-empirical sulfur model has been used to estimate the impact of alternative emission control strategies on sulfur deposition (EPRI, 1991), and SOMS has been used to examine the health impact of NO_x control from power plants (Levy, 1999).

In principle, we could use semi-empirical models to mediate between complex comprehensive models and complex reality: a simple model that can mimic certain responses in the real system can be used to probe the same responses in the comprehensive model. This type of comparison will allow us to bypass the futile exercise of directly comparing model estimates with observations (See Hanna et al., 1998).

ACKNOWLEDGEMENTS

I would like to thank Sven-Erik Gryning and Frank Schiermeier of the ITM organizing committee for providing me this opportunity to "air" my views on air quality modeling.

REFERENCES

Barad, M. L., 1958, *Project Prairie Grass, a Field Program in Diffusion*. Vol. 1, Geophysics Research Paper No. 59, Air Force Cambridge Research Center, Bedford, MA.

Cimorelli A. J., et al., 1996, Current progress in the AERMIC model development program. Proceedings of the 89th Annual Meeting of the Air and Waste Management Association, 96-TP24B.04, AWMA, Pittsburgh, PA.

Deardorff, J. W. and Willis, G. E., 1975, A parameterization of diffusion into the mixed layer. *J. Appl. Meteor.* 14, 1451-1458.

Deardorff, J. W., 1972, Numerical investigation of neutral and unstable boundary layers. J. Atmos. Sci., 29:91-115.

Dennis, R. and Russell, A., 2000, NARSTO critical review of photochemical models and modeling. *Atmos. Environ.* 34, 2283-2324.

Du, S., and Venkatram A., 1998, The effect of streamwise diffusion on ground-level concentrations. Atmos. *Environ.* 32:1955-1961.

Eliassen, A., and Saltbones, J., 1975, Decay and transformation rates of SO2, as estimated from emission data, trajectories and measured concentrations. *Atmos. Environ.* 9:429-429.

EPRI, 1991, Analysis of Alternative SO_2 Reduction Strategies. EPRI report EN/GS-7132, Electric Power Research Institute, 3412 Hillview Avenue, Palo Alto, CA 94304.

Fisher B. E. A., 1975, The long-range transport of sulfur dioxide. Atmos. Environ. 9:1063-1070.

Gryning, S., van Ulden, A.P. and Larsen, S.E., 1983, Dispersion from a continuous ground-level source investigated by a K model. Quart. *J. Roy. Meteor. Soc.* 109: 357-366.

Hanna et al., 1998, Monte Carlo estimates of uncertainties in predictions by a photochemical model (UAM-IV) due to uncertainties in put variables. Atmos. Environ. 32:3619-3628.

Haugen, D.A. (Editor): Workshop on Micrometeorology. American Meteorological Society, Boston, MA.

Horst, T. W., 1979, Lagrangian similarity modeling of vertical diffusion from a ground source. *J. Appl. Meteor.* 18:333-340.

Irwin J. S., 1999, Effects of concentration fluctuations on statistical fluctuations of centerline concentration estimates by atmospheric dispersion models. Sixth International Conference on Harmonisation within Atmospheric Dispersion Modelling for Regulatory Applications, October 11-14, 1999, Rouen, France.

Jeffries, H. E., and Tonnesen, G., 1994, A comparison of two photochemical reaction mechanisms using mass balance and process analysis. *Atmos. Environ.* 28:2991-3003.

Kumar, N., Russell, A.M., Tesche, T.W., and McNally D. E., 1994, Evaluation of CALGRID using two different ozone episodes and comparison to UAM results. *Atmos. Environ.* 28:2823-2845.

Lamb, R. G., 1978: A numerical simulation of dispersion from an elevated point source in the convective boundary layer. *Atmos. Environ.* 12:1297-1304.

Lamb, R. G., 1981, Diffusion in the convective boundary layer. In a Short Course in Turbulence and Air Pollution Modeling, 21-25 September, 1981. D. Reidel, Dordrecht, Holland.

Levy J. et al., 1999, Development of a new damage function model for power plants: Methodology and Applications. *Environ. Sci. Technol.* 33:4364-4372.

Li et al., 1998, Effects of uncertainty in meteorological inputs on O_3 concentration, O_3 production efficiency, and O_3 sensitivity to emissions reduction in regional acid deposition modeling. Preprints of the 10th Joint Conference on the Applications of Air Pollution Meteorology with the Air and Waste Management Association, Paper 9A.14, 1998, Phoenix, Arizona, AMS, Boston, 529-533.

Luhar, A. and Britter, R.E., 1989, A random walk model for diffusion in inhomogenous turbulence in a convective boundary layer. *Atmos. Environ.* 23:1911-1924.

Meng, Z., Dabdub, D. and Seinfeld, J.H., 1997, Chemical coupling between atmospheric ozone and particulate matter. *Science* 277:116-119.

Nordhaus, W. D., 1994, Managing the Global Commons. MIT Press, 213 pages.

Pasquill, F., 1961, The estimation of dispersion of windborne material. Met. Mag., 90:33-41.

Perry S. G., 1992: CTDMPLUS, A dispersion model for sources in complex topography. Part I: Technical formulations. *J. Appl. Meteor.* 31: 633-645.

Reynolds, S. D., Seinfeld, J. H., and Roth, P. M., 1973, Mathematical modeling of photochemical air pollution: Formulation of the model. *Atmos. Environ.* 7:1033-1061.

Russell, A., and Dennis R., 2000: NARSTO critical review of photochemical models and modeling. *Atmos. Environ.* 34:2283-2324.

Sawford, B.L., 1985, Lagrangian statistical simulation of concentration mean and fluctuation fields. *J. Clim. Appl.* Met. 24:1152-1161.

Seigneur C. et al., 1983, On the treatment of point source emissions in urban air quality modeling. *Atmos. Environ*:17:1655-1676.

Sillman, S., 1995, The use of NO_y, H_2O_2, and HNO_3 as indicators for ozone-NO_x-hydrocarbons sensitivity in urban locations. *J. Geophys. Res.* 100:14175-14188.

Snyder W. H. et al., 1985, The structure of the strongly stratified flow over hills: Dividing streamline concept. *J. Fluid Mech.* 152: 249-288.

Thomson, D.J., 1987, Criteria for the selection of stochastic models of particle trajectories in turbulent flows. *J. Fluid Mech.* 180, 529-556.

Venkatram, A. 1982, A semi-empirical method to compute concentration associated with surface releases in the stable boundary layer, *Atmos. Environ.* 16, 245-248.

Venkatram, A., 1983, On dispersion in the convective boundary layer. *Atmos. Environ.* 17:529-533.

Venkatram, A., Kamran, K., Karamchandani, P., and Avitzur, T., 1990, Probing the acid deposition system with a semi-empirical model: the role of oxidant limitation, Atmos. Environ. 24A: 125-131.

Venkatram, A., 1992, Vertical dispersion of ground-level releases in the surface boundary layer. *Atmos. Envrion.* 26A:947-949.

Venkatram, A. 1993, Estimates of maximum ground-level concentration in the convective boundary layer—The error in using the Gaussian distribution. *Atmos. Environ.* 27(A):2187-2191.

Venkatram, A., Karamchandani, P., Pai, P., and Goldstein, R., 1994, The development and application of a Simplified Ozone Modeling System (SOMS*), Atmospheric Environment.* 28:22:3665-3678.

Venkatram, A., 1996, An examination of the Pasquill-Gifford-Turner dispersion scheme. Atmospheric Environment, 30:1283-1290.

Venkatram, A., Du S., Hariharan, R., Carter, W., and Goldstein, R., 1998, The concept of species age in photochemical modeling, *Atmos. Environ.* 32:3403-3413.

Venkatram, A., et al., 1998, AERMOD's simplified algorithm for dispersion in complex terrain. *Proceedings of the 10th Joint Conference on the Applications of Air Pollution Meteorology with the AWMA*, AMS, Boston, MA.

Vukovich et al., 2000, On performing long-term predictions of ozone using the SOMS model. Submitted to *Atmos. Environ.*

Weil, J.C., and Furth, W. F., 1981, A simplified numerical model of dispersion from elevated sources in the convective boundary layer. *Fifth Symposium on Turbulence, Diffusion and Air Pollution*, 91-3 March, 1981, Atlanta, GA, AMS, Boston.

DISCUSSION

D. STEYN: Henk Tennekes, in one of his papers on mixed layer depth, concluded that the flux ratio was a fine example of a simple, sophisticated model which he contrasted with naive, complex models. It would seem to me that you are moving towards simple, sophisticated models (which contain assumptions that capture the essence of the phenomenon) for regional ozone pollution. I wonder if it is possible to capture the essence of such a multifactional, nonlinear phenomenon in a simple model.

A. VENKATRAM: I suppose that it is an act of faith to assume that the essentials of a complicated system can be captured in a simple model. But the evidence that I have presented in my paper indicates that my faith is well placed. As I have stressed several times in my paper, such a simple model is NOT an empirical fit to data or results from a comprehensive model. A simple model is one that contains variables that the modeler deems essential to the system at hand, but it is a model, nonetheless, in that it is consistent with the conservation laws. The simple model is only designed to mimic chosen features of the system at hand, which means that it cannot be criticized because it does treat certain features that the more comprehensive model describes. Clearly, developing such a simple model requires a degree a creativity that is not required in developing a more complete comprehensive model. The adequacy of such a model is determined primarily by its ability to describe observations or results from a comprehensive model. I have shown that you can develop simple models to describe features of complex systems such as those associated with dispersion in complex terrain and the formation of ozone. I need to point out that simple models are not substitutes for comprehensive models, but should be considered to be vital supplementary aids to understanding the system at hand. In an ideal world, simple and comprehensive models should be developed hand in hand.

B. E. A. FISHER: There is a relationship between error in prediction and complexity (or number of input parameters). It has been argued that there is an optimum position (minimum error). Do you think that the position of this optimum has changed in recent years and will move in the future?

A. VENKATRAM: While such a point of minimum error is possible in principle, there is no way of finding it in practice. Again, in principle, this optimum should move in response to better measurements leading to more accurate inputs. However, I can think of no way of finding this optimum in a systematic manner. I am also not suggesting that the simple model represents any sort of an optimum. A simple model based on a small number of variables is one way of understanding the essentials of a complex system. This understanding is likely to complement that from a comprehensive model.

E. GENIKHOVICH: One of the reasons of high scatter on the scatter plots is that concentration fields are highly irregular and comparison of data paired in space and time is based on the use of wrong metrics. One should not ask wrong questions from the model; otherwise stupid answers could be obtained. I think that development of new methods of comparison and validation of the models is of major importance.

A. VENKATRAM: We do need better methods for comparing model estimates with corresponding observations. The development of better and bigger comprehensive models is not likely to decrease the scatter between model estimates and observations. The scatter might actually increase because of unavoidable errors in model inputs. Under these circumstances, it might be more meaningful to compare the distribution of model estimates with those of observations. It might also be possible to use a simple model to mediate between observations and the results from comprehensive models. Specifically, we can fit the simple model to both observations and complex model results and check whether the two sets of fitted model parameters are comparable. In this case, the simple model is used to compare the underlying patterns in the two sets of data: model results and observations.

D. W. BYUN: Venky, it is a very inspiring speech. You pointed out the value of simple models over the comprehensive models. However, the way I view individual modules representing different physical and chemical processes are themselves simpler models. How could you distinguish these types of models based on the term "simple model"?

A. VENKATRAM: I am not suggesting that there is a "unique" simple model for every system. The point here is that it is useful to understand complex systems using "simple" models that incorporate variables that the modeler believes to be most important. In principle, we can construct a hierarchy of simple models that explain different aspects of the system at hand. The understanding derived from such models will complement that obtained from comprehensive models, which attempt to include all the relevant processes in as much detail as possible. We need such simple models because the responses from comprehensive models are often difficult to interpret.

A. ELIASSEN: Why do the semi-empirical models often give better results than the comprehensive model?

A. VENKATRAM: I am speculating here. There are two possible reasons that simple models may at times provide better results than comprehensive models. One is that unavoidable errors in the large number of inputs required by comprehensive models are likely to lead to major errors in outputs even if the physics is correct. The second reason is that errors in the parameterizations of the innumerable processes included in comprehensive models might reinforce each other rendering the outputs useless. On the other hand, simple models can reduce these types of errors by keeping the number of inputs and process parameterizations to a minimum. I am not suggesting that simple models replace comprehensive models. My point is that understanding from simple models can complement that from comprehensive models.

A. ELIASSEN: Can we believe the comprehensive models response to emissions changes when we know that the mixing is incorrectly handled?

A. VENKATRAM: I think not. It is difficult to believe results from comprehensive models when mixing between chemical species is treated incorrectly. This is precisely why we need simpler semi-empirical models to provide alternative understanding of the system.

J. WEIL: How can improved measurement capabilities of meteorology (remote sensing, etc.) and species be integrated with models to produce better results? Do you feel this is important and should we be pursing this more vigorously to improve future models?

A. VENKATRAM: In principle, better measurement techniques should lead to better model inputs and thus to better model results. However, I am more than a little concerned that, in recent years, such thinking has led to large, expensive, field studies involving multiple aircraft and high tech measurement techniques. The cost of the field study is proportional to the comprehensiveness of the model that is supposedly being evaluated with the data being collected. The development of large comprehensive models accompanied by expensive field studies has become a self-perpetuating process. Such a process, which is often primarily of economic value, has no place for simple models and the understanding derived from them.

INVERSE TRANSPORT MODELING OF NON-CO_2 GREENHOUSE GAS EMISSIONS OF EUROPE

A.T. Vermeulen[1], M. van Loon[2], P.J.H. Builtjes[2] and J.W. Erisman[1]

[1]ECN
PO Box 1, 1755 ZG Petten, Nederland
[2]TNO-MEP
PO Box 342, 7300 AH Apeldoorn, Nederland

INTRODUCTION

The relatively inert behavior of the most important non-CO_2 greenhouse gases (CH_4, N_2O, HFC's and SF_6) on a timescale of several days makes them attractive candidates to derive their emissions from observations using inverse transport models. To enable verification of the rather uncertain emission inventories of these gases, as for example required under the Kyoto protocol, a combination of a cost effective concentration measurement program with inverse modeling seems worthwhile.
Through the sensitivity analysis presented in this paper the possibilities for and design of a monitoring and modeling strategy is derived that would allow for monitoring of the emissions of the most important non-CO_2 greenhouse gases of the individual European countries from year to year.
Depending on the model error and the characteristics of the different compounds (atmospheric lifetime, background concentration, temporal and spatial variability of the sources) the maximum accuracy that can be obtained will be discussed.
As a real-life example, results will be presented for the case of methane. Using concentration records for several stations in The Netherlands and NW-Europe in the period 1990-1997, emission estimates are derived using an inverse Lagrangian and an Eulerian transport model. The consequences of the use of different background concentrations, mixing layer height information and a priori emission information for the systematic errors and the uncertainty in the inverted emissions are also presented

Inversion techniques at different scales

Global Scale The smoothing of emissions by atmospheric transport processes limits to a large extent the information contained in measurement data to relatively low spatial and temporal resolution. One of the analysis and solution techniques that can be used to derive the emissions from observations and model results is the Truncated Singular Value Decomposition (TSVD).

SVD can be used to identify the components of solutions that can be resolved by the combination of model and measured data. Brown (1995) concludes that the components of the methane source distribution that are most sensitive to uncertainties in the observations can be neglected (by truncation) and that the truncated singular value decomposition solution can be interpreted as a stable estimate of the unknown (methane) emission distribution. Using TSVD, Brown derived worldwide methane emission estimates for 18 latitudinal bands of 10° width.

The SVD inversion technique has also been deployed by Bousquet et al. (1999a,b) for CO_2 on the global scale, using the TM2 model (Heimann, 1995) and CO_2 observations of the Globalview-CO_2 dataset. A global uptake by land and ocean surfaces was determined with the main uptake at higher latitudes. In tropical Asia a high net source of CO_2 was found. One of the conclusions of Bousquet et al. (1999a,b) is that more measurements are needed to obtain a higher spatial resolution of the calculated fluxes.

In Fan et al. (1998,1999) the global distribution of CO_2 is also derived by using the TSVD technique. Quite large differences in the derived fluxes occur when compared to the fluxes derived by Bousquet et al. (1999a,b). The influence of local disturbances at continental stations was found by Bousquet et al. to have a large impact on the calculated fluxes and they tried to minimise this effect.

Continental Scale. In Biraud (1999) measurements at Macehead (Ireland) are used to derive emissions of the major greenhouse gases in a 'model-free' approach using Rn^{222} measurements over the period of 1996-97. The fluxes derived are in good agreement with the inventories, except for CFC_{11}, of which not enough events with high concentrations occurred in the time-frame of the calculations.

In Stohl (1998) an overview is given of the possibilities of trajectory models to derive emission estimates. The inverse modelling can take place by using statistical methods, like for example Seibert (1997) or the overlay method (Vermeulen et al., 1997) or through source-receptor relationships and inverse calculations. Wotawa and Kröger (1997) show that in the case of a perfect model and measurements the statistical method gives good results for NO_x. Stohl (1996) also used backtrajectories to derive the sources of sulphate, using the EMEP measurement network with a modified statistical method after Seibert, resulting in better use of information in the data and less averaging out like in the unmodified statistical methods, the improved field also better resembles the spatial pattern of current inventories.

By using a simple trajectory model and SVD, Eisma et al. (1995) were able to derive the emissions of $^{14}CH_4$ by nuclear installations in NW-Europe. The same method with an improved version of the transport model was also used to derive the emissions of methane of NW-Europe (Vermeulen et al, 1999).

Simmonds et al. (1996) derived estimates of the emissions of continental Europe of non-CO_2 greenhouse gases using a trajectory model based on concentration data of Mace Head, Ireland. In the concentration data the influence of emission reductions of certain CFC's caused by the Montreal protocol can be seen clearly. They concluded that the CFC_{11} emission was reduced by a factor of 2.5 in the time from 1987 to 1994. In the same time the emissions of CFC_{12} were found to be reduced by a factor of 4.

In Van Loon et al. (1999) and Van Loon (1999), Kalman filtering techniques are used to assimilate measurements of ozone into an atmospheric transport chemistry model. In these simulations the uncertainty in the emissions are considered to be (part of) the explanation of the difference between model and measurements. In the two studies mentioned, the computed emission correction factors are a spin-off and only a rough indication, but by a modification of the data assimilation technique, it is possible to obtain more reliable emission estimates

Local scale. Potosnak et al. (1999) showed that linear modeling of tower measurements of CO_2 combined with tracer elements like CO and C_2H_2 can be used to derive the local influence of antropogenic sources and the relative importance of biological cycle. Using tower measurements of N_2O concentrations during nights with a strong inversion of the mixing layer, Hensen et al. (1999) were able to derive nightly emission rates of N_2O for an area around the tower up to 200 km by 200 km. Airplane measurements allow to measure the local fluxes as a function of surface type, but these measurements are inherently limited in accuracy, the number of elements that can be measured and the time-scale of the measurements. Platform or tower based measurements are more precise, can easily cover long time ranges and can measure more components, but are of course fixed in position and determine therefore only the fluxes of a few surface types reliably. By using high towers well in the mixing layer one can optimize the area for which information is received and one then also minimises the influence of the very local sources. By also measuring at lower altitudes one can easily determine to which degree the higher measurements are influenced by local sources.

METHODS

The COMET (CO_2 MEthane Transport) model is a simple Lagrangian model that can be ise for both predictive and inverse modeling purposes. COMET uses backward trajectories. The calculations described in this paper were performed using trajectory and mixing layer height data provided by EMEP MSC-W in Oslo. The 2D 96 hour backward trajectories were calculated from analyzed wind fields for a height of 975 mbar. In the calculations shown in this paper data on mixing layer height provided by EMEP has been used, which was derived from a combination of weather model data and balloon observations.

In the COMET model two vertical layers are distinguished, a mixing layer and a reservoir layer. The initial methane concentration at the start of each trajectory was taken from the weekly averages of the calculated concentrations of the TM2 (Heimann, 1995) GCM for 1994. A more detailed description of the COMET model can be found in Vermeulen et al. (1997, 1999).

In forward mode the COMET model retrieves per time step the emissions for the grid cells under the current circular source area and calculates the concentration changes and isotopic composition for the modeled components in the column of air with the current mixing layer height. Emission data is retrieved from emission inventories per source category on a regular grid, available are an early ECN, the LOTOS (Builtjes, 1992) and the high resolution METDAT database (Berdowski et al., 1998)

The LOTOS (Long Term Ozone Simulation) model is an Eulerian grid model that uses a 70x70 equidistant grid, covering the domain [10W-60E]×[35N-70N]. This domain is large enough to be able to simulate ozone concentrations in Europe for both episodic and long term calculations. In the vertical direction there are 3 layers: the mixing layer and two layers above it. A detailed description of the LOTOS model can be found in Van Loon, Segers & Builtjes (2000), also in this volume.

Initial and boundary conditions (for each month) from the TNO-Isaksen 2D model. For the modeling of CH_4 concentrations a slightly modified version of the LOTOS model was produced, in which also the necessary bookkeeping was introduced that is needed for the inverse calculations.

Measurements are described as a function of the parameters that need to be determined by the LOTOS or the COMET model. In our calculations no a priori information on the unknown emissions is given and the set of equations is directly solved by (Truncated) Singular Value Decomposition (Press et al., 1992). The transport model is used to determine the contribution of each of the source areas under evaluation to the measured concentration. The measured excess of methane (X) above the background concentration (C_{back}) for each observed concentration (C, 1..t) is the sum of the product of all emission strengths (E) times the contributions (c) of the respective source areas (1..i), as expressed in eq. 1. This simple linear model is only valid for non-reactive and non-depositing components, otherwise unresolvable a-linearities will occur. The c matrix is delivered by the models, the X vector is derived from the measurements and the SVD technique is applied to obtain the best fit for the emission vector E between observation and model. The problem is often overdetermined, this means that conflicting information may be in the dataset and that the solution is uncertain. It could also be that emissions of certain areas contribute little to the total excess concentrations (much less than the measurement precision), but that a variation with very strong factors of the emission strengths of these areas leads to a marginally better solution. (T)SVD allows to identify those areas by objective criteria and to exclude their influence from the solution, without to much cost for the final fit between data and observations.

$$
\begin{aligned}
X_1 &= C_1 - C_{back_{1t}} = \sum_{i=1}^{n} E_i . c_{i,1} \\
&\vdots \\
X_t &= C_t - C_{back_{1t}} = \sum_{i=1}^{n} E_i . c_{i,t} \qquad (1) \\
&so \\
\begin{pmatrix} X_1 \\ \vdots \\ X_t \end{pmatrix} &= \begin{pmatrix} E_1 \\ \vdots \\ E_i \end{pmatrix} . \begin{pmatrix} c_{1,1} & \cdots & c_{1,i} \\ \vdots & \ddots & \vdots \\ c_{t,1} & \cdots & c_{t,i} \end{pmatrix}
\end{aligned}
$$

To estimate the error in the calculations of the inverse model a statistical Monte Carlo routine is applied. In this method the model data is distorted with a user defined random factor (here 50% is assumed), corresponding with an estimated accumulated model error. This accumulated error represents the combined effects of all possible error sources. On this distorted matrix the SVD matrix inversion is applied again. This is repeated several times with newly generated random errors on the matrix elements. The standard deviation in the calculated emissions per area gives a measure of the uncertainty of the result for that area.

In the inverse model the emissions are calculated for aggregate areas on the more coarse LOTOS (Builtjes, 1992) 1/2° by 1° grid (roughly equivalent to a 60 by 60 km grid). Close to the Netherlands a fine resolution is taken starting with aggregates of single cells and farther away from the Netherlands the aggregates contain more cells. The aggregate source areas also resemble in shape the expected emission or country pattern.

RESULTS

COMET and CH4. Inversions were carried out using both the methane concentration data for the Cabauw high tower site in the period from 1993-1997 and concentrations in 1996 at four stations in Europe (Cabauw, Petten, Heidelberg and London). The four European stations were carefully intercalibrated (Nisbet et al, 1997). The c matrix was calculated using the Lagrangian COMET model. The resulting emissions per country for both inversion runs are shown in table 1. This table also shows existing emission inventories. Also the relative uncertainty in the country totals as given by the Monte Carlo error analysis is shown. When only one year of data for one station is used the solution is very unstable (not shown here). The long record from 1993-1997 at Cabauw allows for a quite stable solution, while the use of concentration data for four stations allows a fairly stable result with only one year of data. The absolute deviations between inventory data and the calculated emissions fall within the uncertainty range of the inventories, which estimated to be 30-50%. The emissions derived for the North Sea continental shelf (CS) could not be resolved using the 1 year data of the 4 stations. The transport model indicates that only 25% of the excess concentrations in Cabauw originates from the Netherlands. So the Cabauw concentration signal contains quite some information of the sources outside the Netherlands, covering an area of most of NW-Europe.

Table 1. Emission estimates from inventories and inverse COMET in Tg $CH_4.yr^{-1}$

Country	Lotos [1]	EDGAR [2]	METDAT [3]	COMET Cabauw		COMET 4 stations	
				93-97	Err%	1996	Err%
NL	0.85	1.0	0.88	0.76	1	1.5	4
D	3.9	6.2	5.5	3.8	1	3.2	4
B	0.65	0.59	0.66	0.67	1	0.41	4
GB	2.5	3.9	4.6	5.9	1	7.7	2
EI	0.57	0.65	0.85	1.5	1	0.43	14
CS	0.26	incl.	0.36	0.54	3	<0	23

[1]Builtjes et al., 1992 [2] Olivier et al., 1994 [3]Berdowski et al., 1998.

The inverse model was used to derive simulated concentration data based on the relations between constructed emission data and the relations as derived by the inverse model between emission strength per emission area and the response in excess concentration at the arrival point. When this constructed concentration data is fed to the inversion routine this should lead to an exact representation of the emissions used to construct this concentration data, and this was indeed found to be the case. The Monte Carlo error analysis then allows to estimate the relation between the uncertainty in the calculated emissions and the uncertainties in concentration measurements and the model errors.

COMET and other greenhouse gases. As a theoretical exercise, emission fields were constructed also for nitrous oxide and HFC's, where the nitrous oxides were assumed constant above land and freshwater at a value of the average current estimate for Europe, and zero above sea areas. The HFC emission was taken as zero over the whole area except for five point sources of different sizes in west and south-east France, Ireland, south Germany, Italy and Denmark. The five sources represent hypothetical production facilities with relatively large sources and the diffuse emission of all the other sources is thought to be relatively small per unit area and not leading to detectable changes in concentration above the base signal. With estimated precisions of the concentration measurements for these compounds and different model errors the Monte Carlo error analysis yielded the results depicted in table 3. Trajectory and mixing layer data was used for 1993-1997 for arrival point Cabauw. For methane the METDAT database was used as the a priori emission database. The error in methane concentration measurements was set at 5 ppb. The error in nitrous oxide concentrations was set at 1 ppb. The error in HFC concentration was set at 0.5%.

From table 2 we can conclude that the use of longer time series allows to reduce the error in the emission estimates caused by model and measurement imperfections. Reducing the model imperfection from 50% to 10% reduces the uncertainties in the emissions with a factor of two. Going from 1 to 4 years of data also reduces the uncertainty with a factor of 2. The combined effect allows for theoretical uncertainties in the emission rates of about 5-10%. Uncertainties are largest for nitrous oxided because of its high background concentration compared to the size of the variation in concentration signal caused by the sources.

Table 2. Estimated maximum precision (%) of non-CO_2 greenhouse gas emissions estimates using COMET inverse mode

Model error	50%		10%	
	4 year	1 year	4 year	1 year
CH_4	10	20	5	10
N_2O	30	unstable	15	unstable
HFC's	10	30	<5	10

LOTOS and CH_4. The LOTOS model was used to predict the concentrations in 1994 of methane at four locations at the Netherlands where measurements have taken place, i.e. Cabauw, Kollumerwaard, Delft and Arnhem in the framework of the 2nd Dutch NRP on Climate Change. Also the data calculated for Mace Head were used. The hourly concentration data were averaged to 24 hour mean values. The correlation coefficient (r^2) between measured and calculated concentrations is then 0.41. Which is higher than the correlation found for the COMET model for 6 hour values; when the COMET results are also averaged to 24 hour values the r^2 is 0.39. The LOTOS model used the same aggregate source area definition as was used for the COMET inverse calculations. The LOTOS source-response matrix c was used in the same SVD matrix inversion routine as deployed for the COMET model. The resulting emission fields when used to predict the concentrations at the measurement locations lead to significantly larger explained variation, with r^2=0.67. The calculated update factors per country for a truncation value for 20% of the maximum eigenvalue are given in table 4 together with the uncertainty indicated by the Monte Carlo error analysis calculated as the weighted mean error in the aggregate source areas determining the country total emission.

COMET and LOTOS for other greenhouse gases. In order to test the differences between the two quite different dispersion models COMET and LOTOS on forward and inverse calculations test runs were performed on a compound X, with a rather uniform emission pattern like the one that is expected for N_2O. Both models were provided with the emission data base on the LOTOS grid and a forward run was performed to calculate the expected concentrations for the base year of 1994.

The concentrations calculated by both models in forward mode differ considerably (r=0.14; see figure 1), although the main features of the concentration as a function of time are quite similar. The forward mode calculated time series of the concentrations at Cabauw were then used as input of the SVD inversion routine using the two source-response matrices of the respective models. When the concentrations calculated by one model were combined with the corresponding source-receptor matrix of that model nearly 100% of the variation in the concentrations could be explained, as was to be expected, and the uncertainty calculated by the Monte Carlo error analysis is 1-5% for the main regions around the receptorpoint Cabauw. Then also the calculated emissions are within 0-15% of the original input ones. When the LOTOS predicted concentrations were inverted using the COMET source receptor matrix then only 51% of the variation can be explained, the emissions per country then have a calculated Monte Carlo uncertainty of 10-40%. The calculated emissions are then within 30% of the original input ones.

In a last test the LOTOS model was used to generate source-receptor matrices for the compound X at five imaginary locations over an (rectangular) area covering The Netherlands in a cross-like formation, with one point in the middle and four stations around this central station at equal distances, in the corners of the rectangular area. The sum of the source contributions in each of the receptor points is not equal to the simulated concentrations (the "measurements"), although both the measurements and the source-receptor matrices were generated by the same model. This is due to non-linearities in the (transport) modeling. On average, the sum is equal to the measurements, but at some points in time the deviation can be as high as 10%. The standard deviation of the percentual difference between sum and measurements is 2.2%. Hence a substantial model error is present in this experiment. Moreover, the spatial distribution of the emissions of compound X is rather flat in the neighbourhood of The Netherlands, which may further complicate the inversion.

Table 3. Update factors calculated by TSVD for the country total emissions of methane for the optimal solution of the LOTOS source response matrix for 1994 and five measurement stations in The Netherlands.

Country	Update factor	%Err
Germany	0.55	27
France	1.76	8
Italy	1.50	4
Netherlands	2.00	9
Belgium	5.40	4
Great-Britain	0.65	106
Ireland	0.46	25
Austria	2.21	10
Switzerland	2.46	14
Czech Rep.	1.82	11
North Sea CS	0.85	3
Finland	0.85	2
Poland	0.43	14
Hungary	1.27	21

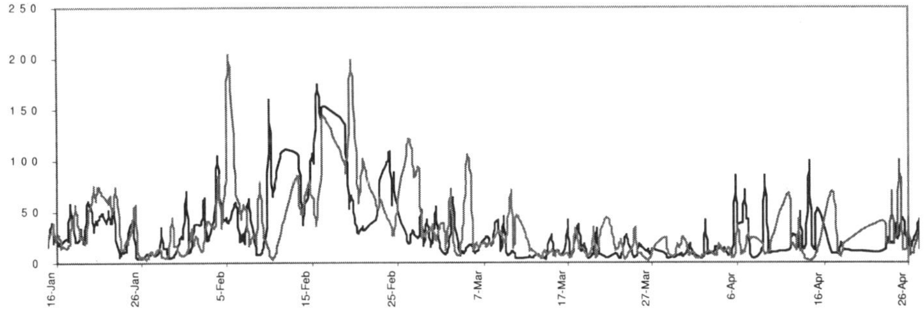

Figure 1. Forward mode calculated concentrations of compound X at Cabauw using COMET and LOTOS.

Inversions were carried out for the 24 hour averaged data of the year 1994 using only the central station and using all five stations. In both cases a nearly perfect solution is found that resolves almost 100% of the variation in the data. The Monte Carlo analysis show however that in the case of only data from the central station the mean relative standard deviation of the emissions in the aggregate areas is about 25%. When the data of all five stations is used this is reduced to 7%. Note that these percentages do not represent the error in the emission estimate, but only the quality of the inversion itself.

In Table 4 the relative emissions compared to the input emissions are compared for the two runs. For most areas the inverse calculated emissions are within a 10-15% range around the input ones. Only for source areas with a very low contribution the inverse calculated emissions can deviate much more (e.g. Ireland and Switzerland). The highest quality is reached for the 5 stations run. The ground level run is marginally better than the 200m run. The SVD inverse emissions tend to be underestimated when the input data is more sparse as it is in the 1 station case.

Table 4. Relative emissions (%) per country (compared to original emissions) of compound X calculated by SVD inversion per country for three source receptor matrices, one with data for five stations, one for 1 central station at ground level and one for 1 central station at 200m agl.

%	Input source receptor matrix	
Country	5 stations	1 middle point
Germany	105	86
France	89	96
Netherlands	85	80
Belgium	101	100
Great-Britain	72	73
Ireland	17	24
Denmark	48	44
Austria	98	130
Switzerland	94	55
Czech Rep.	100	121
Bulgarije	100	97
Roemenië	100	97
Finland	103	96
Poland	96	75
Hungary	99	95
Average emission	87	84

CONCLUSIONS

In general, it can be concluded that with the (T)SVD technique emissions can be estimated with high accuracy, provided that the modeled source-receptor matrices are of high quality as well. In the last experiment the model error was 0-10% resulting into emission estimates that deviate 0-15% from the original input ones. For real world data the (T)SVD inversion method allows to derive satisfying results from imperfect models and concentration data. In order to improve the accuracy of the resulting emission fields more accurate models and more measurement data will (at more stations within the source areas and with longer tie series) be needed. The tools described in this paper allow to develop a feeling of where to go in order to achieve this at cost-effective way.

Acknowledgements

The authors would like to express their thanks to Erik Berge from EMEP MSC-W (Oslo, Norway), for providing the trajectory and mixing layer height data and to Sander Houweling from IMAU for providing the TM2 background methane concentrations and Mike Sanderson from Cambridge University for providing the TOMCAT concentration data.

This work was funded by the European Union (since 1994, EU contract EV5V-CT94-0413) and the Dutch Ministry of Economic Affairs (Contract 53477) with additional financial support from the National Research Program on Global Air Pollution and Climate Change (since 1993, projects 852097 and 950421), and by the ministry of VROM.

REFERENCES

Biraud S., Ciais P., Ramonet M., Simmonds P., Kazan V., Monfray P., European Greenhouse Gas Emissions Estimated from Continuous Atmospheric Measurement Records at Mace Head, Ireland, Submitted to *Atm Env.* (1999).

Brown M., Deduction of Emissions of Source Gases Using an Objective Inversion Algorithm and a Chemical Transport Model, *JGR*, 98, 12639-12660 (1993).

Brown M., The Singular Value Decomposition Method Applied to the Deduction of the Emissions and the Isotopic Composition of Atmospheric Methane, *JGR*, 100, D6, 11425-11446 (1995).

Bousquet P., Ciais P., Peylin P., Ramonet M., Monfray P., Inverse Modelling of Annual Atmospheric CO_2 Sources and Sinks. Part 1: Method and Control Inversion, Submitted to *JGR*, (1999a).

Bousquet P., Ciais P., Peylin P., Ramonet M., Monfray P., Inverse Modelling of Annual Atmospheric CO_2 Sources and Sinks. Part 2: Sensitivity Study, Submitted to *JGR*. (1999b).

Eisma R., Vermeulen A.T., Van den Borg K., $^{14}CH_4$ Emissions from Nuclear Power Plants in Northwestern Europe, *Radiocarbon* 37, 2, 475-483 (1995).

Fan S-M., Gloor M., Mahlman J., Pacala S., Sarmiento J.L., Takahashi T., Tans P.P., A Large Terrestial Carbon Sink in North America Implied by Atmospheric and Oceanic CO_2 Data and Models, *Science*, 282, 442-446 (1998).

Fan S-M., Sarmiento J.L., Gloor M., Pacala S.W., On the Use of Regularization Techniques in the Inverse Modelling of Atmospheric Carbon Dioxide, *JGR* 104, D17, 21503-21512, (1999).

Heimann M, *The global atmospheric tracer model TM2,* Hamburg, Max Planck Institut für Metorologie (1995).

Hensen A., Dieguez Villar A., Vermeulen A.T., Emission Estimates Based on Ambient Concentrations Measured at a 200m High Tower in the Netherlands 1995-1997. In: Ham J. van, et al (eds), *Proc. 2nd Symp. on Non-CO_2 Greenhouse Gases*, Dordrecht, Kluwer Ac. Publ. (1999).

Leclerc L.Y., Thurtell G.W., Footprint Prediction of Scalar Fluxes Using a Markovian Analysis, *Bound Layer Met* 50, 374-393 (1990).

Van Loon M., Builtjes P.J.H., Roemer M.G.M., *Reactive Trace Gas Data Assimilation,* BCRS report USP-2 99-12 (1999)

Van Loon M., 1999. Data Assimilation of Ozone in the Atmospheric Transport Chemistry Model LOTOS, Paper to appear in *Environmental Modelling and Software*.

Seibert P., Inverse Modelling Based on Trajectory-derived Source Receptor Relationships, In: *Proc. 22nd NATO/CCMS Int. meeting air poll. modelling and its application*, Clermont Ferrand (1997).

Simmonds P.G., Derwent R.G., McCulloch A., O'Doherty S., Gaudry A., Long-term Trends in Concentrations of Halocarbons and Radiatively Active Trace Gases in Atlantic and European Air Masses Monitored at Mace Head, Ireland from 1987-1994, *Atm Env* 30, 23, 4041-4063 (1996).

Stohl A., Trajectory Statistics - a New Method to Establish Source-receptor Relationships of Air Pollutants and its Application to the Transport of Particulate Sulphate in Europe, *Atm Env* 30, 4, 579-587 (1996).

Stohl A., 1998. Computation, accuracy and apllications of trajectories - a review and bibliography, Atm Env 32, 947-966.

Vermeulen A.T.., Beemsterboer B, Van den Bulk W.C.M. et al, *Validation of methane emission inventories for NW-Europe*, Petten, ECN, report ECN-C--96-088 (1997).

Vermeulen A.T., Eisma R., Hensen A., Slanina J., Transport Model Calculations of NW-European Methane Emissions. *Env Sci & Policy* 2, 315-324 (1999).

Wotowa G., Kröger H., Testing the Ability of Trajectory Statistics to Reproduce Emission Inventories of Air Pollutants in Cases of Negligible Measurement and Transport Errors, *Atm Env* 33, 3037-3043 (1999).

DISCUSSION

P. SEIBERT: How long were the trajectories?

A. VERMEULEN: The trajectories used in this study were 2D 96 hour backward trajectories as provided by EMEP, based on the windfields of the DNMI weather model.

P. SEIBERT: It is of course very good that you have done the error analysis, but I assume that you have used independent random disturbances in your Monte Carlo analysis. However, real errors both in observations and in the model will have a structure in space and time. If a station is bad, it means that it is locally disturbed which can be felt always as a specific meteorological situation. If meteorology is wrong, it will be so in a whole area and over some time. Also, 50% error can be easily exceeded in the source-receptor relationships, the error can be orders of magnitude. Thus the error estimates are probably overly optimistic.

A. VERMEULEN: The Monte Carlo analysis gives an indication of the stability of the inversion solution, depending on assumed random model errors. The error analysis presented does not deal with the absolute precision of the calculated emissions, as systematic errors in measurements and model are difficult to assess. I think that an average random model error of 50% is possible even with this simple type of transport model. The error analysis illustrates that with more stations and longer time series a more stable emission estimate can be reached. To reach minimum absolute errors the model should be improved as much as possible regarding e.g. mixing layer height, trajectory positions etc, as indicated in the presentation. If the error percentages presented are mistaken as the real uncertainties in the emission estimates, then they are indeed overly optimistic.

P. SEIBERT: A number of people have presented work on inversion, and everybody has used different mathematical notation and even verbal nomenclature, which makes it more difficult to communicate and to understand for other people. I suggest that we adopt the unified notation introduced in the recently published AGU Monographs on Inverse Modelling in the future.

A. VERMEULEN: I cannot speak for the other presenters of inversion work here, but I agree. The notation I used was taken from the SVD method description, which is much older than this recent monograph.

IN SITU DIAGNOSTIC OR NESTED PROGNOSTIC METEOROLOGICAL MODELS TO DRIVE DISPERSION SIMULATIONS IN COMPLEX AREA: A COMPARISON IN A REAL APPLICATION

S. Finardi [1], G. Tinarelli [1], A. Nanni [2], D. Anfossi [3], E. Ferrero [3,4], S. Trini Castelli [4]

[1]CESI S.p.A. Ambiente, via Reggio Emilia 39, 20090 Segrate (MI), Italy
[2]ENVITECH S.r.l., c.so d. Vittoria 2/f, 28100 Novara, Italy
[3]CNR, Istituto di Cosmogeofisica, Corso Fiume 4, 10133 Torino, Italy
[4]Dipartimento di Scienze e Tecnologie Avanzate, Corso Borsalino 54, Alessandria, Italy

INTRODUCTION

In order to assess the air quality impact of industrial emissions over complex area, several different approaches can be used to supply a dispersion model with a satisfactory description of the atmospheric mean flow and a characterization of the turbulence features. On the one hand, large dataset coming from ground level measurements and vertical soundings of wind and temperatures over the interested area can be directly integrated using diagnostic reconstruction tools over complex terrain. On the other hand, a more restricted subset of measurements, describing the atmospheric circulation at large scale, can be used to define initial and boundary conditions for a prognostic non-hydrostatic meteorological model at different nested grid resolutions. Processes directly simulated by this type of model should be potentially capable of reconstructing the mesoscale to local circulation features induced by the resolved inhomogeneities, such as the presence of topography or land/sea discontinuities, without the need of installing and maintaining a complex local measuring network. The aim of this work is to compare the two mentioned modeling approaches through a real application over a very complex area, simulating the dispersion of a plume emitted from a big thermal power plant (TPP). The site of interest, located along the Ligurian coast in northern Italy, is characterized by the occurrence of complex flow configurations due to the presence of the land/sea interface and inland mountains, inducing strong modifications to the synoptic circulation and superposing local breeze cycles. These characteristics are clearly hard to be completely reproduced by numerical models. What we can expect is a good reproduction of the main flow features by the diagnostic tools in the proximity of the measuring points and a lack of this method in the region not covered by the available data. On the other hand, a prognostic tool can directly reconstruct the flow

Air Pollution Modeling and Its Application XIV, Edited by
Gryning and Schiermeier, Kluwer Academic/Plenum Publishers, New York, 2001

features using conservation equations, but capturing only the features modeled by the physical representation on its resolved scale, such as the topography or surface characteristics, which can be inadequate in some circumstances. In this work, the first approach makes use of the diagnostic mass-consistent model MINERVE (Aria tech, 1995) to reconstruct the three dimensional flow field during episodes selected from an intensive field campaign that provides ground level measurements and vertical soundings at different positions. The meteorological fields are used to drive dispersion simulations of the TPP emissions using the Lagrangian particle dispersion model SPRAY (Tinarelli et al., 1994, 2000). In the second approach, the prognostic meteorological code RAMS (Pielke et al., 1992) has been instead used to simulate the synoptic to local circulation on the same period, driving the same dispersion code. Results from the two modelling systems, MINERVE+SPRAY and RAMS+SPRAY, are also compared against the data collected by both the meteorological and air quality local network. Advantages and shortcomings of each of the two systems are also discussed, in order to give an idea of the performances reachable by the different approaches in real applications.

SITE AND DATA DESCRIPTION

The investigated site is located on the north-western Mediterranean coast of Italy, not far from the urban area of Savona, nearly 50 km west of Genoa. It is characterised by complex terrain with mountains rising over 1000 metres few kilometres inland and many valleys along a direction orthogonal to the coastline, where the mean flow and the emitted plumes can be channelled. Different industries are located in the area, the major one is the coal/oil fired TPP operated by ENEL (the larger Italian electricity company) made up by 4 units for a total power of 1320 MW, with two 200 metres high stacks. Intensive meteorological and air quality field campaigns have been performed during summer (July and August 1997) and winter (January, February and March 1998), in order to capture episodes characterising the site specific seasonal features. Meteo/chemical surface stations, a sonic anemometer, two SODARs, meteorological and ozone soundings (pilot balloon, airsonde and thethersonde) have been used during these campaigns. A total number of 20 meteorological and 19 air quality stations have then been considered. A view of the area representing the model target domain for dispersion simulations and the measuring station locations is illustrated in Fig 1. Data from the field campaign have been completed by the large scale meteorological fields from European Centre for Medium range Weather Forecast (ECMWF) analyses, available every six hours. Emission data from the TPP were available with hourly frequency. The complexity of the atmospheric circulation in this coastal site, mainly related to the topographic features and to slope/valley flow is clearly depicted by the analysis of its vertical structure given by the available measurements. This vertical structure shows the presence of three distinct vertical layers, as shown in Fig. 2 depicting vertical profiles of the mean wind measured by Pilot balloon and radio-sounding launched close to the emission point. The first one starting from the ground level up to 300-500m where the flow has features similar to the surface wind, hence directly influenced by surface characteristics. The second layer, located roughly between 500 and 1500 m, shows characteristics weakly related with the surface, being probably influenced by mesoscale topography features. In the upper layer, usually located above 1500-2000 m, the flow tends to be directly connected with the synoptic circulation. This general structure of the flow, emerging above a strong temporal and spatial variability that moves the boundaries between different layers in a quite complicated manner, was observed in both fair and perturbed weather conditions. In particular during fair weather, cyclic rotations of lower level winds have been observed both in the summer and winter periods covered by the field campaign. Even if different pollutant sources (industries, car traffic, house heating,..) should be considered on this area, the simulations have been limited to the SO_2 emitted by the TPP, being Sulphur Dioxide an industrial pollution tracer gas weakly influenced by road traffic and house heating.

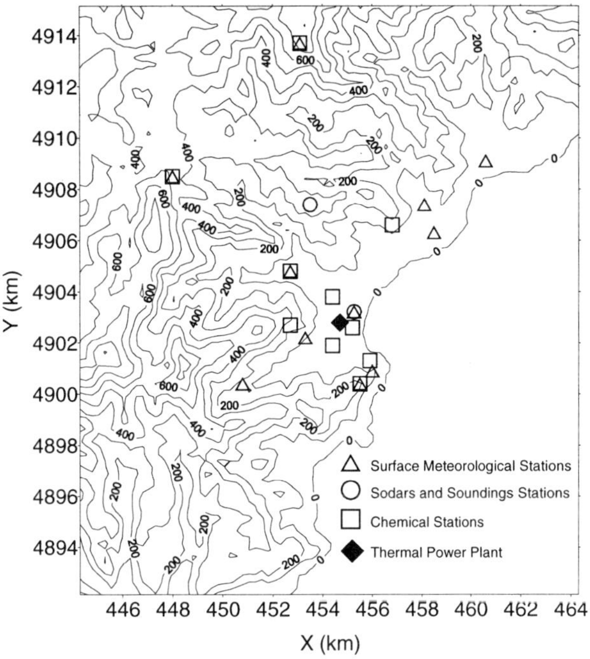

Fig 1. Meteorological and chemical stations location

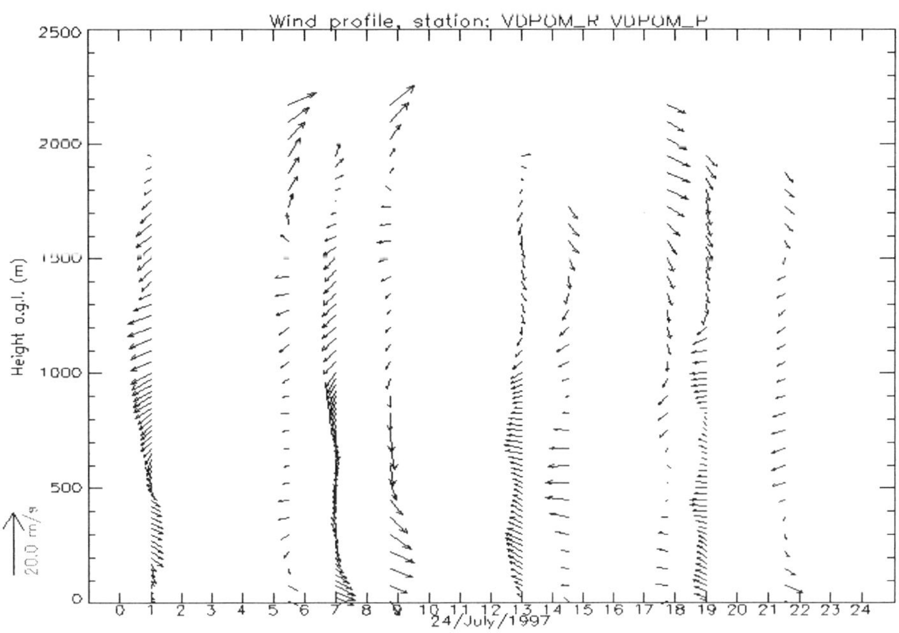

Fig 2. Pilot balloon and radio-sounding wind profiles on 24th July 1997

BRIEF MODEL PRESENTATION

The 3D mean wind and temperature fields have been diagnostically reconstructed applying the mass-consistent model MINERVE, which integrates the different wind measurements and analysis data. All the available vertical profiles (ECMWF analyses, local soundings and SODAR profiles) and selected surface observations have been used to initialise the model. The target computational domain, spanning 20 x 23 km^2 (Fig. 1), had an horizontal resolution of 250 m and variable vertical resolution with the lowest level at 10 m. The entire ensemble of diagnostic meteorological simulations covered 10 short term periods, representative of typical local circulation during the different seasons (Finardi et al., 1999). Each period lasts 1 to 3 days for a total simulation time interval of 20 days. In order to compare different methods, a selected summer episode (23-25/07/1997) has been simulated applying the non-hydrostatic prognostic model RAMS. A two way nesting technique has been employed to describe both the synoptic driving flow and local scale phenomena. The geographical location of the computational grid system is illustrated in Fig. 3.

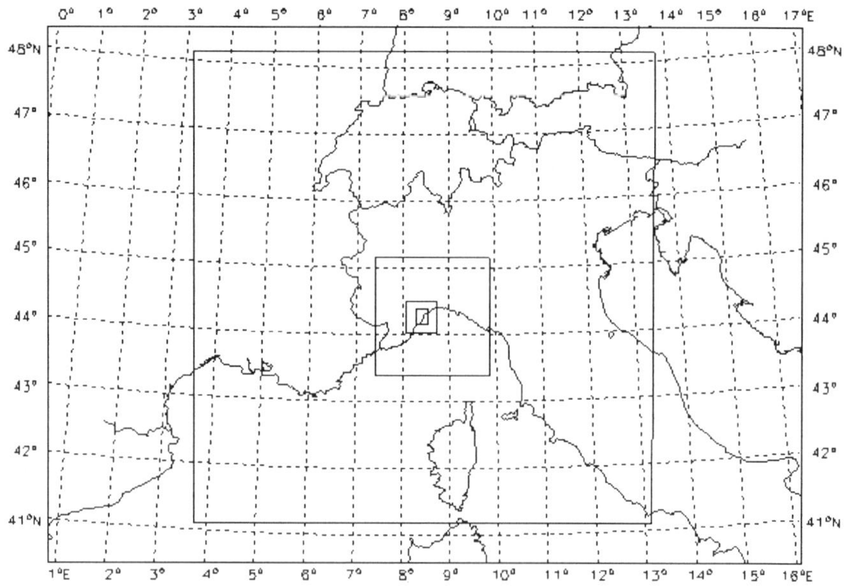

Fig 3. RAMS model nested computational domains

The main characteristics of the different meshes are described in Tab. 1. The previously mentioned dispersion simulation target area (Fig. 1) is enclosed in the central part of the finest grid.

Table 1. RAMS computational parameters

Grid number	1	2	3
Horizontal resolution	16 km	4 km	1 km
Grid points (nx*ny*nz)	50x50x29	50x50x29	54x54x33
Δt	30 sec	15 sec	5 sec
Δz min.	100 m	100 m	50 m
Domain top	13.9 km	13.9 km	13.9 km
Turbulence Closure	Mellor-Yamada	Mellor-Yamada	Mellor Yamada

RAMS has been initialised using the ECMWF 0.5 degrees resolution analysis fields, synoptic and local observations. ECMWF and WMO surface observations have been used to drive boundary conditions with a time resolution of 6 hours. The SPRAY Lagrangian particle dispersion code has been driven by the mean wind fields generated by both MINERVE and RAMS meteorological processors. Diagnostically simulated turbulence fields have been defined using a built-in parameterisation scheme based on the Monin-Obukhov similarity theory and the evaluation of the surface energy budget, taking into account local meteorological data and land use description. In the prognostic case, turbulence fields have been generated by the interface code MIRS (Trini Castelli and Anfossi, 1997). This module tries to use the information given by the closure parameterisation schemes given by RAMS transforming them into the fields needed by SPRAY. In the present simulations, PBL heights were computed from the profile of gradient Richardson number *Ri* according to Maryon and Buckland (1994), velocity variances $\sigma^2(i=1,2,3)$ were computed either from Hanna (HS) scheme (1982) or from RAMS according to the Mellor and Yamada (MY) closure scheme (1982). The decorrelation timescale τ_{L_i} were calculated from the σ^2 and K values and the third moment of the vertical velocity was estimated from the Chiba expressions (1978). SPRAY simulations started at 07:00 UTC of 23 September 1997 and ended at 23:00 UTC of 25 September 1997. Time step was set equal to 5 s and 15 particles were emitted every time step. A cell size of 250 x 250 x 30 m^3 was used to compute hourly averaged ground level concentrations.

RESULTS AND DISCUSSION

Comparing between themselves and with the available measurements the wind fields resulting from the two approaches, some differences can be outlined. Fig. 4 shows the vertical wind profiles obtained on 24th July 1997 over the same point indicated in Fig. 2, using the diagnostic MINERVE tool.

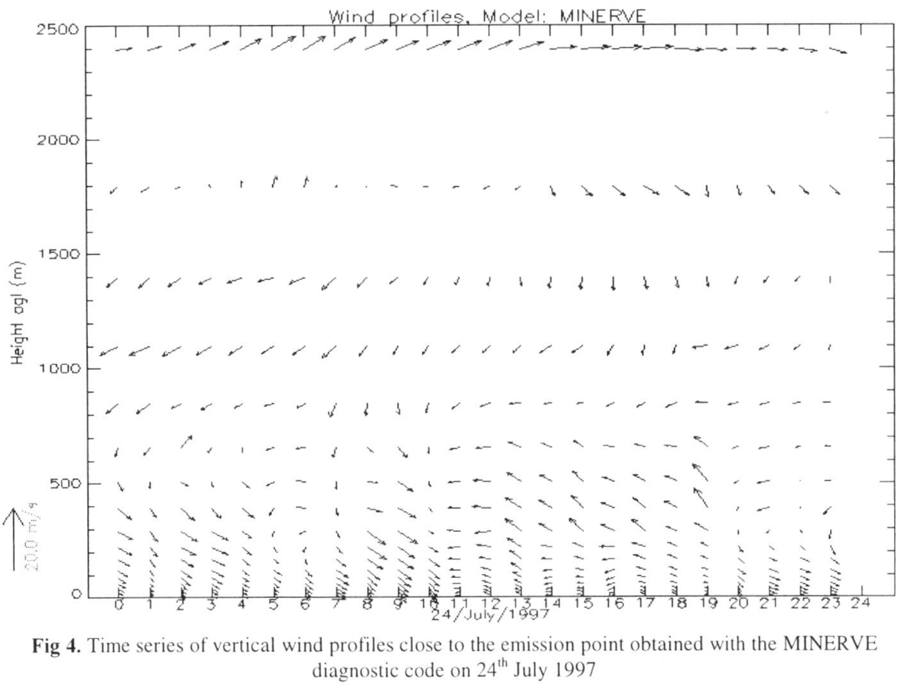

Fig 4. Time series of vertical wind profiles close to the emission point obtained with the MINERVE diagnostic code on 24th July 1997

In this case the sample is taken close to a measured profile given as input to the model, and the comparison with experimental data shows a satisfactory reproduction of the main features previously described, such as the day/night flow reversal and the vertical layers breakdown. Fig 5 shows the profiles obtained in the same positions applying the RAMS code, which doesn't directly use of the same measurement. In this case, the grid 3 with an horizontal resolution of 1 km has been considered. Winds are generally stronger and there is a clear tendency to generate a breeze signal close to the ground, correctly reversing around 10:00 in the morning and at 20:00 in the evening. There are instead some problems to reproduce all the spatial and temporal details shown by the measurements. During a night-time period, in the layer just above the simulated land breeze top level there are probably channelling effects generated by the resolved topography, causing a strong flow reversal not clearly present in the available data. In general, the main discrepancies are located in the layer between 750 m and 1500 m, where the simulated flow characteristics are dominated by the results obtained in the coarser grids and only slightly modified by grid 3. Figs 6a,b show horizontal wind cross-sections produced by the models in the layer closer to the surface during nocturnal land breeze conditions. RAMS computational domain has been zoomed to show the same area considered by MINERVE simulations.

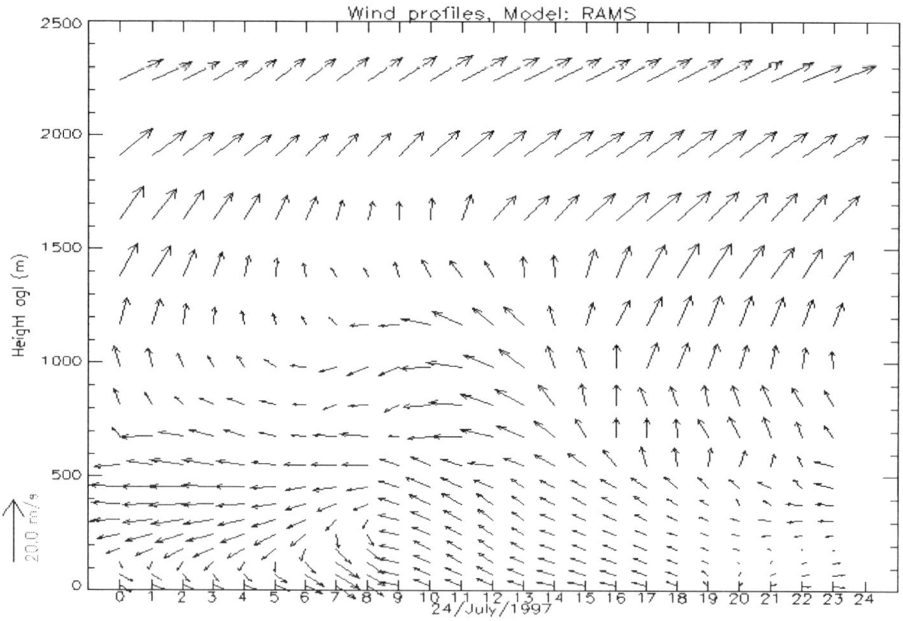

Fig 5. Time series of vertical wind profiles close to the emission point obtained with the RAMS prognostic code on 24th July 1997

It can be observed that the general wind pattern is similar in the region around the TPP, even if MINERVE produces lower wind speeds. On the contrary, in the eastern, north-eastern and south-western part of the domain (poorly covered by local measurements) the two wind fields substantially differ. RAMS shows a general tendency to produce more structure in the flow patterns, taking directly into account the temperature contrasts generated by the surface inhomogeneities.

As regards dispersion simulations, the two modelling systems show results in general of comparable quality, with discrepancies due to the different approaches adopted to generate the mean flow and turbulence driving fields. Moreover it is to keep into account that topography and wind field resolution was 250 m for MINERVE and 1 km for RAMS.

Small scale topographic details were therefore missing in the prognostic meteorological simulation. Taking into account the entire ensemble of flow and dispersion simulations (Finardi et al., 1999) the MINERVE+SPRAY modelling system gave rather satisfactory results, showing discrepancies far from the emission, in locations not covered by meteorological measurements. Comparing diagnostic and prognostic approaches, a particular attention has been therefore paid to these conditions.

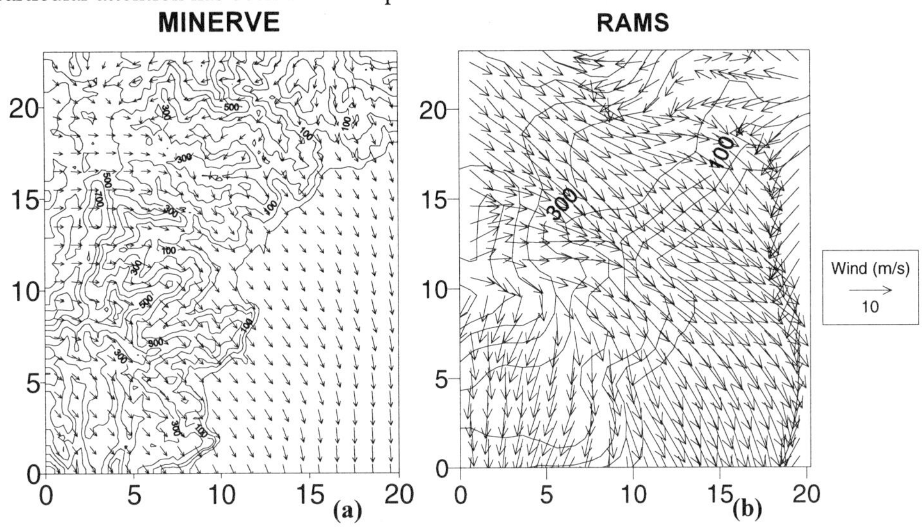

Fig 6. Wind fields produced by MINERVE (left) at about 10m a.g.l. and RAMS (right) at about 20 m a.g.l. on 24th July 1997 06:00

As an example, figs. 7a,b depict the SO_2 g.l.c. trends (full dots) recorded at the Bocca d'Orso chemical station, located 8.7 km to the north-west of the TPP, at an height of 530 m asl. Full triangles indicate MINERVE+SPRAY (Fig. 7a) and RAMS-MY+SPRAY (Fig. 7b) results respectively, while full diamonds represent RAMS-HS+SPRAY (Fig. 7b).

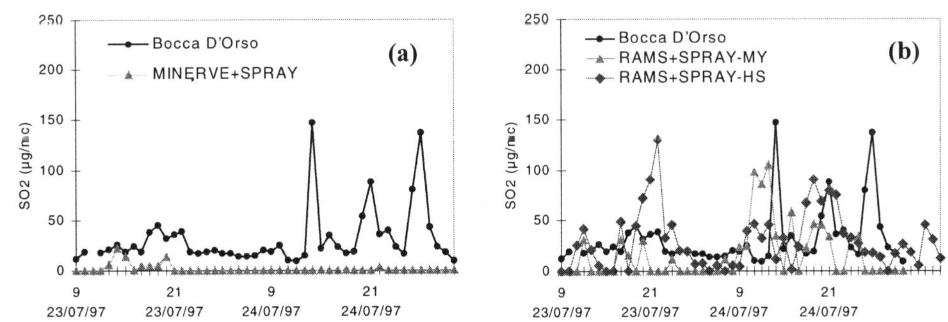

Fig 7 Comparison between computed and measured g.l.c. time series at the Bocca d'Orso station. a) MINERVE+SPRAY b) RAMS+SPRAY

RAMS (both HS and MY) +SPRAY seems to better capture the g.l.c trend, indicating a good reliability in a region not covered by local measurements. The described discrepancies in the 1000–1500 m layer flow generated by RAMS, are likely not to influence the dispersion process in this case since the PBL was generally smaller and the plume mainly traveled below 1000 m. The overall results obtained with the prognostic system seem to be comparable with those obtained using an expensive measuring network. A particular care needs to be paid to the choice and tuning of parameters, such as the higher grid resolution or the turbulence parameterization needed to generate dispersion coefficients. In this respect, a further investigation has to be done in order to better understand the mechanisms involved in each of the two presented methods. This includes a more detailed analysis of the different scale interactions in presence of complex topography and of the uncertainties in each of the two approaches. A further possibility to investigate involves the integration of the two considered modelling methods through the application of four dimensional data assimilation techniques.

REFERENCES

Aria Technologies., 1995, Note du Principe du Modele MINERVE 4.0. Report ARIA 95.008.

Chiba, O., 1978, Stabilty dependence of the vertical wind velocity skewness in the atmospheric surface layer, *J. Meteorol. Soc. Japan,* **56**, 140–142 .

Finardi S., Tinarelli G., Nanni A., Brusasca G. and Carboni G., 1999, Evaluation of a 3D Flow and Pollutant Dispersion Modelling System to Estimate Climatological Ground Level Concentrations in Complex Coastal Sites, *6th International Conference on Harmonization within Atmospheric Dispersion Modelling for Regulatory Purposes*, Rouen, France, October 11-14, 1999.

Hanna S.R., 1982, Applications in air pollution modeling. In: *Atmospheric Turbulence and Air Pollution Modelling* (edited by F.T.M. Nieuwstadt and H. Van Dop), Reidel, Dordrecht, Chapter 7

Maryon, R.H., and Buckland, A.T., 1994, Diffusion in a Lagrangian multiple particle model: a sensivity study, *Atmospheric Environment,* **28**, 2019-2030

Mellor G.T. Yamada T., 1982, *Reviews of geophysical and space physics*, **20**, 851-875

Pielke R.A., Cotton W.R., Walko R.L., Tremback C.J., Lyons W.A., Grasso L.D., Nicholls M.E., Moran M.D., Wesley D.A., Lee T.J., Copeland J.H. 1992, A comprehensive meteorological modeling system – RAMS, *Meteorology and Atmospheric Physics*, **49**, 69-91

Tinarelli G., Anfossi D., Brusasca G., Ferrero E., Giostra U., Morselli M.G., Moussafir J., Tampieri F., Trombetti F., (1994), Lagrangian Particle Simulation of Tracer Dispersion in the Lee of a schematic Two-Dimensional Hill, *Journal of Applied Meteorology* **33**, pp. 744-756

Tinarelli G., Anfossi D., Bider M., Ferrero E., Trini Castelli S., 2000, A new high performance version of the Lagrangian particle dispersion model SPRAY, some case studies, *Air Pollution Modelling and its Applications XIII*, S.E. Gryning and E. Batchvarova eds., Plenum Press, New York, 23, in press

Trini Castelli S., Anfossi D.,1997, Intercomparison of 3D turbulence parametrizations for dispersion models in complex terrain derived from a circulation model, *Il Nuovo Cimento C*, **20**, 287-313

DISCUSSION

A. STOHL: Isn't the fine resolution of your diagnostic wind field model artificial, since you have so little information, especially at the upper levels?

S. FINARDI: It is true that the fine resolution used (250 metres) is useless for the upper layers because we had only two wind profiles located few kilometres apart. But the fine resolution was needed to resolve the surface observations, that were located few hundred metres from one another.

A. VENKATRAM: Your diagnostic model, which conserves mass, does not preserve the winds at the observation points. So even if you conserve mass you might lose valuable information contained in the observations. Have you considered using some simple interpolation schemes to patch the observations with results from the prognostic model?

S. FINARDI: The mass consistent model results have been checked to ensure that only slight modification could be forced on the observed winds. At this aim no stability forcing has been imposed on the wind adjustment method. Until now the diagnostic and prognostic models have been applied independently so the prognostic results have not been used to improve the diagnostic model performance. I believe that in the near future four dimensional data assimilation techniques could be used to take advantage of local observation in prognostic modelling and substitute diagnostic tools to reconstruct local wind fields.

FIRST RESULTS FROM OPERATIONAL TESTING OF THE U.S. EPA MODELS 3 / COMMUNITY MULTISCALE MODEL FOR AIR QUALITY (CMAQ)

J. R. Arnold[1] and Robin L. Dennis[1]

Atmospheric Sciences Modeling Division
National Oceanic and Atmospheric Administration / Air Resources Laboratory
[1]on assignment to the National Exposure Research Laboratory
U.S. Environmental Protection Agency / Office of Research and Development
Research Triangle Park, NC 27711 USA

INTRODUCTION

The U.S. Environmental Protection Agency's Office of Research and Development (USEPA / ORD) recently released its new comprehensive modeling system, the Models 3 / Community Multiscale Model for Air Quality (CMAQ). The system consists of CMAQ, the urban-to-regional scale Eulerian air quality model designed for one-atmosphere assessments for multiple pollutants including ozone (O_3) and other oxidants, aerosols and particulate matter (PM_{10}, $PM_{2.5}$), and acid / nutrient deposition to ecosystems, and Models 3, the computational framework surrounding and supporting CMAQ. The science subsystems making up CMAQ include the MM5 meteorological model, the Models 3 Emissions Processing and Projection System (MEPPS), and the CMAQ chemical transport model (CCTM). More detailed descriptions of Models 3 / CMAQ including theoretical formulations, chemical and physical process parameterizations and numerical algorithms are found in Byun and Ching (1999).

CMAQ's flexible and extensible modular construction allows for exchange of process modules like the chemical mechanisms. Two other examples of the model's flexibility are reported elsewhere in this volume: Byun and Pleim (2000), and Pleim and Byun (2000). In this paper we report results of our evaluation of the model's operational performance for O_3 over a 14 day period (5–18 July 1995) with three one-way nested model resolutions (36km, 12km, 4km) centered over urban centers of the northeast U.S. and using two chemical mechanisms currently available with CMAQ, Carbon Bond 4 (CB4), and the Regional Acid Deposition Model mechanism version 2 (RADM2).

BASIS OF THE TESTING

We previously described a revised model evaluation methodology (Arnold *et al.*, 1998) that distinguishes several types of AQM testing. Two components of that methodology are operational testing with model-to-observations species comparisons and diagnostic testing to help explain the operational predictive performance. Operational performance is determined with direct comparison of model-predicted species against ambient observations of those species, generally at the surface and most often only for the few species of regulatory interest. Diagnostic testing has the objective of providing information to better understand operational performance and to account for anomalies using process-level explanations. These often involve using mass balance techniques, ratios of species other than those of regulatory interest, and process rate and reaction rate information not typically stored by the model. Additional information on instrumenting AQMs for process-level analysis and diagnostic testing is found in Dennis *et al.* (2000). Extensive diagnostic testing has been planned and is underway for CMAQ, but this paper presents only operational testing results.

In our evaluation of CMAQ we judge model performance and behavior over a number of specific attributes including: ability to reproduce the 1 hour peak and 3 hour and 8 hour average daily maximum O_3 mixing ratios $[O_3]$; and ability to reproduce hourly mixing ratios of O_3 and oxides of nitrogen ($NO+NO2=NO_X$) in early morning and overnight periods of O_3 deposition and titration by NO_X. Results for the 8 hour average maximum $[O_3]$ (shown here) are quite similar to those for the 1 hour peak and 3 hour average (not shown), although the model generally fits the 1 hour peaks better than the multi-hour averages.

The observed 8 hour average maximum $[O_3]$ is calculated as the highest 8 hour average occurring between the hours 0900–2000 EST from monitors reporting at least 75% data completeness for the highest 8 hours. There are more than 600 sites reporting O_3 for any hour on any day in the largest CMAQ-modeled domain and this number is reduced somewhat by our time windows and data completeness requirements. Nevertheless, this data loss is less than that incurred by imposing an arbitrary $[O_3]$ threshold at, say, 40 or 60 parts per billion (ppb) as has been recommended in the past (USEPA, 1991).

Statistical comparisons of the model-predicted 8 hour average in the single model cell corresponding to the location of the monitor reporting the 8 hour average maximum $[O_3]$ are computed in this way:

- normalized bias: hourly prediction-to-observation bias normalized to observations using the form

$$NBias = \frac{1}{N_T}\sum_{i=1}^{N}\sum_{j=1}^{Hi}\frac{ModelConc(i,j)-ObsConc(i,j)}{ObsConc(i,j)}\ X\ 100\%$$

- normalized gross error: hourly prediction-to-observation gross error normalized to observations using the form

$$NGrossErr = \frac{1}{N_T}\sum_{i=1}^{N}\sum_{j=1}^{Hi}\frac{|\ ModelConc(i,j)-ObsConc(i,j)\ |}{ObsConc(i,j)}\ X\ 100\%$$

where:

N = number of monitor sites

H_i = number of prediction-observation pairs for monitor site i

N_T = total number of prediction-observation hour pairs in the 8 hour average for monitors in the defined area

ModelConc(i , j) = predicted mixing ratio in the model grid cell corresponding to monitor site i for hour j

ObsConc(i , j) = observed mixing ratio at monitor site i for hour j

Bias and error statistics for the 36km model are substantially similar in magnitude and direction to those at 12km; however, the 12km results do demonstrate better model-to-observed fits nearly everywhere nearly every day. Likewise the 4km results often demonstrate improved fits over the 12km, although results for the 4km are somewhat more variable. Hence, in this report we present 12km results for all statistical measures and refer to differences seen at other resolutions where they are significant.

Model Configurations and Modeled Domains

For these evaluation model runs MM5 v2.10 was configured in nonhydrostatic mode, with 30 layers, and with analysis nudging of winds but not temperature or moisture within the PBL in the 108km, 36km, and 12km (but not 4km) grids.

The CCTM was configured with 21 layers compressed from the MM5 30, RADM-type clouds and aqueous chemistry, full Integrated Reaction Rate (IRR) and Integrated Process Rate (IPR) outputs, and the

SMVGear solver. The CB4 mechanism with aerosol and aqueous updates (CB4_AE_AQ) includes 96 reactions with 45 chemical variables including a 1-product isoprene reaction set; the RADM2 mechanism with aerosol and aqueous chemistry (RADM2_CIS4_AE_AQ) includes 200 reactions with 65 chemical variables including a 4-product isoprene reaction set.

The 36km modeled domain contains 7,047 horizontal cells, 87 columns by 81 rows, and extends east from central Texas to off-shore of the continental U.S., and north from the southern tip of Florida into Nova Scotia, Canada. The nested 12km domain covers New England and the mid-Atlantic U.S. from southern Virginia north to central Maine, and east from central Michigan out over the Atlantic with 6,975 horizontal cells, 93 columns by 75 rows. The nested 4km domain is centered southeast of Long Island, NY and extends from south of Washington, DC to north of Boston, MA and east from central Pennsylvania out over the Atlantic with 21,168 horizontal cells, 144 columns by 147 rows.

Subdomains and airsheds with similar photochemical regimes were defined within the modeled domains by analysis of 1995 surface observations. This showed, for example, that what is commonly referred to as the Ozone Transport Region (OTR) in the highly urbanized northeast U.S. can usefully be divided at 40.50° latitude since the temporal behavior of O_3 and other pollutants is different in the OTR (North) than in the OTR (South). We also defined a subdomain along the New England coast (NE Seaboard) where terrain-influenced meteorology often affects $[O_3]$ differently than farther inland. Within these subdomains we also constructed airsheds of urban-centered photochemical production for cities including Washington, DC; Baltimore, MD; Philadelphia, PA; New York City, NY; and Boston, MA. We defined a remainder subdomain (NE Regional) for sites outside the urban airsheds and northeast corridor.

RESULTS

Figure 1 illustrates the wide range of observed peak 8 hour $[O_3]$ in the regional subdomains (1a-1b) and the urban airsheds (1c-1d) by showing peak values and mean values observed each day in each subdomain and airshed. The range of peak and mean $[O_3]$ in any region or city varies widely over the 14 days of the evaluation period. In many cases the peak observed $[O_3]$ is more than 3 standard deviations greater than the mean, and for several days the difference is 50% or more. On 8 July, for example, the NE Regional peak is a factor of two greater than its mean. Note also that the NE Regional observed $[O_3]$ is greater on some days than the observed peak in the OTRs and in most urban airsheds.

Performance across days: Comparing Figure 1 with Figure 2 shows that CMAQ's performance varies with the range of peak observed $[O_3]$ in that the model, using either CB4 or RADM2 chemistry, demonstrates lower bias and error on days and in regions with generally tighter mean-to-peak ranges and consistently lower bias and error on the days with the highest peak $[O_3]$, 10-15 July. The NE Seaboard subdomain and the Boston, MA airshed where the peak oberved $[O_3]$ is very low, perhaps due to particular local influences such as increased cloud cover and complex wind patterns along the North Atlantic land-ocean interface, remain more difficult for the model to fit well.

Performance across grid resolutions: Figure 3 shows the time series of observed and predicted mixing ratios for O_3 (3a), NO_X (3b), and formaldehyde (HCHO) (3c) at one 36km cell and one 12km cell nested within it in the Washington, DC airshed for 14 July. CMAQ predictions at 36km (lighter lines in 3a-b) and at 12km (darker lines in 3a-b) are shown with CB4 and RADM2 chemistries for each; observations are shown for all reporting monitors in cells at both the 36km (lighter symbols) and 12km (darker symbols) resolution, rather than averaged into a cell mean value for each resolution as was done for the statistical measures described above. With individual monitor values plotted separately in Figures 3a-b, large subgrid scale variation in the observed $[O_3]$ and $[NO_X]$, *i.e.*, the difference between values reported by multiple monitors sited in the same model cell, is apparent. (There are no co-located observations for HCHO here.) This range of subgrid scale variation is significant when considering model performance since we cannot expect the model to correctly represent effects present at scales smaller than its resolved grid.

Substantial improvement in the 12km model (darker lines) over the 36km (lighter lines) for predicting the timing and shape of the $[O_3]$ peak can be seen in Figure 3a. Note that since the smaller 12km cell is

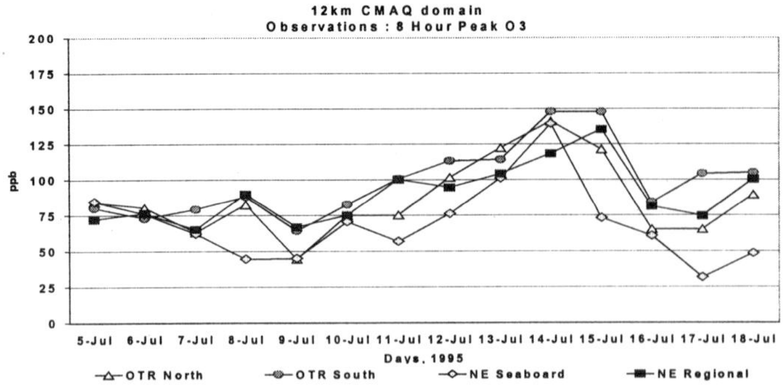

(a)

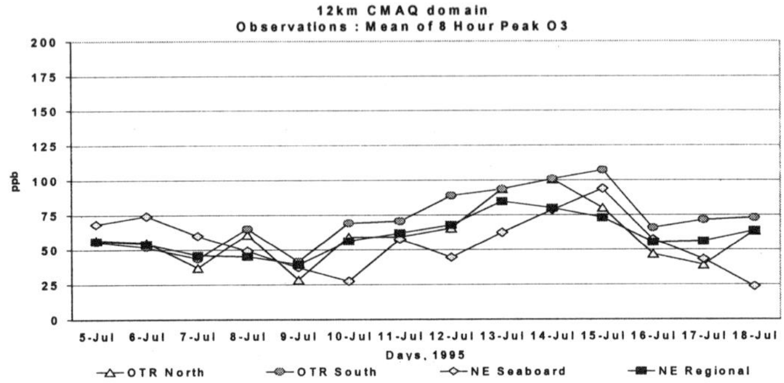

(b)

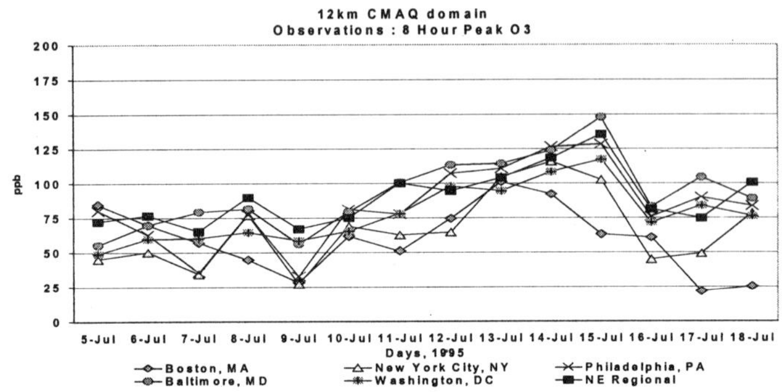

(c)

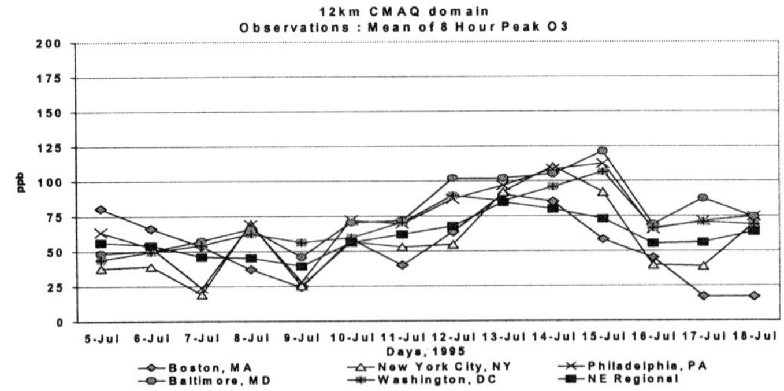

(d)

Figure 1(a-d). Peak (1a, 1c) and mean (1b, 1d) of the highest 8 hour average observed $[O_3]$ (ppb) in the CMAQ evaluation northeast subdomains (1a-b) and cities (1c-d).

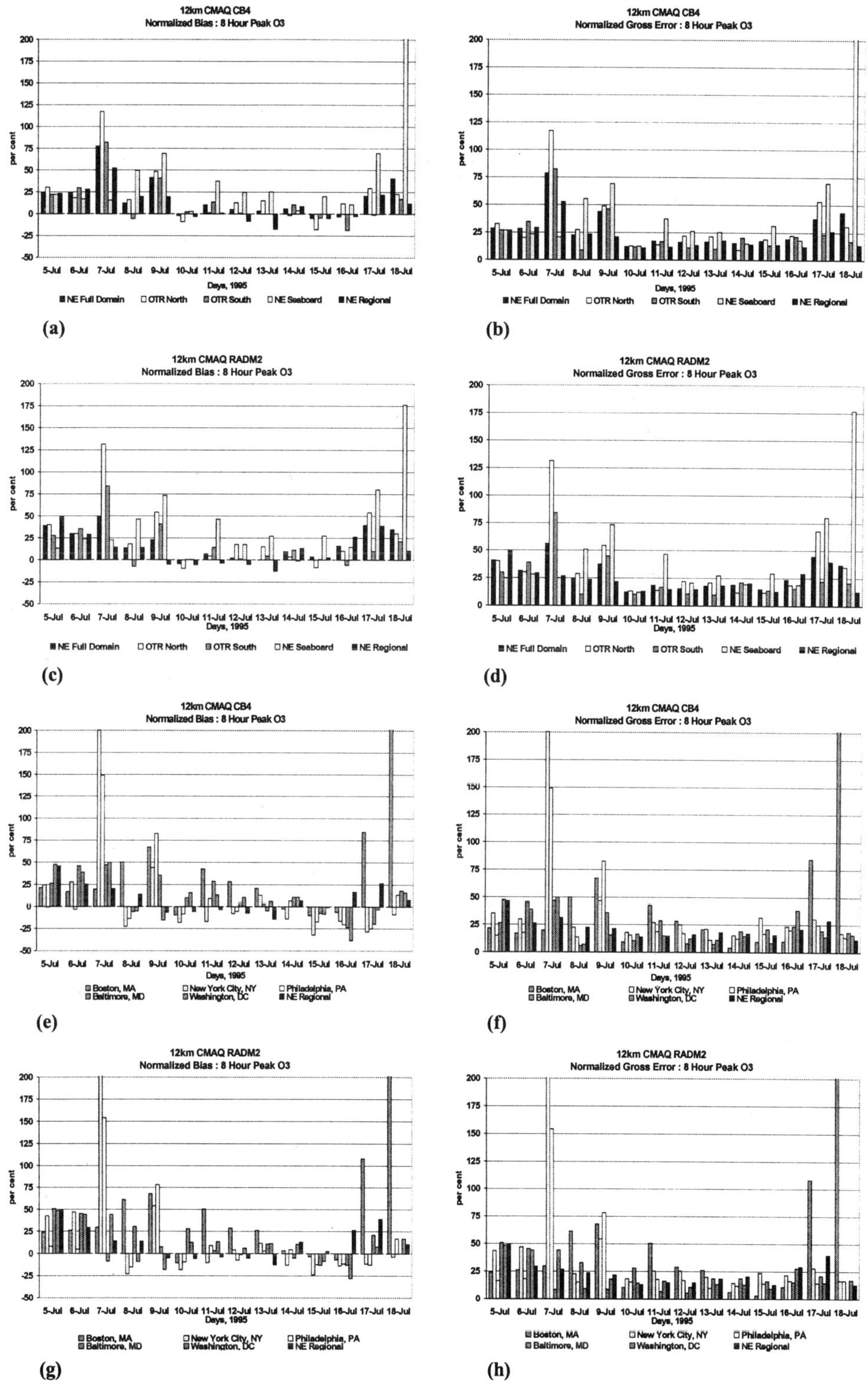

Figure 2(a-h). Bias and gross error (per cent) for 12km CMAQ runs, CB4 (2a, 2b, 2e, 2f) and RADM2 (2c, 2d, 2g, 2h), in the CMAQ evaluation northeast subdomains (2a-d) and cities (2e-h).

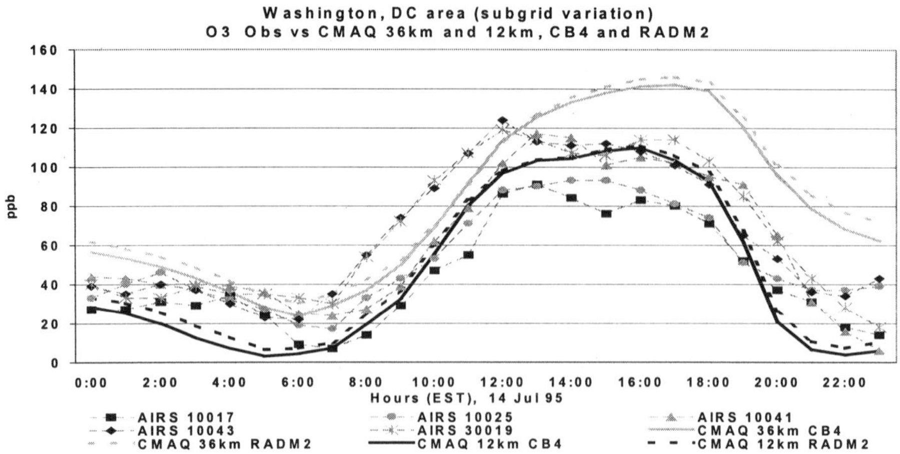

(a)

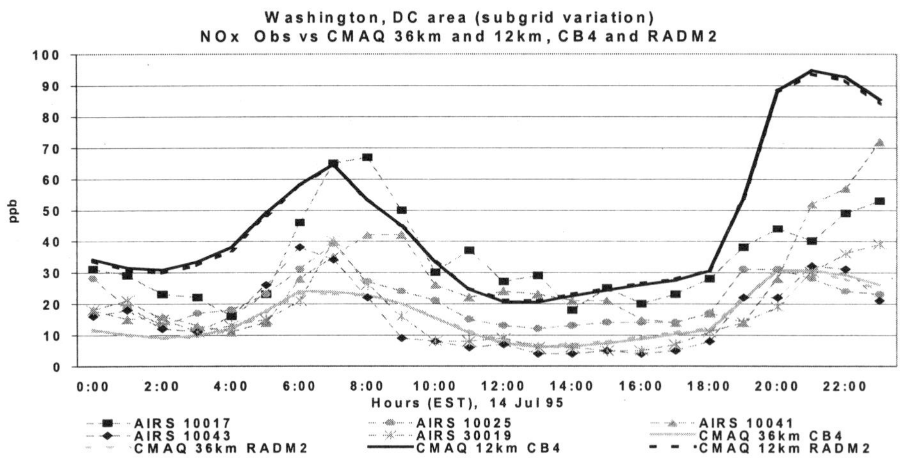

(b)

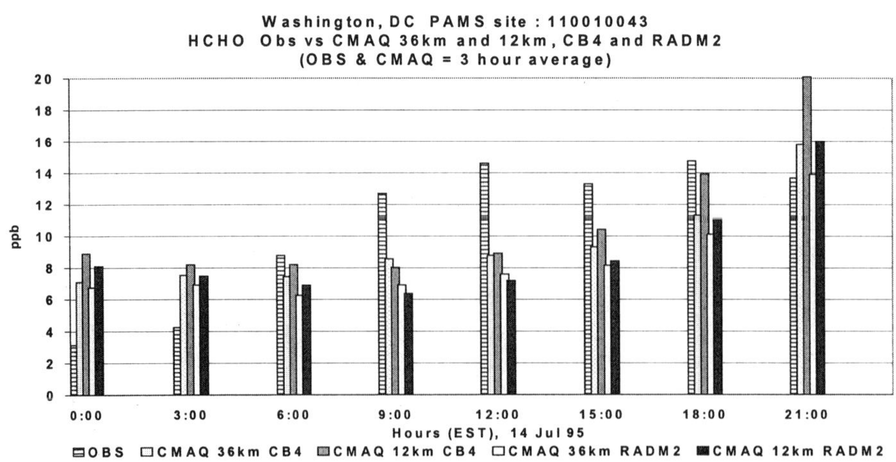

(c)

Figure 3(a-c). O_3 (3a), NO_X (3b), and HCHO (3c) model and observed time series (ppb) at one CAMQ 36km cell (lighter lines and symbols) and one nested 12km cell (darker lines and symbols) in Washington, DC.

nested within the 36km, only monitors AIRS 10017 and AIRS 10043 (darker symbols) are located in the 12km cell shown here. Considering this, the CMAQ 12km predictions for peak [O_3] fall well-within the range bounded by these two observations while the 36km predictions fall outside the maximum [O_3] observed anywhere in that larger cell.

The 12km model (darker lines) predicts significantly less O_3 overnight than does the 36km (lighter lines) owing in large part to the higher [NO_X] predicted overnight in the 12km. The NO_X time series from either model resolution might be considered within the range of observations for that resolution cell as Figure 3b shows, excepting the overnight period at 12km. The 12km holds significantly more NO_X overnight than does the 36km, which accounts for the corresponding 12km nighttime O_3 underprediction. The 12km predictions fit well the peak [NO_X] from one monitor in the 12km cell (darker symbols) at the time of the morning rush hour NO_X injection and track the fall-off through the rest of the day, but are high relative to all observations as the photochemistry slows down in the late afternoon. Note that observations at all sites in the larger 36km cell (lighter symbols) confirm that NO_X should increase during this time, so the model's direction and general timing appear correct.

Performance across chemical mechanisms: Figure 2 shows that [O_3] bias and error results for CB4 (2a, 2b, 2e, 2f) are quite similar to those obtained with RADM2 (2c, 2d, 2g, 2h) for most days in most regions and cities. Different chemistries make very little difference to the predicted [O_3] and [NO_X] at either resolution as Figures 3a-b show. However, relative to CB4 at both resolutions RADM2 does seem to hold more O_3 overnight and accumulates O_3 slightly faster during the mid-morning rise to end with only very slightly higher [O_3] at time of the afternoon peak on this day at this site. Though it is not strictly uniform across all sites every day, this insensitivity of [O_3] and [NO_X] to selection of the chemical mechanism and the behavior of the mechanisms relative to one another has for the most part been confirmed at other urban sites on this and other evaluation days (not shown here).

Performance for HCHO: Comparisons to HCHO observations constitute a special test for the model since the chemical mechanisms were not originally designed or fitted to predict HCHO. Nonetheless, given the important role of HCHO in initiation and propagation of the radicals key to O_3 production (see Dennis *et al.*, 2000), we would hope to find a reasonable degree of congruity between model behavior for HCHO and for O_3. In fact, as Figure 3c shows, CMAQ predictions of [HCHO], both 36km and 12km, with CB4 and with RADM2, demonstrate a fairly good fit to the 3-hour averaged observed [HCHO] reported for this Washington, DC site. For HCHO, though, unlike for O_3, the 12km resolution does not always produce a better fit tan the 36km. However, only for the value at 2100 EST is the 12km prediction significantly worse than the 36km: 20.1 ppb (12km) *vs.* 15.8 ppb (36km) *vs.* 13.7 ppb (observed). Note also that the 12km overprediction is greater for CB4 than for RADM2, as is the predicted [HCHO] in general. This is most likely due to differences in the treatment of HCHO in the different mechanisms whereby a greater number of oxidation steps produce the HCHO in CB4's lumped structure approach while less HCHO is produced in the more explicit lumped species chemistry of RADM2. Fits to observed [HCHO] on other days and at additional sites (not shown here) confirm the trend illustrated here and produce for the most part a reasonable degree of congruity between model behavior for HCHO and O_3.

CONCLUSIONS

The relatively long time period and large geographic extent of our evaluation modeling series allows for examination of model performance over a wide range of temporal and spatial scales. As described above, the observed [O_3] varies widely across the 14 days in all modeled subdomains and airsheds. CMAQ does quite well on high [O_3] days, but less-well on lower [O_3] days and on days with larger mean-to-maximum ranges. A candidate explanation we are currently testing is that this difference in O_3 is tied to the model's inability to predict NO_X correctly for these low O_3 days. This explanation makes chemical sense and has been borne-out by our preliminary investigations. Consequently, our confidence that the model is behaving correctly in a way we can account for is increased.

Operational evaluation of the ozone prediction by AQMs has before now concentrated nearly exclusively on short episodes in large urban centers where $[O_3]$ are high and variation is low. However, predicting for more moderate O_3 days will be necessary for the proposed new USEPA 8-hour O_3 standard (USEPA, 1997) and for predicting oxidant fields related to one-atmosphere issues such as regional haze and particulates. Hence it is important to continue working to understand model behavior for this type of day.

When looking across model grid resolutions we see that the 36km grid is likely too coarse for most city and some regional model applications. The 12km model is an improvement nearly everywhere nearly every day as measured by the bias and error statistics. Statistics for the 4km model (not shown) demonstrate a tendency for improved fits over the12km, but this is not so ubiquitous as for the 12km over the 36km. We see strong potential in moving to the 4km resolution, but modeling at 4km is not a trivial step down from 12km. We are currently investigating known differences in the realizations of the meteorology fields between the 12km and the 4km which we believe are degrading model performance in some locations at the finer resolution. Emissions fields for the 4km model also bear closer scrutiny since they must sometimes be interpolated down from inventories at larger scales.

Improvements across grid resolutions measured with bias and error statistics for large areas are not uniformly seen when plotting time series of the different model resolutions against observations at a given site. This most likely indicates significant problems of monitor siting and representativeness for a model cell. Concerns over monitor siting also appear well-justified by the subgrid variation plots shown above.

We are encouraged that the CCTM appears to behave in a manner we can explain and that supporting evaluation elements like [HCHO] time series are consistent with our results for O_3 and NO_X. We do see some anomalies, however, and are continuing to check the model in greater detail with process analysis-based diagnostic probes and testing of additional species and species ratios across the days and regions.

Our overall conclusion from these first results in the operational evaluation of CMAQ is that the model is for the most part functioning in a way we can account for both where model fits to the observations are good and where they are not as good. Furthermore, we find that CMAQ is operating well-within the range of results from other large Eulerian AQMs in the U.S. for the high O_3 days where model evaluation has traditionally been focused. Hence we find no reason not to use the model and encourage its wide-spread use.

This paper has been reviewed in accordance with the U.S. Environmental Protection Agency's peer and administrative review policies and approved for publication. Mention of trade names or commercial products does not constitute endorsements or recommendation for use.

REFERENCES

Arnold, J. R., R. L. Dennis, and G. S. Tonnesen, 1998, Advanced techniques for evaluating Eulerian air quality models: Background and methodology. *Preprints from the 10th Joint Conference on the Applications of Air Pollution Meteorology with the AWMA*, Phoenix, AZ, American Meteorological Society, Boston, MA.

Byun, D. W. and J. K. S. Ching, eds., 1999, *Science Algorithms of the EPA Models-3 Community Multi-scale Air Quality (CMAQ) Modeling System*, U.S. EPA / National Exposure Research Laboratory, Research Triangle Park, NC.

Byun, D. W. and J. E. Pleim, 2000, Sensitivity of ozone and aerosol predictions to the transport algorithms in the Models-3 Community Multi-scale Air Quality (CMAQ) Modeling System, *Proceedings of the Millennium NATO / CCMS International Technical Meeting on Air Pollution Modelling and Its Application*, Boulder, CO, Plenum Press, New York, NY.

Pleim, J. E. and D. W. Byun, 2000, Application of a new land-surface, dry deposition, and PBL model in the Models-3 Community Multi-scale Air Quality (CMAQ) Model System, *Proceedings of the Millennium NATO / CCMS International Technical Meeting on Air Pollution Modelling and Its Application*, Boulder, CO, Plenum Press, New York, NY.

Dennis, R. L., J. R. Arnold, and G. S. Tonnesen, 2000, Probing the shores of ignorance, in: *Forecasting Environmental Change*, M. B. Beck, ed., Elsevier Press, London, UK.

U.S. Environmental Protection Agency, 1991, *Guideline for Regulatory Application of the Urban Airshed Model*, U.S. EPA Report No. EPA-450/4-91-013, U.S. EPA, Office of Air Quality Planning and Standards, Research Triangle Park, NC.

U.S. Environmental Protection Agency, 1997, *National Ambient Air Quality Standards for Ozone*, 62FR38856.

UNCERTAINTY IN DISPERSION FORECASTING USING METEOROLOGICAL ENSEMBLES

Martin J. Leach and H.-N. Chin

Atmospheric Science Division
Lawrence Livermore National Laboratory
Livermore CA 94550

INTRODUCTION

An approach for quantifying meteorological uncertainty has been via development of an ensemble of forecasts from slightly perturbed initial conditions (Sivillo et al., 1997) to predict the time evolution of the probability density function of atmospheric variables (Mullen and Baumhefner, 1994). We create an ensemble of forecasts by varying the initial (and boundary) conditions for the COAMPS meteorological model. The variations in the initial conditions must be consistent with analysis error. Optimally, the range of initial conditions would encompass the "true" atmospheric state, but which is never actually known. Our method for creating varying initial conditions is to use different global data sets to derive the necessary data. We use two models from the National Weather Service (the AVN and ETA models) and one from the Navy (the NOGAPS model). In addition to those data sets we perturb the data from those models, using a normally distributed random number at each grid point in the COAMPS model. We perturb the (u,v) wind components, the temperature and the moisture. The size of the perturbation is determined by the variability within that variable field. The forecasts, with a total of six simulations, are run for 48 hours. We then use the output from the COAMPS model to drive a Lagrangian dispersion model (LODI) for simulated releases. The results from a simulated release from hour 33 are shown in Figure 1. The center of the domain is Oakland airport and the basic on-shore wind is from the southwest. In three of the simulations, the plume goes over the top of the hills to the northeast, and in the other three the plume hugs the coastline and goes around those hills. The two solutions reflect a dependence on the Froude number, a ratio of the Kinetic energy to Potential energy. Higher Kinetic energy flow (Higher Froude number) flow goes over the top of the mountain, while lower Kinetic energy flow goes around the hills.

MODEL DESCRIPTIONS

A. COAMPS (The Coupled Ocean/Atmosphere Mesoscale Prediction System)
The Coupled Atmosphere/Ocean Mesoscale Prediction System (COAMPS) model (Hodur, 1997) was developed at the Naval Research Laboratory. COAMPS has been used at resolutions as small as 2 km to study the role of complex topography in generating mesoscale circulation (Doyle, 1997). The model has been adapted for use in the Atmospheric Science Division at LLNL for both research and operational use. The model is a fully, non-hydrostatic model with several options for turbulence parameterization, cloud processes and radiative transfer. We have recently modified the code to include an urban canopy parameterization (Brown and Williams, 1998), based on Yamada's (1982) forest canopy parameterization and includes modification of the TKE and mean momentum

Air Pollution Modeling and Its Application XIV, Edited by
Gryning and Schiermeier, Kluwer Academic/Plenum Publishers, New York, 2001

equations, modification of radiative transfer, and an anthropogenic heat source. COAMPS is parallelized for both shared memory (OpenMP) and distributed memory (MPI) architecture.

B.LODI (Livermore Operational Dispersion Integrator. A Lagrangian Particle Model) The dispersion model, LODI, (Nasstrom et al.,2000) simulates the processes of advection, turbulent diffusion, radioactive decay and first order chemical reaction, wet and dry deposition and plume rise. The model uses a Lagrangian, stochastic Monte Carlo method, and is capable of simulations with complex terrain. Similarly to COAMPS, the model is parallelized for both shared memory and distributed memory architecture.

EXPERIMENTAL DESIGN

COAMPS is a regional atmospheric model, requiring initial conditions and boundary conditions from larger-scale models. To create a forecast ensemble, we derived the initial and boundary conditions from 3 models, the Navy's global NOGAPS model, the National Weather Service's (NWS) global Aviation model and the NWS regional Eta model. Using the 3 models, we performed three, 48 hour forecasts using COAMPS. We increased the size of our ensemble by adding small random perturbations to the initial and boundary conditions from the large scale model. Perturbations were only added to the horizontal velocity components (u,v) and to the temperature. All other variables were unperturbed. The size of the random perturbation was determined by the variance in the field itself. The COAMPS forecast used two grids, one nested within the other. Oakland airport is the center of the domain in both nests, and is also the release point for the simulated tracer.

FORECAST DISPERSION PATTERN

Wind forecasts from the COAMPS meteorological model are used to drive the LODI dispersion model. The Results of a three-hour dispersion model simulation, from hour 33 to hour 36 from the COAMPS forecast, are shown in Figure 1. The initial time of the forecast is August 23 at 0000 UTC (August 22, 1700 local time). There is a very distinctive bifurcation in the solutions; three of the plumes travel towards the northeast over the hills, while three of the plumes travel northwest, staying over the bay and hugging the coastal hills. In general the plumes that go over the hills are narrower, while those that go along the shoreline are wider. This indicates stronger winds for the plumes going over the hills. The exception is for the AVN model without perturbation, as a result of less strong wind and a weaker stable layer.

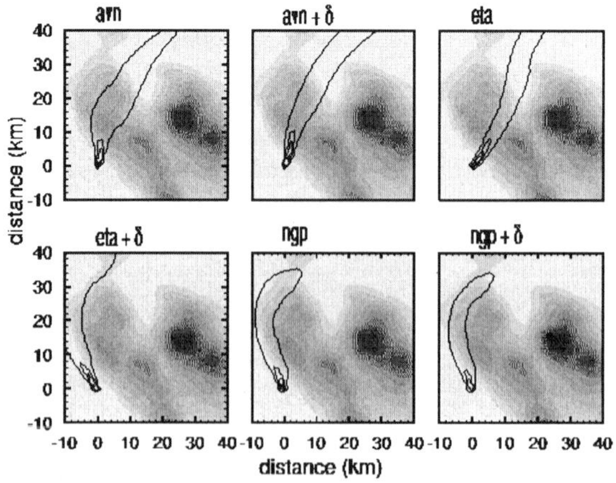

Figure 1. The dispersion patterns for three hour predictions from hour 33 to hour 36. The runs are labeled with the large scale model used to generate initial conditions, including the runs with the random perturbation imposed (+δ). The underlying terrain is represented in grayscale.

FORECAST WIND FIELD

Representative wind fields that drive the forecasts are shown in Figure 2. These winds are from hour 36 in the forecast, valid at 1200 UTC on August 25, or 0500 local time and are from the fourth level from the ground in the COAMPS model, approximately 140 meters above thesurface. Qualitatively, all of the forecasts look similar, an indication that the forecasts have not diverged significantly to this point in the simulations. There is strong westerly flow in the western portion of the domain that turns towards the north (southerly winds) near the center of the upper-half of the domain. On closer inspection of the wind fields however, it is apparent that in the three cases where the plume goes over the top of the hills, the wind shift to southerly winds is farther to the east than in the three cases where the plume stays along the shoreline. Stronger westerly flow in the southern portion of the domain is apparent in the three cases where the flow traverses over the hills. An anti-cyclonic turn of the winds to westerly flow in the northeast section of the domain that is more pronounced in the three cases where the plumes did not traverse the hills but stayed along the shoreline.

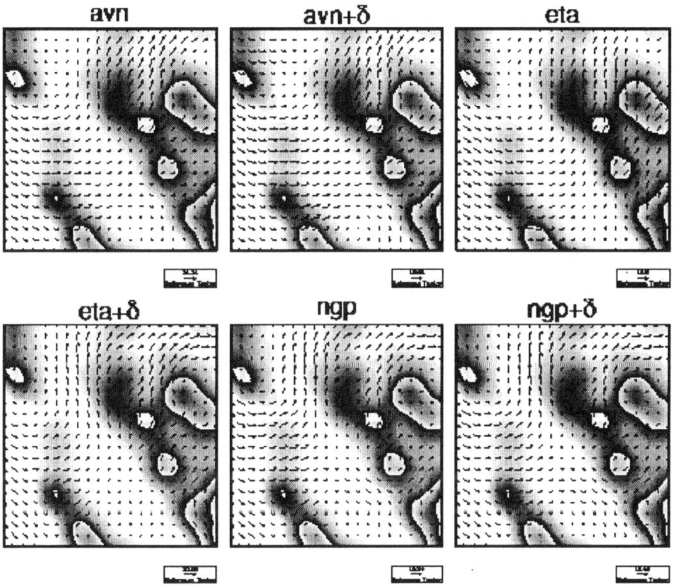

Figure 2. The wind fields at hour 36 for the six forecasts, labeling of the forecasts as in Figure 1.

Froude Number

The mesoscale influence of underlying terrain is evident in the simulations in a way that also shows the uncertainty of the meteorological forecasts. The influence is felt through a blocking effect. As air starts to move up the side of a hill, the air loses kinetic energy and gains potential energy. If the air gets to the top of the hill, before losing all of its kinetic energy, then it moves across the top of the hill. Conversely, if the air loses all of its kinetic energy before it reaches the hilltop, then it will not go over the hill, but rather will move around it. The well-known Froude number is a measure of this effect. The meteorological definition of the Froude number is

$$Fr = U/Nh,$$

Where Fr is the Froude number, U is the wind speed, N is the Brunt-Vaisala frequency, a measure of atmospheric stability and h is the height of the blocking hill. The meteorological Froude number includes the atmospheric stability effect through N. Higher values of N reflect greater atmospheric stratification and therefore more stable air would require greater wind speeds to go over the top of the hill. Table I contains the relevant values for the six cases illustrated in the figures above. The numbers in the tables were calculated using 9 grid points (3x3) centered on the hill. The wind speed and stability were calculated using the lowest 6 levels from the COAMPS model, which is approximately the lowest 330 meters above ground level. It is clear from the values in the table that the Froude number accurately diagnosis whether the air, and therefore the plume, goes over the hills or goes around. In the three cases going over the hill, the Froude number is greater the 1, indicating higher wind speeds, or lower stability. When the Froude number is less than 1, the plume did not go over the hill, but stayed along the shoreline.

Table I. The Froude number, mean wind velocity (U) in the lowest 330 meters and the Brunt-Vaisala (N) frequency for the six forecasts.

Forecast	Froude Number	U	N
AVN	1.63	16.8	0.031
AVN+δ	2.06	21.5	0.032
ETA	3.17	24.7	0.024
ETA+δ	0.98	11.7	0.036
NGP	0.67	9.0	0.040
NGP+δ	0.67	8.6	0.039

One-hour LODI simulations were produced from every 12 hours throughout the 48-hour COAMPS forecast. The results are plotted in figure 3, for the COAMPS runs using AVN as for the initial and boundary conditions. It is clear that diurnal forcing exists. A similar diurnal feature exists in all runs (not shown). A strong onshore wind, consistent with a sea-breeze, exists for the afternoon 24 hour and 48 hour forecasts in all cases. This case was a very warm day in August with temperatures in the Livermore Valley and Central Valley exceeding 100 F. At night, the sea-breeze circulation relaxes, synoptic forcing dominates and more southerly winds are most normal. At all hours in the forecast, it appears that more than one solution exists and is dependent on terrain forcing. As early as 12 hours, differences in the plume forecasts are evident, with the bifurcation that is detailed in the 36 hours forecasts already apparent. Daytime plumes tend to ride westerly winds, but which gap in the terrain that they traverse depends on the forecast. For instance at 24 hours, the NGP and AVN runs are very similar, by 48 hours they are very different with the NGP run much farther to the south. The ETA model run maintains the strongest circulation and synoptic forcing. It is least affected by the terrain induced local forcing and hence variability in the plume forecasts is much less.

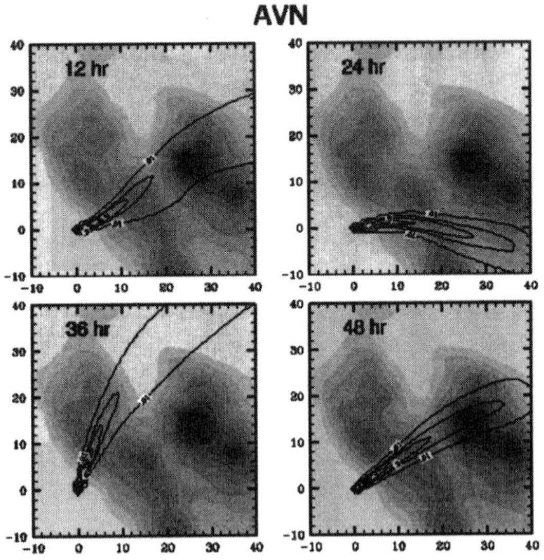

Figure 3. Dispersion patterns using winds from COAMPS forecasts, with the Aviation model as the large scale forcing. The underlying terrain is represented in gray-scale.

PROBABILITY

The probability that the concentration will exceed 0.01 of the maximum concentration amount (Figure 4) demonstrates the uncertainty within the forecasts and that the uncertainty is not simply sub-grid scale due to turbulence fluctuations. The probability is analyzed using the concentration calculations from all six model runs (e.g. if two simulations have concentrations exceeding 0.01 of the maximum value, the probability would be 0.33). Local forcing, and its interaction with synoptic scale forcing, as well as diurnal forcing contributes to the wide spread in the probability distribution, and therefore the uncertainty in the dispersion forecasts. Small changes in the synoptic conditions and/or small changes in the timing of wind shifts can lead to large changes in the small-scale solutions. It is clear from the probability plots, that a unique solution (or plume path) does not exist most of the time.

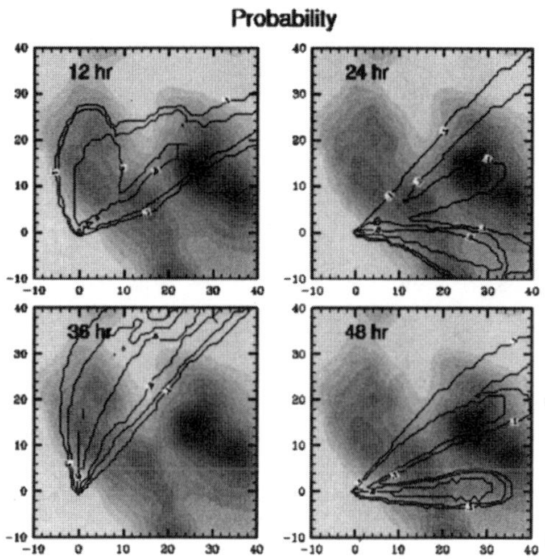

Figure 4. The probability contours of exceeding 0.01 of maximum concentration value at the four times.

REFERENCES

Brown, M. and Williams, M., 1998, An urban canopy parameterization for mesoscale meteorological models, *AMS 2nd Urban Env. Symp.*, Albuquerque, NM.

Doyle, J.D. 1997, The Influence of Mesoscale Orography on a Coastal Jet and Rainband, *Mo. Wea. Rev.* 125:1465-1488.

Hodur, R., 1997, The Naval Research Laboratory's Coupled Ocean/Atmosphere Mesoscale Prediction System (COAMPS), *Mo. Wea. Rev.* 125:1414-1430.

Mullen, S.L. and Baumhefner, D.P. 1994, Monte Carlo simulations of explosive cyclogenesis. *Mo. Wea. Rev.* 122:1548-1567.

Nasstrom, J., Sugiyama, G., Leone, Jr., J.M. and Ermak, D.L., 2000, A Real-Time Dispersion Modeling System, *AMS 11th Joint Conference on the Applications of Air Pollution Metoerology*, Long Beach, CA.

Sivillo, J.K., Ahlquistm J.E., Toth, Z., 1997, An Ensemble Forecasting Primer, *Wea. And Forecast.* 12:809-818.

DISCUSSION

R. BORNSTEIN: What variables were perturbed and what were the magnitudes of the perturbations?

M. LEACH: We varied the horizontal (u,v) components of the wind, as well as the temperatures. We used a normally distributed, random number generator to create the perturbations, with the half-width of the normal distribution being 2-4 m/s for the wind and 1.5-3.0 K for temperature.

COMPARISON OF TWO SAMPLING PROCEDURES FOR THE STATISTICAL EVALUATION OF THE PERFORMANCE OF ATMOSPHERIC DISPERSION MODELS IN ESTIMATING CENTERLINE CONCENTRATION VALUES

John S. Irwin[1]

Atmospheric Sciences Modeling Division (MD-14), Air Resources Laboratory, National Oceanic and Atmospheric Administration
Research Triangle Park, North Carolina 27711, USA

INTRODUCTION

Most operational atmospheric simulation models are deterministic. They provide estimates of the average time- and space-variations in the conditions (e.g., a mesoscale meteorological model), or they provide estimates of the average effects of such time- and space-variations (e.g., a dispersion model). The observations used to test the performance of these models are individual realizations (which can be envisioned as coming from some ideal ensemble), and are affected by stochastic variations unaccounted for within the simulation model. If we believe this, then it makes sense to ask the models to replicate average patterns seen in the observations, but it does not make sense to ask the models to replicate the effects of stochastic variations unaccounted for within the simulation model (maxima, minima, or total variance as seen in the observations, etc.). Thus, we are faced with two basic tasks in order to develop a performance evaluation strategy: Step 1) we should analyze the observations to provide average patterns for comparison with modeled patterns, and Step 2) we should account for the uncertainties inherent in Step 1 so we can tell whether differences seen in a comparison of performance of several models are statistically significant.

Within the American Society for Testing and Materials (ASTM), a standard guide[2] is being developed to address these two steps to foster development of objective statistical procedures for comparing air quality dispersion simulation modeling results with tracer field data. Cox and Tikvart (1990) provided a statistical evaluation procedure, which recommended the comparison of the upper percentile values of the cumulative frequency distribution of observed ground-level concentration values, using a metric called the Robust Highest

[1] On assignment to the Office of Air Quality Planning and Standards, U.S. Environmental Protection Agency.

[2] A review copy of the draft guide Z6849Z and example evaluation procedure (FORTRAN program called ASTM90) is available from the author upon request.

Air Pollution Modeling and Its Application XIV, Edited by
Gryning and Schiermeier, Kluwer Academic/Plenum Publishers, New York, 2001

Concentration (RHC). In contrast to a direct comparison of observations and modeling results, the draft ASTM Guide recommends comparisons be made of some average characteristic of the observed and modeled concentration pattern. To illustrate how this might work, an example procedure (ASTM90[2]) is provided for evaluating performance of models to estimate the average maximum ground-level centerline concentration. In the example procedure, evaluation data having similar meteorological and physical conditions are grouped together. Grouping the data together provides a way for us to compute an average centerline maximum concentration for the group for comparison with what the model simulates (on average) for the same group. Comparison of these averages over a series of groups provides a means for assessing modeling skill (Step 1). In order to place confidence bounds on conclusions reached, bootstrap resampling was used by Cox and Tikvart, and also is used in the example procedure ASTM90 (Step 2).

In this discussion, we will investigate the sensitivity of the evaluation results to how data are selected for averaging within each group. We also investigate the tradeoffs of bootstrap sampling within groups as opposed to bootstrap sampling across all groups.

GROUPING DATA

The observed concentrations, can be envisioned as (Venkatram, 1988):

$$C_o = C_o(\alpha,\beta) = \overline{C_o}(\alpha) + c(\Delta c) + c(\alpha,\beta) \qquad \textbf{(1)}$$

where α are the model input parameters, β are the variables needed to describe the unresolved transport and dispersion processes, the over-bar represents an average over all possible values of β for the specified set of model input parameters α; $c(\Delta c)$ represents the effects of measurement uncertainty, and $c(\alpha,\beta)$ represents our ignorance of β (unresolved deterministic processes and stochastic fluctuations). Since $\overline{C_o}(\alpha)$ is an average over all β, it is only a function of α. The modeled concentrations, C_m , can be envisioned as:

$$C_m = \overline{C_o}(\alpha) + d(\Delta\alpha) + f(\alpha) \qquad \textbf{(2)}$$

where $d(\Delta\alpha)$ represents the effects of uncertainty in specifying the model inputs, and $f(\alpha)$ represents the effects of errors in the model formulations. To perform an evaluation of modeling skill, we separately average the observations and modeling results over a series of nonoverlapping limited-ranges of α. Averaging the observations provides an empirical estimate of what the model is attempting to simulate, $\overline{C_o}(\alpha)$. A comparison of the respective averages over a series of α-groups provides an empirical estimate of the combined deterministic error associated with input uncertainty and formulation errors. These averages can consist of any definable feature or characteristic in the concentration pattern (lateral extent, centerline concentration maximum, variance of centerline concentrations, etc.). This process is not without problems. The variance due to natural variability is of order of the magnitude of the ensemble averages (Hanna, 1993), hence small sample sizes in the groups will lead to large uncertainties in the estimates of the ensemble averages. The variance due to input uncertainty could be quite large, see Irwin et al. (1987), hence small sample sizes in the groups will lead to large uncertainties in the estimates of the deterministic error in each group. Grouping data together for analysis requires large data sets, of which there are few.

From an inspection of data grouped for analysis even with large data sets, "problems" can be seen. Table 1 list the observed (raw) maximum concentration seen at selected distances downwind from the release at the EPRI Kincaid SF6 tracer field study (Bowne et al., 1883) for May 7, 1980. If we sort the results into groups based on ranges of Zi/L and distance downwind, then for -25 > Zi/L > -100 hours 11 through 16 might be grouped together. For the 2 and 3 km arcs, the maximum concentration occasionally varies from one hour to the next by 25% or more. I attribute this to a combination of variations in plume rise and convective

Table 1. Observed maximum concentration divided by emission rate (ppt s/g) for the EPRI Kincaid SF6 field experiment for May 7, 1980. Zi is the estimated mixing height and L is the Monin-Obukhov length.

Zi (m)	950	1253	1548	2250	2450	2506	2528	2540	2546
Zi/L	-39	-38	-57	-79	-60	-44	-37	-27	-5
	Hour (LST)								
Distance (km)	10	11	12	13	14	15	16	17	18
2				10	6	9	7	12	4
3			8	8	9	3[a]	8	6	
5	19	22	9	5	7	8	8	6	5
6.7	14	11	6	2	2	2	2	3	3
10	12	12	7	4	2	1	2	2	3
14		7		3			1	1	2

[a] Only the left side of the plume was sampled at this distance.

looping. Further downwind at the 5 and 6.7 km arcs, maximum concentrations are much higher for hours 10 and 11 in comparison to the rest of the hours. The estimated mixing heights by late afternoon are 2 times higher than that estimated for hours 10 and 11, which accounts for some but not all of the differences seen. We can not verify by post analysis the explanation presented, but the discussion does highlight that sorting the data into groups will be subjective. One must allow that some within group variations may be stochastic (like plume rise variations and looping), whereas some within group variations might be mitigated by refinement in the grouping criteria (although one must retain a reasonable number of cases in each group)

In spite of the problems noted, this process does avoid making physically inappropriate comparisons in determining modeling skill. From (1) and (2), we can conclude that the modeled and observed concentrations (unaveraged) have different sources of variance. Therefore, unless you happen to be working with one of the very few models that attempts to simulate the total variability, any agreement in the upper (or lower) percentile values of the respective cumulative frequency distribution can be attributed to happenstance – that is, the effects of $d(\Delta\alpha)$ and $f(\alpha)$ are making up for the lack of any characterization of $c(\alpha,\beta)$ in the modeling. So it is concluded that grouping the data for determination of average patterns has merit, but it comes at the price of developing grouping and analysis strategies.

CENTERLINE CONCENTRATION

In the discussion that follows, one needs to be aware of the logic within the example procedure, ASTM90, in the draft Guide. The procedure employs a robust scheme for combining observed concentrations along arcs within a group for analysis, by using the computed center of mass from each arc as a common reference point so that all the arcs can be superimposed. Once combined in this manner, a lateral dispersion can be computed for all the results in the group. This group lateral dispersion, σ_y, can then be used to define for each experiment which receptors are close (within $\pm 0.67\sigma_y$) to the center of mass. The statistical properties derived from concentration values from these receptors are considered within ASTM90 as being representative of centerline concentration values.

Olesen (1999) observed that within group variations not only affect the maximum concentration values (e.g. the discussion of Table 1), but also the lateral extent of the plumes in contact with the surface. These variations in lateral extent may cause concentration values to be selected as being 'representative of centerline conditions', using the within $\pm 0.67\sigma_y$ criteria, when in fact they are too far from the center of mass to be judged such. One of the effects of selecting concentration values that are unrepresentative of centerline conditions, is that the average of these values will be lower than it otherwise should be. Presumably, those experiments with comprehensive sampling networks and relatively smooth concentration profiles, e.g., Project Prairie Grass, Barad (1958) and Haugen (1959), would be least affected by this concern, whereas experiments with less complete sampling and 'noisy' concentration profiles, e.g., EPRI Kincaid SF6 trials, would be most affected by this concern.

To test this, the sampling scheme was altered so that one can specify that the n values closest to the center of mass be selected from each arc, assuming they are within $\pm 0.67\sigma_y$, where n can range from 1 to the total number within the selection range for each arc. As an intuitive guide to measure the sensitivity of the computed group maximum to how many values are selected from each arc, I used the normalized mean squared error (NMSE) computed between the computed group maximum and the average of the maximum observed value from each arc within the group. Figure 1 depicted the results obtained for three tracer field studies. The EPRI Indianapolis (Murrey and Bowne, 1988) lateral concentration profiles typically have 5 to 10 receptors, with relatively smooth concentration profiles (likely typical for dispersion of moderately buoyant effluent from a relatively short stack). Most of the arcs only have 1 or 2 receptors within the selection range, hence the NMSE shows little variation beyond n=2. The Prairie Grass lateral concentration profiles typically have 25 receptors. Bifurcating plumes are common and the 'noise' seen superimposed upon the lateral concentration profiles is seen to increase as downwind distance increases from 50 to 800 m (likely typical for dispersion from a ground-level release in flat terrain). With from 2 to 10 receptors within the selection range, there is little sensitivity seen in the results as n increases from 1 to 3, and 'noise' seen in the profiles contributes to larger scatter in the bootstrap sampling results, as seen in the spread between the 25th and 75th percentiles shown in Figure 1. The Kincaid lateral concentration profiles typically have 20 receptors, but the spacing between the receptors is uneven and the 'noise' superimposed upon the lateral profiles is quite large (likely typical for dispersion of strongly buoyant effluent from a tall stack). Typically, there are 2 to 4 receptors within the selection range. The NMSE values rapidly approach the value computed when all possible values are selected, once n is greater than 3. For all three experiments, the NMSE is at a minimum when n is 1 or 2, thus confirming Olesen's concerns that too many 'unrepresentative' concentration values may be selected if all values are used that meet the initial selection criteria. To insure good sampling, n should be set to as large a value as possible. To insure representative values of centerline conditions, n should be set to provide

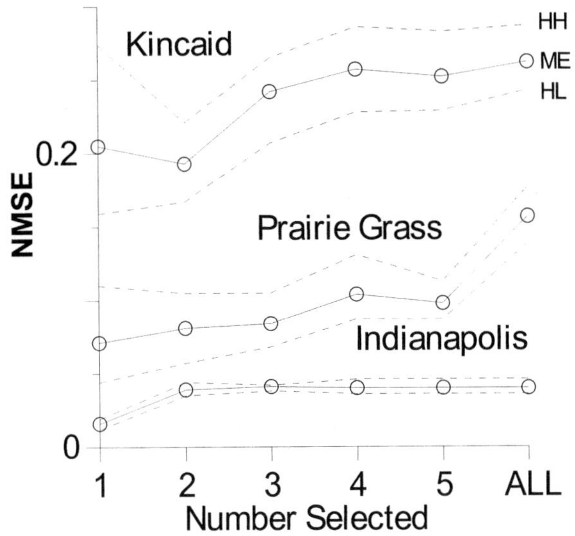

Figure 1. Sensitivity of the Normalized Mean Squared Error between the computed maximum (from selected values) versus the observed arc maximum for three tracer field studies. HH and HL are upper and lower 'hinge' values (robust estimates of 25th and 75th percentiles), and MED is the median, derived from 100 boot samples. Zi/L ranges were used to group the Kincaid and Indianapolis trials along each arc into 23 and and 36 groups, respectively, for analysis. 1/L ranges were used to group the Prairie Grass trials along each arc into 35 groups (contact the author for listings of which arcs are in each group).

a minimum NMSE between the computed maximum and the observed arc maximum. It is concluded from these results, that n can be set to 1 for Indianapolis, 2 for Kincaid, and 3 for Prairie Grass.

BOOTSTRAP RESAMPLING SCHEMES

To place confidence bounds on conclusions reached, we use bootstrap resampling to estimate the mean and variance associated with the comparison measures. In the initial draft of the ASTM Guide, we choose to compare the overall performance of several models using the absolute fractional bias, $AFB_m(\alpha)=2|\overline{C}_m(\alpha)-\overline{C}_o(\alpha)|/[\overline{C}_m(\alpha)+\overline{C}_o(\alpha)]$, and its standard deviation, $\sigma_m(\alpha)$, between the modeled $\overline{C}_m(\alpha)$ and observed $\overline{C}_o(\alpha)$ average centerline concentration for each group α for each model m. $\overline{C}_m(\alpha)$, $\overline{C}_o(\alpha)$, $AFB_m(\alpha)$ and $\sigma_m(\alpha)$ were computed via bootstrap resampling, where pairs of concentration values were selected from all the values within the 'selection range' from each arc, and each arc within a group was treated as being equaling likely during the resampling (Irwin and Rosu, 1998). A grand average, AFB_m, and standard deviation, σ_m, was computed over all groups for each model using inverse variance weighting.

In order to determine which of several models is 'closest' on average (including effects of both bias and scatter) to the average observed centerline concentration, a Model Comparison Measure was computed as $MCM_m = AFB_m + 2\,\sigma_m$. The model with the lowest value for MCM is denoted as the 'base' model. The question that then is asked is whether the values of AFB_m for the other models are significantly different from the value found for the 'base' model, AFB_b. To answer this question, we use a t-test for a difference between two means, computed as: $t = |AFB_m - AFB_b|/\sqrt{\sigma_m^2 + \sigma_b^2}$. For comparisons involving large numbers of cases, if t is greater than 1.96, we can reject the null hypothesis that the two averages are the same with 95% confidence. By successively using the t-test on each model, we can define those models whose performance is significantly poorer than the performance of the base model. Note, if we have many models (say k) to test, i.e. multiple comparisons, we could use the Bonferroni Method, and take the significance level for each t-test to be .05/k rather than the usual .05.

Using a normalized comparison allows pooling results together in which the uncertainties may scale by the magnitude of the group averages. Using an absolute value statistic avoids being misled by offsetting bias between groups, where an overestimation in one might be offset by an underestimation in another. Pooling the results together using inverse variance weighting devalues groups in which the computed standard deviation of the absolute fractional bias is large in comparison to that computed in other groups. A drawback to the initial design is that only comparison measures that are fully defined within a group can be used (which eliminates use of regression statistics, NMSE, etc.). Another drawback is that the AFB is bounded to be between ±2, which can cause $\sigma_m(\alpha) \rightarrow 0$ if the model bias is large for a group, and such groups will dominate the resulting inverse variance pooled statistics.

Both of the drawbacks just mentioned can be avoided by altering the bootstrap sampling scheme. Instead of applying the bootstrap sampling to each group independently (which we will refer to as 'Design 1'), we apply the bootstrap sampling to all groups, on each pass of the bootstrap sampling (which we will refer to as 'Design 2'). We can envision Design 2 as a five (5) step process. Step 1: Form a replicate sample of observed centerline concentration values and modeled centerline concentration values for each model for Group 1. The rules for developing a replicate sample are the same as in Design 1. Namely, each arc is treated as being equally likely; we select a 'pair' of observed centerline values and we concurrently sample the modeling results from arcs selected (see Irwin and Rosu, 1998). Step 2: Compute the average observed centerline concentration for Group 1, and modeled average centerline concentration for each model for Group 1, and store these values for later reference. Step 3: Repeat Steps 1 and 2 for each of the groups. We now have for each group a set of sample averages of observed and modeled centerline concentration values for every group. Step 4: Compute any (or several) comparison statistic of choice, e.g., linear regression of observed versus modeled averages, NMSE, Willmott-d value (Willmott, 1982), and store results for later reference. Step 4: Repeat Steps 1 through 3 for however many bootstrap samples deemed necessary. Irwin and Rosu (1998) outlined a procedure for assessing how many samples might be needed, but 500 samples appears to be more than sufficient in analyses currently completed. We now have results from say 500 bootstrap samples. Step 5: Summarize and analyze the results obtained, which could include characterization of the distribution of values obtained for the NMSE values, computations of averages and standard deviations of the regression results, or one might choose to list the results out for post analysis.

In Design 1 in order to determine which model was most often closest in agreement with the observed average centerline concentration, we used a Model Comparison Measure, $MCM_m = AFB_m + 2\,\sigma_m$, where AFB_m (a measure of bias) and σ_m (a measure of scatter or dispersion) are pooled statistics computed over all groups. In Design 2, we can use the NMSE to provide a composite measure of the effects of both bias and scatter. The model with the smallest value of NMSE can be the 'base' model. We can test whether the results for each of the other models, $NMSE_m$, is significantly different from $NMSE_b$ using the saved bootstrap results, as:

$$\Delta^i_{mb} = (NMSE_{mi} - NMSE_{bi}), \ ; \ \Delta_{mb} = \sum_{i=1}^{n_b} \Delta^i_{mb}/n_b, \ ; \ \sigma_{mb}{}^2 = \sum_{i=1}^{n_b} (\Delta^i_{mb} - \Delta_{mb})^2/n_b \quad \textbf{(3)}$$

where n_b is the number of bootstrap replicate samples, and the t-value to test whether Δ_{mb} is possibly zero is $t = \Delta_{mb}/\sigma_{mb}$. If t is greater than 1.96, we can conclude with 95% confidence that the *NMSE* values computed for model *m* and *b* are significantly different. We can apply this analysis process, (3), to the bootstrap results for any of the comparison statistics, and thereby test to see if a comparison statistic determined for a model is really significantly different from that determined for the other models.

Table 2. Summary of comparison statistics for three plume dispersion models for three tracer dispersion field studies.

	AERMOD	ISC	HPDM
		Project Prairie Grass (*n* = 3)	
NMSE	1.03^b	1.16	4.61*
Slope	1.15^b	0.92*	0.35*
Intercept	0.32^b	0.43	0.27
R^2	0.74^b	0.63	0.29*
Willmott-d	0.89^b	0.87	0.63*
		EPRI Kincaid (*n*=2)	
NMSE	0.67*	1.61*	0.26^b
Slope	0.24*	0.70	0.71^b
Intercept	2.20	0.13*	2.77^b
R^2	0.17	0.10*	0.32^b
Willmott-d	0.57	0.47*	0.72^b
		EPRI Indianapolis (*n* = 1)	
NMSE	0.38	0.28	0.25^b
Slope	0.32*	0.61	0.68^b
Intercept	27.07*	22.70	14.48^b
R^2	0.13*	0.26	0.34^b
Willmott-d	0.58*	0.69	0.74^b

An asterisk (*) indicates that the value is significantly different with 95% confidence from the base model's value. The model with the lowest value for the NMSE is chosen as the 'base' model (superscript **b**) for all significance test. ***n*** is the number of observations selected (as discussed in the previous section).

Table 2 presents the comparison statistics generated using modeled centerline concentrations from three models for three tracer field studies. Irwin (1998) provides the details on how the modeling results were generated. The AERMOD's meteorological data was used to drive the ISC and HPDM models (hence we would expect different performance by these two models, if they were run using their own meteorological preprocessor's data). It would appear that the NMSE statistic does provide a robust assessment of overall model performance. This is seen in that the model seen to have the lowest value for the NMSE, is also generally the model whose linear regression results provide a slope nearest one, an intercept nearest zero, the highest regression coefficient (R^2) ; and the Willmott d-value is closest to one.

CONCLUSIONS

Based on tests using sampling results from three tracer field studies of dispersion, it was ascertained that the criteria for selection of observed centerline concentration values needed to be restricted to the one to three values that are nearest to the computed center of mass and are within $\pm 0.67\sigma_y$. The redesigned bootstrap sampling provides sample averages of observed and modeled centerline concentration values from each group on each pass of the bootstrap sampling. This allows computation of any number of standard statistics for comparison measures, with the added benefit of being able to test whether differences seen are statistically significant. The redesigned bootstrap sampling procedure was tested using modeling results from three models for three tracer field studies. The NMSE statistic is seen to provide a robust assessment of which model's values are nearest on average to that observed, and it is seen that other comparison measures are in reasonable accord with the conclusions one might reach based on the NMSE.

DISCLAIMER

The information in this document has been funded in part by the United States Environmental Protection Agency under an Interagency Agreement (DW 13937039-01-06) to the National Oceanic and Atmospheric Administration. It has been subjected to Agency review for approval for presentation. Mention of trade names or commercial products does not constitute endorsement or recommendation for use.

REFERENCES

Barad, M.L. (Editor), 1958, Project Prairie Grass, A Field Program In Diffusion. *Geophysical Research Paper*, No. 59, Vol I and II, Report AFCRC-TR-58-235, Air Force Cambridge Research Center, 439 pp.

Bowne, N.E., Londergan, R.J., Murray, D.R., and Borenstein, H.S., 1983, Overview, Results, and Conclusions for the EPRI Plume Model Validation and Development Project: Plains Site. EPRI EA-3074, Project 1616-1, Electric Power Research Institute, Palo Alto, CA. 234 pp.

Cox, W.M. and Tikvart, J.A., 1990, A statistical procedure for determining the best performing air quality simulation model. *Atmos. Environ.*, Vol 24A, p 2387-2395.

Hanna, S.R., 1993, Uncertainties in air quality model predictions. *Bound.-Layer Meteor.* Vol 62, p 3-20.

Haugen, D.A. (Editor), 1959, Project Prairie Grass, A Field Program In Diffusion. *Geophysical Research Papers*, No. 59, Vol III, AFCRC-TR-58-235, Air Force Cambridge Research Center, 673 pp.

Irwin, J.S., 1998, Statistical evaluation of atmospheric dispersion model. *Int. J. of Environ. and Poll.* (to be published, 8 pages).

Irwin, J.S. and Rosu, M.-R., 1998, Comments on a draft practice for statistical evaluation of atmospheric dispersion models. *Proceedings of the 10th Joint Conference on the Application of Air Pollution Meteorology with the A&WMA*. Phoenix, AZ., p 6-10.

Irwin, J.S., Rao, S.T., Petersen, W.B., and Turner, D.B., 1987), Relating error bounds for maximum concentration predictions to diffusion meteorology uncertainty. *Atmos. Environ*. Vol21, p 1927-1937.

Murray, D.R., and Bowne, N.e., 1988, Urban Power Plant Plume Studies, EPRI Report No. EA-5468, Research Project 2736-1, Electric Power Research Institute, Palo Alto, CA.

Olesen, H.R., 1999, Model validation kit - recent developments. *Inter. J. of Environ. and Poll.*, (to be published).

Venkatram, A., 1988, *Topics in Applied Modeling. Lectures on Air Pollution Modeling*. A. Venkatram and J.C. Wyngaard (editors). American Meteorological Society, Boston, MA. P 267-324.

Willmott, C.J., 1982, Some comments on the evaluation of model performance. *Bull. of Amer. Met. Soc.*, (63)11:1309-1313.

DISCUSSION

G. SCHAYES: You presented a methodology applied to a small scale problem for analyzing the data. In this case, the underlying physics is rather well known and the procedure gives interesting results. Do you think that the methodology can be applied to the more comprehensive models used for e.g. ozone prediction, where the general complexity is much higher?

J. S. IRWIN: The philosophy applies to all models of environmental processes, not just air dispersion models. Some models may attempt to characterize the uncertainty in the estimates (resulting from input uncertainty and model biases) and some models may attempt to characterize the natural variability (unaccounted for in the model simulation). But all models of environmental processes provide estimates of the ensemble average result for the conditions specified. Thus in principle, the evaluation of such models must involve solving the problem that the models provide estimates of the ensemble averages and the observations are individual realizations from some (poorly defined) ensembles. If the variations between realizations within an ensemble are small, then any one realization provides a good estimate of the ensemble average. In meteorological modeling and in air quality modeling, the within ensemble variations are typically on the order of the magnitude of the ensemble average. Hence, it is a poor assumption to believe that a direct comparison of the observations and the modeling results is a useful exercise. If you have sufficient data to form groups of data, then the methodology I have outlined should work. If you have a few observations (individual realizations), taken from almost as many ensembles (i.e., over a wide range of conditions), it may be dubious whether these data are useful for statistical performance evaluation exercises. There is always the issue that the data available from a field study may be insufficient for their use in statistical performance evaluations, not because of incomplete auxiliary measurements, but because there are insufficient cases (or realizations) to define the magnitude and variation of the ensemble averages as conditions change.

PREDICTING NO_x CONCENTRATION IN ALPINE VALLEYS USING APPLIED DISPERSION MODELING

Peter de Haan

INFRAS, Muehlemattstrasse 45, 3007 Bern, Switzerland;
peter.dehaan@infras.ch

INTRODUCTION

This paper presents an applied method to model the dispersion of pollutants emitted in the basin of Alpine valleys. Frequency distributions of wind speed and direction, mixing height, and PGT stability class are analyzed for three meteorological stations located in Alpine valley basins. An applied Gaussian plume model is used to produce "dispersion matrices" (i.e., transfer functions), which give the annually averaged concentration impact per grid cell for a source of unit strength located in the center of the matrix. Dispersion matrices are computed for different source configurations and different local climatologics. These dispersion matrices are then used to disperse extensive inventories of NO_x emissions for the whole of Switzerland.

The local wind conditions in alpine valleys are often determined either by the topographic forcing of the mesoscale circulation, or by thermally driven valley breezes. As a result, the circulation exhibits a complex pattern, which is neither homogeneous nor stationary, thus preventing the use of classical Gaussian approaches for modeling pollutant dispersion and the resulting ground-level concentrations.

When modeling pollutant concentrations for entire states (Switzerland has a land surface area of approx. 41 000 square kilometers) using emission inventories, commonly applied dispersion models have to be used. Such dispersion models assume homogeneous and stationary conditions, and often take into account only one generic climatology, as the main goal is the prediction of the annual mean concentration only. This approach worked well in Switzerland for the prediction of annually averaged NO_x concentration throughout the country (SAEFL 1997), but NO_x concentrations from transit highway emissions in alpine valleys were clearly underestimated.

Air Pollution Modeling and Its Application XIV, Edited by
Gryning and Schiermeier, Kluwer Academic/Plenum Publishers, New York, 2001

A method has therefore been derived to better take into account the mean local flow patterns in alpine valleys. The main result are so-called dispersion matrices reflecting local climatology for alpine locations, for use within the Swiss PolluMap model, which is described in the next section. For a generic Alpine climatology and for each out of a set of several source configurations (traffic line source, elevated stacks, residential heatings, etc.), a distinct dispersion matrix is used.

THE EMPIRICAL SWISS „POLLUMAP" DISPERSION MODEL

The empirical PolluMap dispersion model (Heldstab and Künzle 1997) has been applied successfully for regional air pollution management purposes, especially for the present and future situation of the NO_2 and NO_x concentrations (SAEFL 1997). Inputs for the model are detailed emission inventories and actual measurements of concentrations for validation purposes.

The PolluMap dispersion model uses annually averaged dispersion matrices which are applied to each cell of the different emission inventories. The spatial resolution of the dispersion matrices differs. For NO_x modeling, grid cells of 200 m x 200 m are used up to a distance of approx. 8 km away from the source. Concentration impacts caused by transport over more than 8 km is not modeled explicitly; instead, the parameterization of the NO_x background concentration is adjusted such that it accounts for this. The PolluMap modeling approach is also used to model PM_{10} concentrations for the whole of Switzerland. Here, a spatial resolution of 200m x 200m is applied up to 4 km away from the source, and a 1 km^2 resolution up to a spatial extension of 200 km (because of the significant long-range transport of fine particulate matter).

Emissions having distinct source characteristics are grouped into different inventories, i.e. urban and extra-urban emissions, and different source heights. For example, urban and extra-urban traffic sources, residential heating sources, and four different industrial stack heights are distinguished. The dispersion matrices are derived from a simple Gaussian plume dispersion model and reflect the annually averaged ground concentration impact of a point source with specific source characteristics onto each of the neighboring grid cells.

To improve the concentration forecast in alpine valleys, where the dispersion matrices currently being used underestimate the persisting channeling of the flow within the valley, a set of dispersion matrices representing local climatologies in alpine valleys, is used. For that part of Switzerland not being part of the alpine mountain ridge, a single set of dispersion matrices representative for that part of Switzerland is still being used.

ANALYSIS OF HOURLY METEOROLOGICAL DATA

65% of the land surface of Switzerland belongs to the Alp mountains. The main residential and commercial areas are located in the densely populated so-called Swiss

Central Plateau, between the Jura and the Alps mountain ridges. Compared to the Alps, the Swiss Plateau might be regarded as being rather „flat", although it cannot be compared to really flat terrain like the central U.S. Therefore, we would expect that differences in the meteorology between Alpine and Swiss Plateau sites are present, and that these differences are the reason for the underprediction in the Alps of the first model version.

We therefore compared hourly met. data from Plateau and Alpine sites. The five stations used to represent the Swiss Plateau cover its entire range. All stations are operated by the Swiss national meteorological service on a regular basis. A wide range of parameters can be obtained with an hourly resolution. The three Alpine stations used are situated in those valleys where the major transit routes are.

Figure 1 shows the distribution of wind speeds. The amount of low wind conditions is roughly the same for Alpine and Swiss Plateau sites. The Alpine sites do experience a somewhat higher amount of high wind episodes caused by Föhn conditions. But the average wind speed of the 5 Plateau and the 3 Alpine sites is very close to each other.

The distribution of the well-known Pasquill-Gifford-Turner stability classes, depicted in **Figure 2**, is very similar as well, a result not expected a priori. As one would expect, the amount of very stable conditions is high in Alpine regions. However, it is equally high in Plateau sites. Very unstable conditions are rarely observed in Switzerland at all.

Mixing heights, shown in **Figure 3**, show no significant difference between Plateau and Alpine sites either. The mixing height is often rather low; inversions occur on a regular basis. This is, however, a common feature of both the Alpine regions and of the Swiss Plateau. In the case of inversions of the temperature profile at ground level, the mixing height is not set at zero, but rather set as the level up to where mixing occurs.

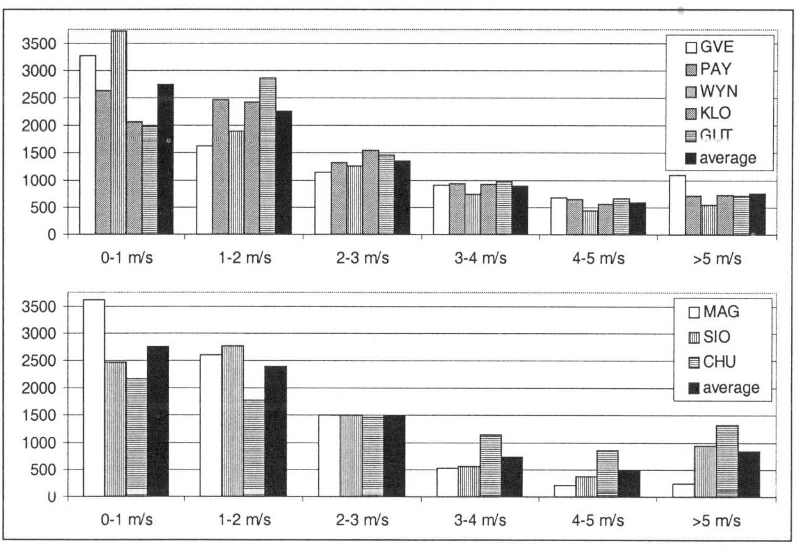

Figure 1. Distribution of hourly wind speed at five Swiss Plateau sites (top) and three Alpine sites (bottom). Data for 1998, 8760 hours.

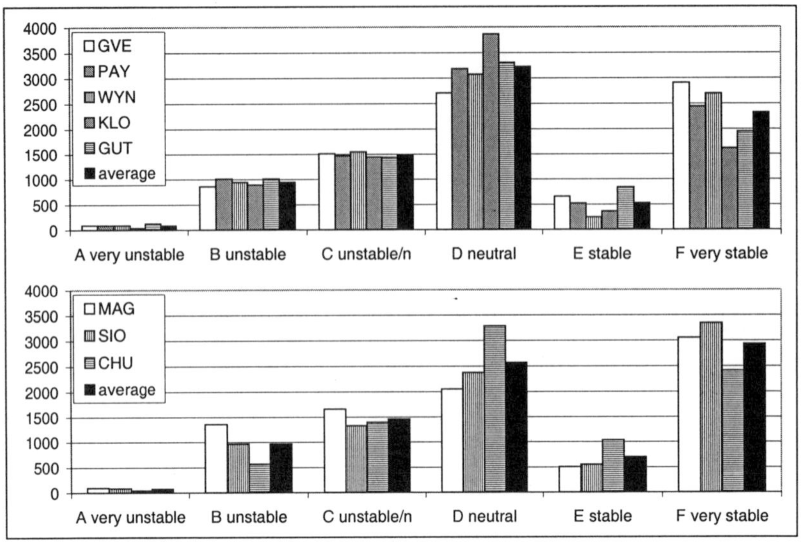

Figure 2. As **Figure 1**, but for PGT stability classes.

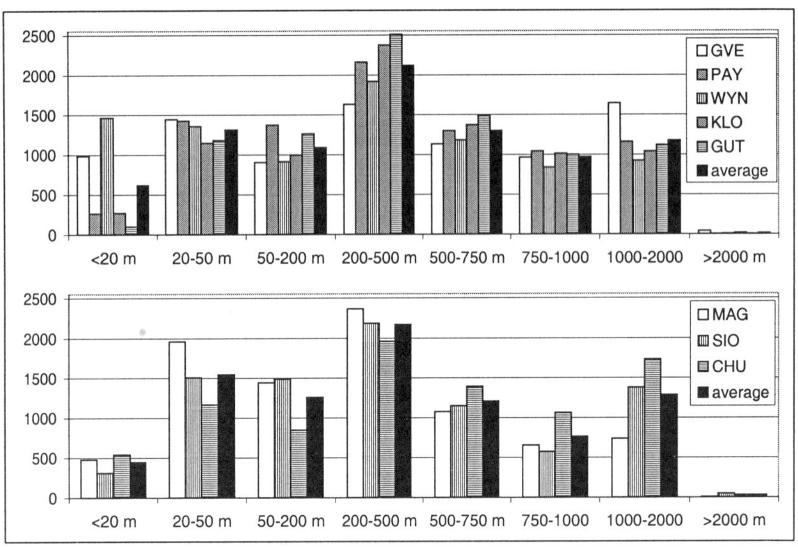

Figure 3. As **Figure 1**, but for mixing height.

COMPUTING DISPERSION MATRICES

A software package has been developed for the computation of dispersion matrices. Such dispersion matrices give the annually averaged concentration impact to each cell of the matrix for a source of unit emission strength, located in the center of the matrix (i.e., the grid). For each type of source (urban, non-urban, ground-level and elevated source, etc.), and for each local dispersion climatology, a different dispersion matrix results.

The Gaussian plume dispersion model being used employs the stability class definitions from the German regulatory model TA-Luft. It assumes homogeneity and heterogenity throughout the modeling domain. Only meteorology for the source location itself is used. The software package enable to compute nested dispersion matrices as well (i.e., a 1 km^2 grid, with a finer 200 m x 200 m grid nested in the center).

A total of six mirror sources are placed beneath the ground and above of the mixing height to preserve mass conservation. To compute the dispersion matrixes, the dispersion of a point source with unit emission strength is computed for a stability class distribution. For 36 different wind flow directions of 10° each, the dispersion is then modeled for different wind speed classes. The dispersion matrices are then computed as the weighted average, and hence give the annually averaged concentration of a unit source.

For medium-range transport distances, the concentration impact at the center of the grid cell is assumed to be representative for the entire grid cell. The local range dispersion in the vicinity of the source location uses a much finer grid spacing within each of the 200 m x 200 m grid cells, in order to obtain the correct cell averaged concentration impact. Different sampling grids for point, line and area sources ensure a correct estimation of the cell-averaged concentration impact even for grid cells in the vicinity of the source location.

RESULTING DISPERSION MATRICES

The main result from the present study is shown in Figure 4 (dispersion matrix for Plateau meteorology) and Figure 5 (Alpine meteorology). The Gaussian model described in the previous section has been used, and all hourly met. data from all sites (i.e., 3 times 8760 hours for the Alpine matrix, and 5 times 8760 hours for the Plateau version). The dispersion matrices correspond to a source emitting 1 ton per year of NO_x. Time series are used to reflect the daily, weekly and seasonal cycle of the emission strength.

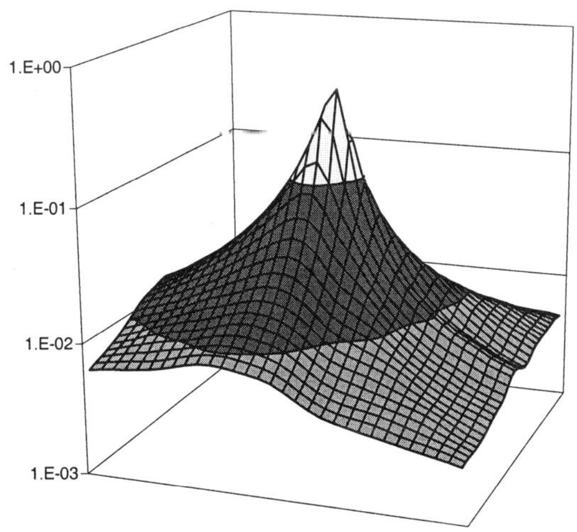

Figure 4: Dispersion matrix with Plateau meteorology for a source located in the center, 20 m above ground (emission: 1 t/a of NO_x). Horizontal axes extend over a square of 6 km by 6 km, each cell is 200 m x 200 m. Vertical axis in µg/m^3.

The main outcome is that no pronounced difference in concentration impact can be seen. Both the near-source peak concentration as well as the decrease of the concentration for increasing distances are comparable. The Alpine matrix is different only because of the more pronounced channeling of the wind directions. This means that for a single point source, the total impact on its surroundings is comparable to a Plateau site. This indicates that the Plateau has to be regarded as a complex terrain region as well. The underprediction of NO_x concentrations in the vicinity of Alpine transit routes, as observed in the first model version, does not occur with the new dispersion matrices anyway. The more pronounced channeling of the wind leads, because of superposition, to higher concentrations in the vicinity of line sources, as long as the line source runs in the same direction as the valley itself.

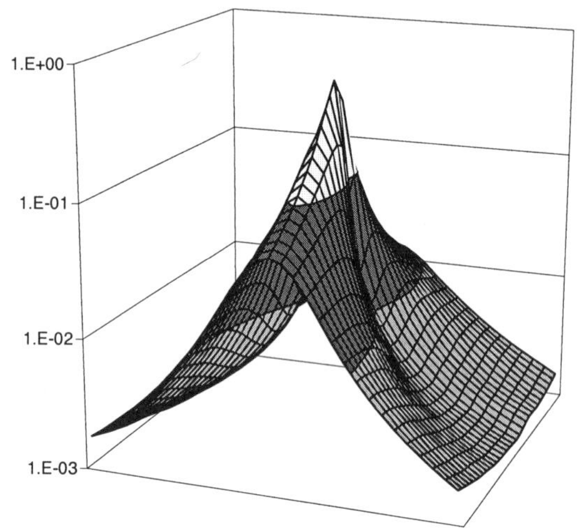

Figure 5: As Figure 4, but for the generic Alpine meteorology.

Acknowledgements—The present work was partly financed by the Swiss Agency of Environment, Forest and Landscape (SAEFL).

REFERENCES

Heldstab, J., and Künzle, T. (1997): Empirical air pollution modeling on the regional scale. Presented at the 4th Symp. Transport and Air Pollution, June 9–13, 1997, Avignon, France

Heldstab, J., de Haan, P., Künzle, T., and Filliger, P. (1999): PM10 map and population exposure for Switzerland. Preprints of 6th Int. Conf. on Harmonization within dispersion modeling for regulatory purposes, 11–14 October, Rouen, France. To be published in Int. J. Environment Pollution

SAEFL 1997: NO2-Immission in der Schweiz 1990-2010. Swiss Agengy for Environment, Forest and Landscape (SAEFL), SRU 289, Bern, Switzerland

DISCUSSION

P. SEIBERT: Do you really believe that a source in an Alpine valley has the same impact on an annual basis as the same source in an open plain?

P. de HAAN: We do not compare our Alpine sites with an open plain site as it may be found in e.g. the central United States. Instead, we compare with a site in the so-called Swiss Central Plateau. Although the Central Plateau from a Swiss perspective may be considered a "plain" site, it should be regarded as complex terrain as well. For example, a clear channeling of the wind direction can be observed in the Plateau, induced by the Jura and Alps mountain ridges limiting the plateau. Comparing a "well-ventilated" Alpine site with a site in the Swiss Plateau, we indeed conclude that the impact on an annual basis is comparable. With "well-ventilated" I mean that no accumulation of pollutants due to inversions lasting more than, say, 12 hours occur. All Swiss Alpine valleys where transit routes are present are "well-ventilated" in this sense. A comparison of the hourly meteorological data of Alpine and of Plateau sites reveals that no significant differences can be observed in the distribution of the mixing heights, the wind speeds or the PGT stability classes. Only the wind direction is even more clearly channeled in the case of the Alpine valleys. This supports are findings that, on an annual basis, a source in an Alpine valley has the same impact as the same source in the Swiss Plateau. Due to the more pronounced channeling of the wind direction, the impact in the Alpine valley will be more concentrated along the valley axis.

MM5 SIMULATION OF THE METEOROLOGICAL CONDITIONS DURING A SOUTH COAST OZONE STUDY (SCOS'97) EPISODE

Dimitra Boucouvala*+, Robert Bornstein*, Douglas Miller°, Jim Wilkinson*@

*Department of Meteorology, San Jose State University, San Jose, CA, USA
+Department of Applied Physics, University of Athens, Greece
°Department of Meteorology, Naval Postgraduate School, Monterey, CA, USA
@Department of Civil and Environmental Engineering, Georgia Institute of Technology, Atlanta, GA, USA

INTRODUCTION

The California Air Resources Board (CARB) sponsored the massive observational South Coast Ozone Study (SCOS'97) in the Los Angeles Basin during the Summer of 1997. One study objective was determination (by data analysis and MM5 simulation) of the role of the downward fumigation of nocturnal residual layer ozone to produce localized early morning surface ozone maxima. The residual ozone layer forms aloft as the inland directed sea breeze flow is deflected by the inland mountain slopes to produce a reversed flow aloft. This flow stagnates in the late afternoon, as the sea to land horizontal temperature gradient dissipates.

The period of 4-7 August was selected for analysis, as it showed high surface ozone values. Analysis has shown that surface ozone values peaked on the coastal side of the San Bernadino Mountains (Fig.1) at about 180 ppb at about 2200 UTC (1500 Pacific Daylight Time, PDT) on 5 August. A secondary maxima (that could have resulted from fumigation) of about 120 ppb was also seen in the same area at about 1100 PDT on the same morning.

A unique feature of SCOS'97 is the richness of its upper air meteorological measurements. Meteorological data from about 259 surface sites case were received from CARB, where they had undergone preliminary QA tests. While data from both the 12 rawinsonde (every six hours) sites and the 25 continuous profiler (mean values only) sites also came from CARB, profiler turbulence data came directly from NOAA/Boulder.

Analysis of surface and PBL meteorological data for the period of 4-7 August by Bornstein and Boucouvala. (2000) showed an inland moving sea breeze front on the 5th, as the large scale flow had become more easterly on that day. The observed location of the front coincided with the zone of maximum observed afternoon ozone. The current paper thus reports preliminary results from an MM5 simulation of the sea breeze front for the period analyzed in the previous observational paper.

Air Pollution Modeling and Its Application XIV, Edited by
Gryning and Schiermeier, Kluwer Academic/Plenum Publishers, New York, 2001

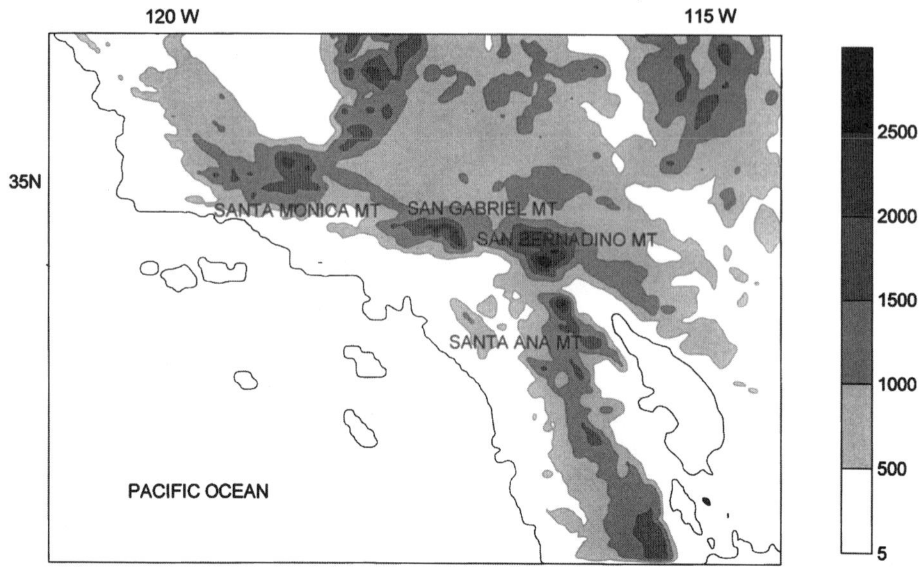

Fig. 1. Topography (in meters) of inner MM5 computational domain.

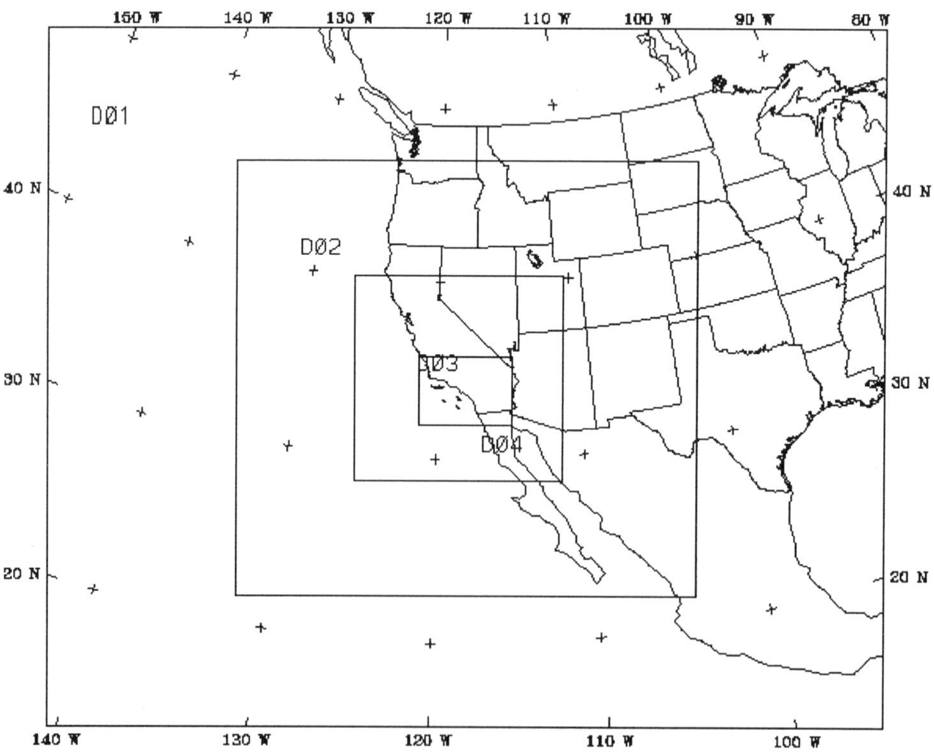

Fig. 2. Four computational domains used in MM5 simulation.

MM5 SIMULATION

Version 2.12+ of MM5 was used to simulate this SCOS97 episode. This version (received directly from PSU) contains the Marine Boundary Layer Initialization (MBLI) scheme of Leidner et al. (1999), not in any NCAR MM5 release. The scheme (used during the first 13 hours of the simulation) establishes a realistic MBL in data-sparse marine regions, a process that normally requires about 48 hours of simulated time.

The quadruple nested Lambert conformal-projection configuration possessed an outermost domain (Domain 1), centered at 34.5N, 118W (Fig. 2), with a horizontal resolution of 135 km, and 41 x 33 horizontal grid points in the E-W and N-S directions, respectively. The domain was chosen to include a large upwind fetch that would include the Pacific High. The second domain, has a grid resolution of 45 km and dimensions 67 x 61; the third a grid length of 15 km and dimensions 91 x 85; and the inner domain (Domain 4) has a grid length of 5 km and dimensions 121 x 85. The new 25 category U. S. Geological Survey (USGS) global land use distribution has been used with a 10 min resolution for terrain and land-use in Domains 1 and 2, 5 min in Domain3, and 30 sec in Domain 4.

NCAR/NCEP global analyses (resolution of 2.5 x 2.5 degrees) were used to provide the initial (1200 UTC on 2 August) and boundary conditions, the latter at a frequency of 12 hours. Improvements were made to these low-resolution analyses by the assimilation (into the coarse-domain gridded analyses) of local upper air and surface data. Initial conditions for the other three domains were obtained by interpolation of values from Domain 1. The simulation used 30 vertical levels, with increasing resolution near the surface, and was carried out for five days.

Other options used included: Dudhia (1989) simple ice for explicit moisture; Kain and Fritch (1993) cumulus parameterization in outer three domains (none used in inner domain), Gayno and Seaman (Gayno 1994) 1.5 order PBL (makes Version 2.12 + different from the Version 2.12 NCAR release); and one-way continuous nesting (one run, but no feedback from an inner nest to its outer domain). For this simulation, model results were only saved at three-hour intervals.

Statistical evaluation of the model results was carried out using the Model Performance Evaluation, Analysis, and Plotting Software (MAPS) system package, developed for urban- and regional-scale meteorological, emissions, and photochemical model evaluations (Tesche 1994). MAPS embodies a variety of the statistical and graphical model testing methods for photochemical and meteorological models recommended by various agencies. Statistics used included a domain wide Wilmont (1981) index of agreement (between observations and modeled values), which ranges from zero (no agreement) to unity (total agreement).

RESULTS

NWS weather charts show only weak synoptic forcing throughout the test period, with a strong 500 hPa ridge over the southwestern U. S. At 1200 UTC on the 5th, however, the weak local coastal anticyclone over the South Cast Air Basin strengthened, which changed the 500 hPa flow direction over the Basin from southwesterly (onshore) to easterly (offshore). At the surface, the offshore Pacific High intensified throughout the period, while the thermal low over the southern part of the state weakened. By the 5th, this produced a relatively light and variable flow over the Basin.

Near-surface temperatures

As described in Bornstein and Boucouvala (2000), by 1200 UTC on the 4th, the smoothed/interpolated observed 1.5 m temperatures showed an average coastal to inland difference of about 8 C. The approximately 10 C difference at 1800 UTC and the corres-

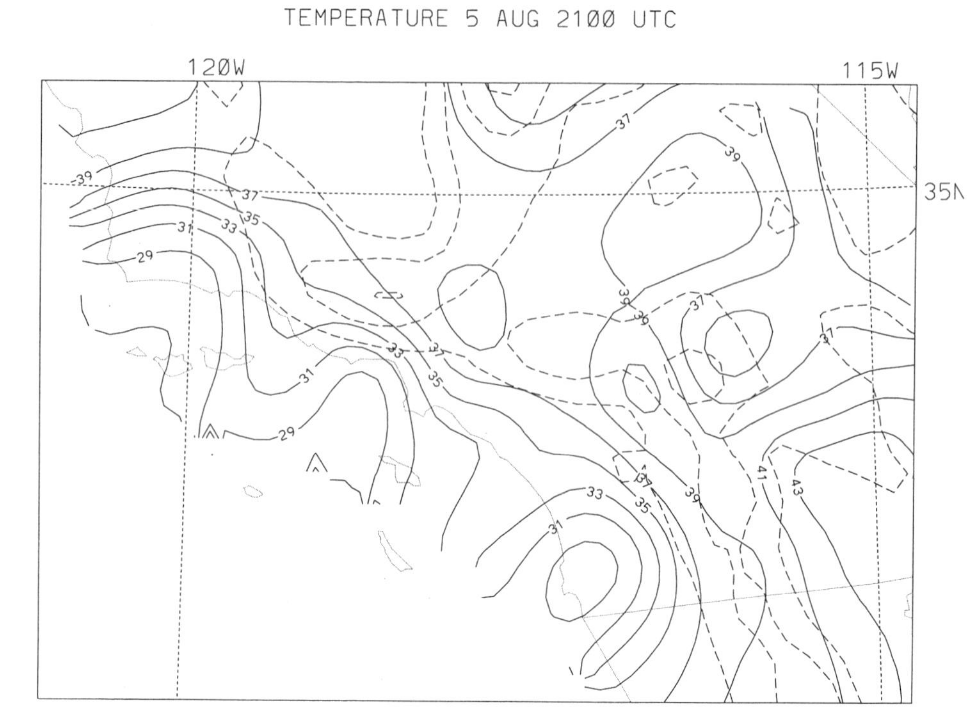

Fig. 3. Domain 4 topographic height contours (dashed lines at 500 m interval), and observed 1.5 m, smoothed, interpolated, over-land temperatures (solid lines in C) at 2100 UTC on 5 August.

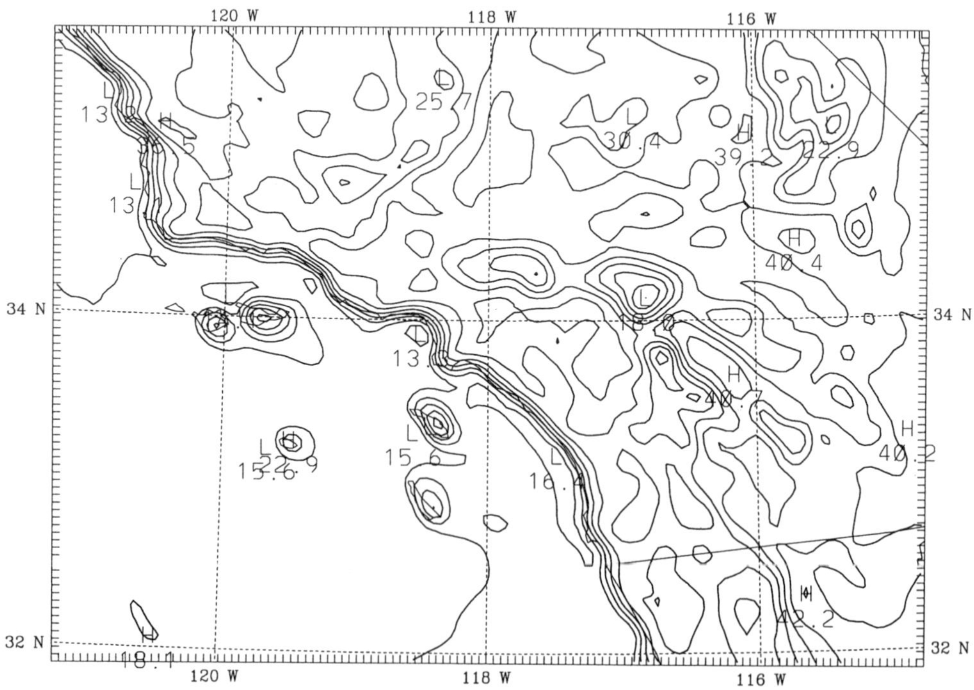

Fig. 4. Domain 4 predicted 20 m temperatures (with a 3 C contour interval) on 2100UTC on 5 August.

ponding 12 C difference at 2100 UTC both reflect a stronger gradient along the coast. Note, that as only a few island observational sites are available, exact coastal gradients are hard to determine.

Throughout the 5th, while all areas had warmed by about 3 C (consistent with NWS charts), spatial patterns were similar to those during corresponding hours on the day before. Coastal-inland gradients were also about the same as on the day before (3). Note that the 2100 UTC distribution is shown, as the ozone maximum occurred one hour after this time on this day.

Throughout the daytime hours of the 6th, temperatures over the entire basin had cooled to almost the values that had existed on the 4th. In addition, coastal gradients were about the same as those on the two preceding days.

Modeled first level (20 m) daytime surface temperature patterns (cool on the coast, warm on the coastal plane, cool on the mountain peaks, and hot in inland areas) on the 4th at 2100 UTC showed generally good agreement with observed values, accounting for the height difference between the two fields. Modeled coastal gradients were larger than in the observations, as they do not lack values over sea areas (as discussed above). Modeled temperature patterns at 2100 UTC on the 5th (Fig. 4) are similar to those on the previous day.

This result is expected, as this simulation does not include analysis forcing, in which the observed large scale warming would be incorporated into the time varying temperature boundary conditions. Domain averaged temperature differences (between observed and simulated near surface temperatures) calculated by the MAPS statistical package as a function of time (Fig. 5) thus show largest differences on the 5th, with its observed large scale warming.

The figure also shows a general under-prediction of daytime temperatures and an over-prediction of nighttime values. These biases should be minimized when (as is being done) MM5 will calculate 1.5 m values in addition to the 20 m values now used.

Near-surface wind flow

The observed nighttime near-surface (10 m) flow pattern at 1200 UTC on the 4th shows a complex combination of flows. While some downslope flow was seen, no clearly defined land breeze flow existed. By 1600 UTC, surface wind speeds increased and had become organized into an onshore sea breeze flow. The inland opposing flow was also more organized and at higher speeds. By 2400 UTC, the inland penetration of the sea breeze generally extended to the ridgeline of the inland high elevation areas. By 0300 UTC on the 5th, the sea breeze had retreated significantly back to the coast.

The nighttime surface flow pattern at 1200 UTC on the 5th showed a similar (to that at the same time on the 4th) complex combination of low speed flows. Nighttime speeds were somewhat faster and a bit more organized on the 5th. By 1600 UTC, surface wind speeds have again become organized into an onshore sea breeze flow, but with less penetration than 24 hours previous. This could be due to the stronger opposing offshore flow on this day in inland areas, due to the change in the upper level flow direction. By 2100 UTC (Fig.6), the inland penetration of the sea breeze again generally extended to the ridge line of the inland high elevation areas, as it had on the 4th. By 0300 UTC on the 6th, the sea breeze had again retreated significantly back to the coast.

The predicted wind flow field on 2100 UTC on 5th shows that the simulation has reproduced the main wind speed and direction flow features of the inland penetration of the various sections of sea breeze frontal convergence zone (Fig. 7). The time variation of the index of agreement statistic (Fig. 8) shows this agreement in a quantitative manner, as its value is close to 0.75 throughout the simulation. This value should increase when model wind predictions are also saved at the same level as the observations (i.e., 10 m).

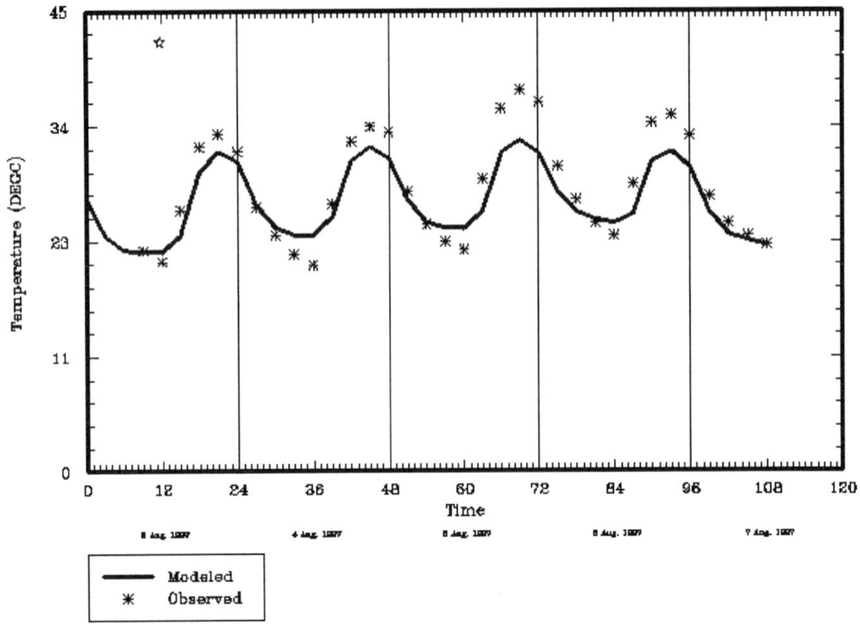

Fig. 5. Domain 4 averaged 20 m predicted (solid line) and observed (asterisks) temperatures(C), where 00 hours is 0000 UTC on 3 August.

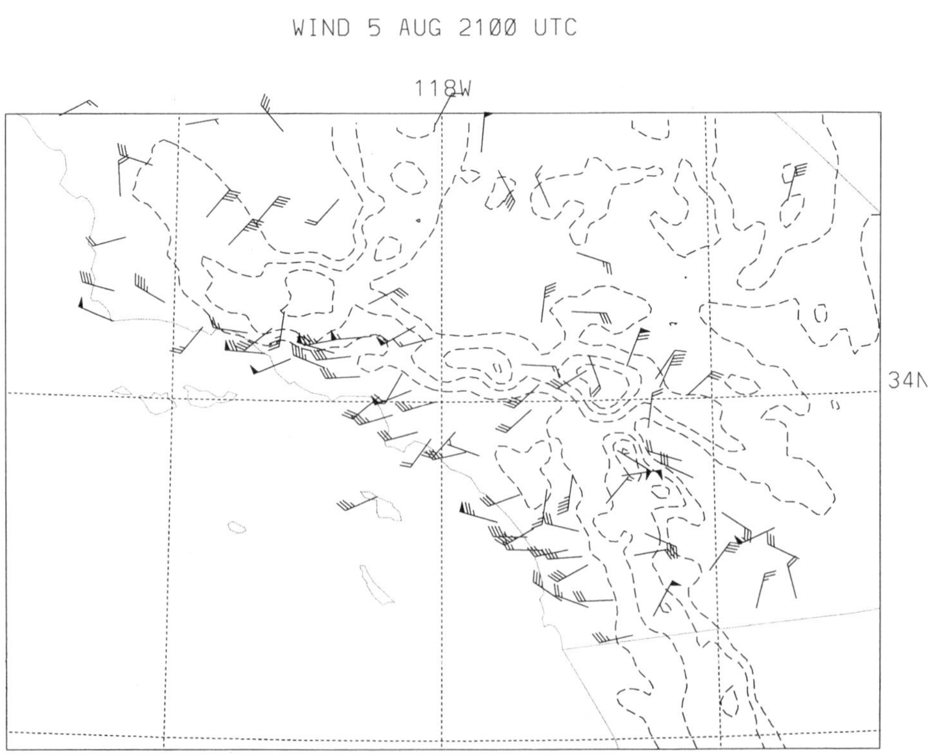

Fig. 6. Domain 4 topographic height contours (dashed lines at 500 m interval), and observed 10 m wind vectors on 2100UTC on 5 August, where a full barb is 1 m/s and a flag 5 m/s.

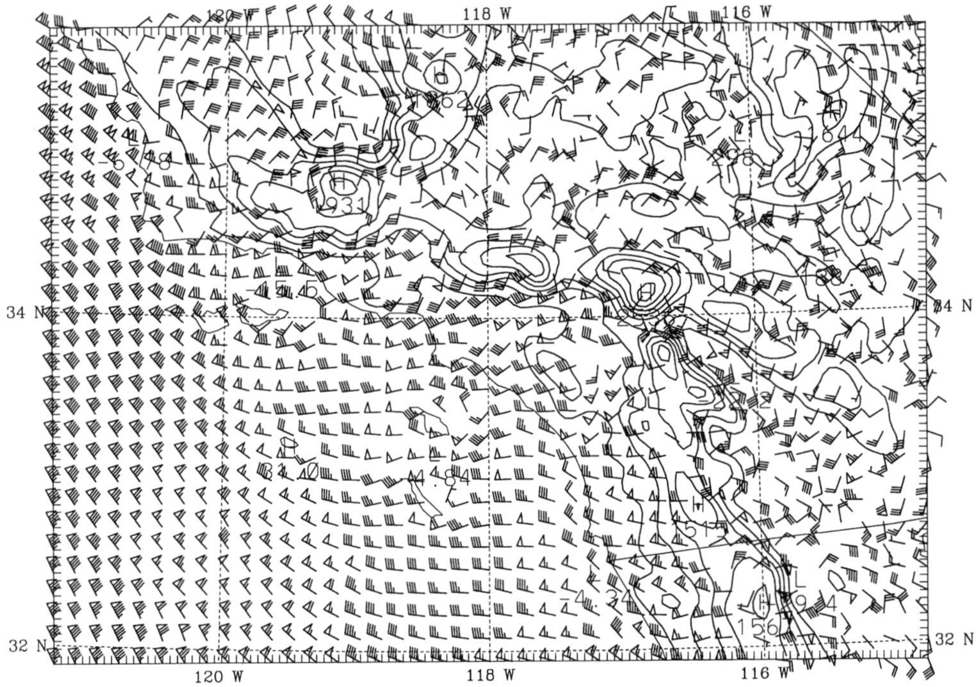

Fig. 7. Domain 4 topographic height contours (solid lines at 300 m interval), and predicted 20 m wind vectors on 2100UTC on 5 August, where a full barb is 1 m/s and a flag 5 m/s.

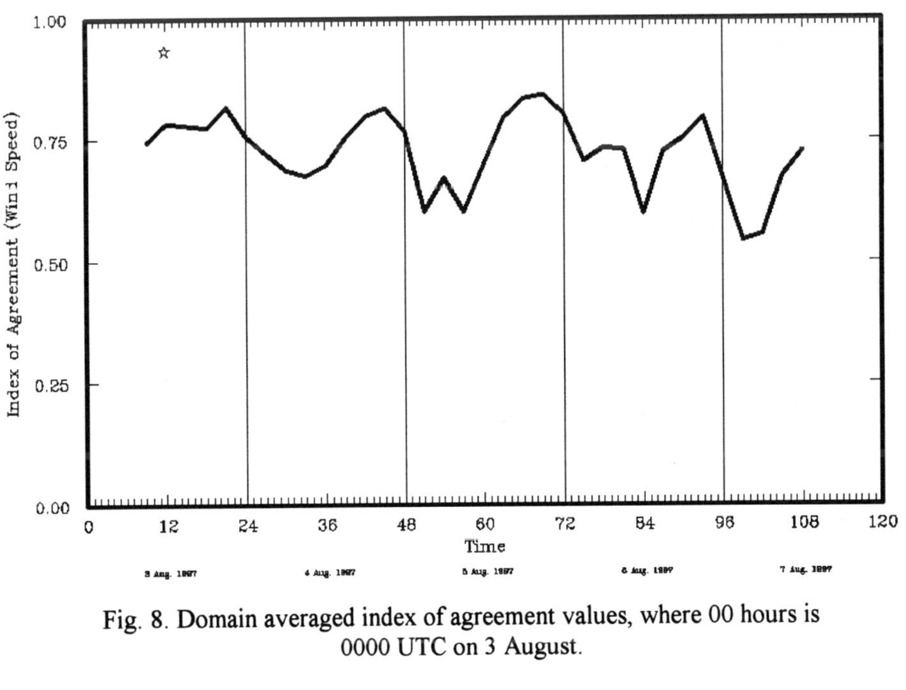

Fig. 8. Domain averaged index of agreement values, where 00 hours is 0000 UTC on 3 August.

CONCLUSION

A five day MM5 simulation of meteorological conditions during an important SCOS'97 ozone episode showed that the model accurately reproduced most of the observed features associated with a inland moving sea breeze front, i.e., its movement speed and inland penetration. Statistical comparisons between near-surface observed and predicted meteorological values (using MAPS) showed reasonable agreement, with differences in temperature, however, largest on the day with the strongest large scale forcing. This effect was not reproduced in the current simulation, as it did not include (model) analysis nudging.

Future evaluation plans include: saving hourly model values for MAPS comparisons, scaling predicted values to observational levels (to reduce biases in MAPS analyses), comparison of predicted and observed upper level values, and compilation of MAPS statistics on various sub-domains (e.g., coastal, mountain, and desert areas). Future modeling simulations will also sequentially include (model) analysis nudging and the use of the SCOS'97 observations via FDDA. Analysis nudging will allow for the incorporation of large-scale meteorological trends (like the subsidence warming on 5 August), while the FDDA will allow for the incorporation of local (sub-grid) flow features into MM5. Both of these additions should further reduce the differences between observed and simulated fields. Results from the final simulations will be used by CARB modelers as input for photochemical air quality model simulations.

***Acknowledgements*:** This research was sponsored under a contract from the California Air Resources Board (CARB). The authors would like to thank Bruce Jackson of CARB for the SCOS'97 data, and Nelson Seaman of PSU for assisting with MM5 Version 2.12+.

REFERENCES

Bornstein, R., and D. Boucouvala, 2000, Preliminary Meteorological Analysis of a 1997 Southern California ozone study (SCOS'97) episode, *11th Joint AMS/AWMA Conf. on Applications of Air Pollution Meteor.*, 9-14 Jan, Long Beach: .70.

Dudhia, J., 1989, Numerical study of convection observed during winter monsoon experiment using a mesoscale two-dimensional model, *J. Atmos. Sci.* 46:3707.

Gayno, G., 1994, Development of a higher-order, fog-producing boundary layer model suitable for use in numerical weather prediction,. *M. S. Thesis*, Dept. of Meteorology, The Penn. State University, University Park.

Kain, J., and J. Fritsch, 1993, Convective parameterization for mesoscale models: The Kain-Fritch scheme, *The Representation of Cumulus Convection in Numerical Models,* K. Emanuel and D. Raymond Eds., Amer. Meteor. Soc.

Leidner, S., D. Stauffer, and N. Seaman, 1999, Improving short-term numerical weather Prediction in the California coastal zone by dynamic initialization of the marine boundary layer, *Mon. Wea. Rev.*, submitted.

Tesche, T., 1994, Evaluation procedures for regional emissions, meteorological, and Photochemical models, *86th Annual Meeting of the Air and Waste Management Association,* 14-18 June, Denver.

Wilmont, C., 1981, On the validation of models, *Phys. Geogr.* 2:168.

POSTER SESSION

MODELLING PHOTOCHEMICAL AIR POLLUTION IN SÃO PAULO, BRAZIL

Ana G. Ulke[1] and M. Fátima Andrade[2]

[1]Dept. of Atmospheric Sciences, Faculty of Sciences, University of Buenos Aires, Buenos Aires, Argentina
[2]Department of Atmospheric Sciences, Institute of Astronomy and Geophysics, University of São Paulo, São Paulo, Brazil

A photochemical airshed model (CIT model, Harley et al., 1993) has been used to study the transformation, transport and removal of the pollutants that cause the air quality deterioration in São Paulo, Brazil. The region is located near the ocean in a complicated topography. The urban core is one of the megacities of the world and the Metropolitan Area of São Paulo is the largest industrialized region in Latin America. Approximately 90% of the ozone precursors are emitted to the atmosphere by the vehicular fleet. The federal imposed limits on ozone and nitrogen oxides have been violated several times every year (Massambani and Andrade, 1994).

Meteorological and air quality field measurements were available to perform model simulations over the period 16-19 February 1989. During this four-day interval, no track of fronts was observed over the region, which was under the influence of a high-pressure system. The flow pattern showed the typical development of local circulations (Silva Días and Machado, 1997). A multiday simulation was carried out to avoid the influence of inaccuracies on initial and boundary conditions. The modeling domain, centered in 23.6° S, 46.5° W, covered a 300x135 km^2 area, with 5-km resolution. The vertical extent was 1100m, divided in five cells with varying depth. Meteorological and air quality fields were generated based on measurements, applying objective analysis procedures. The wind field on afternoon hours of the final day of the simulation along with the topography is depicted in **Figure 1**.

Different emission scenarios were considered in order to analyze uncertainties in the emission pattern. There is a lack of accurate information related to the emissions in São Paulo. The 1989 actual fleet scenario (AF) considered the official emission inventory (CETESB, 1990). Alternative scenarios duplicated the nitrogen oxides contributions (NOx2) and the hydrocarbon emissions (HC2). **Figures 2** to **4** present the predicted ground level ozone concentrations for Feb. 19, 15 LT. The urban core of São Paulo exhibited inhibition of ozone formation due to high nitrogen oxides levels, while the suburban area showed larger ozone concentrations, the product of photochemistry and transport. The ozone levels predicted by the NOx2 scenario were the smallest. The HC2 scenario led to the greatest downwind ozone concentrations. This alternative resulted also in a satisfactory agreement with observed values. The application of the model demonstrated the need of an accurate emission inventory for improved predictions of the pollutant concentrations.

An alternative approach for the turbulence parameterization was introduced in the model. The scheme gives a continuous transition between the different regimes in the atmospheric boundary layer (Ulke, 2000). **Figures 5** and **6** show the evolution of the vertical distributions of ozone during the two final days of the period, obtained with the original and the alternative schemes. The selected grid point (X=27.5, Y=15) represents a typical urban location. The alternative parameterization led to greater concentration levels and less uniform distributions.

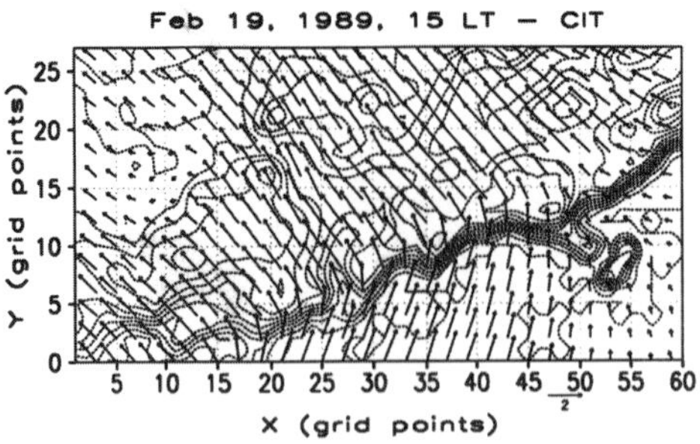

Fig.1 Surface wind field (in ms^{-1}).

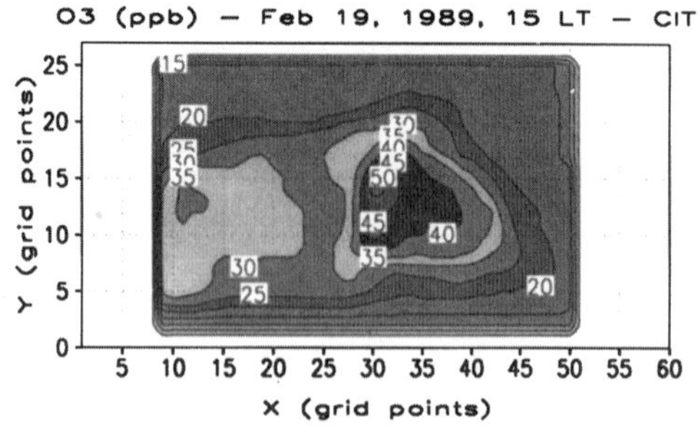

Fig.2 Ozone surface concentration field (AF scenario).

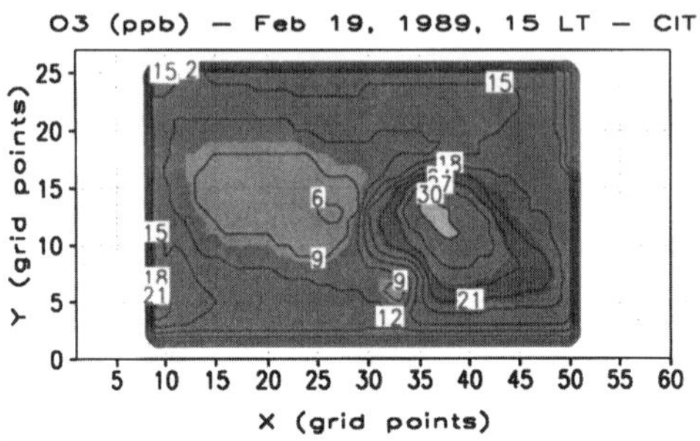

Fig.3 Ozone surface concentration field (NOx2 scenario).

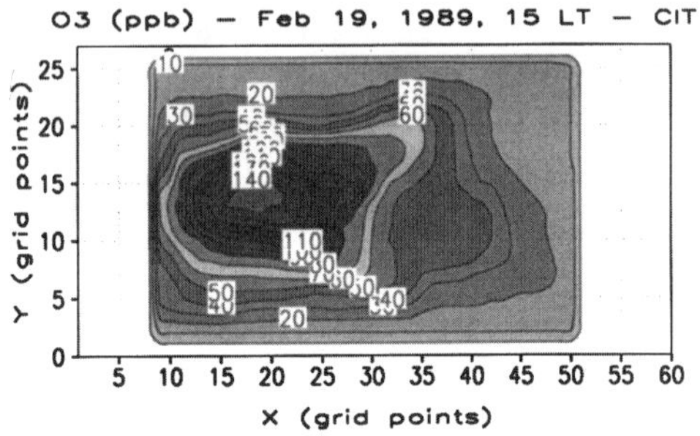

Fig.4 Ozone surface concentration field (HC2 scenario).

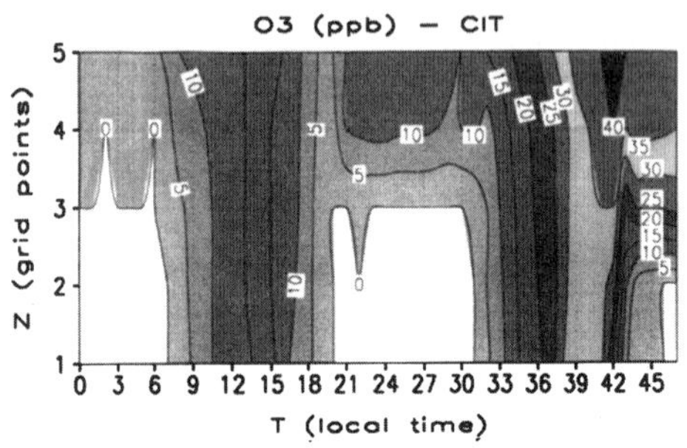

Fig.5 Ozone evolution for Feb.18 and 19 and the original turbulence scheme (AF scenario).

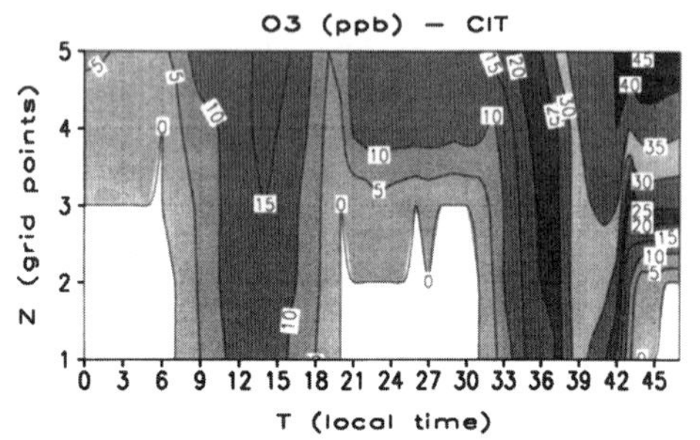

Fig.6 Ozone evolution for Feb.18 and 19 and the alternative turbulence scheme (AF scenario).

Acknowledgments
A.G.Ulke wishes to thank the financial support of FAPESP (Fundacão de Amparo a Pesquisa do Estado de São Paulo) for her stay as visiting professor at the University of São Paulo.

REFERENCES

CETESB, 1990, Relatório de Qualidade do Ar no Estado de São Paulo 1989. Companhia de Tecnologia de Saneamento Ambiental, *Relatórios/CETESB* ISSN 0103-4103, São Paulo (in Portuguese).

Harlcy, R.A., Russell, A.G., McRae, G.J., Cass, G.R., Seinfeld, J.H. , 1993, Photochemical modeling of the Southern California air quality study. *Environmental Science and Technology* 27 (2), 378-388.

Massambani, O. and Andrade, M.F., 1994, Seasonal behavior of tropospheric ozone in the São Paulo (Brazil) Metropolitan Area, Atmospheric Environment 28 (19), 3165-3169.

Silva Días, M.A.F. and Jaschke Machado, A. 1997, The rol of local circulations in summertime convective development and nocturnal fog in São Paulo, Brazil. *Boundary-Layer Meteorology* 82, 135-157.

Ulke, A.G. 2000, New turbulent parameterization for a dispersion model in the atmospheric boundary layer, *Atmospheric Environment* 34 (7), 1029-1042.

ENVIRONMENTAL IMPACT OF BULGARIAN NNP 'KOZLODUY'

Anna Tzenkova, Julia Ivancheva, Dimiter Syrakov

National Institute of Meteorology and Hydrology - Bulgarian Academy of Sciences
Tzarigradsko Shaussee 66 blvd., Sofia 1784, Bulgaria

INTRODUCTION

The Kozlodui Nuclear Power Station (NPS) produces about 2/3 of the electric energy needed by Bulgaria. It is one of the main obstacles for the country's acquisition into the European Community (EC). These facts necessitated the elaboration of a comprehensive estimation of the NPS's impact on the environment. In the present work are presented some of the obtained results for this impact on air quality

METHOD

To calculate the average fields of the concentration and the deposition of radioactive pollutants, versions of the Gaussean plume model were used which account for the specificity of the pollutants and the meteorological input. The aerosol specific processes were described using a parameterization scheme suggested by Galperin [1].

Under normal conditions of operation, the blocks of the Kozlodui NPS emit three types of pollutants: long-living aerosols (LLA), radioactive noble gases (RNG), and radioactive iodine (RI). Three subtasks were calculated: 1) for the years 1997 and 1998; 2) for the interval 1994-1996; 3) for the interval 1987 1989. The input meteorological information for the years 1997 and 1998 was the data recorded by the automatic meteorological station included in the NPS monitoring system. The data used for the interval 1994-1996 was recorded by the Oriahovo climatic station. The data for the interval 1987-1989 was taken from the climatic handbooks for Bulgaria.

The emission data is shown on Table 1. From it, it is quite evident that the emissions from blocks 1 to 4 are much greater compared to the emissions from 5 and 6.

RESULTS

The obtained results reveal that, in 1997, the average yearly field of LLA in the atmosphere reaches a maximum of 1,05 $\mu Bq/m^2$. During the two years, on average, nearly the whole 30 km area stays within the limit of 10 $\mu Bq/m^2$. The maximum values of the diurnal deposition are accordingly 1100 $\mu Bq/m^2$ for 1998, and 3400 $\mu Bq/m^2$ for 1997.

The maximum pollution with RNG for the years 1997 and 1998 in the region of the NPS is about 0,2 $\mu Bq/m^2$. Twelve km away from the source, the maximum concentrations are about 0,04 μBq/m^2, and 30 km away from the NPS they are within the range 9,01-0,015 μBq/m^2. Three km away from the source, in the west direction, is the highest concentration of iodine - 4,46 μBq/m^2, while the lowest one - 1,27 μBq/m^2, is in the north direction.

Table 1. Emission of radioactive pollutants from the Kozlodui NPS

Year	Blocks 1 and 2	Blocks 3 and 4	Block 5	Block 6	Av.conc. 3
Long Living Aerosols (LLA) in [MBq/m^2]					
1988	310,8			-	-
1989	1321,0	740,0	1221,0	-	-
1994	566,0	945,0	150,8	60,9	253,5
1995	689,0	402,0	129,2	83,5	130,8
1996	1144,3	466,3	76,3	56,6	132,4
1997	1071,0	481,0	100,0	59,0	128,0
1998	405,0	488,0	71,4	60,0	142,6
Radioactive Noble Gases (RNG) in [GBq/m^2]					
1988	605357			-	-
1989	124801	114108	526806	-	-
1994	95651	129602	16494	7013	17146
1995	46400	164640	29982	5453	22602
1996	117230	222510	15781	5989	29040
1997	86510	79080	10613	4974	22283
1998	101330	92950	14991	5236	18997
Radioactive ^{131} iodine in [MBq/m^2]					
1988	7580			-	-
1989	1073,0	1665,	1369,0	-	-
1994	752,0	801,0	158,0	126,3	301,6
1995	411,0	413,0	182,8	108,8	355,3
1996	526,2	1055,7	86,8	81,5	210,8
1997	1439,0	828,0	223,0	81,0	166,0
1998	5500,7	4336,6	127,6	22,2	7,2

CONCLUSIONS

For the years 1997 and 1998, the average seasonal fields of the concentration of LLAs were also calculated. Winter appeared to be the season with lowest concentrations, the maximums being accordingly 0,5 μBq/m^2 for 1997, and 0,8 μBq/m^2 for 1998. Higher values were recorded during the transitional seasons, the maximum in fall for both years being about 2,0 μBq/m^2, and in the spring of 1997 - 3,0 μBq/m^2. During all years studied, the maximum values recorded in the roundabouts of the NPS were lower than the maximum concentrations allowed, The level of pollution farther away complied with the sanitary-hygienic requirements.

As a result of the model study it was ascertained that the operation of the "Kozlodui" Nuclear Power Station does not result in the formation of radioactive pollutants with concentration greater than allowed by the sanitary-hygienic norms. The obtained results are comparable to the data from the background monitoring of the country's radioactive air pollution. The radiation loading of the population, calculated based on the above concentrations falls within the prescribed sanitary-hygienic limits.

REFERENCES

Galperin M. and A. Maslyaev (1996): A tentative version of airborne transport POP model; Description and preliminary results of sensitivity studies, EMEP/MSC-E report 3/96, June 1996, Moscow, Russia.

A FIRST ORDER LAGRANGIAN STOCHASTIC MODEL FOR LONG RANGE TRANSPORT

Réal D'Amours[1], Serge Trudel[1], Thomas K. Flesch[2] and John D. Wilson[2]

[1]Canadian Meteorological Centre, Dorval Canada
[2]University of Alberta, Edmonton, Canada

MODEL FORMULATION

The dispersion model is based on a 3-D first order Langevin equation for the turbulent component of the wind. Turbulence is assumed to be quasi-stationary. Horizontal gradients in turbulence statistics, as well as the covariances among of the three turbulent wind components, are neglected. Turbulent velocity variances are obtained by partitioning the turbulent kinetic energy (TKE) given by CMC's operational Global Environmental Multiscale (GEM) model (Coté et all, 1998), depending on local thermal stability. The Lagrangian velocity time scale appearing in the Langevin equation is expressed in terms of the turbulent velocity variances and of the TKE dissipation rate, which is estimated from the TKE and a lengthscale calculated according to Mailhot and Benoit (1982), from the boundary layer height and stability, both provided by GEM. The synoptic components are obtained directly from GEM. The vertical profiles of wind and temperature within the lowest GEM model layer, defined as the surface layer, are diagnosed using Delage (1988) formulation. Particles trajectories are obtained by the time integration the 3-D wind.

SIMULATION OF THE OCTOBER 94 ETEX RELEASE

The dispersion model was used to simulate the October release of ETEX (van Dop and Nodop, 1998). Meteorological fields are obtained from series of 3 and 6 hours forecasts by the GEM model in regional mode, at a resolution of .about 15km over Europe. These forecasts were initialized with the operational CMC global objective analysis available for that period. Figure 1 shows near surface 3hour average concentrations, valid 57 hours after the start of the release, resulting from a simulation with 96000 particles, using a time step of .75s. In simulating an inert tracer gas, the particles are reflected when impacting the ground surface. In this simulation the reflection level is at 3m.

The displayed concentration field was calculated on a polar stereographic grid of 25km resolution using the average residence time, during the given 3-hour interval, of individual particles within a 100m thick, grid cell based, volume. All the available observations for the period are also plotted for comparison. It is apparent the model is able to simulate well the dispersion of the cloud. More results are given on the poster presentation.

BACKWARD SIMULATIONS

An important feature of the model is the ability to run in a "backward" mode, tracking particles backward in time, from a receptor location. In this configuration the backward Langevin equation developed by Flesch, Wilson and Yee (1995) is used. Assuming a surface-area source, particles touchdown locations and touchdown velocities can be used to infer fields of source potential for a given receptor. By combining information on source potentials from different receptors, it is possible to retrieve information on the real source. Figure 2 illustrates a simple approach to delimit the areas of high likelihood for the source location. What is shown is a composite of fields of the number of touchdowns registered within one grid element of a .25° latitude longitude grid, for particles emitted from receptor points, as they travel backward in time. A distinct time integrated touchdown field is obtained for each receptor points, and the composite is simply a biased geometric mean of these touchdown fields, at each grid cell. In the present case, the model was run backward, for 60 hours, from a set of 7 receptors with measured concentrations on Oct 26 00UTC. It can easily be seen that the real ETEX release point (SRCE on the figure) is located in the area where high values of the composite are found. In the poster paper a method to qualitatively determine the location and intensity of the release is presented.

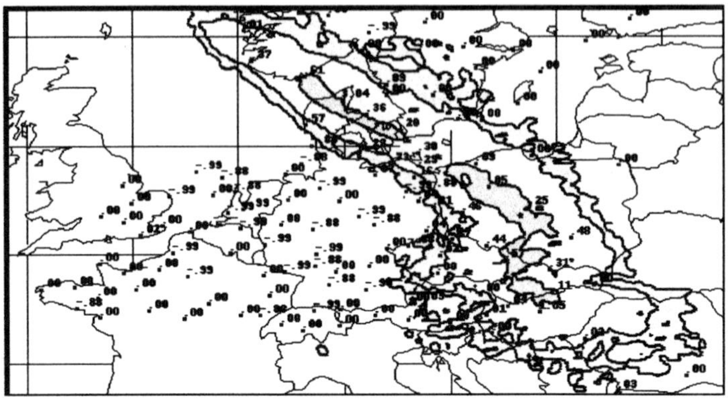

Figure 1 simulated concentrations for the October ETEX release, valid on the 26 00 UTC. Contours are at .01, 0.1 and 1.0 ng m^{-3}.

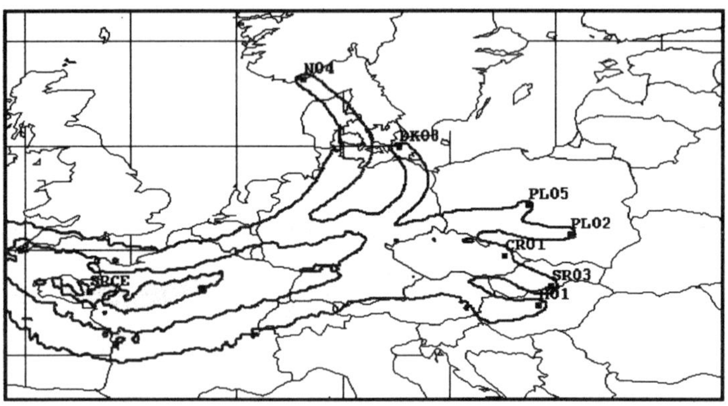

Figure 2 Composite of time integrated touchdown fields for a set of 7 stations which were measuring concentrations on Oct. 26 00UTC. The model was run backward for 60 hours.

REFERENCES

Côté, J., Gravel, S. , Méthot, A., Patoine, A., Roch, M., and Staniforth, A., 1998, The operational CMC-MRB global multiscale environmental (GEM) model, *Mon. Wea. Rev.*, 126 :1373.

Delage, Y., 1988, A parameterization of the stable atmospheric boundary layer, *Bound.-Layer Meterol.*, 43 :365.

Flesch, T.K., Wilson, and D.J., Yee, E., 1995, Backward-time Lagrangian stochastic dispersion models and their application to estimate gaseous emissions, *J. App. Meteorol.*, 34 :1320.

Mailhot, J. and Benoit, R., 1982, A finite-element model of the atmospheric boundary laer suitable for use with numerical weather prediction models, *J. Atmos. Sci.*, 39 :2249.

van Dop, H., and Nodop, K., 1998, ETEX, a European tracer experiment, *Atmos. Env.*, 32 :4089

DISPERSION OF POLLUTANTS UNDER TRAFFIC INDUCED FLOW AND TURBULENCE IN TWO STREET CANYONS FORMING AN INTERSECTION

Jiri Pospisil, Jaroslav Katolicky and Miroslav Jicha

Brno University of Technology, Faculty of Mechanical Engineering,
Institute of Power Engineering, Technicka 2, 61669 Brno, Czech republic

INTRODUCTION

To correctly predict pollutant dispersion around traffic constructions like road tunnels, large city crossroads and/or street canyons, relations between sources and receptors must be established. In this procedure, traffic plays a significant role. Moving vehicles enhance both micro- and large-scale mixing processes in the environment by inducing additional turbulence and partly by entraining masses of air in the direction of vehicle movement. In this study authors focus on traffic and its impact on the pollutants dispersion in two street canyons that form a perpendicular intersection. A model based on Eulerian - Lagrangian approach to moving objects has been developed (Jicha et al., 2000) and integrated into a commercial CFD code StarCD.

DESCRIPTION OF WORK AND RESULTS

The intersection is formed with two street canyons intersecting perpendicularly. Each street has the same aspect ratio (width to the height) of 1.27. In both street canyons two way traffic was set in total four traffic lanes. In the N-S street canyon, traffic with the speed of 50 km/hour and with three different rates was considered namely 360, 720 and 1440 cars/hour/lane. In the other street that forms the intersection, stand-by traffic was assumed only with sources of emissions. Three different wind speeds, namely 4, 7 and 12 m/s at the altitude of 300 m were considered. The direction of wind was assumed in 9 angles covering the total semicircle of 180°. Concentrations are calculated as a passive scalar representing smoke and are normalized with the smoke yield rate generated by one car.

Two of side walls of the solution domain, namely west and south were assigned inlet boundary conditions with specified inlet wind speed, the opposite side walls were assigned pressure boundaries. Inlets and outlets associated with street canyon cross sections were assigned cyclic boundary conditions meaning that the intersection is a part of an infinite structure of the identical crossroads. This also means that concentrations are entrained into the N-S street canyon by cars coming from the south.

In the Figure 1 there are results for the traffic rate of 360 cars/hour/lane and wind speed of 4 m/s. Concentration field is shown at the level of 2m above the ground. From the results it can be seen that due to cyclic boundaries the highest level of concentrations is always in the street into which the wind brings pollutants from "cyclic neighbor". Thus the maximum concentrations are in the N-S canyon for the wind coming from the south and/or north (180° and 360°, respectively) and in the W-E canyon for the wind coming from the west and/or east (270° and 90°, respectively). The lowest concentrations are in the street canyons, which are subject to cross wind. In this case the cross wind forms a vortex in the cross street direction that helps ventilating the street and carries pollutants upwards and away from the street.

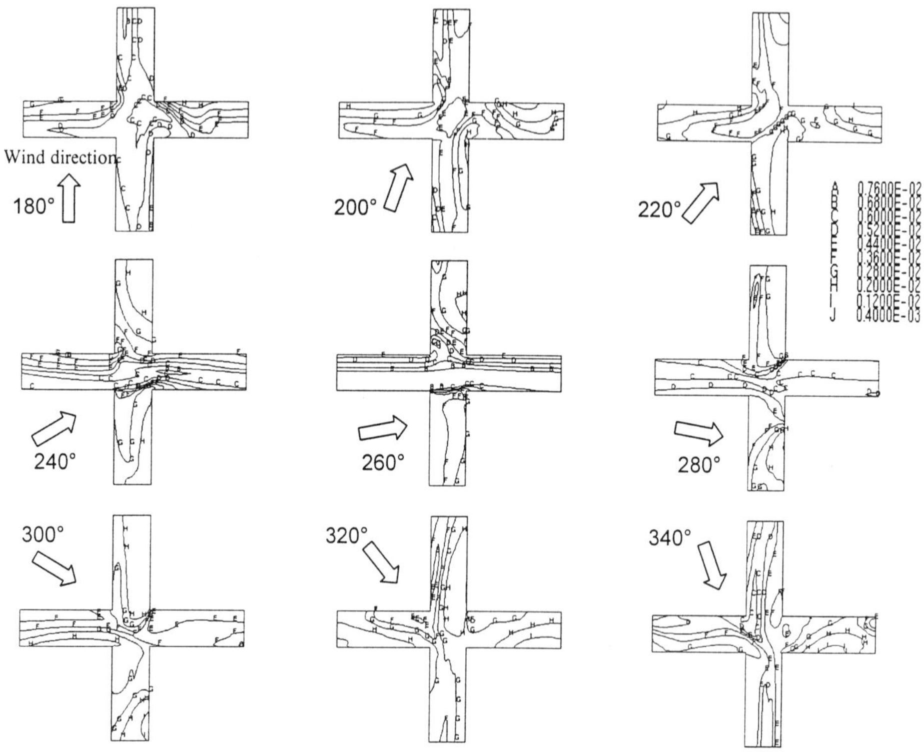

Figure 1. Concentration field at the level of 2m above the ground for traffic rate of 360 cars/hour/lane and wind speed of 4 m/s

ACKNOWLEDGMENT

This work is a part of Eurotrac-2 subproject SATURN and was financially supported by the Ministry of Education of the Czech republic under the grant OE32 /EU1489 and by the Brno University of Technology Development Fund 262100001.

REFERENCES

Jicha, M., Pospisil, J., Katolicky, J., 2000, Dispersion of pollutants in street canyon under traffic induced flow and turbulence, Journal of Environmental Monitoring and Assessment (to appear)

Katolicky, J., Jicha, M., 1999, Pollution dispersion in street canyon under traffic induced flow and turbulence, Proc. of 2nd International Conference on Urban Air Quality, paper 69, Madrid, Spain

EXTENSION OF THE FAST SPECTRAL LINCOM MODEL TO FLOW OVER COMPLEX TERRAIN WITH THERMAL STRATIFICATION

F.N. Dunkerley,[1] J. Moreno Santabarbara[1], T. Mikkelsen[1] and I.H. Griffiths[2]

[1]Risø National Laboratory, Denmark
[2]DERA, Porton Down, UK

INTRODUCTION

LINCOM is a fast linearized and spectral wind flow model. It is designed to generate mean wind field predictions rapidly and is in operational use as wind field driver for fast real-time atmospheric dispersion models like RIMPUFF in assessment of emergency response and decision support. The model has successfully been evaluated in situations involving complex terrain and variable surface roughness under neutral conditions. Recently, an extension of the code to handle non-neutral thermal stratification, proposed by Santabarbara et al. (1995), has been implemented. Both the prescribed vertical temperature field and diffusivity of LINCOM-T strongly affect its solutions. Extensive analysis has been to establish a set of operational inputs for the LINCOM-T based on stability, wind speed and terrain conditions.

POTENTIAL TEMPERATURE FIELD

Thermal effects are introduced by specifying a fixed distribution for the potential temperature perturbation from the background (θ). These are then included in the governing equations as a pre-determined buoyancy term and not as a flow variable for which an additional equation would be required. A vertical profile is postulated for θ in terrain following co-ordinates and then transformed into the cartesian frame such that

$$\theta(x,y) = \tau\, e^{-(z-h(x,y))/\lambda} \qquad (1)$$

The form of equation (1) requires that a temperature lengthscale (λ) and temperature perturbation on flat ground on the upwind boundary of the model domain (τ) must be specified. We choose to calculate these parameters using the Monin-Obukhov lengthscale (L_{mo}), as this measure of stability can now often be obtained from meso-scale models or experimental data, and the background wind speed (U) which is used to drive the model.

Scaling arguments are applied separately to stable and unstable conditions to derive suitable approximate equations. Figure 1 shows λ as a function of L_{mo}. The predicted velocity perturbations in fact respond linearly to τ but have a more complex dependence on λ.

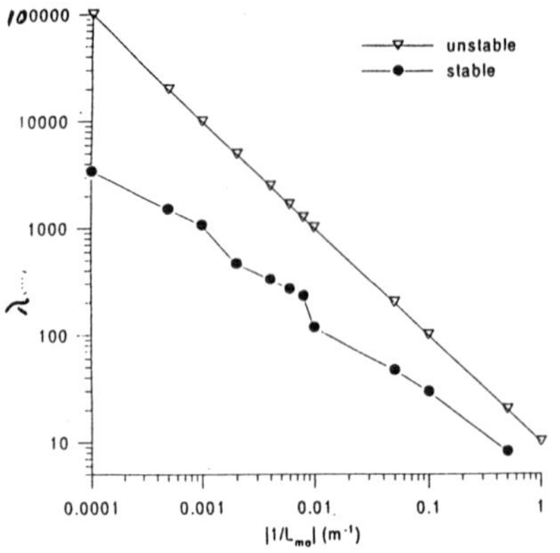

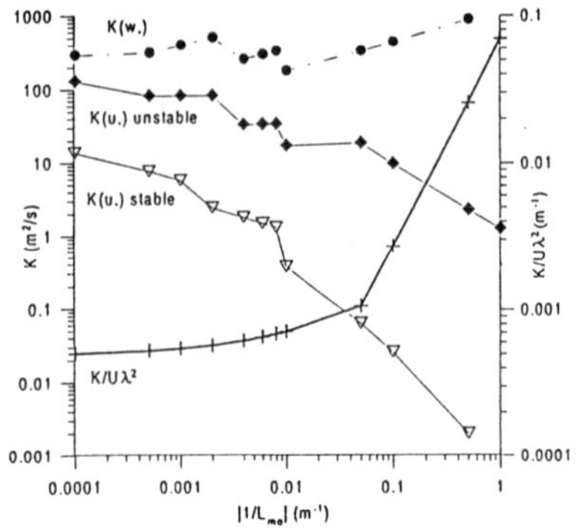

Figure 1.
Diffusivity (K) and $K/U\lambda^2$ as a function of L_{mo}.

Figure 2.
Temperature lengthscale (λ) as a function of L_{mo}.

DIFFUSIVITY MODEL

In the same way as U, the diffusivity (K) enters the model equations as a constant and then must be evaluated in the model solutions at a representative scale height. A number of models for K have been tried in LINCOM-T. Figure 2 shows the relationship between K and $1/(|L_{mo}|)$ for three unstable and one stable form, each defined as the product of a velocity scale and a lengthscale. For $K(w_*)$ the lengthscale of interest is the mixing height; for all other cases we use λ. U has also been allowed to vary with L_{mo}.

It can be seen from Figures 1 and 2 that, under stable conditions, K decreases as $1/L_{mo}$ increases, whereas, for $L_{mo} < 0$, because of the formulation for λ, diffusivity does not increase towards the convective limit as would be expected. However, it is the ratio $K/U\lambda^2$ (solid line), which determines the shape of the perturbed flow field, and this does increase with increasing stability.

RESULTS

Velocity perturbations have been obtained using the above models for potential temperature and diffusivity in LINCOM-T over an idealised Gaussian hill for a range of L_{mo} values. These show the expected flow patterns and reasonable magnitudes are predicted. A limited evaluation has also been undertaken using field data for which L_{mo} values are available (MADONA, Cionco et al199? and WALNUT II, Dunkerley 1999).

REFERENCES

Cionco, R.M.; aufm Kampe, W.; Biltoft, C.; Byers, J.H.; Collins, C.G.; Higgs, T.J.; Hin, A.R.T.; Johansson, P.E.; Jones, C.D.; Jørgensen, H.E.; Kimber, J.F.; Mikkelsen, T.; Nyren, K.; Ride, D.J.; Robson, R.; Santabarbara, J.M.; Streicher, J.; Thykier-Nielsen, S.; Raden H. van; Weber, H., An overview of MADONA: A multinational field study of high-resolution meteorology and diffusion over complex terrain. Bull. Am. Meteorol. Soc. (1999) **80** , 5-19.

Dunkerley, F.N., (1999), Wind flow over complex terrain, *PhD Thesis, UMIST, Manchester.*

Santabarbara,JM; Hernandez,JF; Calbo,J; Mikkelsen,T; Baldasano,JM (1995): Wind flow over moderately complex terrain: A comparison of three recent models and full scale observations. *21st International Technical Meeting on Air Pollution and its Applications. 6-10 November 1995, Baltimore, Maryland, USA.*

PARAMETERIZATION OF WET DEPOSITION OF RADIONUCLIDES

Alexander Baklanov and Jens Havskov Sørensen

Meteorological Research Division, Danish Meteorological Institute (DMI),
Lyngbyvej 100, Copenhagen DK-2100, Denmark

Processes removing radionuclides from the atmosphere, and their interaction with the Earth's surface, are very important for modeling atmospheric transport and consequences of nuclear accidental releases. However, in most long-range air pollution models the parameterizations used for the removal processes do not consider effects of nuclide particle size.

This paper suggests some parameterizations of the removal processes by wet deposition of radionuclides to the surface, and their realization on example of the Danish Emergency Response Model of the Atmosphere (DERMA). DERMA, a 3-D Lagrangian long-range dispersion model using a puff diffusion parameterization (Sørensen, 1998), is developed at DMI for nuclear emergency preparedness purposes. For nuclear emergency response and for post-accidental analysis, new versions of the DERMA model including different complexities of the deposition parameterizations are suggested.

It is shown that the washout coefficient, Λ_w, strongly depends on the particle size (the so-called 'Greenfield gap'), although in most models the washout coefficient does not depend on particle radius. Therefore, a revised formulation of the wet deposition parameterization for particles of different size is suggested. For accidental releases from nuclear power plants, particles with radius $r < 0.5$ μm can play an important role in long-range transport and dose forming. Therefore the new formulation suggested for the washout coefficient, in contrast to formulations in most other long-range transport models, includes the mechanism of Brownian capture, and covers a gap of the washout parameterization for small particles.

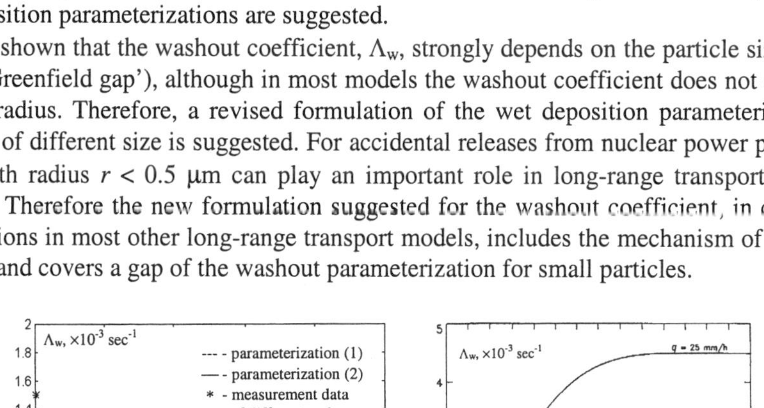

Fig. 1. Dependencies of the washout coefficient, Λ_w, on the particle radius, r, for $q = 5$ mm/h (left) and on the rainrate, q, according to (1) (right).

As the first approximation, we suggest a revised formulation of Näslund & Holmström (1993) for Λ_w (sec^{-1}) avoiding zero values for $r < 0.5$ µm:

$$\begin{aligned} &\Lambda_w(r,q) = a_0\, q^{0.79}, && r < 1.4\ \mu\text{m}, \\ &\Lambda_w(r,q) = (b_0 + b_1 r + b_2 r^2 + b_3 r^3)\, f(q), && 1.4\ \mu\text{m} \le r < 10\ \mu\text{m}, \qquad (1) \\ &\Lambda_w(r,q) = f(q), && r \ge 10\ \mu\text{m}, \end{aligned}$$

where $f(q) = a_1 q + a_2 q^2$, r - particle radius (m), q - rainrate (mm/h), $a_0 = 8.4\cdot10^{-5}$, $a_1 = 2.7\cdot10^{-4}$, $a_2 = -3.618\cdot10^{-6}$, $b_0 = -0.1483$, $b_1 = 322013.3$, $b_2 = -3.0062\cdot10^{10}$, $b_3 = 9.34458\cdot10^{14}$.

For a post-accidental analysis version of DERMA, a more sophisticated formulation for Λ_w, including the mechanism of Brownian capture according to Slinn's formulation, is used:

$$\Lambda_w = \frac{q}{2a_m}\left[\frac{4}{Pe}\left(1 + 0.4\,\mathrm{Re}^{1/2}\, Sc^{1/3}\right) + \frac{4r}{a_m}\left(\frac{r}{a_m} + \frac{(1 + 2\mu_w r/\mu_a a_m)}{\left(1 + \mathrm{Re}^{-1/2}\, \mu_w r / \mu_a a_m\right)}\right) + \left(\frac{St - St_*}{St - St_* + 2/3}\right)^{\frac{3}{2}}\right], \qquad (2)$$

where a_m is the volume-mean raindrop radius, Pe, Re, Sc, St, St_* are the Peclet, Reynolds, Schmidt, Stokes, and critical Stokes numbers, respectively, and μ_w and μ_a are the dynamic viscosities of water and air (for definitions and references see Baklanov (1999)). This parameterization needs to be verified versus empirical data.

The rainout coefficient, Λ_r, for the dynamic precipitation is set equal to the washout coefficient. For the convective precipitation, Λ_r is greater and does not strongly depend on particle size, so the Marion formulations for Λ_r and for snow scavenging are used. The dry deposition velocity, v_d, is parameterized by the resistance approach. For large particles ($r > 1$ µm), the gravitation settling effect is included in v_d through the gravitation settling velocity, v_g, which is simulated by a combination of Stokes law ($r < 1.75$ µm) with the Cunningham correction for small particles ($r < 0.5$ µm) and an iterative procedure for $r > 1.75$ µm.

Earlier comparisons of simulations by the DERMA model versus the ETEX experiment involving passive tracers gave very good results (Graziani et al., 1998). In order to verify the deposition parameterizations, DERMA simulations for the Chernobyl accident and the Algeciras accidental ^{137}Cs release in Spain on 30 May,1998 are made taking into account the deposition processes by different approaches. Comparisons of simulation results for Algeciras with measurement data obtained during nine days after the release are presented in Figure 2.

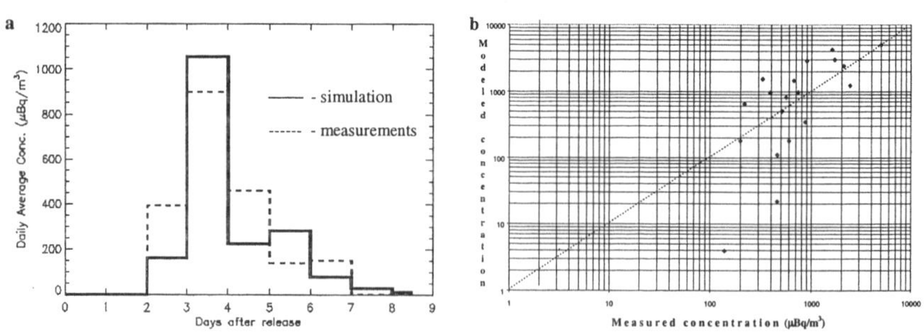

Fig. 2. Daily average ^{137}Cs concentration (µBq/m^3) measured at JRC, Ispra, Italy (a), and at other European stations (b) in comparison with DERMA simulations for the Algeciras accident in Spain on 30 May, 1998.

References

Baklanov, A. 1999: Parameterisation of the deposition processes and radioactive decay: A review and some preliminary results by the DERMA model. DMI Scientific Report, No. 99-4, Copenhagen, Denmark.

Graziani, G., W. Klug & S. Mosca: Editors. 1998: Real-time long-range dispersion model evaluation of the ETEX first release. Joint Research Centre, EU, Luxemburg, EUR 17754 EN.

Näslund, E. and Holmström, H. 1993: Inclusion of a three-dimensional washout coefficient in ADPIC. Report UCRL -ID-114054, Lawrence Livermore National Laboratory, California, USA.

Sørensen, J.H., 1998: Sensitivity of the DERMA Long-Range Gaussian Dispersion Model to Meteorological Input and Diffusion Parameters, *Atmos. Environ.* *32*, 4195-4206.

THE FINNISH OPERATIONAL EMERGENCY MODEL FRAMEWORK

Ilkka.Valkama[1], Pilvi Siljamo[1] and Mikko Ilvonen[2]

[1] Finnish Meteorological Institute
P.O. Box 503
Vuorikatu 24, FIN-00100 Helsinki, Finland
[2] VTT Energy, Nuclear Energy
P.O. Box 1604
FIN-02044 VTT, Finland

INTRODUCTION

This paper presents the current Finnish medium/long range dispersion and dose assessment model framework for emergency applications, called the SILAM Modelling Framework. SILAM is a joint project of Finnish Meteorological Institute (FMI, meteorological assessment) and Technical Research Centre of Finland (VTT, radiological assessment). FMI is a national organisation responsible for the dispersion forecasts during nuclear and chemical accidents as well as other emergencies requiring response actions. For these purposes FMI maintains several short range, regional and long range dispersion models. The SILAM framework system has been operationally tested in several national and international emergency exercises. A dedicated graphical user interface (GUI) is under development.
While primarily designed for real-time assessment of accidents at nuclear installations, SILAM is under modification to be able to handle nuclear explosions, as well. This requires a new 3D description of the source term, based on realistic particle size distributions and nuclide inventories, to be defined.

DOSE ASSESSENMET

SILAM model output can contain instantenous [Bq/m^3] or time-integrated air concentration [Bq s/m^3], deposition [Bq/m^2] fields of dispersed radioactive nuclides, external dose rates [Sv/s] and external and internal radiation doses [Sv]. Radioactive chain decay with generation of daughter nuclides can be calculated for every result point (time) with two optional models : a fast model for chains of three nuclides, using analytical solutions, and a more comprehensive model for arbitrary decay chains having any number of nuclides. Decay chains are described by a group of differential equations of first order. Currently 135 different chains are readily defined, the longest of them having 37 nuclides. The data base contains 496 nuclides. Migration of deposited nuclides is also included.

External dose pathways include direct irradiation from cloud and fallout (gamma and beta), internal pathways include inhaled and ingested radionuclides. In the external pathways, doses and dose rates are directly calculated for any result time point with freely adjustable accuracy of dose integrals, whereas in the internal pathways tabulated dose conversion factors are used and results interpolated. Direct external gamma dose from the cloud is calculated by a fast and accurate method that describes the vertical profiles of concentration in successive layers of the atmosphere. In most pathways, up to 23 human target organs can be considered. For most nuclides the number of dose and dose rate conversion factors in SILAM nuclide data base is 80.

DISPERSION MODELS

The SILAM framework can support various types of dispersion models; Gaussian, Lagrangian, Eulerian etc. The main model is a meso-scale Lagrangian multi-particle dispersion model. The particles, each carrying a part of the total radioactivity and mass, are advected by three-dimensional large scale wind field and dispersed by small scale turbulent eddies. The turbulent dispersion is simulated by random-walk techniques, where the required turbulent velocity statistics, Lagrangian time scales, energy dissipation rates etc. are based on the atmospheric boundary layer similarity theory. The dry deposition is computed by the resistance analogy approach and the wet removal by the scavenging coefficient approach, with sub-cloud and in-cloud scavenging included. All required meteorological data is taken from FMI's current numerical weather prediction model (HIRLAM NWP).

An automated SILAM routine has recently been implemented in the HIRLAM NWP routines. Twice a day (at 00 and 12 UTC) fresh 48 hour simulations for the Finnish nuclear power plants and for 3-5 nearest foreign ones are being run. The output (concentration fields) is written in Vis5D-format, and resulting animations are made available to duty forecasters at the FMI as well as to the Finnish Nuclear Safety Authority (STUK). The output can also contain Vis5D-visualizations of any meteorological quantity available, or derivable, from HIRLAM, such as boundary layer height or cloud water content.

The full description of random-walk turbulent dispersion is very time-consuming task. For faster assessments the SILAM framework includes a 3D particle trajectory model. Our earlier experience with trajectory models has shown that relatively few trajectories can adequatelty describe even as compex an release as the Chernobyl accident (Valkama et al., 1995; Pöllanen et al., 1996). This approach also allows a standard Gaussian puff formulation, similar to the Puff-Particle Model (de Haan and Rotach, 1998) to be used. Recent simulations with the SILAM model (e.g. the Algeciras case in 1998) suggest that on the average less than 200 trajectories are required to describe the same distribution of puff centers as a full scale random-walk simulation of 1000-2000 particles.

REFERENCES

P. De Haan and M. W. Rotach, A novel approach to atmospheric dispersion modelling: the puff-particle model *Q. J. R. Meteorol. Soc.*, vol. 124, pp. 2771-2792 (1998)

Pöllänen R., I. Valkama & H. Toivonen. Transport of radioactive particles from the Chernobyl accident. *Atmospheric Environment*, vol 31, no 21, p 3575 – 3590 (1996)

Valkama I., M. Salonoja, H. Toivonen, J. Lahtinen & R. Pöllänen, 1995. Transport of radioactive gases and particles in the Chernobyl accident :Comparison of environmental measurements and dispersion calculations. *Proceedings of The International Symposium on Environmental Impact of Radioactive Releases*, Vienna, Austria, IAEA-SM-339/69 (1995).

COMBINATION OF DIFFERENT PROCEDURES FOR SURFACE OZONE FORECAST

E. Reimer[1)], G. Wiegand[2)], J. Flemming[1)], M. Dlabka[1)]

[1)] Freie Universität Berlin
Institut für Meteorologie
Carl-Heinrich-Becker-Weg 6-10
12165 Berlin
Germany

[2)] IVU Gesellschaft für Informatik, Verkehrs- und
Umweltplanung mbH, Freiburg

INTRODUCTION

The main object of this scientific project was the development and comparison of different procedures to forecast daily maxima and a three day trend of surface ozone over the Federal Republic of Germany. Especially ozone concentrations higher than the threshold values 180 or 240• g are of particular interest.

PROCEDURE

Investigations were started to develop, evaluate and compare different procedures with regard to their applicability. The basic idea is that a weighted combination of specific approaches might be more successful than one scheme alone.

The approaches are:

- the Eulerian chemical transport model REM3 with three complex chemical codes and with a grid size of about 30km. The model area covers West and Central Europe and uses the Phoxa emission database. The meteorological forecast is given by the Europe Model from German Weather Service. REM3 describes the broad scale transport of chemical constituents and the general formation of episodes with high ozone concentrations. The forecast runs up to three days, so giving also trend information.
- statistical analysis of 300 local time series of ozone and several meteorological parameters by combining hierarchical cluster analysis and regression methods.

- alternative description of the local and highly nonlinear processes of ozone generation by fuzzy and neurofuzzy concepts. Such systems can deal with poorly defined parameter distributions to some extent and with qualitative sets of rules determined by fuzzy cluster analysis and expert advice. Different concepts were developed to get rulebases by use of special cluster algorithms. In this context the neurofuzzy system combines controlled or uncontrolled learning procedures with the determination of the fuzzy system rulebases.REM3 forecasts of ozone concentration maxima and their difference from day to day are used within the cluster procedure.
- forecast of Großwetterlagen in combination to the local distributions of surface ozone observations to get an objective coupling between pressure pattern over Middle Europe and local measurements.

All the separate tools mentioned above are practically combined within a realtime system, so that for each region and episode a combination of the different procedures is determined. On the other hand as a minimum one of these procedures might be usable, if data trouble occures within the daily applications.

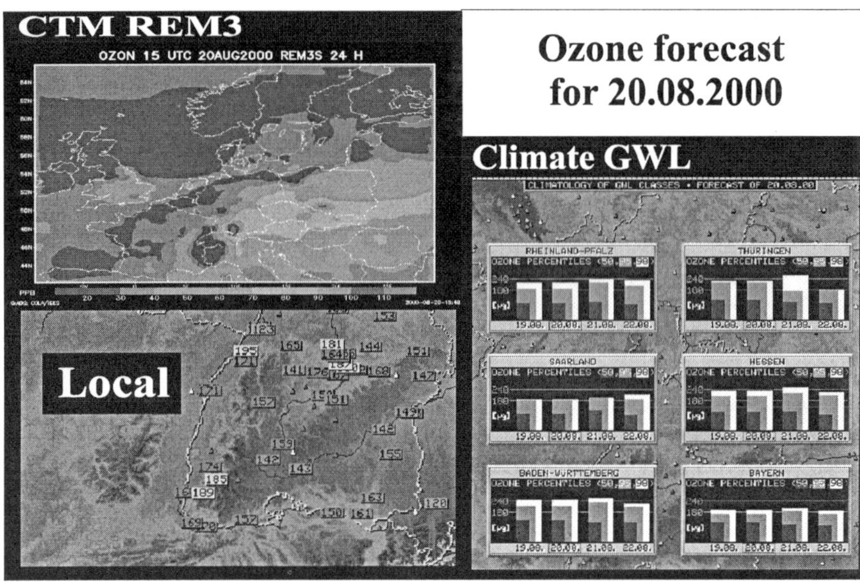

RESULTS AND CONCLUSIONS

The system supports both a better local forecast of low level ozone concentrations with mean absolute errors of 5 to 10 ppm and a better insight into the daily weather formation and chemical processes related to that.
The daily forecasts are presented via internet: http://trumf.fu-berlin.de .

Acknowledgments. The project was funded by German Environmental Agency UBA.

Reimer, E., and G. Wiegand, J. Flemming, W. Enke, K. Berendorf, M. Dlabka, W. Weiss, R. Stern, Erstellung einer Kurzfristprognose für das Smogfrühwarnsystem. Final Report, UBA F&E-Nr. 29543817, FU Berlin

DISPERSION MODELLING WITHIN THE EUROPEAN COMMUNITY'S NEW AIR QUALITY FRAMEWORK DIRECTIVE - THE GERMAN APPROACH

A. Graff[1] and R. Stern[2]

[1]Umweltbundesamt, Postfach 330022, D-14191 Berlin, Germany
[2] Free University Berlin, Institute of Meteorology,
Carl-Heinrich-Becker-Weg 6-10, D-12165 Berlin, Germany

INTRODUCTION

The Member States of the European Community (EC) have to assess air quality under the Council Directive 96/62/EC on ambient air quality assessment and management (framework directive, FWD) throughout their territory and in accordance with the existing resp. upcoming daughter directives. Besides direct measurements, the use of other techniques like air quality modelling is intended.

Among others, the FWD's aim is to assess the ambient air quality in the EC Member States on the basis of common methods and criteria according to Article (Art.) 1. Furthermore Art. 2(4) states that "assessment shall mean any method used to measure, calculate , predict or estimate the level of a pollutant in the ambient air". Thirdly, "The measurements provided for may be supplemented by modelling techniques to provide an adequate level of information on ambient air quality." (Art. 6(2), par. 5), what is also valid for Article 5.

In the context of the framework directive measurements are mandatory for certain criteria. The importance of modelling techniques is stressed the more the ambient air concentration is below a level lower than the limit value (upper assessment threshold) over a representative period. Modelling techniques are considered the only assessment tool for ambient air concentrations below the lower. Thus the EC framework directive recommends/prescribes air quality modelling for the first time in an European regulation explicitly.

DEMANDS ON AIR QUALITY MODELLING

The various requirements concerning air quality assessment depending on the concept of limit values and thresholds shall be covered by a consistent modelling approach including all species addressed in the daughter directives simultaneously.

Air Pollution Modeling and Its Application XIV, Edited by
Gryning and Schiermeier, Kluwer Academic/Plenum Publishers, New York, 2001

The criteria for applying models according to Art. 6 (3) already enclose the demands concerning Art. 5 and Art. 6 (4). Hence, no further differentiation in the modelling approach is necessary, e.g. the technique of 'objective estimation' becomes dispensable.

In zones and agglomerations in which the levels of one or more pollutants are higher than the limit value plus the margin of tolerance, Member States shall take measures to ensure that a plan or programme is prepared or implemented for attaining the limit value within a certain time span (Art. 8). A quantitative analysis of the causes for exceedances in ambient air concentrations requires air quality modelling that takes emissions sources into account explicitly. Air quality models are the only tools to assess the effectiveness of control strategies.

CONCEPT FOR AIR QUALITY MODELLING

The situation of ambient air concentration is related to various spatial scales depending on the species under consideration. For example, whereas elevated ozone concentrations occur on greater areas, high concentrations of NO_2 are measured in street canyons predominantly. Thus, an appropriate modelling approach has to be applied to every spatial scale in question. The spatial resolution is scale dependent. The models are linked by (one way) nesting. Thus, an area covering picture of ambient air concentrations according to the requirements of the framework directive will be produced for every spatial scale. The scales to be considered and the according models applied are:

The *regional scale* (European wide) for calculating the national background concentrations influenced by advection with **R**egional **E**ulerian **M**odel **3** (REM3) has a spatial resolution of ca. ¼° lat. by ½°lon. (30 x 30 km^2).

These concentrations simultaneously serve as background for the zones to be studied by downscaling to *urban scale* resolution of 2 x 2 km^2 at least with CALGRID.

Urban scale modelling interacts with the *micro scale* that is resolved with **M**ICRO-**C**AL**G**RID (MCG) for city quarters or with the **C**anyon **P**lume **B**ox Model (CPB) for street canyons respectively as a link to 'hot spot' related air quality problems.

Measurements will be introduced into this air quality assessment with data available routinely for validating the modelling results and for data assimilation.

The development of an integrated modelling system that fulfils all requirements of the FWD is under way in an ongoing R&D project No.299 43 246 under the research plan of the Ministry of Environment of Germany and commissioned by German Umweltbundesamt and is carried out at the Free University of Berlin. In this context the integrated models will be upgraded to the latest state-of-the-art concerning chemistry, aerosol modelling and numerical methods. The emissions database is going to be adjusted according to the current available information. The meteorological database will be improved with the aim to derive time dependent landuse data.

EXPERIMENTAL INVESTIGATION OF THE TRACE GAS COMPOSITION OF A CITY PLUME

M. Möllmann-Coers[1], K. Mannschreck[2], Th. Schmitz[2] and D. Klemp[2]

[1] Research Centre Jülich, Department for Safety and Radiation Protection (ASS)
P.O. Box 19 13
52425 Jülich, Germany

[2] Research Centre Jülich, Institute for Chemistry of the Polluted Atmosphere (ICG-2)
P.O. Box 19 13
52425 Jülich, Germany

INTRODUCTION

The uncertainties in the results of chemistry transport models (CTM) e.g. for the regional distribution of ozone and photooxidants during photosmog episodes are not only caused by the incompleteness of model physics or by shortcomings of the numerical methods applied. The uncertainties of the source terms of the trace gases also influence the CTM model results. Reliable source data with a high resolution in time and space based on measurements are normally not available. Therefore, the necessary data are usually provided by numerical emission inventories, which are based on estimated emission factors from traffic, industry, and residence heating. The assumptions of these emission inventories, however, have to be verified to provide a basis for the estimation of the magnitude of the uncertainties in the results of the CTM models with respect to the emission input data.
The verification of numerical emission inventories requires independently determined emission data. From these data the quality of the numerically predicted inventories of the known trace gases can be evaluated by comparison. Beside the absolute emission rates the contribution of different sources can be proved.

CONCEPT AND EXPERIMENTAL SITE

Therefore in 1998 field campaigns were designed and conducted in Augsburg, Germany, a medium-sized city (area: 150 km^2, 270 000 inhabitants). Augsburg is sited in a rural environment without major industrial sources in the southwest. The two five-week campaigns in March and October were focused on the investigation of the city plume under southwesterly conditions. Long-term measurements and the source tracer ratio method were applied to investigate the trace gas composition of the city plume and the absolute emission rates of trace gases in Augsburg. For the long-term measurements the ICG-2 laboratory (Pätz at al., 2000) was sited northeast of Augsburg.
The tracer technique is described by Zeuner and Möllmann-Coers (1994). The source tracer ratio

method uses the measured concentration ratio of the tracer to carbon monoxide (CO) downwind of the city and the known tracer emission rate in the city to calculate the city's total CO emission rate. Furthermore, using the concentration ratios of VOC to CO, and NO_x to CO, which result from the ICG-2 laboratory measurements, the absolute emission rates of VOC and NO_x can be calculated from the total CO emission rate determined experimentally. The tracer experiments were carried out in the day time (12 a.m.-3 p.m. local time) under southwest conditions.

RESULTS

A total of six tracer experiments were carried out, two in March and four in October. The experiments no. 2, 3, 5, and 6 could be considered in the analysis. Table 1 shows the results.

Table 1. Absolute emission rates

Exp. Date	1998-03-27	1998-10-10	1998-10-22	1998-10-23
CO Emission Rate [t/h]	3	1.8	1	0.9
NO_x Emission Rate [t/h]	0.35	0.25	0.24	0.22
VOC Emission Rate [t/h]	0.30	0.20	0.17	0.08

The results of the tracer experiments in October are in agreement with the results of mass balance experiments conducted simultaneously (Slemr, 1999; Corsmeier, 1999).
The long-term measurements reveal significant differences between the mean $_{VOCi}$/VOC_{sum} ratios in March and October. The percentages of C2- and C3-VOC are higher in March. These compounds are partly emitted from residental heating (Kühwein, 2000) which played a greater role in March than in October due to lower temperatures. The percentages of most alkanes, toluene and ethyltoluene are higher in October. This could be due to higher solvent evaporation. Furthermore, the comparison of the VOC distribution between the Augsburg measurements and typical road traffic measurements shows good agreement for most trace gases. The composition of measured VOC of the city plume is dominated by traffic emissions.

SUMMARY AND CONCLUSIONS

The combination of both methods, the source tracer ratio and the long-term measurements, permits a detailed qualitative and quantitative analysis of the emissions of a city.
Total CO and NO_x emission rates tend to be higher in spring than in autumn.
The relative contributions of light hydrocarbons are higher in March than in October, the behavior of the heavier fraction of VOC is vice versa.
The measured VOC split in March and October shows that the city plume of Augsburg is dominated by traffic emissions.
The results provide a good data base for a detailed analysis of modeled emission inventories.

REFERENCES

Corsmeier U., 1999, Karlsruhe Research Centre, Private Communication

Kühlwein J., 2000, IER Stuttgart, Private Communication

Mannschreck K. et al., VOC – Reference List – A Database for Atmospheric Research, Conference Proceedings: *Transport and Transformation of Pollutants in the Troposphere*, Garmisch-Partenkirchen, 27.3.-31.3.2000 VOC Database available at http://www.fz-juelich.de/icg/icg2/forschung/VOCREF/

Pätz W. et al., 2000, Measurements of Trace Gases and Photolysis Frequencies During SLOPE96 and a Coarse Estimate of the Local OH Concentration From HNO_3 Formation, *J. Geophys. Res.* ***105***, NO. D1, pp. 1563-1583

Slemr F., 1999; IFU Garmisch-Partenkirchen, Private Communication

Zeuner G. and Möllmann-Coers M., 1994, Dispersion Experiments in Complex Terrain, *Meteorol. Zeitschrift, N.F.* **3**, pp. 107-110

UTILIZATION OF NUMERICAL MODELS AS A TOOL FOR ANALYZING OZONE PRODUCTION FROM TRANSPORTATION SOURCES

D. O. Ranmar[1], M. Luria[2], J. Kaplan[3] and Y. Mahrer[1]

[1]The Department of Soil & Water Sciences, The Hebrew University of Jerusalem, Rehovot 76100 Israel.
[2]Environmental Science Division, the School Applied Sciences and Technology, the Hebrew University of Jerusalem.
[3]The Department of Geography, the Hebrew University of Jerusalem.

Air pollution, with ozone generating processes, is among the most significant ecological impacts of the modern age network transportation. The necessity for protection and preservation of our ecosystem linked to the ongoing need for a robust and functional traffic infrastructure calls for the development and implementation of a modeling system targeting the various factors involved in transportation-to-ozone formation linkage. These include atmospheric dynamics, geographic location of traffic networks, traffic flow, traffic-related pollution emission, and overall development strategies, which are intimately linked to transportation development.

The concept of the work presented is based on the utilization and integration of interdisciplinary models covering, the dynamics of the atmosphere, the characteristics of traffic flow, and the nature of the pollution emission, dispersion and chemistry. The Regional Atmospheric Modeling System (RAMS) was the multipurpose 3D numerical prediction model used to simulate the mesoscale atmospheric variables, which in turn initiated a 3D transport/diffusion mode (TDM). The pollutants' sources location for the TDM is based on the EMME/2 transportation model, which provides the links or region level traffic flows by speed, time and location. The EMME/2 transportation model provided the data of the present (year 1995) and future (year 2020) transportation infrastructure. These data coupled with emission factors obtained from road tunnel measurements, provided the NO_x and the VOC transportation oriented emission factors. The models were applied to a specified coastal area of Israel and the West Bank (Fig. 1), and aimed at understanding the impact of the transportation-system on the spatial evolution and distribution of ozone. Measurements at different sites in central Israel showed a trend of increasing ozone levels with increasing inland distance (from the sea shore eastward due to NW wind) and pointed to the Tel-Aviv metropolis as the origin for the "ingredients" feeding the photochemical transformations. The Gaza strip was addressed as the Palestinian "center of gravity" for the ozone precursors, being their largest coastal metropolitan. The simulation results were compared with summer time flight and ground measurements, where high ozone levels were recorded. The sequential integrated models accurately simulated the spatial and temporal overlap between the primary transportation–oriented pollutants, NO_x and VOC, and the measured secondary pollutant, ozone, indicating an inland transport due to West-NorthWest winds (Fig. 2). The presented work will discuss the application/utilization of integrated interdisciplinary models as well as the implication of the results obtained.

Figure 1. The simulation area, indicating the Tel Aviv metropolis and Gaza-Strip traffic sources, and traffic roads.

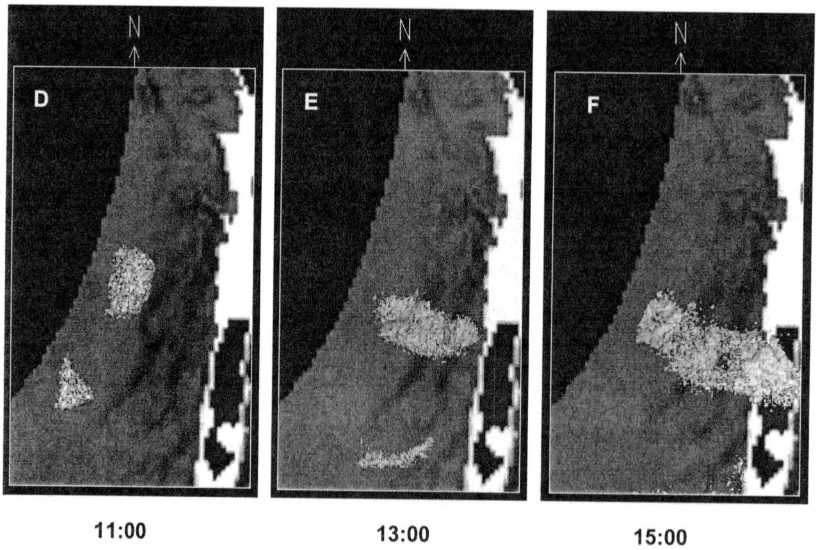

Figure 2. Top view of the particles location originating from the Tel-Aviv metropolitan and the Gaza-Strip transportation sources at different hours (MLT).

PHOTOCHEMICAL MODELING OF A WINTERTIME EPISODE AND HONO'S ROLE

F. Ghezzi[1], G. Maffeis[2], A. Febo[3], M. Tamponi[4]

[1] Via Turati, 11, 20091 Bresso (Mi) [fabio.ghezzi@effeesse.it]
[2] TerrAria s.r.l., Milano
[3] C.N.R., Istituto Inquinamento Atmosferico, Roma
[4] Enviroware s.r.l., Agrate Br. (Milano)

INTRODUCTION

Some field campaigns, among which the one of March 1994 in Milan (FORMONA project, Febo et al., 1999) have shown the impact of wintertime photochemical smog phenomena in urban areas. Ozone and formaldehyde winter concentrations alone, at reduced temperature and radiation, do not seem to be able to explain NO_2 strong increase. During wintertime in fact, NO_2 110 ppb threshold overcoming is not a rare event, and for instance during FORMONA campaign, Verziere NO_2 peak was over 130ppb.

For these reasons one of highest O_x episodes during FORMONA campaign, 5th and 8th of March 1994, has been simulated with the model chain CALMET-CALGRID (Ghezzi, 1999).

DISCUSSION AND RESULTS

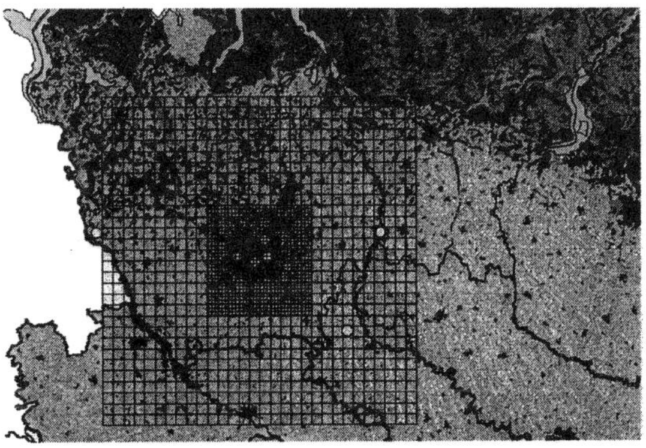

Figure 1: Milan area and the two nested domains.

Milan and its metropolitan area include about 30 towns with a total extent of about 500 km^2 and more than 2.000.000 inhabitants. In this area anthropogenic emissions clearly prevail (e.g. 20.000 NO_x tons/year) (Maffeis, 1999).

O_x was used to define the oxidation state of the atmosphere independently of the season. In wintertime in fact, O_x often presents a significant morning increase, mainly in the form of NO_2. In 1993-1998, NO_2 annual mean values, in Milan, always exceed 40 $\mu g/m^3$, NO_2 98° percentile 200 $\mu g/m^3$.

A one-way nesting with two domains (respectively of 22.5x22.5 km and 67.5x67.5 km, both grids with 900 cells) centered on Milan urban area has been used to reach the needed high spatial resolution (Fig 1). The effect of emissions, mixing height and vertical diffusivity parameterization uncertainties has been evaluated on primary pollutants daily trend. Some of these criticisms have been solved obtaining a better description of O_3 and NO_2. The final model results present a good agreement with experimental data for both the secondary pollutants (Fig

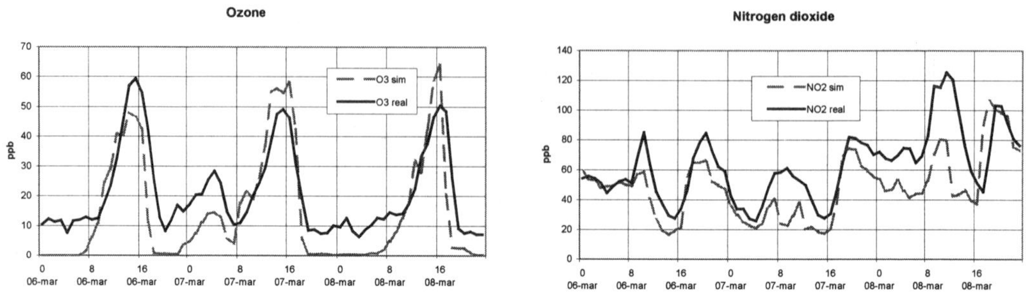

Figure 2: Measured and simulated trends.

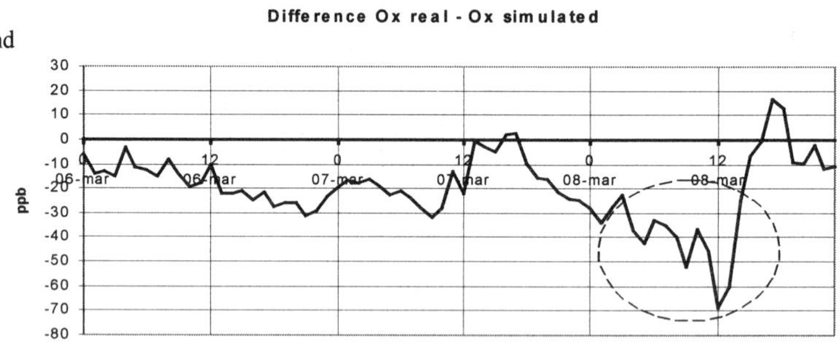

2). R values between observed and simulated trends are equal to 0.87 for ozone and 0.71 for NO_2. The last one, however, decreases to 0.32 considering only the 8th of March: simulated Ox and experimental data reaches this day the "peak difference" of about 70 ppb (Fig 2). SAPRC90 chemical mechanism seems to be not able to reproduce the huge and sudden increase of O_x (11 a.m. peak value). Probably an additional source of OH radicals is needed to better explain the intense wintertime photochemical episodes.

CONCLUSIONS

A model chain was implemented to study photochemical wintertime smog. Milan urban area application gave good results for secondary pollutants O_3 and NO_2, showing the advantages of nesting technique on a small domain and some criticisms linked to hmix parameterization. The possible effect of the heterogeneous creation of HONO is at the moment under investigation.

REFERENCES

Febo A., Perrino C., Allegrini I., 1996, Measurement of nitrous acid in Milan, Italy, by doas and diffusion denuders, *Atmos. Environ.*, 30, 3599-3609.

Febo A., 1999, Formation and occurrence of nitrous acid in the atmosphere (FORMONA), Final Report.

Ghezzi F., 1999, Studio e modellazione dell'inquinamento fotochimico invernale: applicazione all'area metropolitana Milanese, Graduation Thesys.

Maffeis G., Longoni M., De Martini A., Tamponi M., 1999 Emissions estimate of ozone precursors during PIPAPO Campaign, in Photochemical Oxidants and Aerosols in Lombardy Region, Milan, Italy, June 1999.

LAGRANGIAN PARTICLE SIMULATION OF AN EPA WIND TUNNEL TRACER EXPERIMENT IN A SCHEMATIC TWO-DIMENSIONAL VALLEY

E. Ferrero[1], D. Anfossi[2], G. Tinarelli[3], S. Trini Castelli[1,2],

[1] Univ. del Piemonte Orientale, Dip. Scienze e Tec. Av., Alessandria, Italy
[2] Istituto di Cosmogeofisica, CNR, Torino, Italy
[3] ENEL SRI, Area Ambiente, Segrate (MI), Italy

INTRODUCTION

Recently, our group developed a complete 3-D model system aimed to simulate atmospheric pollutant dispersion in complex terrain. The system includes the RAMS model (Pielke et al. 1992) providing the flow field, the Lagrangian Stochastic Models SPRAY (Tinarelli et al., in press; Ferrero and Anfossi 1998) computing the dispersion, and the interface code MIRS (Trini Castelli and Anfossi, 1997). Turbulent quantities can be calculated in MIRS by parameterisations from literature or by the turbulent kinetic energy (E) field, directly extracted by RAMS meteorological output.This model system is applied to the wind tunnel EPA-RUSVAL tracer experiment (Khurshudyan et al. 1990) in which a neutral flow is reproduced on a 2-D cross-wind valley and the source is placed near the bottom of the valley. In the version of RAMS here used, we have replaced the Mellor-Yamada 2.5 turbulence closure with both E-l and E-ε closures, where *l* is the length scale and ε the mean rate of E dissipation. In a previous work (Trini Castelli et al. 1999) we showed that these last reproduce the flow and the turbulent fields in this wind tunnel experiment better than the Mellor-Yamada 2.5 model, originally implemented in RAMS. The influence of these different turbulent closures on the dispersion process is tested by comparing the SPRAY simulations with the observed tracer concentration fields. We focus our attention on the comparison between the simulations carried out using the Hanna (1982) parameterisation and the turbulent field supplied by the E-l or E-ε model.

The turbulent quantities needed as input to SPRAY are: the three standard deviations of the wind fluctuation components (σ_i) and the three Lagrangian time scale (T_{Li}). The turbulent model gives the σ_i, whilst the (T_{Li}) are obtained from the relationships

$$T_{L_i} = \left(K_{m_i}\right) / \left(\sigma_i^2\right) \qquad (1)$$

where K_m is the diffusion coefficient of momentum, in the case of E-l model, or from

$$T_{L_i} = \left(2\sigma_i^2\right) / \left(c_0 \varepsilon\right) \qquad (2)$$

Air Pollution Modeling and Its Application XIV, Edited by
Gryning and Schiermeier, Kluwer Academic/Plenum Publishers, New York, 2001

in the case of E-ε model, where C_0 is chosen according to Anfossi et al. 2000.

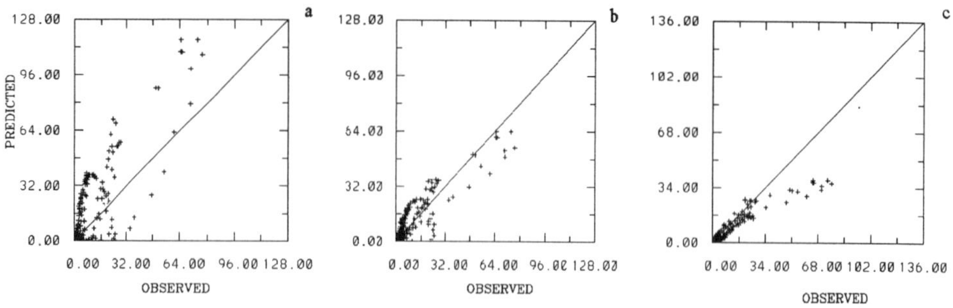

RESULTS

The scatter plots, referring to three out of the simulations performed by our modelling system, are shown in the Figures. In the first case (a), the turbulence model used is E-l, σ_i and T_{Li} are prescribed following Hanna (1982) parameterisation. In the second one (b), the turbulence model used is E-l, σ_i are those generated by this model and T_{Li} are calculated using (1). The last simulation (c) is carried out using E-ε model and σ_i generated by itself, T_{Li} are obtained from (2).

It can be seen that the second and third simulations are preferable. In particular, in the second one the high concentration values are better reproduced whereas the third one shows a rather good fit of the low and medium values.

REFERENCES

Anfossi D., Degrazia G., Ferrero E., Gryning S.E., Morselli M.G:, Trini Castelli S., in press, Estimation of the Lagrangian structure function constant C0 from surface layer wind data, *Boundary Layer Meteorology*,

Ferrero E. and Anfossi D., 199, Sensitivity analysis of Lagrangian stochastic models for CBL with different PDF's and turbulence parameterizations. *Air Pollution Modelling and its Applications XII*, Edited by S-E. Gryning and N. Chaumerliac, Plenum Press, 673-680.

Hanna S.R., 1982, Application in air pollution modeling. in F.T.M. Nieuwstadt and H. van Dop, (Editors) *Atmospheric Turbulence and Air Pollution Modelling*, Beidel, Dordrecht, Ch. 7.

Khurshudyan L.H., Snyder W.H., Nekrasov I.V., Lawson R.E., Thompson R.S, Schiermeier F.A., 1990, Flow and Dispersion of Pollutants within Two-Dimensional Valleys, *Summary Report on Joint Soviet-American Study, EPA REPORT* No. 600/3-90/025

Pielke R.A., W.R. Cotton, R.L. Walko, C.J.Tremback, W.A.Lyons, L.D.Grasso, M.E. Nicholls, M.D. Moran, D.A. Wesley, T.J. Lee and J.H. Copeland, 1992, A Comprehensive Meteorological Modeling System - RAMS, *Meteorology and Atmospheric Physics*, 49, 69-91

Tinarelli G., Anfossi D., Bider M., Ferrero E. and Trini Castelli S., in press., A new high performance version of the Lagrangian particle dispersion model SPRAY, some case studies *Air Pollution Modelling and its Applications XIII*, S.E. Gryning and E. Batchvarova eds., Plenum Press, New York, 23

Trini Castelli S., Anfossi D., 1997, Intercomparison of 3-D Turbulence Parametrizations as input to 3-D dispersion Lagrangian particle models in complex terrain, *Il Nuovo Cimento*, 20 C,N. 3, 287-313,

Trini Castelli S., Ferrero E., Anfossi D., Ying R.: 1999 'Comparison of turbulence closure models over a schematic valley in a neutral boundary layer', Proceeding of the *13th Symposium on Boundary Layers and Turbulence - 79th AMS Annual Meeting*, 601-604

THE DETERMINATION OF THE MIXING HEIGHT: AN EXTENDED VERSION OF THE SAFE_AIR DISPERSION MODEL

Emilia Georgieva[1*], Elisa Canepa[2], and Corrado F. Ratto[2]

[1]Institute of Geophysics, Bulgarian Academy of Sciences, Sofia, Bulgaria
[2]INFM (National Institute for the Physics of Matter)
Department of Physics, University of Genova, Italy

INTRODUCTION

The determination of the mixing height is both a very interesting scientific issue in itself and a crucial aspect as far as the assessment of the dispersion of pollutant in the low atmosphere is concerned.

We implemented a meteorological pre-processor able to build, using different expressions depending on the atmospheric stratification conditions, a 2D mixing height field after calculating the 2D sensible heat flux field in the entire simulation domain. Such pre-processor has been used to provide the input mixing heights to the SAFE_AIR (Simulation of Air pollution From Emissions _ Above Inhomogeneous Regions) model: for a description see, for example, Canepa et al. (1998), and http://aix.meng.auth.gr/database/bin/show_long?SAFE_AIR.

MIXING HEIGHT CALCULATION

The developed pre-processor produces hourly averaged values of the mixing height, using observed or simulated wind speeds, surface temperatures, cloudiness, available upper-air profiles and site specific characteristics such as roughness length, albedo and moisture availability.

The pre-processor is based on the energy balance method to determine the sensible heat flux at a reference point. In diurnal conditions, the scheme by Holtslag and van Ulden (1983) and van Ulden and Holtslag (1985) has been used. In nocturnal conditions, we have adopted a semi-empirical approach as in the recent versions of both the CALMET (Scire et al., 2000) and AERMOD (Cimorelli et al., 1998) codes. Using correction factors depending on the surface roughness the sensible heat flux reference value is extended to each grid point.

*This research activity performed at the Department of Physics, University of Genova (Italy), was sponsored by the International Centre for Theoretical Physics (ICTP), Trieste (Italy), within the Program for Training and Research in Italian Laboratories.

After calculating the sensible heat flux field in the entire simulation domain, the mixing height can be computed as a 2D field using different formulae for stable (night-time) and convective (day-time) conditions. Day-time is defined by an upward (positive) sensible heat flux, night-time by a downward (negative) one. For the parameterisation in night-time conditions the user can select between different formulations, among them: the solution proposed by COST-710 (1998) of the Nieuwstadt (1981) formula, and the Venkatram (1980) expression. The pre-processor uses a slab model for the growth of the mixing height during day-time conditions as proposed by Batchvarova and Gryning (1991). According to this model, the mixing layer growth is initially proportional to the friction velocity, with mechanical production of turbulent kinetic energy being the controlling mechanism; the importance of mechanical production diminishes gradually as the production of convective turbulence becomes important.

We applied the implemented pre-processor to calculate the mixing height to some of the KINCAID (Illinois, USA) experiments (Bowne and Londergan, 1983) that represent relatively homogeneous situations (flat terrain). The comparison between the daily temporal evolution of the measured and simulated mixing height gave good agreement in the considered cases.

The SAFE_AIR model was applied in the region of Ilo (Peru). SAFE_AIR simulates transport and diffusion of airborne pollutants at local-to-regional scale under non-stationary and inhomogeneous conditions using Gaussian segments and/or puffs driven by a 3D wind field reconstructed by the model itself. The Ilo region is a complex coastal area where large SO_2 emissions from a copper smelter plant affect the local air quality and a tracer project was designed to collect diffusion data in the area (Wilkerson et al., 1996). In this application we used SAFE_AIR coupled with the implemented algorithm obtaining a much more realistic time evolution of the mixing height than that deduced by the meteorological pre-processor (WINDS) currently inserted in the model. The effect of these more reliable mixing heights is to improve the SAFE_AIR performance concerning the simulation of the tracer concentrations.

REFERENCES

Batchvarova, E., and Gryning, S.E., 1991, Applied model for the growth of the daytime mixed layer, *Bound-lay. Meteorol.* 56:261.

Bowne, N.E., and Londergan, R.J., 1983, *Overview, Results, and Conclusions for the EPRI Plume Model Validation and development Project: Plains Site*, EPRI report EA-3074.

Canepa, E., Modesti, F., and Ratto, C.F., 1998, About the present version of the dispersion code SAFE_AIR, *J. Wind Eng. Ind. Aerod.* 74-76:305.

Cimorelli, A., Perry, S., Venkatram, A., Weil, J., Paine, R., Wilson, R., Lee, R., and Peters, W., 1998, *AERMOD - Description of Model Formulation,* U.S. EPA.

COST Action 710, 1998, *Harmonization of the Pre-processing of Meteorological Data for Atmospheric Dispersion Models Final Report*, B. Fisher, J. Erbrink, S. Finardi et al., eds., EC Belgium.

Holtslag, A.A.M., and van Ulden, A.P., 1983, A simple scheme for daytime estimates of the surface fluxes from routine weather data, *J. Clim. Appl. Meteorol.* 22:517.

Nieuwstadt, F.T.M., 1981, The steady state height and resistance laws of the nocturnal boundary layer: theory compared with Cabauw observations, *Bound-lay. Meteorol.* 20:3.

Scire, J.S., Robe, F.R., Fernau, M.E., Yamartino, R.J., 2000, *A User's Guide for the CALMET Meteorological Model*, Eart Tech Inc., Concord.

Van Ulden, A.P., and Holtslag A.A.M., 1985, Estimation of atmospheric boundary layer parameters for diffusion applications, *J. Clim. Appl. Meteorol.* 24:1196.

Venkatram, A., 1980, Estimating the Monin-Obukhov length in the stable boundary layer parameters for diffusion applications, *Bound-lay. Meteorol.* 19:481.

Wilkerson, D.R., Risch, L.E., Conger, L.E., and Hauze, W.J., 1996, *Atmospheric Tracer Experiments conducted for Southern Peru Limited February-April 1996*, Final Report. Project No. TRC NAWC 19344-0100, Report No. AQ 96-12, August 28, 1996, Salt Lake City.

COMPREHENSIVE ACID DEPOSITION MODEL AND ITS APPLICATION TO EPISODIC ACID DEPOSITION IN EAST ASIA

Seung-Bum Kim and Tae-Young Lee

Laboratory for Atmospheric Modeling Research/Department of Atmospheric Sciences,
Yonsei University, Seoul 120-749, South Korea

INTRODUCTION

A comprehensive model for air quality and pollutant deposition (will be referred as CoMAP hereafter) has been developed to simulate and predict air quality and pollutant deposition over a large to mesoscale area. The most important features of CoMAP include that 1) it is an on-line 3-dimensional Eulerian model and 2) it treats the effects of both the grid-resolved clouds and subgrid scale convective clouds in a rigorous manner considering microphysics. This paper introduces major features of CoMAP and preliminary results of CoMAP application to the simulation of air quality and pollutant deposition over the eastern Asia region.

MODEL

In CoMAP, both gaseous and aqueous concentrations are predicted. CoMAP consists of modules for transport, gaseous chemistry, cloud effects and aqueous chemistry, and dry deposition (Fig. 1). It is an on-line model directly coupled to CSU RAMS (Pielke et al. 1992). Transport of gas-phase and aqueous-phase pollutants is calculated using a three-dimensional Eulerian model. RADM2 chemistry model is employed to treat the gaseous chemistry. Dry deposition is also treated using the method of RADM2. The new cloud-chemistry model consists of grid-resolved cloud and subgrid scale cloud schemes. The subgrid cloud scheme treats the effects of precipitating and non-precipitating convective clouds and it is based on Walcek and Taylor (1986). The scheme for grid-resolved cloud is derived from Wang and Chang (1993) and considers 6 water substances (cloud water, rain, pristine crystals, snow, aggregates, graupel). All the cloud parameters and microphysical information are updated by the CSU RAMS at every time step.

RESULTS

The CoMAP has been applied to the simulation of episodic air quality and pollutant deposition over eastern Asia during 00 UTC 28 – 00 UTC 31 March 1996 when an extratropical cyclone passed and produced significant precipitation over the area. Both air quality and wet deposition appear to be significantly affected by the introduction of new cloud chemistry model according to the preliminary results. As an example, amounts of sulfate and nitrate wet deposition from CoMAP are shown in Fig. 2, along with those from a model which is identical to CoMAP except that RADM2.1 cloud chemistry scheme is used (referred as CADM in the figure). The pattern of wet deposition amount from CADM tends to follow the pattern of rainfall distribution. On the other hand, such strong correlation between rainfall and wet deposition amounts is not found for the wet deposition from CoMAP. The results are being investigated and yet to be verified.

ACKNOWLEDGEMENTS

This research has been supported by the Ministry of Environment, Korea.

REFERENCES

Pielke, R. A., W. R. Cotton, R. L. Walko, C. J. Tremback, M. E. Nicholls, M. D. Moran, D. A. Wesley, T. J. Lee, and J. H. Copeland, 1992 : A comprehensive meteorological modeling system - RAMS. *Meteorol. Atmos. Phys.*, **49**, 69-91.

Walcek, C. J., and G. R. Taylor, 1986 : A theoretical method for computing vertical distributions of acidity and sulfate production within growing cumulus clouds. *J. Atmos. Sci.*, **43**, 339-355.

Wang, C., and J. S. Chang, 1993 : A three-dimensional numerical model of cloud dynamics, microphysics, and chemistry : 1. Concepts and formulation. *J. Geophys. Res.*, **99**, 14827-14844.

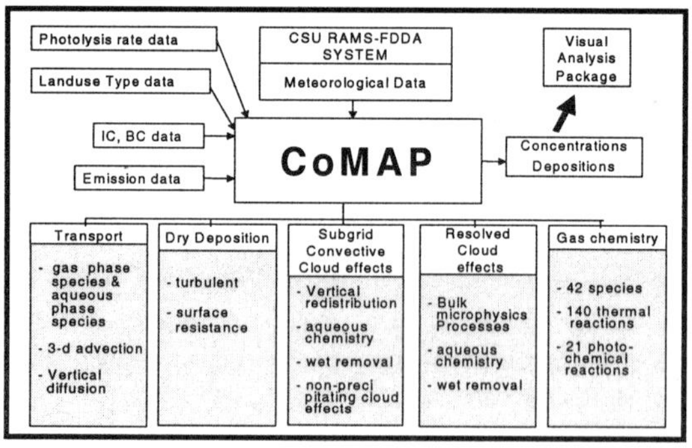

Figure 1. Basic framework of CoMAP.

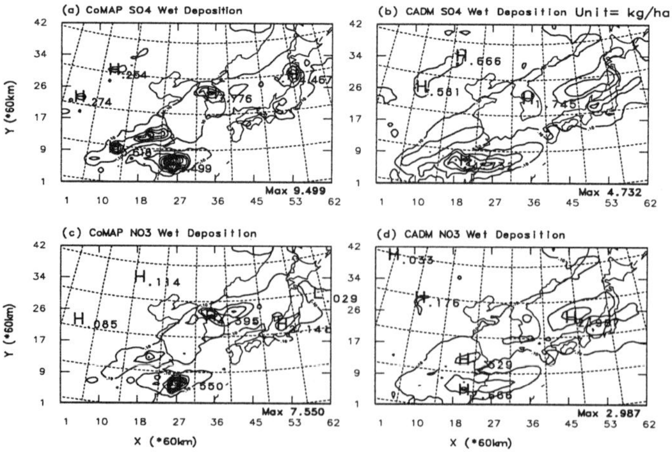

Figure 2. Simulated wet deposition of sulfate and nitrate from CoMAP (a and c, respectively) and CADM (b and d, respectively) for 00 UTC 28 – 00 UTC 31 March 1996.

EVALUATION OF POSSIBLE AIR POLLUTION TRANSPORT FROM IGNALINA NUCLEAR POWER STATION IN REGIONAL SCALE

D.Perkauskas

Institute of physics
Savanoriu 231
2028 Vilnius
Lithuania

INTRODUCTION

The HYSLIT4 (Hybrid Single-Particle Lagrangian Integrated Trajectory) model was used to evaluate possible air pollution transport from Lithuanian Ignalina nuclear power station in regional scale. This model is designed for quick response to atmospheric emergencies, diagnostic case studies, or climatological analyses using previously gridded meteorological data.

MODEL

The model calculation method is a hybrid between Eulerian and Lagrangian approaches. Advection and diffusion calculations are made in a Lagrangian framework while concentrations are calculated on a fixed grid. Without the additional dispersion modules, Hysplit computes the advection of a single pollutant particle, or simply it's trajectory.

Air concentrations Lagrangian model can compute through either of two assumptions. In a puff model, the source is simulated by releasing pollutant puffs at regular intervals over the duration of the release. The puff with its fraction of pollutant mass is advected according the trajectory of its center position while the size of the puff expands in time to account for the dispersive nature of a turbulent atmosphere. Air concentration is then calculated at grid points by assuming that the concentration within the puff has a defined spatial distribution. In a particle model the source can be simulated by releasing many particles over duration of the release. Additionally to the advective motion of each particle,

Air Pollution Modeling and Its Application XIV, Edited by
Gryning and Schiermeier, Kluwer Academic/Plenum Publishers, New York, 2001

a random component to the motion is added at each step according to the atmospheric turbulence at that time. Air concentrations are calculated by summing the mass of all the particles in grid cell.

Although the model can be configured to compute dispersion using three-dimensional puff methods or three-dimensional particle dispersion, only the combination (particle-vertical, puff-horizontal) approach was used. Complete description of the model as well as stability, horizontal and vertical mixing evaluation is presented in[1,2] by Draxler.

RESULTS AND CONCLUSIONS

Analysing the air masses trajectories the earlier performed job[3] was taken into account as well as calculations performed for 1998-1999 using the HYSLIT4. The 1, 6, 11, 16, 21, 26 days of every month with the time period of 120 hours (typical synoptical period 3-5 days) and 150, 500, 1000m start heights were used for the model. The results shows that the 34% of air masses trajectories are of N-NW-W , 36% - S-SW, 8% - E-NE, 10% - SE-S directions and can spread over the some thousands of kilometers. But another 12% of all situations can be evaluated as very slow moving air masses and these situations are most dangerous for the Baltic states. So, the main possile influence of accidental release from Ignalina nuclear power station can be obtained in W,NW and SW directions achieving thousands of kilometers in horizontal direction.

The calculations of pollutant – ^{137}C concentrations were performed using assumption that during 1 hour 1 unit of mass was emited at 150m height. with horizontal spacing of mesh 0.1 deg and the same starting days as for air masses trajectories but only for the first 24 hours. The results shows that within lower 1000m layer of atmosphere the pollutant concentration can achieve 10^{-12} mass unit /m^3 and Ural mountains in the horizontal direction. All calculations were performed using US NOAA Air Resources Laboratory Cray computer (http://www.arl.noaa.gov/ready).

REFERENCES

1. R.R.Draxler and G.D.Hess. An Overview of the HYSPLIT_4 modelling system for trajectories, dispersion and deposition, *Aust. Met. Mag.*, 47:295 (1998).
2. R.R.Draxler and G.D.Hess, *Description of the HYSPLIT_4 modelling system*, NOAA Technical Memorandum ERL ARL-224, (1997).
3. D.Syrakov , M.Kolarova, D.Perkauskas., K.Senuta and A.Mikelinskien• . Model of long-range pollutant transport and acidity of precipitation for Baltic region, in: *Air pollution modeling and it's application X*, Plenum Press, New York and London, (1994).

INFLUENCE OF NON-LINEAR EFFECTS ON SOURCE-RECEPTOR MATRICES AND THEIR POLICY ORIENTED APPLICATIONS TO EMISSION REDUCTION IN EUROPE

Jerzy Bartnicki

Norwegian Meteorological Institute,
P.O. Box 43 Blindern, N-0313 Oslo, Norway

INTRODUCTION

The source-receptor matrices provide the important connection between the emissions and depositions of a pollutant over different time and spatial scales. They play a very important role in the air pollution policy, especially in developing strategies for the sulphur and nitrogen emission reduction in Europe. The country to grid and country to country source receptor matrices for sulphur, oxidized nitrogen and reduced nitrogen are computed annually (Bartnicki, 1999) within the framework of EMEP (Co-operative programme for monitoring and evaluation of the long-range transmission of air pollutants in Europe). There exist two major problems when computing source-receptor matrices with the Eulerian model. The first one is caused by the non-linearities in the model chemistry. The second is created by the non-linearities imposed by the numerical method used for the advection equation. In order to investigate the nonlinear effects on annual source-receptor matrices, a numerical experiment was designed and performed. The main question to be answered by the numerical experiment was if the relations between a single European emitter country and all individual European countries treated as receptors are linear.

NUMERICAL EXPERIMENT

In the experiment, Germany was chosen as an emitter of SO_2, NO_2 and ammonia for the entire year 1997. For each compound the latest version of the EMEP Eulerian Acid Deposition model (Olendrzynski 1999; Olendrzynski *et al.*, 2000) was run, first with the complete emissions and then 10 times with the emissions reduced by 10%, from 100% to 10%. Depositions of oxidized sulphur, oxidized nitrogen and reduced nitrogen to all individual European countries were computed for each run. The departure from linearity was computed, for each run and for each receptor, in percent of the deposition to the receptor.

RESULTS AND CONCLUSIONS

Computed deposition reduction in the receptor country can be presented as a function of emission reduction in Germany. An example for Portugal as receptor is given in Figure 1. The dotted line with in Figure 1 represents the model results and a straight line represents a perfect linear relationship between emission and deposition.

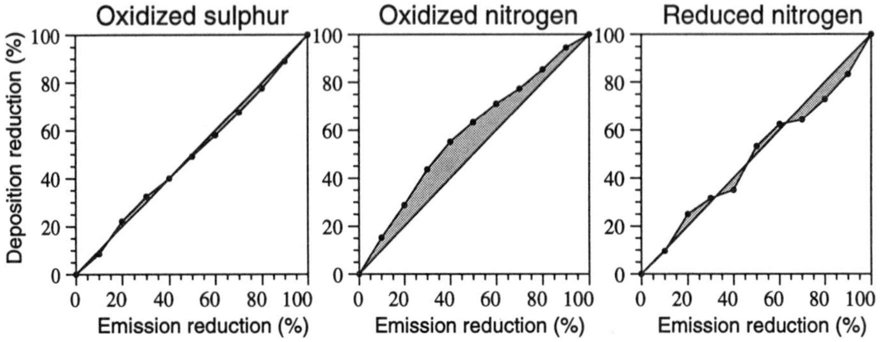

Figure 1. Annual 1997 deposition reduction in Portugal as a function of emission reduction in Germany of oxidized sulphur, oxidized nitrogen and reduced nitrogen, respectively.

In case of Portugal as a receptor, departure from linearity is largest (in absolute terms) for oxidized nitrogen (maximum 15.1%) and smallest for oxidized sulphur (maximum 2.5%). Similar results to those presented in Figure 1 have been examined for other European countries as receptors. For oxidized sulphur the largest (in absolute terms) departure (8%) can be noticed for Cyprus asa receptor. For oxidized nitrogen and reduced nitrogen maximum departures, 32% and 123.8%, occur for Italy as a receptor. For all European countries as receptors, large departures from the linar relationship between deposition and emission can be noticed for oxidized nitrogen (on average 11.7% in absolute terms) and for reduced nitrogen (on average 11.3% in absolute terms). For oxidized sulphur, departures are relatively small, on average 1.1% in absolute terms. For all compounds, the departure from linearity increases with the distance between the emitter and receptor. It is also larger for receptors with relatively small depositions. The results presented here are limited to one emitter only and experiments with more emitters will be performed later. However, these results give an important indication of the error introduced in the source-receptor matrices by the non-linear effects and they should be taken into account in the policy applications.

REFERENCES

Bartnicki J. (1999) Computing source-receptor matrices with the EMEP Eulerian Acid Deposition Model. EMEP/MSC-W note 5/99. Meteorological Synthesizing Centre - West. Norwegian Meteorological Institute, Oslo, Norway.

Olendrzynski K. (1999) Operational EMEP Eulerian Acid Deposition Model. EMEP/MSC-W note 4/99. Meteorological Synthesizing Centre - West. Norwegian Meteorological Institute, Oslo, Norway.

Olendrzynski K., Berge E. and J. Bartnicki (2000) EMEP Eulerian acid deposition model and its applications. *European Journal of Operational Research* **122**, 426-439.

ESTIMATION OF THE INFLUENCE OF THE SOURCES ON AIR POLLUTION AND ITS PERTUBATIONS IN GIVEN REGION USING THE ADJOINT EQUATIONS

Gdaly S. Rivin, Polina V. Voronina

Laboratory of the Modelling of Atmospheric Processes
Institute of Computational Technologies SB RAS,
6, Ac. Lavrentjev Ave., Novosibirsk, 630090 Russia

This work is the continuation of series of the investigations carried out by the authors last years (Rivin, Voronina, 1998a,b; 1999).

The natural and antropogeneous sources and its modifications define a qualitative and quantitative state of air pollutions. The environmental change prognosis consists of the correct description of air pollutions transport, the realization of computational experiment for obtaining the expert estimations with the help of meteorological data observation and ejections in atmosphere, finding and evaluation the influence area for given region.

Usually the solution of this problem is based on the direct simulation with help of the transport equation. This method needs many numerical experiments especially for the evaluation of the contribution of the different industry regions.

We used the technique of adjoint equations based on the G.I.Marchuk method (Marchuk, 1992).

The adoint equation of the transport aerosols we write in the next form:

$$-\frac{\partial \varphi_i^*}{\partial t} - u\frac{\partial \varphi_i^*}{\partial x} - v\frac{\partial \varphi_i^*}{\partial y} - (\tau - \tau_g)\frac{\partial \varphi_i^*}{\partial p} =$$
$$= \frac{\partial}{\partial x}K_x\frac{\partial \varphi_i^*}{\partial x} + \frac{\partial}{\partial y}K_y\frac{\partial \varphi_i^*}{\partial y} + \frac{\partial}{\partial p}K_p\frac{\partial \varphi_i^*}{\partial p} + f_i^*,$$

We have choosen four influence areas and have considered their contributions into the air basin of the “protected” region of the North Siberia. We have used the NCEP/NCAR Reanalysis data in the numerical experiments. On fig. 1 the solution of the adjoint equation is presented for the 21st of December, 1996, for 10 days.

This work is supported by the Foundation of Fundamental Research of the Russian Federation under Grant 98-05-65302 and Grant for Scientific Schools 00-15-98543.

REFERENCES

Marchuk G.I., 1992, *The adjoint equations and analysis of complicated systems,* M: Nauka,. 335 p. (in Russian).

Rivin G.S., and Voronina, P.V., 1998a, The transport of the aerosol in the atmosphere: the imitative experiments, *Optics of atmosphere and ocean,* **11**, 741 – 746.

Rivin G.S., and Voronina P.V., P.V., 1998b, Transport of aerosols in Siberian region: imitative experiments, *Research activities in atmospheric and oceanic modelling*, WMO/TD-No. 865, 5.42 – 5.43.

Rivin G.S., and Voronina, P.V., 1999, Transport of aerosols in Siberian region: the solution of the adjont equation, *Research activities in atmospheric and oceanic modelling,* WMO/TD-No. 942, 5.44 – 5.45.

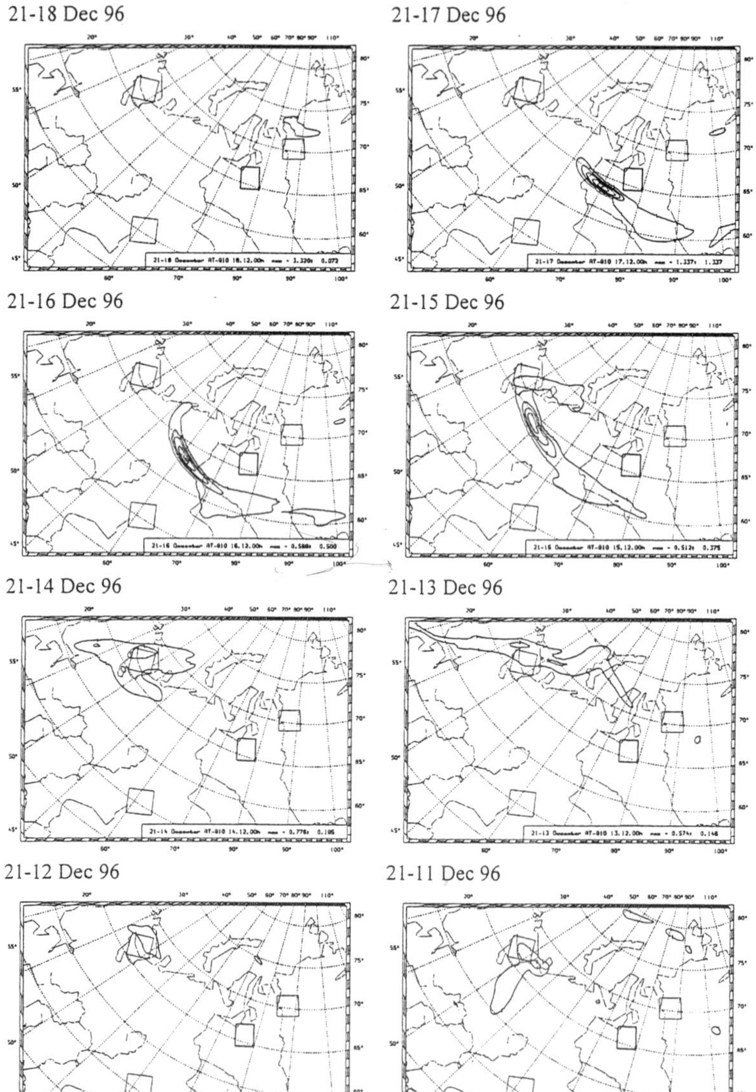

Fig. 1. The isolines of the sensitiveness function for the "protected" region with the Reanalysis data.

OZONE MODELING OF THE BARCELONA AREA: ANALYSIS OF THE INVOLVED TRANSPORT PROCESSES

Isabel Toll, Cecilia Soriano and José M. Baldasano

Laboratory of Environmental Modeling
Department of Engineering Projects, Universitat Politècnica de Catalunya (UPC)
Avd. Diagonal 647, planta 10, 08028 Barcelona, SPAIN
Contact Email: baldasano@pe.upc.es

INTRODUCTION

The city of Barcelona (northeastern corner of Spain) and its surrounding area can reach high levels of O_3 in summertime. An ozone episode that took place between the 3rd and 5th of August of 1990 was simulated with the non-hydrostatic meteorological model MEMO and the photochemical model MARS (Moussiopoulos, 1994). A highly disaggregated emission inventory was also developed taking into account both anthropogenic and biogenic emissions.

The combination of mesoscale circulations (such as sea and land breeze, convection cells and topographic injections) and local emissions strongly influence the production and spatial distribution of ozone in the region. It has been observed how ozone is formed over vegetated areas, where BVOCs (Biogenic VOC) are emitted, due to the inland advection of NO_x from traffic emissions with the sea breeze flow. Orographic injections from the surrounding mountains cause the formation of elevated ozone layers. These elevated layers are dispersed according to the synoptic-scale flow, which is responsible of the orientation of the surface-height decoupled layers

TRANSPORT PROCESSES AND FORMATION OF OZONE

An inventory of gaseous emissions from biogenic and anthropogenic sources has been has been developed for the area of Catalonia (LEM/EIM model), and has been applied in this work to obtain estimation of emissions on the area of Barcelona. BVOCs emissions were estimated taking into account local vegetation data (land-use distribution and biomass factors) and meteorological conditions (surface air temperature and solar radiation) together with emissions factors for native Mediterranean species and cultures (Gómez and Baldasano, 1999). Calculated BVOC emissions were distinguished as isoprene, monoterpene or other BVOCs. Emissions of VOC, NO_x and CO from road traffic, industries, petrol stations, airport traffic, maritime traffic, port tanks and vegetation were calculated in the studied domain (an 80x80 km^2 region with a grid resolution of 2x2 km^2 centered in the city of Barcelona).

CO and VOC emissions take place in the city of Barcelona and are closely related to traffic (emissions of those pollutants are higher at low speeds as occurring in cities). Some important highway junctions in the read network of the area also show important emissions of these pollutants. High emissions of NO_x are found all over the road network, but within the city of Barcelona NO_x are similar to those in the rest of the roads since emissions of this pollutant decrease with decreasing speeds, compensating thus the higher number of vehicles in the city. Finally, The highest BVOC emissions were found in the sclerophyllous forests and shrublands of the coastal and pre-coastal mountain ranges, and also in certain shrubland areas and coniferous forests located in the interior depression. Monoterpenes were the most emitted BVOC throughout the day but during the hours of highest solar intensity and highest air temperature (between 14 and 16 LST) isoprene emissions exceeded those of monoterpenes.

Numerical simulations with the MEMO/MARS modeling system showed that the combination of important BVOC emissions with advection of NO_x due to mesoscale circulations occurring in the region can cause O_3 to form in locations far from the main pollution areas in the region. In Barcelona, during the afternoon hours, sea breeze inflow transports NO_x emissions due to traffic in the city and surrounding road network towards the interior, where high BVOC emissions are taking place during these hours due to the presence of vegetation and the high temperature and solar radiation. The two precursors meet and result in a formation of O_3 and a concentration maximum. Simulations have showed that part of the O_3 found inland is due to the sea breeze and the synoptic wind, introducing significant concentrations through the eastern boundary of the domain. Thus, another conclusion of the study is the importance role of pollutant transport from outside the modeling domain by the geostrophic wind.

The situation during the evening and early morning hours is different. During the early hours of the day, until 9 LST, O_3 is depleted where NO is being emitted, mainly by road traffic both in the urban areas and at the main communication routes. Meanwhile, the land breeze transports this low O_3 concentration air mass towards the Mediterranean Sea.

It is from 13 LST that topographic injection of O_3 becomes significant. The reinforcement of the sea breeze by the upslope winds in the coastal mountain range injects surface air masses, loaded with O_3, towards upper layers of the PBL (reaching a height of around 1300 m ASL). Between 15 and 16 LST vertical mixing in the PBL and the sea breeze circulatory cell transport surface O_3 to upper air layers. From 20 LST until the end of the day the depletion of O_3 where NO is emitted decouples ground O_3 concentrations from those in upper air layers. These elevated layers are transported by the synoptic wind (from the northwest in the case under consideration) towards the southwest.

All these features we have described are shown in the plots of the poster we present, and also in the publications we cite below (Toll, and Baldasano, 1999,2000).

REFERENCES

Gómez O. and J.M. Baldasano (1999) *Biogenic VOC Emission Inventory for Catalonia, Spain.* Proceedings of EUROTRAC Symposium'98. Editors P.M. Borrell and P. Borrell. WITpress, Southampton, pp. 109-115.

Moussiopoulos, N. (1994). *The EUMAC Zooming Model. Model structure and applications,* ed. N. Moussiopoulos, EUROTRAC, Garmisch-Partenkirchen, pp. 1-266.

Toll, I., and Baldasano, J.M. (1999). Photochemical Modeling of the Barcelona Area under Weak Pressure Synoptic Conditions. Atmopheric Environment In: Air Pollution VI. Computational Mechanics Publications.

Toll, I., and Baldasano, J.M. (2000) . Modeling of Photochemical Air Pollution in the Barcelona Area with Highly Disaggregated Anthropogenic and Biogenic Emissions. Atmospheric Environment (*in press*)

MODELING STUDY OF THE RELATIONSHIP BETWEEN PHOTOCHEMICAL OZONE AND ITS PRECURSOR EMISSION OF NITROGEN OXIDES AND HYDROCARBONS IN NORTHERN TAIWAN

Ling-feng HSIAO,[1] Zifa WANG,[2] and Koung-Ying LIU[1]

1Department of Atmospheric Science, Chinese Culture University, HwaKang, Yangmingshan, Taipei, Taiwan

2 Frontier Research System for Global Change, Institute for Global Change Research, 1-18-16, Hamamatu-Cho, Minato-ku, Tokyo 105-0013 Japan

1 INTRODUCTION

Taipei, an well-known densely populated mega-city with heavy traffic, is the rapidly expanding capital city of Taiwan, located in a basin surrounded by mountains on three sides and by two river valleys reaching northwesterly and northeasterly to the Pacific Ocean. Although the monitoring records of recent years show the continuous improvement of air quality, a high level of ozone is still a serious problem, the averaged ozone concentration increased by 10.1% from 1994 to 1998. Photochemical ozone is formed from nitrogen oxides and VOCs through non-linear chemical reactions under special weather conditions. The local scale circulations and airflow would be very complex in this region due to the very complex terrain and the special location. The combination of special meteorological conditions and high emissions in the area may result in very high concentrations of photochemically produced pollutants. The relationship between photochemical ozone and its precursor emission of nitrogen oxides and hydrocarbons in northern Taiwan were investigated based on data analysis and simulation with an air pollution modeling prediction system (APOPS). The use of such a modeling system is needed as northern Taiwan has great changes in topography and pollution episodes are often influenced by local thermal flows.

2 MODEL DESCRIPTION

The basic processes relevant to modeling the urban air pollution problems such as meteorology, dispersion, emission, and chemistry are solved in the APOPS on practically the same grid at the same time. A meteorological model (MMPMS) solves the three-dimensional, non-hydrostatic equations of fluid dynamics and thermodynamics over complex terrain and incorporates the key physical processes that control the boundary layer behavior (Sha et al., 1993). A tracer transport model (GATAM) computes the advection, diffusion, emission, deposition, and chemical process in which the comprehensive gas-phase chemistry mechanisms are considered. The gas-phase chemical reaction scheme used in the present work is a slightly modified version of the CBM-IV (Gery et al., 1989; He and Huang, 1992). The scheme is addressed specially for the accurate modeling of ozone which is produced photochemically from the mixture of nitrogen oxides and hydrocarbons, and it is therefore the best suited for use when ozone is the pollutant in focus. The modified mechanism includes 34 species and 86 photochemical, inorganic and organic reactions.

The domain is divided into 71×51 regular grids in horizontal direction with 3km resolution, covering Northern Taiwan from 120.83°E to 122.4°E and 24.63°N to 25.8°N. Vertically, the model uses 40-layer terrain following coordinate system, with the model top being 7400m above the sea level. Vertical grid spacing is increased gradually from 10m at the surface to 300m at the top. The emission inventory used here is interpolated from the original data provided by the Environmental Protection Administration (EPA) of Taiwan.

The flow patterns typical of the mountain-valley winds and sea/land breezes under stagnant synoptic weather conditions in Northern Taiwan area are replicated well in APOPS. The general characteristics of surface distributions of ozone, SO_2, NO_X and aerosols have been simulated and assessed. The overall agreement between predictions and observations for this case shows that the current model system can be utilized to study the air pollution in urban scale due to photochemical oxidants and other pollutants in the complex terrain regions (Wang et al. 2000).

3 RESULTS

Daily maximum hourly pollutant concentrations of 25 stations between 1996 and 1998 covering northern Taiwan were analyzed as related to the precursors. The observed NOx concentration decreased and ozone levels increased in these five years, which is probably caused by the changed NMHC/NOx ratio. On the base of analysis, many high-ozone episodes were selected to find the quantitative relationship between emission intensity of nitrogen oxides and non-methane hydrocarbons and ozone concentrations covering Northern Taiwan by APOPS simulation. Assuming 50% decrease in NMHC and/or NOx emissions intensity, respectively, simulated results show the air pollution trends afforded by the model have good agreement with observations. The NMHC/NOx ratio is an important parameter to control the formation of ozone. Regarding the sensitivity of NOX and NMHC emissions show that the ozone in north Taiwan is NMHC limited for these cases. In the northern Taiwan, ozone concentrations are more sensitive to the reduction of NMHC emissions.

REFERENCE

Gery, M. W., Whitten, G.Z., Killus, J.P., and Dodge, M.C. 1989, A photochemical kinetics mechanism for urban and regional scale computer modeling. *J. Geophys. Res.* 94: 12 925-12 956.

He, D. and Huang, M. Y. 1992, A photochemical model for regional air quality simulation. *J. Environ. Sci.* 12:192-198.

Sha, W., Kawamura, T. and Ueda, H., 1993, A numerical study of nocturnal sea breezes: Prefrontal gravity waves in the compensating flow and inland penetration of the sea-breeze cutoff vortex. *J. Atmos. Sci.* 50:1076-1088.

Wang, Z., Sha, W., and Ueda, H., 2000, Numerical modeling of pollutant transport and chemistry during a high-ozone event in Northern Taiwan, *Tellus*, (in press).

NUMERICAL AND PHYSICAL MODELING OF URBAN STREET CANYON DISPERSION

Cheng-Hsin Chang, Robert N. Meroney

Fluid Mechanics and Wind Engineering Program
Colorado State University, Fort Collins, CO 80523, U.S.A

INTRODUCTION

The flow patterns which develop around individual buildings govern the wind forces on the building and the distribution of pollution about the building and in its wake. The superposition and interaction of flow patterns associated with adjacent buildings govern the final distribution of facade pressures and the movement of pollutants in urban and industrial complexes. Street canyon depth and width, intersection locations, canyon orientation to dominate wind directions and building geometries will determine peak pollution incidents (Oke, 1988; Theurer, 1995).

WIND TUNNEL SIMULATION OF FLOW ABOUT URBAN STREET CANYON

This study uses as a basic building shape the Wind Engineering Research Field Laboratory building (WERFL) at Texas Tech University, Lubbock. A plastic model of the WERFL structure was constructed to a 1:50 scale and instrumented with multiple pressure and concentration sampling ports. A large number of "dummy" models of similar dimensions were constructed of plastic foam to represent surrounding buildings. These buildings were arranged in various symmetric configurations with different separation distances, and placed in the Industrial Wind Tunnel of the FDDL. Typical building patterns are noted in Figure 1 and associated arrangement patterns are listed in Table 1.

Air Pollution Modeling and Its Application XIV, Edited by
Gryning and Schiermeier, Kluwer Academic/Plenum Publishers, New York, 2001

particle size distribution. Each source was run with 5 years of meteorological data for four stations: Lake Charles, LA; Pittsburgh, PA; Salem, OR; Tucson, AZ. The sites were selected to provide a diversity of climatological regimes. A polar grid of receptors along 16 evenly spaced radials at distances ranging from the edge of the source to several kilometers was used.

We compared the results of various combinations of sampling rates with the results from using the full meteorological database. The basic SCIM approach worked best for meteorological stations with frequent precipitation (e.g. Salem), while wet deposition at sites with infrequent precipitation (e.g. Tucson) was generally underestimated. The inclusion of a higher "wet" sampling frequency with a fairly low frequency for the "regular" sampling improved the results at all sites. Point sources required a higher sampling frequency (regular = every 25th hour; wet = every 8th hour) than area sources (regular =193; wet = 8) to achieve similar results. We also made model runs varying the start hour of both the "regular" and "wet" sampling to characterize the variability of the results. Figure 1 illustrates the results for an example site and indicates that the sampling introduces little bias. The lowest scatter was observed for higher sampling rates and at locations with frequent precipitation. Ratios of annual values paired in space were calculated and frequency distributions were developed to assess the ability of the sampling scheme to reproduce the spatial pattern of impact . An example plot is shown in Figure 2. These plots showed that the enhanced SCIM methodology reproduced the spatial pattern of deposition.

DISCLAIMER

This paper has been reviewed in accordance with the U.S. Environmental Protection Agency's peer and administrative review policies and approved for presentation and publication. Mention of trade names or commercial products does not constitute endorsement or recommendation for use.

REFERENCES

Koch, R.C. and S.D. Thayer, 1974. Validity of the multiple-source gaussian-plume diffusion model using hourly estimates of input; also, Sensitivity analysis of the multiple-source gaussian plume urban diffusion model in Proceedings of the fifth Meeting of Expert Panel on Air Pollution Modeling NATO Committee on the Challenges of Modern Society, Roskilde Denmark.

USEPA, 1995. User's Guide for the Industrial Source Complex (ISC3) Dispersion Models- Volume II - Description of Model Algorithms. EPA-454/B-95-003b, U.S. Environmental Protection Agency, Research Triangle Park, North Carolina, 27711.

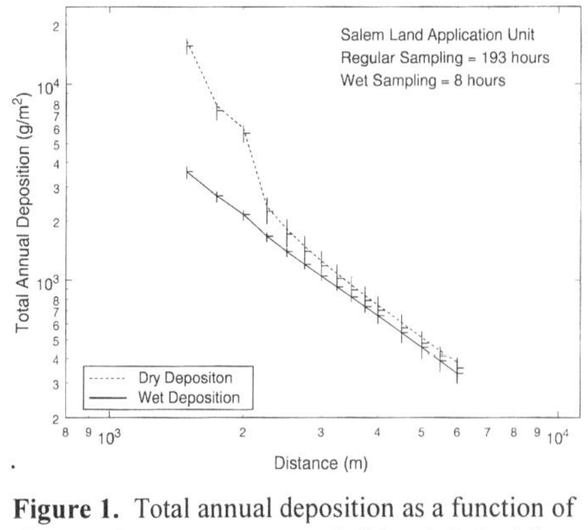

Figure 1. Total annual deposition as a function of distance for an area source. Solid and dashed lines are NoSCIM results. Bars represent replicate samples using SCIM.

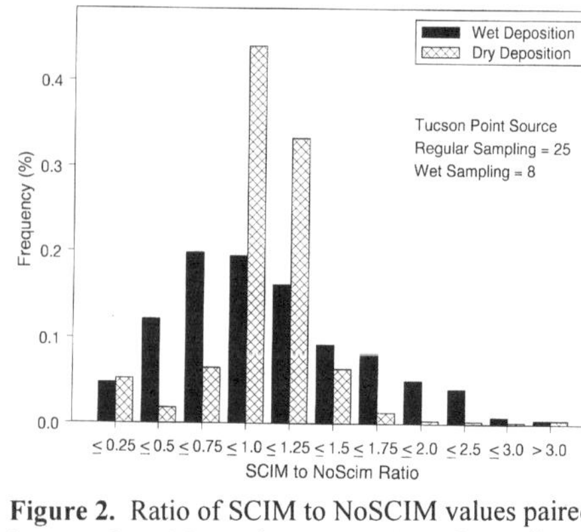

Figure 2. Ratio of SCIM to NoSCIM values paired in space for a 35 m point source. Data from all replicate samples are included.

MODELS-3/CMAQ APPLICATIONS WHICH ILLUSTRATE CAPABILITY AND FUNCTIONALITY

S.K. LeDuc[1], K.L. Schere[1], J.M. Godowitch[1]

Atmospheric Sciences Modeling Division, Air Resources Laboratory, National Oceanic and Atmospheric Administration, Research Triangle Park, NC 27711

and G.L. Gipson

Human Exposure & Atmospheric Sciences Division, National Exposure Research Laboratory, U.S. Environmental Protection Agency, Research Triangle Park, NC 27711

INTRODUCTION

The Models-3/CMAQ developed by the U.S. Environmental Protection Agency (USEPA) is a third generation multiscale, multi-pollutant air quality modeling system within a high-level, object-oriented computer framework (Models-3). It has been available to the scientific community since July 1998 with annual releases since that date. The system was designed to be flexible and modular. The science has been documented (USEPA, 1999a) and a user guide prepared for the framework (USEPA, 1999b). Current users of Models-3/CMAQ have tried various options in their applications, some of which will be shared at the First Annual Models-3 Workshop to be held in June 2000 (http://www.epa.gov/asmdnerl/models3).

The Models-3/CMAQ has considerable capability at this time. The user may select from multiple chemical mechanisms which are provided as well as edit or revise those mechanisms. Two chemical mechanisms have been evaluated thus far in Models-3/CMAQ: RADM2 and Carbon Bond IV. Emissions for these mechanisms are generated using the Models-3 Emission Processing and Projection System (MEPPS). MEPPS can be executed using the framework "Study Planner" (USEPA, 1999b) which directs and monitors multiple execution steps. MEPPS which invokes both SAS[RN] and ARC/INFO™ can also be executed outside the framework. Emission control strategies can be developed within the MEPPS or can be created outside MEPPS by adjusting emission files. The latter approach was used to simulate uniform emission reductions as described on the poster. A Plume-in-Grid capability is also available in Models-3/CMAQ which contains a Lagrangian reactive plume model to simulate subgrid scale pollutant plumes from major point sources within an Eulerian grid. Plume concentrations are integrated into the concentration in a grid cell after the plume segment reaches model grid size.

Open-source visualization tools, Vis5D (http://www.ssec.wisc.edu/~billh/vis5d.html), PAVE (http://envpro.ncsc.org/OTAGDC/pave_letter.html) and Data Explorer (DX), (http://www.research.ibm.com/dci/software.html) are integrated into the Models-3/CMAQ and are used for analyzing and displaying the results. Evaluations of the accuracy and patterns of the simulations have been done using SAS.

The Models-3 framework requires either a Sun workstation, Windows NT on a PC, or Silicon Graphics workstation. The framework Study Planner allows for model simulations to be conducted on computer platforms other than the one running the framework. The science codes will execute on other computing platforms and model execution can be performed on various platforms by using scripts without the Models-3 framework. However, the user must manually make changes in the script that would otherwise be made by the framework. Cray computers, both vector and parallel, have been successfully used with the standalone scripts as well as with the Study Planner in the system framework.

RESULTS AND DISCUSSION

The poster illustrates how Models-3 and its visualization tools can be used to examine and display results of model simulations. The uniform emissions reductions illustrate differences in pollutant concentrations that arise from a simulated reduction of 50% in NO_X emissions from the July 1995 base case in the eastern U.S.

Differences between surface ozone concentrations for the base case and the 50% reduction in NO_X are shown as PAVE generated tileplots; isosurfaces for concentrations of O_3, NO_X and isoprene are shown with Vis5D; and Plume-in-Grid results are shown with DX. Multi-pollutant relationships, effects of controls and the non-linear processes which define the chemistry are discussed on the poster. Time series of pollutants are shown for some urban areas along with the three dimensional structure of ozone and precursors.

DISCLAIMER

This paper has been reviewed in accordance with the U.S. Environmental Protection Agency's peer and administrative review policies and approved for presentation and publication. Mention of trade names or commercial products does not constitute endorsement or recommendation for use.

REFERENCES

USEPA, 1999a, *Science algorithms of the EPA Models-3 Community Multiscale Air Quality (CMAQ) modeling system. Part I: Chapters 1–8* and *Part II: Chapters 9-18*, D.W. Byun and J.K.S. Ching (Eds.), EPA-600/R-99/030, National Exposure Research Laboratory, Research Triangle Park, NC 27711.

USEPA, 1999b, *User Manual for the EPA Third-Generation Air Quality Modeling System (Models-3 Version 3.0)*, EPA-600/R-99/055, National Exposure Research Laboratory, Research Triangle Park, NC 27711.

1. On assignment to Atmospheric Modeling Division, National Exposure Research Laboratory, U.S. Environmental Protection Agency, Research Triangle Park, NC 27711.

A PROPOSAL TO COMPILE PAST ATMOSPHERIC DISPERSION FIELD EXPERIMENT DATA FOR EASY ACCESS AND MODEL EVALUATION

Jeffery T. McQueen, Roland R. Draxler [1]

NOAA Air Resources Laboratory
1315 East West Highway
Silver Spring, MD 20910

INTRODUCTION

The long-range transport of pollutants in the atmosphere has received considerable attention in recent years, corresponding with a comparable number of models to address these issues. In particular there has been consistent emphasis on nuclear reactor accidents since Chernobyl by the International Atomic Energy Agency (IAEA) and the World Meteorological Organization (WMO) through the organization of WMO's Regional Specialized Meteorological Centers (RSMCs). The European Tracer Experiment (ETEX) symposium evaluated 47 different models. The introduction of more fuel efficient jet aircraft engines (running at higher temperatures) has made modern commercial aircraft very sensitive to volcanic ash from eruptions. Volcanic Ash Advisory Centres (VAAC), designated by the International Civil Aviation Organization (ICAO), use transport and dispersion models to forecast the location of volcanic ash, so that aircraft may be warned and rerouted around the affected areas.

Most, if not all, atmospheric transport models (ATM) are linked to one or more operational meteorological forecast models, or a particular meteorological archive which is then used by the ATM. The data predicted by the meteorological models are routinely evaluated and compared against common performance standards. Although similar verification standards exist for the ATMs, there is a sense in the dispersion modeling community that there is very little data available to perform these evaluations. Perhaps this is because these data are not as easily obtained and are usually limited to single events or controlled (and expensive) field experiments that are not available on a routine basis. However there are many isolated experiments, some

Air Pollution Modeling and Its Application XIV, Edited by
Gryning and Schiermeier, Kluwer Academic/Plenum Publishers, New York, 2001

of which are controlled tracer releases and others from the sampling of tracers of opportunity. The problem with many of these data sets is that some are now decades old, reports are difficult to locate, if available the data are in various formats, and not all experiments archived the corresponding meteorological data.

The recent completion of meteorological re-analysis projects at several international meteorological centers, provides an opportunity to link high quality modern meteorological data with each of these dispersion experiment data sets. The concept would be to create a set of CD ROMS, containing experimental data, relevant reports, meteorological data, and statistical analysis and display software, all in a common non-proprietary form. This new common data base would permit the modeling community to conduct sensitivity and verification studies with considerable less preparation and effort than is now required. In addition, each modeling center or research group could produce results for each experiment that then could be compared with results from all other groups through participation in a model verification symposium. The objectives of this report are to stimulate research laboratories that might be interested to do the work and perhaps identify potential sources of funding.

DATA STANDARDS

There are three primary data sources of interest: 1) information about the pollutant release, 2) meteorological data used to calculate the pollutant's transport and dispersion, and 3) measured pollutant values that can be compared with model results. Each of these should be in a common format so that any experiment simulation can be quickly setup and results produced. Depending upon the type of experiment, information or measurements provided, there should be certain minimum requirements for an experiment's inclusion into the data base.

EXPERIMENTAL DATA BASES

Only experiments in which the transport distances from source to the majority of thes amplers are in excess of 200 km are considered for inclusion into the data base. This is the range at which there is a transition from the Planetary Boundary Layer's control of dispersion to the larger-scale 2-D synoptic influence on dispersion.

It is necessary to establish guidelines for which type of data should be included in such an archive: 1)controlled experiments in which the pollutant release rate is known, 2) accidental ones in which it is unknown (and always will be), such as volcanic eruptions, and 3) pure transport experiments, such as balloon releases. The controlled group would include such events as the Chernobyl accident, in which the source term was reconstructed after the fact. Controlled experiments provide quantitative concentration or deposition data, however each experiment has different limitations, usually related to how many samples could be analyzed. Experiments with detailed spatial and temporal resolution are usually limited to a few cases. Experiments that cover many events usually have either low temporal or spatial sampling resolution. Each experiment may serve a different model verification or development purpose.

REGIONAL AIR POLLUTION ORIGINATING FROM OIL-REFINERY FIRES UNDER WAR CONDITIONS

Z. Vukmirovic[1], L. Lazic[2], I. Tosic[2] and M. Unkasevic[2]

[1]Environmental Studies, Alternative Academic Educational Network, 11000 Belgrade, Masarikova 5/XVI, Yugoslavia
[2]Institute of Meteorology, Belgrade University,11001 Belgrade, Yugoslavia

Practically simultaneous release of smoke plumes from the oil refineries in Novi Sad and Pancevo was occurred with total burning rate of 2,000 t h^{-1} during the first 12 h after bombing at midnight between 17 and 18 April 1999 . Using the same methodology applied in the case of the Kuwait oil smoke plume (Johnson et al., 1991), an average emission of carbon particles in overlapping plumes (Melas et al., 1999) is estimated as 65 t h^{-1}.

This study addresses simulation of the regional air pollution originating from oil-refinery fires from Pancevo and Novi Sad incidents and its properties as reflected on typical trajectories. Trajectory calculations are implemented in the Eta Model. A realistic real data simulation of a regional air pollution case is achieved using the model with a 56 km horizontal resolution and 32 layers in the vertical. The synoptic case during 18-21 April 1999 was characterized by an upper trough located in the northern part of area of interest. There is a small gradient field in the lower levels. That synoptic case produces a light variable wind near surface, and much intensive westerly wind in the upper levels. Numerical simulation by the Eta model is initialized at 0000 UTC 18 April using ECMWF (European Center for Medium-Range Weather Forecasts) analysis as the initial. Initial fields of geopotential height, wind components and specific humidity, at ten standard pressure levels, are obtained by bilinear interpolation from the ECMWF operational initialized analyses. Forward trajectories during the oil-refinery fires episodes based on the 72 h Eta model forecast with a horizontal grid resolution of 0.5° x 0.5° are shown in Fig. 1. To improve visualization of the vertical structure for the air pollution, forward trajectories are calculated starting from the middle of the eta layers. According to the trajectories, the pollutant puff was picked up over the area of oil-refinery fires and moved eastward over Romania, Bulgaria, Moldavia, Ukraine and Black Sea. This long-range transport was occurred around level 500 hPa. The lower level trajectories from Pancevo indicate pollutant transport in short-regional and local scales towards the Belgrade area in the first day. Under oxidizing conditions in plume, a significant fraction of Hg(II) from the petrochemical plant in Pancevo might be adsorbed to elemental carbon particles (Seigner et al., 1998). Relatively high concentrations of soot and polycyclic aromatic hydrocarbons (PAHs) are predicted in the air and compared with the available measurements in Novi Sad, Pancevo and Belgrade.

Air Pollution Modeling and Its Application XIV, Edited by
Gryning and Schiermeier, Kluwer Academic/Plenum Publishers, New York, 2001

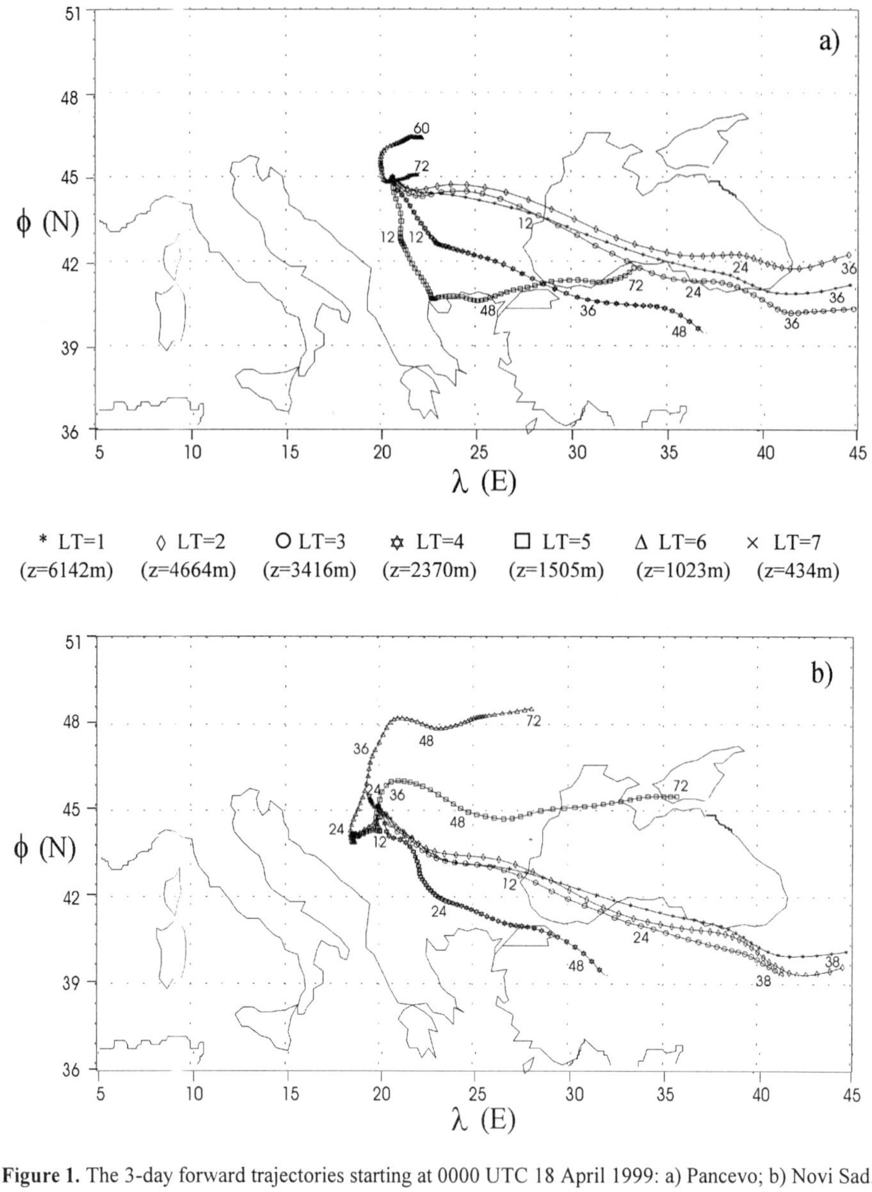

Figure 1. The 3-day forward trajectories starting at 0000 UTC 18 April 1999: a) Pancevo; b) Novi Sad.

REFERENCES

Johnson, D.W., Kilsby, C.G., McKenna, D.S., Saunders, R.W., Jenkins, G. J., Smith, F.D., and Foot, J.S., 1991, Airborne observations of the physical and chemical characteristics of the Kuwait oil smoke plume, *Nature* 353. 617:621.

Melas, D., Zerefos, C., and Raspomanikis, S., 1999, Contribution to the FOCUS report from the Greek delegation, FOCUS Assessment Mission 2 to the Federal Republic of Yugoslavia, Ecology, 18. July-13. August 1999, Executive Summary & Final Report, Bern, October 1999.

Seigner, C., Abeck, H., Chia, G., Reinhard, M., Bloom, N. S., Prestbo, E., and Saxena, P., 1998, Mercury adsorption to elemental carbon (soot) particles and atmospheric particulate matter, *Atmospheric Environment* 32. 2649:2657.

NESTING OF AN OPERATIONAL AIR QUALITY MODEL (OPANA) INTO A GLOBAL METEOROLOGICAL SPECTRAL MODEL (RSM/NOAA): PRELIMINARY RESULTS

Roberto San José, Ignacio Salas, José Ismael Peña, Alejandro Martín, Juan Luis Pérez, Ana Belén Carpintero[1] and Rosa María González[2]

[1]Environmental Software and Modelling Group, Computer Science School, Technical University of Madrid (Spain), Campus de Montegancedo, Boadilla del Monte 28660-Madrid (Spain).
[2]Department of Geophysics and Meteorology, Complutense University of Madrid, Ciudad Universitaria 28040-Madrid (Spain)

INTRODUCTION

The importance of synoptic meteorological fields on mesoscale air quality forecasting is recognised. Also long range transport are also becoming and important key topic in air quality modelling since it is becoming more apparent that long range transport can play an important role to understand some air pollution episodes (high concentrations on stable turbulent conditions, sporadic high concentrations during very unstable conditions, etc.

In this contribution we use the so-called the Regional Spectral Model (RSM) from NCEP (National Centre for Environmental Prediction, USA) to update the meteorological boundary conditions of an operational air quality model (OPANA) which uses a non-hydrostatic mesoscale meteorological model (REMEST) and a chemical model (CHEMA). The RSM is a limited area atmospheric numerical model system which uses spectral computation to solve the fluid equations. Spectral computations seems to have higher accuracy in terms of gradients and spectral interpolation. The difficult to do limited area modelling with spectral method is overcome by the time-dependent perturbation method. We run a Global version of the Spectral Model to generate the input files to run an hydrostatic version over the Iberian Peninsula 1215 x 1215 km. The OPANA model incorporates as initial meteorological forecasting soundings the forecasting sounding produced by the RSM nesting version over the Madrid domain. The OPANA model has run over 80 x 100 km with 5 km spatial resolution and the results are compared with observational station data over the Madrid Municipality air pollution network.

Air Pollution Modeling and Its Application XIV, Edited by
Gryning and Schiermeier, Kluwer Academic/Plenum Publishers, New York, 2001

EXPERIMENT

The experiment presented in this contribution is dated on February, 29, 2000 and it is performed over 120 hours simulation (this is Tuesday to Saturday). The Global model is initialised at 0h00 on February, 29, 2000 by using the Analysis Global Data Set in NCEP. The model is run with T126 resolution (aprox. 105 km). The RSM model over the Iberian Peninsula is run with 27 km spatial resolution. There is an operational version in http://artico.lma.fi.upm.es. The Madrid model domain is 80 x 100 km and we select the 4 centred locations in the four blocks that the domain can be divided regularly. These locations are chosen to incorporate the vertical forecasted meteorological soundings provided by the RSM at those locations during the 120 hours simulation. The AVN/MRF are global operational weather models which run with different spatial resolutions (111 km and 191 km). We currently use the vertical forecasted meteorological soundings provided by these models to compare with RSM results. Figure 1 shows the comparison between observed data and RSM initialisation and AVN/MRF initialisation of OPANA modelling system.

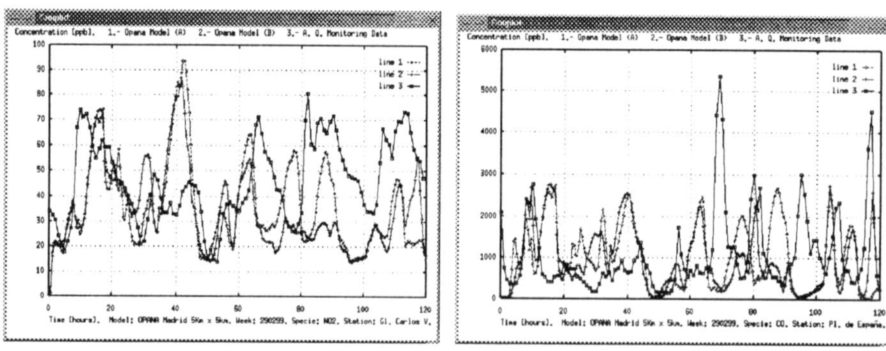

Figure 1.- NO2 temporal pattern at Carlos V station and CO temporal pattern at Glorieta Quevedo station of the Madrid Municipality Network

The results show that peak values are improved when using RSM model although additional nesting in OPANA model improves the comparison with observed data (not presented in this contribution). The spatial resolution of the synoptic model seems to play an important role in air quality modelling.

Acknowledgements

We would like to thank to NCEP (NOAA) for the RSM model and particularly to Dr. Hann-Ming Henry Juang and Dr. Steve Brenner from the Israel Oceanographic and Limnological Research Center. Also, we would like to thank to Madrid Municiaplity for Air Quality Network Dataset and SICE S.A. for technical support.

References

Juang, H.-M.H. and M. Kanamitsu, 1994: The NMC nested regional spectral model. Mon. Wea. Rev., 122, 3-26.

San José R., Prieto J.F., Castellanos N, Arranz J.M. Sensitivity Study of dry deposition fluxes in ANA air quality model over Madrid mesoscale area. Measuring and Modelling in Environmental Pollution. Ed. Computational Machanics Publications. Ed. R. San José and C.A. Brebbia. (1997). pp.119-130.

PARTICIPANTS

The Millennium (24th) NATO/CCMS International Technical Meeting on Air Pollution Modelling and Its Application. Boulder, Colorado, May 15-19, 2000

ARGENTINA

Ulke A. G.

University of Buenos Aires
Department of Atmospheric Sciences
Pabellon II - Piso II, 1428 Ciudad Universitaria
Buenos Aires
E-mail: ulke@at.fcen.uba.ar

AUSTRALIA

Cope M.

CSIRO Atmospheric Research
Private Bag No. 1
Aspendale
3195 Melbourne Victoria
E-mail: martin.cope@dar.csiro.au

AUSTRIA

Seibert P.

University of Agricultural Sciences
Institute of Meteorology and Physics
Türkenschanzstr. 18
1180 Vienna
E-mail: seibert@boku.ac.at

BELGIUM

Kretzschmar J. G.

VITO
Flemish Institute of Technological Research
Boeretang 200
2400 Mol
E-mail: jan.kretzschmar@vito.be

Mensink C.

VITO
Flemish Institute for Technological Research
Boeretang 200
2400 Mol
E-mail: mensinkc@vito.be

Schayes G. H. | University of Louvain
2 Chemin du Cyclotron
1348 Louvain-la-Neuve
E-mail: schayes@astr.ucl.ac.be

van der Auwera L. | Royal Meteorolgical Institute
Ringlaan 3
1180 Brussels
E-mail: vdal@oma.be

BULGARIA

Batchvarova E. A. | National Institute of Meteorology and Hydrology
66 Tzarigradsko Chaussee
1784 Sofia
E-mail: ekaterina.batchvarova@meteo.bg

Syrakov D. | National Institute of Meteorology and Hydrology
66 Tzarigradsko Chaussee
1784 Sofia
E-mail: dimiter.syrakov@meteo.bg

CANADA

Ainslie B. | University of British Columbia
Department of Geography
Atmopsheric Sciences Program
1984 West Mall
Vancouver, British Columbia V6T 1Z2
E-mail: bainslie@eos.ubc.ca

D'Amours R. | Canadian Meteorological Centre
2121 Trans-Canada Highway
Dorval, Quebec H9P 1J3
E-mail: real.d'amours@ec.gc.ca

Fox D. | Environment Canada
Room 200, 4999-98 Ave
Edmonton, Alberta T6B 2X3
E-mail: dave.fox@ec.gc.ca

Gong S. | MSC
Environment Canada
4905 Dufferin Street
Toronto, Ontario M3H 5T4
E-mail: sunling.gong@ec.gc.ca

Rucker M. | University of British Columbia
Department of Earth and Ocean Science
Vancouver, British Columbia V6T 1Z2
E-mail: rucker@geog.ubc.ca

Steyn D. G. | University of British Columbia
Department of Earth and Ocean Sciences
Vancouver, British Columbia V6T 1Z2
E-mail: dsteyn@eos.ubc.ca

CROATIA

Vidic S. — Meteorological and Hydrological Service of Croatia
Gric 3
10000 Zagreb
E-mail: vidic@cirus.dhz.hr

CZECH REPUBLIK

Bubnik J. — Czech Hydrometeorological Institute
Na Sabatce 17
143 06 Prague 4
E-mail: bubnik@chmi.cz

Jicha M. — Brno Universtiy of Technology
Faculty of Mechanical Engineering
Technicka 2
61669 Brno
E-mail: jicha@dt.fme.vutbr.cz

Pospisil J. — Brno University of Technology
Faculty of Mechanical Engineering
Institute of Power Engineering
Technicka 2
61669 Brno
E-mail: pospisil@dt.fme.vutbr.cz

DENMARK

Baklanov A. — Danish Meteorological Institute
Meteorological Research Division
Lyngbyvej 100
2100 Copenhagen Ø
E-mail: alb@dmi.dk

Geernaert G. L. — National Environmental Research Institute
P.O. Box 358
4000 Roskilde
E-mail: glg@dmu.dk

Gross A. — Danish Meteorological Institute
Meteorological Research Department
Lyngbyvej 100
2100 Copenhagen Ø
E-mail: agr@dmi.dk

Gryning S.-E. — Risø National Lababoratory
Department of Wind Energy and Atmospheric Physics
4000 Roskilde
E-mail: sven-erik.gryning@risoe.dk

Sørensen J. H. — Danish Meteorological Institute
Lyngbyvej 100
2100 Copenhagen Ø
E-mail: jhs@dmi.dk

FINLAND

Karppinen A.	Finnish Meteorological Institute Sahaajankatu 20 E 00810 Helsinki E-mail: ari.karppinen@fmi.fi
Valkama I.	Finnish Meteorological Institute Sahaajankatu 20E 00810 Helsinki E-mail: ilkka.valkama@fmi.fi

FRANCE

Bouzom M.	SCEM/OSAS. METEO-FRANCE 42, Avenue G. Coriolis 31057 Toulouse E-mail: michel.bouzom@meteo.fr
Cautenet G.	Laboratoire de Meteorologie Physique CNRS & Universite Blaise Pascal 24, Avenue des Landais 63177 Cedex Aubiere E-mail: g.cautenet@opgc.univ-bpclermont.fr
Cautenet S.	Laboratoire de Meteorologie Physique CNRS & Universite Blaise Pascal 24, Avenue des Landais 63177 Cedex Aubiere E-mail: s.cautenet@opgc.univ-bpclermont.fr
Chaumerliac N.	Laboratoire de Meteorologie Physique CNRS & Universite Blaise Pascal 24, Avenue des Landais 63177 Aubiere Cedex E-mail: chaumerl@opgc.univ-bpclermont.fr
Leriche M.	Laboratorie de Meteorologie Physique CNRS & Universite Blaise Pascal 24, Avenue des Landais 63177 Aubiere Cedex E-mail: leriche@opgc.univ-bpclermont.fr
Pradelle F.	Laboratoire de Meteorologie Physique CNRS & Universite Blaise Pascal 24, Avenue des Landais 63177 Cedex Aubiere E-mail: f.pradelle@opgc.univ-bpclermont.fr

GERMANY

Ebel A.	University of Cologne EURAD Project Achener Strasse 201-209 50931 Cologne E-mail: eb@eurad.uni-koeln.de

Graff A.	Federal Environmental Agency Bismarckplatz 1 14193 Berlin E-mail: arno.graff@uba.de
Hellmuth O.	Institute for Tropospheric Research Permoserstrasse 15 04303 Leipzig E-mail: olaf@tropos.de
Klug W.	Professor emeritus Mittermayerweg 21 64289 Darmstadt E-mail: dg5q@hrzpub.tu-darmstadt.de
Möllmann-Coers M.	Forschungszentrum Jülich Postfach 1913 52425 Jülich E-mail: moe@fz-juelich.de
Müller F.	Meteorological Institute University of Hamburg Bundesstrasse 55 20146 Hamburg E-mail: frank.mueller@dkrz.de
Nester K.	Institut fur Meteorlogie und Klimaforschung Forschungszentrum Karlsruhe Postfach 3640 76021 Karlsruhe E-mail: klaus.nester@imk.fzk.de
Petersen G.	GKSS Research Centre Max-Planck-Strasse 21502 Geestacht E-mail: petersen@gkss.de
Renner E.	Institute for Tropospheric Research Permoserstrasse. 15 04303 Leipzig E-mail: renner@tropos.de
Stohl A.	Universität München Lehrstuhle für Bioklimatologie und Imm.Forschung Ludwig-Maximillians, Am Hochanger 13 85354 Freising-Weihenstephan E-mail: as@atmos1.met.forst.uni-muenchen.de
Thielen H.	GRS Waste Management Division Schwertnergasse 1 50667 Cologne E-mail: thi@grs.de

Wolke R. | Institute for Tropospheric Research
Permoserstrasse 15
04303 Leipzig
E-mail: wolke@tropos.de

GREECE

Kallos G. B. | University of Athens
Department of Physics
Bldg. Phys-V
15784 Athens
E-mail: kallos@mg.uoa.gr

ISRAEL

Ranmar D. O. | Hebrew University of Jerusalem
The Faculty of Agricultural, Food &
Environmental Quality Sciences
76100 Rehovot
E-mail: ranmar@agri.huji.ac.il

ITALY

Anfossi D. | Istituto di Cosmogeofisica CNR
Corso Fiume 4
10133 Torino
E-mail: anfossi@to.infn.it

Baldi M. | IFA-CNR
Via. Fosso del Cavaliere 100
00133 Roma
E-mail: baldi@atmos.ifa.rm.cnr.it

Canepa E. | INFM
University of Genova
Department of Physics
Via Dodecaneso 33
16146 Genova
E-mail: canepae@fisica.unige.it

Deserti M. | ARPA-SMR
Viale Silvani, 6
40122 Bologna
E-mail: m.deserti@smr.arpa.emr.it

Domenichetti M. | Regione Lombardia
Via Stresa 26
20100 Milano
E-mail: marcello_domenichetti@regione.lombardia.it

Finardi S. | CESI
Via Reggio Emilia 39
20090 Segrate
E-mail: sfinardi@cesi.it

Maffeis G.	Terraria SRL Via Zarotto 6 20124 Milano E-mail: g.maffeis@terraria.com
Marchesi L.	A.R.P.A. Lombardia Via Fara 20 20124 Milano E-mail: Luca_Marchesi@regione.lombardia.it
Rizza U.	ISIATA/CNR Strada Per Arnesano 73100 Lecce E-mail: umberto.rizza@isiata.le.cnr.it
Rotondaro G.	Regione Lombardia Via Stresa 24 20100 Milano E-mail: Giuseppe_Rotondaro@regione.lombardia.it
Suppan P.	European Commission Via Fermi 1 21020 Ispra E-mail: peter.suppan@jrc.it
Trini-Castelli S.	Istituto Di Cosmogeofisica CNR Corso Fiume 4 10133 Torino E-mail: Trini@to.infn.it

JAPAN

Kitada T.	Toyohashi University of Technology Tempaku-cho Department of Ecological Engineering 441-8580 Toyohashi E-mail: kitada@earth.eco.tut.ac.jp
Sugata S.	National Institute for Environmental Studies 16-2 Onogawa 305-0053 Tsukuba Ibaraki E-mail: sugatas@nies.go.jp
Uno I.	Research Institute for Applied Mechanics Kyushu University Kasuga Park 6-1 816-8580 Kasuga E-mail: iuno@riam.kyushu-u.ac.jp
Wang Z.	Frontier Research for Global Change IGCR Yokohama 3173-25 Showa-machi Kanazawa-ku 236-0001Yokohama E-mail: zifawang@frontier.esto.or.jp

KOREA

Hwang D.-I. Yonsei University
Department of Atmospheric Sciences
134 ShinChon, Seodaemoon-ku
120-749 Seoul
E-mail: dolf@atmos.yonsei.ac.kr

Lee J.-B. Yonsei University
Department of Atmospheric Sciences
134 ShinChon, Seodaemoon-ku
120-749 Seoul
E-mail: ljb@atmos.yonsei.ac.kr

Lee T.-Y. Yonsei University
Department of Atmospheric Sciences
134 ShinChon, Seodaemoon-ku
120-749 Seoul
E-mail: lty@atmos.yonsei.ac.kr

Park S.-H. Yonsei University
Department of Atmospheric Sciences
134 ShinChon, Seodaemoon-ku
120-749 Seoul
E-mail: psh@atmos.yonsei.ac.kr

LITHUANIA

Perkauskas D. Institute of Physics
Savanoriu 231
2028 Vilnius
E-mail: donatasp@ktl.mii.lt

NORWAY

Bartnicki J. Norwegian Meteorological Institute
PO Box 43 Blindern
0313 Oslo
E-mail: jerzy.bartnicki@dnmi.no

Eliassen A. Norwegian Meteorological Institute
P.O. Box 43 Blindern
0313 Oslo
E-mail: Anton.Eliassen@dnmi.no

Iversen T. Department of Geophysics
University of Oslo
P.O.Box 1022 Blindern
N-0315 Oslo,
E-mail: trond.iversen@geofysikk.uio.no

PORTUGAL

Borrego C. University of Aveiro
Deparment of Environment and Planning
3810-193 Aveiro
E-mail: borrego@ua.pt

RUSSIA

Genikhovich E.	Main Geophysical Observatory 7 Karbysheva Street 194021 St. Petersburg E-mail: ego@main.mgo.rssi.ru
Rivin G.	Institute of Computational Technologies SB RAS 6, Ac. Lavrentjev avenue 630090 Novosibirsk E-mail: rivin@ict.nsc.ru

SERBIA

Vukmirovic Z.	Alternative Academic Educational Network Environmental Studies Masarikova 5/XVI 11000 Belgrade E-mail: zorkav@eunet.yu

SLOVENIA

Boznar M.	Jozef Stefan Institute Jamova 39 1000 Ljubljana E-mail: marija.boznar@ijs.si
Mlakar P.	Jozef Stefan Institute Jamova 39 1000 Ljubljana E-mail: primoz.mlakar@ijs.si

SPAIN

Baldasano J. M.	Lab. de Modelizacion Ambiental Universitat Politecnica de Catalunya (UPC) Avda. Diagonal 647 Planta 10 08028 Barcelona E-mail: baldasano@pe.upc.es
San Jose R.	Technical University of Madrid Campus de Montegancedo Boadilla Del Monte 28660 Madrid E-mail: roberto@fi.upm.es
Soriano C.	Lab. de Modelizacion Ambiental Universitat Politecnica de Catalunya (UPC) Avda. Diagonal 647 Planta 10 08028 Barcelona E-mail: soriano@pe.upc.es

SWEDEN

Ekman A.	MISU, Arrhenius Laboratory 10691 Stockholm E-mail: annica@misu.su.se

Johansson P.-E.

FOA
90182 Umeå
E-mail: johansson@ume.foa.se

Persson C.

SMHI
60176 Norrköping
E-mail: christer.persson@smhi.se

SWITZERLAND

de Haan P.

INFRAS
Mühlemattstrasse 45
3007 Bern
E-mail: peter.dehaan@infras.ch

THE NETHERLANDS

Builtjes P. J. H.

TNO-MEP
P.O. Box 342
7300 Apeldoorn
E-mail: p.j.h.builtjes@mep.tno.nl

van Dop H.

IMAU
Utrecht University
PO Box 80005
3508 TA Utrecht
E-mail: h.vandop@phys.uu.nl

van Jaarsveld J. A.

National Institute of Public Health & Environment
PO Box 1
3720 BA Bilthoven
E-mail: hans.van.jaarsveld@rivm.nl

Vermeulen A. T.

ECN
P.O. Box 1
1755 LE Petten
E-mail: a.vermeulen@ecn.nl

UNITED KINGDOM

Donnell E.

University of Reading
Department of Meteorology
PO Box 243
RG6 6BB Reading Berks.
E-mail: e.a.donnell@reading.ac.uk

Fisher B.

University of Greenwich
Medway Campus
ME4 4AW Chatham Kent
E-mail: fb04@gre.ac.uk

Fournier N.

CEH and University of Edinburgh
King's Building, Department of Meteorology
EH9 3JZ Edinburgh
E-mail: nicolas@met.ed.ac.uk

Warren R. | Imperial College
TH Huxley School of Environment
RSM Building, Prince Consort Road
SW7 2BP London
E-mail: r.warren@ic.ac.uk

UNITED STATES

Arnold J. R.
NOAA/USEPA
PO Box 23193
Atmospheric Modeling Division
Seattle, Washington 98102-0493
E-mail: jra@hpcc.epa.gov

Bach W. D.
Army Research Office
P.O. Box 12211
Research Triangle Park, North Carolina 27709
E-mail: bach@aro-emh1.army.mil

Bornstein R. D.
San Jose State University
Department of Meteorology
San Jose, California 95192
E-mail: pblmodel@hotmail.com

Briggs K.
Colorado Department of Health
APCD
4300 S. Cherry Cir. Drive
Denver, Colorado 80222
E-mail: kevin.briggs@state.co.us

Byun D. W.
US EPA/NOAA
MD-80
Atmospheric Modeling Division
Research Triangle Park, North Carolina 27711
E-mail: bdx@hpcc.epa.gov

Carmichael G. R.
University of Iowa
Regional Environment Research
204 Iowa Advanced Technology Laboratory
Iowa City, Iowa 52242
E-mail: gcarmich@icaen.uiowa.edu

Chang C.-h.
Colorado State University
Civil Engineering
Fort Collins, Colorado 80523
E-mail: cc527330@engr.colostate.edu

Daescu D.
University of Iowa
CGRER
528 Hawkeye Dr.
Iowa City, Iowa 52246
E-mail: ddaescu@cgrer.uiowa.edu

Daye R. L.
EPA REGION VII
21607 NE 178th St.
Holt, Missouri 64048
E-mail: daye.richard@epa.gov

Delle Monache L. — San Jose State University
Meteorology Department
One Washington Square
San Jose, California 95112
E-mail: monache@metsun1.sjsu.edu

Dennis R. L. — U.S. EPA/NOAA
MD-80
Atmospheric Modeling Division
Research Triangle Park, North Carolina 27711
E-mail: rdennis@hpcc.epa.gov

Freedman F. R. — Stanford University
Department of Civil & Environmental. Engineering
Terman Engineering Center
Stanford, California 94305
E-mail: freedman@stanford.edu

Gay B. W. — US EPA
MD-80
National Exposure Research Lababoratory
Research Triangle Park, North Carolina 27711
E-mail: gay.bruce@epa.gov

Guttikunda S. — University of Iowa
204 Iowa Advanced Technology Laboratory
Iowa City, Iowa 52242
E-mail: sguttiku@cgrer.uiowa.edu

Hanna S. R. — George Mason University
CSI Mail Stop 5C3
103 Science and Technology
FairFax, Virginia 22030-4444
E-mail: shanna@gmu.edu

Hendrickson B. — UCAR
Boulder, Colorado 80307-3000
E-mail: hendrick@ucar.edu

Hogrefe C. — SUNY
Department of Earth and Atmospheric Sciences
1400 Washington Avenue
Albany, New York 12222
E-mail: chogrefe@air.dec.state.ny.us

Irwin J. S. — US EPA/NOAA
MD-14
Atmospheric Modeling Division
Research Triangle Park, North Carolina 27711
E-mail: irwin.john@epa.gov

Kumar N. — Electric Power Research Institute
3412 Hillview Avenue
Palo Alto, California 94304
E-mail: nkumar@epri.com

Lavery T. F.	Environmental Science & Engineering, Inc. P.O. Box 1703 Gainesville, Florida 32602 E-mail: tflavery@esemail.com
Leach M. J.	Lawrence Livermore National Laboratory Atmospheric Sciences Division PO Box 808, L-103 Livermore, California 94551 E-mail: mleach@llnl.gov
Leduc S. K.	US EPA/NOAA MD-80 Atmospheric Modeling Division Research Triangle Park, North Carolina 27711 E-mail: leduc@hpcc.epa.gov
Machovec C. M.	Colorado Department of Health 4300 Cherry Creek Drive, S APCD-TS-B1 Denver, Colorado 80222-1530 E-mail: chuck.machovec@state.co.us
Mahrt L.	Oregon State University COAS Corvallis, OR 97331 E-mail: mahrt@oce.orst.edu
McQueen J. T.	NOAA/ARL 1315 East-West Highway, R/E/AR Silver Springs, Maryland 20910 E-mail: jeff.mcqueen@noaa.gov
Mendoza-Dominiguez A.	Georgia Institute of Technology 200 Bobby Dodd Way Atlanta, Georgia 30332-0512 E-mail: albert@themis.ce.gatech.edu
Meroney R. N.	Colorado State University Civil Engineering Fort Collins, Colorado 80523 E-mail: meroney@engr.colostate.edu
Noonan N.	US EPA Headquarters 1300 Pensylvania Avenue, NW Washington, DC 20004 E-mail: noonan.norine@epa.gov
Odman T	Georgia Institute of Technology School of Civil and Environmental Engineering Atlanta, Georgia 30332-0512 E-mail: talat.odman@ce.gatech.edu
O'Reilly P. M.	UCAR/COMET P.O. Box 3000 Boulder, Colorado 80307 E-mail: oreilly@comet.ucar.edu

Petersen W. B.	US EPA/NOAA MD-80 Research Triangle Park, North Carolina 27711 E-mail: petersen.william@epa.gov
Pleim J. E.	US EPA/NOAA MD-80 Atmospheric Modeling Division Research Triangle Park, North Carolina 27711 E-mail: pleim@hpcc.epa.gov
Rao S. T.	NY State Department of Environmental Conservation University at Albany 50 Wolf Road, Rm. 198 Albany, New York 12233-3259 E-mail: strao@dec.state.ny.us
Reynolds N. D.	The Boeing Company Mail Code K95-36 4200 Southeast Blvd. Wichita, Kansas 67210-1618 E-mail: nathaniel.d.reynolds@boeing.com
Russell A. G.	Georgia Institute of Technology 200 Bobby Dodd Way Atlanta, Georgia 30332-0512 E-mail: trussell@themis.ce.gatech.edu
Schiermeier F. A.	US EPA/NOAA MD-80 Atmospheric Modeling Division Research Triangle Park, North Carolina 27711 E-mail: francis.schiermeier@noaa.gov
Schiffman Y. M.	American Meteorological Society 1200 New York Avenue NW, Suite 410 Washington, D.C. 20005 E-mail: YaleSchiffman@dc.ametsoc.org
Stevens D. E.	Lawrence Livermore National Laboratory Earth & Environmental Sciences Directorate L-103 7000 East Avenue Livermore, California 94550 E-mail: stevens9@llnl.gov
Tonnesen G.	University of California College of Engineering Center for Environmental Research and Technology Riverside, California 92521-0425 E-mail: tonnesen@cert.ucr.edu
Tremback C. J.	Mission Research Corporation P.O. Box 466 Fort Collins, Colorado 80522 E-mail: tremback@aster.com

Uliasz M.	Colorado State University Department of Atmospheric Science Fort Collins, Colorado 80523 E-mail: marek@atmos.colostate.edu
Venkatram A.	University of California College of Engineering Riverside, California 92521-0425 E-mail: venky@engr.ucr.edu
Weil J. C.	University of Colorado CIRES CB 216 Boulder, Colorado 80309 E-mail: weil@ncar.ucar.edu
Yamartino R. J.	Consulting Scientist 191 E. Grand Avenue #206 Old Orchard Beach, Maine 04064 E-mail: rjy@maine.rr.com

AUTHOR INDEX

SUBJECT INDEX